Transition metal organometallics for organic synthesis

D1179541

Transition metal organometallics for organic synthesis

(*the late*)
Francis J. McQuillin
School of Chemistry
University of Newcastle-upon-Tyne
Newcastle-upon-Tyne

David G. Parker
ICI Advanced Materials
Wilton
Cleveland

G. Richard Stephenson
Royal Society University Research Fellow
School of Chemical Sciences
University of East Anglia
Norwich

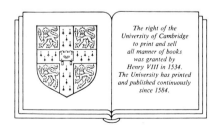

*The right of the
University of Cambridge
to print and sell
all manner of books
was granted by
Henry VIII in 1534.
The University has printed
and published continuously
since 1584.*

CAMBRIDGE UNIVERSITY PRESS

Cambridge
New York   *Port Chester*   *Melbourne*   *Sydney*

Published by the Press Syndicate of the University of Cambridge
The Pitt Building, Trumpington Street, Cambridge CB2 1RP
40 West 20th Street, New York, NY 10011–4211, USA
10 Stamford Road, Oakleigh, Victoria 3166, Australia

© Cambridge University Press 1991

First published 1991

Printed in Great Britain at the University Press, Cambridge

British Library cataloguing in publication data

McQuillin, F. J. (Francis Joseph)
Transition metal organometallics for organic synthesis.
1. Organometallic compounds. Chemical reactions & synthesis
I. Title   II. Parker, David G.   III. Stephenson, G. Richard
547.050459

Library of Congress cataloguing in publication data

McQuillin, F. J.
Transition metal organometallics for organic synthesis/Francis
J. McQuillin, David G. Parker, G. Richard Stephenson.
p.   cm.
1. Organic compounds–Synthesis.   2. Transition metal compounds.
I. Parker, David G.   II. Stephenson, G. Richard.   III. Title.
QD262.M38 1991   547.2–dc20 90–1475 CIP

ISBN 0 521 33353 9 hardback

Contents

Preface

The objective behind this book has been to provide a work that would assist newcomers to the subject with a description of the major new synthetic methods which employ organometallic reactions. After a first draft in 1982, Professor Francis McQuillin was joined in the project by Dr David Parker, and the McQuillin/Parker 'team' developed the theme to include industrial applications and catalytic reactions. This stage was completed in 1983. The untimely death of Professor McQuillin in early 1984 brought progress temporarily to a halt. With the encouragement of Mrs McQuillin, however, the project was quickly reactivated, and Dr Richard Stephenson joined Dr Parker to complete the manuscript. Although the manuscript has been through several stages during its compilation, we have retained and built upon the original McQuillin concept which aimed to provide a compendium of examples to afford insight into new synthetic methods developed over the last 25 years.

Since 1984, Dr Parker and Dr Stephenson have shaped the book into its final form, restructuring the coverage of material and continually updating the examples and references, which now survey the literature up to 1988 with occasional citation of more recent work. The book is illustrated with many examples of synthetic transformations, often taken from target oriented syntheses, and aims to convey an impression of the scope and limitations of the reactions discussed, and to provide a commentary that will form a basis to facilitate the design of new applications for organometallic procedures. Industrial chemists working in the field of organic synthesis, as well as research students and advanced undergraduates following specialised courses, should find this book of use.

The division of material into the fifteen chapters reflects the requirements of a modern treatment of the topic. We start with reactions of alkene π complexes. Ligand rearrangement processes and oxidation reactions, surveyed first (Chapters 2 to 4), are followed by a discussion of nucleophile addition reactions, which occupies much of Chapters 5 to 7. Attention then turns to uses of σ bound species, as nucleophilic reagents (Chapter 8), in carbonyl insertion reactions (Chapter 9), and in hydrogenation and coupling reactions (Chapter 10). In the concluding third of the book, organometallic cycloaddition reactions and metal–carbene chemistry are discussed (Chapters 11 and 12), and, finally, the reactions introduced in the preceding chapters are reviewed in terms of their use in protecting group strategies (Chapter 13), in natural product synthesis (Chapter 14), and heterocyclic synthesis (Chapter 15).

At this time of writing, the rate of appearance of new organometallic research papers in these fields is becoming increasingly frequent. The selection presented here is limited, but we hope, representative. In many ways, during the preparation of this manuscript, the subject under discussion has come of age, and is now widely recognised as a significant and rapidly developing research area. Many industrial applications are appearing. It has been our task to ensure that the style and coverage of this book has grown up with its topic, but also, to strive to retain a perspective of the early days of this subject, and to convey something of the continuing excitement and enthusiasm of the many research workers who have participated in the application of transition metal organometallic chemistry to problems in organic synthesis.

David Parker   Richard Stephenson

Acknowledgements

Assistance from a number of Ph.D. students at the University of East Anglia has helped us during the final preparation of the manuscript for this book. We are grateful to: Miss Michelle Hastings, Miss Wei Dong Meng, Mr David Owen, Miss Roxanne Perret-Gentil, Mr Richard Thomas, and Miss Sarah Williams for their assistance, and to Mrs Angela Saunders for preparation of the art work for this book. GRS thanks the Royal Society for a 1983 University Research Fellowship, during the tenure of which this book was completed. We thank Dr Keith Kelly (ICI) and Mr Donald Smyth, Mr Richard Thomas, and Mr Michael Tinkl (UEA) for assistance with proof reading, and the staff of the ICI Wilton Library (especially Miss Debbi Smith), and the University Library (especially Mrs Juliette Knowles) and the interlibrary loans office at UEA, for obtaining many of the cited references.

The help and encouragement of Mrs Moyra McQuillin, and of our wives and families is gratefully acknowledged.

Abbreviations

Ac	acetyl $COCH_3$
acac	acetylacetonate
ACE	angiotensin-converting enzyme
ACMP	(2-methoxyphenyl)(cyclohexyl)methylphosphine [(118) on p. 132]
AIBN	azoisobutyronitrile
Am	amyl
9-BBN	9-borabicyclo[3.3.1]nonane
bda	benzylideneacetone
BINAP	2,2'-bis(diphenylphosphino)-1,1'-binaphthyl [(45) on p. 34]
bipy	2,2'-bipyridine
Bn	benzyl
BSA	O,N-bis(trimethylsily)acetamide
Bu	butyl
Bz	benzoate
CAMPHOS	1,2,2-trimethyl-cis-1,3-bis(diphenylphosphino)methylcyclopentane
CHIRAPHOS	2,3-bis(diphenylphosphino)butane
COD	cycloocta-1,3-diene
Cp	cyclopentadienyl
CTAB	cetyltrimethylammonium bromide
Cy	cyclohexyl
DABCO	1,4-diazobicyclo[2.2.2]octane
dba	dibenzylideneacetone
DBN	1,5-diazabicyclo[4.3.0]non-5-ene
DBU	1,8-diazabicyclo[5.4.0]undec-7-ene
DDQ	2,3-dichloro-5,6-dicyano-1,4-benzoquinone
d.e.	diastereoisomeric excess
DET	diethyl tartrate
16-DHA	$(-)$-3β-(acetoxy)pregna-5,16-dien-20-one
DIBAL	diisobutylaluminium hydride
DIOP	2,3-isopropylidene-2,3-dihydroxy-1,4-bis(diphenylphosphino)butane [(117) on p. 132]

dipamp	1,2-bis[2-methoxyphenyl)phenylphosphino]ethane
diphos	1,2-bis(diphenylphosphino)ethane (= dppe)
DIPT	diisopropyl tartrate
DME	1,2-dimethoxyethane
DMEU	1,3-dimethyl-2-imidazolidinone
DMAP	4-dimethylaminopyridine
DMF	N,N-dimethylformamide
DMSO	dimethylsulphoxide
DOPA	3-(3,4-dihydroxyphenyl)alanine
dppb	1,4-bis(diphenylphosphino)butane
dppe	1,2-bis(diphenylphosphino)ethane (= diphos)
e.e.	enantiomeric excess
Et	ethyl
Fp	$Fe(CO)_2Cp$
HETEs	hydroxyeicosatetraenoic acids
HMPA	hexamethylphosphoramide
HOMO	highest occupied molecular orbital
hν	photolysis
i	iso
L	any unidentified ligand
LDA	lithium diisopropylamide
LUMO	lowest unoccupied molecular orbital
m	meta
M	any unidentified metal
Magic methyl	FSO_3Me
mcpba	m-chloroperbenzoic acid
Me	methyl
MEM	methoxymethyl
MoOPH	oxodiperoxymolybdenum(pyridine)(hexamethylphosphoramide)
n	normal
NBS	N-bromosuccinimide
NCS	N-chlorosuccinimide
NMMO	N-methylmorpholine-N-oxide
NMP	N-methylpyrrolidinone
n.m.r.	nuclear magnetic resonance
o	ortho
p	para
Pc	phthalocyanin
PCC	pyridinium chlorochromate
PDC	pyridinium dichromate
PEG	polyethylene glycol
PG	prostaglandin
Ph	phenyl

phen	*o*-phenanthrolide
PPA	polyphosphoric acid
PPFA	α-[2-diphenylphosphinoferrocenyl]ethyldimethylamine
Pr	propyl
Py	pyridine
rac	racemic
REDAL	bis-(2-methoxyethoxy)aluminium hydride
RT	room temperature
Sia	CHMeCHMe$_2$
t	tertiary
TBDMS	t-butyldimethylsilyl
TBHP	t-butylhydroperoxide
tcne	tetracyanoethene
TfO	triflate
TFA	trifluoroacetic acid
TFAA	trifluoroacetic anhydride
TG	tetraglyme
THF	tetrahydrofuran
THP	tetrahydropyranyl
TMEDA	tetramethylethenediamine
TMNO	trimethylamine-*N*-oxide
TMS	trimethylsilyl
tol	tolyl
tritox	tri-t-butylmethoxide [Me$_3$)$_3$CO$^-$]
TTFA	thallium tristrifluoroacetate
Ts	tosyl (*p*-tolyl)sulphonyl

Safety bibliography

Although this book does not aim to provide a guide to experimental techniques, it is important that readers should be aware that, like most chemicals, organometallic reagents can present a hazard in use. Metal carbonyls, most especially nickel tetracarbonyl, are a typical example. Even simple solvents and co-solvents such as benzene and HMPA constitute major hazards, and now are strictly controlled in many chemical laboratories. Readers should consult one of the many excellent compilations of hazard data when planning experimental work.

Some examples are:

L. Bretherick, *Handbook of Reactive Chemical Hazards*, Edn 3 revised, Butterworth, 1987.
R. E. Lenga, ed. *The Sigma–Aldrich Library of Chemical Safety Data*, Edn 2, Sigma–Aldrich Corporation, 1987.
N. I. Sax, *Dangerous Properties of Industrial Materials*, Edn 7 revised, Van Nostrand Reinhold, 1988.
Hazard Data Sheets, BDH Ltd, 1989/90.

Computer-based compilations also now exist:

HAZDATA (National Chemical Emergency Centre, Harwell).
Sigma–Aldrich Material Safety Data Sheets (on CD–ROM).

1
Introduction: reactivity properties of coordinated ligands

Transition metal organometallic chemistry can combine organic compounds with transition metal centres in a great variety of ways. Through binding as a ligand to the metal centre, the reactivity properties of the organic compound are often greatly altered. Unstable structures can be stabilised, and stable compounds can be activated in this way. In the context of organic synthesis, these changes in reactivity offer exciting and unconventional opportunities through the provision of new types of reactions to be harnessed for the construction of target molecules. Many different types of synthetic transformations will be encountered later in this book. To make sense of this information it will be helpful to know something of the nature of the bonding between transition metals and their ligands, and of the types of reactivity properties to be expected from structures of this kind. In this first chapter, the common bonding modes that feature later in the book are discussed, and typical mechanisms of organometallic reactions are surveyed. The chapter does not contain comprehensive lists of structural and reaction types, since we are introducing here only a selection of bonding modes and mechanistic steps that relate directly to reactions described in later chapters. This will serve, however, to provide an introduction to the topic that will prepare the reader for reactions of a kind which would otherwise seem surprising to those unfamiliar with transition metal organometallic chemistry. A recent, excellent, and far more extensive survey of structural types can be found elsewhere.[1]

Synthetic chemists turned to organometallic methods in search of special reactivity or selectivity which might not be available by conventional methods. In many cases, these benefits have indeed been demonstrated in good measure, and, through an understanding of the control of the reactions taking place within the coordination sphere of transition metals, a basis for the rationalisation of these effects can be achieved. A number of factors are important:

(a) The availability or creation of free coordination sites (see Section 1.2) at the metal. This permits substrates and reagents to become united by coordination to the metal.

(b) The geometrical restrictions to the orientation of bonding of ligands in many transition complexes. This determines the alignment of reacting species brought together by the metal, so providing control of both chemical selectivity and stereoselectivity.

(c) The use of metals to stabilise charged or otherwise unstable species, or, conversely, to promote reactivity by activation of bonds which would otherwise be inert.

In this chapter we introduce the reader to these effects. The first two sections relate to the identification of coordinatively saturated and unsaturated species. Section 1.3 introduces the topic of oxidation states. The remaining sections in the chapter examine typical reactions of coordinated ligands to identify changes in electron count and oxidation state brought about during each reaction. The final section surveys the bonding and reactivity of organometallic species in slightly greater depth.

1.1 Types of metal–ligand bond

Unsaturated molecules can bind to transition metals in two orientations; they may be 'side-on', or 'end-on'. Some of the first organometallic compounds prepared provided a fine demonstration of these distinctions, although it was some time before the nature of the bonding was adequately understood. An alkene ligand, such as that in Zeise's salt (1),[2] is typical of side-on bonding, while a carbon monoxide ligand, for example, in chromium hexacarbonyl (2),[3] exemplifies the end-on mode. The difference is illustrated

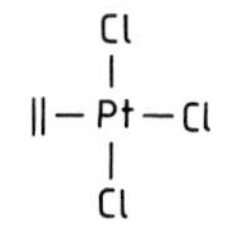

(1) 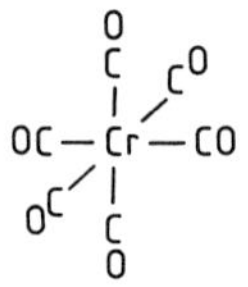   (2)

in Figure 1.1. Location of a metal at the side of the alkene ligand (3) permits an interaction between the π and π^* orbitals of the alkene and d orbitals on the metal, to make new bonding and antibonding orbitals. Side-on bonding to the carbon monoxide π-orbitals also seems reasonable, but in simple complexes, CO ligands prefer the end-on mode shown in structure (4), in which the sp lone pair and π^* orbitals on carbon are directed towards the metal.

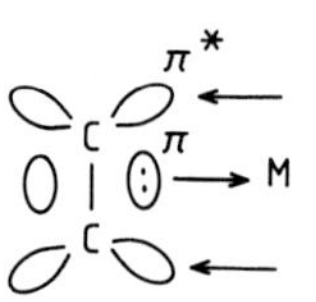

(3)   (4)

side-on alkene complex   end-on carbonyl complex

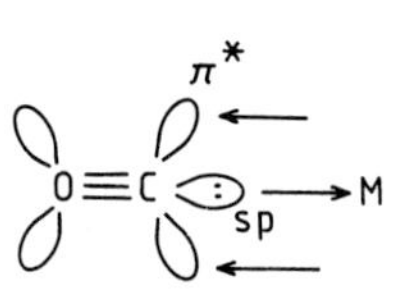

Figure 1.1

Other typical side-on ligands are illustrated in Figure 1.2. The number of carbon atoms π bound to the metal is defined by the superscript of the symbol eta, η, the 'hapto' number, a term coined by Cotton[4] from the Greek word *haptein*, 'to fasten'. Allyl (η^3), diene (η^4), dienyl (η^5), arene or triene (η^6) and trienyl (η^7) ligands all bind the metal to one face of the π system, providing a series of progressively more extensive ligands from alkenes and alkynes (η^2) to η^7 (or even η^8, with lanthanide metals[5]).

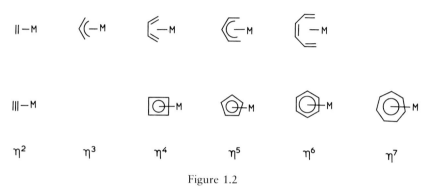

Figure 1.2

Figure 1.3 depicts a selection of end-on ligands. Phosphines (5) resemble CO, as a lone pair is involved with the interaction with the metal, although the second orbital participating in the bonding process is now a phosphorous d orbital, not a π^* orbital. Carbenes (6) and carbynes (7) are similar, with lone pairs and p orbitals on the ligand bonding to the metal.

$$O\equiv C-M \qquad R_3P-M \qquad R_2C=M \qquad RC\equiv M$$

$$(4) \qquad\qquad (5) \qquad\qquad (6) \qquad\qquad (7)$$

Figure 1.3

Binding to transition metal centres is not restricted to unsaturated ligands, though at first, the study of alkyl complexes was difficult because of the ease of β-elimination of hydrogen (see Section 1.4). Alkyl ligands (8) are attached to the metal by a single carbon

$$R-M \qquad\qquad R-OH \qquad\qquad R-SiX_3$$

$$(8) \qquad\qquad (9) \qquad\qquad (10)$$

atom (η^1). The M–C bond is analogous to the O–C bond in an alcohol (9) or the Si–C bond in a silane (10) in that it is a simple σ bond. It is important to note here that for the side-on and end-on ligands discussed earlier, much more complicated bonding is involved, with π as well as σ symmetry interactions (see Section 1.6 and Figures 1.11–1.15). The question of σ and π bonding is introduced in Figure 1.4 and discussed in more detail later in this chapter. In organometallic chemistry, the single line drawn

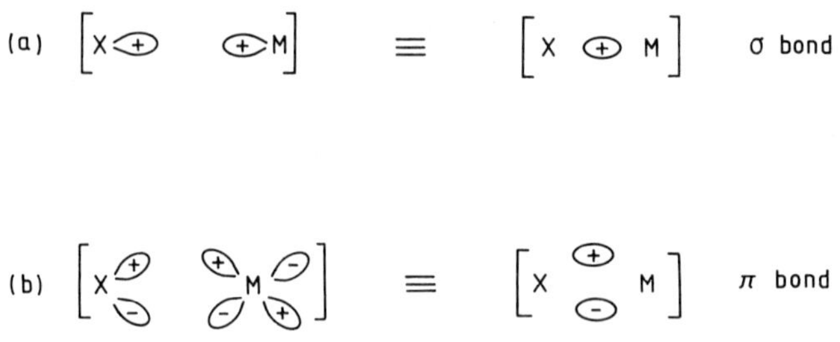

Figure 1.4

between the metal and a ligand can represent a variety of different forms of bonding, and, consequently, care must be taken with interpretation of structural formulae in this field.

1.2 Electron counting: the 18 electron rule

When all the ligands around a metal centre are taken into account, it is possible to make a count of electrons occupying orbitals in this outer valence shell. These electrons originate either from the metal atom or the ligands, but in the complex they occupy the new molecular orbitals formed through the interactions between the metal and its ligands. Once all the metal valence-shell orbitals are used in bonding, further ligands do not usually bind, and the complex can be termed coordinatively saturated. Since there are five d orbitals, three p orbitals and an s orbital available for bonding, a total of nine bonding molecular orbitals can be formed. Each can accommodate two electrons, making a total of 18. This is the origin of the well known 18 electron rule[6] to define coordinative saturation. In fact, the 18 electron rule is only a guide, since complexes are now known in which the total of valence electrons exceeds 18. Furthermore, many complexes are stable with less than 18 electrons in the valence shell. This is notably the case for elements well to the right in the periodic table. In these circumstances the complex can be thought of as coordinatively unsaturated, and can accommodate a further ligand, raising the electron count temporarily to 18. Coordinatively unsaturated complexes of this type are said to have 'free coordination sites', a property that can be very valuable in allowing the approach of a further ligand to initiate reaction at the metal centre. Often relatively unstable complexes with free coordination sites can participate in catalytic cycles, with substrate molecules entering and leaving the coordination sphere of the metal as the cycle proceeds.

There are several counting systems that can be employed to determine the number of electrons in the valence shell. The one recommended here is designed to provide a simple and unambiguous count, by avoiding complications concerning the precise nature of the ligand orbitals. For this purpose any charge on the complex is arbitrarily assigned to the metal, so that, by counting across the periodic table and reducing the total by one for each positive charge on the complex (or increasing by one for each negative charge), a count for the metal is determined. The number of additional electrons contributed by the ligands must now be established on the basis that the ligands are uncharged. For side-on alkene ligands, this is simple. Each p orbital contributes one electron to the ligand π system, and so must similarly contribute one electron to the valence shell of the metal when bound in a π complex. The number of electrons contributed by the ligand is the same as the η value in this method of counting. Thus η^2 alkene ligands bring two electrons with them to add to the metal, η^3 allyls bring three, and η^4, η^5, and η^6 complexes contribute four, five, and six electrons, respectively. Alkynes are normally two electron ligands, but in some situations can contribute four electrons by bringing both bonds into coordination with the metal. With other ligands, for example when heteroatoms are involved in the bonding, it is sufficient to count the total number of electrons in the neutral ligand π system, to discover the number of electrons involved.

For end-on ligands the principle is the same. Carbon monoxide, which offers up a lone pair (two electrons) and an empty π^* orbital (no electrons), can be seen to be a two

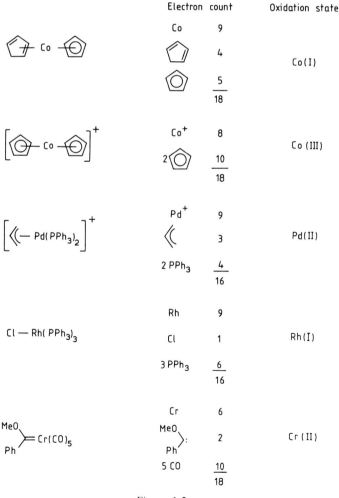

Figure 1.5

electron ligand. The same is true for phosphines (lone pair and an empty d orbital) and for carbenes (an sp^2 and a p orbital, each with one electron, or, alternatively an sp^2 lone pair and an empty p orbital: either way the count is two). Carbyne ligands project an sp orbital and two p orbitals towards the metal (three orbitals and a total of three electrons).

Finally, η^1 complexes must be considered. Here, for alkyl complexes, a single orbital and a single electron is involved. The η^1 alkyl ligand is a one electron ligand. In other situations, a lone pair may form an η^1 interaction. Now two electrons should be counted because the lone pair of the ligand contributes both electrons.

A number of worked examples for side-on, end-on, and η^1 ligands are provided in Figure 1.5.

1.3 Oxidation states

Just as a rather formal system of electron counting was recommended in Section 1.2, since the objective there was to identify free coordination sites, so too, in order to follow oxidation and reduction processes, the formal concept of oxidation state discussed in this section is valuable. In the simple form presented here, the assignment of oxidation states is best used to identify *changes* taking place between products and reactants, or the intermediates in a catalytic cycle. If changes in oxidation state are the main objective of study, it is of paramount importance that the method of assignment used is unambiguous in all situations. To follow organic transformations, this will suffice, but if attempts are to be made to classify reaction types as typical of certain oxidation states of particular metals, then more care is required to avoid anomalies. In this chapter, however, we will concentrate on the simple considerations needed to identify changes in oxidation states.

For this purpose, the most revealing definition of the oxidation state of a central atom in a metal complex is:

Oxidation state: The positive charge remaining on the central atom when all the ligands have been removed *with full coordination shells*.

It will be clear from this definition that the presence of many types of ligand does not influence the oxidation state count. Carbon monoxide and phosphines are *formally* removed with a full coordination shell as neutral species, and so do not alter the charge on the central atom. Similarly alkenes, dienes, and arenes are also neutral 'full coordination shell' species. For alkyl groups, a detachment as a neutral would produce a radical requiring one further electron to complete its shell. From this it can be seen that a σ bound alkyl ligand increases the oxidation state count by one, since formal removal of an anion is required by the rule.

Other ligands present more difficult problems. Allyl and dienyl ligands possess partly filled non-bonding orbitals which require a consistent treatment, and cyclopentadienyl and cyclobutadiene ligands offer tempting analogies between closed shells and aromaticity. In some cases there are several reasonable oxidation states for complexes containing these ligands. For the purposes indicated above, where comparable oxidation state assignments are required, a simple method (Figure 1.6) to determine an unambiguous assignment, can be recommended. When in doubt, divide the ligand π system arbitrarily into alkene links (C=C), and then treat in the normal way the remaining σ bond that arises in the case of π systems with odd numbers of carbon atoms. Following this device, which amounts to considering a canonical form of the true structure, allyl, dienyl, and cyclopentadienyl all increase by one, the oxidation state count at the central atom.

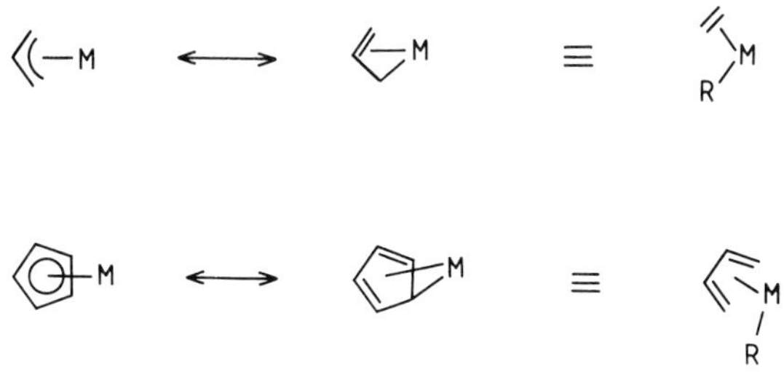

Figure 1.6

Unambiguous determination of oxidation states of organometallic molecules can now be made on this basis. As before, any charge on the complex is arbitrarily assigned to the metal. Ligands are 'removed' as specified by the rule and commentary above, and the charge at the metal adjusted accordingly. The final count provides the oxidation state.

Some examples of oxidation state assignments based on these rules are shown in Figure 1.5. It should be emphasised that the removal of a ligand is an entirely formal device which serves to indicate the oxidation state but does not reflect a real chemical process. Hydride ligands, for example, are normally σ bound (see Chapter 10), and add one charge each to the oxidation state count. In chemical terms, however, despite their name, hydride ligands would commonly leave the metal as protons. Few transition metal hydrides are hydride donors, but many can be deprotonated to yield metal stabilised anions.[7]

1.4. Typical reactions of coordinated ligands

A relatively small number of distinct reaction types form the basis for the interpretation of organometallic mechanisms. A selection of those that are most important are discussed in this section. Some chemical transformations of organometallic compounds arise as a result of just one of these steps, but in other cases, several steps must be combined into a sequence to account for the observed process. Many examples of this are to be found in later chapters, in discussions of both catalytic and stoichiometric processes.

Oxidative addition

If a molecule divides into two portions that become attached as ligands to a metal centre, the electron count and the oxidation state of the metal are both increased by two. For this reason the reaction is termed an oxidative addition; the metal is oxidised during addition of the ligands. The process is illustrated in a generalised form in Scheme 1.1. The two electrons from the R–X bond in (11) both are added to the coordination shell of the metal in (12). The presence of two extra σ bonds in this product accounts for the increase in oxidation state. It is important to note that oxidative addition requires a free coordination site on the metal.

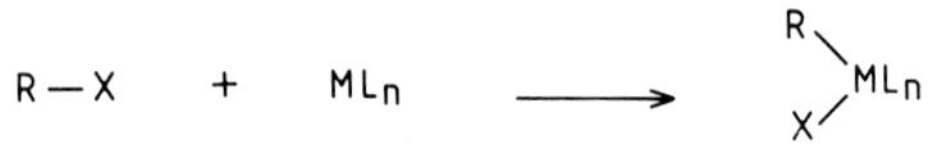

$$R-X \quad + \quad ML_n \quad \longrightarrow \quad \begin{array}{c} R \\ \diagdown \\ \diagup \quad ML_n \\ X \end{array}$$

$$(11) \hspace{8cm} (12)$$

Scheme 1.1

Nucleophilic displacement

An alternative reaction path might result in attachment of only one of the two portions of R–X. This is illustrated by the nucleophilic displacement reaction shown in Scheme 1.2, in which a metal anion displaces X^- to form a metal alkyl bond. Here the electron count in the product (14) is unchanged from that in (13), but the oxidation state is again increased by two. An anion (-1) is replaced by a σ bond $(+1)$. Unlike oxidative addition, this reaction does not change the electron count, but is facilitated by negative charge on the metal.

$$R-X \quad + \quad ML_n^- \quad \longrightarrow \quad R-ML_n \ + \ X^-$$

$$(13) \hspace{6cm} (14)$$

Scheme 1.2

Reductive elimination

Scheme 1.3 depicts a generalised reductive elimination reaction. This is the reverse of oxidative addition, and reduces both the electron count and the oxidation state of the metal. The metal centre is reduced (hence the name) and an organic product is liberated by the covalent union of two ligands.

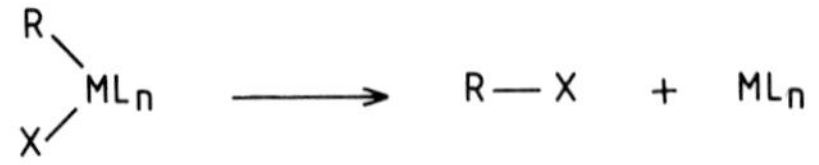

$$\begin{array}{c} R \\ \diagdown \\ \diagup \quad ML_n \\ X \end{array} \quad \longrightarrow \quad R-X \ + \ ML_n$$

Scheme 1.3

β-Elimination

In another process by which organic molecules are commonly detached from a transition metal, hydrogen is removed from a position in the ligand adjacent to the point of attachment of a σ bound metal (i.e. from the '*β*-carbon'). The M–C σ bond in the starting material is replaced by an M–H bond in the product, and an alkene is released. Because the hydrogen is transferred from the *β* position in the ligand, the reaction (illustrated in Scheme 1.4) is called a *β*-elimination. Both (15) and (16) contain σ bonds, so this reaction leaves both the electron count and the oxidation state unchanged. No free coordination site is required, and none is produced.

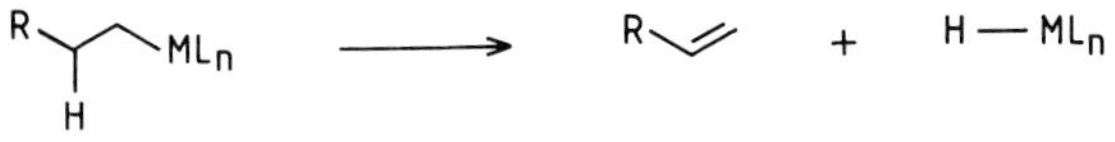

(15)   (16)

Scheme 1.4

Alkyl migration

The best known alkyl migration process is carbonyl insertion (discussed at length in Chapter 9). This reaction is depicted in Scheme 1.5. Again a σ bound ligand is involved, and, as in the *β*-elimination reaction, a new σ bound ligand is introduced into the product. In this case, however, the alkyl group is transferred to a carbonyl ligand (drawn in (17) in a canonical form that emphasises the π bonding between the C=O carbon and the metal). In the product, a metal acyl, the electron count is reduced by two, but the oxidation state is unchanged. This reaction creates a free coordination site at the metal, but is often accompanied by attachment of an additional ligand to maintain coordinative saturation.

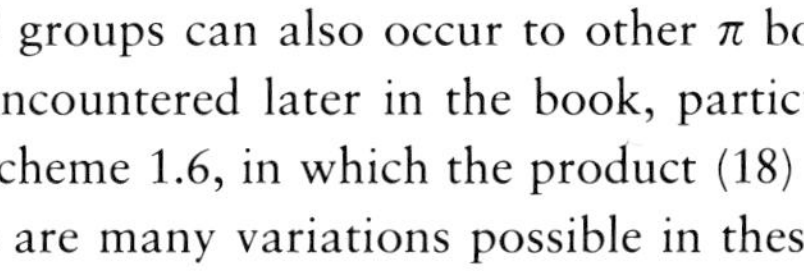

(17)

Scheme 1.5

Migrations of σ bound groups can also occur to other π bound ligands. A number of examples of this are encountered later in the book, particularly in Chapter 10. A typical case is shown in Scheme 1.6, in which the product (18) could now proceed to a *β*-elimination step. There are many variations possible in these reactions.

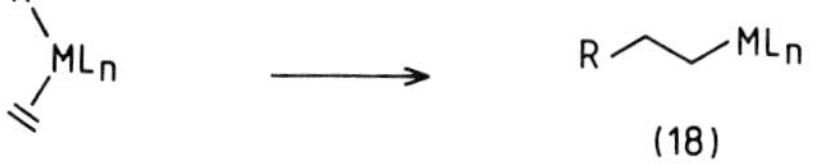

(18)

Scheme 1.6

Nucleophile addition

Another reaction that changes the nature of the organic ligand is shown in Scheme 1.7. Nucleophile addition can occur both with neutral and cationic complexes, but, in either case, the electron count of the metal is unchanged while the oxidation state is decreased. Nucleophile addition to a ligand does not require a free coordination site.

$$Nu^- \quad + \quad ||-ML_n \quad \longrightarrow \quad Nu \diagdown\diagup ML_n^-$$

$$Nu^- \quad + \quad ||-ML_n^+ \quad \longrightarrow \quad Nu \diagdown\diagup ML_n$$

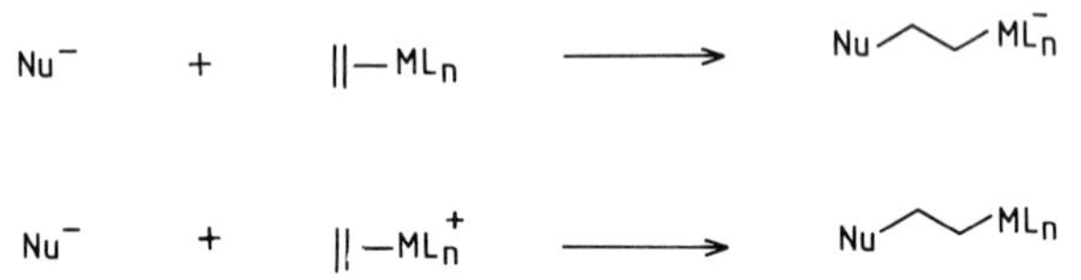

Scheme 1.7

A further possibility, shown in Scheme 1.8, is nucleophile addition to the metal. This does require a free coordination site on the metal, and the electron count on the metal is increased in the product (19). Because a σ bound ligand is introduced, the oxidation state of the metal is not changed despite the fact that the cationic starting material is converted into a neutral product.

$$Nu^- \quad + \quad ML_n^+ \quad \longrightarrow \quad Nu - ML_n$$
$$(19)$$

Scheme 1.8

Electrophile addition

Electrophiles can also add either to a ligand, or to the metal itself. Examples are shown in Scheme 1.9. In the formation of (20), a new σ bond is made to the metal and the electron count is not changed. For (21), the electron count is reduced by two and a free coordination site is formed.

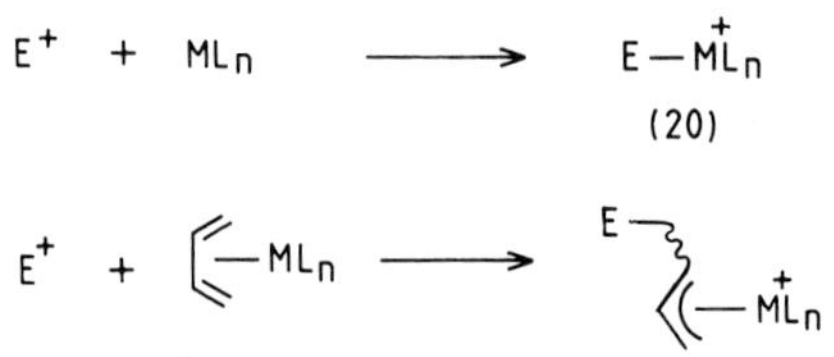

$$E^+ \quad + \quad ML_n \quad \longrightarrow \quad E - \overset{+}{M}L_n$$
$$(20)$$

$$E^+ \quad + \quad \langle -ML_n \quad \longrightarrow \quad E \diagdown \langle - \overset{+}{M}L_n$$
$$(21)$$

Scheme 1.9

Stereochemistry of organometallic reactions

When transition metal organometallic reactions are used in organic synthesis, it is important to consider the stereochemistry of the steps involving changes in the organic ligand undergoing reaction. Alkyl migration, reductive elimination, and oxidative addition reactions typically proceed with retention of configuration of the alkyl group, and *β*-elimination follows a *syn* transition state. These are shown in Scheme 1.10. A more detailed discussion of stereochemistry is reserved for later chapters, where specific examples are examined in depth.

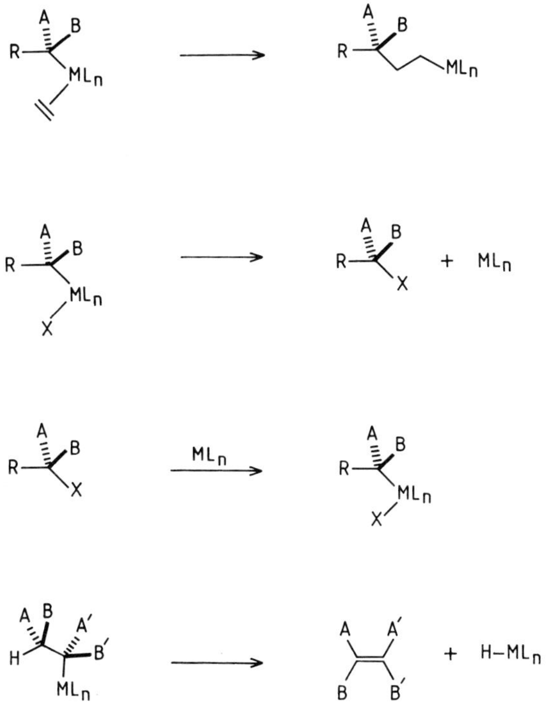

Scheme 1.10

Nucleophile and electrophile addition can introduce substituents to either face of a ligand, depending on the circumstances. Both '*exo*' (22) and '*endo*' (23) products of nucleophile addition are shown to illustrate the possibilities. Direct *exo* nucleophile addition is most common, but *endo* addition is known, particularly when the mechanism of addition involves initial nucleophilic attack at the metal or another ligand. Examples proceeding via attack at the metal are to be found in Chapters 5 and 6.

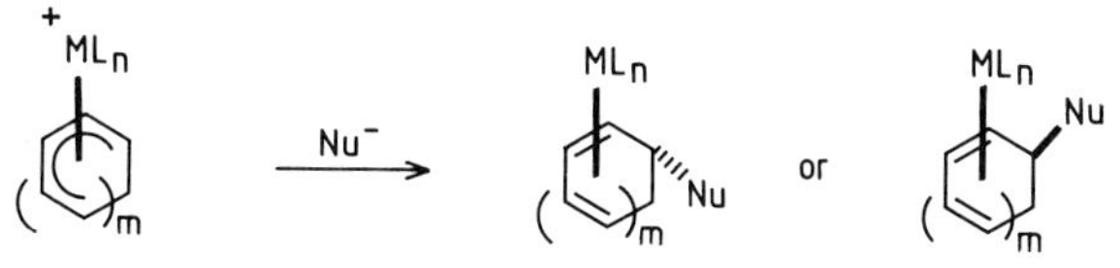

(22)   (23)

Electrophile addition often affords *endo* products. Protonation, in particular,

normally proceeds *endo*, after initial addition to the metal to form the metal hydride
intermediate shown in Scheme 1.11.

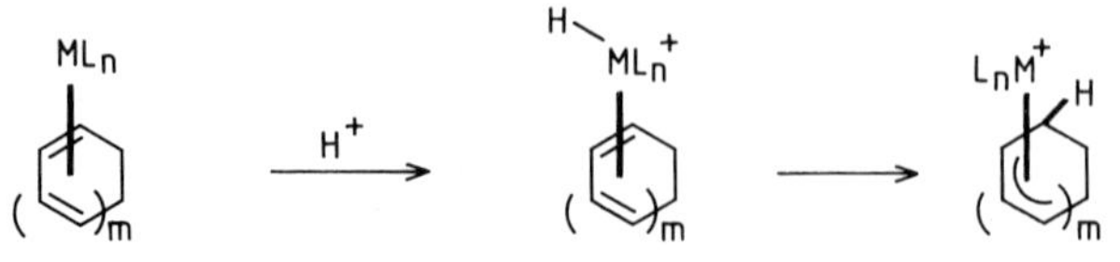

Scheme 1.11

1.5 Agostic bonding and fluxional and non-rigid complexes

Agostic and multicentre bonding

In some cases the bonding in organometallic species may not be as straightforward as
suggested in the examples above. An *endo* hydrogen, for example, may span the ligand
and the metal, being partially bound to each, as shown in (24). Similarly, the *syn*
alignment of β-elimination may present a hydrogen which can again span the gap
between the ligand and the metal.[8] This is shown in (25). Hydrogens bridging in this

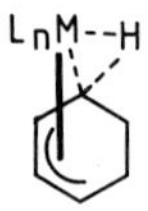

(24) 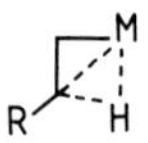   (25)

way have been termed 'agostic',[9] a word drawn from Homer's Greek expression which
translates as 'to clasp oneself'. Examples of this type of structure arise in quite simple
reactions. For example, the reduction of the manganese complex (26), or the
protonation of the cobalt complex (28), result in the products (27)[10] and (29)[11] in which
agostic hydrogens span the metal and carbon centres. An agostic bond is also proposed

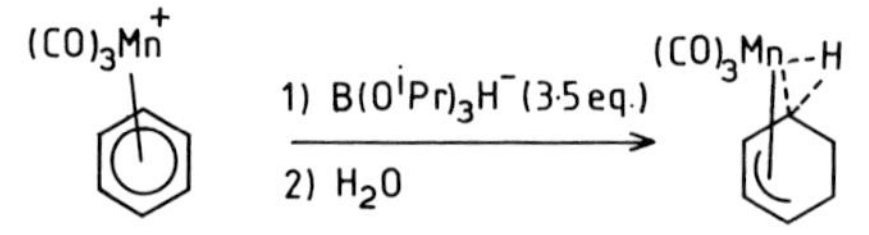

(26)   (27)

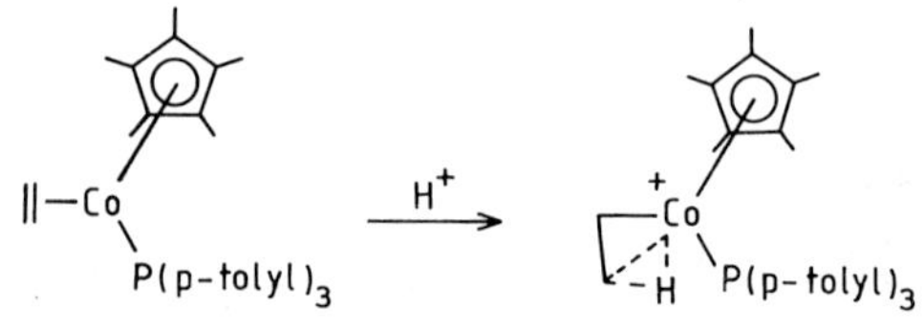

(28)   (29)

for an intermediate in the equilibrium between the η^1 metal alkyl complex (30) and the η^2 alkene π complex (31).[12] This offers a good example of an intermediate stage in β-

(30)   (31)

+ PMe$_3$

elimination. Such situations can also arise in the formation of larger rings, when the circumstances are correct. In complex (32), there is a hydrogen that occupies an intermediate position on the way towards orthometallation.[13]

(32)

In each of these situations, the hydrogen occupies a position in which it is completely bound neither to carbon nor to the metal, but, rather, is partially bound to both centres. Although this situation has been named agostic when a transition metal is involved, it is similar to the phenomenon of three centre bonding by hydrogen, now familiar in the chemistry of boranes. Similar three centre bonding is also found in molecular hydrogen, $M-(H_2)$ complexes described in Chapter 10.

When other atoms bridge between two centres, the situation looks less strange. Bridging carbonyl groups are common in organometallic chemistry,[14] and in some cases bridging can be distorted towards a side-on interaction to one of the metal centres.[15] Anions derived from acyl ligands (see Chapter 9) can also sometimes adopt side-on bonding.[16] Alkenes themselves span several centres when bound to transition metal clusters[17] and in bimetallic complexes such as (33).[18] Even in simple dicobalt alkyne complexes of type (34), discussed in Chapter 5, multicentre bonding provides the best model to understand the interactions between the ligand and the metals. A few examples[19, 20] of reactions involving multicentre bonding are shown in Schemes 1.12 and 1.13. Some of the reactivity properties of ligands spanning bimetallic,[20] and higher

(33)   (34)

nuclearity systems,[17] indicate potential for organic transformations at present unapplied in target oriented organic synthesis.

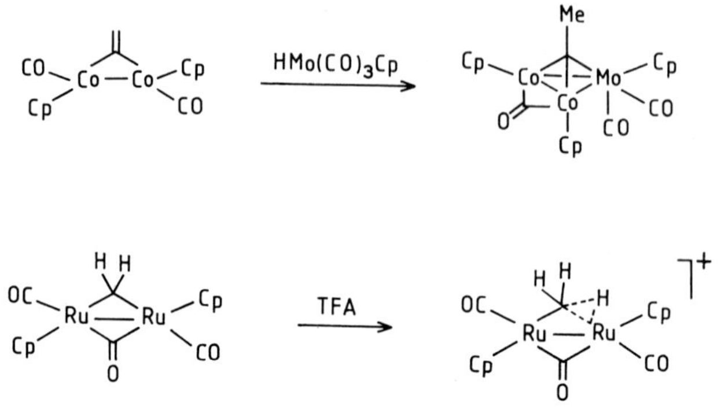

Scheme 1.12

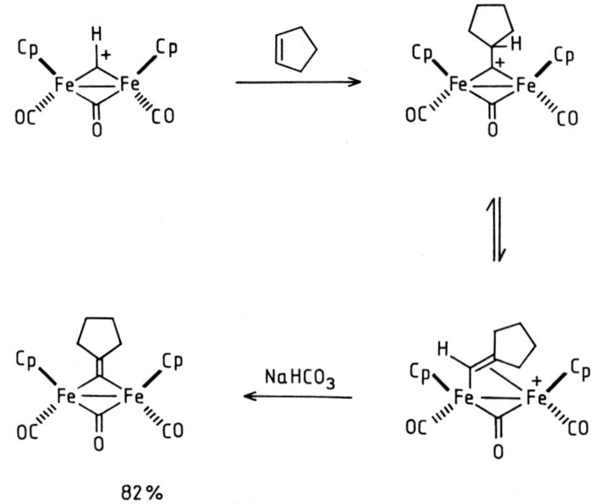

82%

Scheme 1.13

Of all of these cases, it is agostic bonding that has the greatest impact on the description of organometallic mechanisms. Although starting materials and products of a transition metal mediated reaction may be well defined, and plausible mechanisms available, the details of bonding in reaction intermediates participating in the transformation may not always be as simple as might at first be anticipated.

Fluxional and non-rigid structures

In organometallic chemistry, the interconversion of regioisomers can be a facile process. Some compounds interchange rapidly between forms, often showing averaged n.m.r.

spectra at normal temperatures. These complexes are designated non-rigid, and, in cases where all interconverting forms are chemically and structurally equivalent, the phenomenon is called fluxionality.[4,21] Such properties not only complicate the interpretation of n.m.r. spectra of organometallic species, but can also be important in assessment of chemical reactivity, since an apparently simple material can, if non-rigid, adopt many forms in a chemical reaction. The interconverting species may be related by changes in attachment of ligands in bimetallic complexes or in larger clusters,[22] or may reflect a rearrangement of the position of binding of a single metal between different sites on an organic ligand.[21] A classic example of this latter type of reactivity is the fluxionality of η^1-dienyl complexes such as (35), discovered early in the development of organometallic chemistry.[23]

(35)

There are many variants of this type of process. The movement of a π bound metal in a complex, where the unsaturated portion of the ligand is only partially bound to the metal, is of particular importance in organic synthesis, since the metal, and with it the driving force for many chemical reactions, is mobile in these circumstances. Such isomerisations may occur rapidly in non-rigid complexes,[24] or quite slowly in other situations,[25] and are referred to as haptotropic rearrangements[26] when the metal moiety changes its connectivity (η number) to its ligand. The examples of non-rigid, fluxional, and haptotropic rearrangements in iron complexes, shown in Scheme 1.14, have been selected because iron carbonyl complexes will be discussed at length in Chapters 2 and 7. This type of reaction, however, is in no sense unique to iron, and is found with most metals. The examples in Scheme 1.14, however, give an indication of the range of ΔG values for reactions of this type. Comparison of the cyclooctatetraene and cyclo-heptatriene ligands, for example, shows a particularly marked effect.

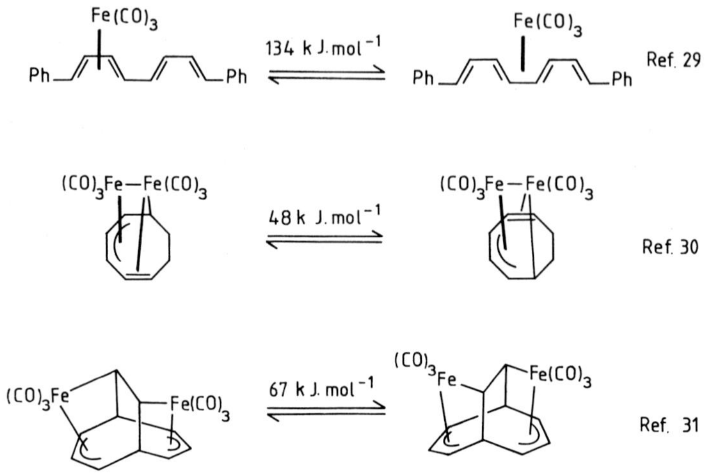

Scheme 1.14

While many organometallic complexes are structurally static and straightforward to deal with, it is important to keep in mind that, in some circumstances, a metal centre in an organometallic synthetic intermediate may not necessarily remain where it is first placed.

An important effect in rearrangements involving changes in η^n for the metal centre arises as a consequence of the creation of a free coordination site during the rearrangement process. A number of reactions discussed in Section 1.4 require a free coordination site, and so can be initiated in this way. Particularly in catalytic processes, the availability of a coordinatively unsaturated intermediate is essential to allow the reaction substrate to interact directly with the metal. Many catalytic reactions employ rhodium or palladium catalysts, which, being on the far right hand side of the periodic table, commonly adopt 16 electron coordination modes and so naturally have a free site to bind a substrate as a ligand. In other situations, π to σ rearrangements create a free coordination site. This can assist the start of a catalytic cycle.[32] A variation on this theme involves the indenyl ligand which is particularly suited to $\eta^5 - \eta^3$ rearrangements. An intermediate η^3 form has recently been intercepted and identified.[33] These processes are illustrated in Figure 1.7.

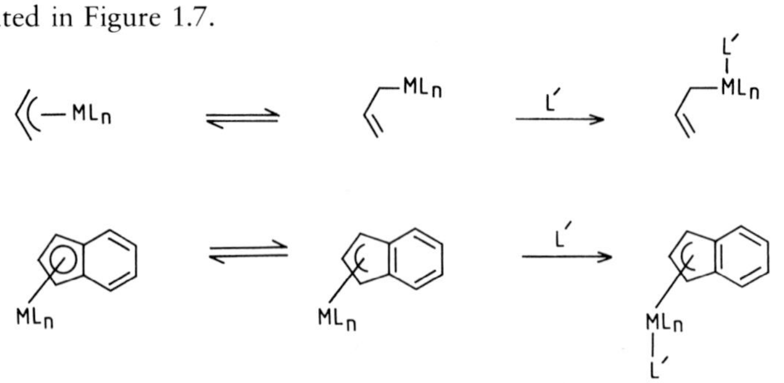

Figure 1.7

Distinction between multicentre bonding and non-rigidity

Non-rigidity refers to a dynamic process, while multicentre bonding exists as a static phenomenon. When compounds interconvert, the reaction can be depicted schematically on an arbitrary reaction coordinate which plots the energy barrier between the two components (Figure 1.8). At the transition state (the top of the barrier), the bonding takes an intermediate form between product and reactant, but this structure is only a transient form, reflecting the situation at the instant of maximum energy. If a complex has multicentre bonding (Figure 1.9), then the intermediate form is stable (or

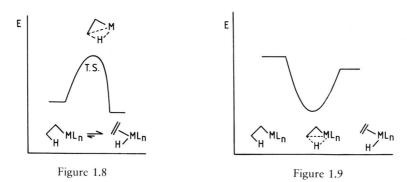

Figure 1.8   Figure 1.9

metastable) with respect to other forms, and is at a minimum on the energy plot. In some reactions, isomers with agostic bonding look reasonable as intermediates in the interconversion, as, for example, in the oxidative addition reactions of H_2 to be discussed in Chapter 10. A reaction involving multicentre bonding in an intermediate is depicted in Figure 1.10. Here the bonding takes on an intermediate form, but this occurs at a local minimum in terms of energy. Since there is an intermediate in the reaction, there are two transition states; one between the starting material and the intermediate, and a second as the reaction proceeds to form the product.

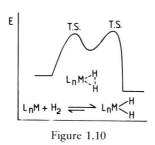

Figure 1.10

At present, organometallic mechanisms are discussed in a rather imprecise fashion by linking series of well known steps in a reasonable way. Only rarely are these mechanisms founded on the firm basis of kinetic and labelling experiments that underlies many purely organic reaction mechanisms. As the detailed investigation of organometallic complexes proceeds, the fine distinctions of bonding in different types of reaction intermediates will become of greater significance. At present, however, such considerations lie beyond the scope of a book intended to assist organic chemists to employ transition metal complexes in organic synthesis. The topic is covered in far greater detail elsewhere.[34] We do, however, require a rather more complete discussion of bonding in organometallics than could be presented in Sections 1.1 and 1.2, to help readers follow discussion of reactivity properties in later chapters. This is provided in the next section.

1.6 The description of bonding in organometallic complexes

In principle, the bonding in transition metal complexes is no more difficult to understand than chemical bonds in purely organic molecules. A chemical bond exists between two atoms when orbitals overlap to form bonding orbitals with energies low enough to make occupation by electrons a reasonable possibility. This is no less true for the tetrahedral array of carbon monoxide ligands around the central nickel atom of nickel tetracarbonyl (36) than for the four hydrogen atoms in methane (37). The

(36)   (37)   (38)

difference lies in the greater complexity of bonding interactions that are available in the case of (36). Even here, the initial concepts are simple. The complete set of s, p, and d orbitals of the nickel atom, and those orbitals of the ligands that are directed towards the central position, are combined to produce a set of bonding orbitals and a set of anti-bonding orbitals. The total of the electrons belonging to the metal outer valence shell and the appropriate orbitals of the ligands (in this case two electrons each for the four carbonyl groups), must then be accommodated in the bonding orbitals of the complex. This simple picture, unfortunately, is also rather uninformative. It does not explain why nickel tetracarbonyl (36) has tetrahedral geometry rather than the square-planar form of Zeise's salt, why titanium alkoxides have polymeric structures with alkoxide ligands spanning two metals, or why trigonal bipyramids are often preferred to square-based pyramids. Indeed computational techniques do not at present permit the a priori calculation of the structure and reactivity properties of even simple molecules by the complete analysis of all interactions between all orbitals available within the coordination sphere of the metal. Instead, bonding in specific regions of the molecule, for example, the Ni–CO bond itself in (36), are tackled in isolation by considering the overlap of selected d orbitals on the metal with appropriate orbitals on the organic ligands. This can provide considerable insight into the nature of the metal ligand bond without impossibly detailed calculations.

Orbital interactions

The distinction between σ and π bonds is familiar to organic chemists and has been introduced in the context of organometallic bonding in Section 1.1. Its importance is just as profound in consideration of the bonding and stability of organometallic complexes as it is in the discussion of the organic chemistry of alkenes and alkynes. To form a chemical bond, interacting orbitals must be of the same symmetry, of similar energies, and directed into the same region in space.[35] The distinction between σ and π bonds, and indeed δ bonds between metal atoms, arises from the different symmetries of orbitals that can combine in bonding combinations. This was illustrated briefly by the examples (a) and (b), shown in Figure 1.4, which depicted two possibilities for the overlap of atomic orbitals on a metal (M) and a ligand (X). Further discussion of this point is appropriate here. In the case of σ bonding, (a), both orbitals are aligned along the internuclear axis and, in the bonding combination, both show the same symmetry label (+). This produces a σ bond symmetrically placed along the M–X axis. Combining the M and X orbitals in the opposite sense (+ combined with −) would produce an anti-bonding (σ^*) orbital. In the case of π bonding, (b), two lobes of an orbital on the metal and two more on the ligand X are shown again in positions that allow an additive combination. Although this is a bonding interaction, the orbital produced is anti-symmetric with M and X lying on a nodal plane, a region where the wave function for the orbital is zero. This forms a π bond, which, like the corresponding (π^*) anti-bonding orbital, is distinct from a σ bond in terms of the symmetry of the orbital. This description of π bonding is equally valid when the lobes of the orbital of the metal atom drawn in Figure 1.1 are part of a metal p or d orbital or when part of a metal hybrid orbital. Similarly for X, the lobes may belong to an atomic orbital (perhaps a d orbital on a phosphine) or be part of the molecular orbital set of the free ligand, as is the case for carbon monoxide. Indeed, the use of hybridisation to obtain a set of appropriate metal atomic orbitals, directed conveniently to suit the ligand geometry, is just as convenient in simplifying the depiction of bonding in transition metal chemistry as it is in forming the simple bonding picture describing methane. The use of hybridisation in this is discussed in more detail shortly.

While the σ bonded framework of a metal complex usually involves orbitals that, in the free ligand, are already occupied, for π bonds the ligand orbitals may be either filled or empty depending on the circumstances. It is enlightening in this context to compare the two organometallic complexes, nickel tetracarbonyl (36) and titanium tetraisopropoxide (38), encountered above. A carbon monoxide ligand in (36) directs a filled sp orbital towards the central atom. An empty π^* orbital belonging to the carbon–oxygen triple bond is also aligned correctly to overlap with orbitals on the metal (see Figure 1.1). The oxygen of the isopropoxide ligand could be envisaged with an sp lone pair but also with p orbitals. Although the π^* orbital of carbon monoxide in (36), and the p orbital in (38), both have the correct symmetry for a π interaction, the π^* orbital in (36), being empty, can accept electrons from the metal, while in (38), the p orbital, which is filled, would add two further electrons to the valence shell. Here we encounter a fundamental difference between these two ligands. The carbon monoxide ligand is a 'π acceptor', while alkoxide ligands are not. Indeed, the latter can donate electrons to the metal when required. The carbon monoxide ligands assist the stabilisation of electron-rich complexes such as (2) or (36) by accepting electrons from the metal, which, in these cases, is in the zero oxidation state. In (38), the metal is in a high oxidation state, Ti(IV), and electron rich alkoxide ligands are appropriate.

The original theories describing bonding in organometallics, particularly the depiction of the alkene–metal bond for (3) in Figure 1.1, are now referred to as the Chatt, Dewar, Duncanson model.[36] This description places heavy emphasis on the donor/acceptor properties of ligands. In the case of an ethene or carbon monoxide complex, a filled ligand orbital can be regarded as interacting with an empty metal orbital, forming the σ bond and transferring electron density to the metal in the process. The π bond originates from interaction between an empty ligand π^* orbital and a filled metal d orbital, and accepts electron density back from the metal into the π system. While this effect can be clearly seen reflected in bond lengths obtained from crystal structure data,[37] or, particularly in the case of carbon monoxide,[38] in the i.r. stretching frequency of the carbonyl group, where weakening of C≡O bonding is consistent with the filling of an anti-bonding orbital, descriptions that can take into account the interactions of a greater range of orbitals provide a more generally applicable model. None the less, the concept of π acceptor ligands is valid and important. Low oxidation state metals often form stable complexes with π acceptor ligands because of the opportunity they provide to lessen the electron density on the central atom.

Interaction and correlation diagrams

Interaction diagrams are now common in organometallic chemistry.[39] These diagrams depict the interactions between orbitals on the metal and the ligand, indicating a set of new molecular orbitals for the complex as a whole. These will then accommodate the valence electrons. The set of metal hybrid orbitals developed by Hoffmann[40] to describe the bonding possibilities of metal ligand fragments are the most widely used for the description of bonding by transition metal moieties.

A simplified form of an interaction diagram using hybrid orbitals for the metal fragment is shown in Figure 1.11. This describes qualitatively the bonding between the common cyclopentadienyl ligand (C_5H_5) and the dicarbonylcobalt fragment.[41] At one side of the diagram (the right hand side in this case) are shown the hybrid orbitals of CoL_2, available to bind to the cyclopentadienyl ligand. These are represented by horizontal lines, accompanied by drawings that illustrate the main features of the corresponding metal hybrid orbital in a way that emphasises symmetry and orientation. On the opposite side of the diagram, the π orbitals of the cyclopentadienyl ligand are

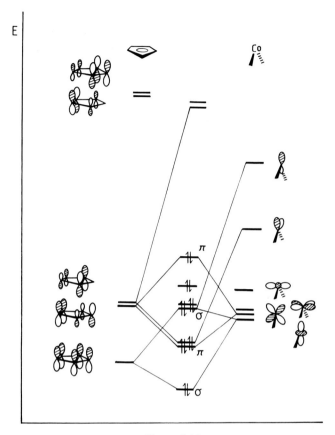

Figure 1.11

portrayed. Note that there are five in all (five p orbitals contribute to this orbital set) and that two, the two anti-bonding orbitals, are a great deal higher in energy than the other three. Possibilities for interactions between all these orbitals must be considered, but, since only orbitals of the same symmetry can interact, a great many combinations can be ruled out at once. The most significant interactions will be between orbitals of the same symmetry and similar energies. Where the energy difference is large, as, for example, with the cyclopentadienyl anti-bonding orbitals, involvement in interaction with bonding orbitals is minimal. The results of these deliberations are depicted in the centre of the diagram, where the molecular orbitals of the complex are shown. The spacing between these levels, shown qualitatively in the figures, can be calculated from the degree of orbital interaction in each case. Note that two of the metal orbitals do not encounter significant interactions with the ligand, and appear at roughly the same energies in the central column. The valence-shell electrons are now added to the diagram; in this case, five from the cyclopentadienyl ligand and nine from cobalt, making a total of 14. From this it can be seen that all the low lying orbitals in the bonding system are filled.

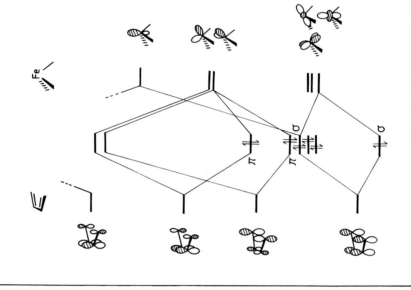

Figure 1.12

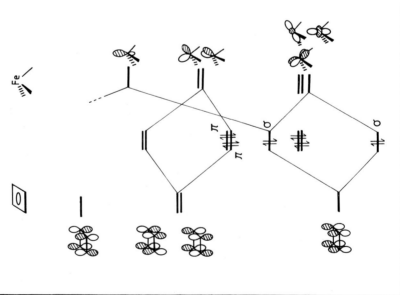

Figure 1.13

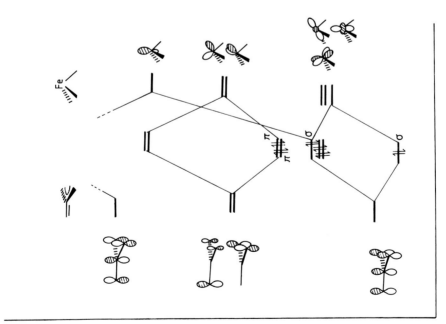

Figure 1.14

Figure 1.15

For a particular metal/ligand combination, the pattern of molecular orbitals obtained will depend on the nature of the orbitals of the ligand. This can be seen clearly in the diagrams in Figures 1.12, 1.13, and 1.14 (not drawn to scale), which show the bonding in three different η^4 tricarbonyliron complexes. Cyclobutadiene,[42] butadiene[43] and trimethylenemethane[44] ligands all form Fe(CO)$_3$ complexes, but have quite different π orbitals. Thus, while the right hand side of each of the interaction diagrams remains constant since it describes the Fe(CO)$_3$ moiety which is common to all three, the orbital sets on the left vary. This results in corresponding changes in the molecular orbitals of the complexes.

When complexes of Cr(CO)$_3$ are considered, different interactions are found since a total of six π orbitals are present on an arene ligand. The interaction diagram for a chromium arene complex is shown in Figure 1.15. The Cr(CO)$_3$ unit has two less electrons than the iron fragments shown in Figures 1.12 to 1.14, but the arene compensates with two more, so that the total number of electrons available to occupy the bonding orbitals remains unchanged. Comparing these figures will show, however, that the orbital sets for the Fe(CO)$_3$ and Cr(CO)$_3$ fragment are the same. In contrast, metals with different geometries, or with different numbers of ligands, will have different orbital sets; a full derivation of the form of these orbitals is set out in an extensive series of papers by Hoffmann,[26,40,45,46] and a recent review.[39] Other molecular orbital treatments,[47] too, can also offer important insights. In cluster compounds, in particular, far greater complexity arises in the procedures for electron counting and description of bonding, and different methods are required.[48] Whatever the nature of the complexes under discussion, however, it is the molecular orbital description of bonding

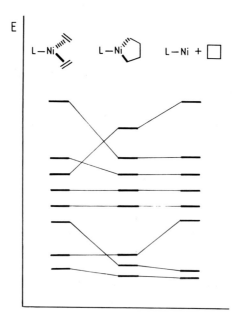

Figure 1.16

that underlies modern thinking about the nature of organometallic complexes and their reactions.

Correlation diagrams show the changes in bonding taking place during a chemical reaction, or sometimes, during a rotational, translational, or conformational change in a molecule. A typical example[49] is shown in Figure 1.16, in which differences between two ethene ligands, a tetramethylene metallocycle, and reductive elimination product, are illustrated and compared. Diagrams of this type can be used to probe the course of chemical reactions. In this case, the cross-over of orbital correlations of the left hand side of the diagram indicates that the coupling of the ethene ligands is not an allowed process. Reductive elimination, on the other hand, would be expected to proceed.

Isostructural, isoelectronic, and isolobal complexes

Terms that facilitate the comparison of organometallic complexes provide convenient devices to make fine distinctions about bonding. 'Isostructural' and 'isoelectronic' are fairly simple concepts, but the recently introduced idea of the 'isolobal' analogy is more difficult to interpret. 'Isostructural complexes' have the same structural form. They may all be octahedral (e.g. $V(CO)_6$, $Cr(CO)_6$, $Mo(CO)_6$, and $W(CO)_6$), for example. 'Isoelectronic' relates to the electron count of an organometallic fragment; $Fe(CO)_3$, $Mn(CO)_2NO$, $CoCp$, and $Co(CO)_3^+$ will all form 18 electron complexes with dienes. These fragments themselves each have 14 valence electrons, and so are 'isoelectronic', a useful definition because of its power to suggest replacement metal/ligand systems for a particular purpose.

The term 'isolobal' relates to the frontier orbitals of the metal ligand fragments. Fragments that have a similar number of frontier orbitals possessing similar symmetry properties and energies, and that contain a similar number of electrons in the fragment, are called isolobal.[46,50] It is important to note that the orbitals need not be exactly the same. The intention of the definition of isolobal is to allow the construction of wide-reaching analogies spanning both organic and organometallic chemistry.[50,51] Thus $Mn(CO)_5^.$ and $CH_3^.$ are isolobal, each having a single frontier orbital with a single electron. Clearly they are neither isostructural, nor isoelectronic, but are, none the less, comparable in terms of their bonding capabilities with other ligands.

1.7 Roles for the metal centre in organic synthesis

In this introductory chapter the reader has been presented with the background information that will be needed to make sense of the great variety of reactions mediated by transition metals. From basic ideas of electron counting, bonding, and organometallic reaction mechanisms, we have progressed to more advanced topics, introducing less common forms of bonding and more modern bonding theories. It should now be clear that ligand interactions with transition metals can determine the orientation between groups, and the types of chemical transformations that may occur. Different metal ligand systems can have quite different properties, and it is reasonable that there should be corresponding differences in the detailed nature of the chemical bonds that they form. It is the diversity of combinations between organic ligands and differing metal ligand systems that opens up a wealth of specific, well controlled, and now increasingly well understood, chemical reactions. The stage is set for a detailed discussion of the uses of organometallic reactions of transition metal complexes in organic synthesis.

2
Isomerisation and rearrangement reactions

2.1. Isomerisation of alkenes

When transition metal complexes interact with organic compounds that are unsaturated or strained, rearrangement reactions frequently result. A simple example of this, at least from the point of view of the organic transformations involved, is the movement of the position of an alkene by hydrogen migrations induced by the metal. Reactions of this type have useful synthetic applications and have been studied extensively over the last 30 years. Some typical examples are shown in Scheme 2.1.

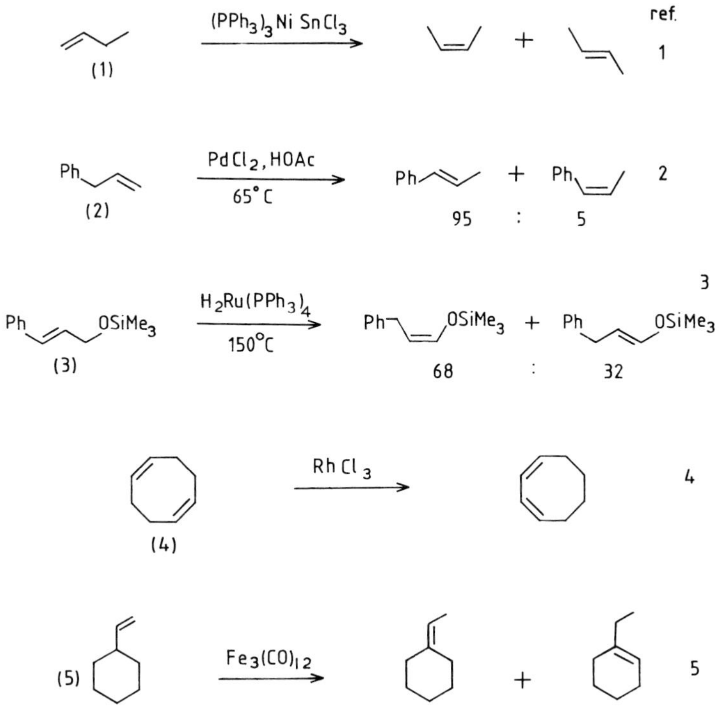

Scheme 2.1

The isomerisations of (1), (2), (3), and (4) make use of complexes with the metal in a positive oxidation state. In the case of (5), the iron carbonyl complex is in the zero oxidation state. In appropriate structures, a terminal alkene is converted into the *cis* and *trans* in-chain isomers. Since initially the proportion of the *cis* form may frequently far exceed the thermodynamic equilibrium amount, the choice of reaction time can be important if stereoselectivity is to be achieved.

2.2 Isomerisation mechanisms

Consideration of the nature of the reagents indicated in Section 2.1 provides a clue to the mechanisms of the processes involved. Two distinct general mechanisms have been proposed depending on the type of catalyst employed as shown in Scheme 2.2. The first, (A), involves addition/elimination reactions of metal hydride complexes.[6] The metal hydride may be present initially in the catalyst as in the isomerisation of (3), or may arise by reactions with, for example, an alcohol solvent, as is the case with (4). Mechanism

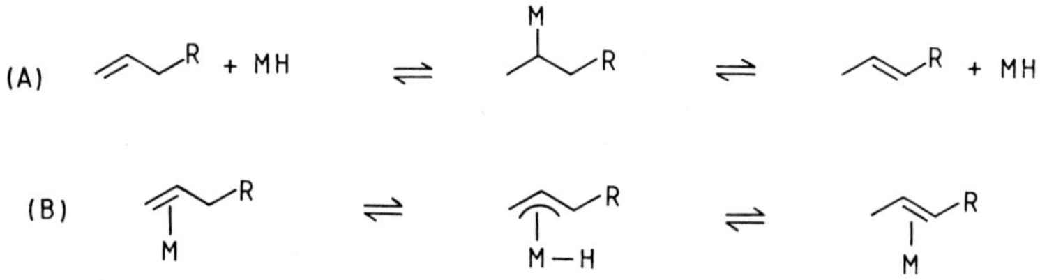

Scheme 2.2

(B) proceeds by coordination of the alkene as a π complex prior to the formation of the M–H bond which arises by hydrogen transfer creating a π-allyl metal hydride intermediate.[5,7,8] Hydrogen transfer to a new position in the ligand accounts for the isomerisation and has been confirmed[9] in some cases by studies using deuterium labelling techniques. In some circumstances the C–H bond adjacent to the alkene interacts with the metal without completing hydrogen transfer.[10] 'Agostic' interactions of this type have been described in Chapter 1 and are discussed again in more detail in Chapter 10. Although superficially similar, there is an important distinction between the two processes. The formation of the allyl intermediate involves a formal increase in oxidation state. Indeed, catalysts operating by mechanism (B) are frequently those starting in low oxidation states. In mechanism (A), on the other hand, the oxidation state remains constant since the M–H σ bond is replaced by a M–C σ bond.

Most isomerisations appear to proceed via mechanism (A), although the π-allyl metal hydride mechanism has been shown to be involved in iron carbonyl catalysed isomerisations,[11] as in the example shown in Scheme 2.3, and has been invoked for some

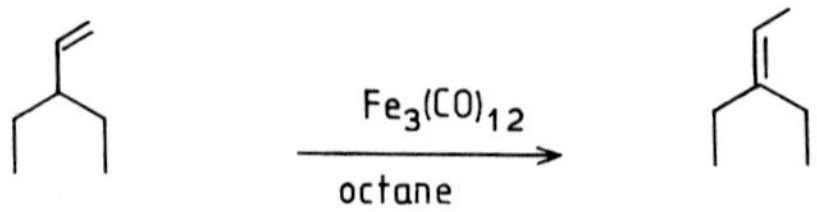

Scheme 2.3

palladium catalysed isomerisations. It should, however, be noted that deactivation of palladium catalysts can arise by formation of stable π-allyl palladium complexes.[12]

2.3 Alkene isomerisation in synthesis

One of the more widely used reagents for alkene isomerisation is rhodium chloride ($RhCl_3 \cdot 3H_2O$) ethanol.[12] The Rh(III) species coordinates the alkene bond and reaction proceeds as in mechanism (A) above. Bertele and Schudel used $RhCl_3 \cdot 3H_2O$, followed by treatment with $Fe(CO)_5$ to convert the β-sinensal skeleton of (6) into all *trans-α-*

sinensal (7).[13] $RhCl_3 \cdot 3H_2O$ has also been used to isomerise *trans-β-farnesene* (8) to *trans,trans-α-farnesene*, (9). The smooth transformation of 3, 7-dimethylocta-1,6-diene (10) to *cis-* and *trans-3,7-dimethylocta-2,6-dienes* (11, 12) was effected by $RhCl_3 \cdot 3H_2O$.

The mole ratio of catalyst to olefin was 1:10.[14] The same isomerisation was observed with acetone as solvent. However, in this case, prolonged reaction yielded the

cycloalkenes (13) and (14). Andrieux, Barton and Patin have illustrated the use of the reagent to effect reactions that were otherwise difficult or impossible.[15] $RhCl_3 \cdot 3H_2O$ in aqueous ethanol/chloroform quantitatively converted the benzylidene chromanone (15)

into 3-benzylchromone (16). Similarly, the benzylidene tetralone (17) gave 2-benzyl-1-naphthalenone (18). A further $RhCl_3 \cdot 3H_2O$ catalysed exocyclic–endocyclic double bond rearrangement allowed the efficient synthesis of 2-substituted cyclopentenones which are useful intermediates in the synthesis of many natural products (Scheme 2.4).

R = Ph : R = $(CH_2)_nCH_2$, n = 2,3

Scheme 2.4

In a similar study[16] of a spirocyclic system, the preferred product was unexpectedly found to be the debenzylated exocyclic alkene (19), a result attributed to $A_{1,2}$ strain in this particular system.

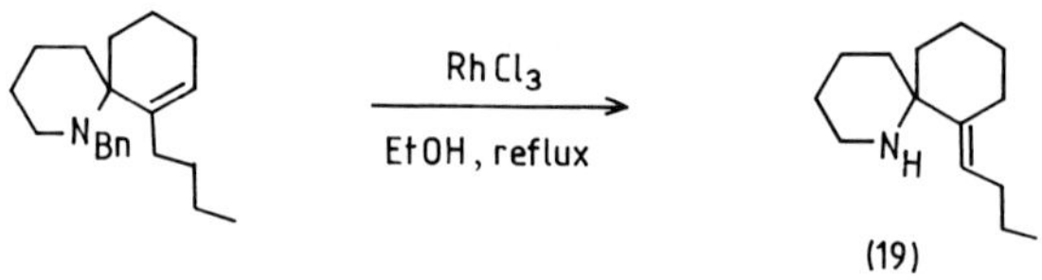

The driving force for such rearrangements is sometimes readily apparent when, for example, remote double bonds are brought into conjugation. Thus treatment of carvone (20) gave the aromatic phenol (21), whilst analogous treatment of the unsaturated imine (22) yielded the substituted aniline (23).[17] In this last instance, an addition of potassium

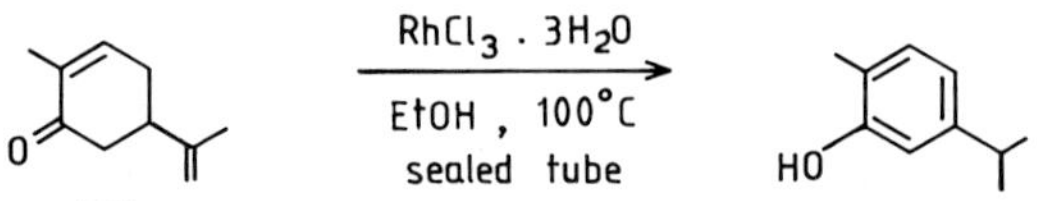

carbonate was found necessary to prevent hydrolysis of the imine to the ketone. In other cases the product may contain a more highly substituted alkene, so providing a thermodynamic driving force to bring the reaction to completion. In another example where, again, a clear thermodynamic driving force can be seen, an alkyne is isomerised to a 1,3-diene. A ruthenium hydride catalyst was used in this case.[18]

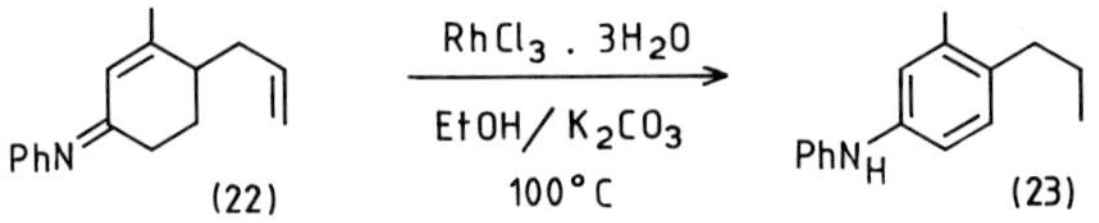

It would be misleading, however, to give the impression that it is the stability of the final alkene that must inevitably determine the outcome of such rearrangements. Work at E.I. du Pont related to adiponitrile formation has explored an example where the thermodynamically most stable alkene is clearly not the major product. In this study of the isomerisation of 3-pentenenitrile, McKinney[19] has shown that the terminal alkene (24) was formed in preference to the conjugated product.

The ratio of the 3- and 4-pentenenitriles, however, never exceeds the equilibrium value suggesting that, while metal complexes leading to the 2-isomer are kinetically inaccessible, the observed formation of the 4-isomer otherwise reflects normal thermodynamic influences. This example brings home the point that the reactivity properties of intermediate complexes can have a profound influence on the selectivity of processes of this type.

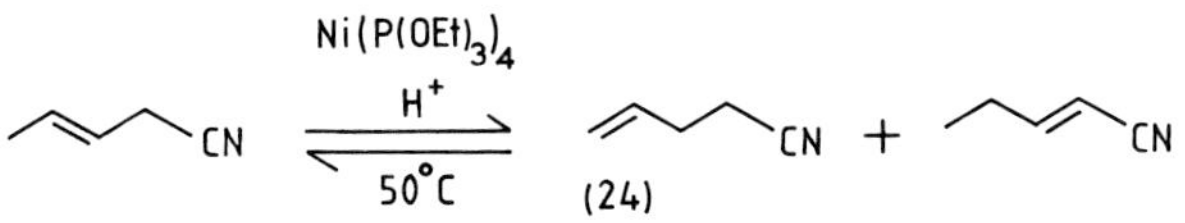

Birch's investigation[20] of the utility of Wilkinson's catalyst, $RhCl(PPh_3)_3$, as an isomerisation catalyst, provides an example where the product obtained is different to that formed by the action of strong base (Scheme 2.5). In many cases, this method effects the important conversion of 1,4-dienes to equilibrium mixtures of 1,3 and 1,4-dienes.

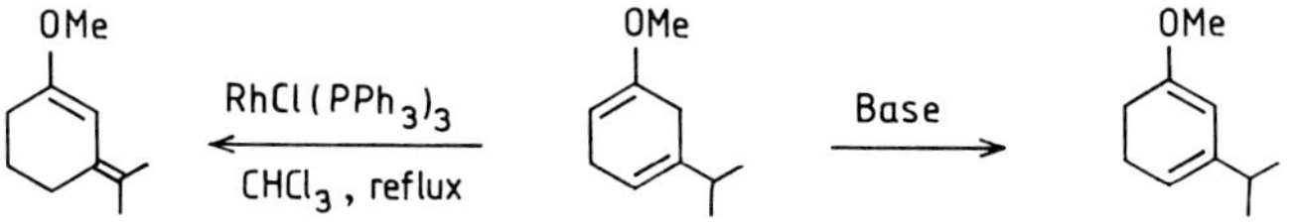

Scheme 2.5

The reactions discussed so far have highlighted transition metal catalysed double bond migrations. Many of these reactions are also accompanied by *cis–trans* isomerisation. A good example where *cis* to *trans* isomerisation is specifically effected involves reaction of the end product of a synthesis of vitamin A acetate (25) which used $PdCl_2(PhCN)_2$ in acetonitrile solution to yield the all *trans* form (26).[21]

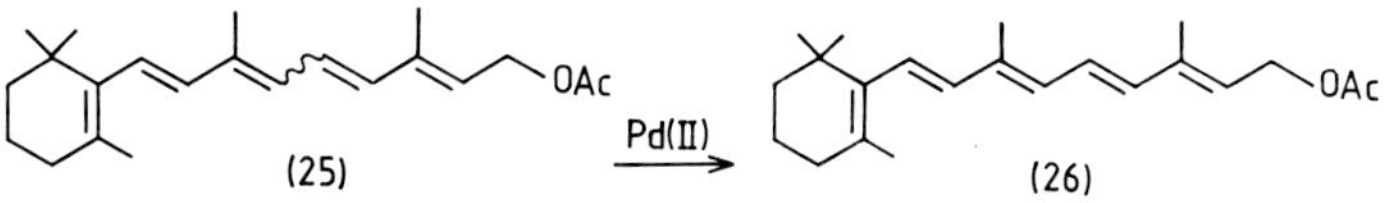

Synthetic applications

The rigours of synthetic work frequently give rise to the requirement to adjust double bond locations at an appropriate stage. Transition metal catalysis may provide a selective way to achieve this, as seen in a synthesis of $(\pm)$-hirsutene in which (27) was conveniently converted into the more fully substituted alkene (28).[22]

Type (B) catalysts have also found synthetic application. Palmisano's synthesis[23] of $(\pm)$-deplancheine used $Fe_2(CO)_9$ to form the α,β-unsaturated ketone (29).

The two types of catalyst can sometimes give different results. By the action of rhodium chloride, ergosterol (30) was transformed over 30 hours at 70 °C to coprostatrienol (31) 25%, ergosterol B1 (32) 30%, and ergosterol B2 (33) 45%.[15] Barton

subsequently found that use of chromium hexacarbonyl promoted the formation of only one product from ergosterol acetate (30, acetate) namely, ergosterol B2 acetate (33, acetate) in 81% yield.[24]

Iron pentacarbonyl is also an effective catalyst for double bond isomerisations. A notable example is provided by the conversion of $(-)$-β-pinene (34) into $(-)$-α-pinene (35)[25] with an optical yield of 97%.

In cases where a second alkene is present in the molecule, the possibility of the formation of a stable $Fe(CO)_3$ complex arises (see Chapter 7). Oxidative removal of the metal as a separate step can render this method useful for the formation of *cis* dienes, so directly exploiting the bonding requirements of the intermediate complexes. Early work by Alper[26] employed this technique for the formation of the new *cis* steroidal diene (36) from 3-methoxycholesta-3,5-diene.

Many 1,4-dienes, obtained by Birch reduction of the corresponding aromatic, have been converted into 1,3-diene complexes. The isomerisation of (37) illustrates[7] this route, as well as indicating the 1,3-hydrogen transfer.

An interesting application of this method is found as a stage in a synthetic entry to compounds in the prostaglandin C series.[27] Reaction of the intermediate (38) with $Fe_3(CO)_{12}$ in DME at 95 °C gave the $Fe(CO)_3$ complex (39) in which the alkene bonds have been brought into conjugation. As a potential aldehyde, (39) could be converted

by a Wittig reaction into (40), which by oxidation (CrO_3) afforded (41). The reaction preceded with concomitant oxidative removal of the $Fe(CO)_3$ residue.

Optically active metal complexes may result from hydrogen migrations of this type. Thus carvone (42) afforded the complex (43) without racemisation. Similar treatment of limonene (44) gave access to an optically active complex of the achiral diene α-terpinene.[28]

(42) $[\alpha]_D$ −67·5° $[\alpha]_D$ −542° (43)

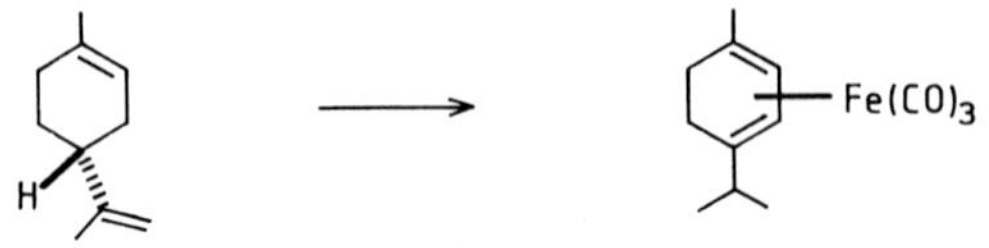

(44) $[\alpha]_D$ +126° $[\alpha]_D$ +12°

Use of type (A) isomerisation catalysts in enantiomer synthesis has also been examined in a process that produces a chiral centre from a prochiral alkene. The asymmetric induction is extremely effective when the rhodium complex of the chiral phosphine BINAP (45) is used. The enamine (46) was formed in > 95 % enantiomeric

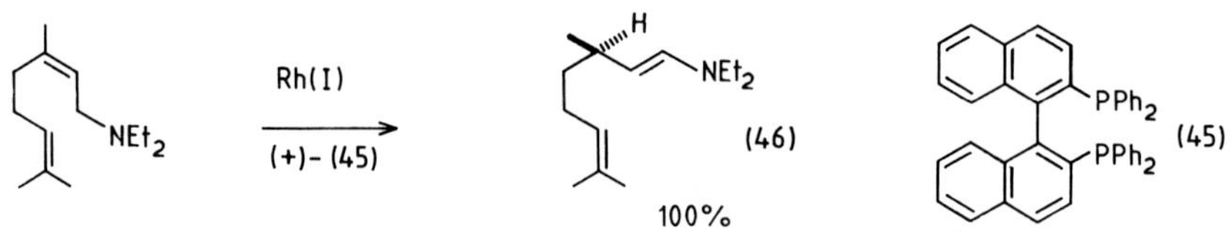

Rh(I)
(+)- (45)

(46)

(45)

100%

excess.[29] This isomerisation forms the basis of a major commercial preparation of menthol by the Takasago Perfumery Company in Japan. Hydrolysis to the aldehyde and cyclisation with zinc bromide produces an alkene that is reduced to menthol. The

(46) 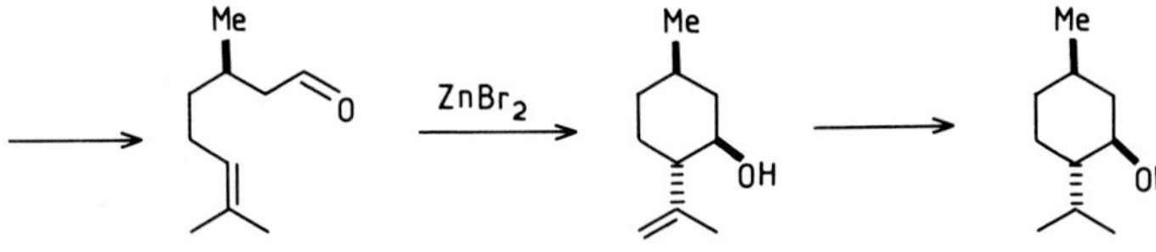 ⟶   ZnBr$_2$ ⟶

isomerisation of the allyl amine can be performed on a 9 tonne scale; one third of the world demand for menthol is currently met by this process.[30]

2.4 Isomerisation of allyl ethers and alcohols

The formation of the synthetically valuable enamine functionality by relocation of a double bond described above is an example of the use of alkene rearrangements to bring about functional group interconversions. The analogous reactions of allyl ethers and alcohols have been far more extensively studied (see Chapter 13).[31] The transformation

(47) 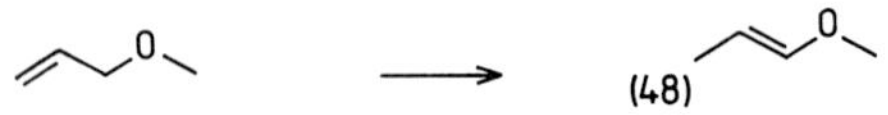   ⟶   (48)

of (47) into (48), which can be effected catalytically by means of RhCl(PPh$_3$)$_3$ is useful as a means of converting an allyl alcohol into the related aldehyde or ketone. Direct conversions of, for example, (49) into (50), or (51) into (52) with metal hydride catalysts are also useful.

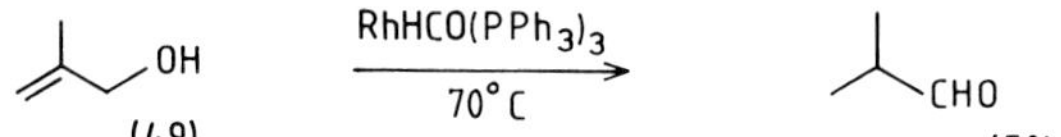

RhHCO(PPh$_3$)$_3$, 70°C

(49) → (50)

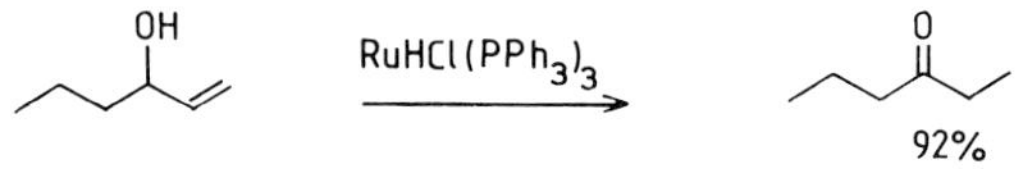

RuHCl(PPh$_3$)$_3$

(51) → (52), 92%

A similar rearrangement of (53) has been effected using a rhodium catalyst bearing the chiral ligand (45). Unlike (46), the diketone produced in this example does not contain a chiral centre. The process does, however, provide a moderately effective kinetic resolution of the readily available racemic starting material to afford (R)–(53), recovered in 91% enantiomeric excess at 72% conversion.[32]

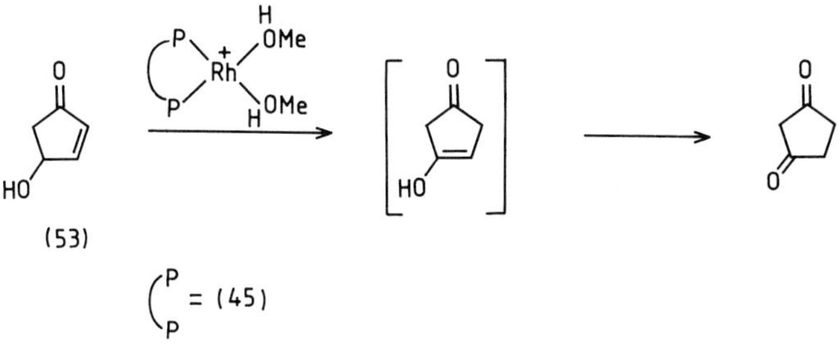

(53) → (45) →

Treatment of phenyl allyl ethers with PdCl$_2$(PhCN)$_2$ in boiling benzene gave the corresponding propenyl ethers.[33] The same reactions, however, were more effectively catalysed by RuCl$_2$(PPh$_3$)$_3$. Ethers (54) and (55) were found resistant to isomerisation. Here, coordination of the metal species to the alkene bond is hindered by the alkyl substituents.

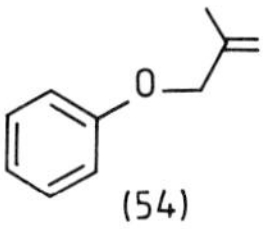

(54) 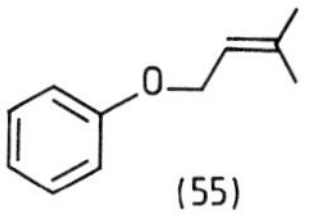   (55)

Isomerisation of allyl alcohols to aldehydes or ketones has also been effected[34] using nickel(0) complexes, e.g. (C$_2$H$_4$)Ni(tri-o-tolylphosphate)$_2$ in the presence of HCl. Examples include the conversions (56) into (57) and (58) into (59).

(56) → (57)    (58) → (59)

A further method[35] which is effective for primary allyl ethers such as (60) or (61) uses an iridium complex, [Ir(cyclooctadiene)-(PMePh$_2$)$_2$]PF$_6$, in the presence of hydrogen in either THF or dioxan.

(60)

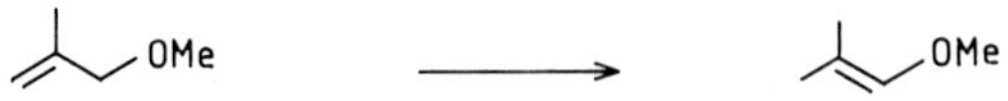

(61)

An alternative method to isomerise an allyl ether to the corresponding propenyl ether has used Fe(CO)$_5$ as the catalyst with UV irradiation.[36] The product (62) was found to

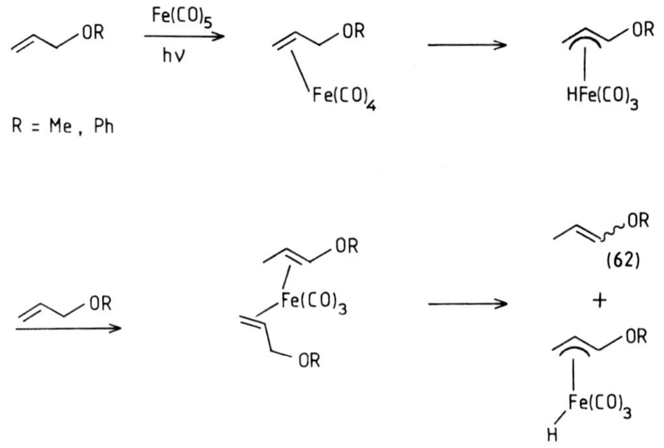

R = Me, Ph

be a mixture of *cis* and *trans* isomers. This is a disadvantage of the iron method compared to a conventional isomerisation of the allyl ether by means of KNH$_2$ on alumina,[36] which is an effective alternative and gives preferentially the *cis*-isomer.

A similar rearrangement using the molybdenum catalyst Mo(N$_2$)$_2$(dppe)$_2$ is also considered to proceed by formation of a π-allyl metal hydride intermediate. A recent study by Fiaud and Aribi-Zouioueche has revealed an interesting stereoelectronic control effect in this reaction. Comparison of the two allylic alcohols (63) and (64), used

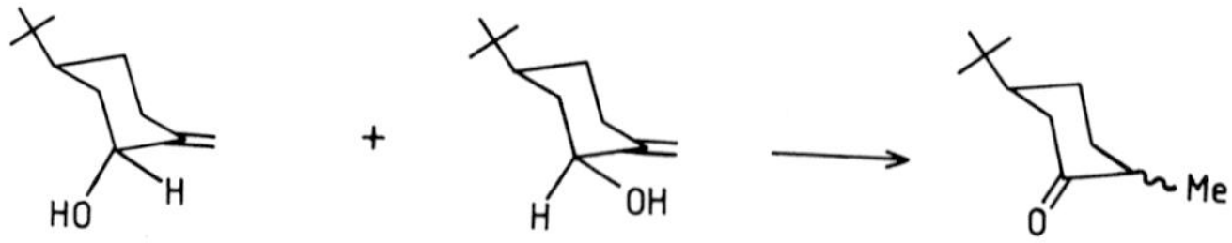

(63)   (64)

as a 3:2 mixture, indicated that (64) was the more reactive diastereoisomer. In (64) the C–H bond which is to be broken is aligned so as to interact with the π orbitals of the alkene. In contrast, the corresponding bond in the isomer (63) has an orientation approaching the plane of the σ bonds of the alkene.[37] This distinction points to a possible explanation for the difference in reactivity of the two isomers in terms of the relative degree of activation of equatorial and axial substituents adjacent to the alkene bond.

2.5 C–C bond rearrangements

So far this chapter has examined isomerisation reactions involving hydrogen migration. Alternative types of reactions are encountered in the many instances where transition metal complexes induce carbon–carbon bond rearrangements. Interesting examples arise with carbocyclic molecules containing strained bonds, which are subject to insertion reactions with a range of transition metal complexes. This is a common occurrence in the chemistry of cyclopropanes. A theoretical study of cyclopropane ring opening by palladium (0)[38] and palladium(II)[39] complexes has led to the conclusion that interaction of the transition metal at one corner of the cyclopropane is more favoured than approach of the metal to an edge of the ring, as might be expected if the C–C bond cleavage is regarded as a simple oxidative addition reaction at the metal centre. Ring opening also occurs in structures containing a cyclobutane ring in a rather strained environment. Insertion may lead to a relatively stable organometallic product, as in examples (a)–(c) in Scheme 2.6 or an unstable intermediate which can rearrange with elimination of the metal moiety, as in Scheme 2.6 (d)–(i). Most of these examples involve the use of $[Rh(CO)_2Cl]_2$. With this reagent, insertion may be followed by transfer of

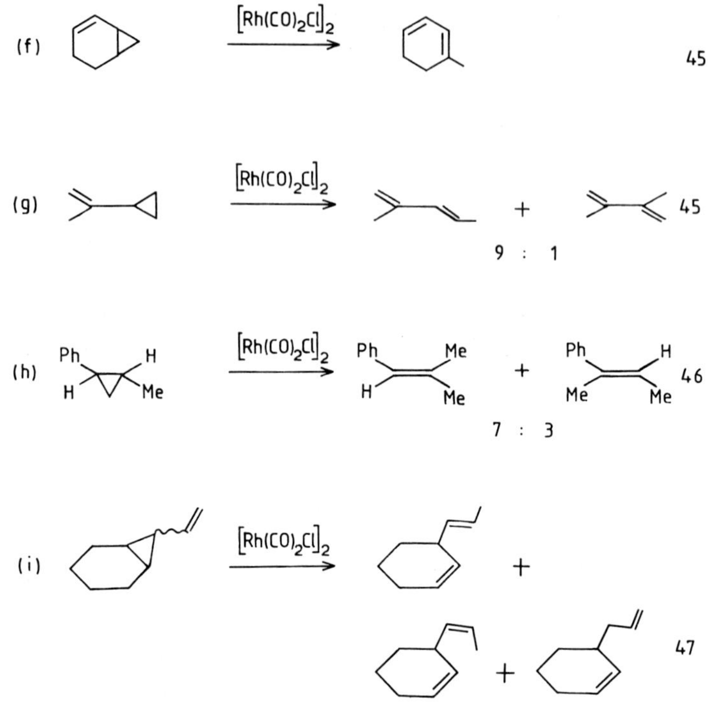

Scheme 2.6

carbon monoxide (see Chapter 9) as in (b) or (c). Reactions (d)–(i) exemplify the general utility of $[Rh(CO)_2Cl]_2$ in inducing rearrangements showing a degree of preference between alternative sites of reaction and the intervention of a hydrogen transfer step leading to the production of the final product by metal elimination. Thus reaction (h) may be rationalised as shown in Scheme (2.7).

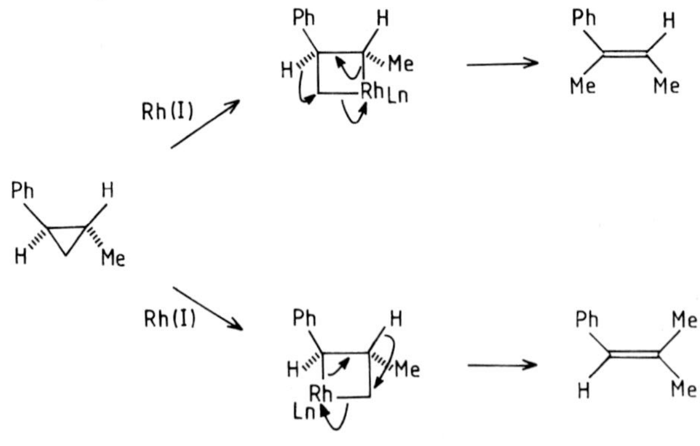

Scheme 2.7

In an example of the rearrangement of a more complex structure, diademane (65) was converted into (66) and thence into (67), on treatment with Rh(I) complexes.[48] We have

already seen examples of CO insertion in (b) and (c) above, using $[Rh(CO)_2Cl]_2$. Similar reactions have been reported for iron carbonyls (Scheme 2.8). The cyclobutane ring in α or β-pinene is subject to carbon monoxide insertion on treatment with iron pentacarbonyl.[49] This reaction clearly involves a more complex rearrangement. This is

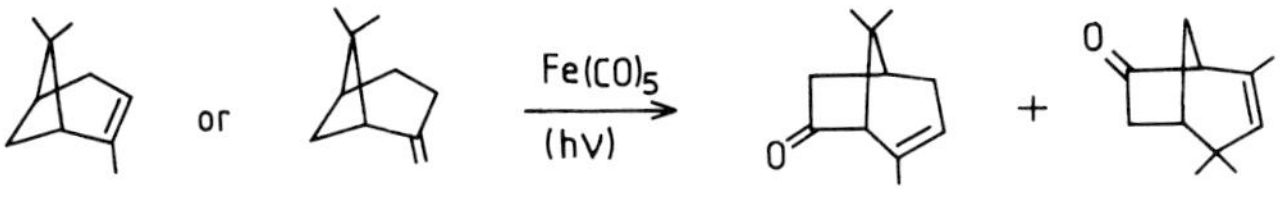

Scheme 2.8

also the case for the rearrangement of apopinene (68) to 1,2,3-trimethylbenzene (69), brought about by treatment with $PdCl_2$ in acetic acid.[50]

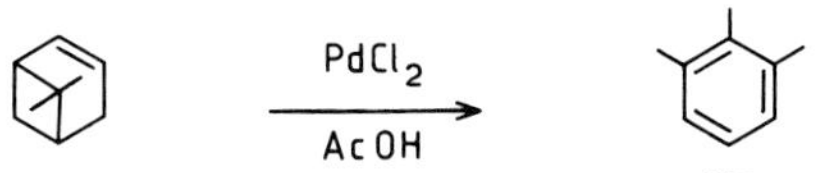

An important and potentially general palladium mediated C–C bond rearrangement occurs with allyl alcohols in which the hydrogen migration discussed in Section 2.4 is blocked by disubstitution as a four-membered ring. Compounds such as (70) have been converted[51] into substituted cyclopentenones by reaction with bis(benzonitrile)-palladium dichloride. The mechanism proposed for this reaction invokes a hydrogen shift as a subsequent step.

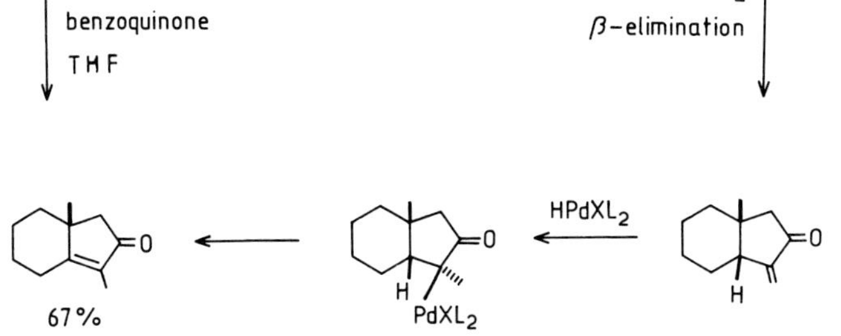

A similar rearrangement occurs when an alkyne is present in the place of an alkene. This is seen in the conversion of (71) to (72). The related ketone (73) also underwent the rearrangement reaction, but the dione product proved unstable during chromatography.[52] The course of this reaction is rather different to the palladium catalysed

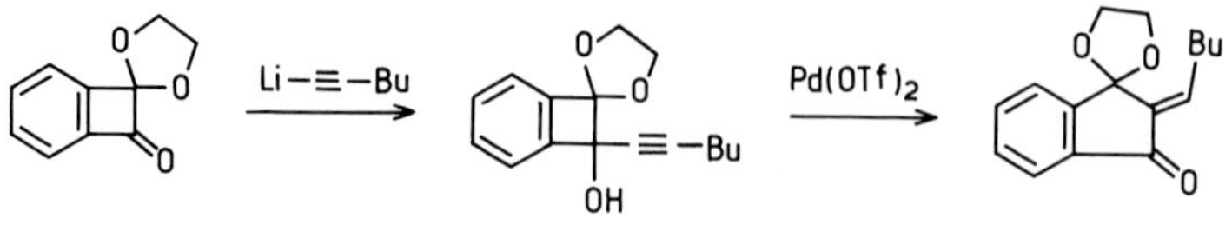

92%   (71)   91%   (72)

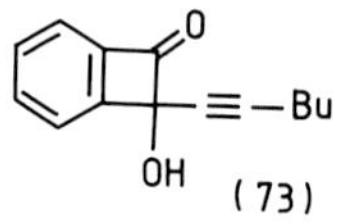

(73)

rearrangements of alkynyl alcohols that will be discussed in Section 2.7. Those reactions do not involve ring expansion and so are interpreted in terms of an initial interaction of the palladium at the alcohol. In the case of (71)–(73), a more likely mechanism involves addition of the palladium to the alkyne, a view supported by the highly stereocontrolled nature of the ring expansion which is consistent with a reaction path that proceeds through the best stabilised cationic intermediate.

A quite different palladium catalysed rearrangement route to cyclopentenones has been described[53] by Rautenstrauch. This intramolecular cyclisation of an allylic acetate to an alkyne appears to involve a concurrent acetate migration. Trapping experiments confirm the involvement of an acetoxycyclopentadiene as an intermediate. This is consistent with the unusual stepwise oxidative addition, followed by reductive elimination of the diene (Scheme 2.9), that is proposed to account for the observed products.

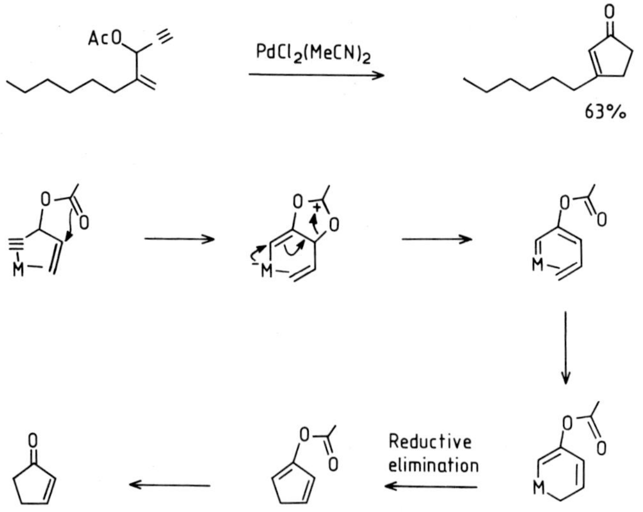

63%

Reductive
elimination

Scheme 2.9

Palladium(II) is also an important catalyst for Cope rearrangements. Since this process was described[54] in 1980, work by Overman and by others has shown the reaction to be generally applicable to a wide variety of substrates including, for example, compounds with alcohol[55] and acyl[56] groups as shown in Scheme 2.10.

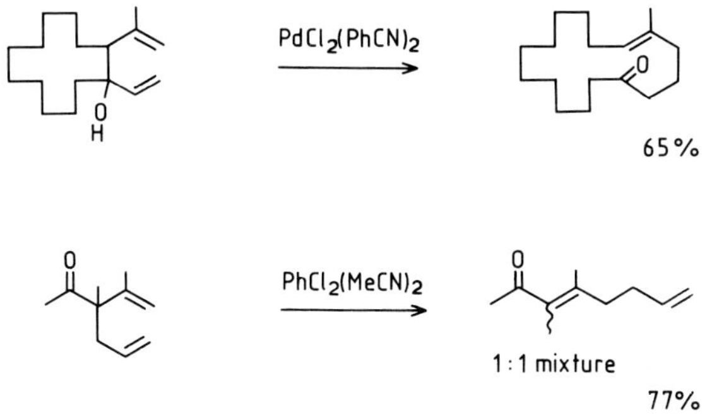

65%

1:1 mixture

77%

Scheme 2.10

Palladium is not alone in its ability to catalyse [3,3]-sigma-tropic rearrangements. Nickel, silver, platinum, mercury, and aluminium are also important catalysts for processes of this type.[57] A mechanism involving an initial cyclisation is favoured. When allylic electron-withdrawing substituents are present, the reaction is halted[58] at the cyclised product (Scheme 2.11). Bosnich[59] has investigated an ostensibly similar

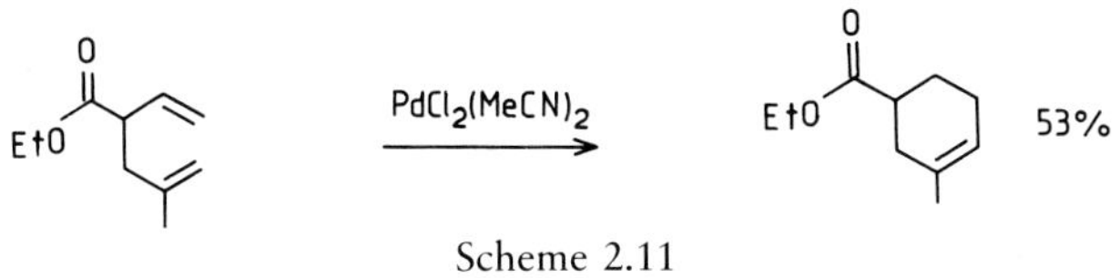

53%

Scheme 2.11

rearrangement of allyl imidates such as (74). The presence of an allylic heteroatom between the double bonds introduces possibilities for the involvement of other mechanisms. Studies using chiral ligands or deuterium labels suggest that, while palladium(II) catalysts operate in a fashion similar to that shown above, palladium (0) catalysis involves a π-allyl intermediate of a type discussed in Chapter 6.

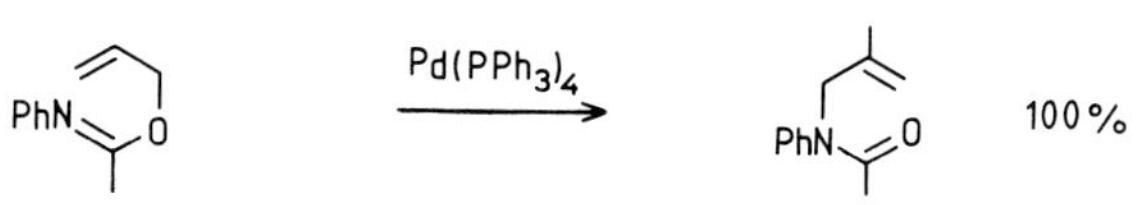

100%

(74)

2.6 Rearrangements of allylic esters

Allylic acetates are well known to undergo 1,3-shift of the acetoxy group in the presence of catalysts, notably $PdCl_2(MeCN)_2$ or similar soluble Pd(II) derivatives. The process is also observed with other allylic esters. Reaction may be accompanied by ester exchange with the medium.[60]

The reaction is brought about by the use of a catalytic amount of $PdCl_2(MeCN)_2$ in THF and gives good yields.[61] Examples include the rearrangement of (75) and (76).

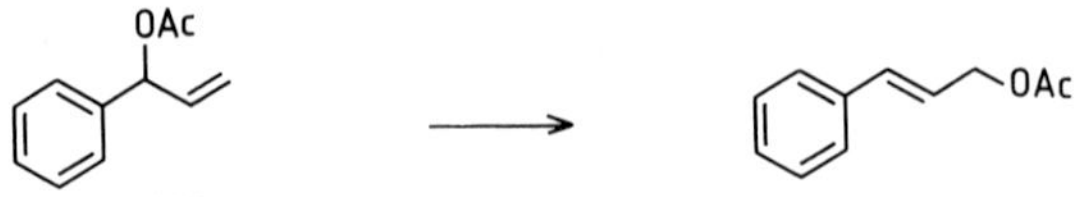

Where there is a second centre of unsaturation in the molecules, as in linalyl acetate (77), stable palladium π-allyl complexes may be formed.[62] Such complexes are useful in their own right, and are discussed in detail in Chapter 6. These complexation reactions are analogous to those leading to η^4 diene complexes described in Section 2.3.

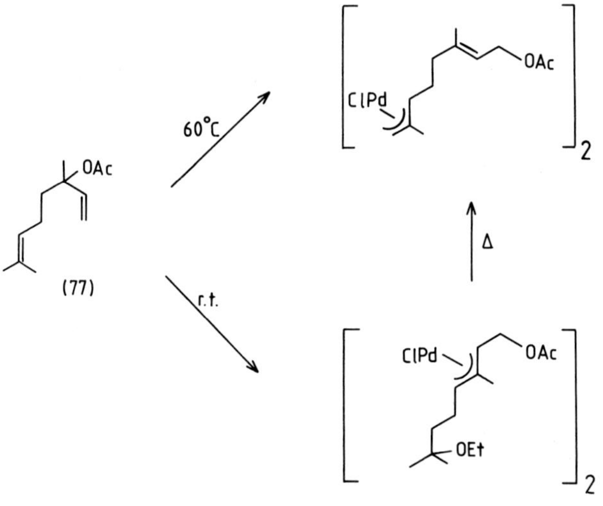

The rearrangement reaction may also be brought about, albeit less effectively, by $Hg(OCOCF_3)_2$. Both the Hg(II)[61] and Pd(II)[62] catalysed reactions appear to proceed via a cyclic intermediate (78) as shown in Scheme 2.12.

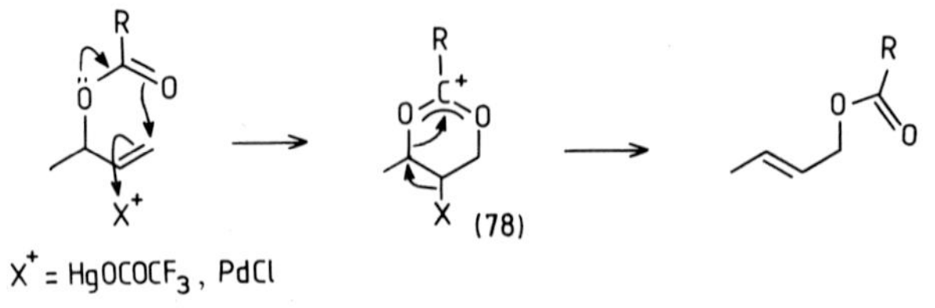

$X^+ = HgOCOCF_3$, PdCl

Scheme 2.12

For examples (75) and (76), there was a thermodynamic driving force in converting a tertiary ester into a primary alcohol derivative, or in achieving conjugation. More generally, an equilibrium mixture is formed, as in the case[61] of (79), where (79) and (80) were formed in a ratio of 59:41. The mechanism and kinetics of this reaction have been explored.[60]

(79)   (80)

A feature of this rearrangement of allylic acetates is the possibility of stereochemical inversion[63] by transfer of acetate to palladium and thence back to carbon. A well studied example of the epimerisation of (81) gave progressively increasing amounts of the isomer (82) after a short period under reflux. The reaction is considered to proceed via the π-

(81)   (82)

(83)   (84)   (82)

allyl cation (83) which binds acetate first at palladium to form (84). Transfer of the acetate ligand to the allyl moiety subsequently produces (82). In isomerisation via a cyclic acetate such as (78) as described above, the ester group will also remain on the same side of the molecule.

The allylic acetate rearrangement shows a high degree of steric preference in favour of formation of the (E)-alkene.[61] Thus, with a catalytic amount of $PdCl_2(MeCN)_2$ in THF, (85) gave (86) with 80% (E)-isomer, and (87) gave (88) with the proportion of (E)-isomer > 97%.

(85)   (86)

(87)   (88)

The intermediate π-allyl complexes involved in these reactions can be intercepted by nucleophile addition under certain circumstances. These synthetically important reactions are discussed in Chapter 6.

2.7 Rearrangement of alkynyl alcohols and esters

Alkynyl alcohols (89, R = H) or esters (89, R = COMe) may be rearranged to the unsaturated aldehyde (90) or the allene (91).[64] Rearrangement of esters has been brought

about by warming in acetic acid solution in the presence of an added silver salt, e.g. silver carbonate, and the procedure has been applied to the preparation of citral (94) from dehydrolynalyl acetate (92, R = Ac) via the allenic acetate (93).[65]

A more direct route to citral from linalool has been developed by Rhône-Poulenc. The idea that the formation of an ester of an inorganic acid could be used to introduce the metal required to catalyse the isomerisation led to the use of alkoxy and siloxy vanadates to promote the rearrangement reaction. This gave rise to an industrial scale process for the isomerisation of (92, R = H) into (94). A good account of this work is given by Chabardes *et al.*[66] With a catalytic amount of tris(triphenylsilyl)vanadate at 50 °C, dehydrolinalool gave neral and citral in a ratio of 1·1:1 in high yield. Further, the vanadate catalyst continued to be effective for rearrangement of many batches of substrate.

The reaction is regarded as occurring by rearrangement of the vanadate ester (95) to the allenic vanadate (96).

The procedure has also been applied[67] in a synthesis of vitamin A by a sequence initiated by a Grignard reaction to form (97). Acid induced allylic rearrangement of (97) yielded (98, R = H) which was acetylated to (98, R = MeCO). Rearrangement using $(Ph_3SiO)_3VO$ in refluxing xylene gave a product shown to be predominantly the (3*Z*)-isomer (99) which could be inverted to the (3*E*)-isomer (100) using piperidine acetate. Hydride reduction and acetylation gave the diacetate (101). Finally, treatment with 60% HBr yielded all (*E*) Vitamin A acetate (102) in good yield after chromatography.

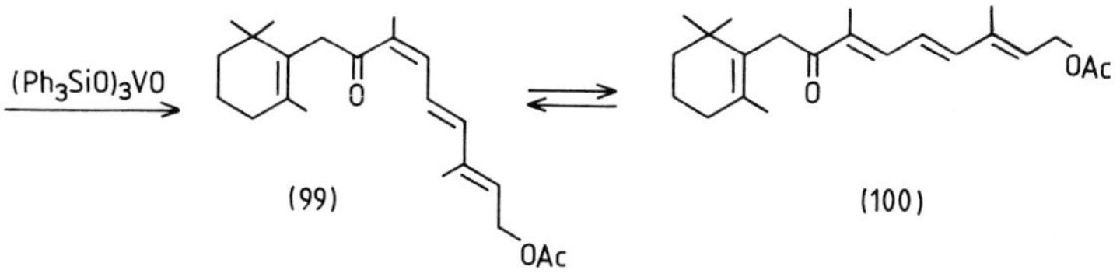

(99) (100)

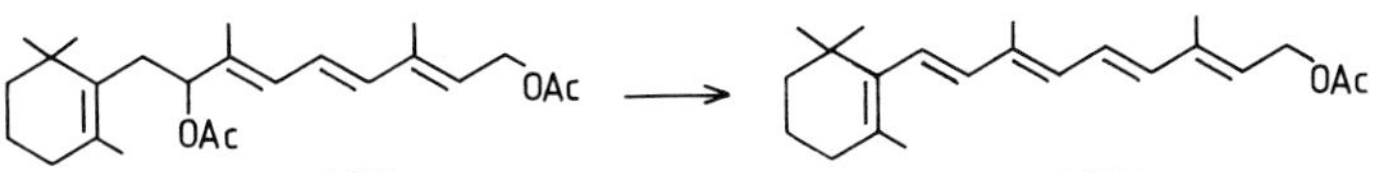

(101) (102)

The rearrangement of some bis-propargyl esters has been effected by means of $[Rh(CO)_2Cl]_2$ in chloroform. An example is the formation of (103). Unfortunately, the

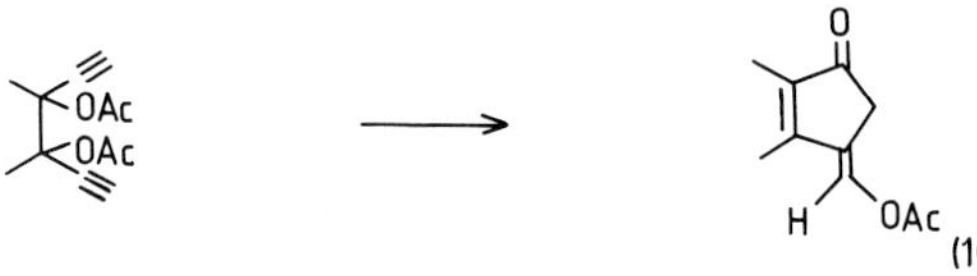

(103)

yield was only modest. A mechanism is proposed in which the acetoxy group is regarded as being eliminated as ketene.[68] In another case, in which the diacetate was flanked by two alkenes, rearrangement proceeded in the normal way, placing a 1,3-diene between the acetoxy groups. Palladium catalysis was used in this instance.[69]

2.8 Rearrangement and deoxygenation of epoxides, oxetanes and peroxides

Reaction of various arene oxides with a catalytic amount of $[Rh(CO)_2Cl]_2$ gave results[70] of the type summarised by the reactions shown in Scheme 2.13.

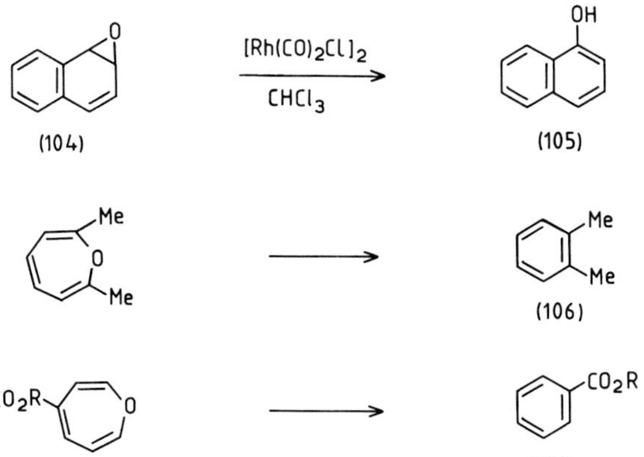

(104) (105) (106) (107)

Scheme 2.13

Reaction of (104) gave some naphthalene in addition to naphthol (105). Reaction of the type exemplified by the conversion of (104) into (105) may be rationalised[70] by the mechanism shown in Scheme 2.14 in which $[Rh(CO)_2Cl]_2$ acts essentially as a Lewis acid.

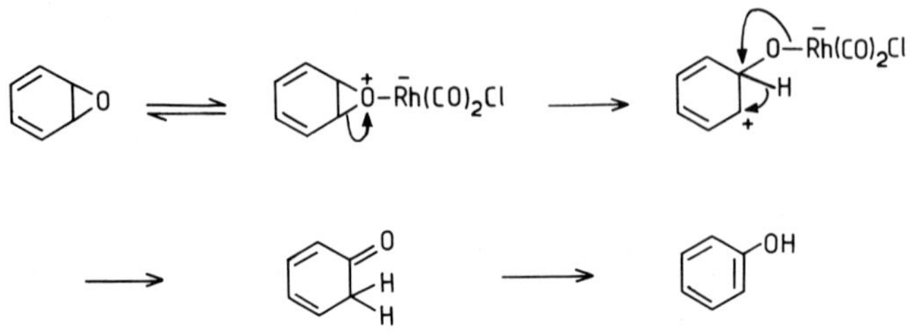

Scheme 2.14

In the deoxygenation process illustrated by the formation of (106) or (107), the oxygen removed is thought to be transferred to rhodium and eliminated as CO_2 via a process indicated in Scheme 2.15. By the use of methanol as the solvent, instead of

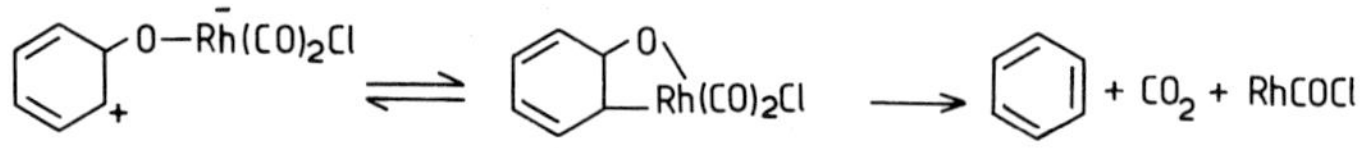

Scheme 2.15

chloroform, a methoxy group may add to the intermediate cation. In practice, the process is almost certainly more complicated than is indicated by the simple mechanism given above.

In related work, Grigg *et al.* found that the cyclooctatetraene epoxide (108) isomerised to (109) on contact with $[Rh(CO)_2Cl]_2$ at $-50\,^{\circ}C$. The product was subsequently converted into (110).[71]

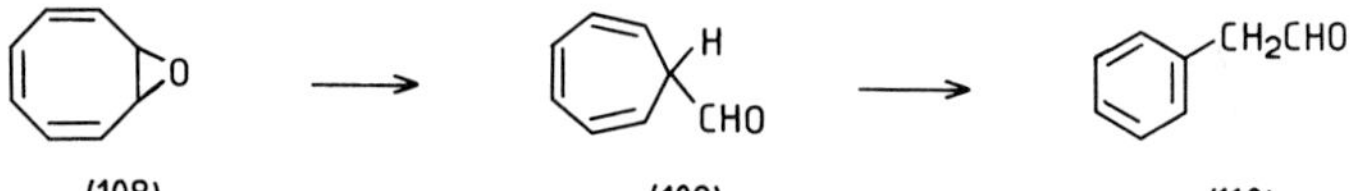

With $[Rh(CO)_2Cl]_2$, the epoxides (111, $R^1 = R^2 = H$; $R^1 = Me$, $R^2 = H$; or $R^1 = H$, $R^2 = Me$) were found[72] to rearrange to the aldehydes (112). This observation provides evidence to suggest that (113) is an intermediate in the process leading to (112, $R^1 = Me$, $R^2 = H$).

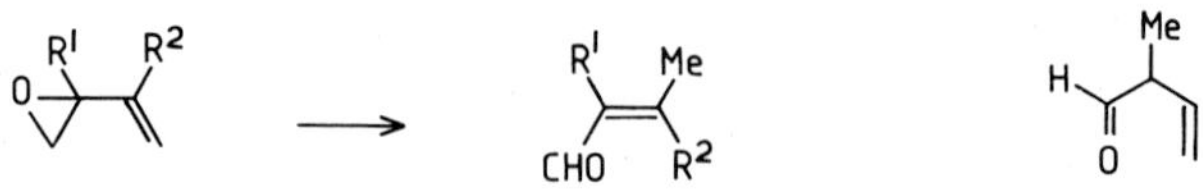

A similar reaction of the allylic epoxide (114), this time catalysed by palladium, has been used in the conversion[73] of glucose into the useful synthetic intermediate (115).

Epoxides derived from allyl alcohols also undergo ring opening catalysed by

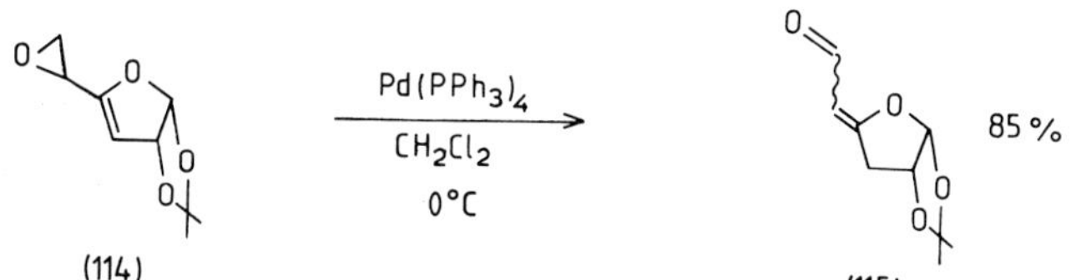

(114) → (115)   Pd(PPh$_3$)$_4$ / CH$_2$Cl$_2$ / 0°C   85%

palladium. The hydrogen shift associated with this process is strongly influenced by substituents such as aryl groups. The alcohol substituent itself does not appear to provide a strong control effect. Thus (116) rearranges to give only (117), while (118) affords a mixture of products.[74] The chemistry of epoxides derived from allyl alcohols, and, in particular, the preparation of such epoxides, is discussed in detail in Chapter 3.

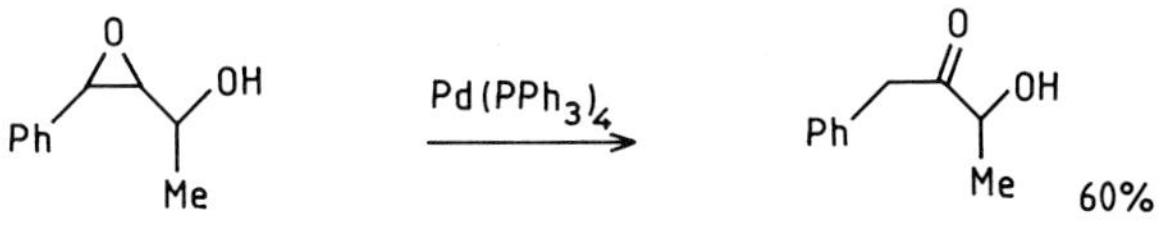

(116) → (117)   Pd(PPh$_3$)$_4$ 60%

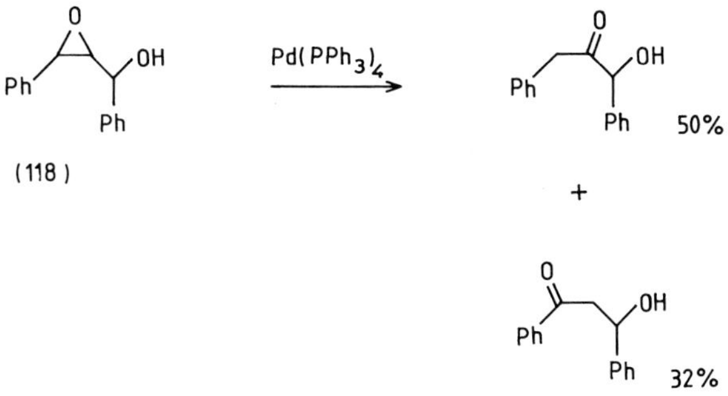

(118) → (50%) + (32%)   Pd(PPh$_3$)$_4$

Under the influence of [Rh(CO)$_2$Cl]$_2$, some oxetanes undergo cleavage, as in the conversion of (119) into (120).[72]

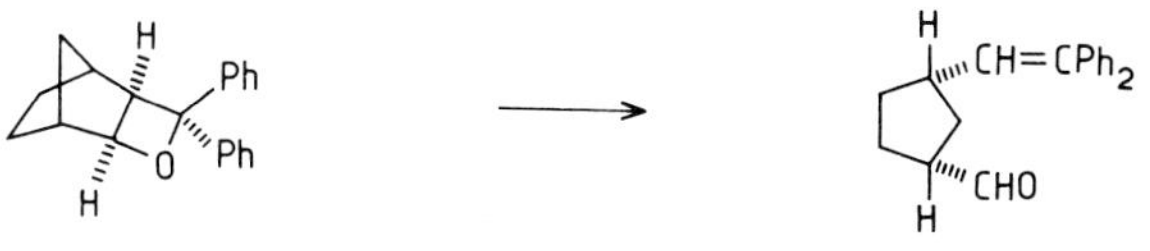

(119) → (120)

A series of β-lactones (121), with a catalytic amount of PdCl$_2$, Pd(OAc)$_2$, or PdCl$_2$(PhCN)$_2$, were shown[75] to isomerise to the dienoic acid (122). The vinyl group was necessary for the success of the elimination reaction. With an alkyl substituent in (121), the result, instead, was polymerisation.

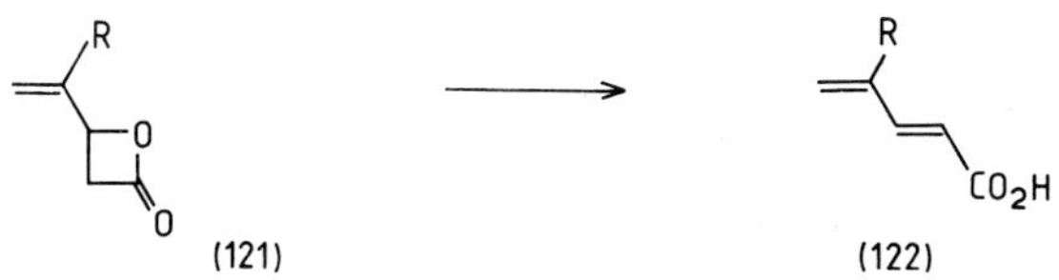

(121) → (122)

R = H, Me

Deoxygenation occurs[76] when 1,4-epoxy-1,4-dihydronaphthalenes are heated with $Fe_2(CO)_9$. A related deoxygenation of the 7-oxanorbornadiene (123) derived from the

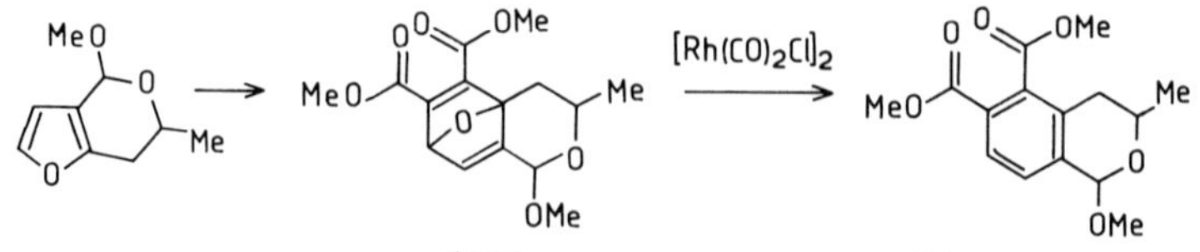

(123)　　　　　49%

corresponding furopyran by a cycloaddition reaction has been employed[77] in a synthesis of isochromans. The $[Rh(CO)_2Cl]_2$ reagent, however, has been found by Hogeveen to isomerise very similar substrates, e.g. (124), to 6-hydroxyfulvenes, e.g. (125).[78]

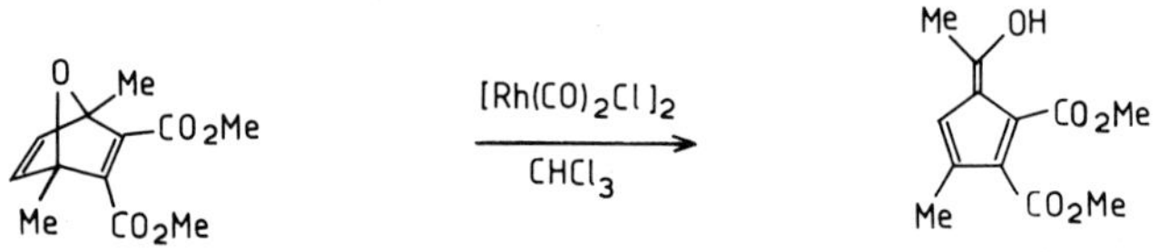

(124)　　　　　(125)

The oxepine (127) could be obtained in the transition metal promoted isomerisation of oxyquadracyclone (126). A different oxepine (128) is obtained by thermal rearrangement of (126).[79] When a trace of methanol was added to the reaction mixture, the 6-hydroxyfulvene (125) was the major product formed.

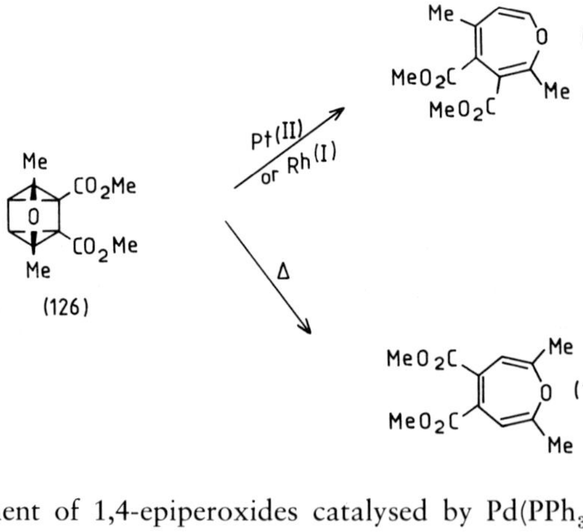

The rearrangement of 1,4-epiperoxides catalysed by $Pd(PPh_3)_4$ is also of interest. Some simpler examples provided precedents for the rearrangement of the prostaglandin endoperoxide (129).[80] A small amount of ring fragmentation products was also found.

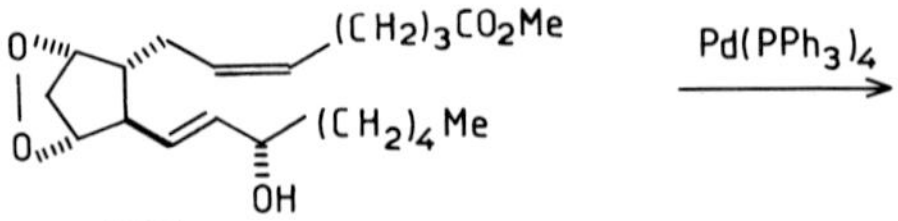

(129)

17%   11%   41%

Other metal catalysts have been examined. A ruthenium catalysed rearrangement of another cyclic peroxide (130) forms an important step establishing stereocontrol in a synthesis of epi-chorismic acid. In this example, the peroxide bridge spans an alkene. The rearrangement produced the bis epoxide (131). This was converted to the alcohol (132), which was obtained as a single diastereoisomer, since the epimeric centre adjacent to the ester was removed in this step.[81] The resolution of (132) has recently been described.[82]

(130)   (131)   (132)

This chapter has drawn attention to a variety of rearrangement processes which are initiated by the interaction of transition metal reagents with the substrate molecules. Through consideration of these examples at the start of this book, the uses of some important types of organometallic reactions have been introduced. These reactions appear again, in subsequent chapters, as stages in transition metal mediated bond forming processes. While rearrangements themselves have applications in synthesis, as seen in the examples given above, the processes involved have perhaps greater significance as an introduction to general reaction types, in particular to the π complexation of alkenes and insertion (oxidative addition) into σ bonds. These are encountered frequently in forthcoming chapters. The discussion of the uses of π and σ bound reagents and intermediates begins in Chapters 3, 4, and 5, which concern a variety of uses of alkene π complexes. Detailed discussion of the chemistry of σ bonds is reserved for the second half of the book.

3

Epoxidation of alkenes

3.1 Epoxidation by V, Mo, W, and Ti

Although a vanadium pentoxide catalysed epoxidation of 1-octene was originally described by Hawkins[1] in 1950, it was to be many years before the value of transition metal epoxidation catalysts was widely appreciated, and reliable and efficient reactions were developed. By the late 1980s, however, the method has become important both in industrial process chemistry and organic synthesis. Examples of the former include the preparation of glycerol by the epoxidation of allyl alcohol (Scheme 3.1),[2] and the molybdenum catalysed preparation of pyrogallol from cyclohexene-3-ol (Scheme 3.2).[3] However, by far the most significant large scale industrial reaction is the Halcon process for manufacture of propene oxide (Scheme 3.3).[4]

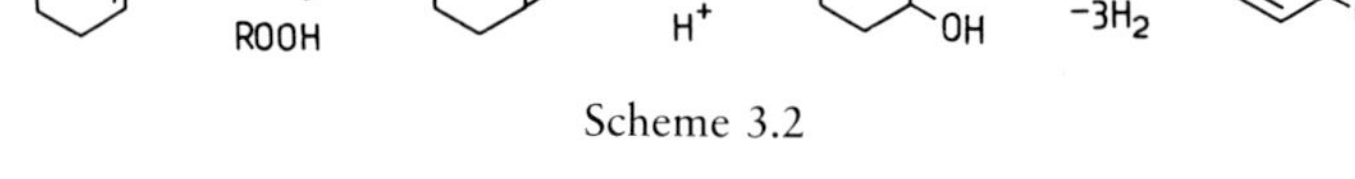

Scheme 3.1

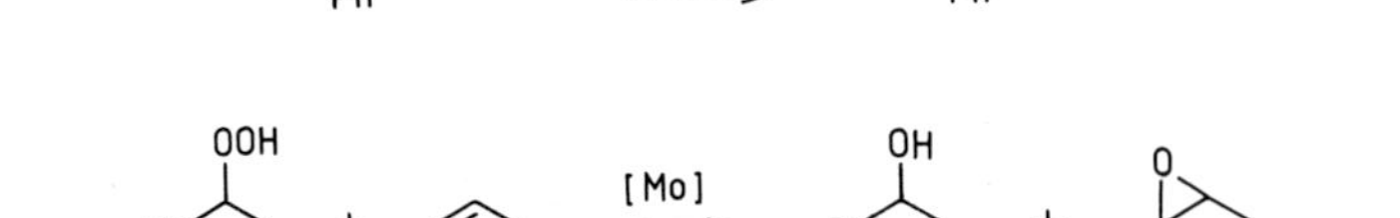

Scheme 3.2

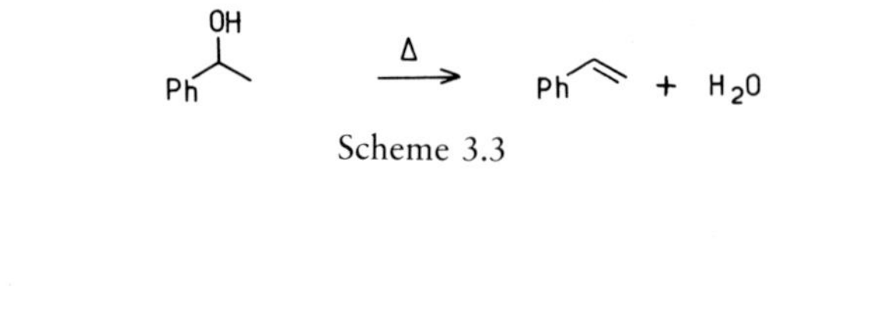

Scheme 3.3

Factors influencing the metal catalysed epoxidation reaction have been examined in great detail.[5] A simple view of the overall process is that coordination of the source of oxygen, typically an alkyl peroxide or hydroperoxide, and the alkene molecule promotes an oxygen transfer that takes place within the coordination sphere of the metal.

The presence of electron donor substituents at the double bond increases reactivity, whereas electron accepting groups are deactivating. The effect of alkyl substitution, increasing the reactivity of the double bond, is clearly indicated by the selective epoxidation of the ring alkene in 4-vinylcyclohexene (1).[6]

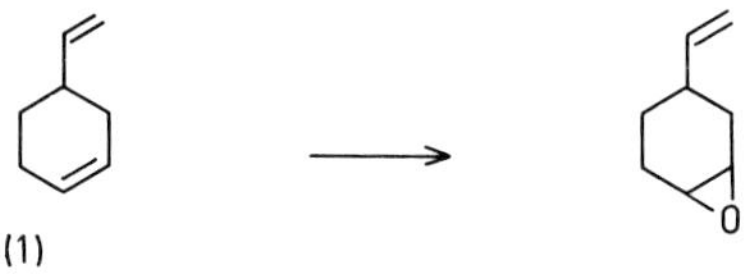

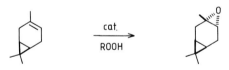

(1)

Although in a number of specific cases, considerable effort has been made to elucidate the mechanistic details of these reactions (see Section 3.4), the general picture is complicated. The range of metal complexes that are capable of acting as epoxidation catalysts is very wide and includes many examples of the same metal in a variety of oxidation states. While catalysts based on molybdenum and tungsten appear suitable for all types of alkenes, vanadium and titanium catalysts are highly selective for allylic alcohols (see below). Other metals such as zirconium, hafnium, niobium, tantalum, and chromium generally form unsatisfactory epoxidation catalysts.[7] A number of comparative studies of a range of catalysts with the same substrate have been performed. For example, in the epoxidation of carene by alkyl peroxides, variation of both selectivity and yield were found when a range of molybdenum catalysts was examined (Scheme 3.4).[8] The best results were obtained with Mo(VI) species. A similar

Scheme 3.4

type of comparative study[9] of molybdenum catalysts has examined the role of ligands in determining catalyst activity for the epoxidation of cyclohexene.

Most workers have chosen alkyl peroxides as the source of oxygen for the epoxidation step. Hydrogen peroxide itself can also be used successfully provided conditions are anhydrous.[10] Catalytic species of great complexity are employed in some cases. The molybdenum complex $[PMo_{12}O_{40}]^{3-}[C_5H_5N^+(CH_2)_{15}CH_3]_3$, prepared from 12-molybdatophosphoric acid and cetylpyridinium chloride, for example, is a highly selective catalyst for the direct epoxidation of allylic alcohols by hydrogen peroxide.[11] Phase transfer conditions[12] for the use of hydrogen peroxide offer considerable practical advantages. In this case the catalyst was derived from Na_2WO_4 and phosphoric acid. The oxidising species that can be used in transition metal catalysed epoxidation processes include dioxygen[13] itself and sodium hypochlorite.[14]

More recent work has provided some good examples where palladium[15] and platinum[16] reagents have been used. The platinum system employed hydrogen peroxide as the oxidising agent. With simple alkenes as the substrates, moderate enantiomeric excess was achieved in asymmetric epoxidation reactions. The palladium catalyst proved selective for strained alkenes, as indicated in the formation of (2), and required O_2, rather than a peroxide, as the source of oxygen.

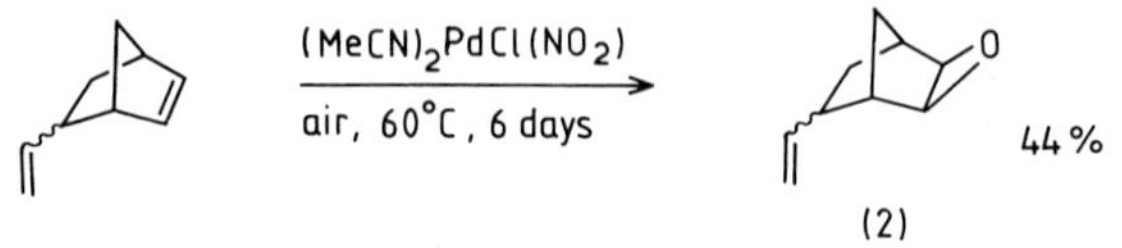

$$(MeCN)_2PdCl(NO_2)$$

air, 60°C, 6 days

44%

(2)

The great number of variables (the catalytic metal, its oxidation state, and the choice of ligands and oxygen sources) will ensure that this area remains a complicated field of study. There are now, however, a number of procedures that have proved to have general applicability. The most important is the Sharpless epoxidation reaction that is discussed in Section 3.6. These methods can offer substantial advantages in terms of selectivity. Examples are described in detail in the following sections.

3.2 Regioselectivity of epoxidation

Selectivity of reaction is a pronounced feature of metal catalysed epoxidations and reflects the sensitivity of the reaction to the nature of the alkene. In the case of molybdenum peroxo complexes, kinetic studies[17] have identified the binding of the alkene to the metal centre as the rate limiting step of the process. As seen above, the reactivity of alkene varies with its nucleophilic character, thus the susceptibility to catalysed epoxidation increases in the order monosubstituted < disubstituted < trisubstituted < tetrasubstituted.[18]

Steric constraints are also significant and lead to *cis*-alkenes displaying higher reactivity than their *trans*- analogues, as is evident from epoxidation of *cis*- and *trans*-1,4-hexadiene[6] shown in Scheme 3.5.

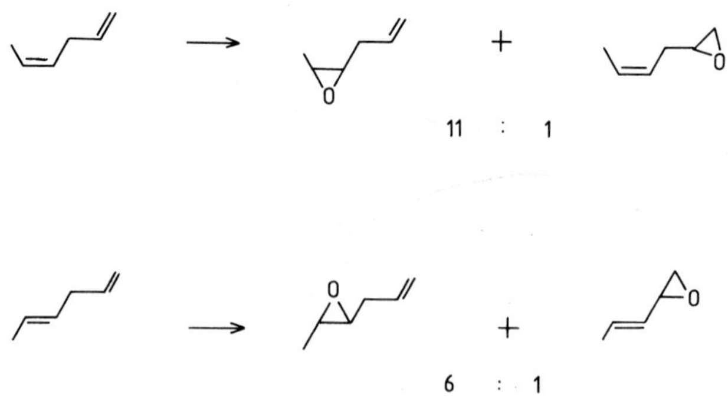

11 : 1

6 : 1

Scheme 3.5

A major control effect (the *syn* effect) is exerted when substituents adjacent to the double bond are capable of direct interaction with the metal catalyst. Some of the most elegant demonstrations of the *syn* effect have come from Sharpless and co-workers[19]

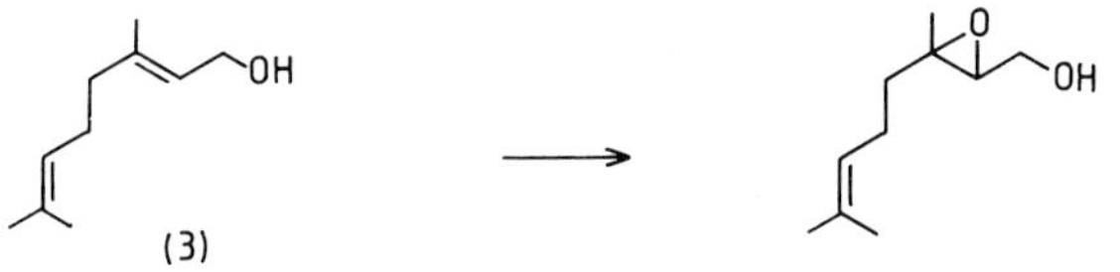

(3)

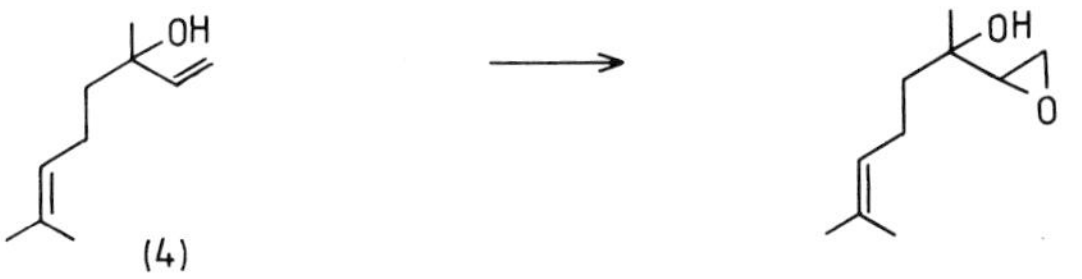

(4)

who made use of t-butylhydroperoxide (TBHP) as oxidant, in the presence of a vanadium complex [VO(acac)$_2$] and the now famous titanium epoxidation catalyst Ti(O^{i}Pr)$_4$/DET discussed in Section 3.6. The *syn* effect has also been observed with molybdenum [Mo(CO)$_6$] catalysts.

The selectivity of this type of reaction is illustrated by the epoxidation of geraniol (3) and linalool (4) using TBHP with VO(acac)$_2$ in benzene at reflux.[19] Useful discrimination between an allylic alcohol and alkene is also seen in the synthesis[20] of (±)-periplanone B (6). Selective epoxidation of (5) is followed by oxidation of the alcohol function and an elimination step.

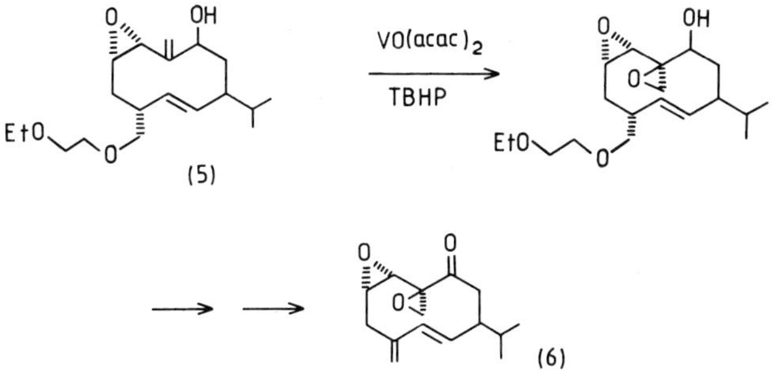

(5)

(6)

A similar instance[21] is to be found in the synthesis of (±)-C$_{18}$-*Cecropia* juvenile hormone (10). The bis allyl alcohol (7), derived from (*E,E*)-farnesol was epoxidised using TBHP in the presence of VO(acac)$_2$. The product (8) was treated with Me$_2$CuLi to form (9).

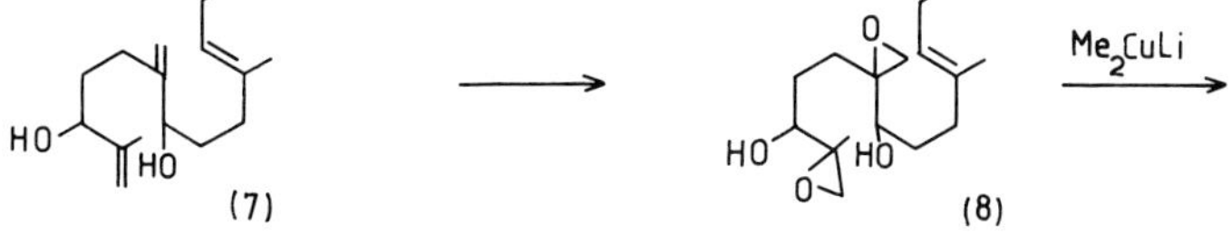

(7)

(8)

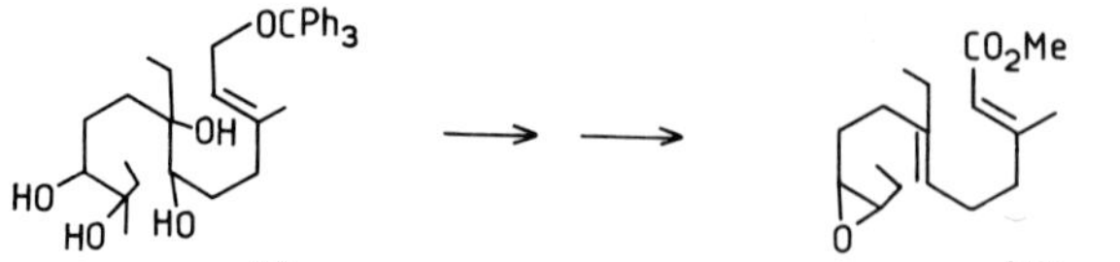

(9)   (10)

A further example[22] is found in a synthesis of $(\pm)$-zoapatanol where an allylic alcohol was again the substrate of an epoxidation at a key point (Scheme 3.6).

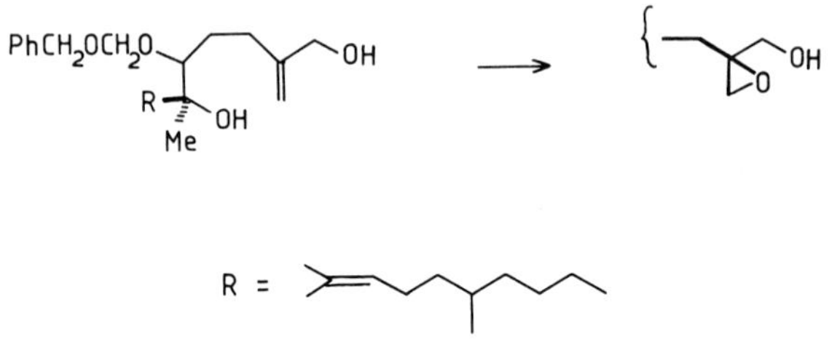

Scheme 3.6

3.3 Stereocontrol of epoxidations

As is the case with the use of peracids, the metal catalysed epoxidation reactions usually proceed stereospecifically when the substrates are hydrocarbon alkenes, as was demonstrated in an early investigation by Indictor and Brill. These workers found that *cis*- and *trans*-4-methyl-2-pentene yielded exclusively the corresponding *cis*- and *trans*-epoxides.[23]

Coordination by the OH group of an allylic alcohol provides[24] a high degree of stereocontrol in addition to the regioselectivity noted above. The examples (11) and (12) illustrate the preference for epoxidation on the same face as the hydroxyl substituent.

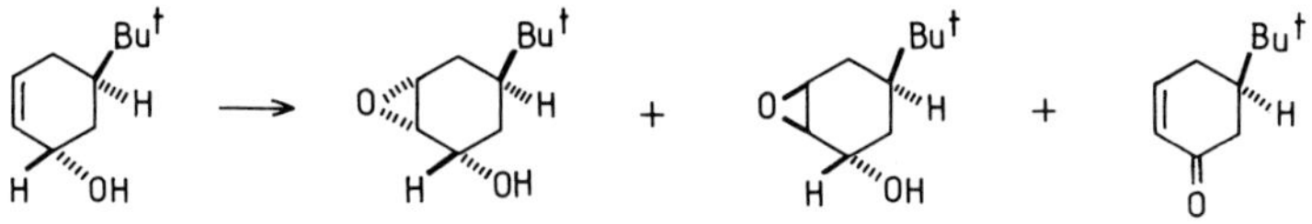

(11)   83%   0.08%   15%

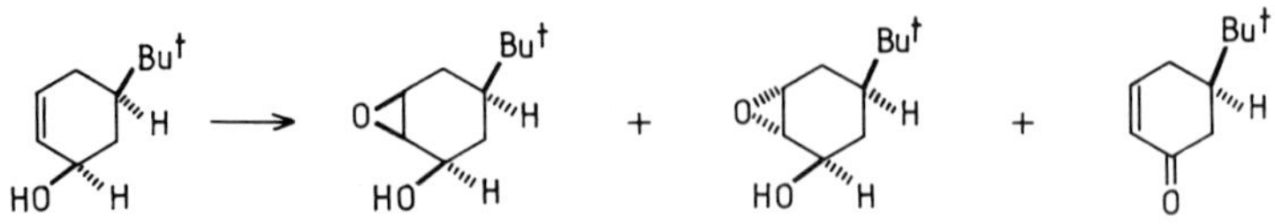

(12)   18 %   1%   74%

These examples also draw attention to a competing oxidation of the allyl alcohol to the enone, which, although not generally important, may become so when the alkene bond is seriously sterically hindered as in (12). The substrate (*E*)-cyclooctenol (13) is exceptional in yielding an epoxide *trans* to the hydroxyl, because of preferential reaction[24] through the structure (14).

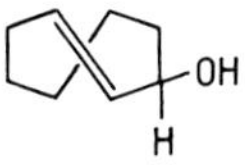

(13) 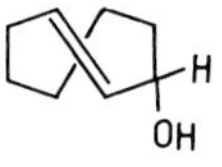   (14)

The epoxidation process can be completely suppressed in favour of alcohol oxidation by control of pH when ammonium molybdate is used to catalyse oxidation by hydrogen peroxide.[25] The presence of alkyl substituents exerts an important influence. This is illustrated by the product ratios shown in Scheme 3.7. The reagent allows the selective

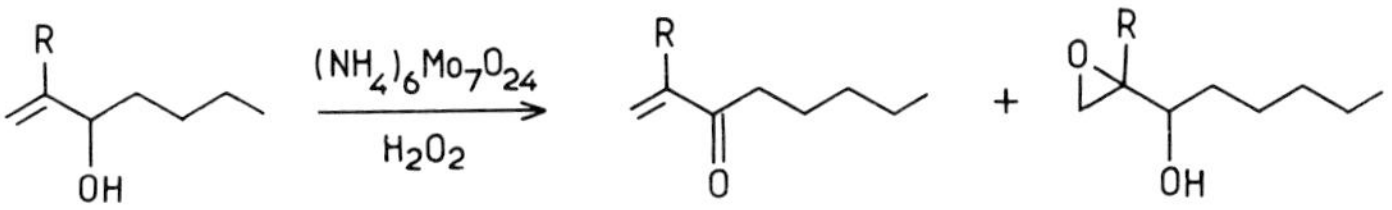

R = H   DMF   83%   –

R = Me   DMF   49%   10%

R = Me   THF/K$_2$CO$_3$ 64%   22%

Scheme 3.7

oxidation of alcohols at more hindered positions, for example, the 17-hydroxyl group in (15).

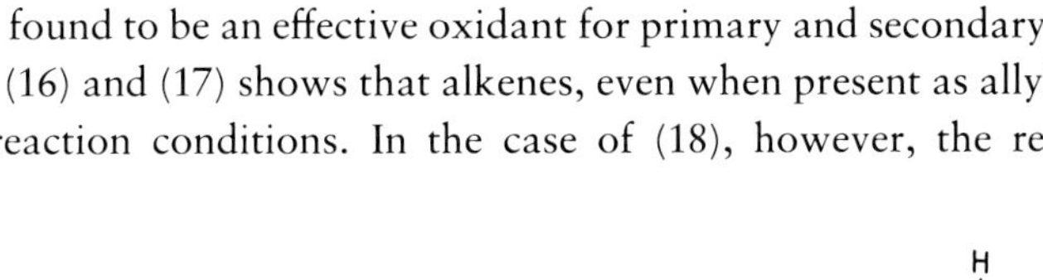

(15)

Another molybdenum catalyst, a molybdenum-picolinate N-oxidoperoxo complex, has also been found to be an effective oxidant for primary and secondary alcohols.[26] The formation of (16) and (17) shows that alkenes, even when present as allylic alcohols, can survive the reaction conditions. In the case of (18), however, the reaction was less selective.

(16)   99%

(17)   99%

(18)   80%

                                                  +

                                                  20%

In practice, oxidation of the allylic alcohol OH group, rather than epoxidation, does not limit the general utility of alcohols as directing groups controlling the site of epoxidation. Many reliable methods are available to avoid such side reactions. The use of allylic alcohols to control epoxidation is a strategy that is now finding widespread application (see also Section 3.5).

In the synthesis of (±)-cerulenin[27] (21) a useful degree of stereoselectivity was achieved in the epoxidation of (19) to generate the intermediate (20), which was converted to an amide with NH_4OH. Oxidation then completed the synthesis of cerulenin.

(19)   VO(acac)$_2$ (20)
                    TBHP

        (i)   (ii)

(21)

(R =)

(i) AgCO$_3$ on celite,   (ii) NH$_4$OH, ether

In the oxidation of (19), the stereoselectivity is presumably induced by the hydroxyl group. In a related instance, the oxidation of hydroxy steroids, a different and more unusual control effect operates. The selectivity indicated in the TBHP/VO(acac)$_2$ epoxidations[28] of (22) and (23) must derive from the relative degree of steric hindrance to approach to the steroid side-chain in the preferred conformation of the molecule.

For (22) and (23), the use of *m*-chloroperbenzoic acid oxidation showed no stereoselectivity of the type found with TBHP/VO(acac)$_2$. The contrasting degree of stereocontrol in these two methods of epoxidation has also been compared for the relatively simple allylic alcohol (24). There is a preference in the TBHP/VO(acac)$_2$ system for production of the *erythro* isomer (25), whilst *m*-chloroperbenzoic acid shows the reversed preference in epoxidation. The proportions of *threo* and *erythro* products depend, however, on the degree of substitution of the alkene bond.[29]

In several acyclic situations, replacement of an appropriate hydrogen by a trimethylsilyl group resulted in a substantial alteration in *erythro* selectivity. Epoxidation of (26) and (28) indicates[30] that either the *erythro* (27) or *threo* (29)

product can be selected, depending on the location of the trimethylsilyl group. Investigations of various substrates show this effect to be of general importance.[30, 31] The mechanistic implications[31] of these results are discussed in Section 3.4.

Homoallylic alcohols also show useful stereocontrol. The reaction is highly sensitive to the stereochemistry of intervening alkyl substituents, as can be seen in the epoxidation of (30). Comparison of the results obtained for the diastereoisomeric substrates (31) and (32) shows that the effect of the alkyl group overrides the directing influence of the alcohol group itself.[32]

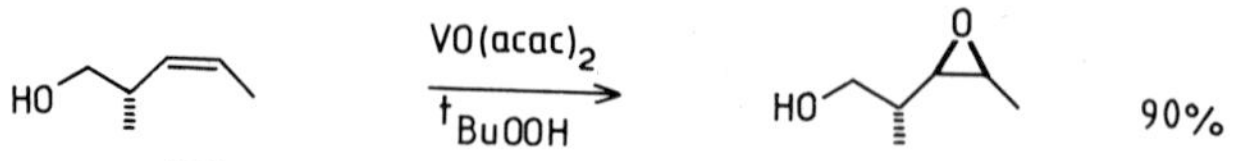

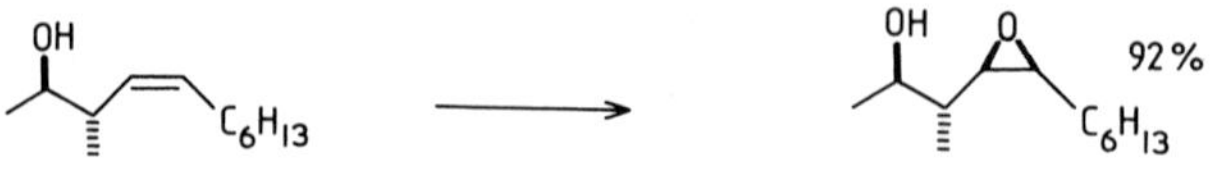

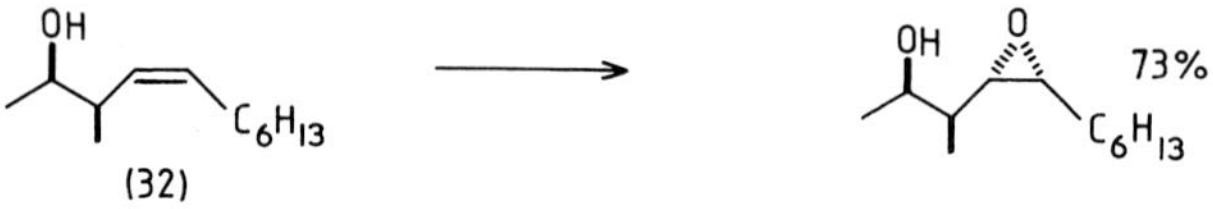

A homoallylic version of the palladium catalysed epoxidation of vinyl silanes, described above, has been used in a similar way to establish chiral centres in an intermediate (33), which was used for the construction of the erythronolide segment (34).[33]

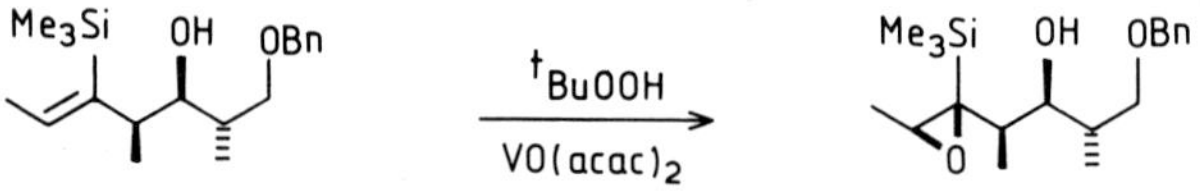

Siloxyl groups can also exert a directing influence. A modified vanadium catalysis procedure, using a silicon Lewis acid, gave rise to considerable stereocontrol in the epoxidations of both allylic and homoallylic ethers (Scheme 3.8).[34]

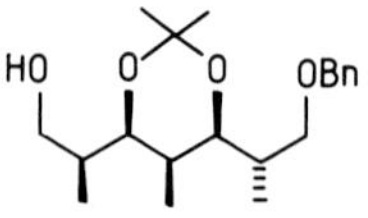

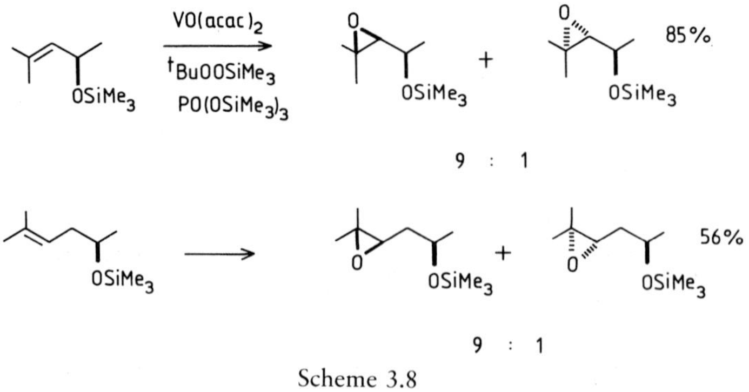

Scheme 3.8

Useful stereoselectivity is achieved even in the presence of a variety of potentially competing functional groups, provided an appropriate catalyst is used. Ti(IV) catalysis proved superior[35] to the use of vanadium for the epoxidation of (35) with t-butylhydroperoxide. Comparison of the stereochemistry of the products with those arising from (30) or (32) indicates the dominance of metal coordination by the OH group. The metal catalysed epoxidation procedure gave complementary results, in this case, to the use of *m*-chloroperbenzoic acid which was highly selective for the formation of (36).

	Ti(IV)	V(V)
Selectivity	10 : 1	3 : 2

3.4 Mechanistic considerations

The details of the interaction of hydroperoxides and alkylperoxides with early transition metals are clearly of great relevance to the discussion of the mechanism of the epoxidation process. Studies in this area by Mimoun[18] have concentrated on the correlation of catalytic activity with reactivity properties of related peroxo and alkylperoxidic complexes. The general structural characteristics of these species are similar, involving[36] a triangular coordination of the two oxygen atoms as in (37). Although not all compounds containing this structural unit are active as epoxidation reagents or catalysts, the involvement of such species in catalytic cycles is well supported by ^{18}O labelling studies.[37,38]

Two distinct types of oxygen transfer reactions appear to be involved in epoxidation processes. These are heterolytic oxygen transfer, and homolytic oxygen transfer. Which one operates depends on the nature of the catalyst. The essential difference between the two is that while heterolytic processes can be considered in terms of movements of pairs of electrons in the making and breaking of bonds, homolytic bond cleavage distributes electrons equally between the atoms involved, so producing radical species as intermediates.

Heterolytic oxygen transfer

Heterolytic oxygen transfer is characterised by complete stereoselective conversion of *cis*-alkenes to *cis*-epoxides and *trans*-alkenes to *trans*-epoxides, in a process that is strongly retarded by σ donor solvents such as DMF. This is interpreted in terms of an initial coordination of the alkene to the metal peroxide complex, followed by an intramolecular rearrangement[39] to form the metallocyclic intermediates[40] shown in Scheme 3.9. The effect of donor solvents is thus a competition with the alkene for the

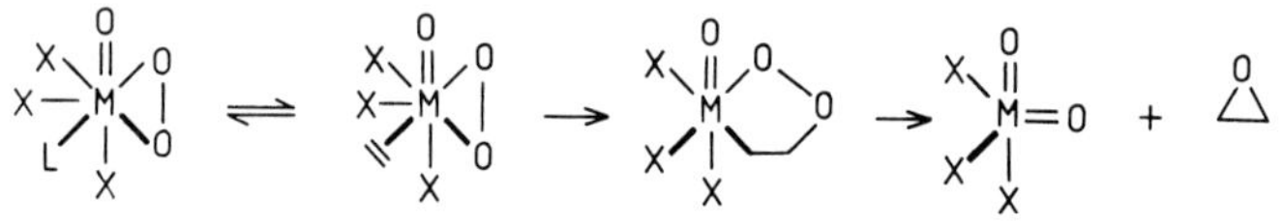

Scheme 3.9

free coordination site of the catalyst. This also accounts for the correlation between the susceptibility to epoxidation and the nucleophilic character of the alkene. This need not, however, be interpreted as evidence for direct attack on the peroxide species.

In the case of alkylperoxidic complexes, ^{18}O labelling studies[37,38] have shown that peroxo complexes are not intermediates. Kinetic investigations of the epoxidation of styrene have been interpreted[41] in terms of intermediate catalyst–alkyl peroxide and catalyst–alcohol complexes. The similarity of reactions using alkyl peroxides to those involving hydrogen peroxide can be accounted for by proposing the related mechanism shown in Scheme 3.10. It is important in both cases that there should be a releasable

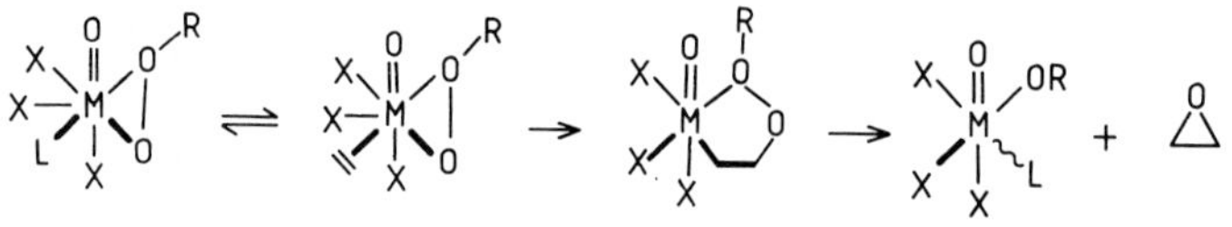

Scheme 3.10

coordination site adjacent to the peroxide group to allow alkene binding to occur at a position where C–O bond formation is possible.[18]

Homolytic oxygen transfer

In contrast to the behaviour described above, some complexes, typically V(V) peroxo complexes, show little stereoselectivity in their reactions with *cis*- or *trans*-alkenes, giving mixtures of *cis*- and *trans*- epoxides in both cases. Little inhibition is observed in these cases in donor solvents, and reactive complexes are also capable of hydroxylating arenes to phenols. These properties are consistent with the involvement of a biradical intermediate such as the V(IV) species (39), formed by homolytic cleavage of a V–O

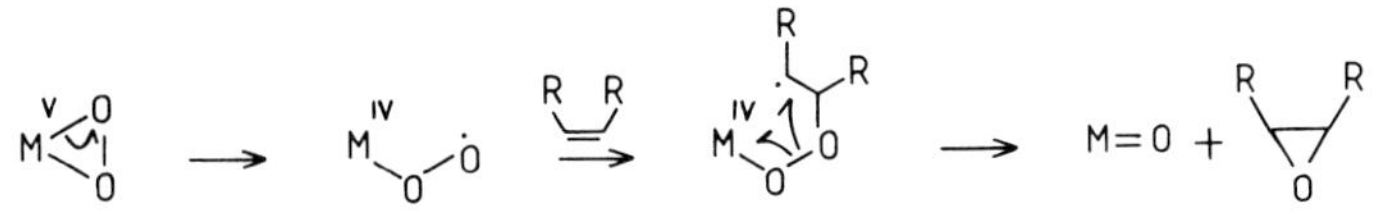

(38)   (39)

bond in (38). Complexes of hydrogen peroxide[42] and TBHP[43] with structures as shown in (38) have been isolated and characterised by X-ray crystallography.

Allylic and homoallylic alcohols

As was seen in Section 3.3, alkenes with nearby alcohol substituents undergo efficient epoxidation with catalysts that are ineffective for the epoxidation of simple alkenes. This can result in the valuable chemoselectivity described above. The best catalysts for the exploitation of this property are vanadium or titanium complexes, such as $VO(acac)_2$, $VO(OR)_3$, $V_2O_3(OR)_4$ or $Ti(OR)_4$, that have no coordination site available for interaction with an alkene, and yet can bind alcohols by displacement of an alkoxide ligand (Scheme 3.11).

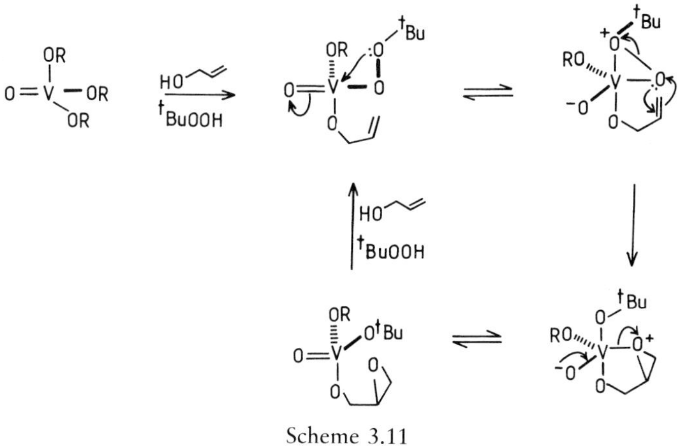

Scheme 3.11

In such cases the alkene is not itself coordinated to the metal and so can directly approach the bound oxygen atom. Reactions of this type have been examined extensively by Sharpless who has proposed[44] a mechanism involving a trigonal bipyramid intermediate with the now familiar triangular binding of the oxygen atoms. Indirect support for the involvement of this intermediate arises from an analysis[31] of relative contribution of *threo* (40) and *erytho* (41) transition states to the reaction path.

(40)   (41)

Far more complicated intermediates have been proposed for titanium catalysed asymmetric epoxidations, see Section 3.6.

It is clear from the mechanism of oxidation by TBHP and VO(acac)$_2$ that other metals capable of coordination both to the hydroperoxide and to an alcohol function may show the same type of catalysis. This has been found[45] to be the case for aluminium, using (^{t}BuO)$_3$Al and TBHP in benzene. Geraniol gave the expected 2,3-epoxide, and the *threo:erythro* ratios of epoxides from other allylic alcohols have been recorded.

Titanium, zirconium, and hafnium epoxidation catalysts have been the subject of a structural investigation by Wolczanski[46] who advanced a mechanism for the epoxidation process which is similar to that illustrated above for vanadium. It is notable, however, that the structure of the titanium catalyst [(tritox)TiMe$_2$]$_2$(μ-OMe)$_2$ although a dimer, is very different from the Sharpless asymmetric epoxidation catalysts discussed in Section 3.6.

3.5 Selective epoxidations in synthesis

Recourse to transition metal catalysed epoxidation is often made in cases where conventional peracid reagents give an excess of an undesired diastereoisomer. For example, in a biomimetic synthesis[47] of 14α-methyl-19-norsteroids, molybdenum catalysis gave the epoxide (42) in 5:1 excess. This was the minor product from treatment[48] with peracid.

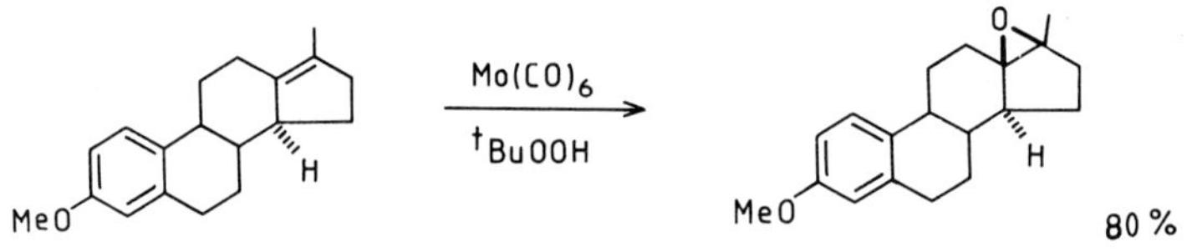

(42)

In cases where alkoxide binding can provide stereoselectivity, metal catalysis is frequently the best approach to follow. In an example related to prostaglandin synthesis, epoxidation of the homoallylic alcohol (43) proceeded[49] with *syn* addition of oxygen. In this cyclic system, the stereochemistry of the OH group dominates stereocontrol, since, unlike the epoxidations of (31) and (32) described in Section 3.3, only one face of the alkene is available for reaction with the metal-bound peroxide once the OH group is coordinated to the metal.

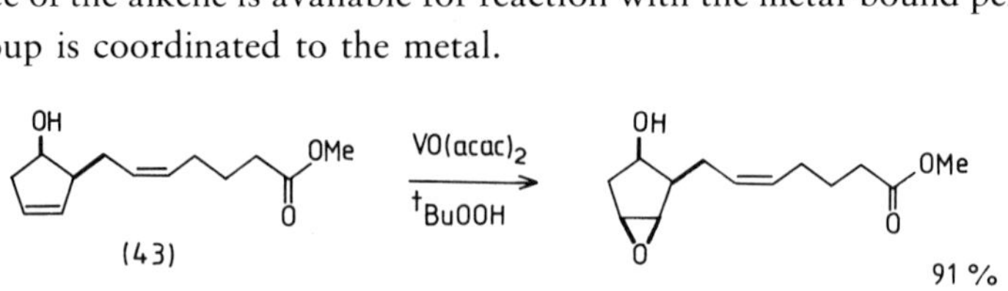

(43)   91%

Another instance is provided[50] by the oxidation of (44) to (45) by means of TBHP/VO(acac)$_2$. This was an early step in a synthesis of a prostacyclin analogue.

An example of selective epoxidation of a bis-homoallylic alcohol is encountered[51] in the case of the trisubstituted alkene (46). Increased steric compression in the

(44)   (45)

conformation (50), compared to (49), led to preferential formation of about 80% of the product (47), rather than (48), by the use of TBHP/VO(acac)$_2$.

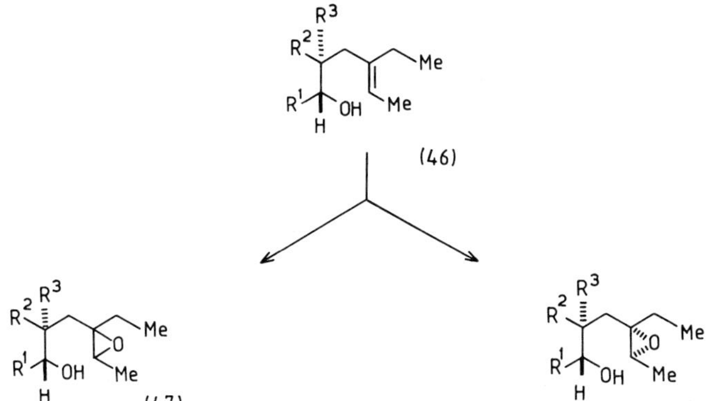

(46)

(47)

(48)

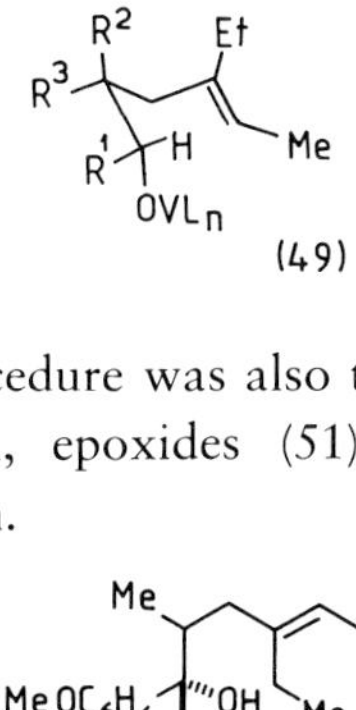

(49)

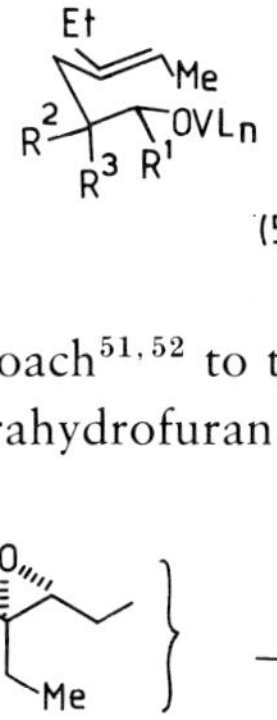

(50)

This procedure was also the basis of a convenient approach[51, 52] to the synthesis of lasalocid A, epoxides (51) or (52) providing the tetrahydrofuran moiety upon acidification.

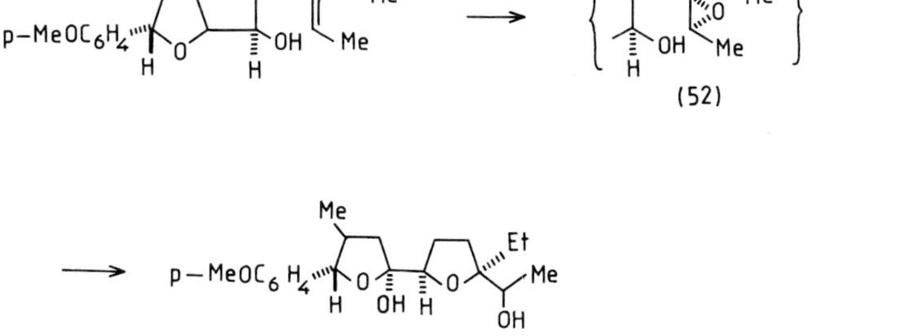

(51)

(52)

A stereocontrolled total synthesis of (−)-maytansinol (54) by Isobe[53] features an epoxidation reaction controlled by the OSiMe$_2$'Bu group of the advanced intermediate (53) that was prepared in resolved form from D-mannose

(53)   ≈ 80%

(54)

The factors contributing to selectivity can be subtle. Transannular reaction, rather than attack at a side-chain alkene, is a feature of a synthesis[54] of cytochalasins in which the intermediate (55) is selectively converted to (56).

(55)   (56)

3.6 Asymmetric epoxidations

The intimate involvement of alkene and peroxide molecules in the coordination sphere of the metal catalyst, described in Section 3.4, has led several groups to examine the possibility of an asymmetric modification of the epoxidation reaction by the inclusion of chiral auxiliaries as additional ligands. Initial investigations by Yamada and Terashima[55] and by Sharpless[56] obtained only moderate optical yields using the molybdenum complex (57) of (−)-N-methyl ephedrine and a vanadium complex of a hydroxamic acid (58) derived from camphor, respectively.

(57)   (58)

The first practical general asymmetric epoxidation procedure emerged[57] from the Sharpless group with the discovery that titanium tetraisopropoxide formed an excellent chiral catalyst in the presence of diethyl tartrate and t-butylhydroperoxide. Typical examples with optical yield in excess of 95% are shown in Scheme 3.12.

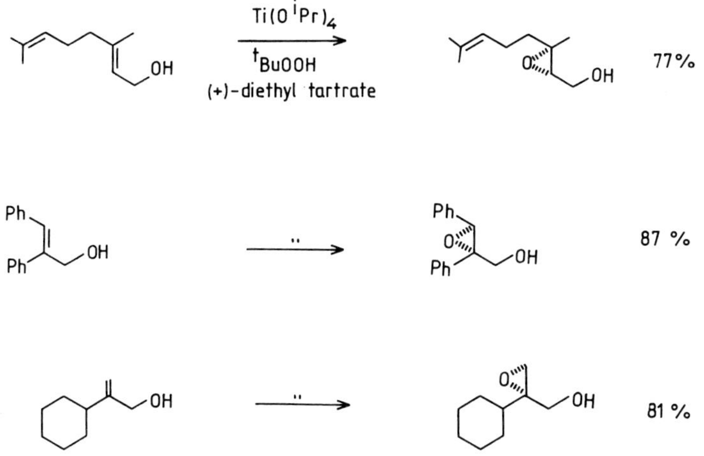

Scheme 3.12

In these reactions the chiral auxiliary, diethyl tartrate, serves to distinguish the enantiofaces of a prochiral alkene, forming the epoxide from the back face of the allyl alcohol in the examples shown. An interesting variation of this procedure employs symmetrical substrates such as (59) which contain two enantiotopic alkenes. In this case two forms of stereodifferentiation occur, the selection of one of these two symmetry-related alkenes, and selectivity between the enantiofaces of the double bond. These types of reactions have been developed by Schreiber[58] who has demonstrated a remarkable feature of processes of this type, in which kinetic resolution increases the enantioselectivity of the reaction. (For conventional examples of kinetic resolution, see pp. 67–8). By allowing the formation of small amounts of bis-epoxides, Schreiber has shown that the enantiomeric excess of the major product increases with time. There is an optimum reaction time for a given substrate, combining remarkable enantioselectivity with a very satisfactory yield. In principle, any required enantiomeric excess could be achieved, although at the expense of yield.

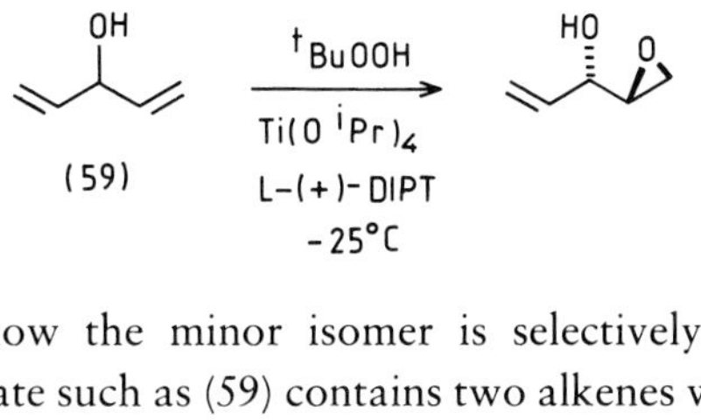

It is simple to see how the minor isomer is selectively removed in the kinetic resolution step. A substrate such as (59) contains two alkenes which react with the chiral catalyst at different rates. The minor product arises from epoxidation of the slower-reacting alkene. In this case, the faster-reacting alkene remains unchanged in the mono-epoxide intermediate. This, then, will undergo the second epoxidation step far more rapidly than its epimer. This effect gives rise to a guaranteed kinetic resolution in these circumstances.[58] This ingenious type of reaction has now been employed in the preparation of intermediates such as (60) and (61) in reaction sequences that are particularly efficient since the symmetry of the molecule allows identical reactions to occur at each end of the intermediates during the preparation of the epoxidation substrates.[59]

(60)

89%

(61)

81%

Asymmetric epoxidation of homoallylic alcohols usually gives products in much lower yield and enantiomeric excess. An example is the formation of (62), an intermediate in the asymmetric synthesis[60] of $(-)$-γ-amino-$\beta(R)$-hydroxybutyric acid (63). An improvement in enantiomeric excess with homoallylic alcohol substrates has

been obtained by the use of a zirconium catalyst with dicyclohexyl-tartramide and TBHP. Enantiomeric excess (e.e.) was somewhat variable depending on the nature of the substrate, but a good example is the formation of (64), which proceeded with 77% e.e. The yield of the reaction, however, was only 25%.[61]

(64)

For the more straightforward asymmetric epoxidation of allylic alcohols, the use of titanium appears to be the key to success, and in the further exploration of this chemistry, similarly excellent optical yields have been obtained[62] by using amino alcohols as the chiral auxiliaries under reaction conditions employing either hydrogen peroxide or alkyl or aryl hydroperoxides as the oxidant.

The original Sharpless tartrate-based procedure has itself been developed extensively and is now widely used as a synthetic entry to resolved epoxides. A particularly important development has been the addition of molecular sieves to the reaction mixture. By establishing reliably anhydrous conditions in this way, the reaction was made truly catalytic, requiring only 5 to 10 mole % tetraisopropoxide and only 6 to 13 mole % of the tartrate ester.[63] Previously, near stoichiomeric amounts of the titanium-tartrate catalyst were frequently required to achieve efficient reactions. A related procedure dramatically reduced reaction times by the addition of a catalytic amount of calcium hydride and silica gel to the Sharpless reagent.[64] Modified in this way, the reagent proved effective for the epoxidation of the gibberellate (65) for which the standard Sharpless reagent was unsatisfactory.[65]

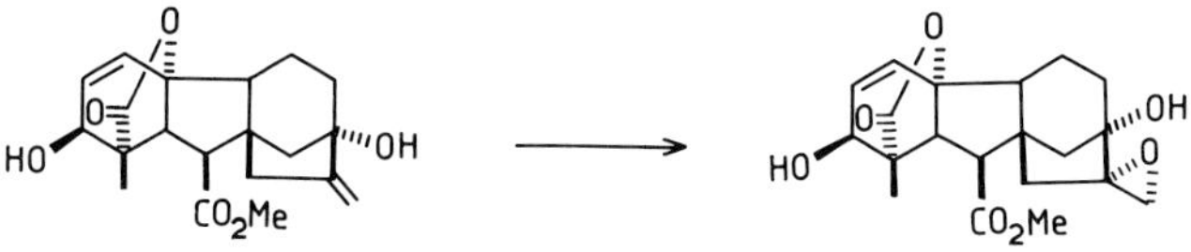

(65)

71%

Scheme 3.13 shows a variant[66] of the method using $TiCl_2(O^iPr)_2$ and tartrate diesters and amides to form chlorodiols derived from the opening of epoxides with the opposite stereochemistry to those obtained under standard conditions. The yield of the direct epoxidation can be improved to 51% by use of $Ti(O^tBu)_4$; but epoxide formation via the chlorodiol is still more efficient.

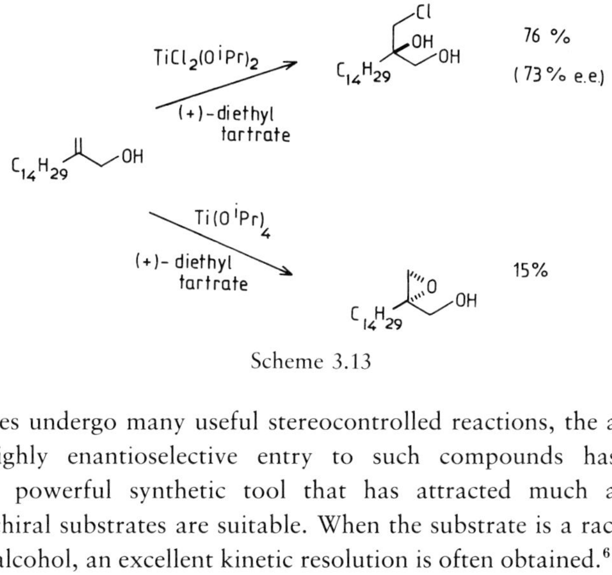

Scheme 3.13

Since epoxides undergo many useful stereocontrolled reactions, the availability of a simple and highly enantioselective entry to such compounds has provided an extraordinarily powerful synthetic tool that has attracted much attention. Both prochiral and chiral substrates are suitable. When the substrate is a racemic sample of a chiral allylic alcohol, an excellent kinetic resolution is often obtained.[67] Depending on the degree of conversion used, an arbitrarily high enantiomeric excess is claimed for the unreacted isomer. An enantioselective version of the epoxidation of vinyl silanes discussed in Section 3.4 provides an example of the use of this type of reaction in the presence of partially protected diols. The process effected a kinetic resolution affording both the epoxide (66) and the allylic alcohol (67) in greater than 99% enantiomeric

excess. The enantiomer of (66) was obtained from (67) by the use of (−)-diethyl
tartrate instead of the (+) isomer.[68]

A further example of a kinetic resolution was used to prepare CF_3 substituted allyl
alcohols in 97 % e.e. and 31 % yield. The reaction was performed at low temperature
(between −78 °C and −20 °C) over a period of 6 days.[69]

Nature of the Sharpless catalyst

Considerable effort has now been expended in the study of the active catalyst species in
the Sharpless asymmetric epoxidation. Kinetic and structural evidence[70] is consistent
with the presence of a dimeric catalyst with C_2 symmetry and consequently, two
stereochemically identical active sites. It has become clear that when titanium
tetraisopropoxide and diethyl tartrate are mixed with t-butylperoxide, a great diversity
of complexes are formed. Materials with a variety of stoichiometries can be identified
in such mixtures.[71] The dimer (68), however, constitutes about 90 % of the mixture, and
also is the most active catalyst of all the many species present. Consequently it is the

dimer (68) which dominates the asymmetric epoxidation process, accounting almost
entirely for the conversion of the allylic alcohol into the epoxide.[72] This dimeric species
has been the subject of a theoretical investigation by Jorgensen, Wheeler, and
Hoffmann, who sought to define the electronic and steric factors that determine the
course of the asymmetric epoxidation step.[73] A further attempt to probe the nature of
the stereocontrol is to be found in the work of Hawkins and Sharpless. Here, the
problem was simplified by the deliberate formation of a monomeric catalyst species (69),
which proved, itself, to be a fairly effective catalyst in asymmetric epoxidations,
producing products with enantiomeric excesses between 40 and 60 %.[74]

Examples of the use of Sharpless asymmetric epoxidation in synthesis

In this concluding part of Section 3.6, we shall examine a number of examples of the application of the Sharpless asymmetric epoxidation in synthesis, and then discuss the effect the development of this reaction is having on the direction of the evolution of synthesis design. These examples have been drawn from the work of many different synthetic chemists, for the Sharpless reaction is one which has been taken up and used by many other, less specialised groups, a fine testament to its simple and robust nature.

First, however, we should consider an example from Sharpless' own work. In his synthesis of the mannosidase inhibitor swainsonine (70), the sequence, asymmetric epoxidation followed by opening of the epoxide by nucleophiles, was used twice at separate stages to introduce the required stereocontrol in the run of four adjacent chiral centres.[75]

The same type of process, epoxidation/nucleophile addition, has provided Oehlschlager and Czyzewska with a convenient route[76] to resolved allenes such as (71), opening the epoxide via cuprate addition to the alkyne. (Further examples of the addition of cuprates to alkynes are given in Chapter 8.)

The epoxidation of (72) provided[77] an enantioselective route to the resolved acetylene (73), a key intermediate in Baker's synthesis of the spiroketal units of avermectins B_{1b} and B_{2b}.

The simple epoxides (74) and (76), available from allyl alcohol by the use of either
(+) or (−) tartrates and cumene hydroperoxide, have been elaborated *in situ* to
produce intermediates (75) and (77) for the synthesis of the β-blocker (78).[78] The route
via (77) provides a convenient and flexible process for the synthesis of a range of
analogues while the process using (75) seems more suitable for large scale preparations.

There are now examples of very large scale applications of the Sharpless epoxidation
process.[72] The epoxide (79) has been produced by Upjohn on a 150 mole scale, and Arco
has developed an even larger scale process (660 mole) for the epoxidation of allyl
alcohol itself.

In a synthesis of an unusual cyclosporine amino acid (80), selective epoxidation of the
allylic alcohol in (81) gave a convenient first step for an enantioselective route. Opening
the epoxide intermediate with cuprate reagents was not selective, the best results being
obtained in the presence of Lewis acids (see Chapter 8). At this point, however, a second
important chiral centre in (80) was introduced.[79]

Another example which combines regioselectivity with asymmetric induction arises
in a synthesis of a polyene natural product from *Laurencia pinnatifida*.[80] Epoxidation

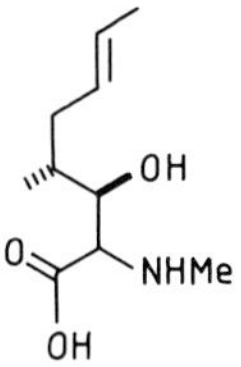

(80)

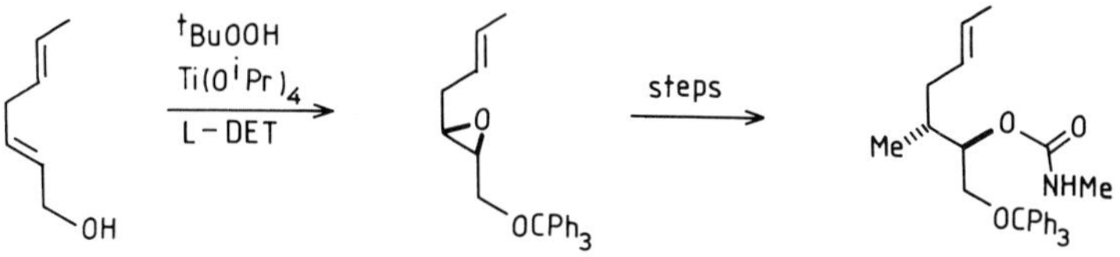

(81)

of a terminal allylic alcohol in a triene intermediate was followed by isomerisation of the α-hydroxy epoxide to move the epoxide group to the end of the carbon chain, where it was required for reaction with an acetylide.

An important fragment (84), used by Nicolaou in his total synthesis of the macrolides amphoteronolide B and amphotericin B, was also obtained by nucleophilic opening of the epoxide. Asymmetric epoxidation, followed by protection, produced the epoxide (82) which contains a C_2 axis of symmetry. In this way, the question of regiocontrol

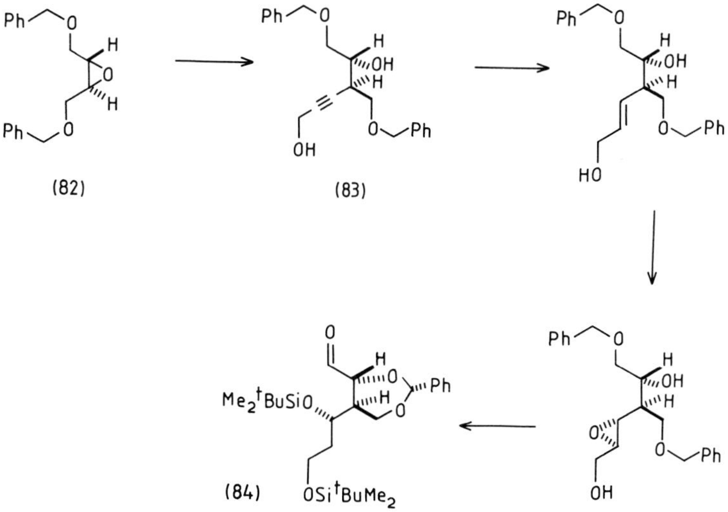

(82)   (83)

(84)

during the opening of the epoxide was avoided. The product of acetylide addition, (83), was elaborated to (84) in a sequence of reactions that employed a further Sharpless epoxidation step and regiocontrolled opening of the epoxide with REDAL.[81]

It is a measure of the great advantage of the allylic alcohol based epoxidation processes that it can now be worthwhile introducing an alcohol control group in a synthesis specifically to obtain the benefit of an easy enantioselective entry. A good example arises in the synthesis of a key chiral portion of another macrolide, roflamycoin, by Lipshutz.[82] Once again, Sharpless epoxidation was followed by nucleophile addition, this time producing the intermediate (85). The alcohol group which served to control the epoxidation step and so establish the two chiral centres in (85) was then removed by reduction of the derived tosylate. Several other epoxide intermediates were used in this synthesis to introduce other oxygen-bearing chiral centres in roflamycoin.

(85)

Because of the remarkable generality of application of the Sharpless asymmetric epoxidation, the development of this reaction has had a profound impact on the design of organic synthesis. Epoxides have gained greatly in popularity as key intermediates and, as the acceptance of the Sharpless reaction has become more widespread, so too has there been an increase in the development of new methods to utilise epoxides in synthesis. In this sense, the Sharpless epoxidation has become one of the truly great, influential reactions of recent years. A number of syntheses have been reported which utilise the epoxide product in a variety of interesting ways. This epoxidation method is a reaction that has come of age, and is now leading the development of synthetic design, rather than serving it. Several examples follow to illustrate this statement.

The bis-epoxide (86) was prepared by a series of two epoxidation steps for utilisation in a cascade cyclisation of interest as a potential route to polycyclic polyethers. Hydrolysis of the ester with pig liver esterase resulted in the formation of (87) in 77% yield.[83]

(86)   (87)

A cyclisation promoted by an epoxide also figured in a synthesis of the epoxy-3',6'-auraptene derivative (88) and its enantiomer. After the initial epoxidation step, it was again necessary to remove the alcohol control group before cyclisation with SnCl$_4$.[84] This somewhat roundabout route again points to the premium now placed on the efficient enantioselective start to the reaction sequence offered by the use of allyl alcohol derived epoxides.

The construction of a building block for the A ring of the anti-tumour compound taxol used a similar cyclisation, but retained the original alcohol group, since this was required in the intermediate (90). The epoxidation step leading to (89) further emphasises the selectivity and tolerance to other functional groups possible with this reagent.[85]

Another anti-tumour compound psorospermin (91) inspired the synthesis of the analogue (92). These target molecules contain epoxides but the route to (92) originally introduced the epoxide unit at a different position using the allylic alcohol control strategy. Subsequent conversion to the mesylate led to the formation of (92) upon removal of the benzyl protecting group.[86] In this case, also, the alcohol control group was put to work again, rather than simply deleted, after the epoxidation step.

(91)

An intramolecular ring opening step, combined with the enantioselective epoxidation of (93), introduced the two chiral centres required in Roush's synthesis of (+)-*erythro*-dihydrosphingosine (94).[87]

Instead of nucleophilic opening of the epoxide ring, an interesting alternative employs an oxidation step, following the epoxidation, to form an aldehyde. Nucleophile addition to the aldehyde then required the epoxide itself to serve as a control group. Use of $C_3H_5SnBu_3$ with a Lewis acid proved moderately selective, giving diastereoisomer ratios of between 6:4 and 9:1, depending on the conditions.[88] Techniques like this, which make further use of the alcohol functionality following the completion of its role in the epoxidation step, are typical of the direction in which these methods are developing.

3.7 Oxidation of ketones to α-hydroxyketones

A rather useful oxidation, the conversion of a ketone to an α-hydroxyketone, may be effected[89] using the peroxy molybdenum reagent $MoO_5.C_5H_5N$ in HMPA (MoOPH) with the enolate of the ketone. A typical example is the conversion of (95) into (96).

(95)   (96) H

The MoOPH oxidation has been used[90] to perform an oxidation at the less strained ring junction of (97) in a synthesis of (±)-patchouli alcohol which used (98) as a key

1) LiN^iPr_2
2) $MoO_5.C_5H_5N$, HMPA

(97)   (98)

intermediate. The same reagent served to introduce an hydroxyl group adjacent to the steroidal ketone in (99).[89]

i) LDA
ii) MoO_5. Py. HMPA

(99)   75%

The MoOPH reagent has now become popular and widely used, following gradual development in the 1970s. More recently, however, an alternative procedure[91] for this transformation has been developed. This approach uses the trimethylsilyl ether of the enolate, which may then be oxidised by a catalytic amount of OsO_4 and *N*-methylmorpholine-*N*-oxide to give, for example, the α-hydroxyketone (100) in high yield. Trapping the enolate as the silylenol ether has a distinct advantage, since

(100)

enolisation may be directed towards the kinetic or thermodynamically preferred enolate. In this way, the site of hydroxylation can be controlled, as illustrated by the two pathways shown in Scheme 3.14.

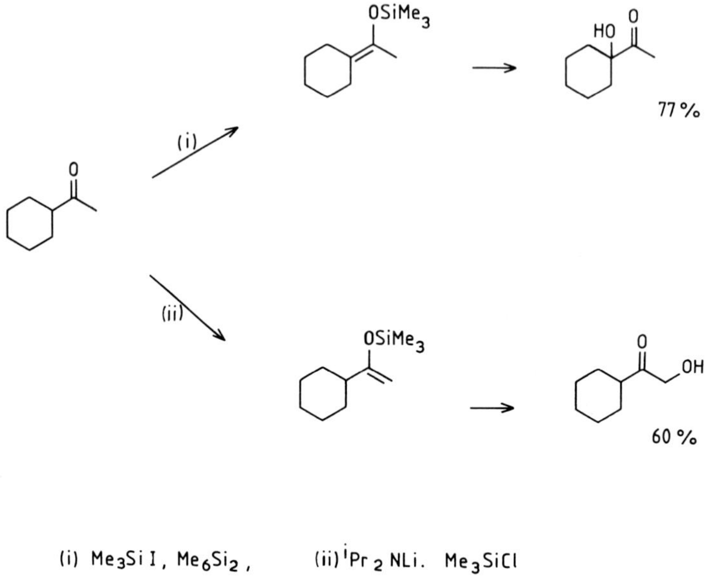
OSiMe3
HO
O
77 %
(i)
O
(ii)
OSiMe3
O
OH
60 %
(i) Me3SiI, Me6Si2,
(ii) iPr2NLi. Me3SiCl

Scheme 3.14

Oxidation of alkenes

4.1 The Wacker process and the oxidation of alkenes

The oxidation of alkenes to carbonyl compounds is an organic transformation of considerable antiquity. It was Phillips, in 1894, who first noted that aqueous palladium(II) chloride is reduced to metal by ethene, and that ethanal is formed.[1]

It was not until the 1950s, however, that the synthetic application of this reaction was developed by Smidt and co-workers in a study that led to the well known Wacker catalytic process.[2] In this procedure, the palladium metal is re-oxidised to palladium(II) by an oxidant with a higher redox potential, e.g. Cu(II) or Fe(III), the reduced species formed being itself re-oxidised by oxygen or air, so creating the catalytic cycle shown in Scheme 4.1. Since that time, the process grew rapidly in importance until in 1976 it

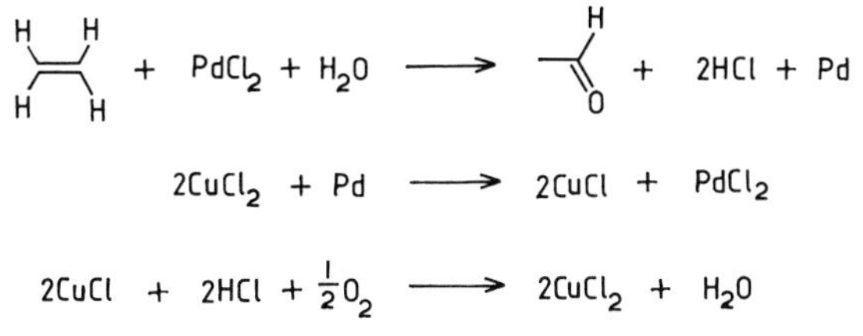

$$2CuCl_2 + Pd \longrightarrow 2CuCl + PdCl_2$$

$$2CuCl + 2HCl + \tfrac{1}{2}O_2 \longrightarrow 2CuCl_2 + H_2O$$

Scheme 4.1

was estimated that 82% of the world's 2·3 megatonne per year plant capacity for ethanal used the Wacker process.[3] More recently, the Monsanto process (see p. 282) has superseded some Wacker operations.

Over the same period it was established that the oxidation of alkenes in this way also offered a promising synthetic technique in laboratory scale synthesis. The reaction of aqueous palladium(II) chloride (or indeed other Pd(II) salts such as nitrate and acetate) with alkenes is general and has been applied to a wide range of substrates.[2,4] Terminal alkenes are converted to methyl ketones. Cyclic alkenes yield cycloalkanones, and

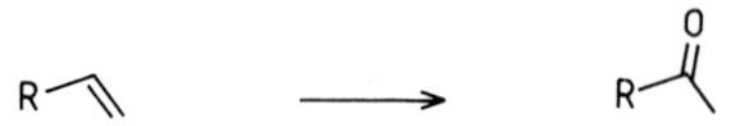

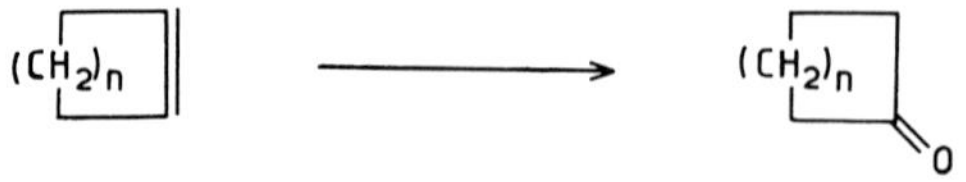

dienes react with isomerisation. For higher molecular weight alkenes (generally those

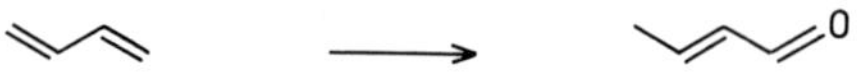

larger than hex-1-ene) a co-solvent is necessary. The solvent most commonly employed is dimethylformamide (DMF).

The direct formation of ketals occurs when alcohols replace water in the reaction. Acrylonitrile was converted[5] to the ketal (1) in high yield with $PdCl_2$ and $CuCl_2$. An

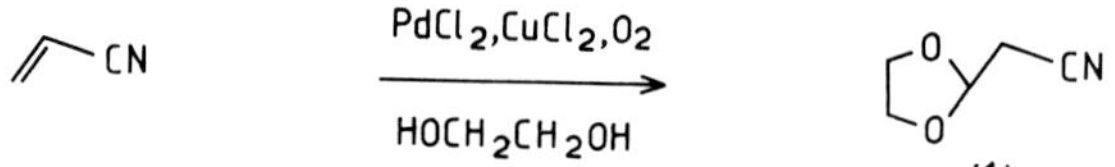

intramolecular version of the same process has been used[6,7] to effect a cyclisation to form endo-brevicomin (2), a reaction sequence discussed in detail in Chapter 14. A similar

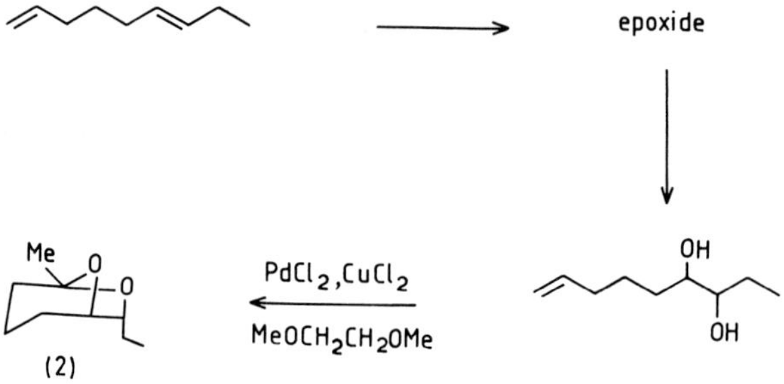

sequence starting from diallyl ethers (3) was also described,[6,7] yielding the trioxobicyclo[3,2,1] series (4).

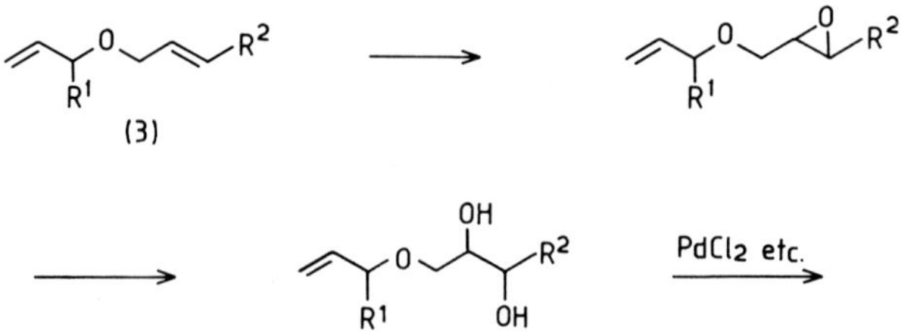

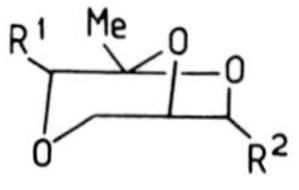

(4) $R^1 = R^2 = H$;
$R^1 = H, R^2 = Me$;
$R^1 = R^2 = Me$

This cyclisation method has been used in the synthesis of (5) and (6), attractant molecules of the Norwegian beetle *Trypodendron lineatum*. Dianion addition to 4-bromo-1-butene followed by borohydride reduction gave a mixture of *erythro* and *threo* products. Consequently a mixture of *exo* and *endo* isomers was produced.[8]

$$(5)\quad R^1 = Me; R^2 = H$$
$$(6)\quad R^1 = H; R^2 = Me$$

The use of a separate oxidising agent to re-oxidise the palladium is an important aspect of these reactions. A great variety of re-oxidising conditions have been examined.[9] Cu(II) salts are effective for this purpose, but since $CuCl_2$ can chlorinate[10] carbonyl compounds, CuCl is commonly used in the presence of oxygen. In this way side reactions are minimised.

Problems are sometimes encountered in the maintenance of the catalytic cycle by re-oxidation of CuCl. If this is carried out in aqueous DMF, for example, the reaction medium is not completely homogeneous and undesirably large amounts of palladium(II) and copper salts may be required. This is particularly notable in the oxidations of long-chain alkenes; sometimes nearly a stoichiometric amount of copper salts is needed to complete the oxidation. Efficient re-oxidation in homogeneous conditions can be achieved with organic oxidants such as benzoquinone[11] or peroxides.[12,13] A recent development is a catalytic use of benzoquinone which is regenerated by an

electrochemical method in which hydroquinone is re-oxidised at the anode of an electrochemical cell. The oxidation product (7) was obtained in this way in 82% yield.[14] In some cases, direct re-oxidation of palladium intermediates by O_2 has been proposed.[15,16] A multistep catalytic system employing $Pd(OAc)_2$ and benzoquinone effects oxidation in chloride-free media.[17]

A second recent development concerns the use of microemulsions to accelerate the oxidation reaction. In this case a stoichiometric amount of benzoquinone was used to effect re-oxidation of the catalyst. Oxidation of hex-1-ene proceeded over three times as fast when performed in a microemulsion obtained by the addition of surfactant materials to the reaction mixture.[18]

In some cases the 're-oxidant' may be intimately involved in the reaction. An investigation[15] of the first asymmetric induction during a Wacker oxidation provided evidence that in this catalytic system, in which Cu(OAc)$_2$ and O$_2$ were used to complete the cycle, direct combination to form (8) was followed by oxygen insertion into the Pd–H bond. In such a case, palladium would remain in the (2+) oxidation state throughout the cycle.

(8)

The C–O bond formation depends on initial coordination of the alkene by the Pd(II) salt (see Section 4.3), and, since coordination is impeded by substitution, terminal alkenes react faster than alkenes with internal double bonds. Similarly, *cis*-alkenes react faster than *trans*-alkenes.

4.2 Selective oxidation of alkenes

The pronounced differences in reactivity of alkenes with different substitution patterns gives rise to useful selectivity when substrates contain more than one alkene. This selectivity is frequently employed to oxidise terminal alkenes in the presence of internal alkenes as in the formation of (10) from (9) and (12) from (11).[9] The diketone (12)

(9)   (10)   59%

(11)   (12)   85%

produced in the last of these oxidations was used as an intermediate in a synthesis of muscone.[19] Another simple example which illustrates selectivity is found in the synthesis of *cis*-jasmone (14).[20]

Such products as (12) and (13) are typical in reflecting the general tendency of these processes to form methyl ketones from terminal alkenes. The preference for oxygen

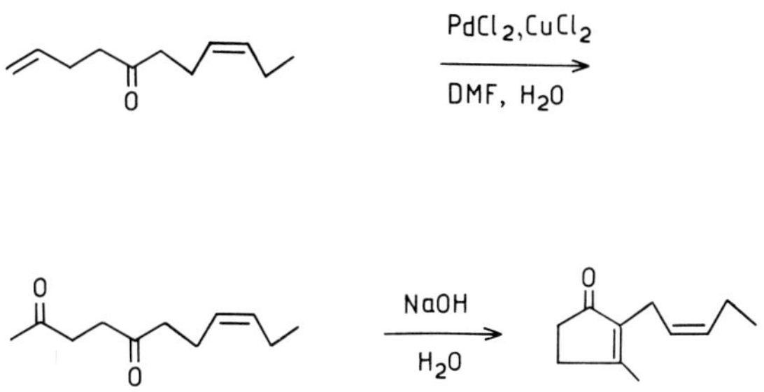

(13) (14)

addition to the inner carbon atom is generally explained in terms of an associated movement of the metal towards the less hindered end of the alkene (see Section 4.3). The presence of an electron withdrawing substituent on the alkene reverses the regiochemistry in the case of alcohol additions. The reaction was originally examined[5] for ethylene glycol and acrylonitrile as described in Section 4.1, but has since been found[21] to be of general importance. A good example is the acetalisation shown in Scheme 4.2.

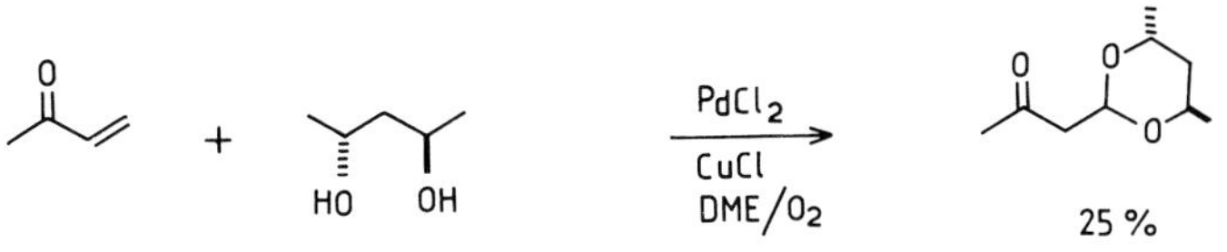

Scheme 4.2

A subsequent mechanistic investigation[22] of reactions of this type using deuterium labelling is discussed in Section 4.3.

Although the process is known, the oxidation of internal alkenes is an extremely slow reaction. Phase transfer catalysts[23] have been employed to increase reactivity. The production of mixtures of regioisomers, as seen in the products from (15) and (16) is,

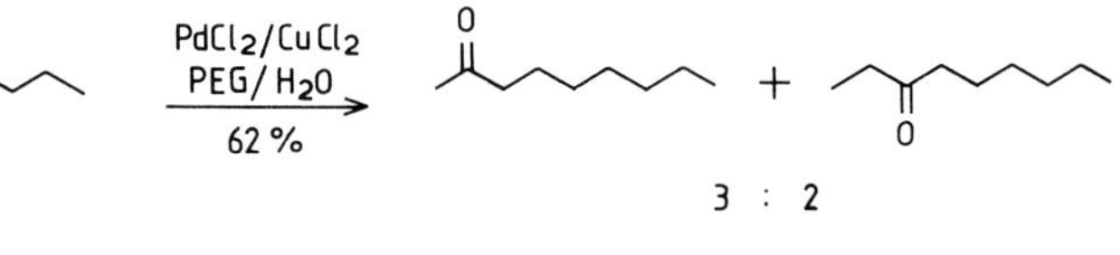

(15)

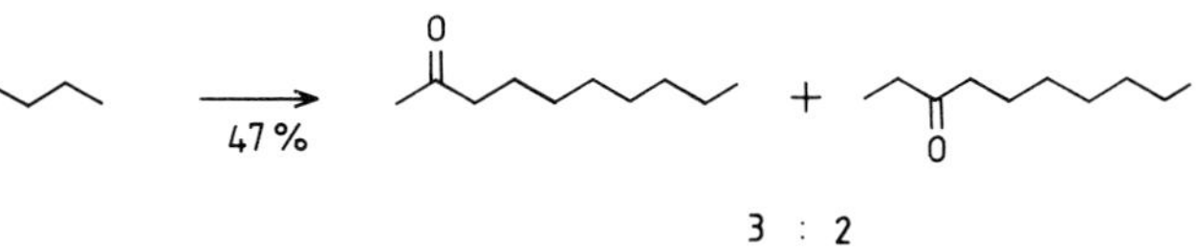

3 : 2

(16)

3 : 2

unfortunately, a typical feature of the reactions of internal alkenes. An exception is the formation[24] of (18) in 73% yield from the γ,δ-unsaturated lactone (17). This was performed under conventional conditions and so required a reaction time of seven days.

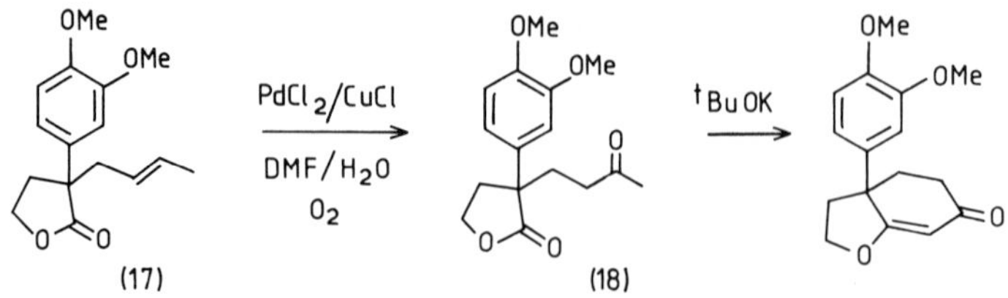

The strong directing influence of electron withdrawing substituents is also apparent in the reactions of internal alkenes and provides reliable regiocontrol in such cases.[9] Typical examples are illustrated in Scheme 4.3. These reactions use hydrogen peroxide

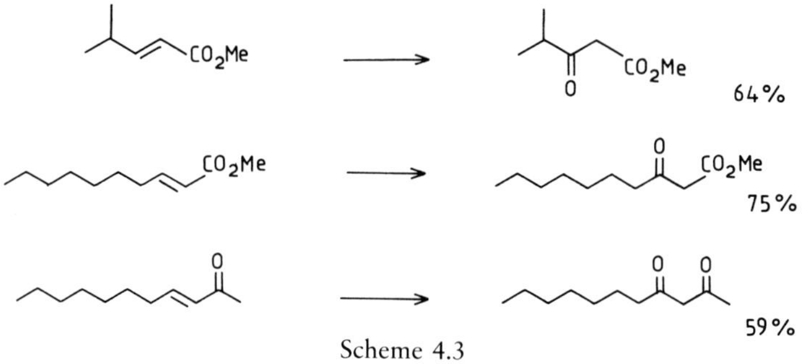

Scheme 4.3

or t-butylhydroperoxide as the re-oxidant and provide an interesting contrast to the epoxidation reactions discussed in Chapter 3. Both the catalyst and the reaction conditions are quite different, the oxidation in this case being performed by disodium tetrachloropalladate in aqueous acetic acid.

Surprisingly regioselective oxidations of β,γ-unsaturated esters can be achieved[25] by the use of $PdCl_2/CuCl/O_2$ in aqueous dioxane or THF. Use of DMF for this reaction is unsatisfactory since π-allyl complexes are formed (see Chapter 6). Regioselectivity is also achieved[26] in the presence of other remote substituents, as found in allyl and homoallyl ethers and acetates. Representative examples of these reactions are given in Scheme 4.4. Products such as (19) undergo facile elimination to form enones. Allyl ethers

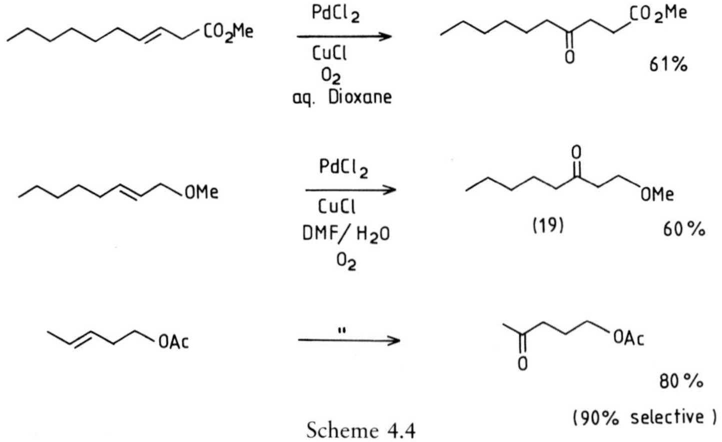

Scheme 4.4

can thus serve as masked vinyl ketones by means of this oxidative procedure (see also Chapter 13 and Section 4.4).

4.3 Mechanism of the oxidation of alkenes by palladium complexes

The original mechanism for the Wacker oxidation and related reactions, as described in Henry's review[27] in 1975, involved a *cis* transfer of OH to a metal-bound alkene. Although the stereochemistry of this process cannot be examined directly, subsequent stereochemical investigations by Åkermark[10] and Stille[28] are more easily explained[29] by a *trans* addition of water in the fashion now well accepted for nucleophilic additions to coordinated alkenes (see Chapter 5). On this basis, the mechanism can be depicted as shown in Scheme 4.5. This process is consistent[29] with the original kinetic evidence that

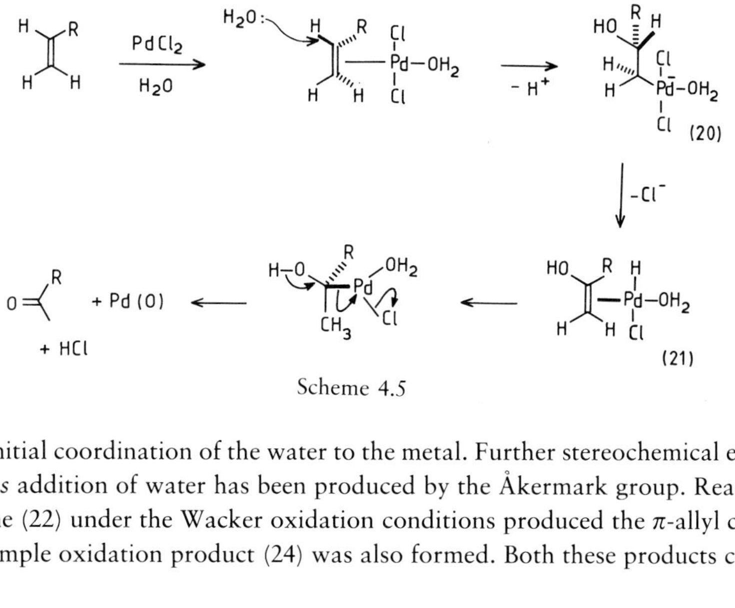

Scheme 4.5

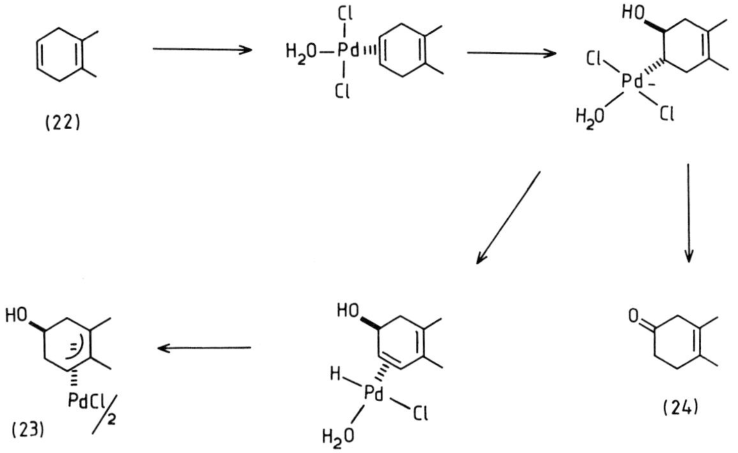

suggested initial coordination of the water to the metal. Further stereochemical evidence for the *trans* addition of water has been produced by the Åkermark group. Reaction of the 1,4-diene (22) under the Wacker oxidation conditions produced the π-allyl complex (23). The simple oxidation product (24) was also formed. Both these products can arise from initial coordination of the less substituted alkene followed by nucleophilic addition of water. There are then two possibilities for β-elimination, which could proceed as in the formation of (21), or from a CH_2 group to form a π bound intermediate which could afford (23) by hydrogen transfer to the ring. The *trans* relative stereochemistry in (23) strongly supports the initial addition of water to the alkene on the face opposite to that bound by the metal.[30] The formation of dimeric palladium chloride complexes such as (23) is discussed in more detail in Chapter 6. A further example of the formation of a π complex under the Wacker oxidation conditions is to be found at the end of Section 4.4. While modification of the reaction for stereochemical investigation points to *trans* addition of water, the possibility of a *cis* mechanism in the original Wacker oxidation of ethene still cannot be discounted.

Reactions of labelled compounds have demonstrated that all hydrogens in the final product come directly from the alkene. This can be explained either in terms of a rapid reaction sequence of β-elimination and hydrogen transfer to the metal via the intermediate (21), or the concerted rearrangement involving a hydrogen shift shown in Scheme 4.6.

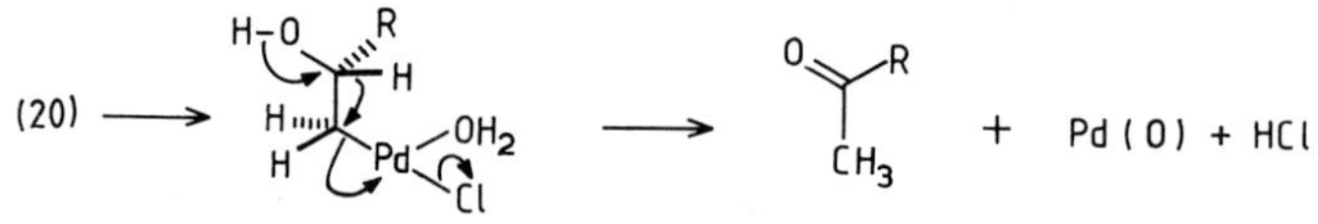

Scheme 4.6

Reactions in alcohol solvents that proceed directly to form acetals presumably follow a related mechanism in which a second alcohol adds to the coordinated alkene (25) formed in the β-elimination step. The final metal complex can now decompose by reductive elimination.

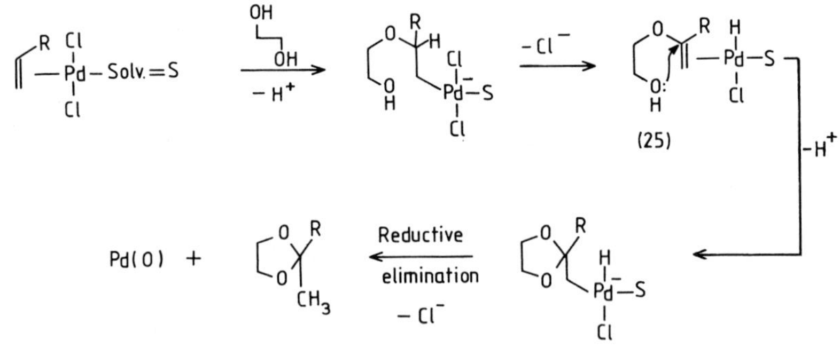

An investigation using labelled compounds has indicated that cyclisation to form the acetal can occur in several ways. The deuterated substrate (26) forms a 1:1 mixture of (27) and (28), indicating loss of deuterium from an intermediate palladium alkene complex.[22] After alcohol addition, β-elimination can initially produce a Pd–D species

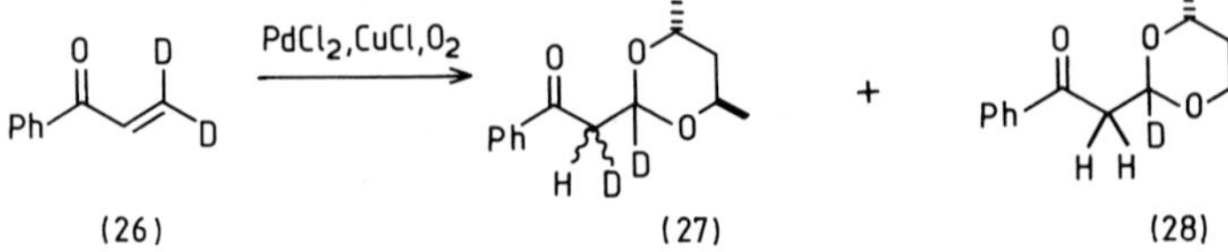

(26)   (27)   (28)

(29). This can equilibrate with the Pd–H species (30) by an addition–elimination process, and if (29) and (30) are in equilibrium with the free ligands, the deuterium will

be lost from (29). Cyclisation, promoted either by palladium, or by acid, could occur at very similar rates for both d_2 and d_1 compounds. This would produce equal amounts of (27) and (28), if the equilibration processes are fast compared to cyclisation. On this basis, an additional refinement must be added to the mechanism previously described for five-membered ring acetals. In the six-membered ring case, the intermediate equivalent to (25) appears to be in equilibrium with the uncomplexed vinyl ether. For $HOCH_2CH_2OH$, however, cyclisation of (25) may be more rapid since a five-membered ring is being formed. The participation of the vinyl ether may not be so significant in these circumstances.

From the point of view of regiocontrol, the important step is the formation of (20). The commonly observed addition of the OH group to the more hindered end of the alkene can be accounted for if steric effects are dominated by the large PdL_3 group which moves selectively towards the more open end of the alkene (the CH_2 end) during the oxidation of terminal alkenes such as (9). Exceptions arise when substituents take over the regiocontrol process by introducing electronic effects through direct interaction with the coordinated π system.

The overall effect of the process is the required interconversion:

$$PdCl_2 + H_2O + RCH{=}CH_2 \rightarrow Pd(0) + 2HCl + RCO{-}CH_3$$

The palladium(0) formed in this reaction must be re-oxidised, as described in Section 4.1, to enable the reaction to proceed in a catalytic manner.

4.4 Synthetic versatility of the Wacker oxidation of alkenes

In terms of synthetic applications, the main value of the Wacker oxidation lies in the introduction of masked ketones (see also Chapter 13) in the form of vinyl groups. Examples given in this section testify to the tolerance of the reaction to the presence of a wide variety of functional groups elsewhere in the molecule. This allows the unmasking of the ketone to be delayed until a late stage in the synthesis. A good example arises in Tsuji's synthesis[31] of (+)-19-nortestosterone (32) in which the acetyl group in (31) is unmasked after the Robinson anellation had been used to form the B ring.

Another route[32] to (32) employed two PdCl$_2$ oxidations following an asymmetric aldol reaction catalysed[33] by L-phenylalanine. A similar approach[34] was also successful for the synthesis of D-homo-4-androstene-3,17-dione.

In a stereocontrolled synthesis of the steroid D ring fragment (33), the ketone required for cyclisation was again introduced by a Wacker oxidation. The subsequent step, a decarbonylation by means of Wilkinson's catalyst, afforded (33),[35] an example of a type of reaction that is discussed fully in Chapter 9.

A further interesting example of the use of the Wacker oxidation is provided by the synthesis[36] of (±)-zearalenone (34) which uses a starting material obtained by a Pd-catalysed telomerisation of butadiene[37] in the presence of acetic acid (see Chapter 10).

The variety of ways available to introduce the vinyl group contributes to the appeal of the Wacker oxidation. A simple example is the alkylation of allyl bromide, as in the synthesis[9, 38,] of the cyclopentenone (35). This approach has also been used in a synthesis

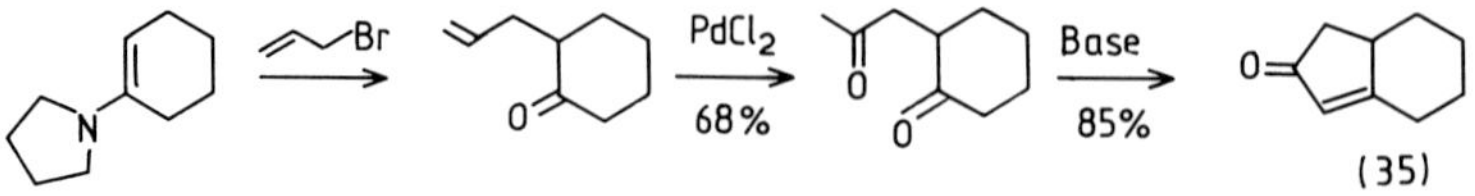

of muscone via the cyclopentenone (36). In this case the three carbon unit was introduced by alkylation of a β-ketoester.[39]

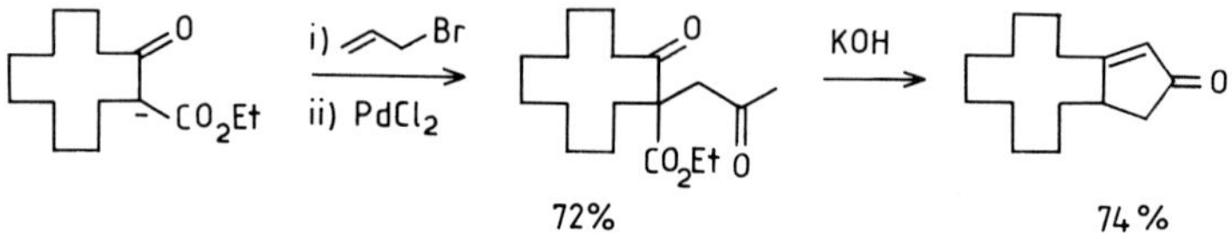

Allylation of an aldehyde as a route to ketoaldehydes required for cyclisation, is successful in a synthesis of cuparenone (37) by Wenkert.[40]

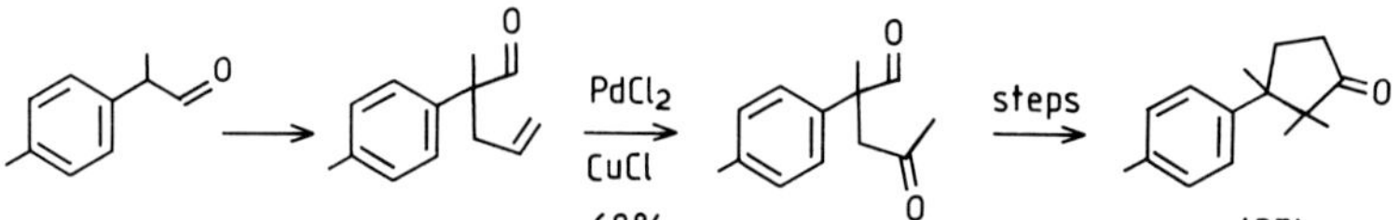

Enolate alkylation with allyl bromide, followed by $PdCl_2$ oxidation, has been used to form (38) in a synthesis of coriolin.[41]

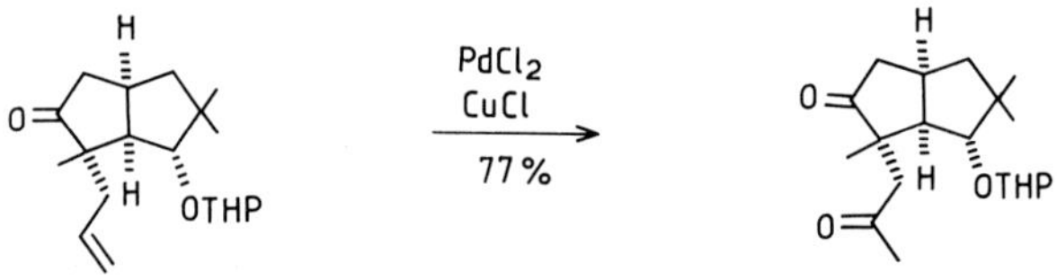

Conjugate addition by lithium divinylcuprate (see Chapter 8) provides another simple entry to vinyl intermediates (Scheme 4.7).[42]

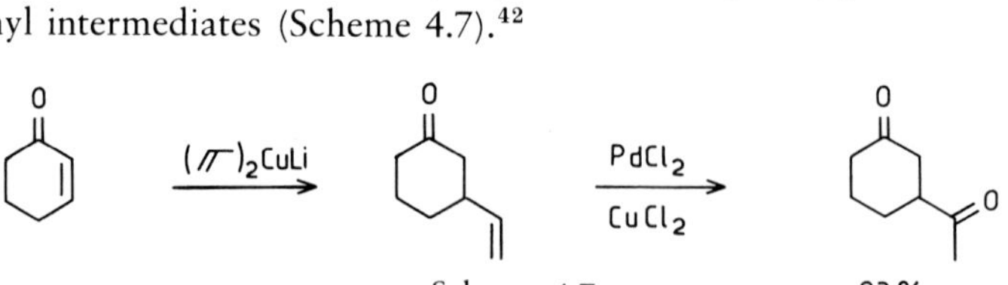

Scheme 4.7

In a dihydrojasmonate synthesis,[43] a Cope rearrangement gave access to the vinyl intermediate (39). A dihydrojasmone synthesis[44] required the selective oxidation of the vinyl group in the presence of an internal vinyl silane.

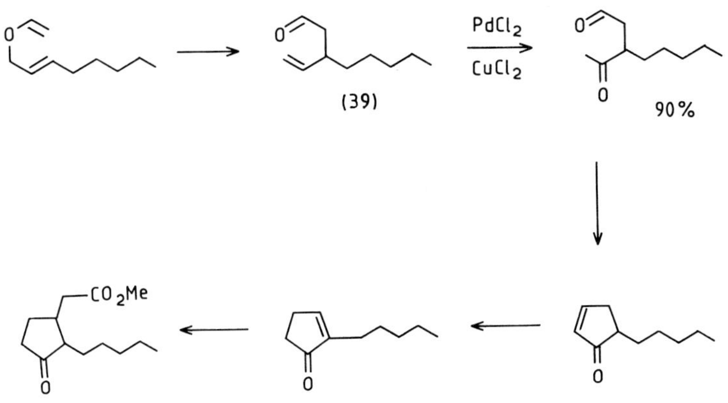

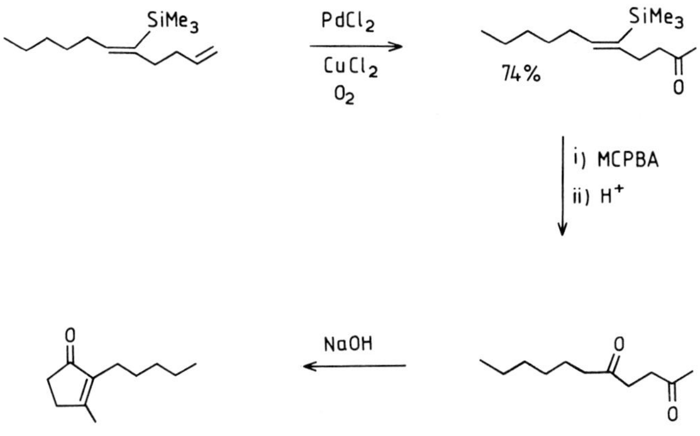

These examples illustrate the wide range of methods that can be used for the introduction of a vinyl group for use as a masked ketone.

An alternative oxidation sequence[45] initiates reaction of the alkene with mercuric acetate in an aqueous medium, and then uses Li_2PdCl_4 to effect a *trans* metallation in which an HgOAc group is replaced by $PdCl_2^-$. In an aqueous medium, for example H_2O/THF, the latter decomposes to the ketone. The process is shown in Scheme 4.8.

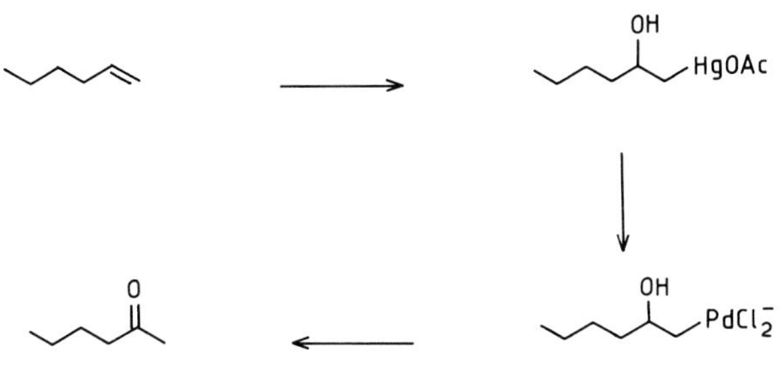

Scheme 4.8

This sequence was applied[46] in the course of a synthesis of (+)-nootkatone from (−)-β-pinene in which Michael addition of allyl silane was used to introduce the vinyl group at the required distance from the carbonyl group in (40).

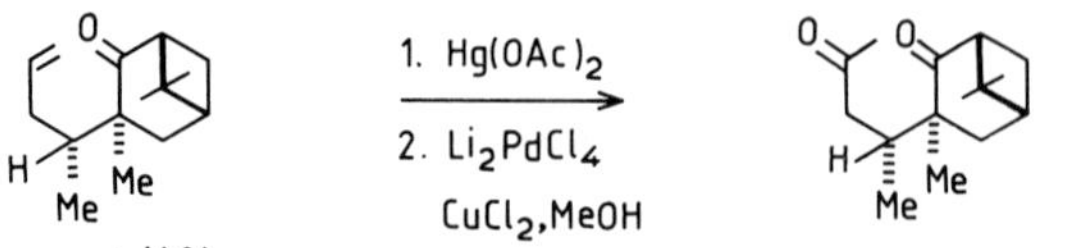

A further variant[47] employed $RhCl_3$ in place of $PdCl_2$, with $FeCl_3$ as reoxidant, replacing $CuCl_2$. This combination, used in aqueous DMF, oxidised[47] (41) to the ketone (42) in 80% yield. With $PdCl_2$, (41) was found[47] to undergo a more complex reaction, in which reduction of the vinyl group and complexation by palladium produced the π-allyl derivative (43). The use of $RhCl_3/FeCl_3$ or $RhCl_3/Cu(ClO_4)_2$ may therefore have

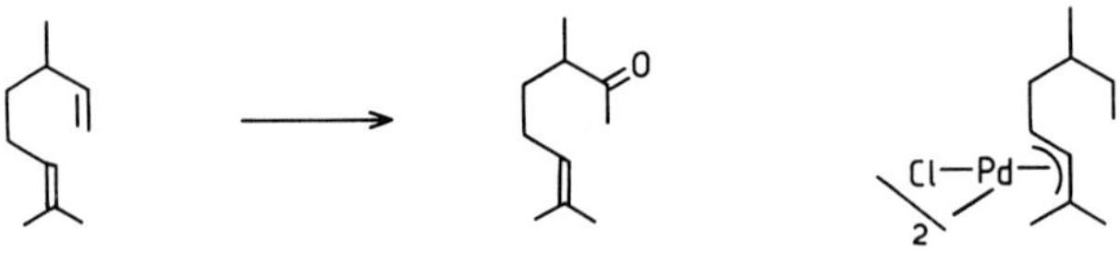

advantages with some dienes. A redox combination of Rh(III) with Rh(I) no doubt replaces the Pd(II) and Pd(0) combination of the palladium process.[48]

4.5 Formation of vinyl acetates and allyl acetates

Oxidation of a terminal alkene to the derived enol acetate can occur when $Pd(OAc)_2$ is used.[49] In this case the reaction sequence leading to the ketone is terminated by the elimination of PdH(OAc) to form Pd(0) and acetic acid, as in the conversion of (44) to (45).

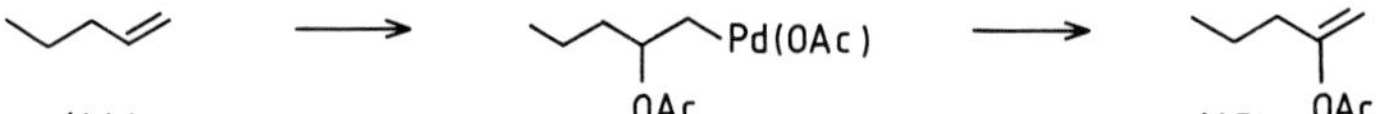

The consequent reduction of palladium to the metal is again accommodated by the addition of a re-oxidant. It has, however, been shown that re-oxidants such as $CuCl_2$ may also react with intermediates in the reaction to form other products[50] in slightly aqueous acetic acid. Indeed, use was made of this reaction by the addition of lithium chloride to determine the steric course of hydroxypalladation.[10] The product is shown in Scheme 4.9.

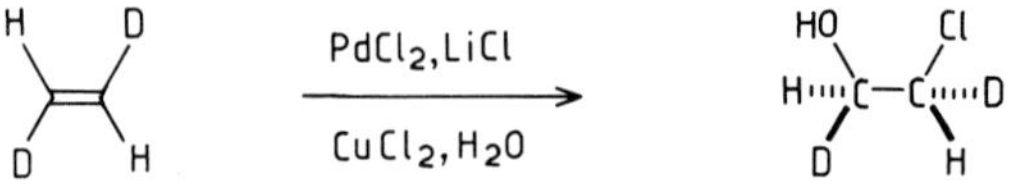

Scheme 4.9

In the case of a non-terminal alkene such as (46), alternative hydrogen atoms are available for elimination. The main product is the enol acetate (47).

Thus under suitable conditions,[51] for example, in water-free acetic acid (HOAc + Ac$_2$O), an alkene may undergo a useful allylic oxidation. A simple instance[52] is provided by the formation of (48) brought about by $[Pd(OAc)_2]_3 \cdot NO_2$.

Benzoquinone serves very efficiently as an oxidant in this reaction, possibly because of its ability to function also as a ligand. When manganese dioxide is used as an additional re-oxidant, products such as (48) can be obtained in high yield.[53]

4.6 Formation of enones

Silyl enol ethers, which are readily formed from aldehydes and ketones,[54] may be conveniently cleaved with Pd(OAc)$_2$ to yield the corresponding α,β-unsaturated carbonyl compounds.[55] The intermediate palladium derivative eliminates HPdOAc, or its equivalent, with the introduction of an alkene bond, as shown in the conversion of (49) to (50).

The simple example, cyclopentanone to cyclopentenone, was the basis for the step (51) to (52) in a synthesis of aphidicolin[56] by Trost and co-workers. Similarly, the

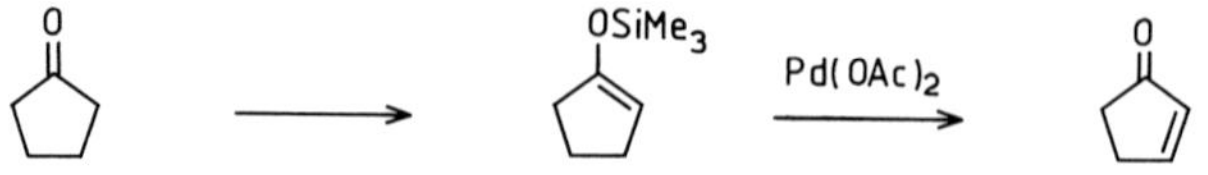

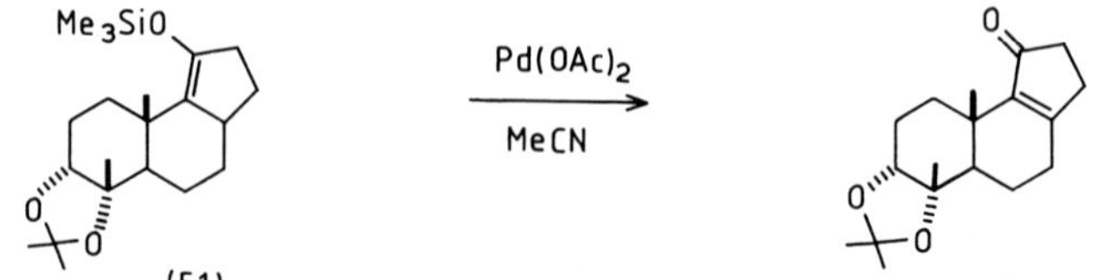

sequence, ketone to trimethylsilylenol ether to α,β-enone, shown in Scheme 4.10, was used as a step in another synthesis of aphidicolin, this time by Ireland's group.[57] Yet

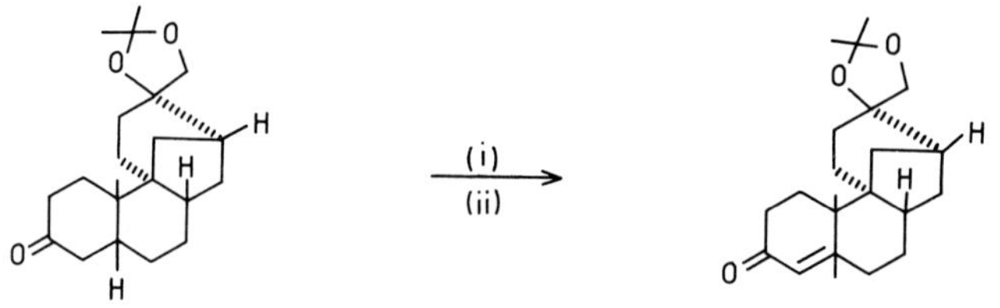

(i) KH, THF, Me$_3$SiCl, Et$_3$N
(ii) Pd(OAc)$_2$, MeCN

Scheme 4.10

another example of this reaction may be seen in a synthesis of $(\pm)$-isabelin where the formation of (53) was effected by Pd(OAc)$_2$.[58]

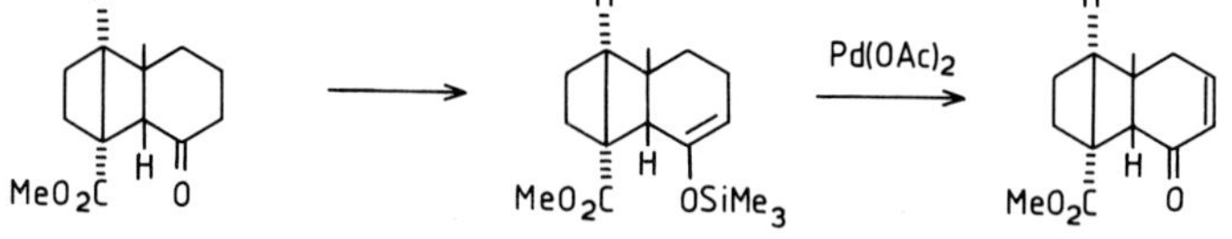

In a process related to the reaction sequence cyclopentanone to cyclopentenone described above, use of RhCl(PPh$_3$)$_3$ with oxygen was found to effect allylic oxidation to the trimethylsilylenol ether of cyclopentanedione (54).[59] In this case the trimethylsilyl ether was not cleaved.

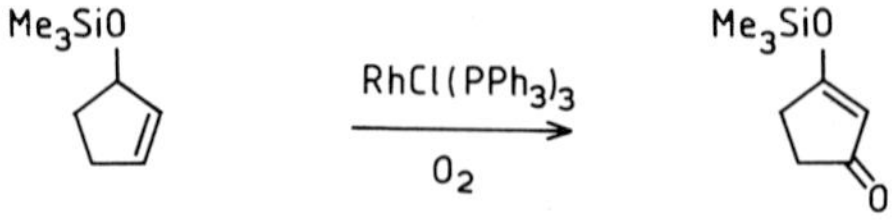

Where there is a second alkene bond in the TMS-enol ether, cyclisation may be induced upon treatment with Pd(OAc)$_2$ in acetonitrile,[60, 61] as shown by the examples in Scheme 4.11.

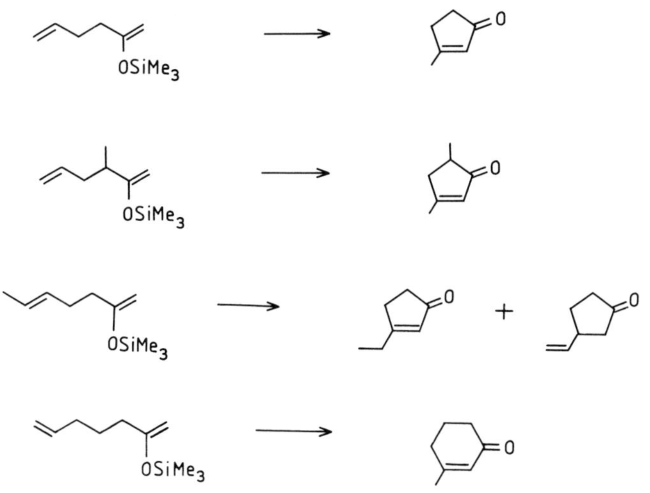

Scheme 4.11

It is proposed[60] that the silylenol ether forms an oxa-π-allyl palladium complex which has two possibilities for further reaction.

The first is an elimination reaction.

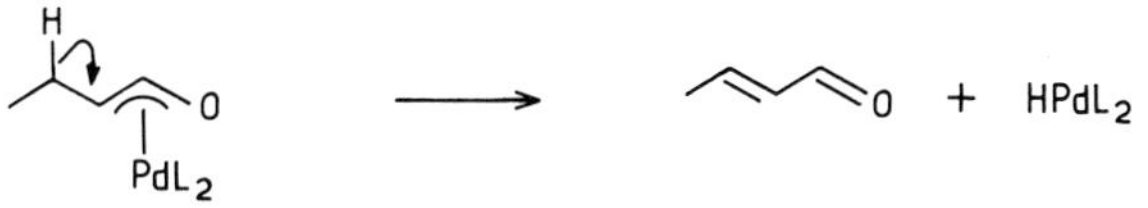

Alternatively, cyclisation may occur.

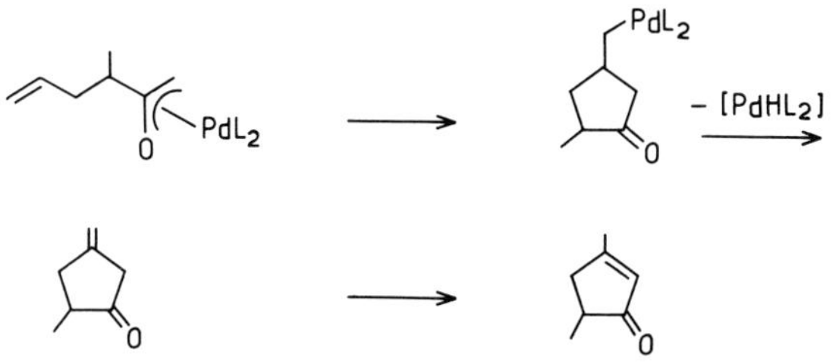

In these reactions palladium is eliminated as the metal, but it has been shown that re-oxidation *in situ* is possible using Pd(OAc)$_2$ in catalytic amount with Cu(OAc)$_2$ in MeCN solution in the presence of oxygen.

An unusual version of the conversion of TMS-enol ethers to enones uses a concurrent reduction of allylic carbonates to CO_2 and propene in the place of the normal oxidation step.[62] A nitrile solvent is important for the success of the method. In other conditions, alkylation via a π-allyl complex occurs (see Chapter 6).

Other silyloxy groups in the substrate are cleaved during the reaction, as seen[63] in the formation of the steroidal enone (55).

Direct conversion of ketones to enones by $PdCl_2$ in t-butanol has been examined for a range of steroid systems. In some cases the conversion was incomplete and in others, selective oxidation, as in the formation of (56), was observed.[64] Some ketones, however, were inert under the reaction conditions and the introduction of the double bond via TMS-enol ethers appears more satisfactory as a general method.

A similar and more useful direct reaction of a carbonyl group is the conversion[65] of β-amino ketones such as (57) to enaminones by $PdCl_2(MeCN)_2$.

Coordinated alkenes and alkynes as synthetic intermediates

5.1 Nucleophile addition to palladium complexes

Although the hydroxypalladation process that forms the basis of the Wacker oxidation, discussed in the previous chapter, is by far the most generally applied example of the addition of nucleophiles to alkenes bound to palladium, this type of reaction is really a special case of the much wider synthetic potential of palladium alkene complexes. Examples of both stoichiometric and catalytic reactions demonstrate the generality of this addition process. While the few catalytic systems that are available are of the most practical use, the majority of investigations have examined stoichiometric complexes, which often provide valuable stereochemical and mechanistic information. Palladium metal is usually deposited from such reactions so recovery is practicable.

The examination of nucleophilic addition of secondary amines to *cis*- and *trans*-but-2-ene catalysed by palladium salts is typical of the early work in this area. Entry of the R_2NH group trans to the metal in these reactions has been established.[1]

The addition[2] of 1-phenylethylamine has been developed[3] as a method to effect an asymmetric induction at the adjacent carbon atom by removal of the metal by oxidation with $Pb(OAc)_4$ (Scheme 5.1).

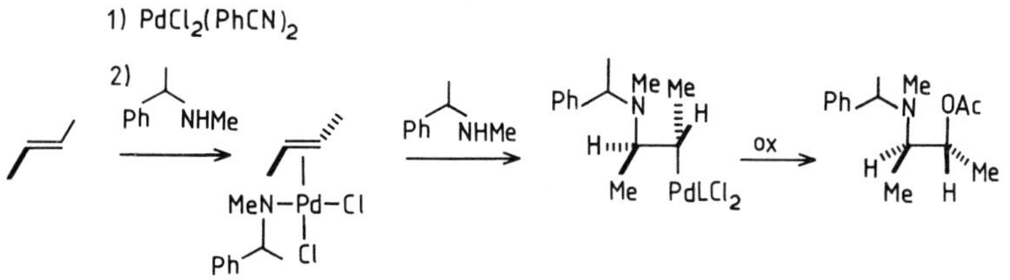

Scheme 5.1

When taken to completion, amine addition requires three equivalents of the amine. No carbon adducts are formed after the addition of the first equivalent, possibly because of an initial cleavage of the dimer to form the intermediate (2). Although the details of the reaction sequence vary from case to case, the overall process generally occurs in three steps, first the cleavage of the chloro dimer (1), followed by *trans* addition by the amine and then cyclisation to form (3).[4]

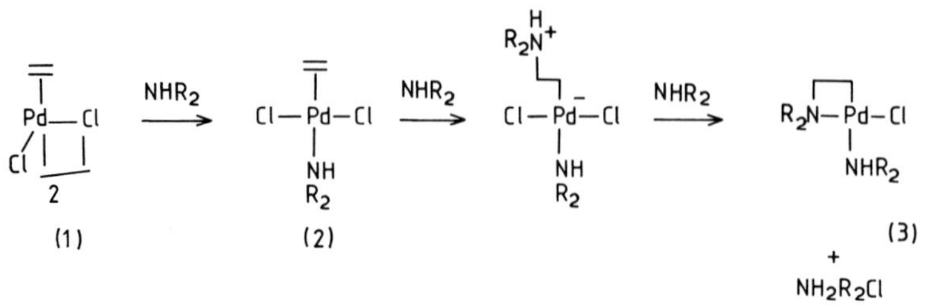

(1)   (2)   (3)

$+$

NH_2R_2Cl

Intramolecular coordination by an amine is very effective in promoting alkylation reactions. Pronounced regiocontrol, attributed to a preferential formation of a five-membered metallocycle, is demonstrated by the carbopalladation reactions of allyl[5] and homoallyl amines (Scheme 5.2).[6] This type of process has been developed by Holton[7] to provide an elegant entry to prostaglandin synthesis (see Chapter 14).

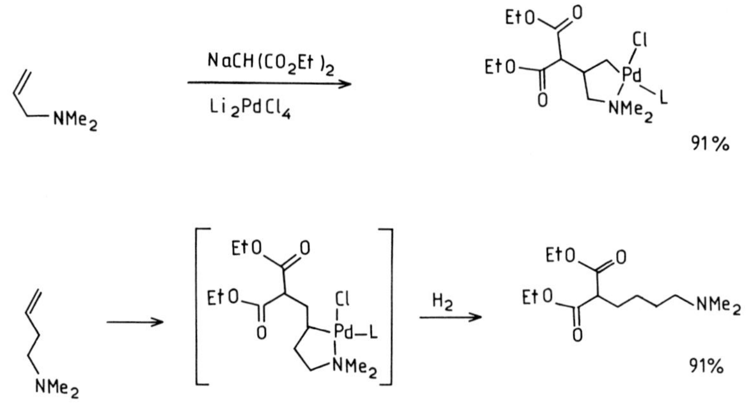

Scheme 5.2

Direct evidence for the *trans* addition of nucleophiles to the bound alkene is found in the reactions of palladium diene complexes where the metal is retained in the product. In the case of (4) the outcome was confirmed by conversion to the known ether (5) with sodium borohydride.[8]

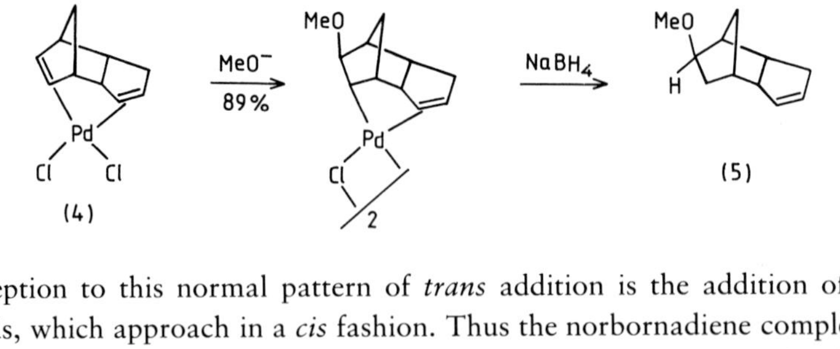

An exception to this normal pattern of *trans* addition is the addition of mercury compounds, which approach in a *cis* fashion. Thus the norbornadiene complex (6), for example, afforded[9] the cyclic product (7). Initial coordination of the nucleophile to the metal is proposed to account for this anomaly. Carbopalladation by mercury

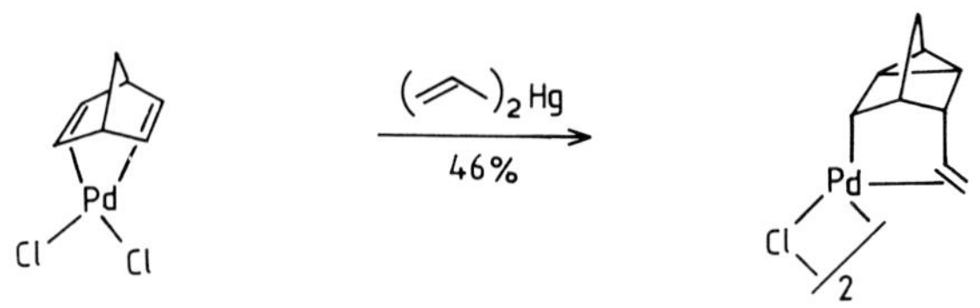

(6)   (7)

compounds has been used by Larock[10] in the synthesis of the interphenylene PGH_2 analogue (8) and its diastereoisomer in 58% yield.

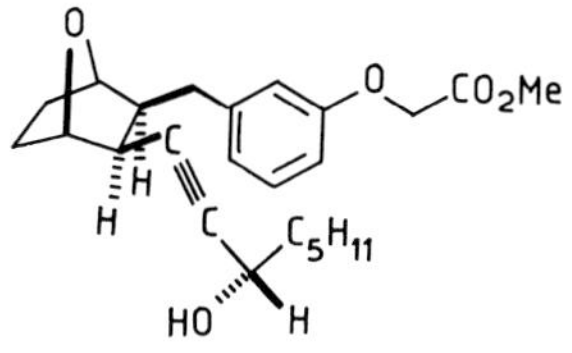

(8)

5.2 Addition–elimination reaction sequences

A β-elimination to form a substituted alkene is an alternative fate for the σ bound complex produced by nucleophile addition to an alkene π complex.

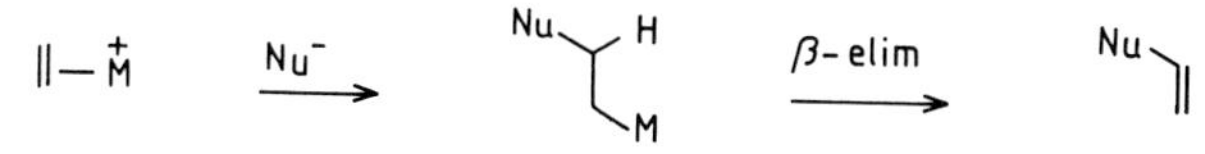

Enolates and lithiated species can substitute alkenes in this way. The use of two equivalents of triethylamine assists alkylation and HMPA is included when basic anions are used.[11] An interesting special case of this process, the nitration of alkenes by $Pd(NO_2)_2(MeCN)_2$ is shown in Scheme 5.3.[12] In these reactions, elaboration of the

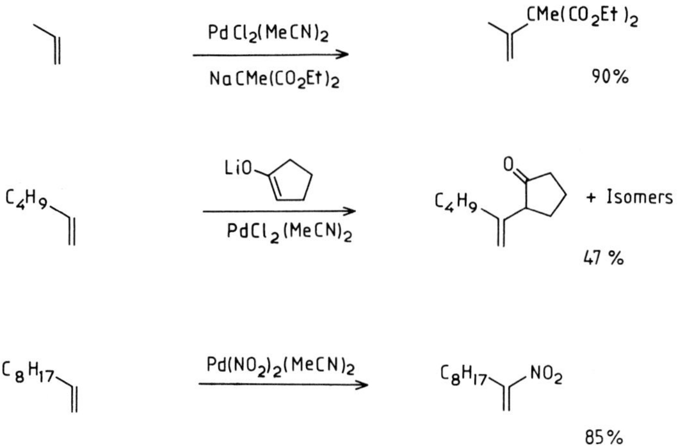

Scheme 5.3

alkene is effected through a process that enables the alkene functional group to be retained in the product. This can be a significant advantage when compared to other nucleophile additions to alkenes (for example, to enones), where the alkene group is consumed unless it is substituted with a sensitive leaving group such as a halide, for use in addition–elimination reactions. The role of β-elimination in re-forming the alkene after nucleophile addition to an alkene complex is discussed further in Chapter 10 where a variety of coupling reactions are compared.

5.3 Cyclisation reactions of palladium alkene complexes

Early in the investigation of cycloocta-1,5-diene complexes, it was found that two sequential nucleophile additions led to cyclised products arising from reductive elimination of the metal.[13] A similar cyclisation occurs under oxidative conditions (Scheme 5.4).[14]

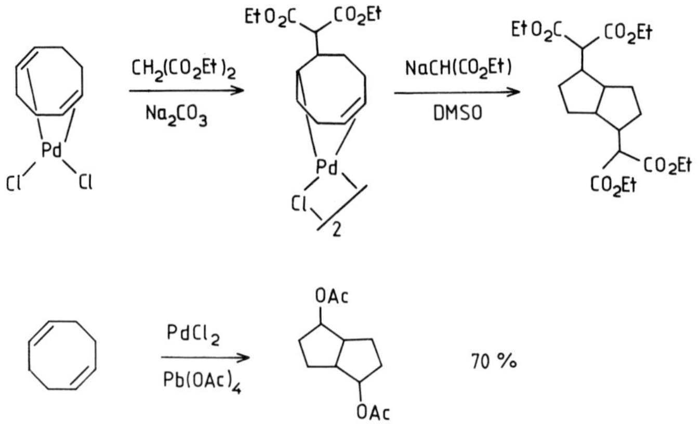

Scheme 5.4

In some cases these reactions are completely regiocontrolled as shown in Scheme 5.5.[15]

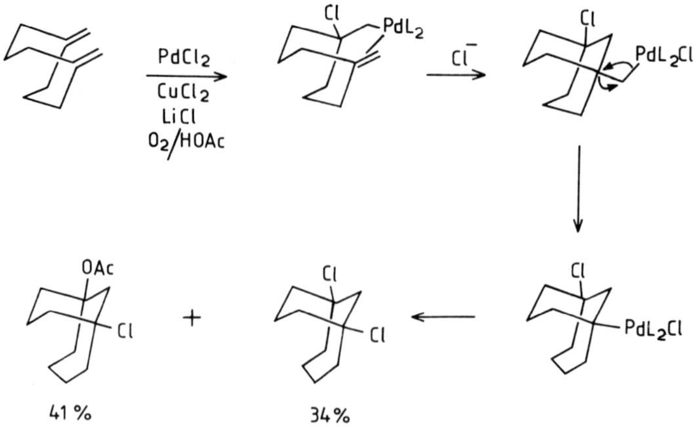

Scheme 5.5

A more obvious source of cyclisation reactions arises when intramolecular nucleophile addition occurs. Indoles are formed in this way from the reaction[16] of *o*-allylanilines with $PdCl_2(MeCN)_2$ and triethylamine. A probable mechanism involves β-elimination and isomerisation as the concluding steps, as shown in Scheme 5.6. Further examples of this type of reaction are discussed in Chapter 15.

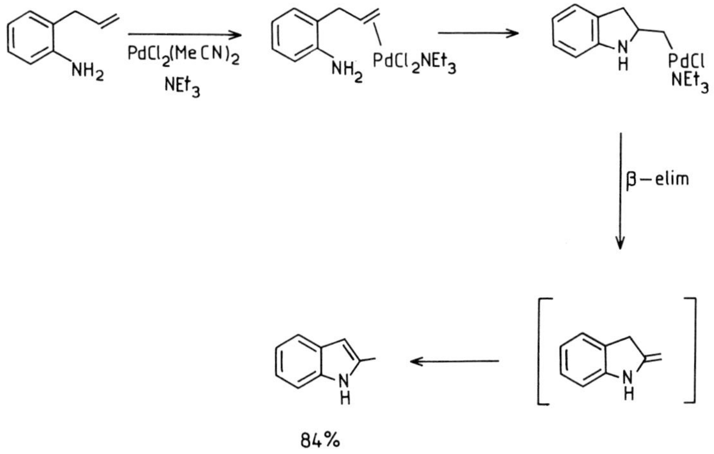

84%

Scheme 5.6

An application[17] of the addition of a phenolic hydroxyl group to an alkene is found in the synthesis of the flavone (9).

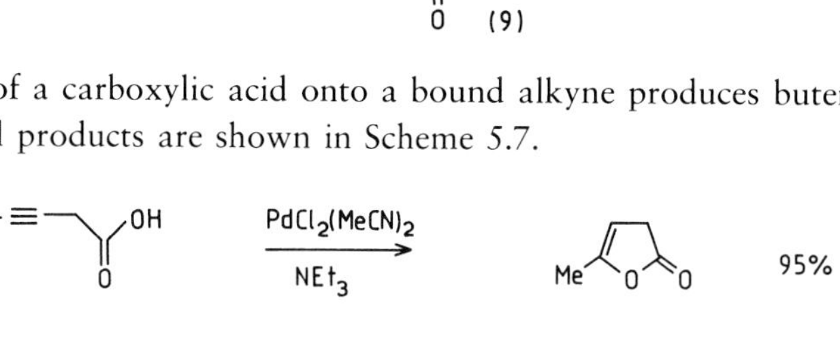

Cyclisation of a carboxylic acid onto a bound alkyne produces butenolides in high yield.[18] Typical products are shown in Scheme 5.7.

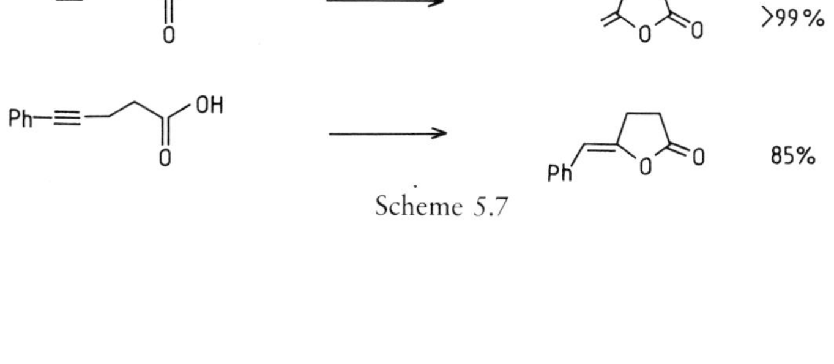

Scheme 5.7

An interesting alkene–alkene addition reaction (dimerisation of 2-*trans*-1-propenyl-4,5-methylenedioxyphenol (10), brought about in the presence of PdCl$_2$ and NaOAc in methanol) led to a synthesis of carpanone (11).[19]

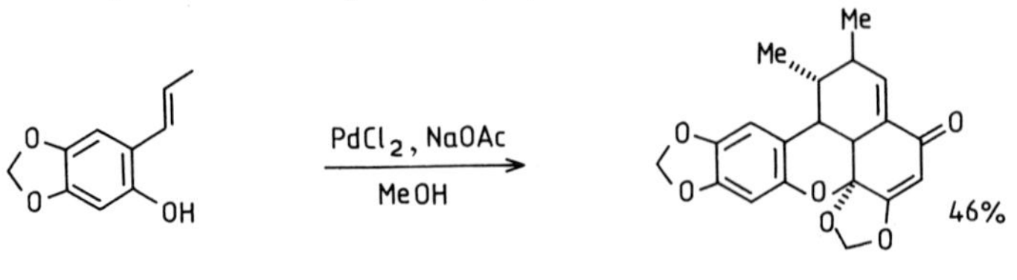

5.4 Nucleophile addition to iron alkene complexes

Since dicarbonyl(cyclopentadienyl)iron (Fp) complexes of alkenes are cationic, they react well with a wide variety of nucleophiles. Following the initial work in this area

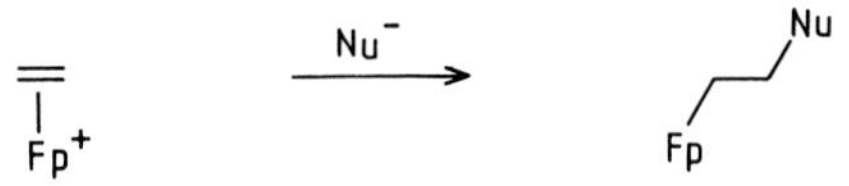

$$Fp = CpFe(CO)_2$$

from the groups of Pauson,[20] Fischer,[21] Green,[22] and Rosenblum,[23, 24] the potential of these reactions in synthesis has been developed to a considerable extent.[25, 26] Alkylation by enolates, enamines and cuprates provide synthetically useful C–C bond forming reactions.[27, 28] Heteroatom nucleophiles such as oxygen, sulphur, nitrogen, and phosphorus can also be used.[29] Some examples are shown in Scheme 5.8.

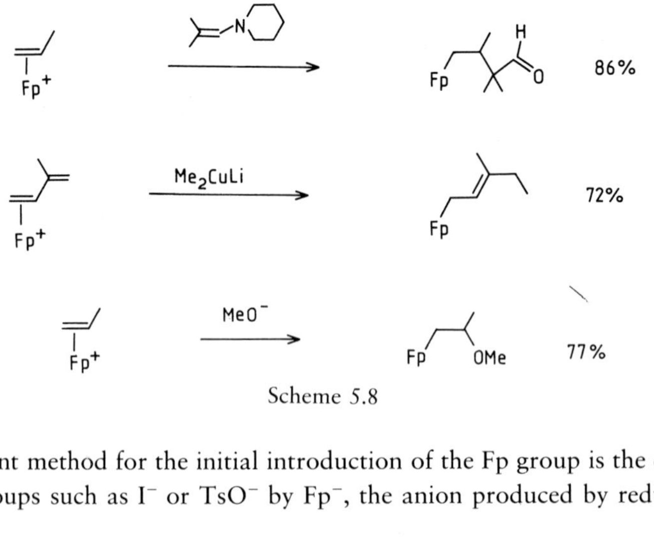

Scheme 5.8

A convenient method for the initial introduction of the Fp group is the displacement of leaving groups such as I$^-$ or TsO$^-$ by Fp$^-$, the anion produced by reduction of the

$Cp_2(CO)_4Fe_2$ dimer. The alkyl complex is then converted to an alkene complex by hydride abstraction with $Ph_3C^+BF_4^-$.

The stereoselectivity and regioselectivity of this process has been examined in detail.[30] Reaction of Fp^- with an epoxide, followed by treatment with acid, has provided[31] an alternative preparation of alkene π-complexes.

A similar approach makes use of nucleophile addition to the aldehyde (12) to form an alkoxide intermediate. Acid treatment is used once again to produce the alkene complex. Both organolithium and Grignard reagents have been used successfully in the presence of the Fp complex.[32]

For the direct complexation of an alkene, Fp^+ can be transferred from the isobutene complex (13).[23] An alternative is a complexation procedure that generates Fp^+ in the presence of the alkene by reaction of FpI with $AgBF_4$ in CH_2Cl_2.[33]

Although carbon nucleophiles generally show poor regiocontrol in additions to the activated double bond of Fp(alkene) cations, it has been found that nucleophile addition to vinyl ether complexes such as (16) is highly regiospecific.[34] The complex (16) may be obtained by reaction of an α-bromoacetal (14) with $NaFe(CO)_2(C_5H_5)$ to yield (15). Protonation with HBF_4 in CH_2Cl_2 at $-78\ °C$ converted (15) into the cationic vinyl ether complex by removal of an OEt group as ethanol. Nucleophile addition occurs at the oxygenated carbon of the vinyl ether. The example shown in Scheme 5.9 arose in a synthesis of isopiperitone in which the metal is finally removed by treatment with Et_4NBr.[35]

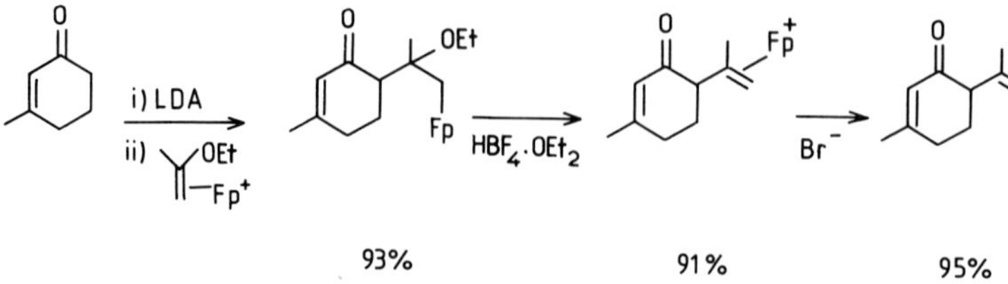

93%   91%   95%

Scheme 5.9

Reactions of this type can show a very high degree of diastereoselectivity. Addition of the enolate derived from cyclohexenone to the methylvinylether complex (17) afforded (18) as a single diastereoisomer. Further elaboration by stereocontrolled reduction of the ketone and cyclisation to the lactone (19) by oxidation of the iron

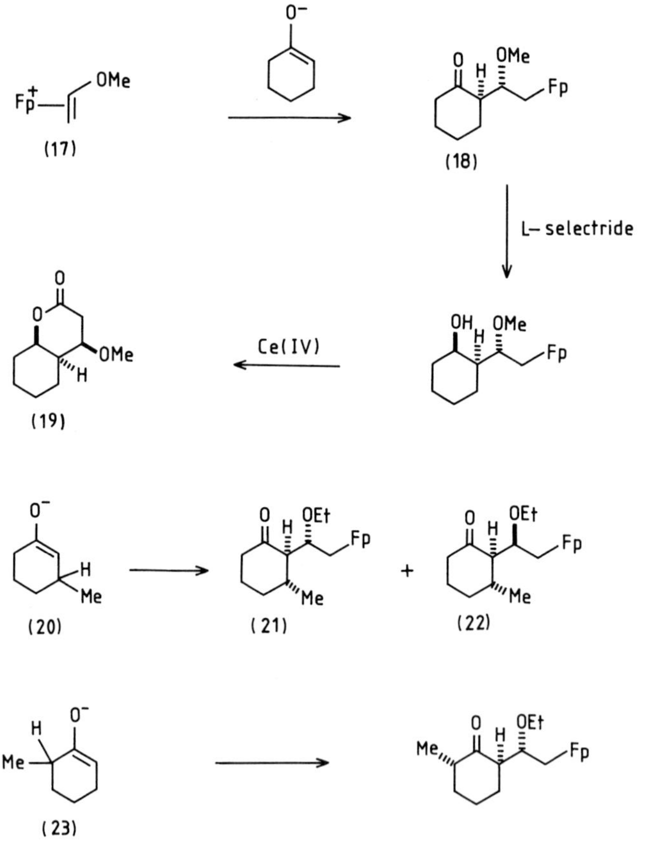

complex (ceric ammonium nitrate) proceeded in 63% overall yield. The lactone product contains a sequence of three adjacent chiral centres which have been introduced in this process with complete stereocontrol. The enolate (20) offers two stereochemically distinct faces for reaction with the metal complex. Reaction with (16) gave a 3:1 mixture of the diastereoisomers (21) and (22). When the chiral centre in the enolate was adjacent to the carbonyl group, stereocontrol was superior. In this case a single diastereoisomer was produced by the reaction of the enolate (23) and (16). In both these reactions, however, diastereoface recognition at the enolate was complete.[36] Reaction sequences of this type rely on the powerful and reliable directing influence of the alkoxy group to pull the nucleophile in towards the site of substitution. The origin of the regiodirecting influence of the alkoxy group has been related to a distortion of the ground state structure of the complexes that makes one of the two competing transition states for alkylation more easily accessible.[37] Other explanations are also reasonable, however, and a more general discussion of effects underlying the reactivity patterns of electrophilic transition metal π complexes is given in Chapter 7.

The functionalised vinyl ether complex (24),[38] obtained from ethyl pyruvate diethylketal, has provided[39] a facile synthesis of α-methylene lactones. In the case of (24), the regiodirecting effects of the OEt group and the ester group are in competition. Although the ester would normally direct nucleophiles to the more distant, unsubstituted, terminus of the alkene complex, the products shown in Scheme 5.10

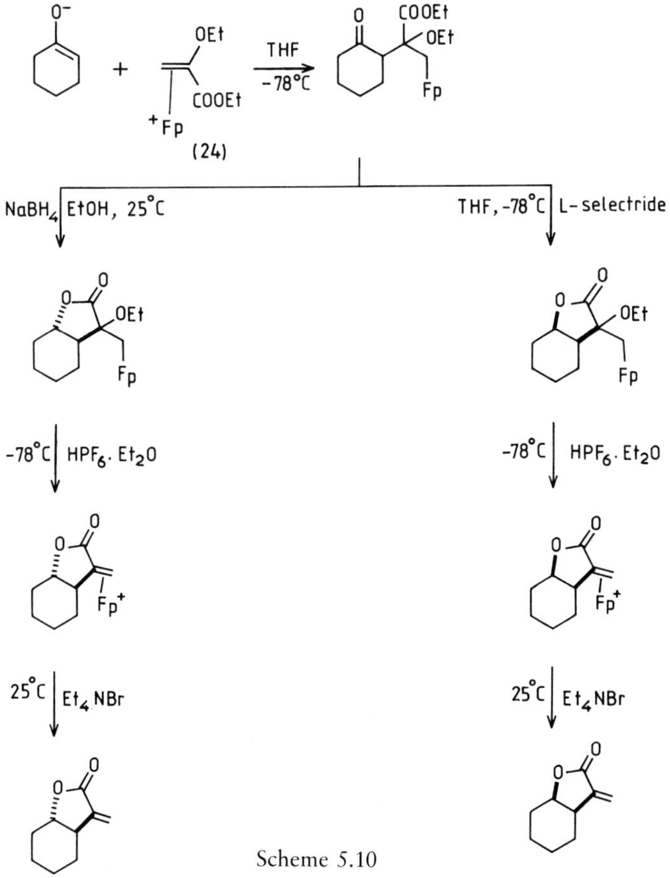

Scheme 5.10

demonstrate that the ether substituent totally dominates the control of this reaction. Several organometallic approaches to α-methylene lactones are compared in Chapter 15.

The dialkoxyalkene complex (25) can promote a sequence of two nucleophile additions to give either *cis* or *trans* alkene products depending on the method used.[40]

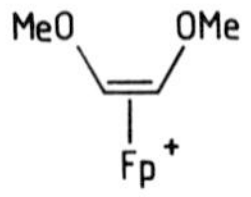

MeO   OMe

Fp⁺

(25)

The achiral complex (25) has been converted into an optically active bisvinyl ether complex by ether exchange with the resolved diol (R,R)-2,3-butanol. In this way the alkene complex (26) can easily be obtained in optically pure form. Reaction with nucleophiles, for example Grignard reagents and enolates, proceeded with high regioselectivity, as seen in the example of the formation of (27). Reduction with sodium

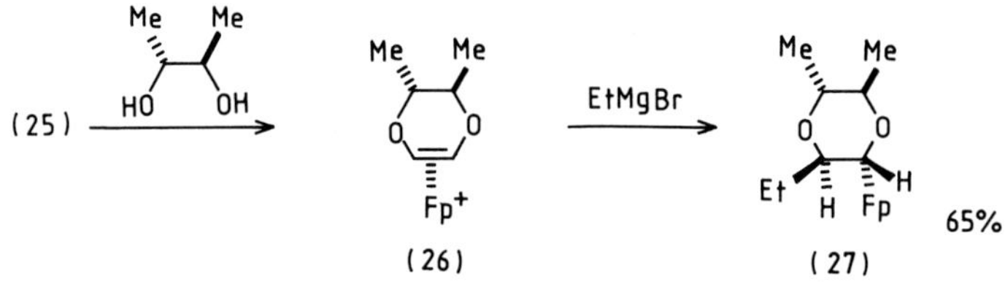

cyanoborohydride was similarly well controlled. Reaction of the product with trimethylsilyl triflate afforded the vinyl ether complex (28). Simple optically active Fp complexes can also be obtained. For example, by reaction of (26) with MeCuCNLi and trimethylsilyl triflate, the complex (29) was produced. This complex can be converted to the *cis* alkene complex (30) which is itself a source of (31) by demethoxylation by reduction and acid treatment.

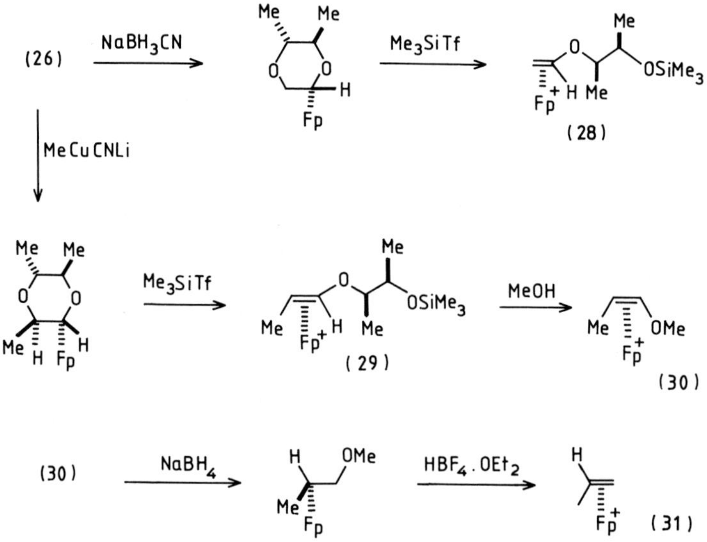

The acyclic optically active alkene complex (28) proved not to be completely optically stable. Epimerisation of the metal complex, however, was slow at room temperature, requiring 19 hours to attain an equilibrium mixture of diastereoisomers. The alkoxy substituent on the alkene is implicated in the epimerisation mechanism. A significant contribution of the structure (32) allows rotation to occur about the formal C–C double bond in the complex. This mechanism is supported by the observation that epimerisation occurs only at C-1, the end of the alkene bearing the alkoxy substituent.[41]

The methyl vinyl ketone complex also showed useful regioselectivity.[28] In this case the metal can be removed from the product by exposure to base, as in the cyclisation examples shown in Scheme 5.11. The outcome of these reactions is the same as a

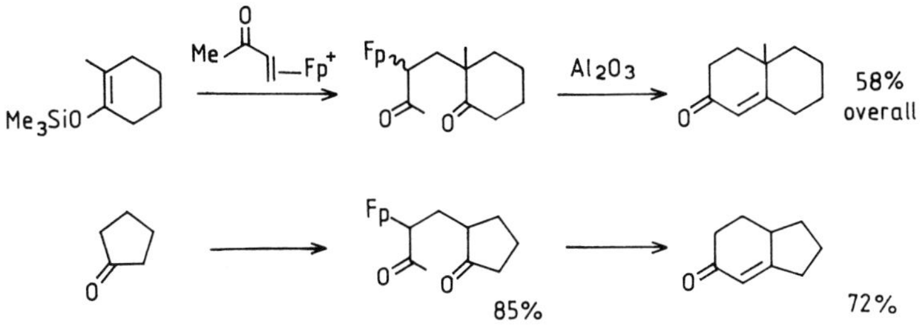

Scheme 5.11

conventional Robinson annellation, but use of the Fp cation complex allowed mild conditions for the initial alkylation. The first example shown was performed at 0 °C., without the need for Lewis acid catalysts. Unfortunately, enone Fp π complexes are not generally available, since binding of Fp^+ at the ketone carbonyl group prevents the formation of many C-bound enone complexes by standard methods. The methyl vinyl ketone complex was obtained from the corresponding epoxide by the method discussed earlier.

The many examples[26] of regiocontrolled alkylations of the type shown above provide useful precedents for the use of these methods in synthesis. The synthesis of β-lactams using this type of reaction is a particularly good example. This application is discussed in Chapter 15. Uses in organic synthesis require versatile methods to remove the metal, after its role is completed. In addition to the methods already discussed, Scheme 5.12 shows several useful decomplexation procedures which are available to introduce valuable substituents such as ester groups, halides and organomercury compounds.[24,42,43] The stereochemistry and degree of stereocontrol of these processes has been studied using deuterated alkenes.[44] Related, optically active, $CpFe(CO)PPh_3$ complexes have also been employed[45] to probe stereoselectivity, which was found to be far from complete in this case.

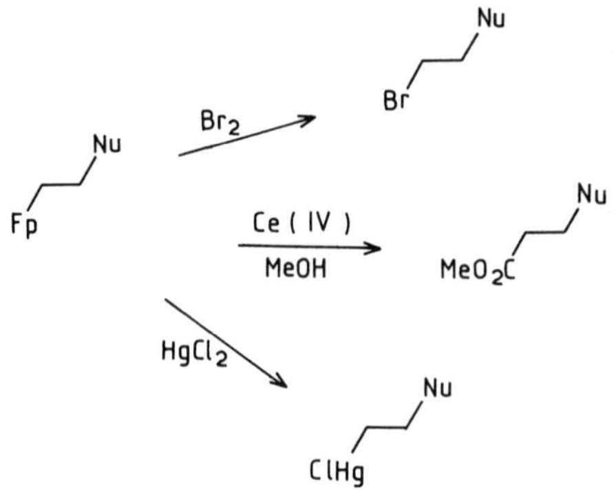

Scheme 5.12

The iron complexes discussed so far in this section have all been positively charged. Naturally this enhances the electrophilic character of the compounds. Positive charge, however, is not essential to render complexes electrophilic. As was seen with the palladium examples in Section 5.1, binding to a transition metal centre is in itself sufficient to promote nucleophile addition. The same is true for iron complexes. The $Fe(CO)_4$ complex (33), upon reaction with malonate anions, produced anionic iron intermediates of a type now familiar from the use of Collman's reagent (see Chapter 9). Decomplexation was effected by protonation and oxidation of the iron complex to produce (34) in 45% yield. Use of an excess of (33) increased the yield to 68%.[46]

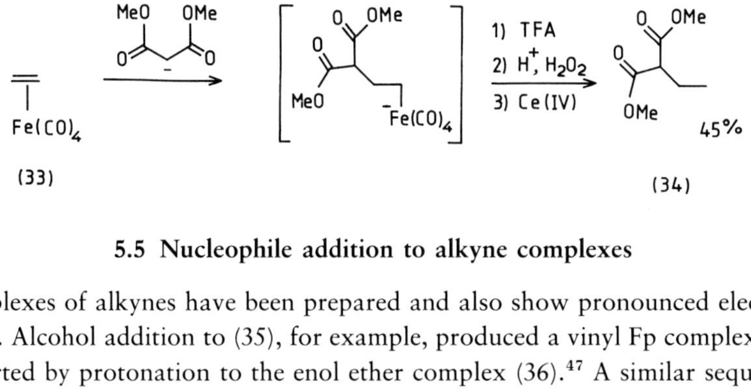

(33)   (34)

5.5 Nucleophile addition to alkyne complexes

Fp^+ complexes of alkynes have been prepared and also show pronounced electrophilic character. Alcohol addition to (35), for example, produced a vinyl Fp complex that can be converted by protonation to the enol ether complex (36).[47] A similar sequence,[47,48]

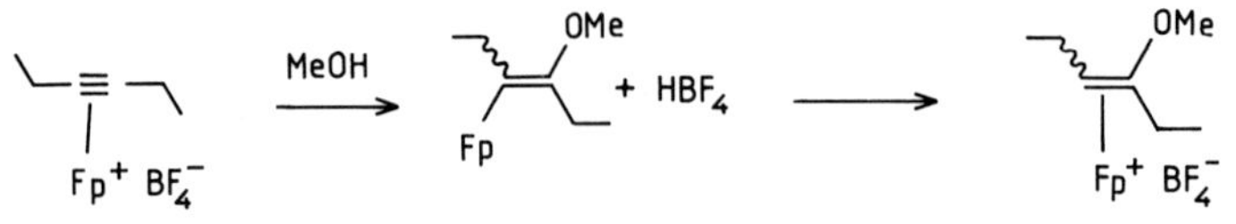

(35)   (36)

involving intramolecular nucleophile additions, resulted in the formation of dihydro-furan complexes such as (37). The same product was obtained from the corresponding allene.[48]

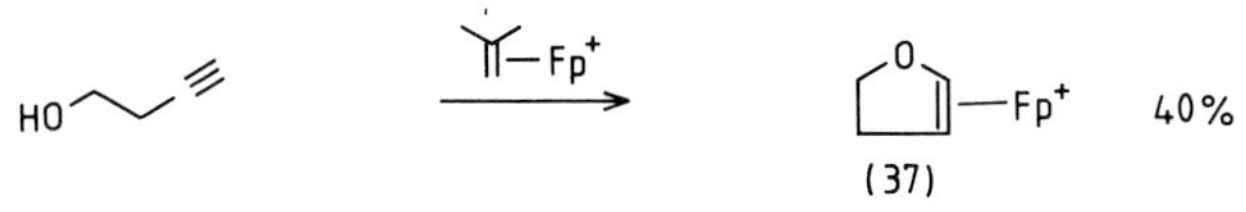

(37)   40%

A reaction in which ring closure is effected by C–O bond formation in a similar way, has been catalysed both by palladium and by rhodium. The most efficient reagent for this purpose was the dimeric rhodium catalyst (38), a reagent that gave both higher yields and far superior regiocontrol to more conventional methods for cyclisation using mercury.[49] The lactone (40), for example, was obtained in 93% yield using (38), compared to 81% $(Pd(PPh_3)_4)$ and 48% $(Rh(PPh_3)_3Cl)$ when other transition metal catalysts were used.

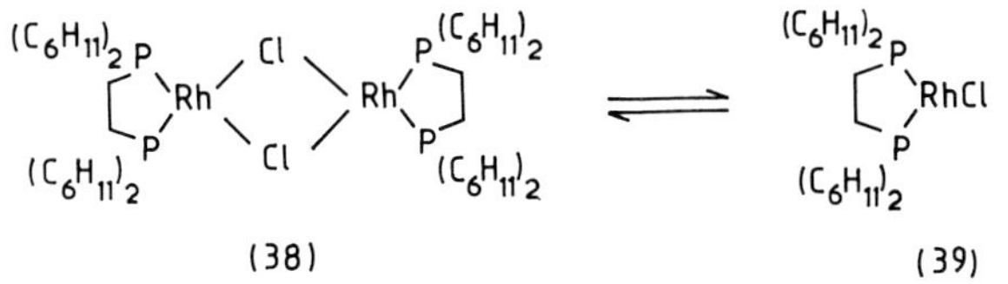

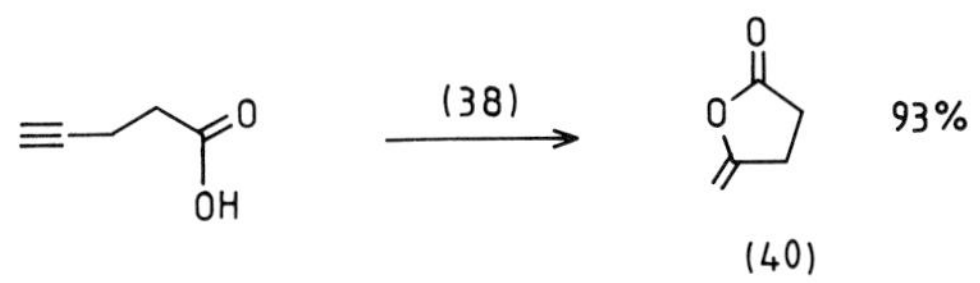

(38)   (39)

(40)   93%

This process is thought to begin by oxidative addition to the OH bond of the acid, followed by transfer of rhodium to the alkyne. Nucleophilic ring closure and reductive

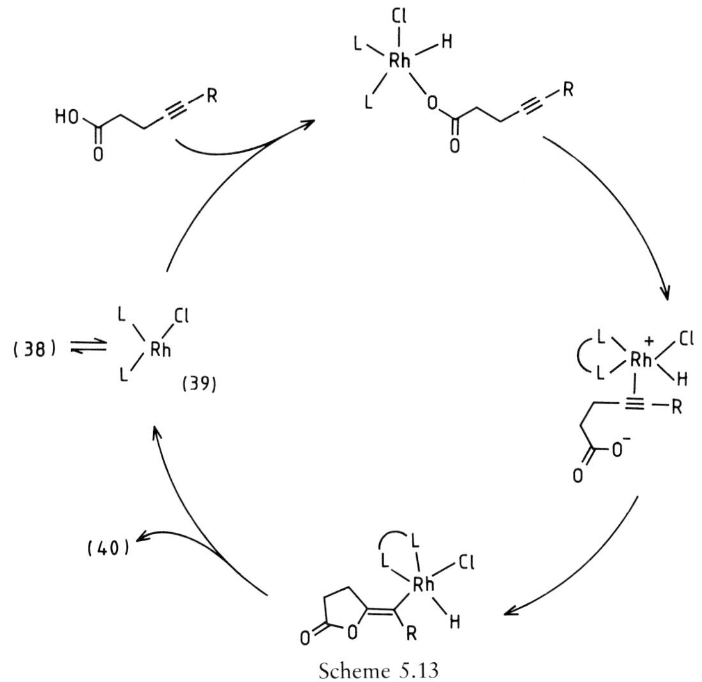

Scheme 5.13

elimination affords the product (e.g. (40)) and reforms the coordinatively unsaturated intermediate (39). The process is shown in Scheme 5.13.

Addition of a range of nucleophiles to $CpFe(CO)PPh_3$ complexes, for example (41), has been examined.[50] Typical cases include cyanide, malonate, and $Ph_2CuCNLi_2$. In Scheme 5.14, two cuprate examples have been chosen to illustrate the way that unsymmetrical alkynes bearing phenyl[50] and ester[51] groups react regioselectively.

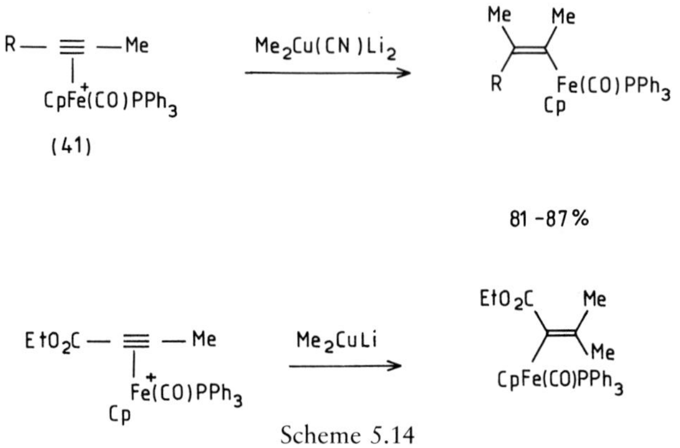

Scheme 5.14

81–87%

Alkyne complexes of iron,[52] platinum[53] and the electron deficient molybdenum complex (42)[54] have also been shown to undergo nucleophile addition.

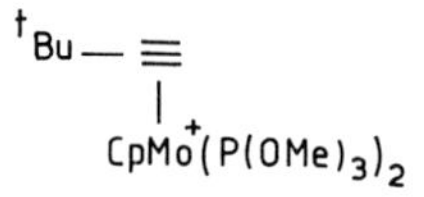

(42)

5.6 Nucleophile addition adjacent to alkyne complexes

Cationic cobalt complexes can be generated from $Co_2(CO)_8$ complexes of propargyl alcohols. The reaction of these species with nucleophiles has been examined extensively by Nicholas who has shown that the process is a useful way to effect propargyl substitution without the competing reactions that are otherwise common in such situations. Mild oxidation serves to remove the metal from the products shown in Scheme 5.15. These are derived from anisole[55], malonate[56] acetylide[57] or allylsilane[58] additions. Many types of nucleophiles can be employed. Both TMS-enol ethers, and ketones themselves, are alkylated efficiently[59] by cationic complexes of this type. This

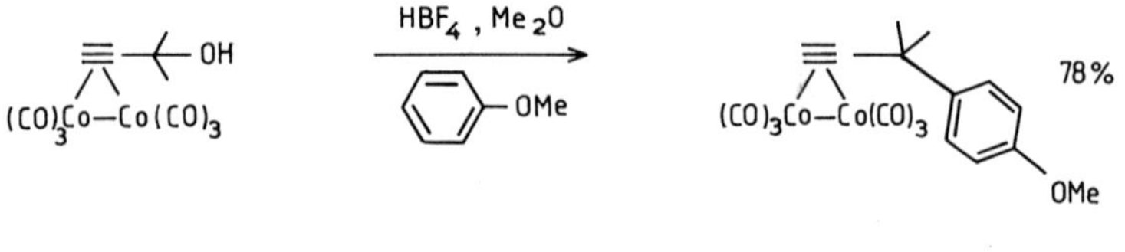

78%

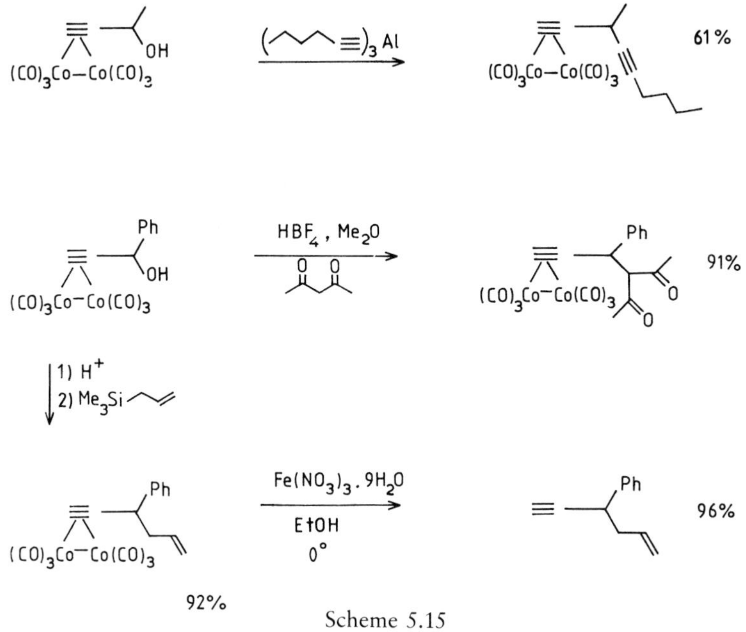

Scheme 5.15

is illustrated by the formation of (43) which was converted to the cyclopentenone (44) by hydration and cyclisation.[60]

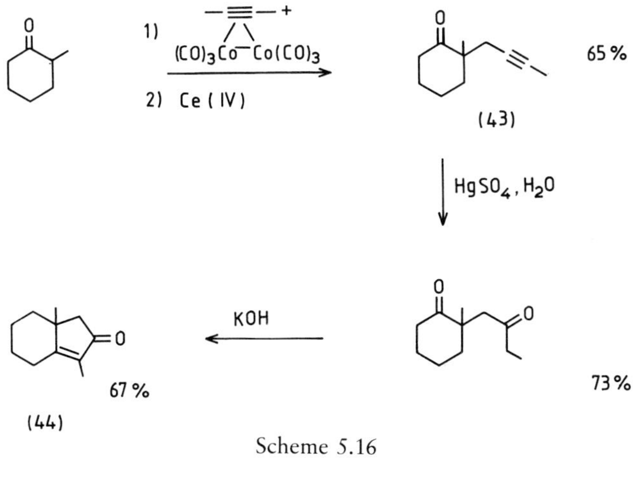

Scheme 5.16

The key to the success of the alkylation of dicobalt complexes of this type lies in the substantial stabilisation of positive charge achieved by the metal. (For other examples of this effect see Chapter 7). Extended cation systems are also stabilised. In most cases selective alkylation at the remote terminus is observed (Scheme 5.16).[61]

This ability to stabilise charged intermediates allows the sequential use of

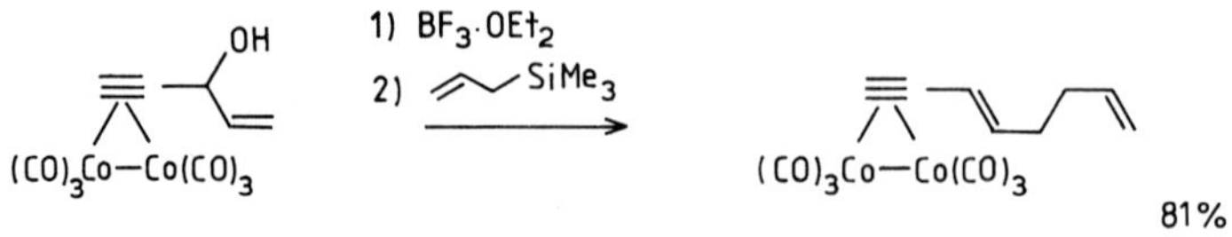

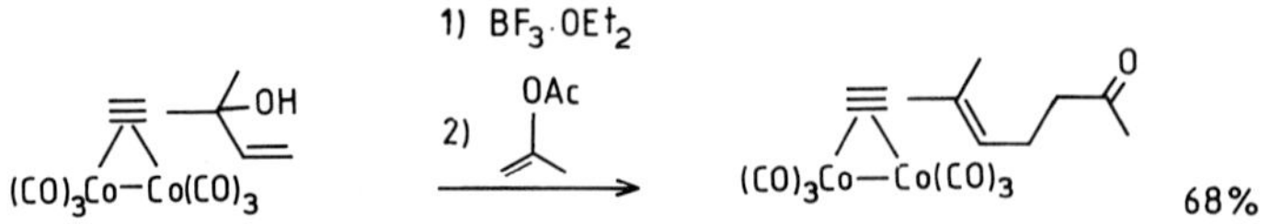

81%

68%

electrophilic and nucleophilic reagents, as is illustrated by the example of the formation of (45).[62] An application of this process in natural product synthesis is described in Chapter 14.

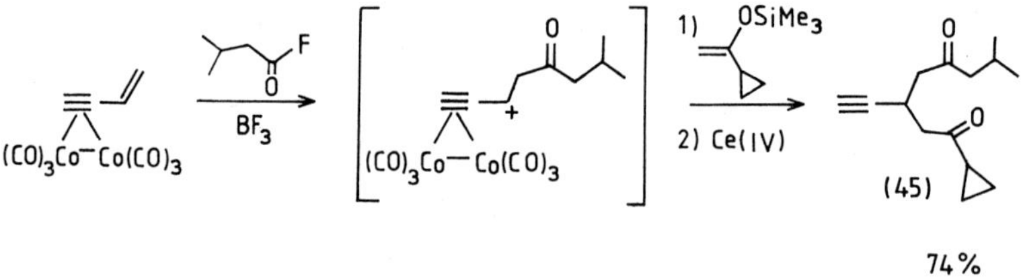

74%

An interesting recent example in which a cyclisation is achieved with good stereocontrol employs the $Co_2(CO)_6$ complex of a propargyl aldehyde derivative in a Lewis acid promoted reaction with an alkoxyallylstannane.[63] A twelve-membered macrocyclic ring is formed in 70% yield in this reaction. Propargyl acetal complexes have also been used as precursors for cationic cobalt alkyne complexes.[64] More examples of cobalt mediated reactions of propargyl cations are discussed in Chapter 9. In these cases, alkylation is combined with a 'Pauson-Khand' cyclisation, thus using the cobalt complex twice within the synthetic sequence.

6
π-Allyl complexes of palladium as synthetic intermediates

Despite the great diversity of the organometallic chemistry of transition metal π complexes, the study of one metal, palladium, has far outstripped the rest in the degree to which its application[1,2] to synthesis has been developed and exploited. This is no less true for π-allyl complexes than was the case for the chemistry of alkenes, discussed in Chapters 4 and 5. Indeed, for coupling reactions too (discussed in Chapter 10), palladium dominates the scene. This chapter deals with the chemistry of π-allyl (η^3-allyl) complexes of palladium, and their reactions with nucleophiles.

Two distinct types of system are important,[1] neutral stoichiometric complexes such as the chloride dimer (1) derived from alkenes, and cationic intermediates of the type (2) which arise in catalytic cycles by displacement of allylic leaving groups. Both types of palladium allyl complex react with nucleophiles under suitable conditions.

6.1 Formation of stoichiometric palladium allyl complexes

The early definitive studies of palladium allyl complexes concerned the preparation and stereochemistry of dimeric palladium chloride complexes.

Palladium π-allyl derivatives are readily available from alkenes by reaction of $PdCl_2$[3] or Na_2PdCl_4[4] under appropriate conditions. In the case of (3), Trost and Strege[5] found that π-allyl dimers, with the PdCl$_{/2}$ complex cis (4) or trans (5) to the ^{t}Bu substituent, are formed in a ratio of approximately 2:3, showing only a modest degree of control.

111

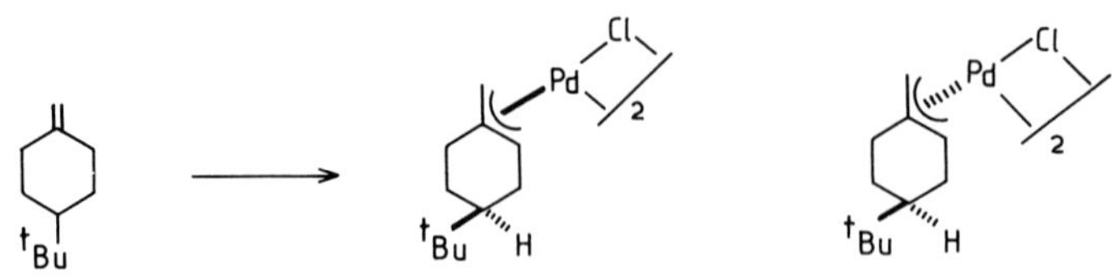

(3)   (4)   (5)

Such complexations can, however, on occasions be highly stereoselective. An early example of this is the formation[6] of a single complex ($[\alpha]_D + 32\cdot8°$) from either α or β-pinene, shown in Scheme 6.1.

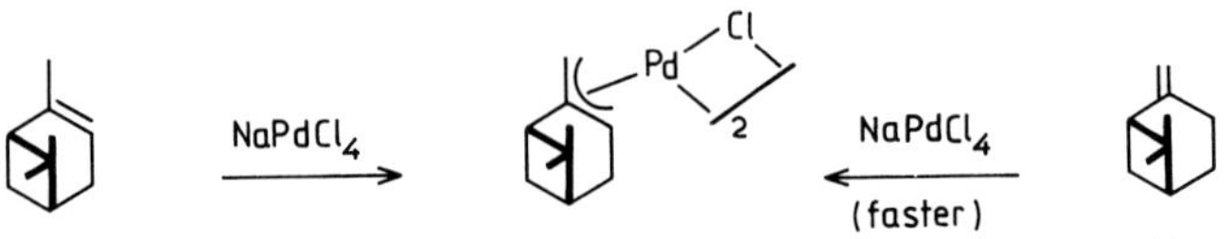

Scheme 6.1

Mixtures of regioisomers may arise in complexations of this type, but it is not unusual to obtain a single product. Some examples[7] are given in Scheme 6.2.

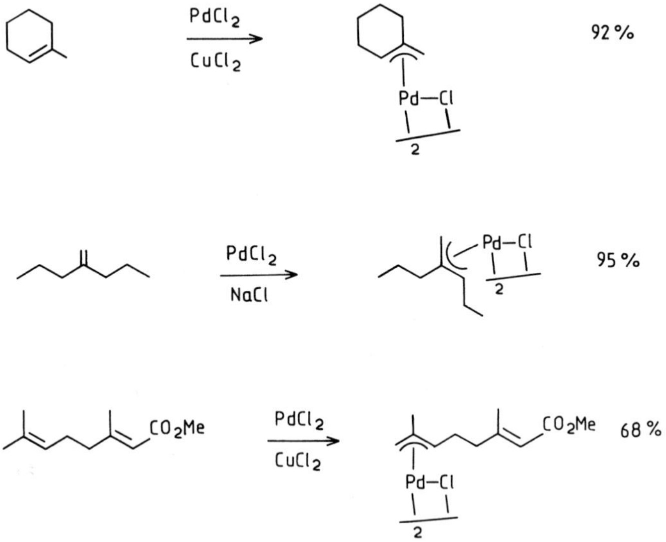

Scheme 6.2

When there is no strong factor to distinguish the alkene bonds, the formation of a Pd π-allyl complex from a non-conjugated diene may yield isomers, though these are commonly distinguishable by n.m.r. examination. As an example, consider *trans*-geranylacetone (6) which gave the π-allyl derivatives (7) and (8) in the ratio 3:2. *Cis*-geranylacetone (9) formed a 2:3 mixture of the same products.[8] Reactions with conjugated dienes can give rise to compounds such as the alcohol (10),[9] obtained from β-myrcene. This reaction is related to the palladium catalysed addition of water to alkenes via the Wacker process discussed in Chapter 4, in that nucleophile addition to an alkene complex has occurred. In this case, however, formation of a π-allyl complex intervenes before the β-elimination step.

In another reaction in which nucleophile addition occurred during the formation of a palladium chloride dimer, the bicyclic methylenecyclopropane (11) underwent ring opening upon complexation to produce (12).[10] Such reactions are of interest from a

mechanistic view-point and have been the subject of considerable investigation.[11] Frequently, however, mixtures of products are formed in complexations proceeding in this way. Despite this difficulty, some organic transformations of these complexes have been described. The dimer (13), for example, undergoes solvolysis in methanol with retention of configuration at the oxygenated chiral centre.[12,13] This product (14) can be converted[13] in methanolic potassium hydroxide to give metal-free alkene products such as (15) in addition to simple alkenes lacking the OMe substituent. This decomplexation process resembles the reverse of reactions forming π-allyl complexes by direct reaction of alkenes with palladium chloride.

A clear distinction must be made between this and decomplexation by elimination with base, which is described in Section 6.9. The formation of alkenes from palladium chloride dimers of type (12) appears quite a general process. When a chloride substituted compound, for example (16), is used, dehalogenation also occurs during the reaction. Although quite complicated mixtures are often obtained, alkenes such as (17) and (18) are typical of the type of product formed.[13]

Palladium π-allyl complexes may be also obtained by other methods that are often more convenient than the reactions of PdCl$_2$ with alkenes used in early studies. One such procedure used Na$_2$PdCl$_4$ in reaction with an allylic halide in methanol under a stream of carbon monoxide.[14] This produced a π-allyl complex in high yield. 6β-Bromocholest-4-en-3-one (19) gave (20) in this way.[15] In this reaction carbon monoxide acts as a reducing agent for the palladium.[16] In some cases direct reaction of allyl bromides with palladium metal can be effected by ultrasound irradiation.[17]

6.2 Alkylation of stoichiometric palladium allyl complexes

The addition of nucleophiles to dimeric Pd(II) complexes such as (1) results in the displacement of the metal as Pd(0) due to the overall transfer of electrons from the nucleophile to the metal.

This process is assisted by the use of coordinating solvents and the presence of phosphine ligands.[5,8,18] Phosphines serve a useful function, splitting Pd(II) dimers into more reactive Pd(II) intermediates of the general types (21) or (22). Similar monomeric adducts of this type, produced from strongly coordinating amines, have been characterised.[4] The nature of these intermediates will vary considerably with the choice of solvent and the quantities of phosphine used.[19] This can have a marked effect on the regiochemistry of the addition reaction. In the presence of HMPA, only (24) was formed from (23). When HMPA was replaced by tri-o-tolyl-phosphine, the regiocontrol was reversed and alkylation occurred at the more hindered position to give an 85:15 mixture of (25) and (24).[5] In this case the metal, bound by bulky ligands, has moved towards the less hindered location as the reaction proceeds.[19] A similar effect has been discussed for alkene complexes in Chapter 4. The most appropriate choice of phosphine is related to the nucleophilicity of the entering group. In general PPh_3, or $Ph_2PCH_2CH_2PPh_2$ are effective with anions such as $^-CH(CO_2Me)_2$ or $^-CH(CO_2Me)PhSO_2$. The effect of acceptor ligands such as phosphines on the properties of electrophilic palladium allyl complexes has been the subject of direct study using ^{13}C n.m.r. spectroscopy in an attempt to obtain information about the relative charges at the termini of the η^3-allyl system.[20] Precise experimental conditions can also be of great importance. Many displacement reactions of Pd π-complexes with carbanion nucleophiles have been carried out in THF or $MeOCH_2CH_2OMe$ as solvent, but use of DMSO can be very advantageous.[21]

Normally nucleophiles add to the less hindered terminus of the allyl system. Trost's route[22] to the ester of the monarch butterfly pheromone (28) provides a good example,

demonstrating that complete regiocontrol is possible during both the formation (see above) and alkylation of the allyl complex which afforded the intermediate (27).

Reaction with the anion (29) has been used to lengthen a terpene chain by one isoprene unit, after the CO_2Me and SO_2Ph groups had been removed. This produced an

intermediate for the alcohol (30). Similar stabilised enolate additions have been used to elaborate a variety of steroids in the B ring.[23]

In some cases alkene stereocontrol can be introduced during the palladium mediated alkylation reaction. An E, Z mixture of 1-substituted butadienes was converted to the

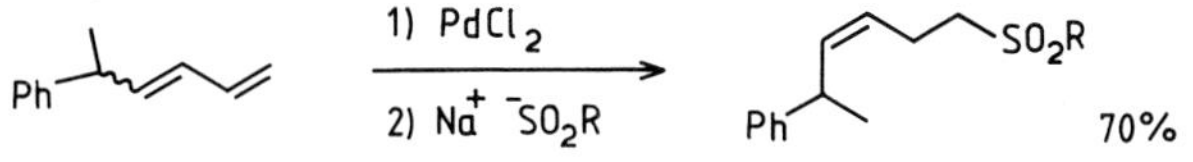

(31)

Z-isomer (31).[24] The formation of a Z-isomer in this reaction is unusual and contrasts with catalytic versions of this process discussed in Chapter 14.

It is useful to note that in alkylation of Pd π-allyl, carbanions stabilised by RSO_2, RSO, or RS groups offer a range of applications in synthesis.[25] As seen above, the $PhSO_2$ or $MeSO_2$ group may be removed by reduction with Li in $EtNH_2$ at -78 °C or with Ca in liquid ammonia, or with sodium amalgam.[25] Similar removal of a sulphide residue has been achieved using Na/Hg and Na_2HPO_4 in methanol,[25] as seen in the preparation of (33) and (34) from (32). When the alkylation product contains a sulphoxide group,

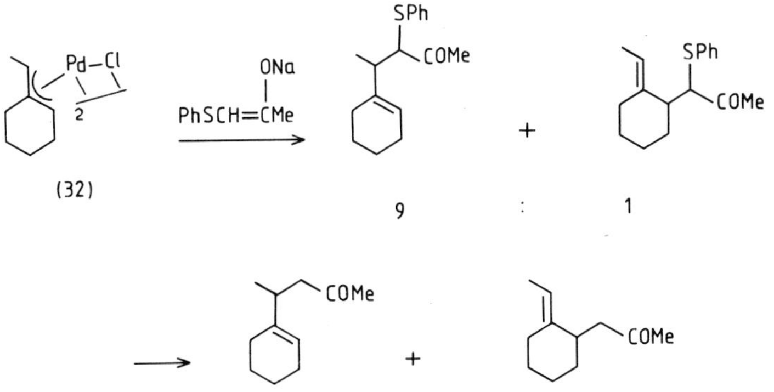

(32)

9 : 1

(33) (34)

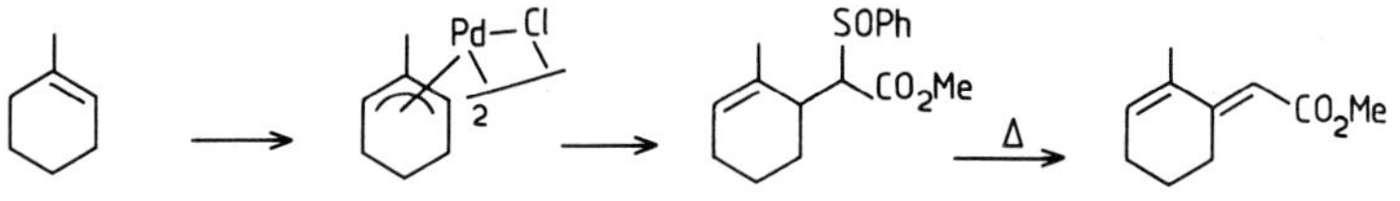

(35)

on the other hand, thermal elimination can be used to yield alkene products such as (35), (36) and (37).

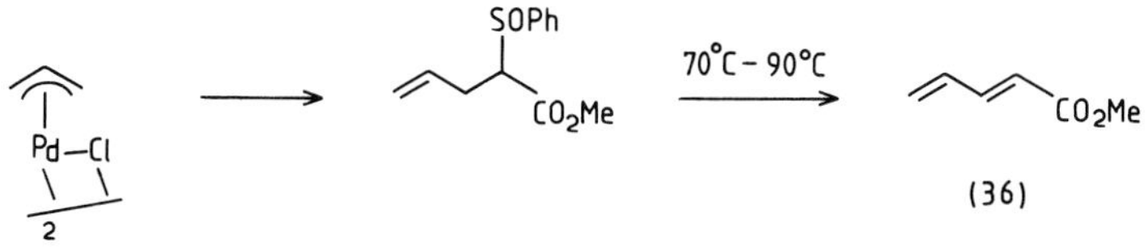

(36)

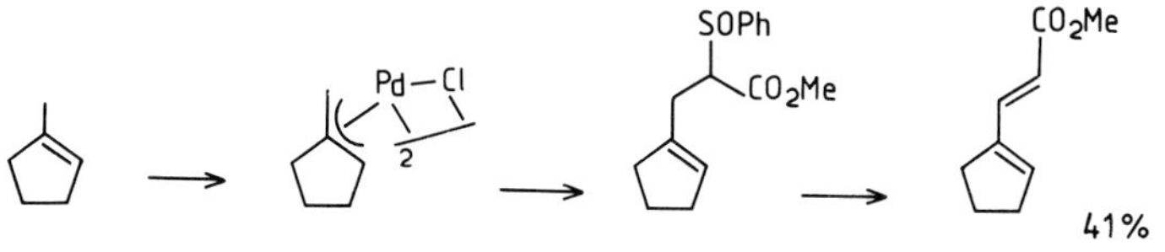

41%

(37)

PhSO$_2$ elimination has been used to produce the double bond required in a synthesis of the macrocyclic antibiotic A26771B (39).[26] Palladium catalysed cyclisation (see Section 6.4) of (38) produced an enol ether which was converted into a mixture of α-hydroxyketones before esterification and elimination to form (39).

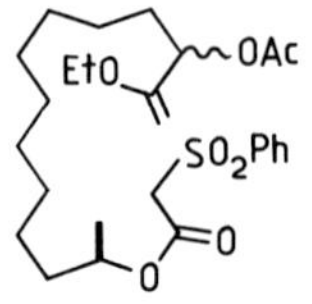

(38) 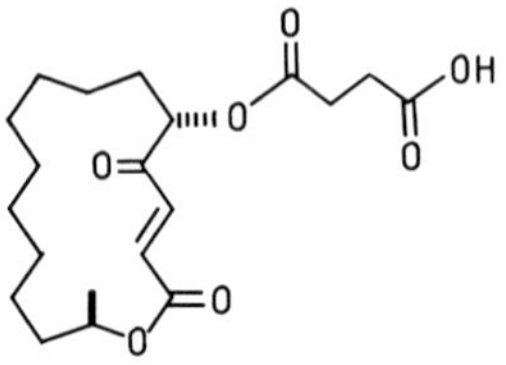   (39)

Among other syntheses which make use of alkylation of a Pd π-allyl system, is an interesting route[27] to vitamin A developed by Marchand and co-workers. The phenylsulphonyl derivative (42) was elaborated from the β-ionol derivative (40) via the bromide (41) (PBr$_3$, C$_5$H$_5$N) and reaction with PhSO$_2$Na in DMF. The sulphone (42) was converted to the sodio derivative and condensed with the Pd π-allyl complex (43).

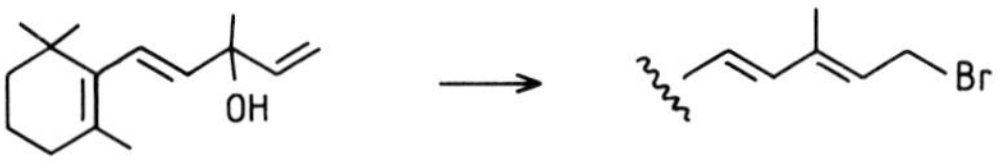

(40)   (41)

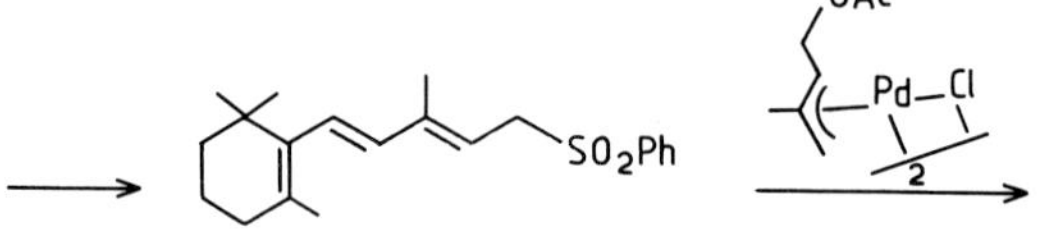

(42)

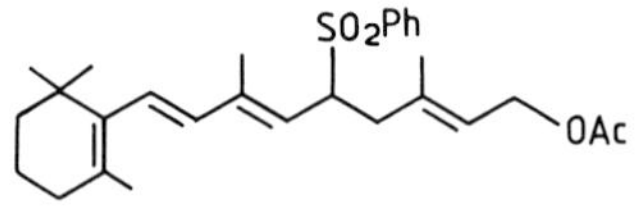

(44)

Elimination of the phenylsulphone residue from (44) using NaOEt in hot ethanol, gave a vitamin A mixture containing 67% of the all-*trans* isomer (45).

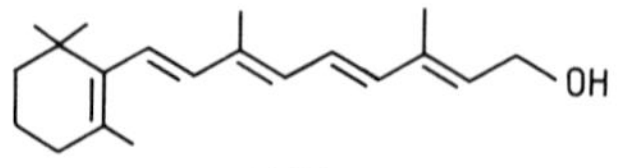

(45)

Examples employing stabilised enolates illustrate a range of related carbanion derivatives which are effective in displacing palladium. The sulphonyl stabilized anions

$[MeSO_2CHCO_2Me]^-$ or $[PhSO_2CHCO_2Me]^-$ appear to be more regioselective than $[CH(CO_2Me)_2]^-$, favouring alkylation at the less substituted allyl terminus.

Substituents on the allyl ligand can exert a strong directing influence on the alkylation reaction. This effect has been used in early work by Jackson[21] to demonstrate the use of (46) as an umpolung equivalent of an extended ester enolate. More recently,

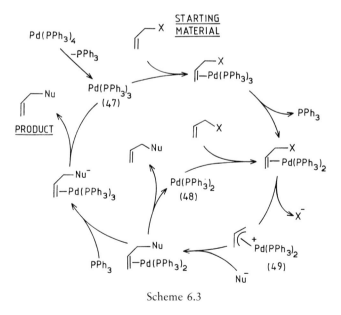

regiodirecting effects have been examined in detail in catalytic systems employing palladium(II) complexes, the topic dealt with in the next section.

6.3 Catalytic uses of palladium allyl complexes

The displacement of an allylic leaving group by a nucleophile is catalysed very effectively by palladium complexes. The main attraction of the method is the complete stereocontrol often obtained in reactions of this type. This aspect is discussed in detail in Section 6.4.

A representation of the catalytic cycle is shown in Scheme 6.3. Reasonable

Scheme 6.3

intermediates can be drawn as either 16 electron (47) or 14 electron (48) complexes, although the initiation reaction will form the 16 electron species (47). In either case, the key cationic palladium allyl complex (49) is implicated in the nucleophile addition step.

The overall process, typically the replacement of an allyl acetate or carbonate by a nucleophile, is indicated in Scheme 6.4. First the leaving group (X^-) is displaced by palladium. The second step is the addition of the nucleophile, which occurs in a fashion analogous to that described in Section 6.2.

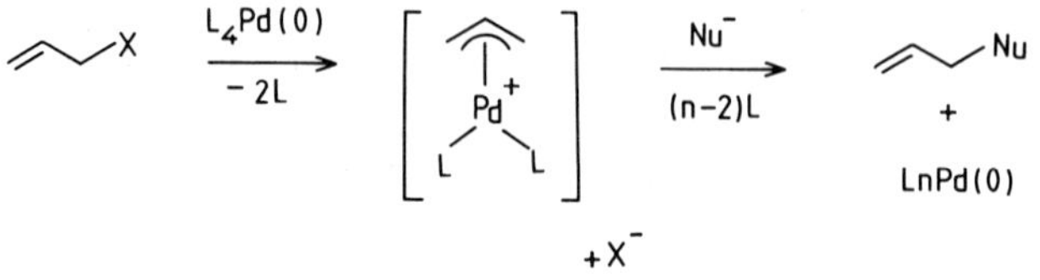

(49)

Scheme 6.4

Although the allyl intermediates in this process contain Pd(II), the palladium complex is added to the reaction mixture in oxidation state (0), typically as $Pd(PPh_3)_4$.[19] When first encountered, this might appear paradoxical. It is, however, exactly what should be expected since the displacement of the leaving group involves the removal of a pair of electrons from the complex. In redox terms, this is the reverse of the nucleophile addition process that was discussed above.

The nature of the nucleophile can have a pronounced effect on the site preference for additions.[28] In reactions of the intermediate complex (50), soft nucleophiles such as malonate gave only the adduct (51) while harder species such as the allyl silane formed only (52). Reaction with typical hard nucleophiles such as alkyl lithium or cuprate reagents resulted in this case in reduction to form (53) rather than addition. Of

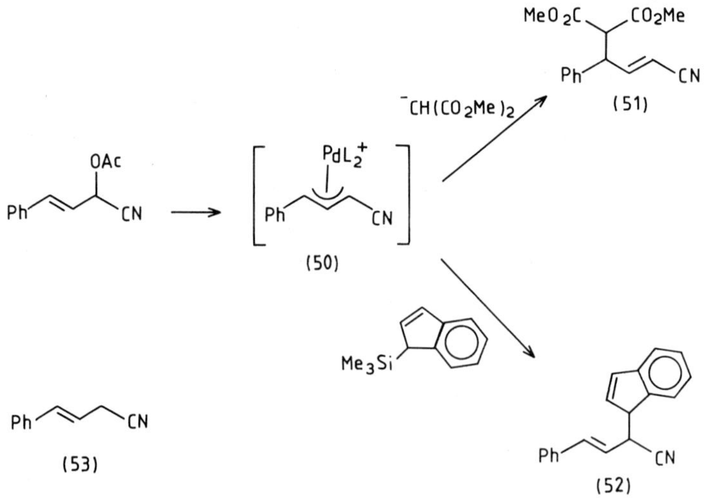

particular importance is the possibility that, in some instances, the nucleophile may be bound initially to the metal before transfer to the allyl ligand. The two paths also have different stereochemical outcomes, and so can usually be distinguished. These mechanisms are discussed in Section 6.4. Another complication in predicting regiocontrol arises when extended π systems are involved. In such cases the positional isomerisation of the metal attachment must be considered.[29]

In either stoichiometric or catalytic systems organic substituents, such as Me_3Si[30] and PhS[31] groups, can have a strong directing effect. The trimethylsilyl group in (54) forces the enolate of methyl acetoacetonate to add to the complex at the far end of the allyl system, overcoming the steric bulk of the pentyl substituent. In (55) a silyl group is placed in competition with a phenyl group. The production of (56) shows the domination of the silyl substituent in this example.[32] Sulphide directing groups have also been used in synthetic work.[31]

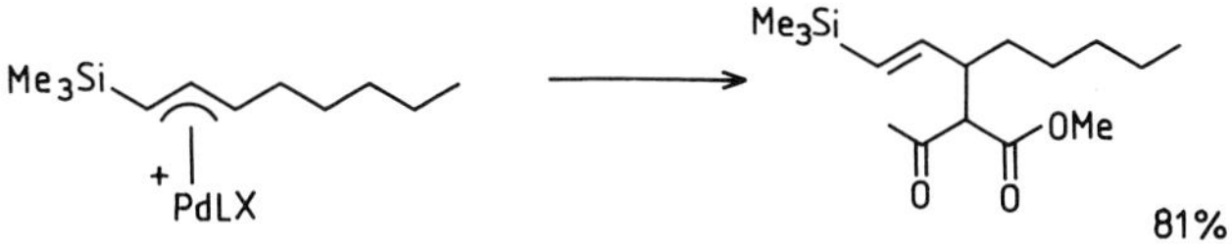

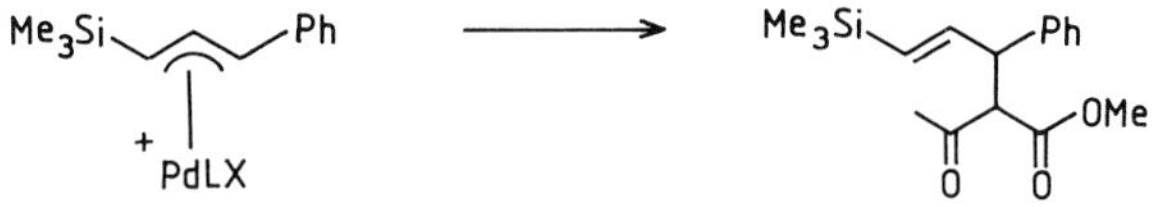

In this way, the vinylsulphide allylic acetate (57) can be used[31] as the precursor for a reaction equivalent to a Michael addition, but with reversed stereocontrol. The outcome is consistent with attachment of palladium to the less hindered face of the double bond in the expected fashion (see Section 6.4). Scheme 6.5 compares palladium catalysed nucleophile addition to (57) with conjugate addition to (58), which produces the *trans* isomer (59). In the stoichiometric complex (46) discussed at the end of Section

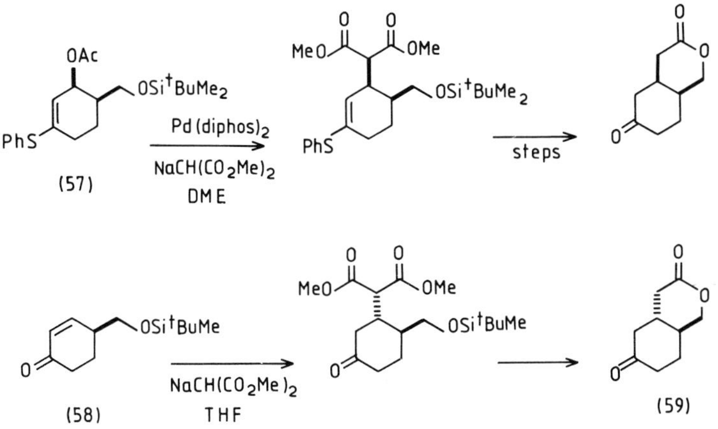

Scheme 6.5

6.2, an electron withdrawing group also sends the nucleophile to the far end of the π system. A similar effect has been obtained in a catalytic reaction using the acetate substrate (60).[33] In this case two allyl complexes could be formed. The product (61), however, is consistent with alkylation of an enol ether–substituted allyl complex at the unsubstituted terminus. The enol ether (61) has been converted into the enone (62),

(60)   (61)   60%   (62)   78%

|||

(63)

leading Trost to describe (60) as a synthetic equivalent of an enone cation synthon (63). This is a catalytic analogue of the equivalence discussed earlier for complex (46). Allylic phosphates have also been examined. These also direct nucleophiles to the unsubstituted terminus.[34] This type of alkylation is not well stereocontrolled. The products (64) and (65), for example, were produced as a 3:2 mixture. A number of alkylations of this type have been examined; the yields vary from about 40 to 90%.

(64)

+

(65)

Even in quite extensive polyenes such as (67), considerable regiocontrol can be obtained, as seen in the addition of the sulphur stabilised enolate, derived from (66), to the unsubstituted terminus of the π system. This product was converted by a palladium catalysed elimination reaction into vitamin A analogues (68).[38] For other examples of palladium catalysed elimination reactions, see Section 6.9.

Alkylation at the less hindered terminus of an unsymmetrically substituted allyl has been used to good effect in a synthesis of humulene,[35] by Yamamoto and co-workers. By oxidation (SeO$_2$) followed by reaction with PBr$_3$, geranyl acetate was converted into (69) which, with the dianion of methyl α-methylacetoacetate and subsequent re-

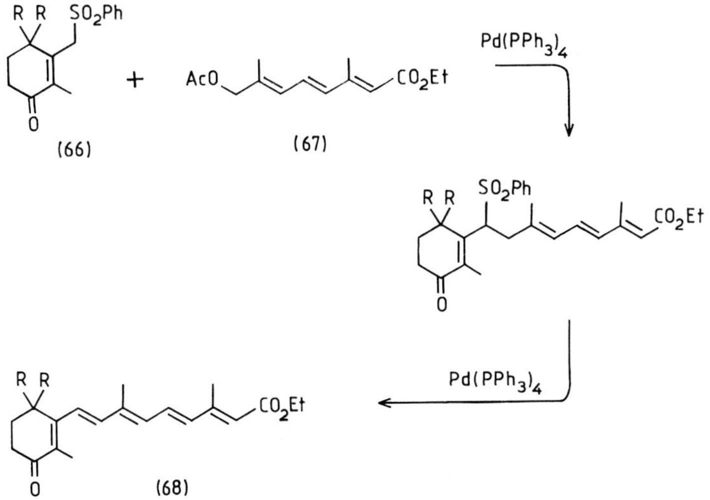

(66) + (67) →

(68)

acetylation with acetic anhydride/pyridine, led to (70). The product (70), as its sodio derivative, was treated with Pd(PPh$_3$)$_4$ in catalytic amount, in the presence of Ph$_2$P(CH$_2$)$_3$PPh$_2$ and HMPA in THF. This effected cyclisation via (71) to give (72). The

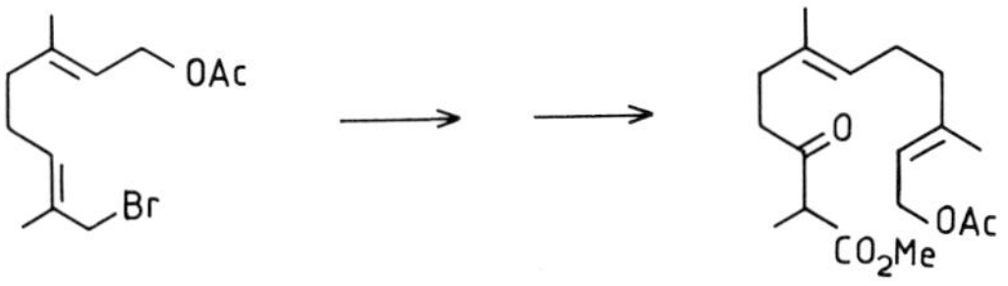

(69) → → (70)

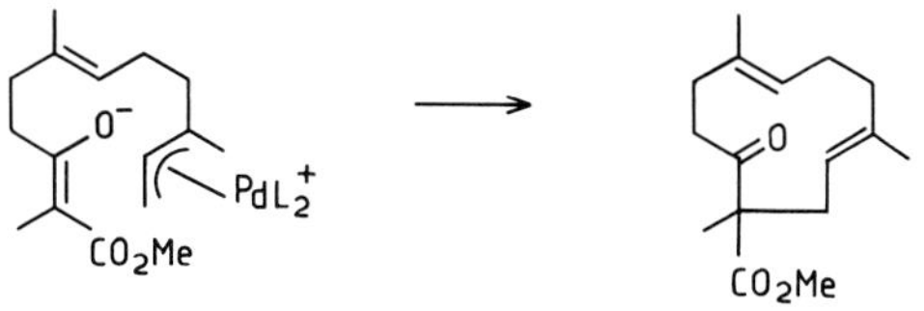

(71) → (72)

use of Ph$_2$P(CH$_2$)$_3$PPh$_2$ very much reduced the formation of dimeric C-32 products. From (72), humulene was obtained by LiAlH$_4$ reduction, tosylation of the primary hydroxyl group, and treatment with KO^{t}Bu to give the oxetane (73). Reaction with Et$_2$AlNMePh converted (73) into (74), which possessed the required alkene stereo-

(73) → (74)

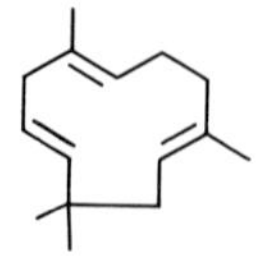

(75)

chemistry. The CH$_2$OH side chain was then modified in a series of steps to produce humulene (75).

The process of cyclisation of (71) to (72), incorporated in the synthesis of humulene, has been extended by Trost *et al.* to the synthesis of other macrocyclic compounds,[36] including (±)-recifeiolide.[37] The intermediate (76) which contains an allylic acetate terminus and also a potential carbanion terminus, was cyclised in the presence of Pd(PPh$_3$)$_4$ to yield (77), which could be modified, (Me$_4$NOAc in HMPA, then Na/Hg and Na$_2$HPO$_4$) to recifeiolide (78).

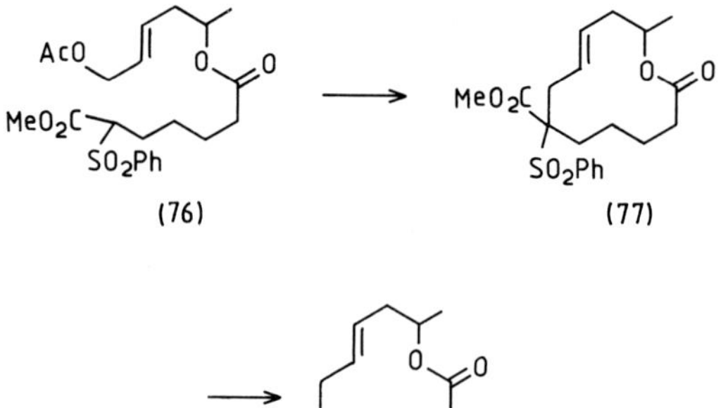

(76)   (77)

(78)

In a related approach aimed at the clinically important erythronolide antibiotics, the cyclisation of (79) by a similar method has been examined. The product has been converted by decarbomethoxylation and *cis*-hydroxylation to the erythronolide analogue (80).[38]

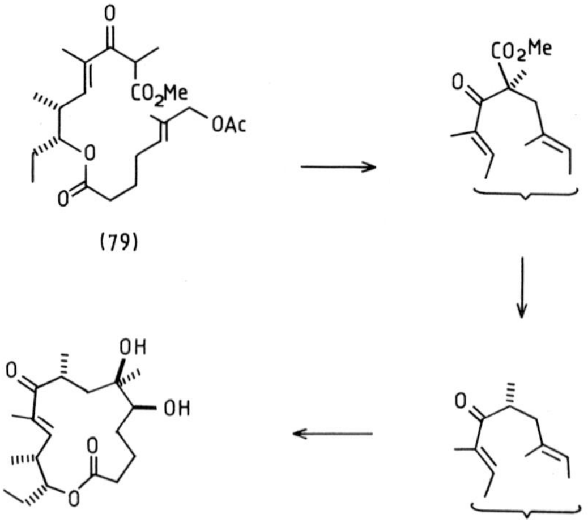

(79)

(80)

Similar cyclisation can provide a convenient route to smaller rings as shown in a synthesis of methyl dihydrojasmonate by Tsuji.[39] Spirocyclisation can also be achieved.[40] Thus (82) was obtained by treatment of the allylic acetate (81) with NaH in the presence of $Pd(PPh_3)_4$ in THF. A shorter chain length can lead to oxygen addition to the allylic intermediate to form, for example, (83). A somewhat similar alcohol addition has been

used to form cyclic and spirocylic ethers.[41] Both spirocyclic and bridged products can be obtained from palladium catalysed intramolecular additions of silyl enol ethers to alkenes.[42] Similar results have recently been obtained with methoxyenol ether precursors. In an application[43] of this reaction, (84) was used to construct the bicyclo[3.3.1]nonene framework of the alkaloid selagine.

6.4 Stereochemistry of reactions of palladium allyl complexes

Displacement of palladium with inversion by the entering nucleophile was demonstrated[44] by Trost and Weber, using the Pd π-allyl complex of 2-ethylidenenorpinane (85), which was found by X-ray analysis to have the stereochemistry shown in (86).

Reaction with $^-CH(CO_2Me)_2$ in THF in the presence of $(Ph_2PCH_2)_2$, followed by decarbomethoxylation, gave a product that was shown to be (87), since oxidation (OsO_4) led to a lactone, recognised as (88) by n.m.r. analysis.

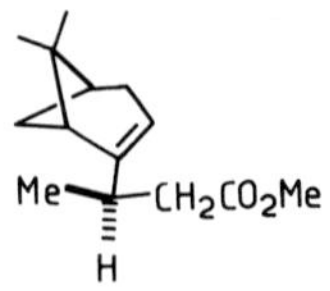

(87) 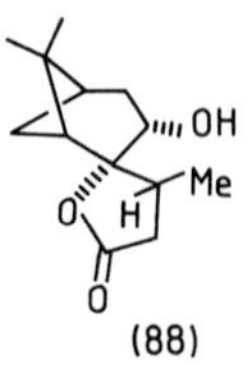   (88)

In a second example[45] which established the stereochemistry of displacement of palladium, the π complex (89) was shown to afford (90) by reaction with $NaCH(CO_2Me)_2$ in DMSO. The yields in these reactions were generally high, although in alcohol solvents (89) showed a tendency to undergo elimination to yield the dienone instead of the required nucleophilic substitution product (see Section 6.9).

(89)   (90)

The stereochemistry of palladium catalysed alkylation is the result of two sequential displacement reactions, and the stereocontrol at each step must be considered. The initial approach of the palladium atom to the substrate occurs from the side opposite to the leaving group. Direct alkylation of the π bound ligand occurs from the side opposite to the metal, as was the case with the stoichiometric examples considered above. The consequence of these two inversions of configuration is thus overall reaction with retention, and this is the usual outcome observed.

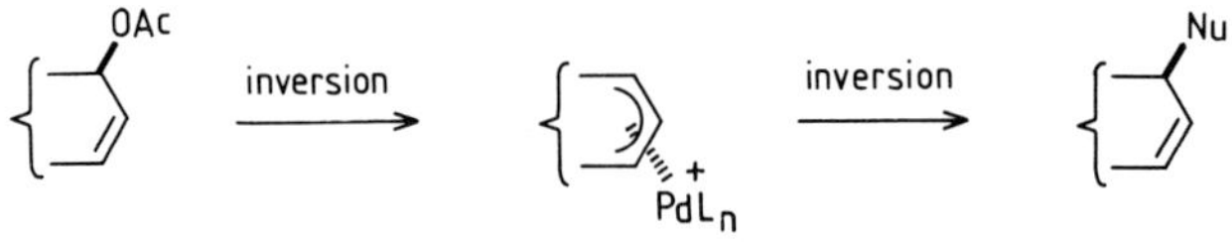

The stereochemical basis of this sequence was determined[46] using the examples shown in Scheme 6.6.

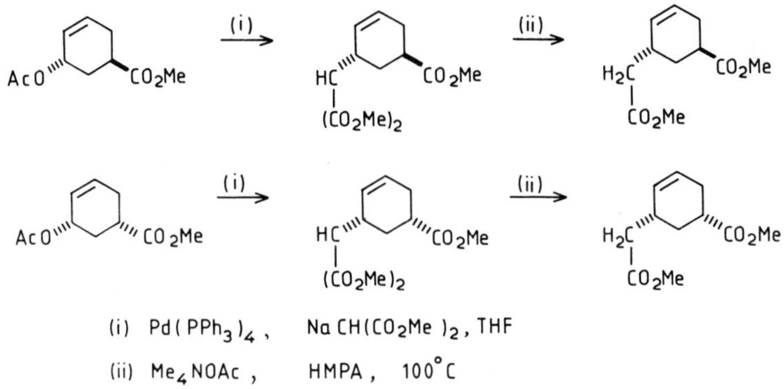

(i) $Pd(PPh_3)_4$, $Na\,CH(CO_2Me)_2$, THF

(ii) Me_4NOAc, HMPA, 100°C

Scheme 6.6

The expected stereochemistry of substitution of the acetoxy group with retention is modified if the substituting agent reacts slowly relative to rearrangement (see Chapter 2) of the acetate. This effect is seen in the formation of (92). When $^-CH(SO_2Ph)_2$ is

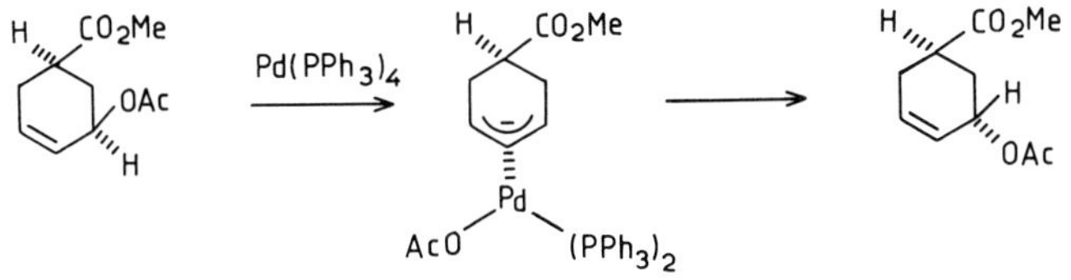

(91)   (92)

employed, the bulk of the nucleophile causes slow alkylation and allows epimerisation of the allylic acetate by AcO^- addition to the metal to form (91) from which acetate could be transferred to the ring. The result is the formation of a mixture of substitution products.[47] This was also observed[47] in a substitution of allylic acetates by amines[48] in the presence of $(Ph_3P)_4Pd$. Substitution with retention was, however, increased if the palladium catalyst was supported on silica gel or polystyrene carrying phosphine groups.[48]

In early examples, the ability, in some cases, of allyl complexes to interconvert with their acetate precursors had often confused attempts at mechanistic investigations in this area. Recent work by Hayashi *et al.*[49] on the alkylation of the optically active allylic acetate (93) with the malonate anion has now verified the stereochemistry of each step in the catalytic cycle which leads to net retention in the overall formation of the product.

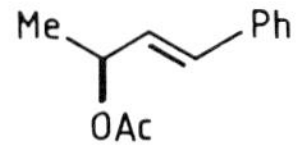

(93)

Although acetate replacement with overall retention of configuration is the most common outcome of palladium catalysed allylic substitution, an alternative mechanism can operate in some circumstances, leading to the formation of products with overall inversion. Just as was seen in Chapter 5 for additions to palladium alkene complexes, so too, with π-allyl complexes, nucleophile addition can also proceed via the metal. This alternative is shown in Scheme 6.7. This realisation has led to a recent attempt to classify

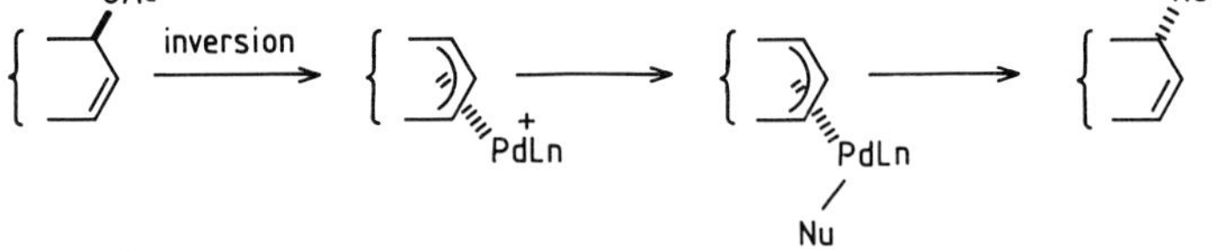

Scheme 6.7

nucleophiles into two categories, those which react by direct addition (for example sodium dimethyl malonate, sodium cyclopentadienide, and morpholide) and those which proceed by addition via the metal (phenylzinc chloride, sodium indenide, and ammonium formate).[50] Some specific examples are given later in this section.

In the case of allylic carbonates, the relative stereochemistry of the starting material has a great effect on the degree of stereocontrol obtained. While the *cis* substrate (94) reacted predominantly with retention, the alkylation of the *trans* substrate (95) was found to be much less stereoselective.[51] In the reaction of the corresponding acetate (96)

with diethylamine, adducts were obtained in a ratio (*cis*:*trans*) of 88:12. This is consistent with prior isomerisation of the starting material. Addition of lithium chloride to the reaction mixtures reversed this product ratio to 15:85, showing a clear effect of halide ion concentration on the reaction.[52] N.m.r. and conductivity studies on complexes derived from norcamphene have revealed considerable interaction of halide ions with palladium complexes in solution.[53]

Most cases where reactions do not follow the normal path of overall retention in displacement are encountered when nucleophiles initially react at the metal itself, rather than the ligand. Reduction by NaBD$_4$ of (97) introduced[54] deuterium with net inversion of the stereochemistry of the allylic substituent. Similarly, acetate transfer from the dimer (98) under a carbon monoxide atmosphere, possibly via the σ bound intermediate (99), occurs to the face of the ligand attached to the metal.[55] Some organometallic nucleophiles also appear to add via the metal. Good examples are the organometallic organozinc[56] and organozirconium[57] reagents (see also below) which add in a *cis* fashion.

It is clear from this evidence that there are two distinct paths that can be followed during the stage in the catalytic cycle at which addition of the nucleophile to the ligand occurs. Path A, direct addition to the ligand, proceeds with inversion at this stage. Path B requires two steps. Addition of the nucleophile at the metal to form (100) leaves the stereochemistry of the allyl ligand unchanged. Subsequent transfer to the ligand places the nucleophile on the same face as that bearing the metal, i.e. retention of configuration at this stage. The stereochemical outcome of these two paths relative to the configuration of the allyl acetate substrate is indicated in Scheme 6.8.

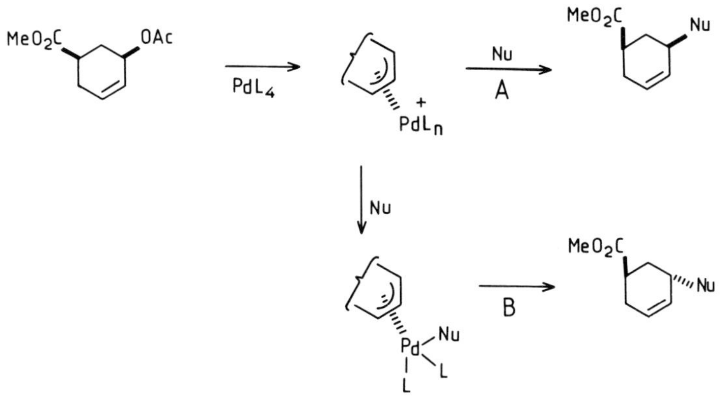

Scheme 6.8

The use of the stereocontrolled alkylation of allyl acetates may be illustrated with an example from Trost's work on the steroid (101).[58] The product (102) could be

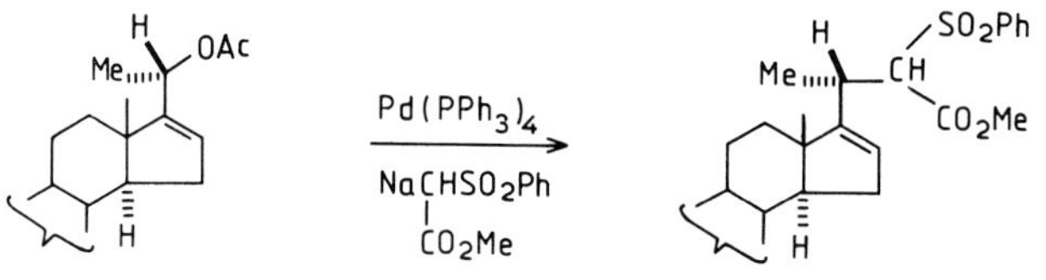

elaborated as shown to introduce the cholesteryl side chain. It is important to observe that (102) was formed with overall retention of stereochemistry at C-20. Displacement of OAc by palladium, and of palladium by the phenylsulphonyl acetate anion, both involved inversions.

The C-20 stereochemistry set up in the conversion of (101) into (102) was reversed in an alternative sequence[58] which started from the alkene (103). Since the Pd π-allyl complex from (103) took up the *syn* stereochemistry, displacement of the metal from (104) with Na[CH(CO$_2$Me)$_2$] or Na[PhSO$_2$CHCO$_2$Me] yielded (105) and thence (106) in which C-20 had the configuration opposite to that found in the natural steroids.

(R=CO$_2$Me or SO$_2$Ph)

In an application of the stereochemical consequences expected when addition of the nucleophile occurs via the metal,[57] a complex analogous to (104) has been used in a synthesis[59] of 25-hydroxycholesterol. Addition of the vinylzirconium reagent (107) and maleic anhydride to the palladium complex (108) resulted, in this case, in the natural stereochemistry at C-20. An allylic acetate displacement similar to the conversion of (101) into (102) also showed this reversal of stereocontrol when a zirconium nucleophile was employed.

Like simpler examples described above, the regiochemistry of this process is sensitive to the catalytic conditions.[60] Use of maleic anhydride in the reaction mixture resulted in the preferential formation of the desired product (109), whereas triphenylphosphine led to a preponderance of the opposite product (110).

Maleic Anhydride — 5 : 1

$Pd(PPh_3)_4 + 4PPh_3$ — 35 : 65

These studies in the field of steroid synthesis demonstrate the approaches available from palladium allyl chemistry to control the stereochemistry of a new chiral centre. Both the nature of the substrate, the method of introduction of the metal, and the mechanism of addition of the nucleophile, can be manipulated in an attempt to achieve the desired stereochemical outcome.

6.5 Asymmetric synthesis using palladium allyl complexes

The stereocontrolled displacement of allylic carbonates has been examined[61,62] as a method of remote asymmetric induction when optically active substrates are used. The new bond formation may occur at a different carbon atom to that originally bearing the acetate group. In lactones, too, chirality transfer can be virtually complete in some cases.[62] Alkylation of the chiral lactone (111), available from glucose, gave a resolved intermediate (112) for the synthesis of the vitamin E side-chain (113).[63]

In another example, using mannose as the starting point, a C_5 side-chain was introduced into the lactone (114).[64]

(114)

In cases where either the allyl acetate itself, or a complex derived from it, is prochiral, an alternative entry to resolved products is available by the inclusion of a chiral auxiliary in the catalytic system. Asymmetric catalysis of this general type is desirable because a catalytic amount of optically active catalyst is used to form a much larger quantity of the resolved product.

An attractive approach is the use of a chiral phosphine to influence the site of alkylation. This influence has been demonstrated in the conversion of (115) into (116) where optical yields of about 20% were observed using (+)-DIOP (117), or (+)-ACMP (118) as phosphine ligands.[65] (−)-Sparteine (119) was also found to be effective, demonstrating that a suitable nitrogen ligand may replace the phosphine.

(115)

(116)

(117)

(118)

(119)

A recent, though also relatively inefficient, asymmetric induction used (+)-DIOP, (−)-CHIRAPHOS and (+)-BINAP to introduce asymmetry by directing double bond stereochemistry to distinguish the sides of a cyclohexane ring in a product obtained from a prochiral allylic acetate. (+)-BINAP gave the best results; about 40% e.e. and 74% yield.[66]

Catalytic reactions employing this principle at first showed similar disappointing optical yields. In the presence of (+)-DIOP, the product (120) was obtained in moderate optical yield (62 % R,R and 38 % S,S isomer). The use of CAMPHOS, derived from camphoric acid, was also explored[67] in this reaction. More recently, asymmetric alkylation procedures employing CHIRAPHOS have been developed to give far better optical yields, now greater than 80 %.[68]

(120)

The mechanism of this asymmetric induction process has been studied.[69] In direct nucleophile addition (i.e. following Path A, Scheme 6.8), the nature of the alkylation reaction is such that, unlike asymmetric epoxidation (see Chapter 3) or asymmetric hydrogenation (see Chapter 10) where both of the reacting components are bound to the metal, only the allyl ligand is held close to the chiral auxiliary by coordination. Different steric effects or activating effects at the termini of the allyl ligand must be responsible for the asymmetric induction. The design of special ligands to create an appropriate chiral environment in the region of approach of the nucleophile has been attempted.[70]

The true basis of the asymmetric induction effect may vary from case to case, and unexpected results are sometimes obtained. DIOP mediated addition reactions to allyl acetates have been examined in an attempt to probe the mechanistic details of the alkylation reaction. In a comparison of reactions of racemic and optically active precursor acetates, it was anticipated that a common symmetrical π-allyl complex would be produced. In the event, however, this proved not to be the case, since different specific rotations were obtained for the products from the two reactions.[71]

An intramolecular version of the asymmetric alkylation process converted the prochiral alkene (121) to a chiral 2,3-dihydrobenzofuran (122). In this case β-pinene was used as a chiral auxiliary. The optical yield of the product was only 12 %.[72]

(121)   (122)

In another asymmetric cyclisation reaction, (123) was converted into the optically active 4-vinyl-2-oxazolidone using the ferrocenyl chiral diphosphine (125, R = $CH(CH_2OH)_2$). Although the chemical yield for this reaction improved to 88 % at lower temperature, the optical yield worsened. The product (124) was formed in 73 % e.e.

when the reaction was performed in THF at reflux. This was reduced to 63% e.e. at 40 °C and only 59% e.e. at 15 °C.[73]

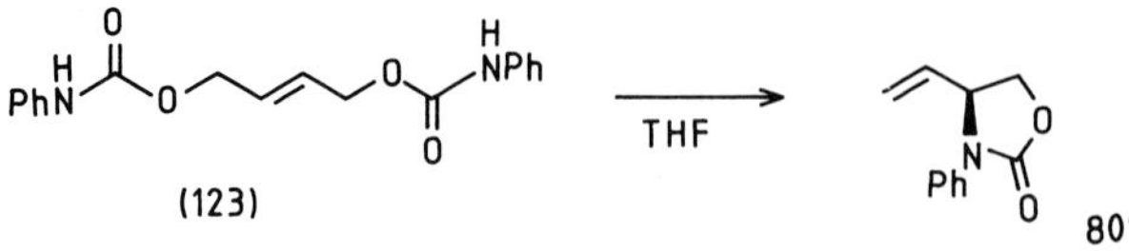

(123)

(124)   80%

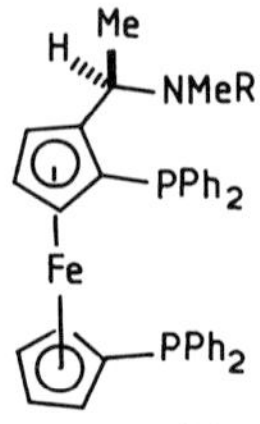

(125)

Good results have also been obtained recently with chiral phosphines of type (125). By addition of malonate, the bis-carbonate (126) has been converted into the vinyl cyclopropane (127) in 67% e.e. Here a sequence of two allylic displacements has been performed. The more simple phosphine (125, R = Me) was used in this case.[73] The ring closure step used here provides an asymmetric induction modification of a chirality transfer cyclopropanation employed by Genêt[74] in a synthesis of (+)-dictyopteren A.

(126)

(127)   24%   67% e.e.

Kinetic resolution has proved to be more effective. Malonate addition to the racemic allylic acetate (128), using the same chiral auxiliary that was employed for the formation of (124), gave the (S) isomer of the adduct (130) in 97% yield and 90% e.e., presumably

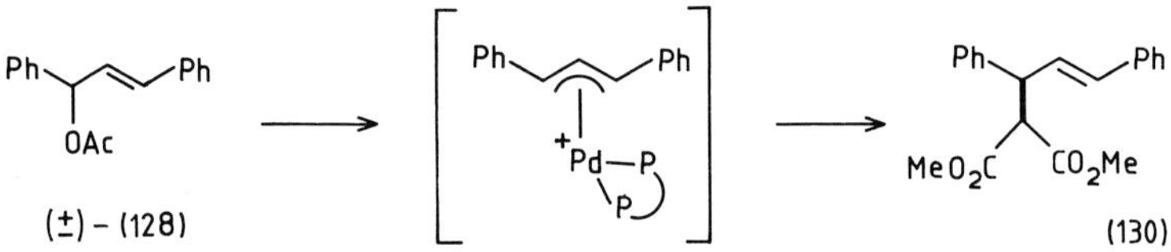

(±) – (128)

(129)

(130)

90% e.e.

by selected addition to the π-allyl intermediate (129).[73] When an unsymmetrically substituted racemic precursor was used, for example (131), a mixture of regioisomers was produced in roughly equal amounts. Both were formed in high optical yield.[75]

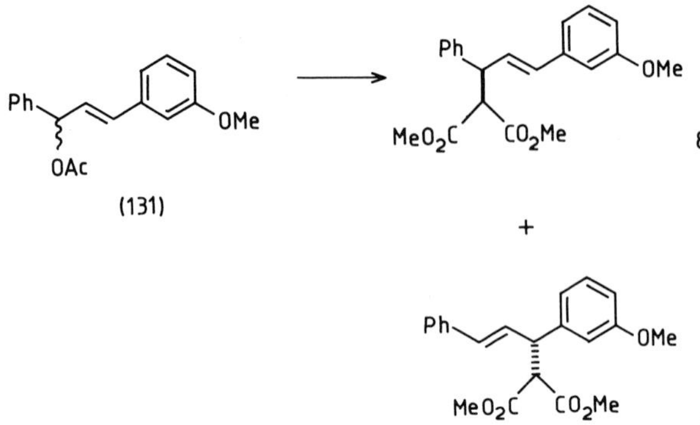

(131)

80% e.e.

+

95% e.e.

6.6 Range of nucleophiles applicable to palladium allyl alkylations

One reason why the reactions of palladium allyl complexes have received such detailed attention from synthetic chemists is the great variety of nucleophiles that are compatible with the method. This provides valuable flexibility when developing a synthetic plan.

The range of synthetically useful nucleophiles extends beyond the use of stabilised enolates to include organoaluminium,[56] zinc[56] and zirconium[57] reagents, and even unstabilised enolates[76] and enolstannanes.[77] Amination reactions are also important and have been examined in some detail. The direct addition of secondary amines has been applied[78] to the conversion of the pyranoside (132), readily available from D-glucal, to amine derivatives in 70 to 90% yield.

The carbocyclic compound aristeromycin (134) has been synthesised by a sequence using two palladium mediated nucleophile additions which established the *cis* relative stereochemistry between the heterocycle and the CH$_2$OH group. Nucleophilic addition of adenine to the epoxide (133) was catalysed by palladium acetate. One equivalent of

butyllithium was used to deprotonate the adenine nucleophile. After conversion of the resulting alcohol to a carbonate, a second nucleophile displacement introduced the additional carbon atom required for the hydroxymethylene group.[79] Another example of the alkylation of heterocyclic bases by transition metal π complexes is to be found in Chapter 7.

A similar enolate stabilised by a nitro group has been used to elaborate (135) into the heptose derivative (136).[80]

Bäckvall has developed the use of amines[81] and sulphonamides[82] for the

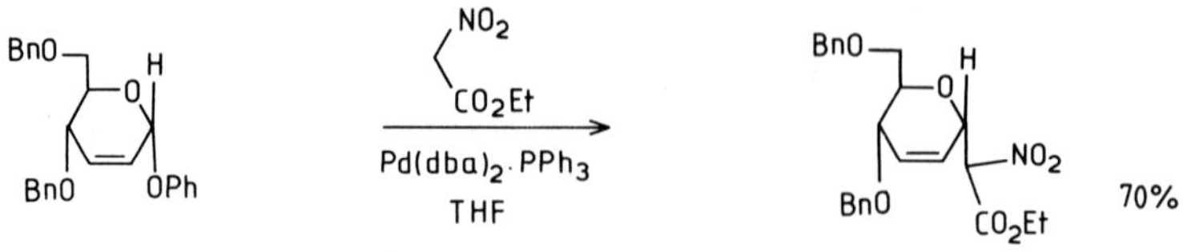

(135)   (136)   70%

stereoselective synthesis of compounds such as (137) and (138) from starting materials that are easily obtained from 1,3-dienes by another palladium catalysed reaction.[83] This

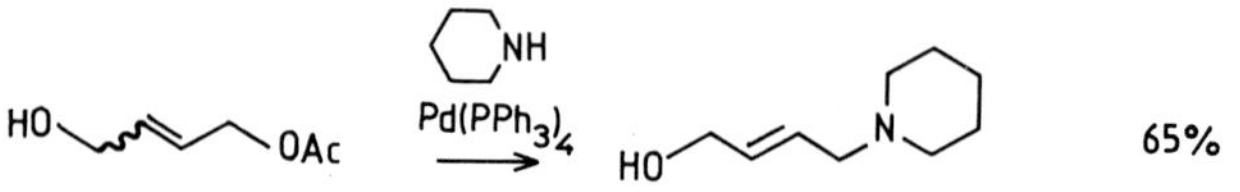

(137)   65%

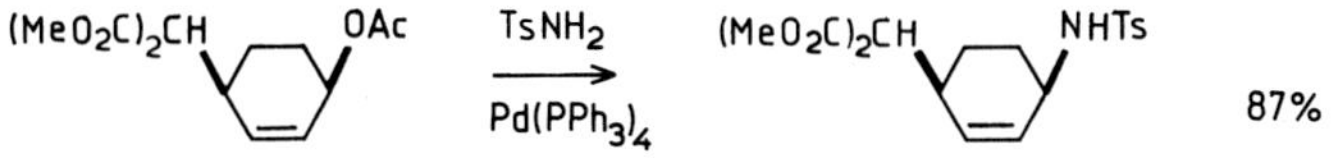

(138)   87%

palladium catalysed 1,4-difunctionalisation of 1,3-dienes has now been extensively examined,[84] and offers an alternative to singlet oxygen addition. Both methods produce a *cis* relationship between the two groups that are introduced.

This type of reaction can be seen in the functionalisation of the oxygenated cycloheptadiene (139). Replacement of the chlorine in the product (140), again using the sulphonamide nucleophile, provides an example which emphasises the complementary stereochemistry of these reactions. Direct displacement of chlorine by sulphonamide proceeded with inversion. In contrast, the use of palladium catalysts to form (141) proceeded with overall retention of configuration.[85]

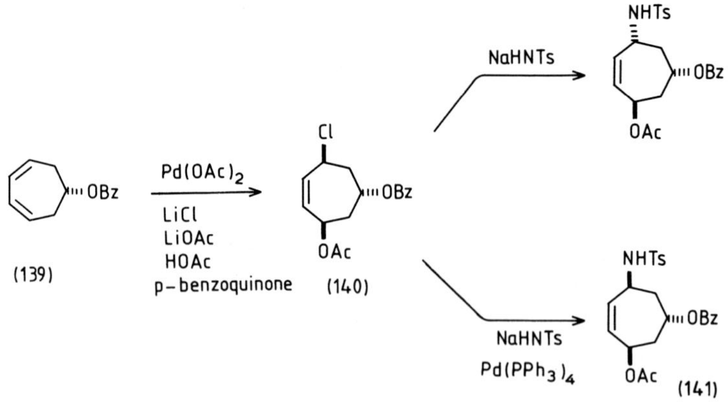

(139)   (140)   (141)

Another application is to be found in the stereoselective synthesis of the antibiotic anticapsin (143). Initial functionalisation of the diene was performed using the same conditions employed for the production of (140). The subsequent steps in the synthesis employ conventional methods. First the displacement of chloride by the stabilised enolate, then decarboxylation and chain extension to introduce the amino acid centre. Epoxidation with *m*-chloroperbenzoic acid was directed by the oxygen substituent on the six-membered ring before oxidation to form the ketone required in the product.[86]

An electrochemical method has been developed to drive the re-oxidation step in the catalytic cycle. This makes the reaction catalytic also in benzoquinone. In acetic acid, a bis-acetate is produced, but when performed with lithium chloride and lithium acetate the product was (142).[87]

Schiff base nucleophiles have been used by Genêt to introduce the amino acid functional group by palladium catalysed displacement of allylic leaving groups. The convenient Schiff base enolate provides a suitable source of the amino acid unit for synthesis, for example, of (144).[88] α-Aminophosphonate groups have also been introduced in this way.[89]

This process has been employed to form the phosphonate ester (145) from butadiene. Reaction of (145) with an aldehyde derived from levogluscosan gave access to the diene (146) with the correct stereochemistry for the lower chain in griseoviridin (147).[90] Other heteroatom nucleophiles to be introduced using palladium catalysis include sulphur,[91] silicon,[92] and tin.[93]

A similar allylic displacement by PPh$_3$ has been used[94] in a Wittig based synthesis of pellitorine. In another study, imides were converted[95] to *N*-allyl derivatives.

Earlier in this chapter, many examples of C–C bond formations using stabilised

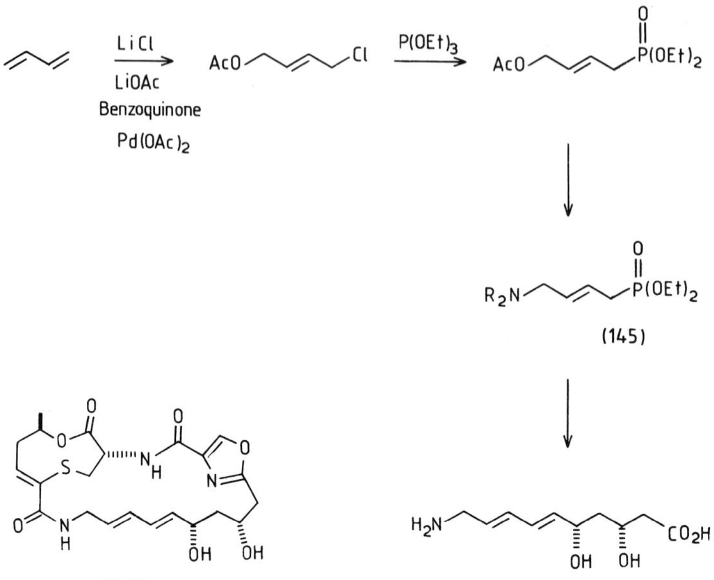

(145)

(147)

(146)

enolates are to be found. The reaction is tolerant of a wide variety of pK_a values. As seen above, the range has been extended toward very low pK_a by studies of the alkylation of α-nitroesters[80] and α-nitroketones.[96] Enolates with high pK_a, however, are also suitable and simple ketone enolates give satisfactory alkylation reactions.[76] An amine substituted enolate, an alternative to the Schiff base enolate discussed above, has been used for the formation of (148) which has been converted to α-amino acids by hydration, hydrolysis and decarboxylation.[97] Analogous syntheses have been based on the catalysed dimerisation of butadiene or of isoprene in the presence of Pd(OAc)$_2$ with $^-$C(NHAc)(CO$_2$Et)$_2$ as the nucleophilic addend, as seen in the formation[97] of (149). Reactions of this type are discussed in detail in Chapter 9.

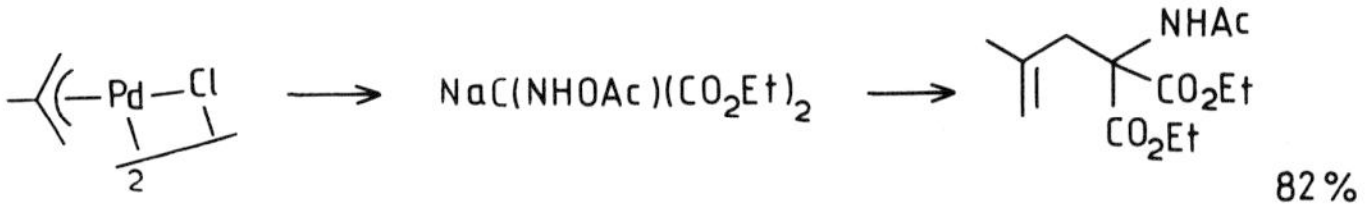

(148)

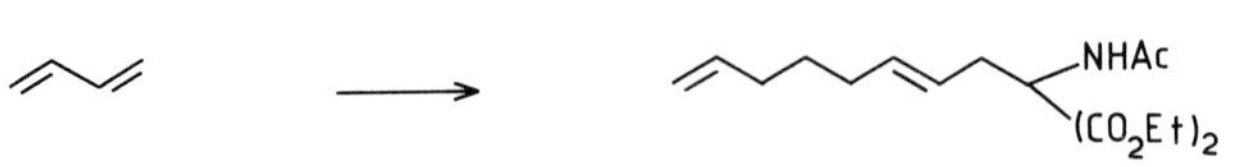

(149)

In another example of the use of a nitro stabilised enolate, Genêt and Grisoni describe an interesting cyclisation reaction leading to ergoline precursors. The best results were obtained using a carbonate leaving group. The 75% yield of (150) was improved to 85% when potassium fluoride on alumina was added as a basic catalyst.[98] Simple

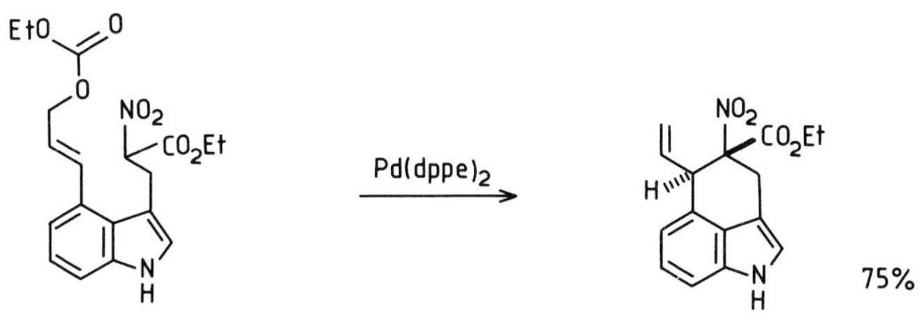

(150)   75%

nitroalkane analogues also cyclise, a reaction that has recently been developed further to effect an asymmetric induction by inclusion of a chiral auxiliary into the catalyst.[99] Enantiomeric excesses of up to 70% have been obtained.

Enolstannanes also give useful alkylation reactions.[77] Allylstannanes have been used in the same way,[100] but with much lower yields. Some examples are given in Scheme 6.9.

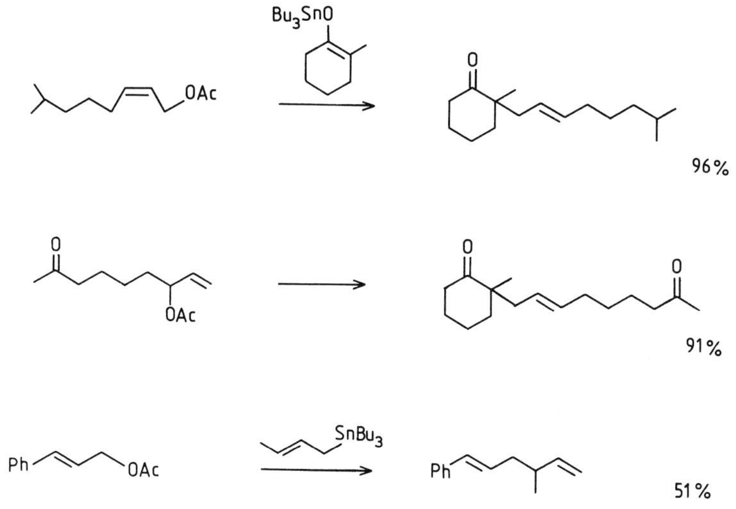

96%

91%

51%

Scheme 6.9

In the case of (151), the product was conveniently cyclised to give a cyclopentenone in what amounts to a five-membered ring version of Robinson annellation.[77]

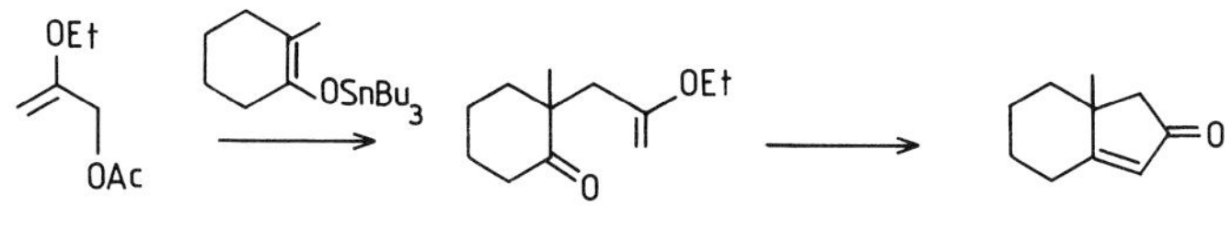

(151)   72%

Other palladium mediated routes to five-membered rings also involve allyl intermediates but proceed as cycloaddition reactions and are discussed in Chapter 11.

In the formation of (152), a stable enol was itself used to effect the alkylation, which took place at one of the carbons on the heterocyclic ring.[101]

(152)   83%

Examples of organometallic nucleophiles include the use of organoaluminium reagents as seen in the conversion of (153) into the triene (154).[102] Grignard reagents have also been used.[103]

(153)   (154)   82%

Further examples using organozirconium chemistry have been discussed in Section 6.4. Some coupling reactions between palladium bound intermediates and organo-metallic nucleophiles discussed in Chapter 10 are also related mechanistically to these reactions.

6.7 Other palladium catalysed allylic displacements

Although displacements of allyl acetates have been most extensively studied, there are clear indications that palladium complexes are able to catalyse the replacement of most allylic leaving groups by nucleophiles. In the examples in previous sections, the displacements of acetates and carbonates, and the opening of epoxides, under conditions using palladium catalysis have been encountered on a number of occasions. The use of carbonates and epoxides in particular, has been one of the main developments in this area in recent years.[104]

Carbonates,[105] dithiocarbonates,[106] silyl ethers,[103] and nitro groups,[107, 108] have all been successfully displaced by nucleophiles using palladium catalysts. Chiral allylic sulphinates have been converted by Pd(PPh$_3$)$_4$ to optically active sulphones with high stereoselectivity at temperatures below those required for the uncatalysed reaction.[109]

In other cases, the leaving group itself contains a nucleophilic species. Treatment of the vinyl substituted enol ether (155) with Pd(PPh$_3$)$_4$ or Pd(dppe)$_2$ results in the formation of a carbocyclic product (156).[110] The reaction has been successfully applied

(155)   (156)   80%

(157)

to the synthesis of an 11-deoxy-PGE intermediate (157).[111] Another application of this cyclisation is the formation[112] of (158), a potentially useful entry into steroid synthesis.

The size of ring that is formed is sensitive to bulk of phosphine ligand employed. The use of polymer bound phosphine again had a particularly pronounced effect, as shown[113] by the reactions of (159), derived from mannose.

Allylic β-ketocarboxylates can also produce C-alkylated products. Here, decarboxylation forms an enolate which reacts with the palladium allyl complex to form monoalkylated derivatives such as (160).[114] Both O-bound[114] and C-bound[115] enolate metal adducts have been proposed as intermediates. Under different conditions α,β-unsaturated ketones are produced.[116,117] Some recent examples of the use of allylic esters include the formation of (161) and of (162).[118]

(162)

Allylic carbonates are also important substrates for reactions of this type.[119] These processes have been used to effect dehydrogenation of alcohols by conversion to allyl carbonates followed by reaction with a phosphine-free palladium catalyst in acetonitrile.[120] A similar alcohol oxidation is described in Section 6.8.

Trost's group have investigated reactions of the vinyl epoxides with palladium complexes.[121] These substrates usually react by the path shown in Scheme 6.10.

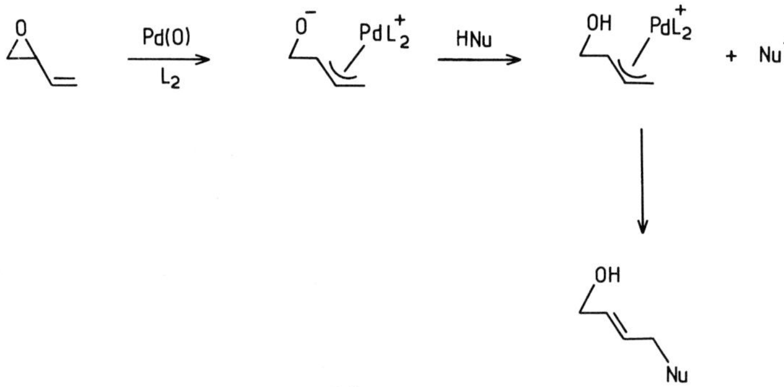

Scheme 6.10

Recently, tin reagents have been used to reverse regiocontrol, effecting alkoxide addition adjacent to the OH group.[122] Another recent development demonstrates the analogous opening of four-membered rings of vinylic oxetane.[123] This provides a process of alkylation in which the potential nucleophile is in the neutral form (HNu). Scheme 6.11

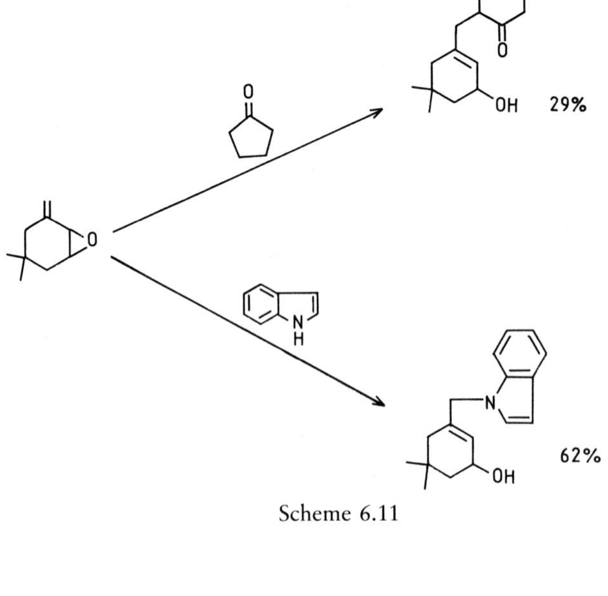

Scheme 6.11

shows an example, typical of many, to illustrate this point. The epoxide opening reaction proceeds by a sequence[121] that involves the usual two inversion steps. Consequently, the nucleophile enters *cis* to the epoxide oxygen. Opening of epoxides by the alkylation of allyl complexes provides excellent stereocontrol at remote positions. The hydroxyl group and the bond formed are separated by four carbon atoms. This principle has been used in Tsuji's group to control stereochemistry relative to the C-15 alcohol of 11-deoxy-PGE$_1$ (163)[124] and at the D-ring alcohol of the steroid derivative (164).[125]

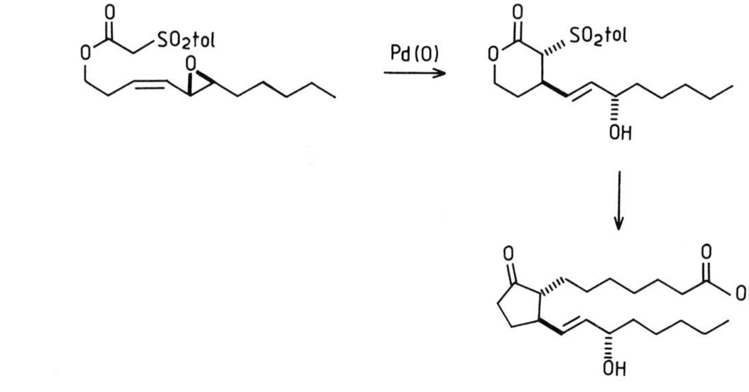

(163)

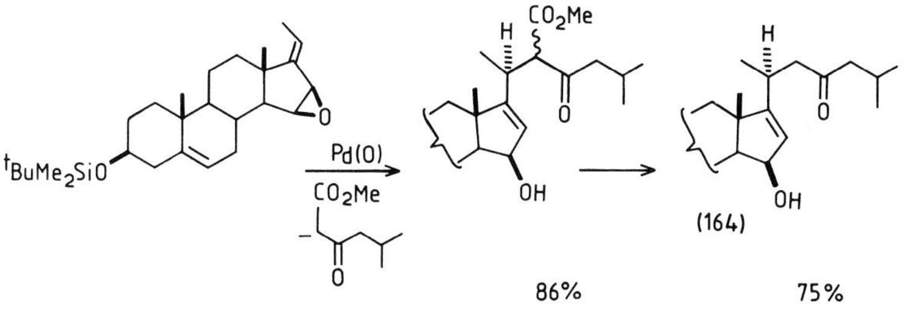

(164)

86%   75%

Epoxide opening has also been used in the formation of the macrocycle (165). A variety of conditions were used to close the 26-membered ring. Pd(PPh$_3$)$_4$ and Pd(OAc)$_2$/reducing agents gave relatively low yields but the newly developed use of a tri-isopropylphosphite ligand[126] improved the yield of (165) to 70–92%.[127]

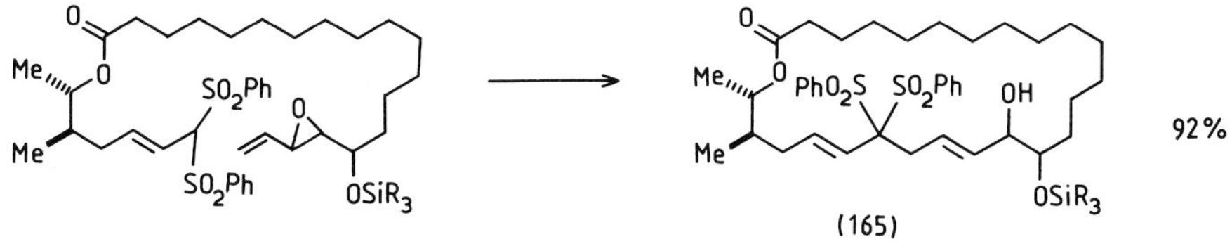

(165)   92%

An interesting palladium catalysed stereoselective reduction of vinyl epoxides such as (166) has been described using ammonium formate and a palladium catalyst.[128] The inversion of stereochemistry at the allylic position suggests hydrogen transfer from the metal to the ligand. Initial coordination of formate to palladium followed by decarboxylation, has been proposed to account for this result.

(166)

Allylic epoxides can also be converted into cyclic carbonates in a catalytic reaction that is thought to proceed by oxygen addition to carbon dioxide following opening of the epoxide by palladium.[129]

Alkynyl epoxides, too, have been used as substrates for palladium catalysed reactions. A variety of cyclised products have been obtained. A typical example is the formation of (167), which was quantitatively converted into the bicyclic product (168). In a similar example, the furan (169) was produced.[130]

Another cyclisation occurs following the opening of an allyl cyclopropane bearing a *gem*-diester group. A wide range of examples have been examined.[131] The reaction of (170) shows that this reaction is not completely stereocontrolled when chiral substitution is present on the cyclopropane. The isomer (171) was used in a synthesis[131] of the cyclopentanoid iridodiol.

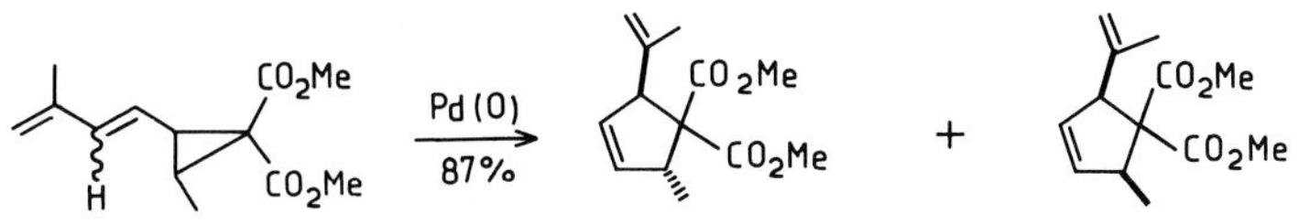

(170)   (171)   3 : 1

Intermolecular addition occurs when an external nucleophile is added to the system. The new enolate in this case was apparently generated *in situ* by proton transfer to the anion formed upon ring opening (Scheme 6.12).[132]

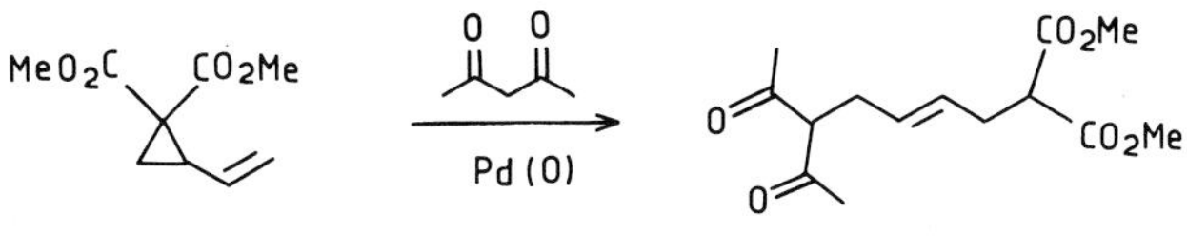

58%

Scheme 6.12

In a further example, electrophilic vinyl cyclopropanes have been subjected to nucleophile addition using Pd(PPh₃)₄ to catalyse the opening of the three-membered ring. The cyclopentene (172) was produced in good yield in this way.[133] By interception

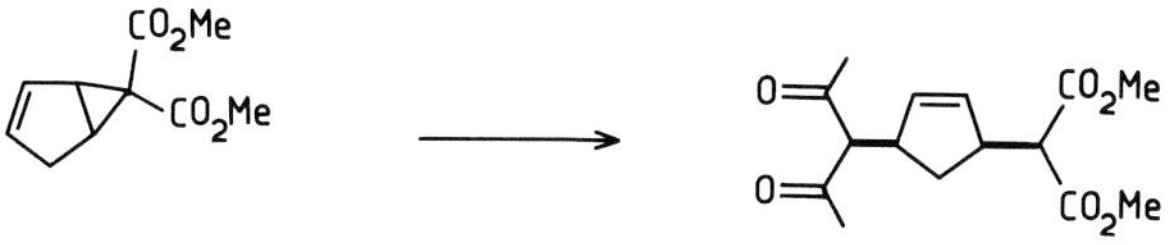

(172)   89%

of the π-allyl intermediate with an aryl isocyanate, amides such as (173) have been produced.[134]

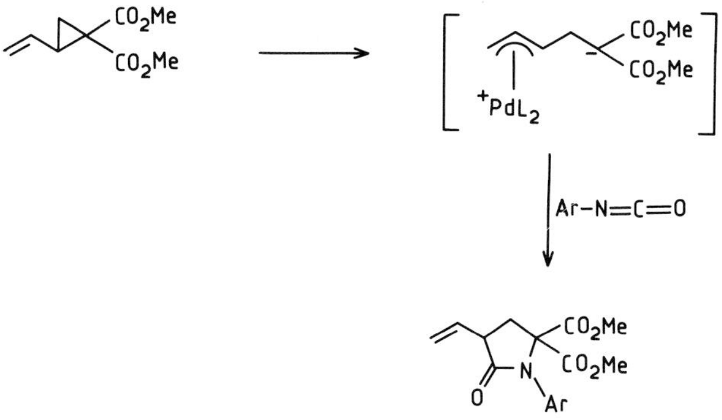

Ar−N=C=O

40−90%

(173)

6.8 Formation of enones

Photolysis of palladium allyl complexes under an oxygen atmosphere has been shown to result in the formation of enone products such as (174) and (175).[135] Another oxidation reaction has been used[136] to prepare the steroidal enone (176).

A catalytic version of the reaction converts allyl ethers into enones by treatment with PdCl$_2$ in air.[137] Another catalytic procedure[138] is a variant of the displacement of allyl carbonates described in Section 6.6. No oxidising reagent is introduced, but propene is a by-product of the reaction, showing that the C$_3$H$_5$ group has been reduced in the reaction during the formation of the aldehyde (177). Ketones are similarly produced from carbonates such as (178).

6.9 Palladium catalysed elimination reactions

Early in the study of palladium chloride dimers it was found that pyrolysis in high vacuum resulted in elimination of palladium and HCl to form dienes.[139] Later, it was discovered that such drastic conditions were unnecessary. Indeed, deprotonation is a common outcome from attempted nucleophile addition. Base treatment of (179), for example, gives the dienone (180) in moderate yield.[140]

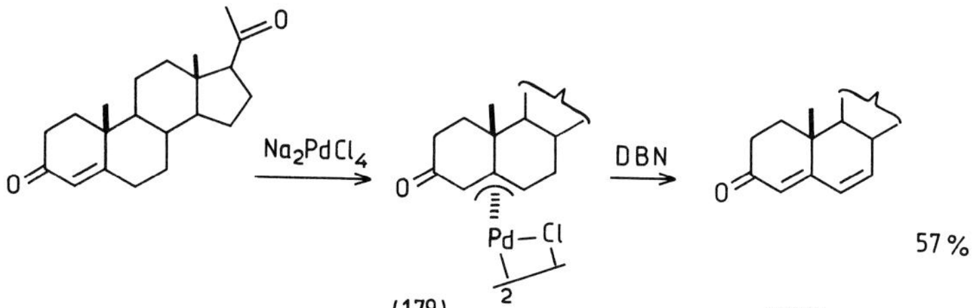

Allylic acetates undergo elimination by treatment with Pd(OAc)$_2$[141] or Pd(PPh$_3$)$_4$, as shown in Scheme 6.13.[142] These reactions show poor stereocontrol. Despite this

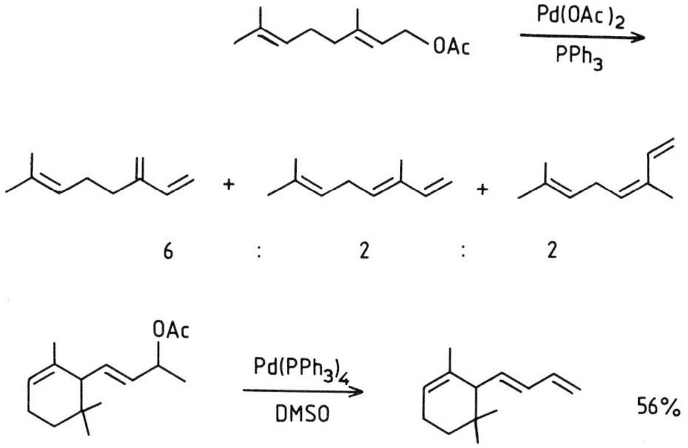

Scheme 6.13

disadvantage, the process has been used effectively by Suzuki[143] in a synthesis of citral (182) from diprenyl ether. An enol ether is produced by deprotonation of the palladium complex (181). Citral is then obtained by consecutive Claisen and Cope rearrangements.

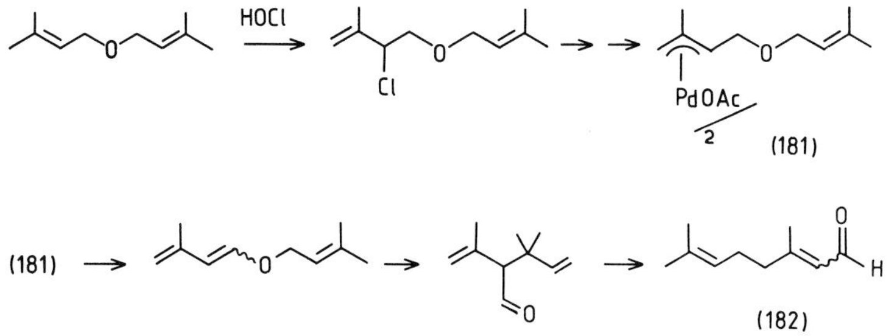

In an elimination reaction which showed better stereocontrol, the insect sex pheromone codlemone (183) was produced by a decarboxylation process displacing both CO_2H and OAc groups from the substrate. The stereochemistry of the double bond originally present in the starting materials was unchanged during the reaction. A similar step in the *cis*-alkene series produced mainly the *Z,E*-isomer contaminated by about 10% of (183).[144]

(183)

+ 22%   Z,E - Isomer

Allyl carbonate based methods have also been used to effect decarboxylation and elimination but by a different mechanism. In this way, the α-methylene lactone (184) was produced in 74% yield.[145] Other methods for the formation of α-methylene lactones are discussed in Chapter 15.

(184)

Use of triethylamine in the reaction mixture has also been shown to promote much higher isomeric purity in the products arising from the elimination of allyl acetates.[146]

The aspects of the chemistry of palladium η^3 π-complexes discussed in this chapter demonstrate all the general points that arise when transition metals are used to render alkenes electrophilic. In the next chapter, these reaction types are encountered again in other allyl complexes, and also for diene, dienyl and arene complexes. The parallels in reaction type for different η values illustrate clear analogies between the chemistry described in Chapters 5, 6, and 7. For diene, dienyl, and arene complexes, however, stoichiometric methods are much more common than catalytic reactions.

7

Further applications of hydrocarbon π complexes as synthetic intermediates

The patterns of reactivity that we have already described for η^2 complexes in Chapter 5 and for η^3 palladium complexes in Chapter 6 are typical of a wide range of hydrocarbon π complexes. A great variety of such complexes have been studied, and, although there are many differences in the details of their properties and uses, they have much in common with one another in general terms.

7.1 Alternative η^3 complexes

Although the development of uses for η^3 complexes has been dominated by work on palladium compounds, practical alternatives using less expensive metals do exist. Early work from the Whitesides group described[1] the alkylation of the tetracarbonyliron complex (1) by a stabilised enolate, in a reaction that showed partial regiocontrol. Birch and Pearson's alkylation of (2) by dibenzylcadmium was somewhat more regioselective.[2]

In a comparison of regiocontrol effects, the *anti* isomer (3) has been found to show consistently better control than the *syn* counterpart (2). Alkylation of (3) with sodium dimethyl malonate produced (4) and (5) in a ratio of 74:26. The corresponding *syn* products were formed as a 57:43 mixture. This trend for better regiocontrol with the *anti* isomer was apparent for a range of nucleophiles. Morpholine, for example, gave an 88:12 mixture with (2), improving to a 94:6 ratio when the substrate was (3). In all these cases, however, attack at the unsubstituted terminus predominated, showing, in common with earlier examples, that nucleophile addition occurred preferentially at the less hindered position,[3] although, as will be seen below, this need not always be the case.

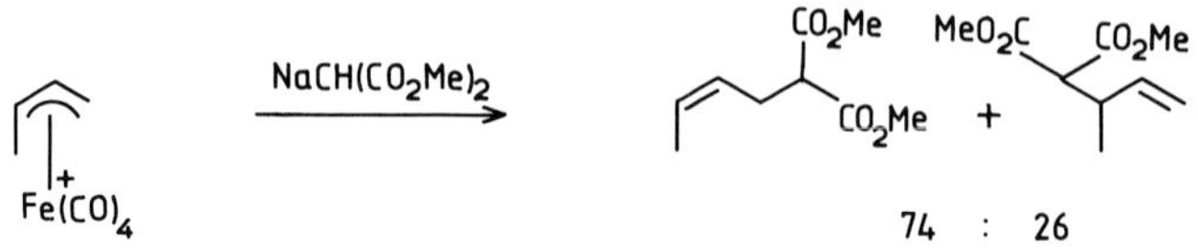

(3)   (4)   (5)

Cationic allylic complexes of $Fe(CO)_4$ have also been shown to react with activated aromatics. Addition of 1,3-dimethoxybenzene to (6) afforded an adduct in 60% yield. The presence of a third OMe group on the aromatic ring in (7) improved the yield of the addition reaction in the case of (6) to 87%. Reactions of this trimethoxyarene with (2) and (3) again proceeded selectively at the unhindered terminus, producing 3:1 and 4:1 mixtures of regioisomers respectively. The phenyl substituted allyl complex (8) gave a single product upon addition of the same nucleophile.[4]

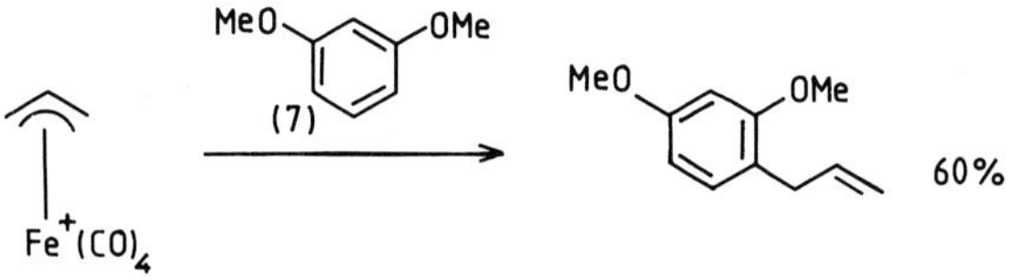

(6)

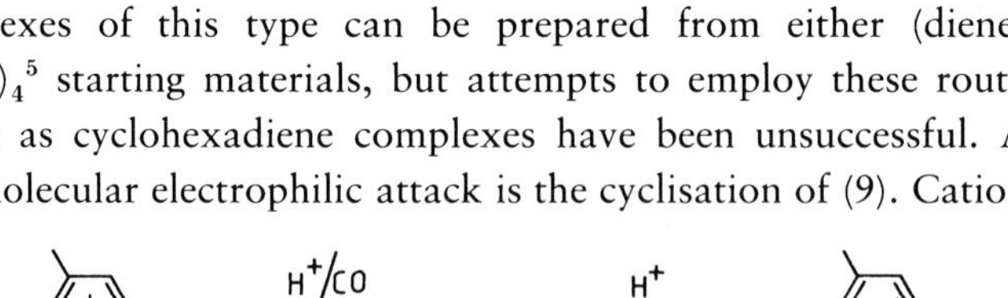

(8)

Allyl complexes of this type can be prepared from either $(diene)Fe(CO)_3$[1] or $(alkene)Fe(CO)_4$[5] starting materials, but attempts to employ these routes with cyclic examples such as cyclohexadiene complexes have been unsuccessful. An interesting case[6] of intramolecular electrophilic attack is the cyclisation of (9). Cationic ruthenium

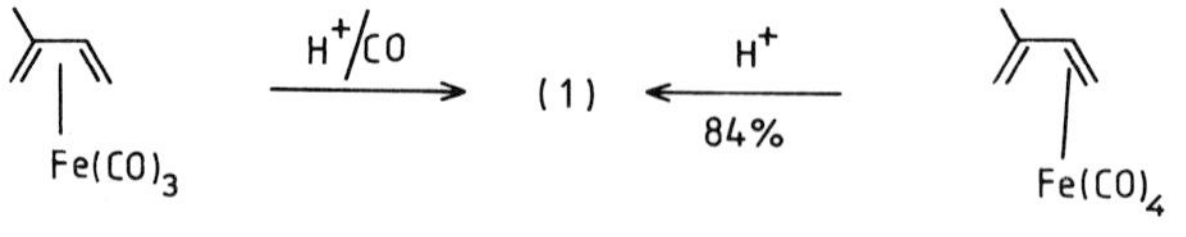

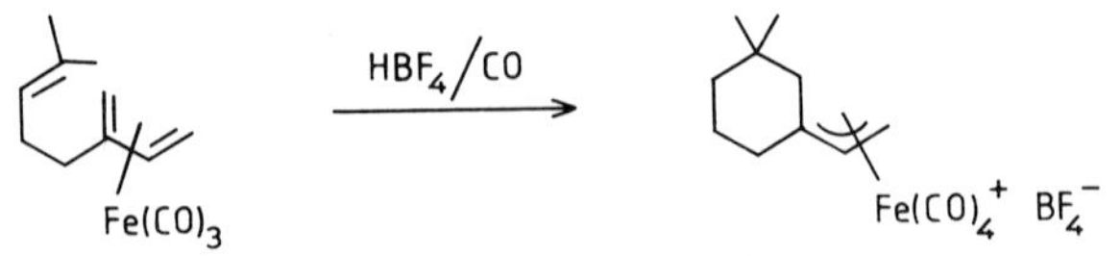

(9)

arene complexes of allyl ligands have also been prepared, and have been examined in reactions with $NaBD_4$.[7]

Displacements of allylic acetates and halides using iron catalysis have been described.[8] Either $Fe(CO)_3NO^-$ or $(\eta^2\text{-crotyl})Fe(CO)_2NO^-$ can be employed. More recently, a similar reaction used[9] commercially available $Fe_2(CO)_9$ as the source of the catalyst. Examples are shown in Scheme 7.1.

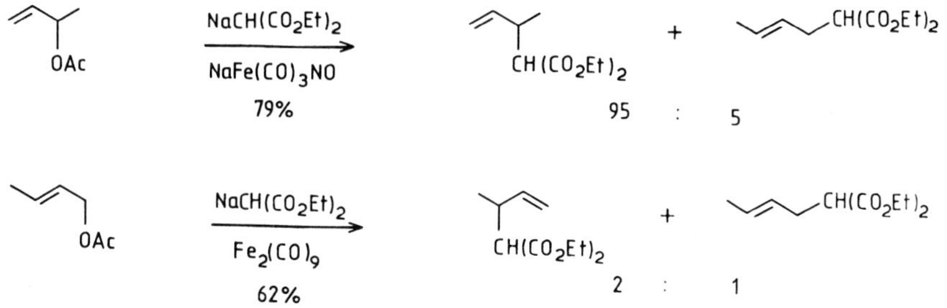

Scheme 7.1

It now seems probable that the reaction catalysed by $Fe_2(CO)_9$ proceeds by an indirect mechanism. On the basis of a comparison of regiochemical information, and of spectroscopic studies of intermediate species produced by addition of malonate nucleophiles to the iron catalyst, a mechanism involving initial binding of malonate to iron to form an η^1 $Fe(CO)_4$ anion has been proposed.[3] A key step in the catalytic cycle is the addition of malonate to the η^3 ligand in the neutral intermediate (10).

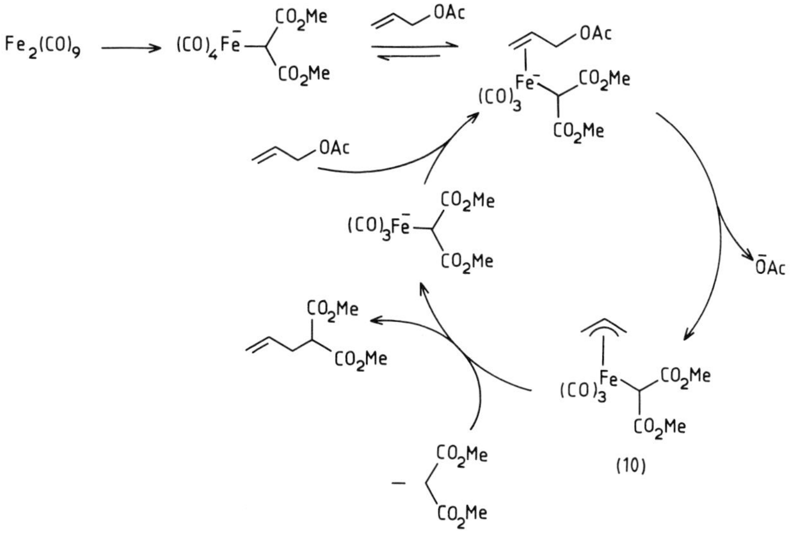

Recently, considerable efforts have been made to find alternatives to palladium for the catalysis of allylic displacements. Rhodium, and nickel,[10] and more importantly, tungsten and molybdenum[11] complexes can be used in this way. An example[12] of a similar reaction of a stoichiometric molybdenum complex has been described. The alkylation of (11) provides a further example of nucleophile addition at the more hindered terminus of an allyl system.[12] The regiocontrol in this type of alkylation has

been the subject of much investigation.[13] Considerable differences in regiocontrol can occur, depending on the catalyst system employed. With $Mo(CO)_4$(bipyridyl) catalyst, (12) produced a 7:3 mixture of regioisomers with the major product again resulting

from addition at the substituted terminus, a complete contrast to the result described above for the cationic stoichiometric complex (8). The use of $Mo(CO)_6$, however, shifted the site of addition back to the unsubstituted end. The substrate (13) provides an example where silyl and ester directing groups are placed in competition (see Section 7.6). With a molybdenum carbonyl/BSA catalyst system the single product (14) was formed.[13]

A comparison of regiocontrol for methyl substituted allylic acetates (15) also shows the variation that is possible depending on the catalyst used.[3] Considerable (9:1) selectivity for the unsubstituted terminus, obtained with $Fe(CO)_3(NO)^-$ as the catalyst, was reversed to more than 3:1 in favour of internal attack with a tungsten catalyst. Palladium catalysts showed no selectivity in this case.

~~OAc	W(CO)$_3$(MeCN)/bipy	24 : 76
"	Fe(CO)$_3$NO$^-$	91 : 9
"	Fe$_2$(CO)$_9$	37 : 63
~~OCOCF$_3$	Fe$_2$(CO)$_9$	52 : 48
~~OSOTol	Pd(PPh$_3$)$_4$	50 : 50
~~OSOTol	Pd(PPh$_3$)$_4$	50 : 50
~~SPh	Mo(CO)$_6$	27 : 73

The acetate (16) also exhibited a reversal of regiocontrol in malonate addition reactions. The molybdenum tetracarbonyl bipyridyl catalyst gave the same result as palladium catalysis, but molybdenum hexacarbonyl showed the reverse selectivity.[13]

Mo(CO)$_4$ bipy	100	:	0
Mo(CO)$_6$	15	:	85
Pd(PPh$_3$)$_4$	100	:	0

Reductive couplings (see Chapter 10) using zinc and the molybdenum hexacarbonyl catalyst produced a mixture of regioisomeric dimers from (12). The major products arose from linkage of an unsubstituted terminus to either the unsubstituted or to the substituted terminus of the second substrate.[14]

47 : 49 : 4

In a substrate with two allylic leaving groups such as (17), further differences between palladium and tungsten catalysis can be seen. Addition of malonate, catalysed by palladium, displaced the primary carbonate group to form (18) which was isolated as a mixture with the cyclisation product arising from a second nucleophilic carbonate displacement to produce a cyclopropane (see Chapter 6). With the tungsten catalyst, however, (19) was the only product.[15]

The stereochemistry of acetate displacement by the molybdenum hexacarbonyl catalyst has been examined for the diastereoisomers (20) and (21). These two isomers were found to give the same ratio of products from the addition of malonate. Trost and Lautens ascribe this result to a rapid interconversion of diastereomeric η^3 allyl complexes. Again, nucleophile addition to the substituted terminus is notable.[12]

Molybdenum complexes can also be used[16] to effect elimination reactions (see Chapter 6) such as the formation of (22). This type of elimination procedure has been used at the Wellcome Research Laboratories in a synthesis of the insecticidal natural product piperovatine.[17]

7.2 η^4-Diene complexes as synthetic intermediates

While a number of useful alkylation reactions of non-conjugated diene complexes were discovered in the early days of organometallic chemistry (see Chapter 5), the development of applications of the chemistry of η^4 complexes of conjugated dienes is only just beginning. Alkylations of organoiron compounds such as (23) in the presence

of carbon monoxide, described[18] by Semmelhack and Herndon in 1983, are of particular note if considered for use in conjunction with the complementary alkylation chemistry of η^5 complexes to be discussed in Section 7.3, since a wide variety of substituted complexes are available as substrates. In these reactions, the metal is removed from the diene complex in a process that introduces both a nucleophile and an electrophile, forming potentially useful products with *trans* stereochemistry. Substrates such as (23), however, suffer from a disadvantage shared with other neutral metal complexes (see, for example, Section 7.4); they react with a relatively restricted range of nucleophiles. Alkyl substituted nitrile and ester anions, and lithiated dithianes have been used successfully.[19]

With simple nucleophiles, such as organolithium and Grignard reagents, there is a tendency[20] for the nucleophile to attack a carbonyl ligand in (23). Reaction with phenyllithium, for example, produced the ketone (24) in 26% yield. Benzaldehyde was also produced.[18] The ketone (24) could arise from addition of a phenyl group to carbon

monoxide, followed by acyl transfer to the ring. More recently, $Fe(CO)_3$ complexes of α,β-unsaturated ketones[21] and $Fe(CO)_4$ complexes of α,β-unsaturated esters[22] have been found to undergo similar acyl transfer reactions in very much better yields. Both organolithium and Grignard reagents can be used. The products (25) and (26) arise from acyl transfer to the β position of the organic ligand. When the reaction with α,β-

(26)

unsaturated ketone complexes is performed under an atmosphere of carbon monoxide, α,β-unsaturated ketene complexes can be formed. These vinylketene complexes react directly with nucleophiles.[23]

Promising alkylation reaction sequences using dienes can be achieved with cationic complexes of $Mo(CO)_2Cp$. Originally observed by Faller *et al.*[24] for the addition of Grignard reagents to the cyclohexadiene complex (27), this reaction has been employed by Pearson and Khan[25] in a synthesis of a fragment (28) with the correct stereochemistry for inclusion in tylosin or magnamycin B.

(27)

(28)

Carboxylic acid substituents in side-chains can be induced to cyclise by addition to the metal complex upon treatment with iodine or NO^+. This produces a lactone, as seen in the formation of (30) from the complex (29), itself obtained by elaboration[26] of cycloheptene. In other circumstances, iodine incorporation occurs, leading to a stereocontrolled synthesis of allyl iodides such as (31). Alternatively, addition of NO^+ can be used to form cationic allyl complexes that can be alkylated[26] a further time (see Section 7.5).

Cationic η^4 complexes indicated in the above examples were obtained by hydride abstraction using Ph_3C^+. An interesting alternative is to be found in work from Green's group[27] where DDQ and HBF_4 are employed for this purpose. These studies also provide useful examples of regiocontrolled nucleophile additions in cases such as (32) where the diene bears substituents. In an acyclic example, a vinyl cuprate addition has been used[27] in the synthesis of the naturally occurring *trans-trans* triene (33).

Interesting η^4-diene iron cations have been prepared by reaction of $C_5Me_5Fe(CO)_2 \cdot THF^+$ salts with a variety of dienes under photolytic conditions. Yields are in the range of 75–85 %.[28] The reaction of the cyclohexadiene complex (34) with a

number of amine nucleophiles was examined and second-order kinetics for nucleophile addition was observed. The rate of addition was relatively insensitive to the nature of the nucleophile. Variations in rates of reactions of Fe(CO)$_3$ dienyl cations, discussed in the next section, are far more pronounced. Ruthenium analogues of this type of complex have also been prepared.[29]

Recent work by Nicholas has demonstrated the use of cationic cobalt diene complexes. Nucleophile addition produces an η^3-allyl product which cyclises with an oxygen centre in the nucleophilic portion upon decomplexation.[30] Tricarbonylcobalt derivatives look promising as a general alternative to the use of molybdenum in electrophilic diene complexes. A further example of an alkylation reaction of this type is discussed in Section 7.6.

7.3 η^5-Dienyl complexes as synthetic intermediates

The development of most synthetically useful methods in this area springs from the pioneering work in the 1960s of Birch and Lewis and co-workers[31,32] employing the chemistry of cationic tricarbonyliron complexes, following the first report of compounds of cyclohexadienyl complexes of this type by Fischer and Fischer[33] in 1960. These highly stable, non-hydroscopic compounds are suitable for reaction with a great diversity of nucleophiles, and have been prepared with a wide variety of substitution patterns.[34,35] Until very recently, they have dominated synthetic applications of η^5 complexes.

Many of the more useful complexes for synthetic applications carry methoxy substituents on the dienyl ligand. Complexes of this type can frequently be obtained by complexation with Fe(CO)$_5$ (see Scheme 7.2) followed by hydride abstraction with Ph$_3$C$^+$ from the resulting neutral complexes. This is a convenient method[36] because unwanted regioisomers can be easily hydrolysed and removed as neutral by-products by solvent extraction. The 2-OMe isomer (36) is stable to hydrolysis and is typical of salts

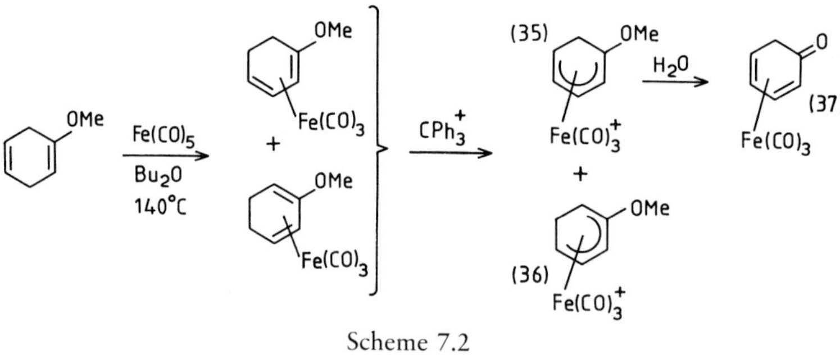

Scheme 7.2

of this type in that it is easily stored for long periods. The 1-isomer (35), however, is quickly hydrolysed in hot water.

The rapid hydrolysis of 1-methoxy substituted salts such as (35), presumably via a hemiacetal, which provided the basis for this separation, has also halted their development as intermediates in their own right, although there are examples of their use as phenylating agents,[37] most probably by nucleophile addition at the site of methoxy substitution.

Recently the first example of an alkylation of a 1-methoxy substituted dienyl complex has been reported. The complex (38) underwent reaction predominantly at the 1-position.[38] Nucleophile additions to the ethyl ether analogue of (35) have also been found to proceed at the substituted terminus of the dienyl system.[39]

Some examples of substituted complexes that can be conveniently obtained are indicated in Scheme 7.3.[40, 41, 42] In some cases, such as the 1-methyl substituted complex (39), the ester (40) or the silyl substituted compound (41), the hydride abstraction process itself is highly selective.

There are, however, a great many substitution patterns that are inaccessible by hydride abstraction. A convenient alternative is acid catalysed demethoxylation.[43] If performed in TFA,[44] the dienyl cation produced always has one terminus of the dienyl

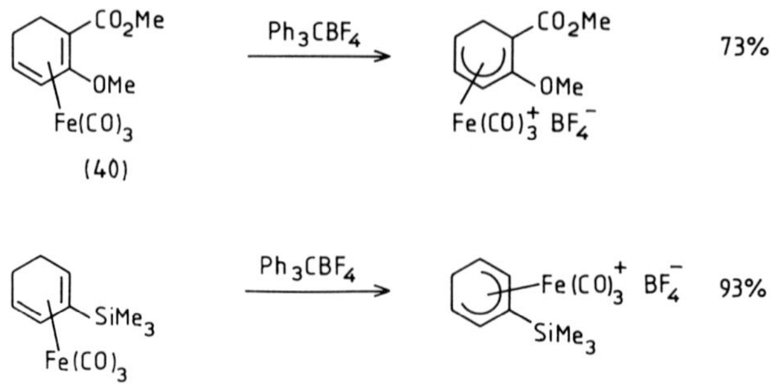

Scheme 7.3

system at a carbon atom that originally bore a methoxy group. This method is particularly valuable when, as in the formation of (42), a mixture of complexes forms

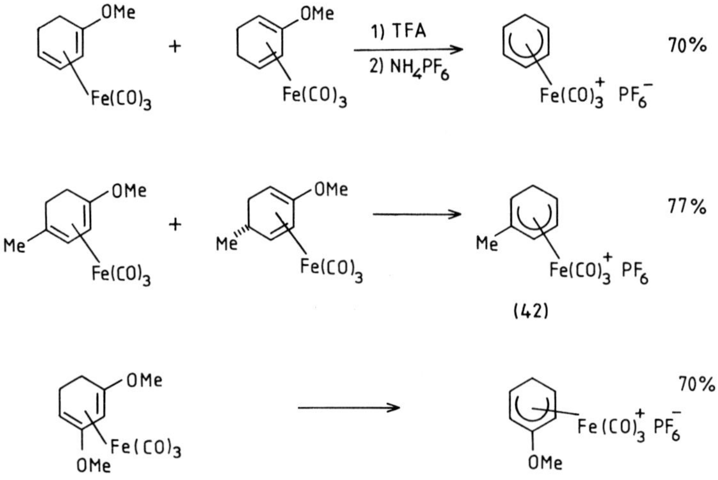

a single product. The reaction has been the subject of a detailed mechanistic study[45] using deuterated acids, as is illustrated, for example, in the formation of the 1-D salt (43).

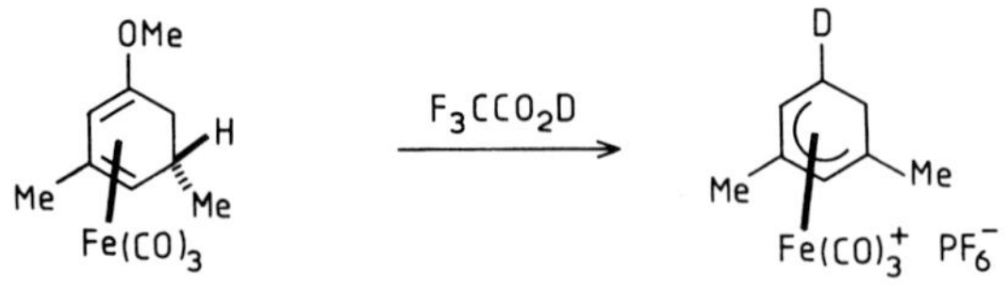

Substituents adjacent to the MeO group can exert a powerful directing influence that has been employed[46] in the formation of the single product (44) from a mixture of complexes obtained from 2,4-dimethylanisole. The same effect leads to the exclusive formation of (46) from (45).[47] Phenyl substituents have also been found to have a powerful effect in the demethoxylation reaction.[48]

Acid catalysed decarbonylation has similarly been used to produce dienyl cations. The 2-substituted complex (47) was obtained in this way. When decarbonylation was placed in competition with demethoxylation, the product (48) was produced.[49] Both these conversions used concentrated sulphuric acid, the conditions originally employed for the normal type of demethoxylation reaction.

A further regioselective method for the formation of specific complexes employs thallium oxidation.[50] Yet another approach involves the protonation of alcohol[51] or alkene[52] substituents in side-chains. This produces only 1-alkyl substituted salts such as (49) and (50).

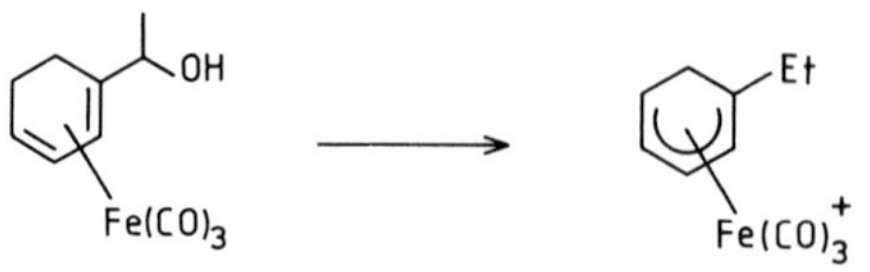

With a range of methods available for the formation of η^5-dienyl cations, a wide variety of different substitution patterns have been produced. This has made available a selection of versatile intermediates for use in organic synthesis. By making many starting materials accessible on a substantial scale, this diversity of methods for the preparation of specifically substituted dienyl complexes has contributed significantly to the utility of these compounds. Complexes bearing alkyl, methoxy and carbomethoxy substituents of the types shown above are finding increasing use in organic synthesis.

Differing substitution can have a pronounced effect on the reactivity of the dienyl complexes. Large differences in the rates of alkylation reactions (see Section 7.6) have been observed,[53] but the range of nucleophiles showing useful reactions appears not to be restricted by this variation unless a substituent is present at the terminus of the dienyl system.

Types of nucleophiles

Examples given in Scheme 7.4 provide an indication of the many types of nucleophiles that have been used successfully with tricarbonyl(η^5-cyclohexadienyl)iron cations. The

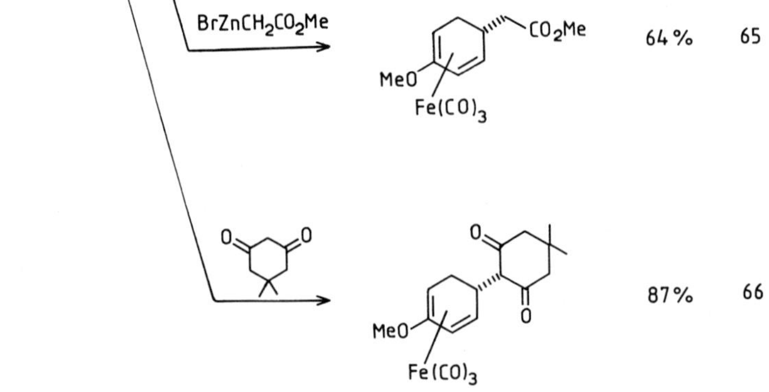

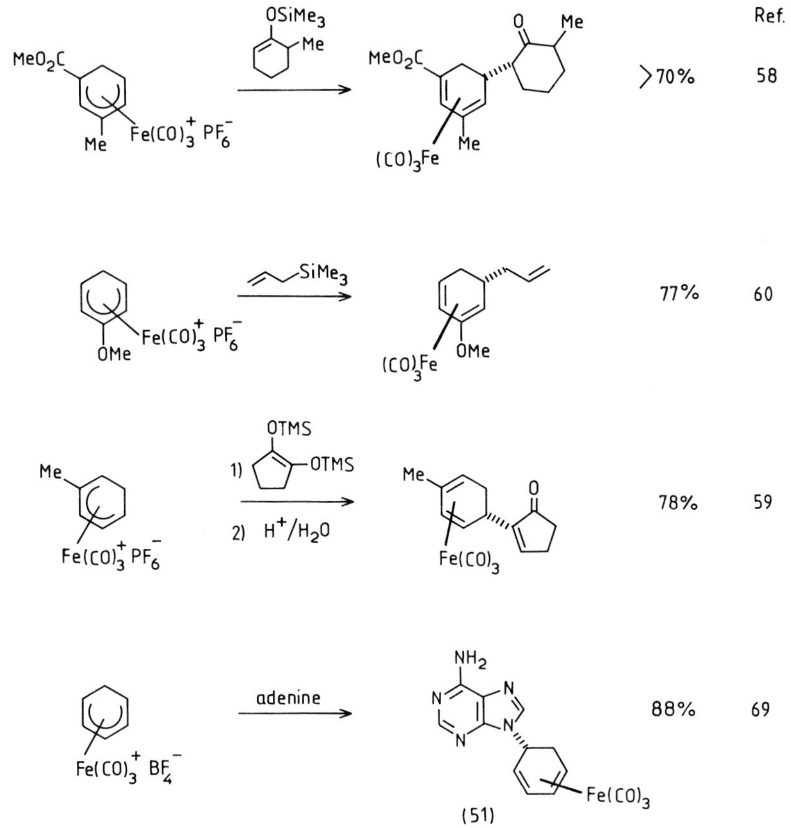

Scheme 7.4

range includes organozinc and cadmium compounds,[2,54] cuprates,[55] organoboron reagents,[56] trimethylsilylenol ethers[57,58,59] and allylsilanes.[60] Alkyllithium species can also be made to add but give unusual regiochemical results.[61] By use of $Fe(CO)_2PPh_3$ complexes rather than the more usual tricarbonyliron complexes, successful addition of Grignard reagents can be achieved.[62,63] The inclusion of phosphine ligands, which are less powerful π acceptor ligands than carbonyl groups, adjusts the charge density in the complex and has been found useful to facilitate other alkylation reactions in difficult situations.[63,64] Despite these advantages, however, it is, at present, the tricarbonyliron group that has been most extensively studied. While ketone enolates themselves are unsatisfactory as nucleophiles,[58] the Reformatsky reagent can be employed[65] successfully. Stabilised enolates are highly effective as nucleophiles with these complexes; the usual procedure[32] is to pre-form the enolate with sodium hydride, but in some cases direct reaction, presumably via the enol, is satisfactory.[66] Sodium cyanide gives reasonable yields in many cases[32,67] but trimethylsilyl cyanide has been shown[68] to be a superior reagent with sensitive compounds, or when a high yield is essential. A great many heteroatom nucleophiles have also been used. The last example[69] in Scheme 7.4 shows an alkylation of adenine to form the interesting adduct (51).

This impressive versatility, probably more than any other single factor, explains the successful use of tricarbonyl(η^5-cyclohexadienyl)iron cations as key intermediates in a number of natural product syntheses and underlies their convenient definition as synthetic equivalents of arene, diene and enone cations.

Removal of the metal: access to dienes, enones and arenes

In general, organic compounds can be obtained from the products of these reactions by removal of the metal by a variety of processes as shown in Scheme 7.5. The type of

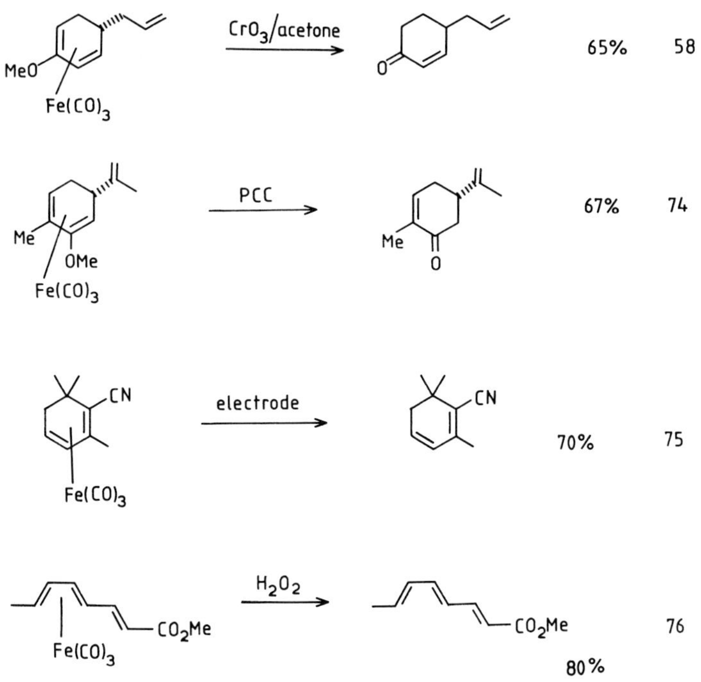

Scheme 7.5

compound obtained depends on the method used. The diene ligand is liberated by the use of one-electron oxidising agents such as Fe(III),[32] Ce(IV),[32] or Cu(II)[70] to oxidise the metal, or by treatment with trimethylamine-N-oxide[71] which attacks a coordinated CO to liberate it as carbon dioxide. Both types of reaction result in the destruction of the metal complex. Enol ethers produced in this way are often hydrolysed *in situ*, and enones are obtained. Pyridinium chlorochromate (PCC) has proved[58,72] to be a particularly convenient reagent for the direct production of enones.

The last two examples in Scheme 7.5 show relatively unusual methods for the liberation of dienes. Oxidation in these cases is performed either by electrochemistry[75] or by use of hydrogen peroxide.[76] Good yields, however, are reported in each case for several examples. Another unusual decomplexation procedure effects concomitant reduction to produce an alkene from the diene complex upon removal of the metal.[77] An example of the application of this method is discussed in Chapter 14.

Dienes produced by decomplexation can also be aromatised by DDQ. An alternative for the formation of arenes is the dehydrogenation of the metal complex itself. Some typical examples of both these methods are shown in Scheme 7.6. These reactions are not as easy or as general as the decomplexation routes to dienes and enones.

Direct palladium catalysed aromatisation during the decomplexation step furnished the cyclopentenone (53), bearing an inter-phenylene analogue of the prostaglandin top chain.[79] In this synthesis, regiocontrolled formation of the dienyl cation (52) by acid catalysed demethoxylation was followed by alkylation[59] with a bis(trimethyl-

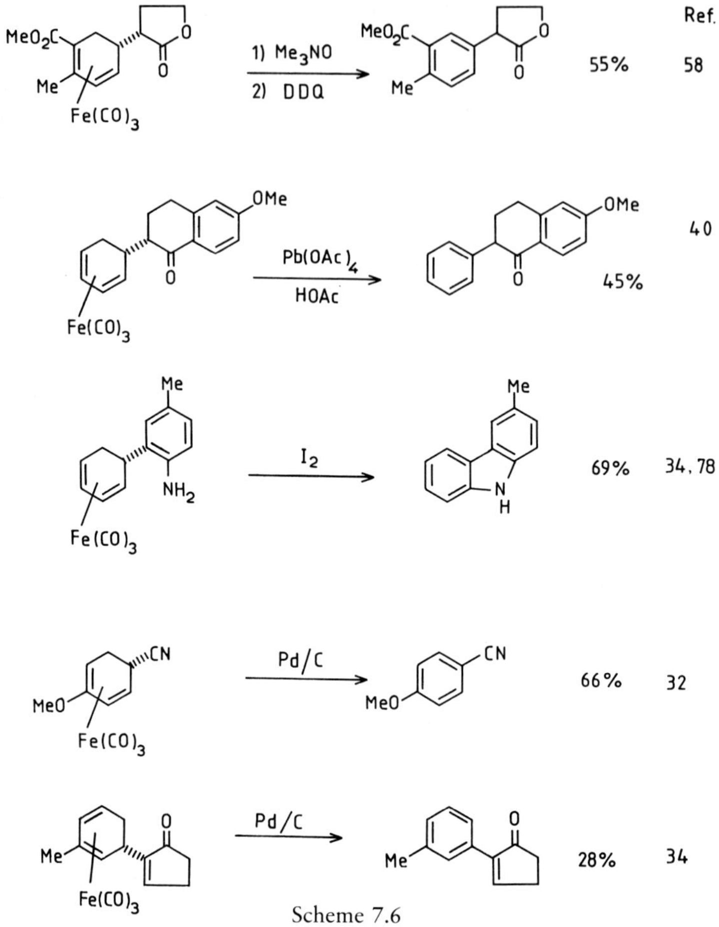

Ref.

55%   58

40

45%

69%   34, 78

66%   32

28%   34

Scheme 7.6

(52)   83%

1) structure

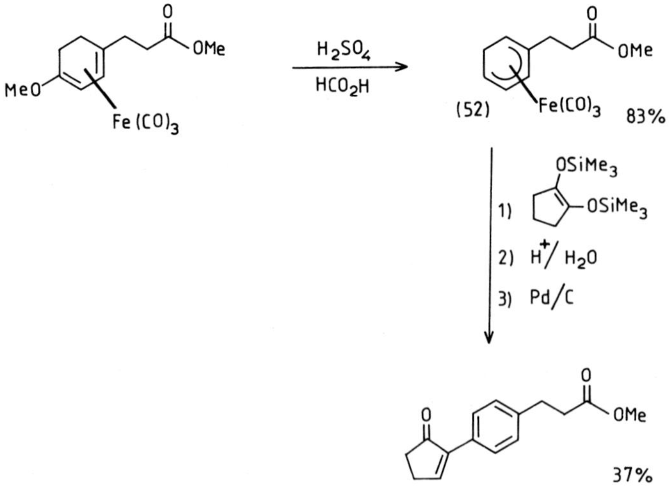

2) H^+/H_2O

3) Pd/C

37%

(53)

silyloxy)alkene to introduce the five-membered ring which was converted into the cyclopentenone by elimination in dilute acid.

Pearson's route to the *Sceletium* alkaloid O-methyljoubertamine (57) illustrates the alternative aromatisation method, decomplexation followed by oxidation with DDQ. Addition of the stabilised enolate (54) to the 2-methoxydienyl complex (36) afforded (55). After functional group transposition in the enone ring, reaction with Me₃NO produced the diene (56) which was then aromatised. Since the iron centre is removed without further use, the formation of (55) as a mixture of diastereoisomers is unimportant. Side-chain homolagation, however, was needed to complete the joubertamine skeleton.[80]

Aromatisation by these rather vigorous methods is not always compatible with sensitive substituents on the metal complex. Both dehydrogenation and DDQ oxidation have been examined for the formation of diaryl ethers, but complications arise through an alternative aromatisation by elimination of phenol. In the case of (58, R = H), DDQ was the only reagent to give any of the desired product, and then, only in low yield. The introduction of the aryloxy group, however, by nucleophilic addition to a cationic dienyl complex was a useful procedure and allowed[81] the efficient preparation of the organometallic intermediate (59) bearing the dibromoalanine portion present in L-thyroxine and analogues such as SK&F L-94901. Alternative organometallic routes to diaryl ethers such as ristocetin A are discussed in Section 7.4.

The examples encountered so far have demonstrated a variety of nucleophile additions to tricarbonyl(cyclohexadienyl)iron cations, and have illustrated the

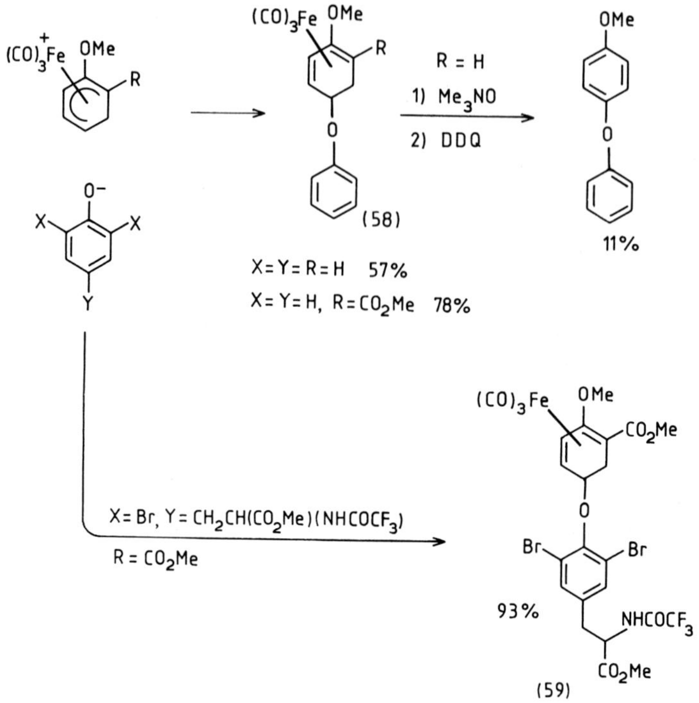

X=Y=R=H   57%
X=Y=H, R=CO₂Me   78%

(58)

11%

(59)

decomplexation of the resulting cyclohexadiene complexes to form organic products. These methods have become quite extensively used to effect alkylation reactions in organic synthesis.

Examples of applications of cyclohexadienyl complexes in routes to dienes and enones

In recent synthetic work, Grieco and Larsen have used dicarbonylphosphine complexes in alkylation reactions with Grignard reagents, to prepare a cyclohexadiene precursor for a synthesis of dihydrocannivonine (61).[62] A Grignard reagent derived from 4-

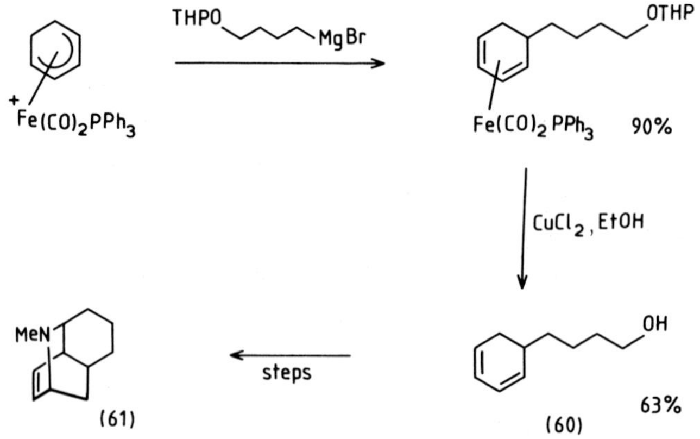

(61)   (60)

bromobutyltetrahydropyranyl ether proved very effective in the alkylation reaction. Oxidation using the copper(II) chloride procedure of Thompson[70] worked well to remove the dicarbonylphosphine metal complex. The tetrahydropyranyl ether protecting group was detached at the same time to produce the cyclohexadiene (60). This was converted by Swern oxidation to the corresponding aldehyde which was transformed by an intramolecular immonium ion cycloaddition reaction into the tricyclic product (61).

Complexes in bicyclic systems have also been examined as synthetic intermediates. A number of useful alkylation reactions, such as the introduction[72] of an angular substituent to form (62) have been observed. In some cases[82] alkylation is blocked by steric effects (see Scheme 7.7).

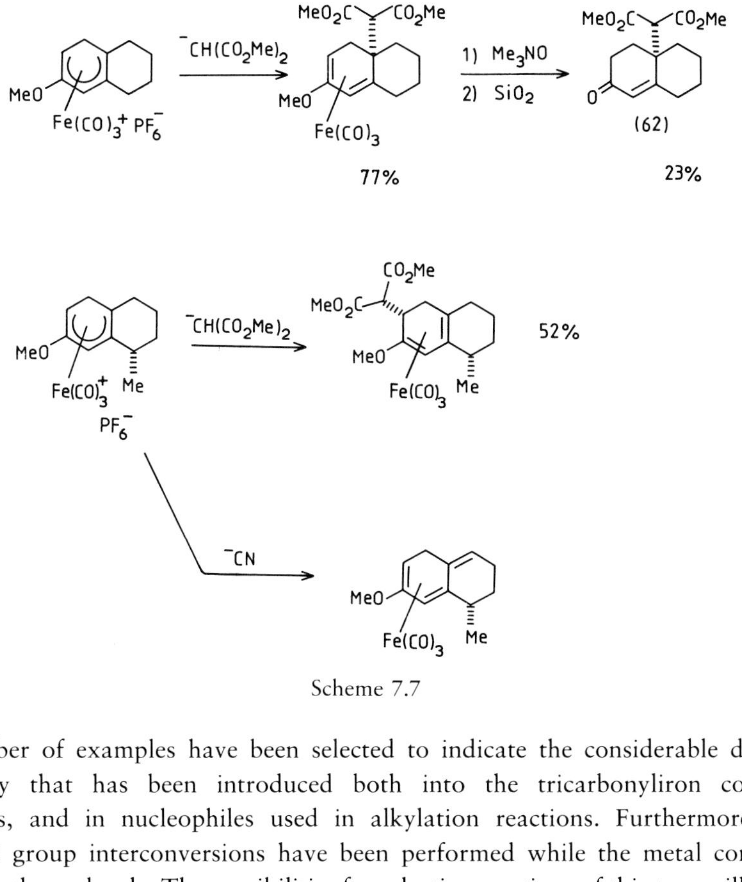

Scheme 7.7

A number of examples have been selected to indicate the considerable degree of complexity that has been introduced both into the tricarbonyliron complexes themselves, and in nucleophiles used in alkylation reactions. Furthermore, many functional group interconversions have been performed while the metal complex is retained in the molecule. The possibilities for selective reactions of this type will become apparent in the following examples, and are described in detail in Chapter 13 when organometallic masking groups are discussed.

In some cases some quite elaborate nucleophiles have been used in reactions with tricarbonyl(η^5-cyclohexadienyl)iron complexes. The examples[83] given here are related to studies of the total synthesis of steroids. The use of stabilised enolates provides the key for the introduction of the CD rings with the concomitant formation of the quaternary centre at the AB ring junction. Initial studies concentrated on routes to D-homosteroids such as (64). More recently the same approach has been applied to the synthesis[84] of the ergostene analogue (65), although the product contained the unusual *cis* CD linkage.

While silylenol ethers will not form quaternary centres in this way, it has recently been discovered that silyl ketene acetals are suitable reagents for efficient alkylation of (63).[85] This result is important in situations where the application of enolates themselves is limited by deprotonation side-reactions. As indicated earlier, trimethylsilyl cyanide has also been found to be a particularly valuable reagent in these circumstances.[36]

Another recent application of the malonate addition procedure can be found in studies related to quassinoid synthesis by Chandler, Mincione, and Parsons. Alkylation of the dienyl cation (63) was used to produce the adduct (66). The metal and the acetal

protecting group were removed at this stage to form (67) which was elaborated by a series of cyclisation reactions employing the two carbonyl groups in the molecule to produce the polycyclic compound (68) which still contained the diester group used in the initial alkylation step.[86]

Another important alkylation using a stabilised enolate is seen in the spirocyclisation leading to products such as (69).[87] A cyclisation of this type has been employed[88] in

studies designed to give access to the ring system of histrionicotoxin congeners, by means of amine addition to the tosylate (70). Other examples[89] of spirocyclisation lead to compounds related to acorane and cedrol (Scheme 7.8), and to tricyclic systems formed by intramolecular addition of an activated aromatic to the cyclohexadienyl complex.

Examples of more elaborate metal complexes include a number of steroidal complexes such as ergosterol[34, 90, 91] and chlorestadiene[92] derivatives. An interesting

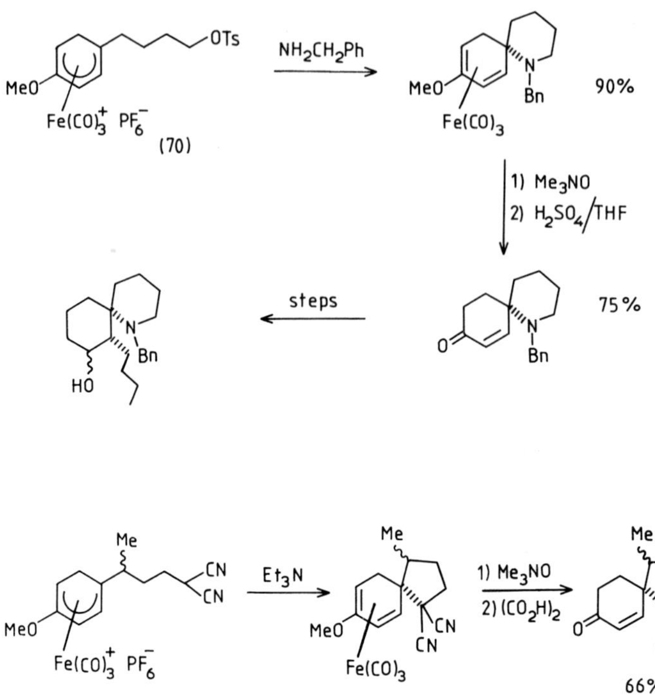

Scheme 7.8

example is to be found[93] in studies by Birch *et al.* of thebaine modification using tricarbonyliron complexes. Electrophilic opening of the oxygen bridge produced the salt (71) that can be alkylated, or induced to rearrange to form a new ring system.

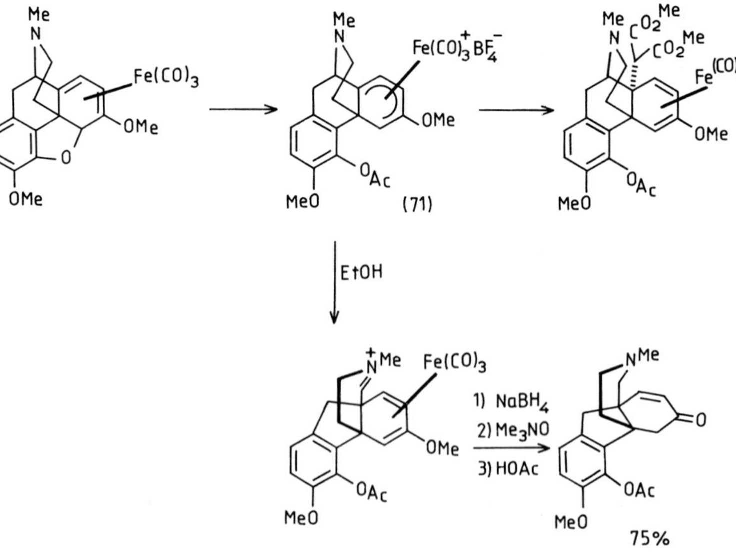

A number of other synthetic applications of tricarbonyl(η^5-cyclohexadienyl)iron cations will be discussed in Section 7.7, and in Chapters 13 and 14.

Acyclic complexes

Acyclic tricarbonyl(η^5-pentadienyl)iron cations have been known for over 25 years since the description of the perchlorate and tetrafluoroborate salts shown in Scheme 7.9, which were made by Mahler and Pettit.[94] Only recently, however, have these types of

Scheme 7.9

complexes received from synthetic chemists the attention they deserve. This change has been stimulated by the growing importance of E,Z 1,3-dienes for use in HETE and lipoxin synthesis. The attraction of pentadienyl complexes is that they contain a natural E,Z stereochemistry that can, in principle, be locked by bonding to a suitable metal centre. This can be seen in the conversion of the diastereoisomers ($-$)-(72) and ($-$)-(73) to the salt (74) which was used in a stereoconvergent construction of the complex ($-$)-(75). This product was formed as a 95:5 mixture with ($-$)-(72) by addition of water to (74).[95]

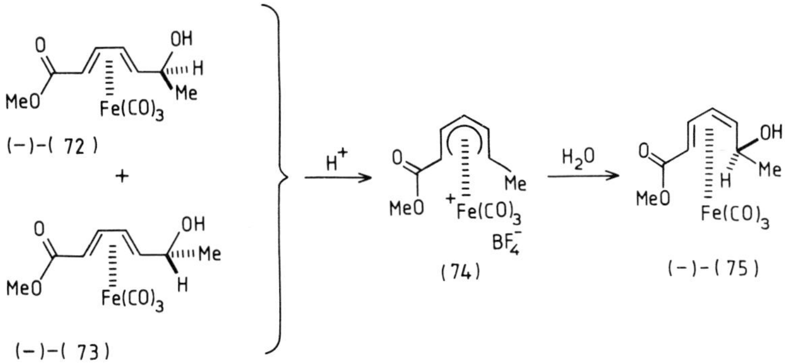

Grée has used a similar approach to obtain the Z linkage in (76). This compound was converted to the aldehyde (77), in which the metal complex distinguishes the faces of the aldehyde group. Addition of methyllithium proved more stereocontrolled than the use of a Grignard reagent, and a 4:1 mixture of (78) and (79) was obtained in this way. The selectivity was reversed with MeTi(O^{i}Pr)$_3$, which gave a 1:10 mixture in 77% yield.[96] Thus both relative stereochemistries in the product are available from the same starting material.

The isomerisation of (77) to the E,E isomer has been examined by n.m.r. at 77–96 °C. The activation energy of isomerisation was sufficiently high to suggest that intermediates of this type have reasonable prospects as stereocontrolled synthetic intermediates if used at moderate temperatures.[96]

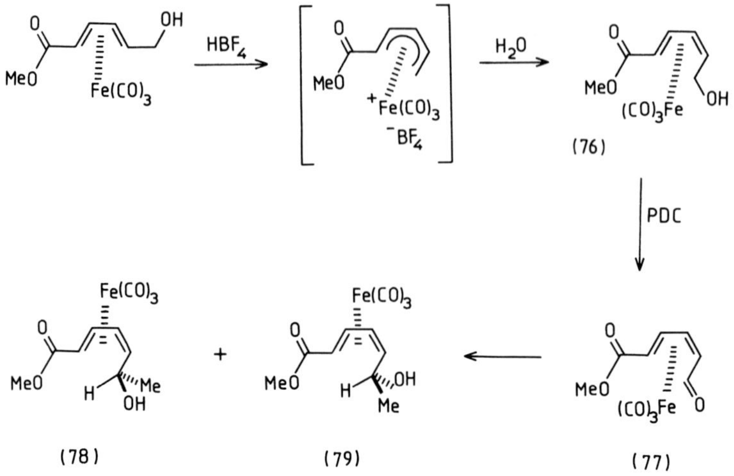

(76)

PDC

(78) + (79) ← (77)

Addition of an acetylide to (80), the *E,E* isomer of (77), produced a 7:3 mixture of (81) and (82). Both selective reduction, and epoxidation of the free alkene have been examined.[97] A mixture of isomeric epoxides (84) and (85) was obtained from (83) by reaction with the Sharpless reagent TBHP-VO(acac)$_2$ (see Chapter 3).

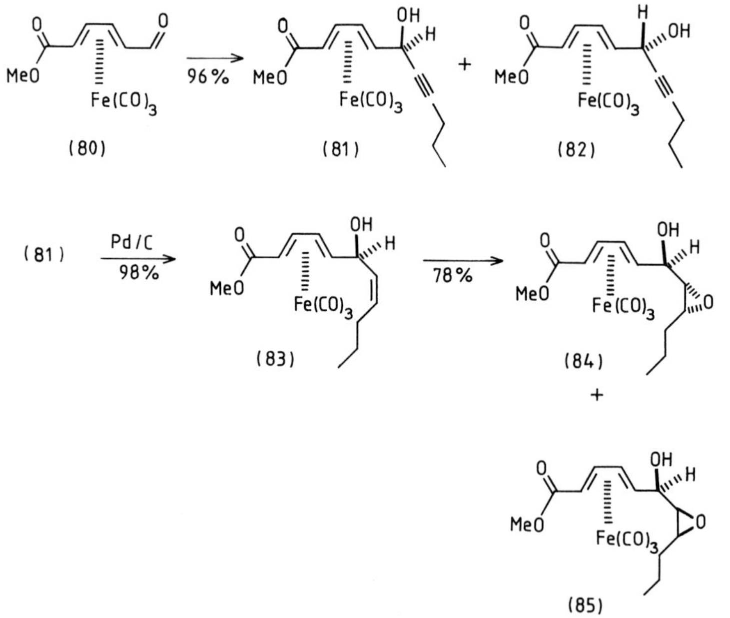

(80) 96% (81) + (82)

(81) Pd/C 98% (83) 78% (84)

+

(85)

Far more selective addition to the aldehyde (80) was encountered in a reaction with an allenylsilane using a Lewis acid catalyst. This method introduces a CH$_2$—C≡C—R group, producing (86) in 65% yield with better than 98% diastereocontrol. Reduction using a nickel catalyst and decomplexation with ceric ammonium nitrate produced an

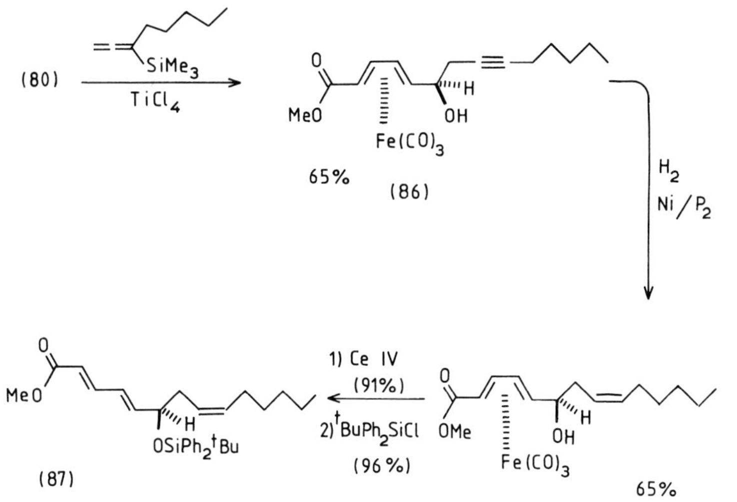

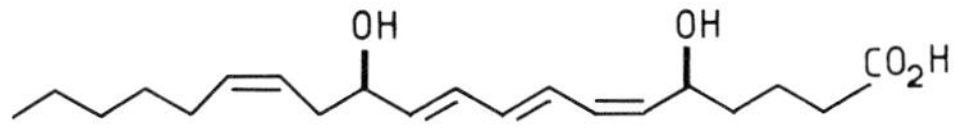

(88)

E,E,Z triene which was protected at oxygen by formation of (87), a key intermediate which incorporates the required functionality for the left hand portion of leucotriene B$_4$ (88).[98]

Stereocontrolled alkylations of the dienyl cations themselves have also been examined in the context of approaches to HETEs.[99] The alkyl substituted dienyl cation (89) is typical of the type of complex under investigation.

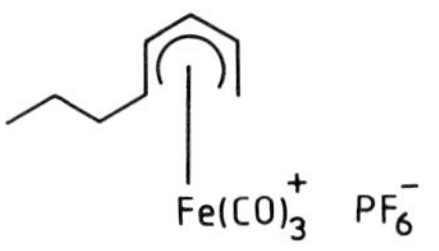

(89)

Stabilised enolates,[99, 100, 101] organocadmium reagents,[2] and activated aromatics[102] have been used in additions to a variety of acyclic cations. In these reactions, care must be taken to be certain of alkene stereochemistry in the products. Both the studies by Bonner, Holder and Powell,[102] and by Semmelhack and Park[101] show E,E isomers as the product. The first of these studies, a kinetic investigation, provided evidence for an initial equilibration of the cisoid dienyl complex with a transoid isomer (see below).[102] The work of Pearson[2, 100] and Donaldson,[99] however, has demonstrated that it is also possible to obtain alkylation products retaining the normal cisoid geometry of the dienyl complex.

Regioselectivity of nucleophile addition is also more complicated in acyclic examples. Selective addition to either the substituted[103, 104] or unsubstituted[2] terminus of the dienyl complex (90) can be achieved, depending on the choice of nucleophile. Often, however,

mixtures are produced.[2, 99] In the case of amine additions,[105] the nature of the product depended on the basicity of the amine, whilst in another comparison, the outcome was determined by the choice of reducing agent.[104] Nucleophile addition at internal positions (see Section 7.6) has also been observed. Reduction of (91) produced roughly equal amounts of η^4 and η^3, η^1 products.[106] The unusual tristrimethylphosphine complex (92) underwent exclusive internal attack in high yield.[107] This result was achieved with

the almost equally unusual nucleophile (93), and so is not directly comparable with other examples. Even in more normal circumstances, though, an unexpected η^3, η^1 complex has been identified by Donaldson as the major product (20:1) of malonate addition to a 1-carbomethoxy substituted pentadienyl cation.[99] Here, the ester group must be implicated in the displacement of C–C bond formation to C-2; Grée's group have shown that a similar nucleophile addition product was the minor component (2:3) from a reaction between the same salt and trimethylphosphite. The major product in this case resulted from normal terminal addition of the nucleophile.[99]

Normally it is not possible to use hydride abstraction to form acyclic dienyl complexes; hydride abstraction only works when the cisoid geometry required in the product is already established. The examples discussed above used acid catalysed removal of an OH group to produce the dienyl system. In the case of (94), hydride

abstraction was satisfactory, since the presence of an *anti* methyl group is guaranteed. The cation (95) was alkylated with malonate to form the *E,Z* product (96) for use in preliminary studies by Pearson[100] directed towards a subunit of erythronolide A. A recent alternative to hydride abstraction has employed carbene complexes to form precursors to acyclic dienyl salts.[101]

The difficulties with hydride abstractions have caused many of the original investigations to concentrate on solvolysis reactions. These early studies make clear the greater complexity of this type of chemistry in acyclic situations, and are useful in the interpretation of stereocontrol and regiocontrol results. Transoid intermediates in the solvolysis reaction,[108] and interconversions of *E,E* and *E,Z* isomers,[109] have been observed by n.m.r., and kinetic data[102,110] and product ratio studies[111] all point to the operation of several reaction paths.

Recently a process which effected alkylation with retention of *E* configuration has been described by Uemura *et al.*[112] This employed nucleophiles in combination with Lewis acids in direct reaction with η^4-(1-substituted-*trans*-pentadienol)Fe(CO)$_3$ derivatives such as (97). Similar results were obtained with triethylaluminium.

Because of the complications of alkylation at different sites and with different stereochemistries, care must be taken in the use of acyclic dienyl cations. Synthetic applications of these complexes are only now beginning to emerge.

Another and more recent series of acyclic complexes, the cross-conjugated cation complexes investigated by Lillya, are even further behind in terms of development for use in synthesis. These structures are considerably less stable than those discussed above, although they appear still to have the η^5 bonding shown in the example (98), as opposed to a coordinatively unsaturated η^3 structure.[113]

$$\underset{\overset{|}{Fe(CO)_3^+}}{}\qquad (98)$$

Role of the Fe(CO)$_3$ control group in synthesis

The examples of cyclic and acyclic cations encountered so far should give a clear impression of the way in which the control influence of the metal can be put to work when the alkylation of electrophilic π complexes is used in organic synthesis (see also Chapter 14). In the case of Fe(CO)$_3$ complexes, these methods have been developed to a considerable extent and the use of these metal complexes has been shown to be compatible with a great many functional groups and chemical reagents.

When complexes are used in this way, the metal is fulfilling two important roles. It provides the source of activation for bond formation and completely controls the stereochemistry of the alkylation reactions. In common with other additions of this type (see Chapters 5 and 6), the nucleophiles add *trans* to the metal. A typical example of this is nitrile addition in which the outcome has been determined[114] by X-ray crystallographic analysis of the product (99). ^{1}H n.m.r. spectra of alkylation products usually provide definitive evidence for the stereochemistry of addition.[115]

$$\underset{Fe(CO)_3^+BF_4^-}{} \xrightarrow{\text{NaCN}} \underset{Fe(CO)_3}{}\overset{\text{CN}}{} \qquad (99)$$

Substituents on the organic ligand are also important since regiocontrol of alkylation arises from their interaction with the complexed π system. Some substituents such as MeO and MeO$_2$C exert a powerful directing influence (see Section 7.6), while with alkyl substituents, as might be expected, steric effects are most important. This can still give valuable, though rather less reliable, regiocontrol. Fortunately, however, many nucleophiles such as stabilised enolates and activated aromatics can be persuaded to give single products, even with poor directing groups like methyl groups as the controlling substituent.

Methods for the introduction of the Fe(CO)₃ group

The synthetically interesting reactions described in this section would lose much of their appeal if the complexes involved were difficult to prepare. Following considerable work in this area, a number of methods for the synthesis of neutral η^4-diene complexes, the precursors to η^5 salts discussed in this section, have been developed. Some examples are given in Scheme 7.10. Fe(CO)₅ can be employed to isomerise 1,4-dienes during

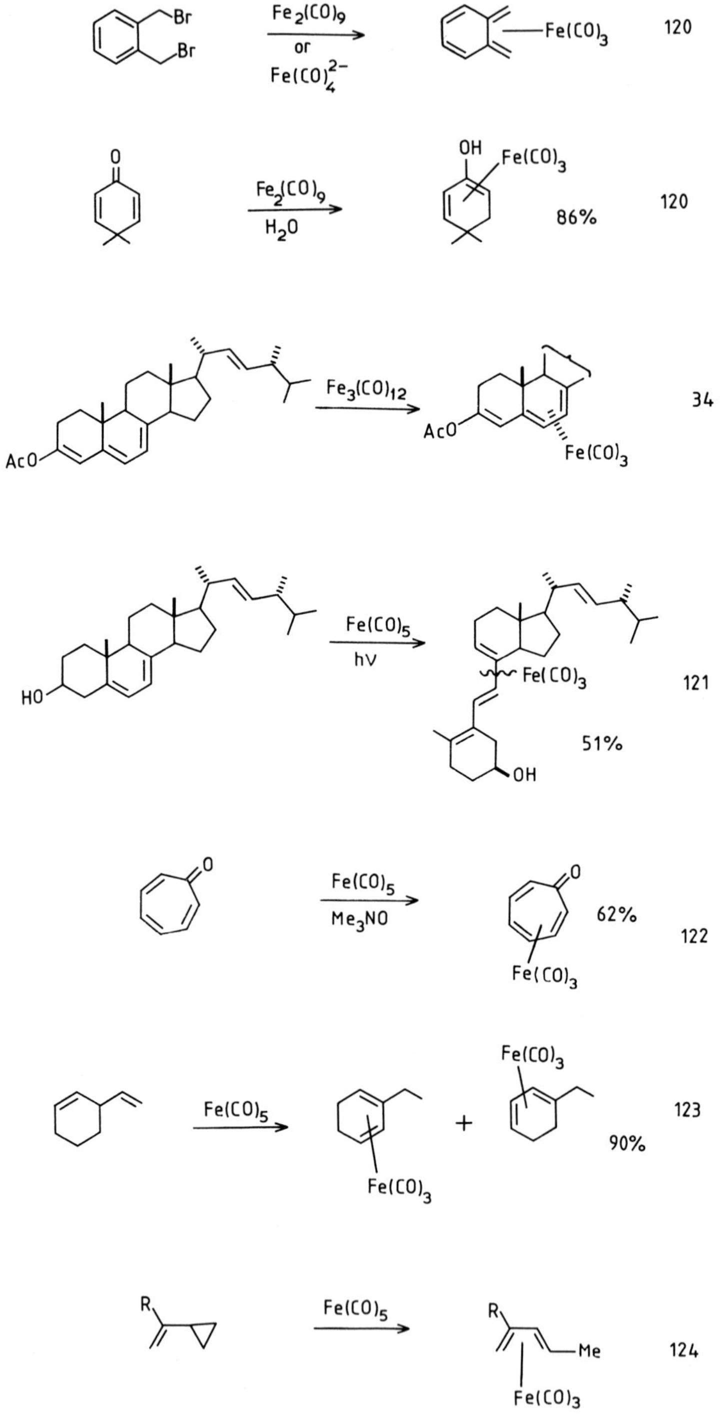

Scheme 7.10

complexation (see Chapter 2) but $Fe_2(CO)_9$ and $Fe_3(CO)_{12}$, which require less vigorous conditions, use 1,3-dienes as substrates. Transfer reagents such as amine[116] and enone[91,117] adducts can sometimes give superior results.

Use of ultrasound[125] irradiation with the $Fe_2(CO)_9$ reagent has also proved particularly advantageous in some cases (Scheme 7.11). This type of ultrasound

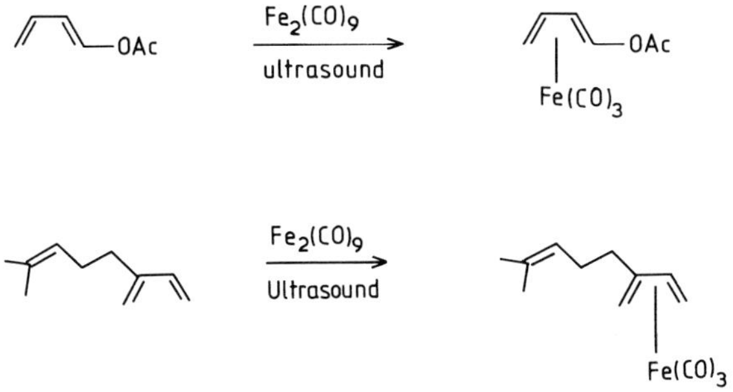

Scheme 7.11

procedure can be performed with the use of a normal ultrasound cleaning bath and so does not require special expensive apparatus.[126] Regiocontrol in many of these complexation reactions is still a problem, however, and an attractive way to introduce the metal moiety initially is by the alkylation of a simple dienyl salt that has a well established preparation.

Alternatives to Fe(CO)$_3$ complexes

There is much to be gained from the development of alternative η^5 complexes for synthetic applications if their properties can be manipulated to achieve reactions that previously were not possible. An example is the replacement[127] of CO by phosphines mentioned above. This can improve the yields of difficult alkylation reactions and also increase the range of nucleophiles that are suitable for use. Vinylmagnesium bromide, for example, can be used[63] successfully in this way to form (100). With tricarbonyliron complexes, Grignard reagents cause electron transfer to the dienyl cation (see Chapter 10) to form dimeric diene complexes.[32]

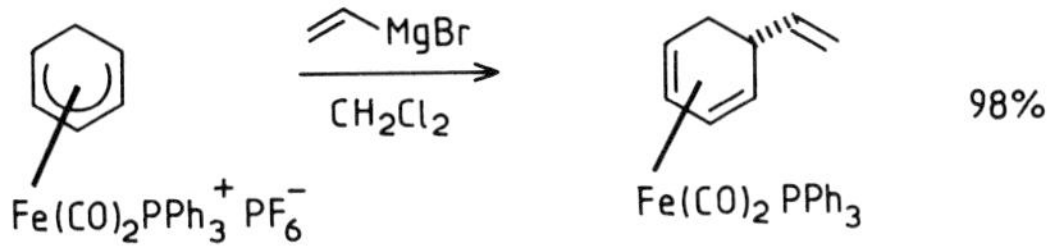

(100)

In other cases, nucleophile additions can be achieved in situations where the reaction would not be expected to proceed at all, as in the addition of a t-butyl group to the cobalt complex (101).[128] The reaction of (101) with methyllithium is also of note. Addition to the cyclopentadienyl ligand would not normally be expected (see Section

7.6). The hydrogen transfer between rings is a good example of an unusual but important reaction.

Another metal ligand system that shows great promise for use with η^5-dienyl cations is the $Mn(CO)_2NO$ fragment which is also isoelectronic with $Fe(CO)_3$. Examples of the reactions of complexes of this type are given in Section 7.5. Although reduced with ease by hydride reducing agents, few alkylation reactions of these cations have been reported and $Mn(CO)(NO)PPh_3^+$ complexes may prove superior. Recent work on $(C_6R_6)Fe^+$ dienyl complexes by Astruc *et al.*,[129] and on $(C_6H_8)RuCO^+$ complexes by Roulet[7] indicates further possibilities for alternatives to $Fe(CO)_3$. The iron analogue $(C_6H_8)FeCO^+$ is also available,[130] but, although reliable regioselective alkylation at the dienyl ligand would be expected (see Section 7.6), these complexes have not yet found synthetic applications.

There is still a challenge to be met in defining criteria for the rational choice of organometallic complexes for use in alkylation reactions. Furthermore, practical intermediates for use in synthesis require a formidable list of attributes. They must be stable and able to be stored for long periods, easily prepared, inexpensive, well controlled, and, ideally, optically stable as well (see Section 7.7). When their role in a synthesis is complete, it must be a simple matter to remove them to afford an organic product. Currently, for dienyl ligands, $Fe(CO)_3$ complexes offer the most general reactions, now reasonably well understood. In cases where difficulty is encountered, however, the possibility that other complexes will have different, and perhaps more advantageous properties, should not be overlooked.

7.4 η^6-Arene complexes as synthetic intermediates

The general features of nucleophile addition to coordinated π systems, described in previous sections, hold true for η^6 compounds and indeed for larger ligands such as those found in η^7-tropylium complexes,[131] although these latter cases have so far found little use in organic synthesis. η^6 Complexes, in particular the chromium arene complexes,[132] already have important applications and have been investigated over a long period of time. Unlike many arene complexes, they possess all the attributes needed for convenient synthetic utilisation: they are easily formed,[133] usually by direct reaction with $Cr(CO)_6$ in acetonitrile, THF, or di-n-butyl ether, they undergo a varied selection of efficient bond-forming reactions (both at the aromatic nucleus and in side-chain positions), and the metal is easily removed by mild oxidation in air.

Nucleophile addition

The addition of nucleophiles to complexes of arenes has received a great deal of attention. The electrophilic character of the aromatic ring is greatly enhanced, even when present in a neutral metal complex such as (102), and requires no further activating substituents. Initial attack by a nucleophile in this case produces an anionic intermediate (103) that can be oxidised with iodine to remove the metal and afford a substituted aromatic product.[134]

If a suitable substituent is present to serve as a leaving group, the anionic intermediate (104) isomerises to place the substituent at an allylic position. The final product is an arene complex in which overall substitution has occurred in the ligand.[134,135] The metal can now be removed in a separate oxidation step. Prompt oxidation following nucleophile addition avoids the regioisomerisation and produces a *meta* substituted chlorobenzene.[136]

A rather different course of reaction can be seen in the addition of the lithiated dithiane to the complex (105). After addition of TFA, the *meta* substituted arene (107)

(105)   (106) 70%   (107)   18%

was produced, indicating attachment of the nucleophile adjacent to the chlorine. The major product (106) was consistent with this regiochemistry, and arose through protonation of intermediate chromium anion species. Applications of this type of protonation reaction are discussed later in this section. Alkylation of (105) by LiCMe$_2$CN and LiCHMeSO$_2$(tol) gave similar results.[137] The nucleophile addition to chromium arene complexes is reversible. It is possible that differences in regiocontrol under different reaction protocols arise from competition between kinetic and thermodynamic processes. A similar replacement of chlorine has been used in the preparation of O-aryl-hydroxylamines.[138]

The cationic iron arene complex (108) was also alkylated by replacement of a chlorine atom. The base used for this alkylation was the mild reagent, potassium fluoride on celite.[139] Displacement of a nitro group from the FeCp$^+$ complex of o-

(108)

chloronitrobenzene has been described.[139] Similar displacements from FeCp$^+$ complexes have been effected by addition of enolates derived from malonate[140] and nitroethane.[141] A Cr(CO)$_3$ analogue of (108) was used to form macrocyclic polyethers by means of two chlorine displacements.[142]

For FeCp, like Cr(CO)$_3$, variants of the displacement reaction are possible. Addition rather than substitution has been effected by the oxidation of intermediate complexes. Oxidation of (109) with DDQ to produce the tri-substituted arene (110) illustrates the

(109)   (110)   72%

way that the elimination of halide from intermediates can be prevented by changing reaction conditions.[143]

Manganese tricarbonyl is iso-electronic with FeCp and so provides another example of cationic arene complexes. Halogen substitution from (111) by phenoxide has been

examined by Pearson's group as a potential entry to poly-aryl ether structures such as thyroxine or more complicated structures related to ristocetin A. This is a synthetic application of a procedure described originally by Pauson in which chlorine was displaced from (111) by phenoxide.[144] Arylation of the phenoxide (112) by (111) gave a 46% yield of the manganese complex (113). Gradual loss of the manganese tricarbonyl unit from this product occurred in acetonitrile at 20 °C. Direct conversion of (111) to the 3-diphenyl ether, without isolation of (113), improved the yield to about 55%.[145] Similar addition of a range of phenoxides to chlorobenzene complexes has also been examined using $Cr(CO)_3$ and $FeCp^+$.[81]

Many examples of addition reactions to substituted arenes have been examined, and factors affecting regiocontrol have been related to molecular orbital descriptions (see Section 7.6). Other factors such as the size of substituents and the reactivity of the nucleophile are also clearly of importance.[146] It is apparent,[136,147] however, that the directing influence of substituents is a major factor to be assessed when synthetic applications are considered.

Methoxy substitution directs strongly to the *meta* position. The alkylation of the substrate (114) indicates that the *para* position is greatly deactivated. Even methyl substituents can exert a strong directing influence, but some *ortho* substitution was possible with less bulky nucleophiles. The alkylation of (115) indicates that the influence

(115)

of a methoxy substituent is greater than that of a methyl substituent. Trimethylsilyl groups ensure *para* substitution. The use of a silyl directing group (easily introduced by metallation, see below) provides[136] a convenient way to increase the range of accessible substitution patterns. The silyl group is removed along with the metal during the oxidation step to form (116).

(116)

The regiocontrol of additions to naphthalene complexes has also been examined. This has been the subject of a recent review.[148] The examples given below illustrate the types of effects that can be expected. The chromium tricarbonyl complex of naphthalene itself, (117), reacts with nucleophiles at the α position. Oxidation of the intermediate anion with iodine produced the 1-substituted naphthalene (118). A good yield, 83%, was obtained when a lithiated dithiane was used as the nucleophile. Lithiated acetonitrile was less efficient than the dimethyl substituted reagent shown, producing the corresponding 1-substituted naphthalene in only 54% yield. Lithium methoxide was worse still, with nucleophile addition proceeding in only 30% yield.

The powerful electron withdrawing properties of the carbonyl ligands on the chromium are important for the efficient nature of nucleophile additions. When trimethylphosphite is used in place of carbon monoxide virtually no addition of nucleophiles was obtained under comparable conditions to those used for the production of (118). PF$_3$ ligands were similar to, though slightly worse than, the carbonyl group in terms of activation.[147,149]

Substituents on the naphthalene ring can produce a powerful directing effect. The methyl-substituted complex (119), in which the chromium is bound to the unsubstituted six-membered ring, underwent nucleophile addition to produce (120) with very high regiocontrol (98:2). The methyl group, it seems, exerts a powerful steric blocking effect at the adjacent carbon atom in the second ring. The methoxy-substituted complex (121) provides an interesting comparison. Here addition at the less hindered position still predominated, although the ratio of products was now 77:23. The two methoxy groups

in (122) exert a blocking effect at both the α positions and in this case nucleophile addition formed the β substituted complex (124) as well as (123).[148,150]

Because nucleophile addition of this type is reversible, competition between kinetic and thermodynamic effects can be observed. At −72 °C, the ratio of (123) to (124) was 38:62. At 0 °C, however, (123) was very much the major product, present in a ratio of 96:4. At an intermediate temperature of −40 °C, the ratio of products was 79:21. At this temperature, the best chemical yield (95%) was obtained. The change in regiocontrol indicates that (124) is the kinetically preferred product but that (123), the α substitution product, was thermodynamically favoured. In all of these reactions, the anionic chromium intermediate was converted to the aromatic products by oxidation with iodine, as shown for the formation of (118). The effect of temperature in selecting thermodynamic or kinetic control was dependent upon the nature of the nucleophile. The exclusive formation of the thermodynamic product (126) was observed for addition of LiCH$_2$CN even at −20 °C. With a lithiated dithiane, however, reaction at 0 °C produced only the kinetic product (125).[148, 151]

The examples above indicate similarities between the reactions of neutral arene π complexes. Although quite a number of nucleophiles show useful addition reactions with neutral chromium complexes, they tend to be similar in character, and the range is limited as indeed it is for the alkylation of most neutral complexes. The reaction fails with a number of nucleophiles of great synthetic importance, including Grignard reagents, cuprates, simple enolates, and also malonate anions.[136]

The range of useful nucleophiles can be extended by the use[152, 153] of cationic complexes. The corresponding Mn(CO)$_3^+$ arene complexes with a variety of substitution patterns have been studied. Following initial examination of malonate additions, it has been found that simple ketone enolates can also be used successfully, as shown in Scheme 7.12. Organometallic nucleophiles such as aryllithiums and related species are also satisfactory reagents.

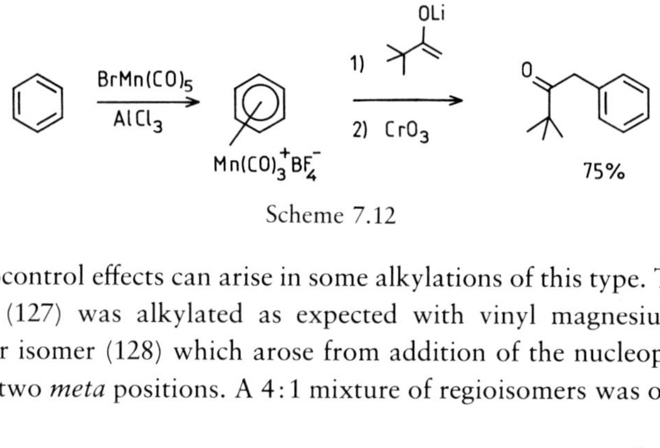

Scheme 7.12

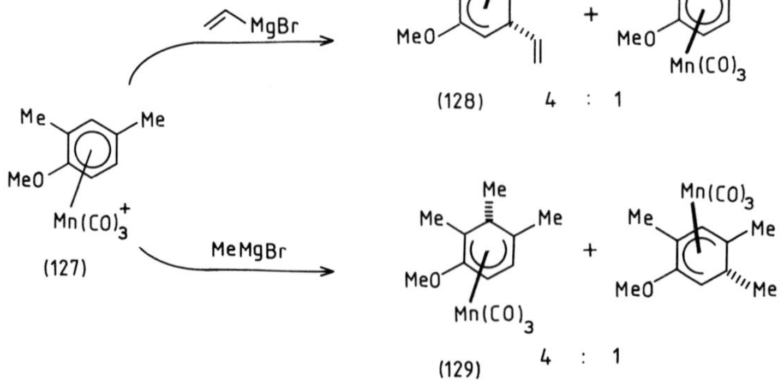

Curious regiocontrol effects can arise in some alkylations of this type. The manganese arene complex (127) was alkylated as expected with vinyl magnesium bromide to produce a major isomer (128) which arose from addition of the nucleophile at the less hindered of the two *meta* positions. A 4:1 mixture of regioisomers was obtained. When the alkylation was performed with methylmagnesium bromide, a marked reversal of regiocontrol was observed. The major product in this case was (129) in which the nucleophile is attached between the two flanking methyl substituents. A possible explanation is that formation of (129) has occurred at the site within (127) where the greatest positive charge is present in the arene ligand. The approach of vinyl Grignard reagents, however, might be more influenced by orbital overlap considerations. It is possible that an intermediate π interaction forms between the nucleophile and the metal complex in this case. Whatever the explanation, it is clear that steric effects are far more significant in the case of the vinyl Grignard reagent. The use of Grignard reagents derived from 1-bromopropene improved the regioselectivity from 4:1 to 9:1 in a reaction that proceeded in 75% yield.[154]

An improved procedure for the preparation of (arene)Mn(CO)$_3$$^+$ complexes by Pearson and Richards[155] may well provide the necessary stimulus for their more widespread use in synthesis. These workers have found that enolate addition to anisole and veratrole complexes shows the same regiocontrol effects as seen for the chromium complexes discussed above. The formation of (130) is typical of this type of reaction.

Semmelhack's group has developed methods that provide alternatives for the exploitation of intermediate anion species of the type (103). Protonation produced a neutral diene complex which, in the case of (131) gave[136] a 5-substituted enone upon oxidation. A slow isomerisation, similar to that discussed earlier, occurred. The adduct (131) was converted into (132) from which the enone (133) could be obtained.

Displacement of methoxide can also be achieved by nucleophile addition to methoxyarene complexes. An example occurring with estradiol derivatives[156] provides an interesting contrast to results obtained in lithiation experiments described below.

Use of methyl iodide as the electrophile has been found[157] to result in carbon monoxide insertion (see Chapter 9) and the formation of *trans* acyl products (134) and (135), a result that is consistent with initial reaction of the alkyl halide at the metal. These processes are related to those described in Section 7.2 for η^4 complexes.

The regiocontrolled alkylation of (136) forms the basis for a synthesis[158] of acorenone sesquiterpenes. The initial alkylation sequence formed the metal-free product (137) which was further elaborated and then complexed a second time. A mixture of stereoisomers resulted, since the benefit of stereocontrol by the metal was lost in this sequence because of the need to repeat the complexation step. The two isomers, however, are both useful products and were converted separately to acorenone (138) and acorenone B by a spirocyclisation that again utilised the control available from the metal complex.

The $Cr(CO)_3$ group exerts a considerable steric influence on adjacent rings, providing a useful control effect that has been largely developed through the work of Jaouen and his group,[159] following an early report by Jackson and Mitchell in 1969.[160] This effect is discussed further in Section 7.7. Some illustrative examples of this type of reaction are given in Scheme 7.13.[159–161] Even when not constrained within a second ring, this type

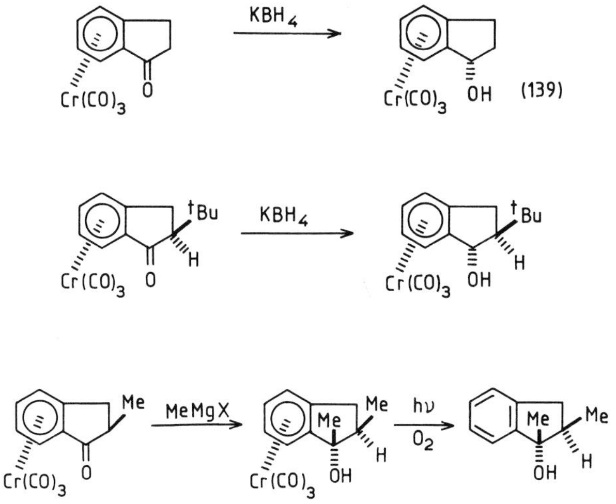

Scheme 7.13

of reaction can show quite pronounced diastereoselectivity.[162] An interesting example from Brocard's group uses the $Cr(CO)_3$ complex of 2-methoxybenzaldehyde.[163]

Delocalisation of negative charge has been examined in studies[161] of enolates formed from indanone complexes. Again, stereocontrolled products were obtained as shown in Scheme 7.14. The last case, (140) in Scheme 7.14, taken from a study by Boudeville and des Abbayes,[164] employs the same principle, but uses an ester to stabilise the enolate.

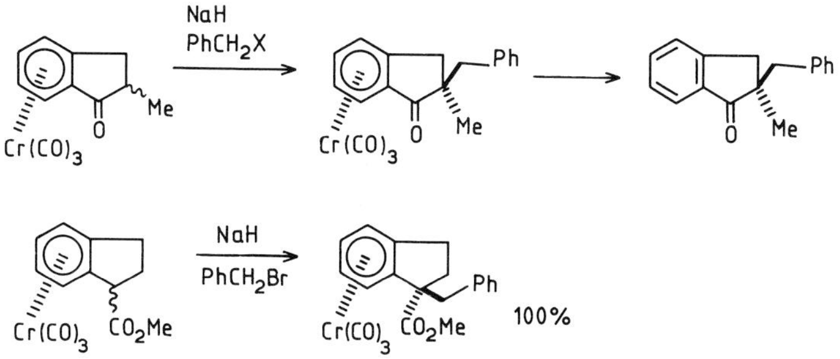

Scheme 7.14

More recently, by means of a hydride addition reaction similar to that used to form (139), Uemura *et al.*[165] were able to selectively reduce the complex (141) in a synthesis of the fish poison (144), in which the *cis* methyl group was introduced via the acetate (142) (benzyl cation stabilisation used in this step is discussed later in this chapter). The

regiodirecting influence of the MeO group now ensured that alkylation of (143) produced the required isomer.

Lithiation of chromium arene complexes

Coordination with $Cr(CO)_3$ induces a pronounced electron deficiency in the arene ring which enhances the kinetic acidity of ring hydrogens. This facilitates both the formation and functionalisation of (145) and its reactions with electrophiles.[166]

Interception of the lithiated species is performed at low temperature to avoid the formation[167] of butyl substituted products. Lithiated species can also be obtained by lithium exchange with iodine and mercury substituted arene complexes.[167] For direct lithiation, the use of a mixed ether/THF solvent system has been recommended.[168] Regiocontrol in substituted examples has been studied.[169] Chelation control favours *ortho* substitution as expected (Scheme 7.15).

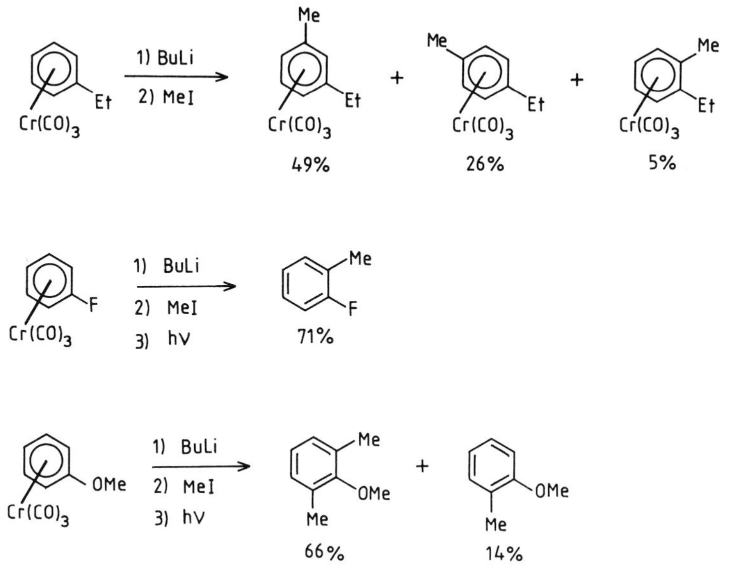

Scheme 7.15

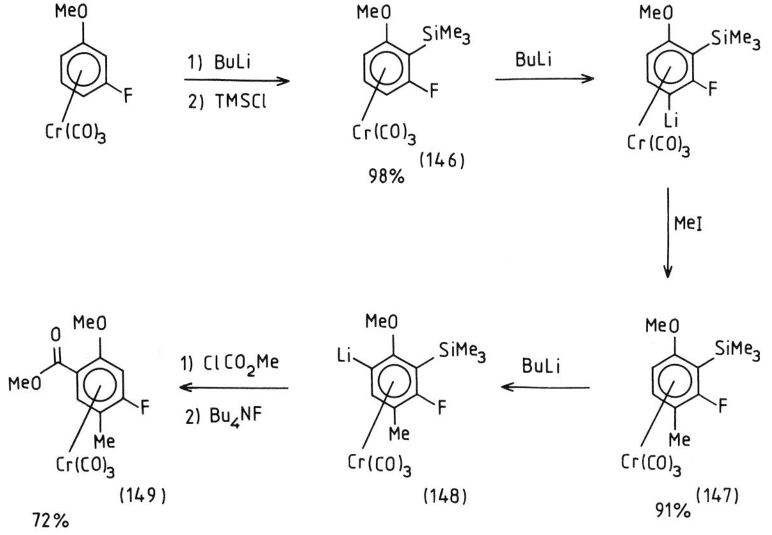

Methoxy and fluorine substituents can be placed in competition with one another. Examination of lithiation of fluoroanisole tricarbonylchromium complexes has indicated that fluorine has by far the more potent directing effect. Lithiation of the complexes of 4-fluoroanisole and 2-fluoroanisole both occurred exclusively adjacent to the fluorine substituent to produce 3-lithiated intermediates. The complex of 3-fluoroanisole underwent lithiation at C-2 between the methoxy and fluorine substituents.[170] For the silyl substituted complex (146), lithiation was again controlled by the fluorine substituent. In (147) the fluorine is flanked by substituent groups. A further lithiation is now directed by the MeO group to produce (148). Reaction with an electrophile, in this case methyl chloroformate, followed by removal of the silyl blocking group with fluoride produced the tetra-substituted arene complex (149).[171]

Steric effects can also be used to modify the directing influence of substituents. The bulky silyl ether in (150) promoted mainly *meta* lithiation. In this case, tBuLi was used.[172] Although lithiation at the *para* position is disfavoured,[173] efficient lithiation of the disubstituted complex (151) was possible. The bulky amine substituent in (152) also promoted *meta* lithiation.

The lithiation of naphthalene derivatives has also been examined. In this case LDA was used to effect lithiation of (153) and (156). Use of a short reaction time afforded a mixture in which (154) was the major product. Prolonged reaction, however, progressively increased the proportion of (155), indicating that equilibration was

-95°C	15min	78 : 22	
-95°C	4·5hrs	48 : 52	
-95°C	10 hrs	39 : 61	34%

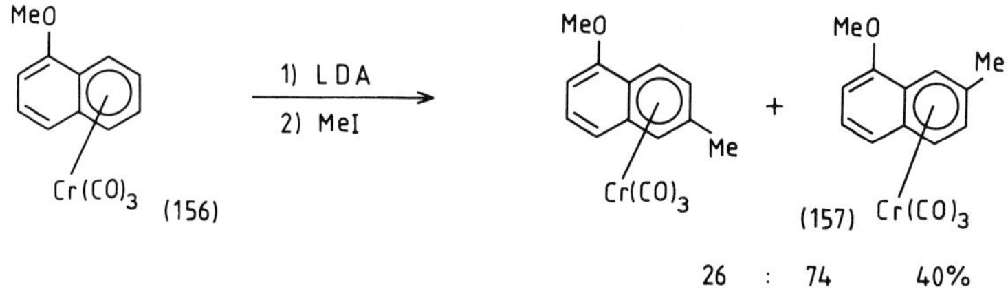

26 : 74   40%

occurring to enhance the proportion of the thermodynamically preferred product. Lithiation of the methoxy naphthalene complex (156) produced complementary regiochemistry to that described earlier (p. 187) for nucleophile addition. Lithiation followed by alkylation with methyl iodine produced a mixture of β isomers in which (157) predominated.[174]

Complete regiocontrol[175] in the lithiation of some tetralol complexes can reverse results obtained for the free ligand (Scheme 7.16). This effect has been ascribed[176] to the configurational preferences (see Section 7.6) of the $Cr(CO)_3$ group.

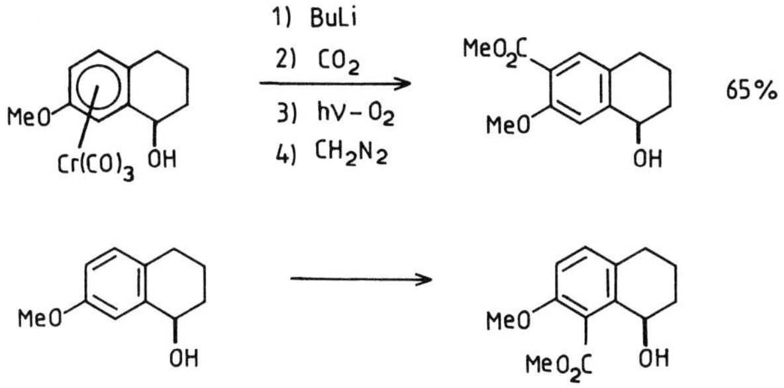

Scheme 7.16

The MoOPH hydroxylation reagent (oxodiperoxymolybdenumhexamethylphosphoramide) (see Chapter 3) has been applied to an oxidation of lithiated chromium arene complexes obtained from the mixture of diastereoisomers (158), the product of direct complexation of the steroid with chromium hexacarbonyl. After a reductive work-up with sodium sulphate and decomplexation in sunlight, the 2-hydroxysteroid derivative (159) was obtained in 65% yield.[177] Hypervalent iodine oxidation of enolates has also been used to introduce OH groups in the presence of $Cr(CO)_3$.[178]

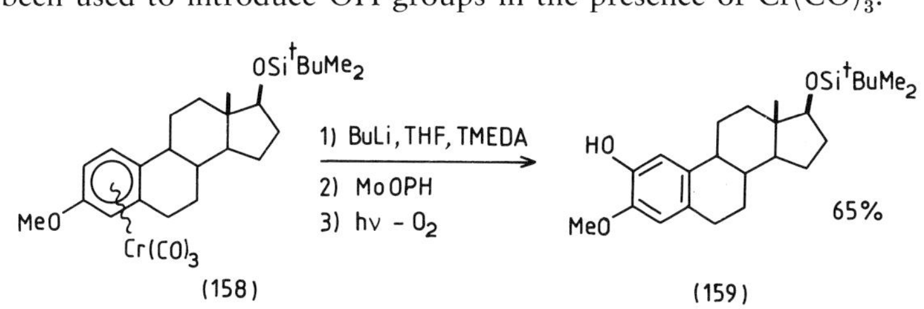

The cases above provide typical examples of the bond forming reactions of lithiated arene complexes in processes that often lead to different regioisomers to those obtainable by nucleophilic substitution. The scope of such alkylation processes has been extended[179]

by the formation of derived cuprate species and the use of palladium catalysis to effect cross-coupling reactions (see Chapters 8 and 10).

The stabilisation of benzyl anions by complexation to $Cr(CO)_3$ is related to the ability of the metal complexes to stabilise the anionic intermediates in nucleophilic substitution, since similar dienyl anion species (see also Section 7.8) make a contribution to the stabilisation process.

This effect is apparent in the much increased acidity of benzylic protons and forms the basis for useful alkylation reactions. Alkylation[180] of the phenylacetate complex (160), for example, is straightforward. The ester group is not essential for successful alkylation, and variation of other ligands at the metal can assist[181] the control of

multiple alkylation. Replacement of CO by CS led to double substitution and inclusion of a phosphite ligand had the reverse effect, due to its much lower π acceptor properties. In a recent example, a toluene $Cr(CO)_3$ complex was elaborated to phenylpropionic acid by deprotonation with lithium amide, alkylation with $BrCH_2CO_2Na$ and removal of the metal with ceric ammonium nitrate.[182]

Stereocontrolled alkylations can be achieved in cyclic systems such as the examples shown in Scheme 7.17.[183,184] A third example illustrates the use of stabilisation of

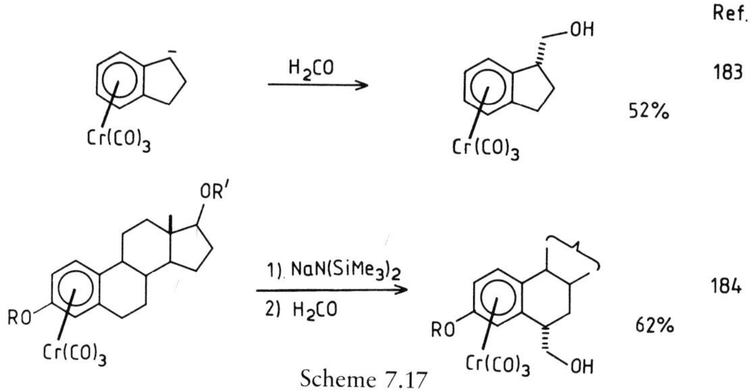

Scheme 7.17

negative charge by the tricarbonyl chromium unit to suppress Wittig rearrangements in anions derived from benzyl ethers. Deprotonation in this case was effected with butyllithium, providing an interesting comparison with the lithiation reactions discussed earlier in this section. Deprotonation occurred at the site of substitution by OMe. Alkylation with methyl iodide then produced (161) as a single diastereoisomer. The

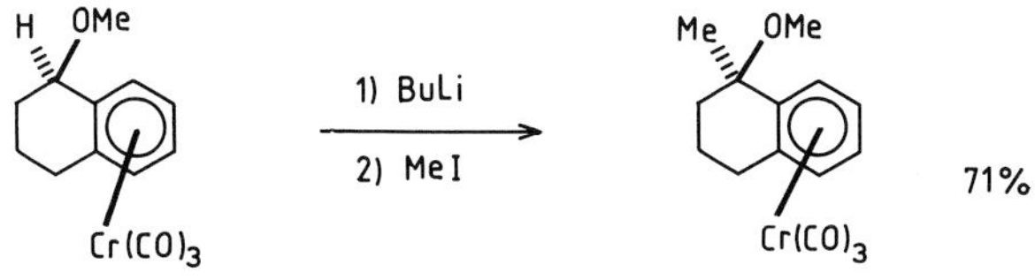

benzyl ether (163) was made from (162) in a similar way. This reaction has also been examined with chiral ethers in which the methyl group was replaced by a chiral secondary alcohol. Diastereoselectivity in these circumstances was relatively small. Diastereoisomer ratios of between 3:2 and 2:1 were obtained. The alkylation of the anion derived from (162) with ethanal was also examined. This produced a 2:1 mixture of diastereoisomers.[185]

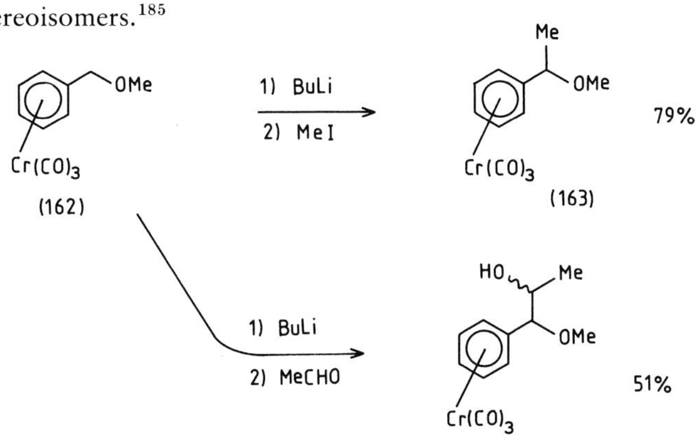

Stereocontrol can be extended to acyclic cases in suitable circumstances. The alkylation[186] of the anion of the (+)-N,N-dimethylamphetamine complex (164) gave only the isomer (166), indicating that the degree of β-substitution is sufficient to induce strong conformational preferences. Oxidation of the anion with the MoOPH reagent similarly gave (+)-N-methylpseudoephedrine as the sole product.

The stereoselectivity in these examples has been ascribed to the involvement of the dimethylamine group in an interaction with lithium during the approach of the butyllithium base to effect deprotonation of (164). Since the dimethylamine substituent is present at a chiral centre, two diastereoisomeric transition states could arise. In one case, the methyl group points away from the chromium arene ring. In the other, it would point towards it; this latter possibility is disfavoured. Examination of the stereochemistry of deprotonation and protonation is consistent with this speculation. Addition of D+ to the anion (165) produced a single diastereoisomer which, when treated again with butyllithium, lost only the deuterium atom, indicating that both deuterium addition and deuterium removal was stereospecific.[187]

Lithiation of tetrahydroisoquinoline complexes has also been examined. The formation of (168) shows that lithiation of (167) occurs α to the chromium complex at

the position remote from nitrogen substitution. In common with other cyclic examples, the production of (168) proceeded with good stereocontrol.[188] The anion intermediate (169) could be oxidised by MoOPH to produce a 4-hydroxy-N-methyl-tetrahydroisoquinoline in about 35% yield. A phenyl group was also introduced via the anion (169). Reaction with the chromium tricarbonyl complex of fluorobenzene, followed by oxidation, afforded (170) in 30% yield. With complexes of this type, although a range of alkylations have been performed, the most efficient procedures for synthetic use have employed alkyl halides. Methyl iodide (89%) and benzyl bromide (71%) gave comparable yields to the formation of (168) described above.[189] Lithiation

(169)   (170)   30%

at the 1-position is possible with the use of t-butyllithium. With (167), a 2:1 mixture of 4- and 1-methyl substituted tetrahydroisoquinoline complexes was obtained after addition of methyl iodide. Lithiation of (168), followed by addition of ethyl iodide, produced the disubstituted complex (171) from which (172) was obtained as a single diastereoisomer.[190]

The use of a silyl blocking group, seen in lithiation studies, has also been applied to control benzyl anion formation. The alkyl silane (173) was easily obtained by silylation of the anion derived from (167). Deprotonation of (173) with t-butyllithium followed by alkylation and desilylation with fluoride, produced the 1-substituted complex (174).[190]

The alkylation of tetrahydroisoquinoline complexes has been applied in the elaboration of the alkaloid canadine (175).[191] The two tetrahydroisoquinoline units in (175) were distinguished during complexation to form (176) as a mixture of diastereoisomers. Reaction of the major isomer with n-butyllithium resulted in lithiation rather than deprotonation at a benzylic position. Once a silyl blocking group had been introduced with trimethylsilyl chloride to form (177), the formation of a benzyl anion could be successfully performed. n-Butyllithium was sufficient for this purpose also, and, in contrast to (167), alkylation with methyl iodide introduced the methyl group adjacent to the nitrogen atom. This reaction, in common with others of this type, was completely stereocontrolled. Desilylation with fluoride followed by decomplexation by oxidation afforded (−)-(8R)-methylcanadine as a single diastereo-isomer. The (8S) isomer was similarly produced from the minor stereoisomer obtained in the complexation reaction.

In a sense, the alkylation of coordinated benzyl anions is similar to the alkylation of

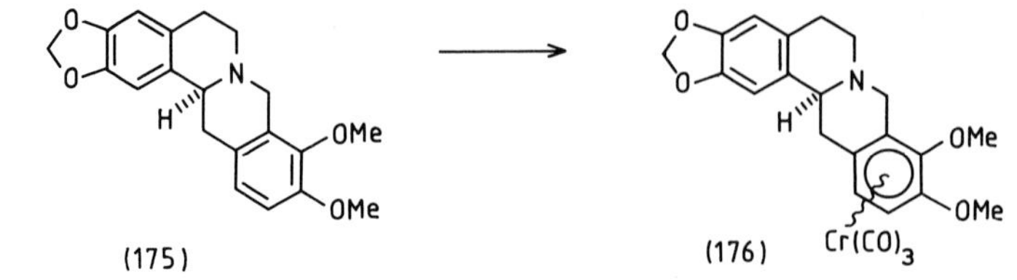

(175)   (176)   Cr(CO)₃

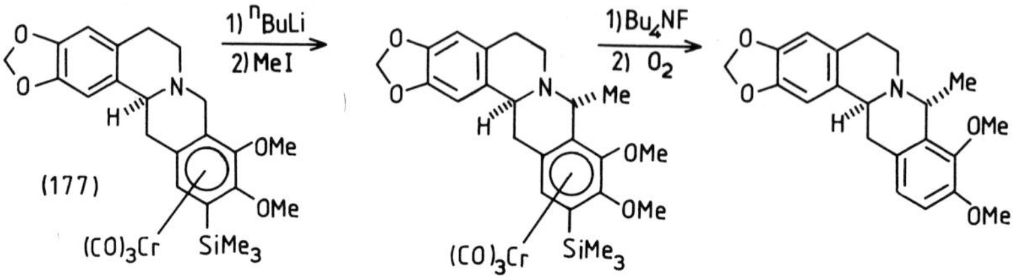

(177)

(CO)₃Cr   SiMe₃   (CO)₃Cr   SiMe₃

enolate anions, with the arene complex, rather than a carbonyl group stabilising negative charge. This analogy has been further developed by Semmelhack *et al.*[192] in reactions reminiscent of the conjugate addition–enolate trapping reactions (see Chapter 8) of enones. In cyclic cases, however, a major difference was observed. Both bond-forming reactions were controlled by the metal centre of (178) and the resulting substitution pattern was *cis*.

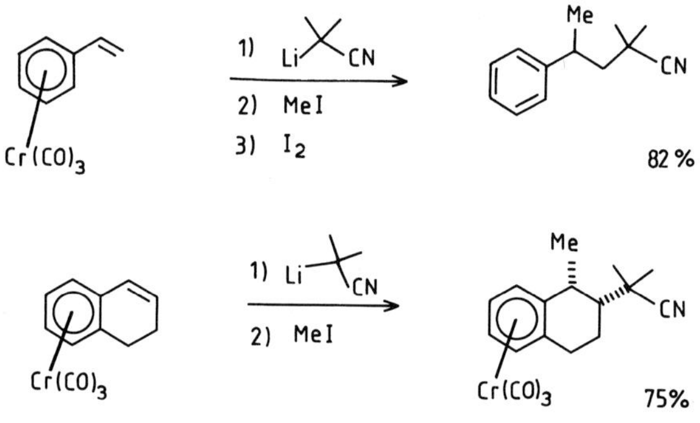

82 %

(178)   75%

The influence of the chromium tricarbonyl moiety on adjacent unsaturated groups is also apparent in the potential required for cathodic reduction, which was very much lower for the metal complexes than in the free ligands. In the case of the stilbene complex (179), the coupling of the radical anions (see Chapter 10) was investigated. A

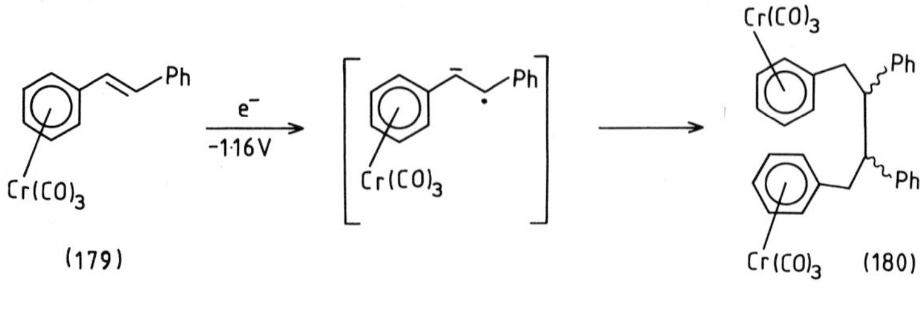

(179)   Cr(CO)₃ (180)

tetraphenyl substituted butane was isolated by decomplexation of the initially formed dimer, presumably (180).[193]

Stabilisation of benzyl cations

Cationic centres adjacent to transition metal complexes experience marked stabilisation. This effect was inferred initially in the case of chromium complexes from the increased ease of solvolysis at such positions. Solvolysis reactions of, for example, side-chain halogens or mesylates are accelerated[194] by 10^5 for (181) and by 10 for (182), in comparison with uncomplexed substances. When there are two adjacent coordinated

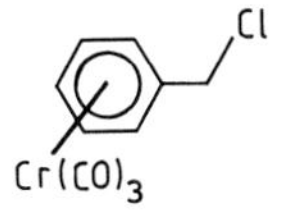

(181)

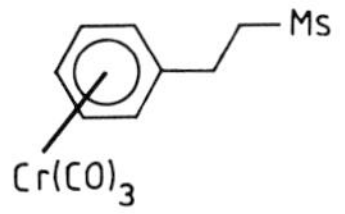

(182)

rings far greater charge stabilisation is achieved, and an insoluble, royal blue cation (183) can be obtained.

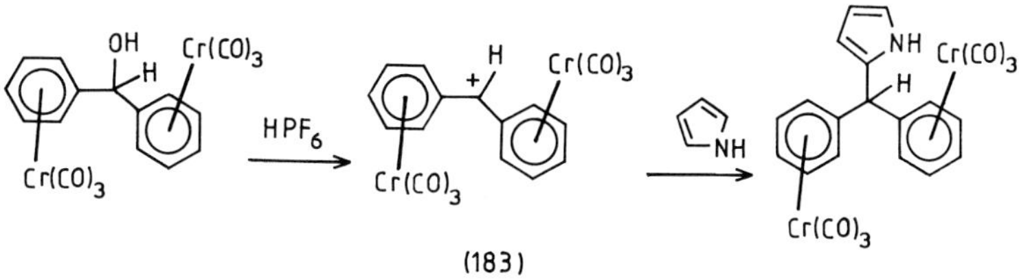

(183)

Solvolysis in cyclic systems is completely stereocontrolled and considerable control is retained even in the acyclic example (184).[195]

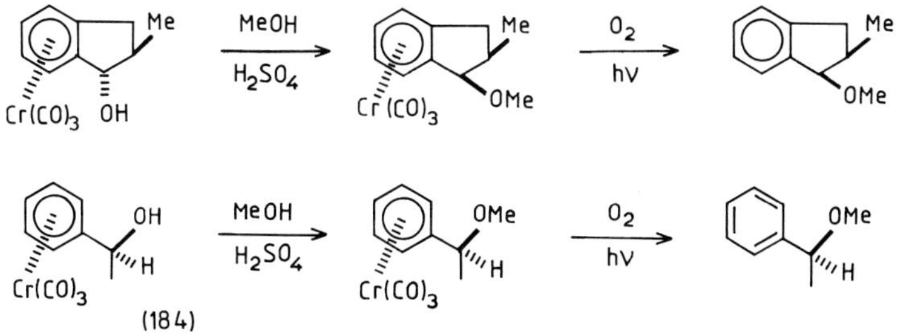

(184)

89% ee

64% ee

Stereocontrolled alkylation reactions employing Lewis acid catalysed reactions of trimethylsilylenol ethers have been developed[196] using this principle. Stereoselectivity was confirmed in this case by the resolution[197] of intermediates to allow the synthesis of optically pure (185) (see Section 7.7).

Temporary attachment of a tricarbonyl chromium moiety offers the possibility of efficient chirality transfer via an intermediate chromium-stabilised carbenium ion, produced by removal of a leaving group from a carbon α to the arene ring. To constitute an effective procedure, side-chain chirality must exert efficient control over the complexation of the arene ring by Cr(CO)$_3$. Since an aromatic ring with two suitable substituents is a prochiral group, attachment of the tricarbonyl chromium moiety distinguishes these two faces producing a new chiral centre. Diastereoselectivity in the complexation step is thus required. Prospects for this type of diastereoselective complexation have been examined for a number of complexes.[198,199,200] Studies by Brochard of the complexation of *ortho* substituted benzyl alcohols illustrates this point. Complexation of the racemic benzylalcohol (186) showed substantial diastereoselection. The stereochemistry of the major product (187) was consistent with control of the approach of the chromium by initial coordination with the OH group. The structures (187) and (188) are drawn to illustrate the steric effects operating in such an approach. In (187) the isopropyl group is directed away from the OMe substituent on the arene ring. Since the starting material is racemic, both enantiomers of (187) are produced, the chromium being directed to the upper face of the arene ring preferentially, in the enantiomeric structure.

The alcohols (187) and (188) show strikingly different properties in solvolysis reactions. Again this is due to the different steric effects that would operate during the formation of the benzyl cation intermediate. When a mixture of the two diastereoisomers was treated with methanolic sulphuric acid, the minor product. (188), was converted to the methyl ether (189). The major isomer (187) was recovered unchanged in 90% yield. As the carbenium ion intermediate (191) is formed, the interaction of the chromium complex with the developing positive charge occurs from the side of the α carbon atom away from the departing OH group. The preferred conformation for this process is shown in the Newman projection (190). This is the opposite configuration to that which provides the best placement of the OH group to direct the initial complexation reaction. This is depicted in the projection (192). Hence it is the minor diastereoisomer which is selectively converted to the ether. These reactions illustrate the power of *ortho* substituents on the aromatic ring to provide considerable stereocontrol even in acyclic situations.[199]

The chiral centre controlling the complexation step need not be at the α position. Complexation of (193) under mild conditions[200] using (naphthalene)Cr(CO)$_3$ produced a 92:8 mixture of the two racemic diastereoisomeric complexes (194) and (195). The bulk of the isopropyl ether in (193) assists the selectivity of the reaction. When a methyl ether was used as the substrate, the ratio of products was reduced to 85:15. In (196) the asymmetric centre is moved one position further down the alkyl chain. Again good stereocontrol was achieved with the naphthalene reagent, producing (197) and (198) as an 89:11 mixture of racemic diastereoisomers. Complexation using the naphthalene complex was effected at 70 °C. These mild conditions are important for the stereoselective complexation. Use of the more usual chromium hexacarbonyl reagent at 130–140 °C produced a 55:45 mixture of (197) and (198).

($\pm$)-(193)   70°C, 88%   ($\pm$)-(194)   92 : 8   ($\pm$)-(195)

($\pm$)-(196)   85%   ($\pm$)-(197)   ($\pm$)-(198)

89 : 11

The best selectivity was obtained with the isopropyl ether analogue of (196). In this case the reaction was 94% diastereoselective and proceeded in 95% yield. The major product from (196), (197), was converted by protection as the benzyl ether and hydrolysis of the acetal to the ketone (199), which was formed as single diastereoisomer. This compound has been used to demonstrate the power of the chromium tricarbonyl moiety in the relay of chiral information originally arising from the asymmetric alcohol substituent. Reduction of (199) produced an alcohol that was acetylated to afford (200). Nucleophilic displacement of the acetoxy group adjacent to the arene complex proceeded with retention of configuration to produce (201) as a single diastereoisomer. An alternative sequence used methyllithium to produce (202) which was reduced, again with retention of configuration at the α position, to afford (203), the epimer of (201). By these two complementary reaction sequences, arene complexes bearing two diastereo-isomeric chiral side chains have been prepared from the same precursor.[200] These reactions have been demonstrated with racemic material but in view of the extensive work on resolution of chromium arene complexes (see Section 7.7) there are interesting possibilities in processes of this type for the enantioselective control of acyclic stereochemistry.

Synthetic versatility

The varied and highly controlled reactivity properties of chromium arene complexes, when considered in combination, offer flexible possibilities for synthetic planning. An alternative route[201] towards hydroxycalmanenes such as (204) that bear a different methoxy substitution pattern to those described above (e.g. 144), can none the less be undertaken. In this case, an electrophilic cyclisation method was employed to form the bicyclic system.

Alkylation with methyllithium proceeded *trans* to the metal. At this stage, lithiation was performed, followed by addition of DMF to place a formyl group at the less hindered of the two carbons α to the OMe group. Reduction of the chiral centre in the saturated ring proceeds with inversion in this case. Although the removal of the leaving group to form the intermediate carbenium ion is most easy when the carbon–oxygen bond points away from the metal complex, this, like several examples seen earlier in this section, shows that when freedom of rotation is removed in a cyclic system, access to a carbenium ion intermediate is still possible even in the less favoured stereoisomer. Reduction occurs in the normal way, at the face of the molecule opposite to the metal. This results in inversion of the stereochemistry of the reduced product. The formyl group is reduced to a methyl group in the same step. Finally, decomplexation produced (204).

A combination of conjugate nucleophile addition and directed lithiation has been used to prepare an intermediate for a synthesis of 11-deoxyanthracyclinone. The dihydronaphthalene complex (205) was acylated with a suitable acyl anion equivalent to produce (206) by nucleophile addition to the uncomplexed alkene in a fashion similar to that observed with (178) earlier. Lithiation, rather than deprotonation from an α-methylene group, was required for the next step. This was achieved using n-butyllithium and TMEDA at −78 °C in THF. Once again, the OMe group served to direct the lithiation to produce (207), which was quenched with 2-formyl-3-methoxy-N,N-diethylbenzamide, forming (208) after decomplexation by exposure to sunlight and hydrolysis with dilute acid and then with base. The B-ring of the anthracyclinone was formed in the normal way by conversion of the phthalide to a benzoic acid derivative which was cyclised with TFA. This produced an anthrone derivative which was oxidised to the anthraquinone and partially demethylated to complete a formal total synthesis.[202]

Directed lithiation followed by a nucleophilic substitution that removes the directing group, provides a method for the synthesis of *ortho* substituted arenes. The fluorobenzene complex (209) was found highly suitable for this purpose[203] and has been described as an equivalent of an aromatic dipolar synthon (Scheme 7.18). The reaction can be applied to one-pot cycloaddition reactions that lead to a variety of heterocycles.

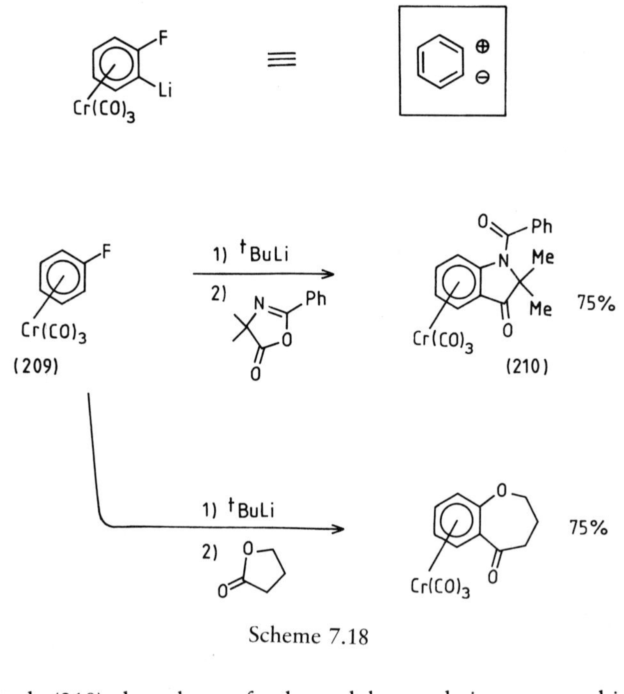

Scheme 7.18

The compound (210) has been further elaborated in steps which owe their stereocontrol to the tricarbonylchromium moiety. Nucleophile addition to the ketone using methylmagnesium iodide produced (211), in which the methyl group has been added *trans* to the chromium centre. Ionic reduction using an alkylsilane and HPF$_6$ also proceeded with complete stereocontrol forming (212) as a single product.[204] The

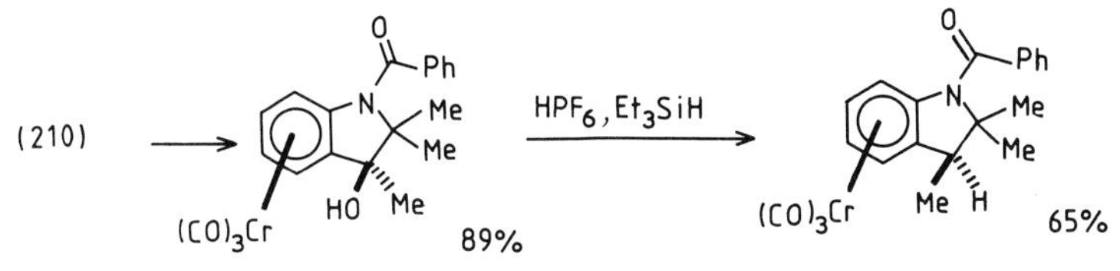

(210) →　　(211)　89%　　$\xrightarrow{HPF_6,Et_3SiH}$　　(212)　65%

inversion of stereochemistry seen here is comparable to that observed in the formation of (204).

In the conversion of (209) into (212), directed lithiation, nucleophilic displacement of fluoride, stereocontrolled reduction and hydride addition to a stabilised carbenium ion, have all been employed to achieve the reaction sequence.

Because of its ability to dominate the course of lithiation in the presence of OMe substituents, the use of fluoride as a directing group for lithiation provides opportunities for highly flexible syntheses of polysubstituted aromatics from fluoroanisole starting materials. The 2-fluoroanisole complex (213) provides a good illustration[171] of the type of flexibility in synthesis that can be obtained. Lithiation of (213) proceeded *ortho* to the fluorine substituent, to produce (214) after addition of the electrophile. Nucleophilic displacement with pyrrolidine afforded the trisubstituted arene complex (215). In the

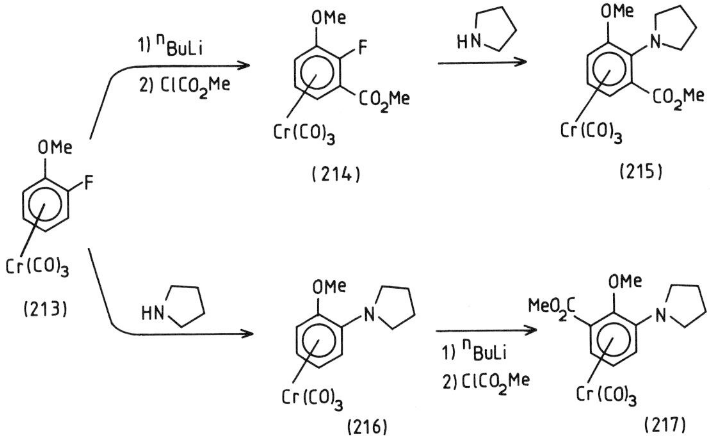

alternative sequence, the order of steps is reversed. First, nucleophilic displacement of fluoride from (213) produced (216). In this disubstituted complex, the OMe substituent is now the more powerful directing group. Accordingly, lithiation and addition of methylchloroformate produced only (217) in which the three substituents appear in a different order. With removal of the chromium in the usual way, such alkylation sequences would provide access to a wide range of substituted aromatics.

From the discussion earlier, it will be clear for 2-, 3-, and 4-fluoroanisole complexes that many different sequences of nucleophile and electrophile addition to the arene ring can be performed in this way.[171] These possibilities are summarised in the anisole aryl cation and anion equivalents shown in Scheme 7.19.

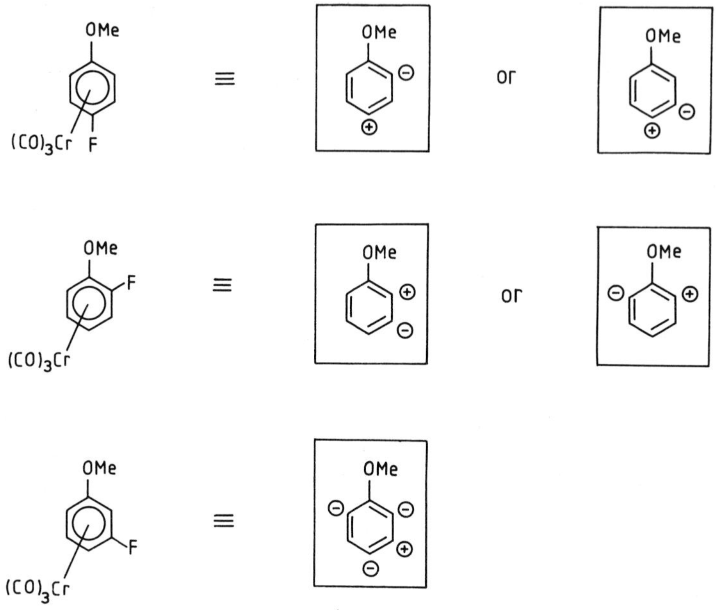

Scheme 7.19

7.5 Sequential nucleophile additions to organometallic π complexes

The ability to use transition metal complexes to promote nucleophile addition to alkenes and polyenes becomes particularly attractive if a series of alkylations can be achieved. Stereocontrol arising from the influence of the metal would then ensure the relative stereochemistry of products obtained from multiple alkylation sequences. This goal has been achieved,[64] for example, with some cycloheptadiene complexes leading to the formation of the *cis* product (218).

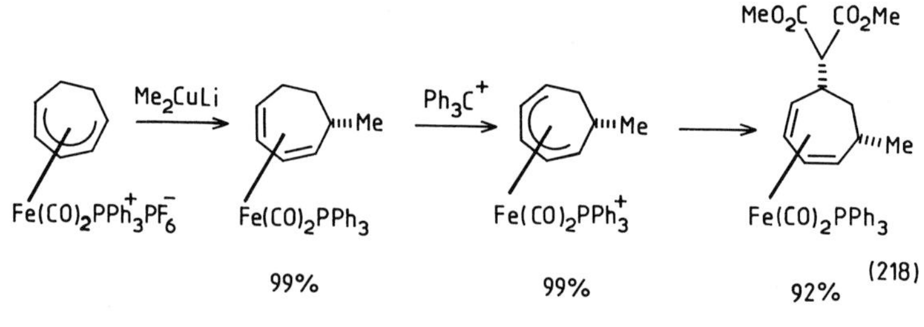

In the widely used cyclohexadiene complexes of $Fe(CO)_3$, the removal of hydride after alkylation is blocked by the presence of the new substituent, and repeated alkylation in this way cannot normally be achieved with Ph_3C^+ reagents. Recent work has tackled this problem in a study of alternative hydride abstraction procedures.[50] Of more immediate promise is the observation that allylic substituents can be introduced intramolecularly by suitable oxidation methods. This has provided a good method to overcome the limitation of the hydride abstraction reaction. Originally manganese dioxide[205] was used, but more recently methods employing ferric chloride on silica gel,[206] or thallium tristrifluoroacetate[207, 208] have been adopted. The method has been demonstrated[63, 208] for the formation of the disubstituted 1,3-diene (219). The use of phosphine complexes in routes to (218) and (219) provided the key to obtaining much better yields at the alkylation steps in the manner discussed earlier in this chapter.

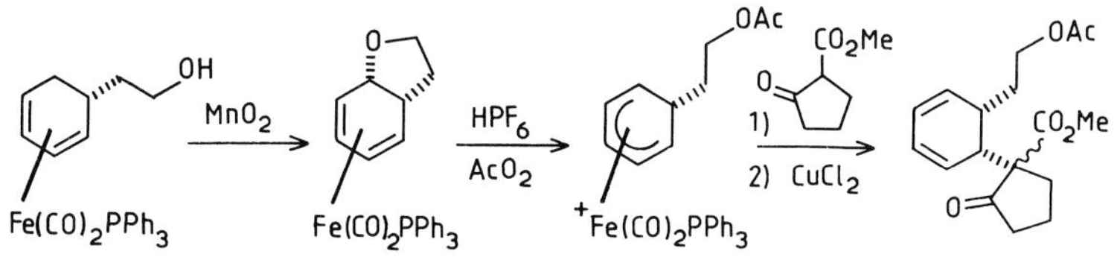

(219)

The cyclisation approach has also been used in the cycloheptadiene series. Formation of (220) was performed in this example using DDQ. Direct conversion of the alcohol to the dienyl cation was unsuccessful. The salt (221) was obtained by acetylation followed by reaction with HBF$_4$ in acetic anhydride.[209]

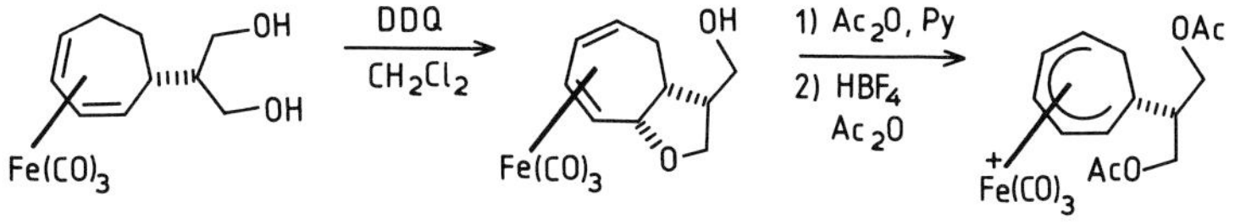

97%   65%

(220)   (221)

A 6-methoxy substituted dienyl complex[210] can also introduce a leaving group for use when a dienyl complex is re-formed after alkylation. The complex (222) was conveniently obtained from cyclohexadiene-5,6-diol which is available by a microbial oxidation of benzene.

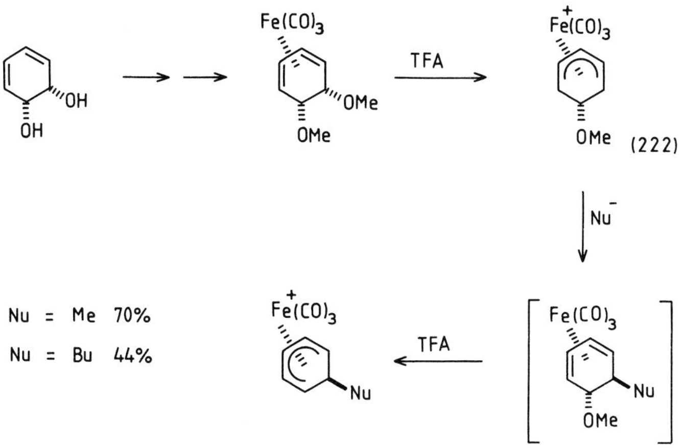

Nu = Me   70%

Nu = Bu   44%

It should be noted that in some exceptional cases hydride abstraction from tricarbonyl(cyclohexadiene)iron complexes can be achieved in the presence of blocking groups using a triphenylcarbenium reagent by performing the reaction at reflux. In the case of the cholestadiene complex (223),[92] initial abstraction from a methylene group in the adjacent ring, followed by a rearrangement similar to that leading to (49) and (50), may account for the result. The effect[211, 212] of the silyl groups in promoting this reaction is harder to explain.

With larger ring sizes,[64] or different ligands in the place of the carbonyl groups,[213] a second hydride abstraction can become possible. In these circumstances, nucleophile addition to cationic metal complexes can be followed by hydride abstraction and a second alkylation. This approach has been used to control relative stereochemistry in a synthesis of a cycloheptenone intermediate to the Prelog–Djerassi lactonic acid that makes use of the multiple alkylation of cycloheptadienyl complexes[64] described earlier. In this case the phosphite complex (224) is converted to the dimethyl complex (225) from which the 4-hydroxycycloheptenone (227) was obtained by peroxidation of (226) and ring opening.[214]

When $(C_6H_6)Fe$ is used in place of $Fe(CO)_3$, the steric limitations on hydride abstraction are also removed and the sequence of two alkylations leads, after decomplexation, to the formation of disubstituted products. Success of hydride abstraction in these cases has been ascribed to the operation of an electron-transfer mechanism.[213, 215]

Multiple alkylation might also be anticipated if dicationic complexes were chosen as starting materials. While yet to find application in synthesis, a dialkylation of this type has been achieved.[216] The arene complex (228) reacted with sodium cyclopentadienide

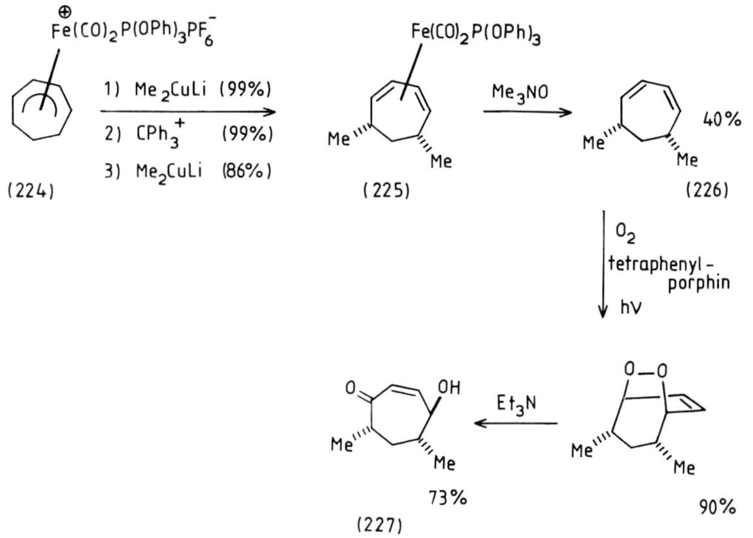

(224) (225) (226) (227)

to give the unstable complex (229) as a mixture with other regioisomers. The major product, however, appeared to arise from addition of the nucleophile at the metal.

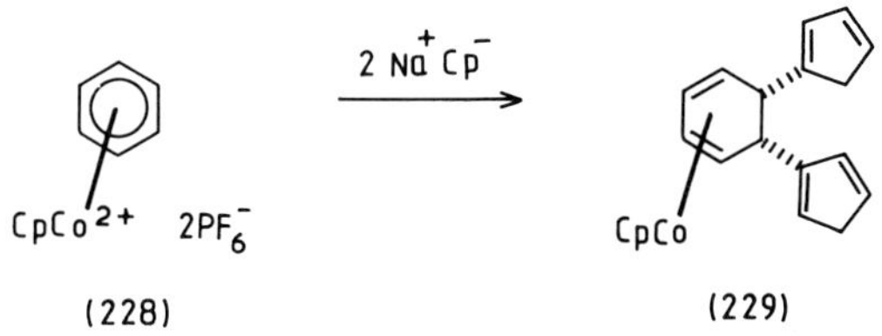

(228) (229)

There are a number of related examples where double hydride additions have been made to dicationic substrates.[217] A similar double addition[218] to the monocation (230) results in the formation of anionic metal complexes, although it was possible to obtain a single final product by protonation, treatment with KH, and removal of the metal by oxidation. Alkylation procedures combining the methods of Birch and Semmelhack (see Sections 7.2 and 7.3) to make double additions to a dienyl system should also offer interesting possibilities.

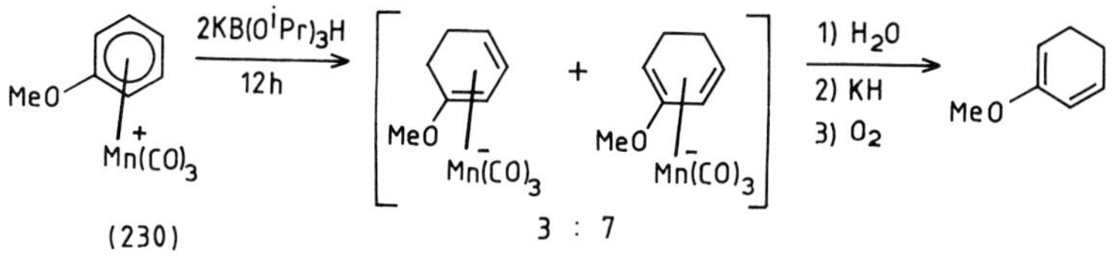

(230)   3 : 7

An alternative approach to the problem has been based by Sweigart and co-workers[152, 219] on the replacement[220] of CO by NO$^+$ following the first alkylation. Reaction of the Mn(CO)$_3$ cation with Grignard reagents provides a good source of neutral dienyl complexes such as (231)[153] that can be converted into cationic dienyl intermediates by reaction with NOBF$_4$ or NOPF$_6$.

(231)   83%   38%

A second alkylation is then possible, but this has not yet been developed as a general procedure. Indeed, there are limitations to double alkylation reactions based on $Mn(CO)_3$; sequences using conversions of $(\eta^5$-dienyl)$Mn(CO)_2PR_3$ to $(\eta^5$-dienyl)$Mn(CO)(NO)PR_3^+$ are more promising.[220] Hydride addition, however, has been achieved with several $Mn(CO)_2NO$ cations. The example of the formation of (232) shows exclusively an unusual mode of addition.[219] *Endo* hydride addition can compete with normal *exo* attack in tricarbonyliron complexes, too, when the reaction takes place at a substituted terminus of the dienyl system.[40]

(232)   3 : 2

A good many other examples of replacement of CO by NO^+ have been studied. Since the studies of nitrosylation by Efraty *et al.*,[221] the method has been used by Bamberg and Bergman[222] for cyclobutadiene complexes. In some cases NO_2^+ was found[221] to be a more effective reagent for the introduction of NO^+. Use of $NOPF_6$ in this latter case gave (233) in only 23% yield. Indeed, the use of $NOPF_6$ for carbonyl replacement may not

40 - 50%

(233)

63%

be of general applicability; reactions with (arene)$Cr(CO)_3$ complexes, for example, proved to be satisfactory only when complexes carried several methyl substituents on the ring.[223]

Multiple nucleophile additions based on these procedures provide attractive methods for use in synthesis. Examples drawn from the work of Faller *et al.*[24] and Pearson *et al.*[224] on cyclohexenyl and cycloheptadienyl chemistry are shown in Scheme 7.20. The formation of (234) illustrates the complementary nature of this approach to the hydride abstraction method leading to (225).

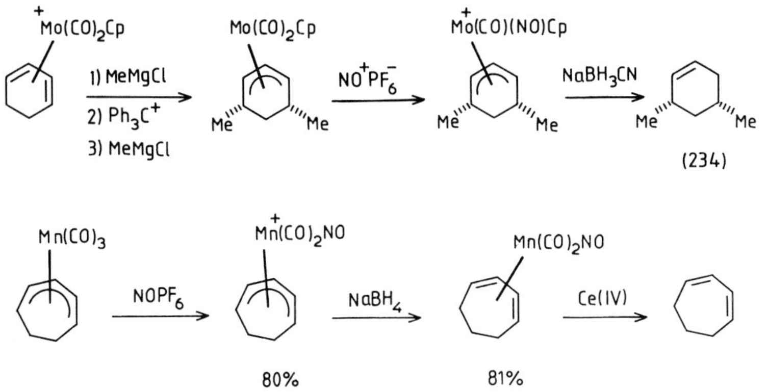

Scheme 7.20

Double alkylation of the cycloheptatriene complex (235) has been performed in this way.[225] The $Mn(CO)_2NO^+$ complex (236) reacted with the enolate of dimethyl malonate to give the adduct (237). In the alkylation of the $Mn(CO)_2NO^+$ complex it is important to avoid electron transfer reactions and hence the replacement of a carbonyl group by a phosphine, as discussed above, can also be advantageous with these complexes.[225]

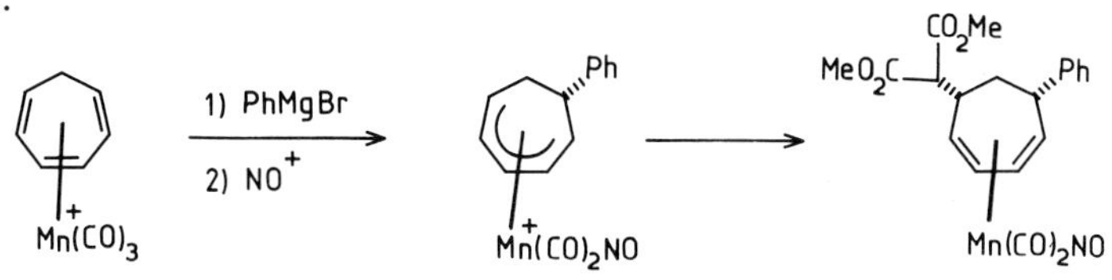

Both the dienyl and diene complexes (236) and (237) were unstable and the nature of the product was confirmed by decomplexation using ceric ammonium nitrate. Complexes bearing NO ligands offer interesting opportunities in organic synthesis, but in some cases can lack the convenient stability and reactivity properties that have proved so advantageous in applications of the more widely studied types of complexes.

7.6 Factors that affect reactivity and selectivity

Previous sections in this chapter, and also Chapters 5 and 6, give many examples of the controlled alkylation reactions of specific ligands. So far, however, no attention has been paid to the state of our understanding of the factors behind such selectivity. In 1978, Davies, Green and Mingos[226] published a series of simple rules, based ultimately on the premise of charge control, and the use of molecular orbital considerations to assess charge distribution. The rules predict which ligand will be the site of alkylation when several π systems are bound to the same metal atom in cationic metal complexes. The rules apply only to reactions proceeding under kinetic control.

Ligands are classified depending on whether an even or odd number of carbon atoms are bound to the metal, and whether the π system is open or closed. The terms 'open' and 'closed', when used in this context, refer to properties of the π system itself, not to whether the ligand is cyclic or acyclic; thus cyclohexadienyl is an 'open' ligand and cyclopentadienyl is closed. A simple statement of the rules for nucleophilic addition is given below:

(1)   Reaction at 'even' ligands is more favoured than at 'odd' ligands
(2)   'Open' ligands are preferred to 'closed' ligands
(3)   Reaction at 'even open' ligands will always occur at a terminus of the π system. Reaction at 'odd open' ligands will occur at the terminus if ML_n^+ is a strongly electron withdrawing group

This leads to a useful guide to relative reactivity which is set out in Figure 7.1.

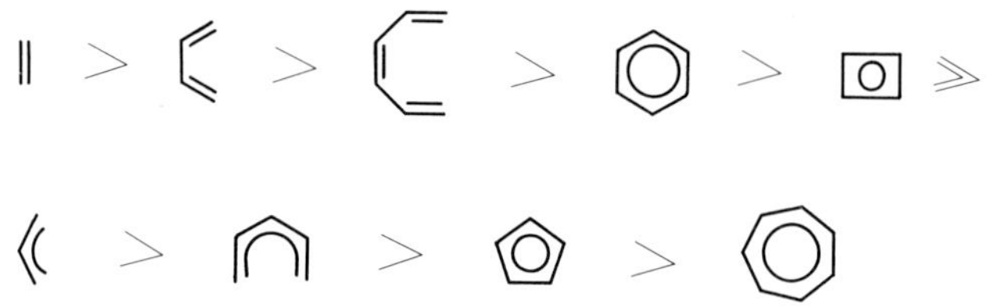

Figure 7.1

Carbon monoxide ligands take no place in this classification, and may be attacked in competition with hydrocarbons. Nucleophiles that are expected to show irreversible additions, however, usually give hydrocarbon adducts. There are many instances where the nature of the nucleophile is of great significance, and in some cases the outcome of alkylation may arise[226] from orbital control (see below). Iodide addition is of particular note since, following an initial addition to a carbonyl group,[227] final attachment of iodide to the metal has been observed.[228]

A second important aspect is the prediction of the site of alkylation within any particular ligand. The third rule from Davies, Green and Mingos[226] provides a guide in 'open' cases. An important point is that the nature of the metal and of other ligands can play an important role in determining regiocontrol when 'odd' ligands are considered.

An interesting comparison has been made between nucleophile addition to neutral η^4 Fe(CO)$_3$ complexes and cationic Co(CO)$_3$ analogues. The former react at an internal position while the latter react at a terminus of the π system, as expected on the basis of rule 3. Extended Hückel calculations were used to probe the origins of this reversal. For the Fe(CO)$_3$ case, a cross-over between internal and terminal attack was predicted if sufficient variation of the nucleophile were possible. Terminal attack for Co(CO)$_3$ was always preferred. This study used a frontier orbital approach in which the energy of the LUMO of the electrophile was scrutinised.[229]

For dienyl ligands, instances can be found for addition at any position bound to the metal, and a full theoretical description must be able to account for these differences, and also the effects of substituents that are described in Table 7.1.[130, 230]

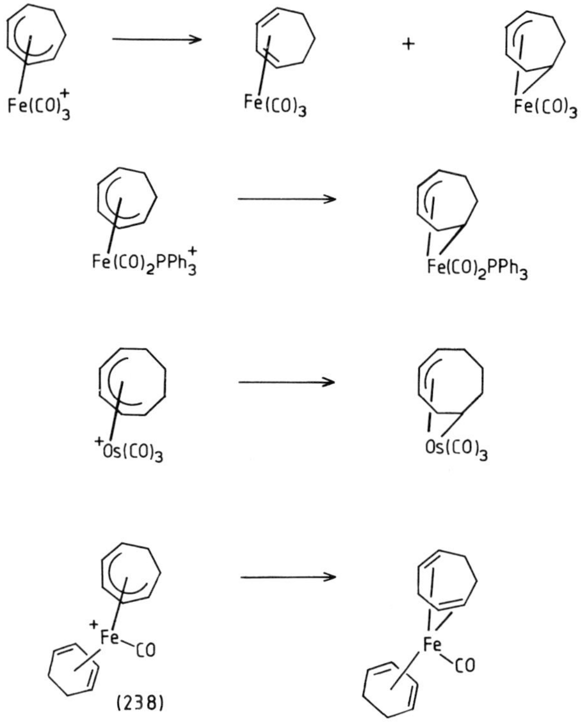

Scheme 7.21

When considered in this light, some synthetically important examples given in this chapter represent special cases, but since the reactivity preferences that they show appear to be general for that particular type of complex, this does not constitute a problem for synthesis design. It is, however, important to be aware that changes in ring size, or the choice of metal or its ligands, can introduce complications in some cases.

The reduction[130] of the η^5-dienyl, η^4-diene complex (238) in Scheme 7.21 is of particular note since it provides one of the few examples that appear to be exceptions to rule 1 (another was seen in the case of a cobalt complex encountered in Section 7.4). A possible explanation for the anomaly with (238), however, is that a different mechanism may be in operation, since hydride may be transferred to the cyclohexadienyl ring after initial addition to the carbonyl group.

The Davies, Green and Mingos rules are based on charge control under kinetic reaction conditions and should not be applied to cases where equilibration results in thermodynamic control, or where the outcome is directly determined by the relative sizes of molecular orbital coefficients rather than by charge density. Apparent contradictions to the rule can, upon detailed consideration, be reasonably accounted for in this way. Two further examples suffice to illustrate these distinctions. Addition[231] of cyanide to (239) proceeds as expected at the η^2-ligand but addition of $C_3H_7S^-$ (which is far more polarisable) at the allyl moiety, requires an explanation based on orbital control. In the case of the azide addition[232] to (241), attack at the alkene may be reversible, and the formation of (240) can be explained by the operation of thermodynamic control.

(239)

(240)   (241)

The effect of substituents on the ligand is also of great significance. A number of attempts have been made[233, 234] to employ the charge control arguments here, and ^{13}C n.m.r. chemical shifts have been used[235] as an indicator of this parameter. More recently, discussion of nucleophile addition to dienyl systems has centred on the interpretation of frontier orbital descriptions with the recognition that LUMO coefficients may be of significance. Substituent directing effects may be considered to operate by distinguishing the orbitally favoured positions through differences in charge density.[236] Interpretation of these influences is not straightforward since complexation can greatly perturb bond orders within π systems, as shown by a series of detailed studies[237] of neutral complexes. No comprehensive molecular orbital treatment of substituted dienyl complexes is available to allow a direct consideration of LUMO coefficients in such cases, although some candidate LUMO orbitals have been identified for 2-OMe substituted complexes.[238] A simple consideration of the effect of substituents on charge stabilisation within the ligand itself, can often provide a qualitative explanation for regiocontrol effects, but fails to explain some of the rate differences discussed later in this section.

With neutral complexes, at any rate in the case of arenes bound to $Cr(CO)_3$, orbital considerations hold sway[239] in explaining apparent anomalies in the directing effect of substituents. An interesting rationale,[240] supported by molecular orbital calculations, is that the conformational preference of the $M(CO)_3$ group influences the relative reactivity of positions on the ring, and that nucleophiles will add at carbons that are eclipsed by carbonyl groups. This can be understood by consideration of the alternating donor–accepter character of the hybrid orbitals[241] of the metal carbonyl fragment, and their interaction with the arene orbitals. Charge distribution in the LUMO at the eclipsed and staggered carbon atoms are shown against structure (242). The significance of these values was tested by calculation of the approach of a model CH_3^- nucleophile.

(242)

The orbital and charge influences caused by substituents operate in concert. Because of this, their relative significance cannot be distinguished. This analysis is based on the assumption that the reaction proceeds most readily through the minimum energy conformation of the carbonyl groups, a situation which cannot be guaranteed for all reactions.

A similar analysis of the control of deprotonation α to tricarbonylchromium arene complexes identified electronic effects as the source of regiocontrol. Preferential attack at a carbon eclipsed by a chromium–carbonyl bond could not account for the observed regiochemistry.[242]

Table 7.1 Regiodirecting effect of substituents

A different basis for the calculation of factors influencing regiocontrol has been adopted by Eisenstein and Hoffmann[243] for a study of additions to coordinated alkenes. Here distortion of metal ligand bonding due to substituents was assessed, and nucleophile addition was predicted at sites where metal ligand bonding is lessened. This approach is related in its philosophy to attempts[233] to identify positions of highest 'free valence' used for other systems.

Although the full picture is by no means clear, some general trends for substituent directing effects in charged 'open' ligands are apparent from examples given earlier in the chapter. Until a more complete theoretical description is possible, however, these can provide nothing more than a rough guide. 1-Carbomethoxy and 2-methoxy substituents direct nucleophiles to the far terminus of the complexed π system. 1-Alkoxy substituents, on the other hand, direct attack to the site of substitution. When compared to the effects of alkyl groups, the directing influence of groups that can interact directly with the π system of the ligand is more powerful. Examples shown in Table 7.1 illustrate a useful empirical pattern of control influences for complexes ranging from η^2 to η^5. Although care must be taken with the interpretation of these trends (some entries arise only from a single example and may not be generally reliable), an interesting pattern emerges.

Pronounced steric effects resulting from the presence of alkyl substituents can, however, be dominant in some cases.[82, 244] The results of some of the rate measurements discussed below provide further insight into the effects of substituents.

Kinetic data

A great many kinetic investigations of nucleophile additions to π complexes have been made, and the body of information available is now sufficient to allow some important generalisations to be made.[245] The variation in rates observed for different metal/ligand systems and ring sizes is extensive, spanning nine orders of magnitude. The most reactive species shown in Figure 7.2 are very powerful alkylating agents, being some 10^4 times more rapid in their reactions than 'magic methyl'. Effects of substituents can cause significant changes in the order of reactivity, altering the relative reactivity of a particular species by several orders of magnitude in some cases.

Linear free energy relationships indicate a fairly early transition state for most nucleophiles. For phosphine addition, for example, rehybridisation from sp² to sp³ is approximately one third complete in the transition state.[246] M—C bond breaking is thus of little importance from this standpoint and, indeed, both free and complexed electrophiles appear to follow a simple, unified reactivity law.[245] The reactivity of nucleophiles also spans an extensive range of orders of magnitude, and is not related to the nature of the electrophilic partner. A further detailed analysis of data of this type has recently been made.[247]

As far as the bound π system is concerned, significant steric effects can be discerned. The relative rates indicated in Figure 7.2 give a measure of overall reactivity differences,

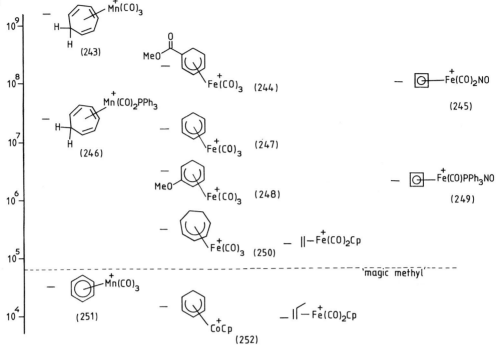

Figure 7.2

combining both electronic and steric influences. Relative rates in this figure are based on $(C_7H_9)Fe(CO)_2I$ at unity.

Several points of major significance emerge, however. The open η^6 ligand in (243) reacts nearly 10^5 times faster than the closed counterpart in the arene complex (251). This corresponds to the qualitative kinetic differences between ligands within the same complex implied in the regioselectivity predictions made using the second of the Davies, Green, and Mingos rules. To explore the analogy with the first rule, however, one must change the metal and/or ligands. The difference between the 'even' ligand in (243) and the 'odd' ligand in (247) lies in the right direction, but changes in the metal/ligands alone (compare, for example, (247) and (252)) can have a far greater effect.

Carbonyl groups are stronger π accepters than phosphine ligands, an effect that remains important even in cationic complexes. Replacement of a carbonyl group by PPh_3 lessens the positive character of the metal centre, reducing the electrophilicity of the complex. The rate differences comparing (243) with (246), or (245) with (249), reflect these changes. The low relative reactivity of (252) compared to (247), mentioned above, thus largely reflects the loss of electron withdrawing power of the carbonyl groups. Practical advantages of phosphine substitution, encountered in Sections 7.3 and 7.5, must in some respects arise from this attenuated reactivity. The ester in (244), an electron withdrawing group, increases reactivity, while the lower rate of reaction of

(248) seems reasonable in view of the electron donation of the OMe substituent. The effects are not so straightforward, however, and must also reflect the nature of the bonding within the complex. The 3-OMe isomer of (248), for example, reacts faster than (247).

Steric effects also have an important influence on relative rates. Dienyls in seven-membered rings (e.g. complex (250)) are typically less reactive at the termini than when in six-membered rings, a distinction ascribed[248] to steric blocking by the methylene group as shown in structure (253). Substituents themselves also have a direct steric effect. A methyl group at the site of expected alkylation can cause a significant reduction in rate, and more pronounced blocking can displace the site of attack or change the nature of the reaction (see Section 7.3). *Exo* substitution at C-6 can also have a great effect.[245, 248]

(253)

The detailed rate expressions reflect special factors for the nucleophile involved. With β-diketones, there is a pre-equilibrium to form the enol, while primary amines, for example, form an ammonium salt that is then deprotonated. The actual addition step, however, in which the nucleophile–substrate bonding is established, is interpreted by Kane-Maguire, Honig, and Sweigart[245] as following simple bimolecular kinetics with no interaction with the metal in most cases. Arene nucleophiles, however, appear to form a π interaction prior to the creation of the σ bond.

The overall picture for alkylation reactions can thus be summarised as an approach of the nucleophile to the face of the ligand opposite to that bound to the metal, a trajectory facilitated by the interaction of the nucleophile lone pair orbital and the LUMO of the metal/ligand orbital set. It is noteworthy that approach to the other face would be accompanied by an unfavourable interaction between filled orbitals. In contrast, the approach of electrophilic reagents to neutral metal complexes is favoured from below because of the presence of this filled orbital.

7.7 Stoichiometric metal π complexes in enantiomer synthesis

The resolution of transition metal complexes has been studied since the early days of organometallic chemistry. As progress was made with the development of synthetically useful alkylations of the type discussed in this chapter, so interest in the use of these highly stereocontrolled reactions in enantiomer synthesis developed. There are three types of chiral transition metal complexes. Some cationic examples are shown in Figure 7.3. Compounds such as (254)[249] are chiral at the metal atom. Compounds like (255)[250] bear ligands that contain chiral centres, and in π complexes (256)[251] of prochiral ligands, binding to the metal distinguishes the faces of a prochiral π system. In this latter

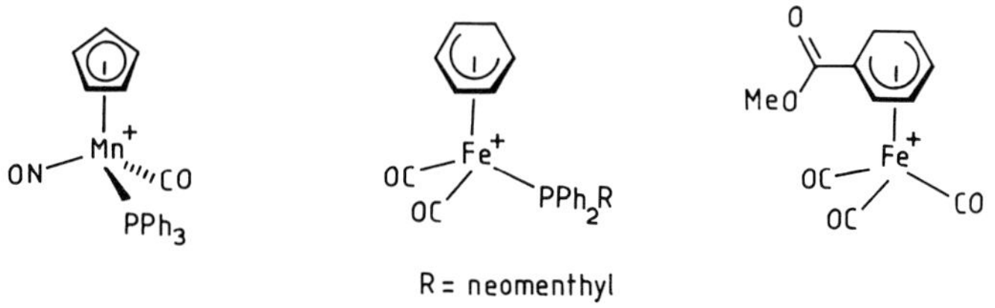

Figure 7.3

case the complex can be said to possess planar chirality. The formation of such complexes as racemates was recognised early in the 1960s and a number of resolutions were performed.[252] More recently a detailed stereochemical analysis has been made.[253] Neutral complexes with chiral metal atoms have found extensive application in asymmetric synthesis by use of reactions elaborating carbonyl insertion products. These topics are described in Chapter 9. This chapter will concentrate on the complexes containing planar chirality.

The resolution of the tricarbonyliron complex (257) by the separation of

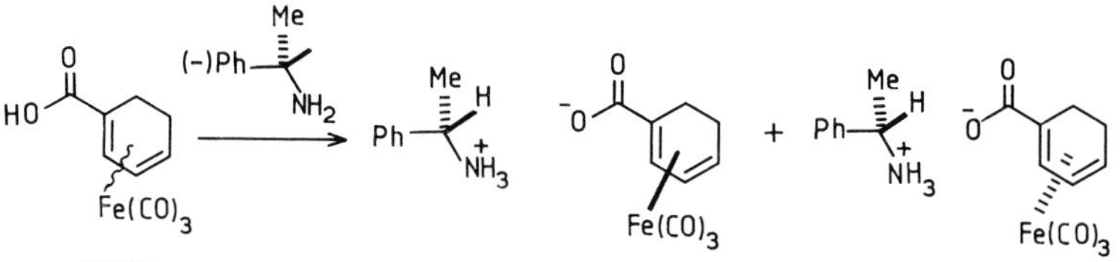

diastereoisomers by crystallisation of $(-)$-1-phenylethylamine salts, was used by Birch's group[254] in work that lead to an enantiomer synthesis[255] of gabaculine (258) by the completely stereocontrolled addition of $H_2NCO_2{}^tBu$ to (256). Since the absolute configuration of (257) had been determined in a series of chemical correlations,[256] the absolute configuration of gabaculine (258) was defined by the synthetic study.[255,256]

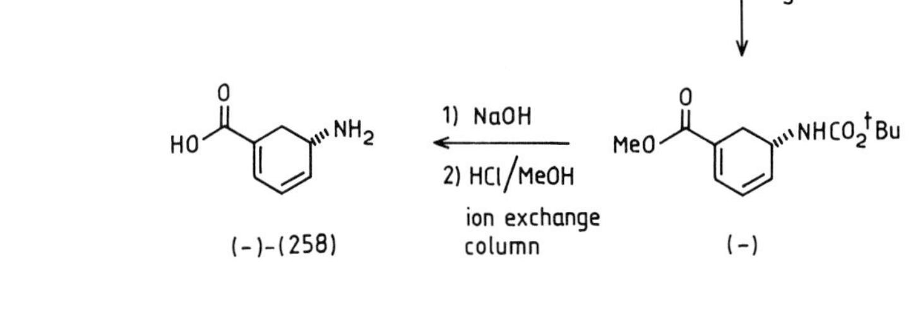

Hydroxide addition has been used in a similar way to begin the conversion of the (+) isomer of (256) into (−)-methyl shikimate (262). The product could be obtained from either enantiomer of (256). Oxidation of an alcohol intermediate (−)-(259) (obtained from (−)-(256)) with CrO₃ in pyridine afforded the dienone complex (+)-(261) which could be reduced by sodium borohydride in the presence of zinc chloride to afford the diastereoisomer of (259) which similarly gave (+)-(260). This was converted to (−)-(262) by reaction with OsO₄ and deprotection with fluoride.[257]

There has now been considerable effort made to study resolution procedures to provide general routes to η^5 salts of this type. Covalent attachment of chiral alcohols[250, 258] amines[250, 259] and phosphines[250, 260] and chromatographic separation of the resulting isomers has been used in a number of cases. The complex (36) has been resolved in this way and used in a synthesis[261] of optically active 4,4-disubstituted cyclohexenones. Alkylation of anionic complexes of this type is discussed further in Section 7.8.

An earlier approach[262] based on the salt (63), obtained in partially resolved form by an asymmetric induction[263] during the complexation of (263), was used to examine conditions for removal of the metal without racemisation. A method has since been developed[264] to give access to optically pure material following a partial induction of asymmetry. A third method[265] for access to fully resolved complexes relies on the complexation of chiral substrates such as the terpenes limonene and carvone, which, upon reaction with Fe(CO)₅, rearranged to give optically active diene complexes.

Originally investigation was limited from a synthetic viewpoint because of a shortage of suitable substrates, but could give optically active complexes of prochiral ligands by a rearrangement of chiral starting materials during complexation (see p. 34).

More recent extension of the use of biologically derived diols described in Section 7.5 offers a method of more general applicability. Microbial oxidation of unsymmetrically substituted (prochiral) aromatics can afford optically active diols. The complex (264), obtained from this material, could be converted into the dienyl cations (265) and (266).

$$E^+ = H^+ \quad 1 : 1 \quad 74\%$$
$$E^+ = Ph_3C^+ \quad 5 : 1 \quad 78\%$$

The major product, (265), was separated from the mixture.[266] The optical purity of these products was established by formation of (267), and conversion via the 2-methyl cation to the malonate adduct (268), which was examined with a chiral shift reagent[267] by n.m.r. spectroscopy.[266]

A number of other techniques, kinetic resolution,[58] asymmetric destruction,[268] or direct separation[269] on chiral columns have been examined but do not yet constitute practical methods for complete resolution on a large scale. Particularly of note is a new method that employs pig liver esterase to hydrolyse esters within organometallic complexes that have proved to be suitable substrates for this enzyme.[270]

Whatever method is used, however, much work is required to determine efficient procedures when a new example is attempted. The elaboration of simple resolved complexes to obtain the required substitution pattern can offer an attractive alternative.

Howell and Thomas[271] have used this approach in a route to (+)-(63) from (+)-(36). The Wittig reaction with formaldehyde, however, proceeded in rather low yield.

(+)-(36)

(+)-(63)

40% overall

Much synthetic work using resolved π complexes has concentrated on the control of stereochemistry in rings adjacent to the site of complexation. This is particularly well developed in organochromium chemistry. The resolution of chromium arene complexes has been extensively examined.[272] Recently, a remarkably simple kinetic resolution has been described.[273] Racemic (269) has been converted into the diastereoisomeric imines (270) and (271) by reaction with valinol. By passage down an alumina column, (270) was

(±)-(269)

(270)

(271)

hydrolysed to (−)-(269) which was eluted with ether. Subsequent elution with CH_2Cl_2/MeOH afforded (271). This was hydrolysed separately to form (+)-(269).

Racemic (269) has also been a substrate in an examination of the kinetic resolution of chromium complexes by yeast reduction. The (−) isomer proved to be the better substrate for the yeast, but (+)-(269) was also reduced to some extent, and the optical purity of the product was about 66%.[274] Asymmetric induction at a position α to the chromium arene complex has also been performed using yeast. In the case of (272), the

(272)

96%
>99% e.e.

metal complex was not, itself, chiral, and an optically pure product was obtained by asymmetric induction.[275]

The alkylation and reduction sequences involving indanone complexes discussed in Section 7.4, have been used[276] for the enantiomer synthesis of disubstituted products such as (273).

There are now many examples, mainly from the work of Jaouen *et al.*,[277] of this type

(+)　　　　　　　　(−)　　　　　　　　(−)-(273)

of process. An interesting application,[278] to be found in a study by Meyer *et al.*, is an attempted Robinson annellation approach in which cyclisation of the intermediate (274) was dominated by deprotonation next to the coordinated ring to give predominantly the ketoalcohol product (275). A small amount of the normal enone product (276) was also obtained. This type of cyclisation was followed exclusively when the diastereoisomer of (274) was cyclised.

(−)　　　　　(−)-(274)　　　　　(−)-(275)　　75%

(−)-(276)　　　　　　(−)

Direct alkylation of cationic chromium complexes have also found successful application in enantiomer synthesis. Even in acyclic cases, considerable selectivity can be achieved in alkylation reactions. Despite the possibility of rotation about the bond to the ring, alkylation of a benzyl cation resulted[279] in the synthesis of (278) in 92% optical purity. An indanol example gave complete stereocontrol. There are two important differences between these two cases. First, free rotation is prevented in the bicyclic complex, and secondly, in this latter case, the metal complex comprises a planar chiral element, while the complex (277) does not.

Alkylation reactions can also be effected by stereocontrolled displacements of

(+)-(277)　　　　　　　　　(−)　　　　　(−)-(278)

(+)-(279) (-)-(280) 80% (-)-(281)

acetates by silylenol ethers and a Lewis acid.[196] This indicates an alternative use of stereodirected oxygen substituents such as that present in the precursor to (273).

The (+) isomer of (279) was converted directly to the acetal (−)-(280) during hydroxylation α to the ketone with PhI(OAc)$_2$. Decomplexation afforded (−)-(281) in 90% yield.[280]

The use of asymmetric inductions in the place of resolution has also been explored in the search of access to optically active arene complexes. Original investigations of asymmetric inductions such as the complexation of (282) gave very low optical yields.[281] The key to successful asymmetric synthesis of chromium complexes was shown by Levine in 1984 to be the removal of free rotation of the chiral auxiliary. Complexation of (283) gave a 50:1 mixture of isomers, and pure (−)-(284) was obtained by crystallisation.[282]

(282) (283) (-)-(284)

Complexation of the optically active amine (−)-(285) produced the chromium complex (−)-(286) in which the arene complex is not a chiral centre. Lithiation, however, occurred selectively at one side of the ring, to produce (−)-(287) as the only product after reaction with trimethylsilyl chloride. When the alkylation was performed with methyl iodide the reaction was less selective, affording a 2:1 mixture of diastereoisomers. This unusual reaction induces asymmetry in the complexed arene, producing an optically active complex with planar chirality.[283]

Substantial transfer of chirality has been achieved by asymmetric inductions at positions remote from the ring. Selectride reduction[284] of (288) gave (289) in 85–90% e.e. The inclusion of an additional methyl group on the arene ring is significant in making the bound ring itself an additional chiral centre. Complete asymmetric induction has been reported in some Grignard additions to (290).[285]

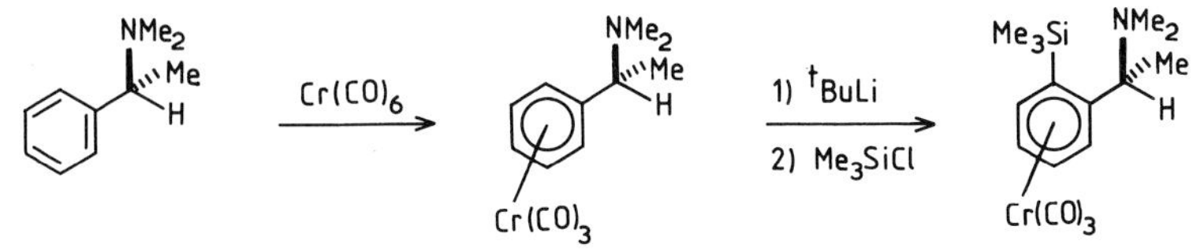

(+)-(285) (+)-(286) (-)-(287)

65% 56%

(288) (289)

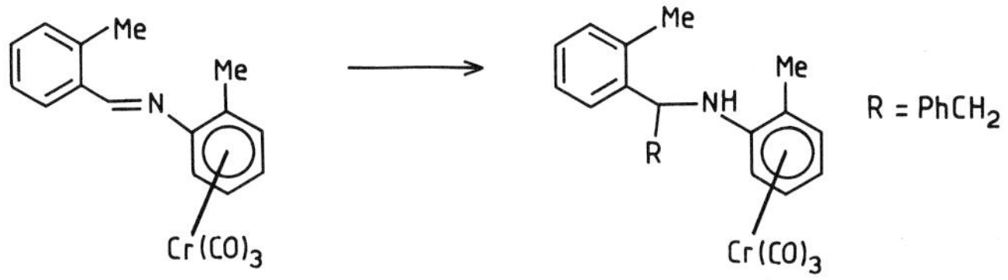

(290)

Iron complexes have been used[286, 287] to control cyclopropanation in enantioselective synthesis of pyrethroid precursors. Two examples show that differences in *cis/trans* stereochemistry of the alkenes obtained by Wittig–Horner reactions, can be relayed to the relative stereochemistry of the products (291) and (292), by control of the face of alkene that is attacked by the diazoalkane.

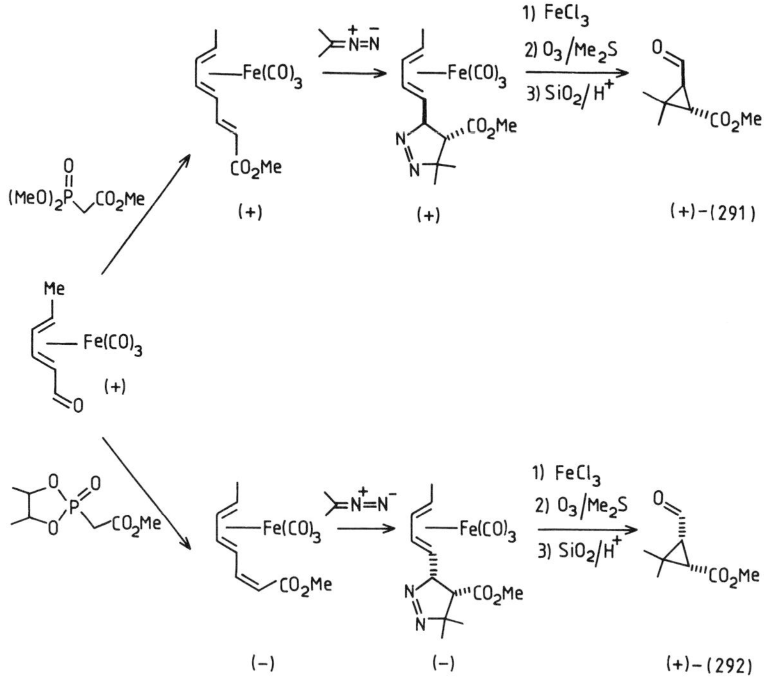

(+)-(291)

(+)-(292)

This type of procedure has recently been used by Grée[288] in an enantioselective synthesis of (−)-verbenalol and (−)-epiverbenalol. The aldehyde (−)-(293)[287] was condensed with Meldrum's acid to produce (−)-(294) which underwent stereocontrolled alkylation with methylmagnesium iodide to produce (−)-(295). Removal of the metal with ferric chloride, and the simplification of the Meldrum's acid moiety to CH_2CO_2H, was followed by a copper catalysed cyclisation of a diazoester (see Chapter 12). This produced a mixture of diastereoisomers which were elaborated into the target molecules. In this synthesis, the organoiron unit relays chirality to the alkene substituent in a very efficient manner.

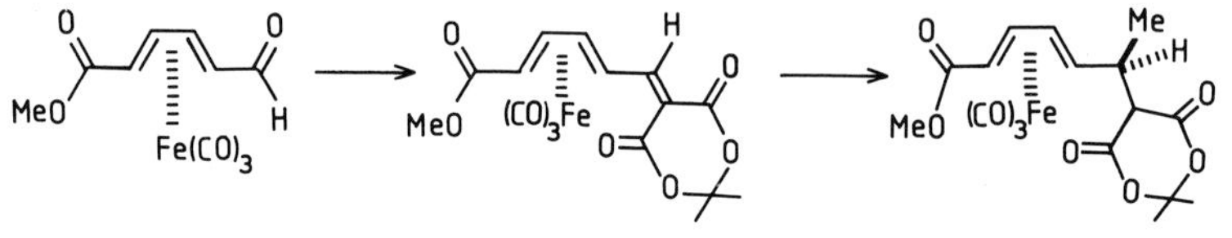

(−)-(293)   (−)-(294)   (−)-(295)

A limit to the efficient control of asymmetric induction is reached at this point along the chain. In hydride addition at a γ-position of a Wittig–Horner product from (±)-(293), stereoselectivity was completely lacking. The products were converted to methyl-3-indolizidines using racemic material. The mixtures of diastereoisomers arising in the reduction step simplify to a mixture of enantiomers when the planar chirality of the metal complex is removed upon decomplexation.[289] Elaboration of the γ position in the side-chain by these methods, however, does not appear an attractive approach for enantiomer synthesis.

The studies described above use optically active metal complexes to give products with high optical purities, but care must be taken when working with acyclic complexes since racemisation can occur by dechellation and bond rotation. This has been observed in photo-induced reactions[290] and in the thermal isomerisation reactions[291] of diene and triene complexes. Similar racemisation can occur in some allyl complexes by π to σ interconversions as shown in Scheme 7.22.[292] This does not always preclude the use of

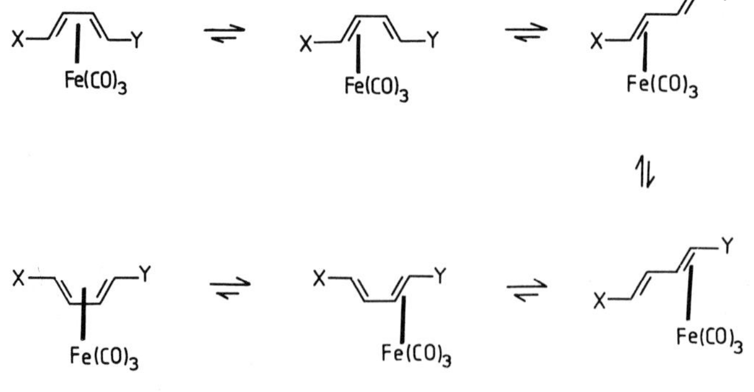

Scheme 7.22

allyl intermediates and many examples with palladium complexes have been described (see Chapter 6).

In some cases, racemisation processes in which the metal exchanges between faces of a cyclic ligand are proposed. These occur in special situations where arene substituents are present. A mechanism has been suggested in which the metal transfers to an arene substituent which can then rotate to return the metal to a different face of the polyene ligand.[293] Another racemisation process occurs by the movement of the position of double bonds. This proceeds via the formation of a symmetrical intermediate such as (296). Here an interesting stereochemical property of planar chirality comes to the fore,

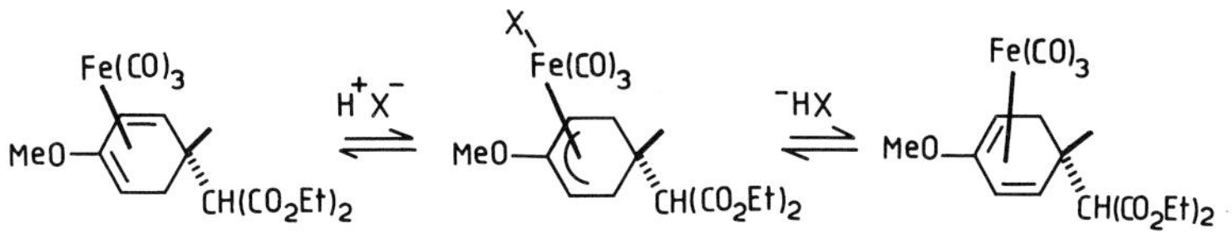

(296)

since in cases where the substitution pattern is correctly chosen, this process forms new regioisomers rather than racemates. It follows from this that, with such systems, when a single regioisomer is obtained from an optically pure starting material, the product will itself be optically pure. This realisation has prompted the resolution of salts such as (297)[294] for the demonstration of this property in enantiomer synthesis.

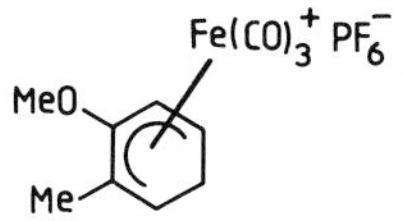

(297)

Investigation of a new type of carbon–carbon bond forming reaction of tricarbonyliron complexes[295] also demonstrated the importance of rearrangement reactions in enatioselective work. Racemisation was encountered in an ene reaction of

(298). The high temperature required for the reaction was sufficient to promote thermal rearrangement[296] of the position of the double bonds in the complex. Introduction of an additional substituent prevented this isomerisation and heating of (+)-(299) produced a single isomer (+)-(300).[297]

The effect of substituents on isomerisation processes of complexes containing planar chirality is well illustrated by the interconversion of (−)-(301) and (+)-(301) which leads to racemisation, and by the formation of (−)-(302).[297]

Chiral ligands also have an important role to play in the control of alkylation of electrophilic π complexes. The influence of chiral ligands on alkylations has been discussed elsewhere in this book (p. 95, pp. 132–4), in the context of the chiral modification catalysts. Two further examples are given here to illustrate asymmetric inductions using stoichiometric complexes to transfer chirality. In the first, activation of the molybdenum allyl complex (303) by NO⁺ gave an electrophilic intermediate that afforded the ketone (304) at near optical purity[298].

Chiral phosphine ligands in the dienyl complex are also capable of a large directing effect forming the diastereoisomer (305) as the major product of cyanide addition.[250]

Nice examples of asymmetric inductions are to be found when chiral nucleophiles are used to alkylate prochiral complexes.[299] Kinetic resolutions have been observed in reactions based on the same principles.[58]

The first seven sections of this chapter, together with Chapters 5 and 6, describe the use of π complexes as electrophiles in organic synthesis. In the range of examples from η^2 to η^6 which have been chosen for detailed examination, synthetically important applications have been developed. From these examples it will have become clear that these many different complexes strongly resemble one another in terms of their reactivity and control properties. Alkylation reactions generally proceed with complete stereocontrol, and the regiodirecting effects of substituents show pronounced similarities between complexes with a range of η values. As methods for the multiple alkylation of electrophilic π complexes are improved, possibilities will arise for the design of syntheses in which control by the transition metal centre plays a central role in the synthetic strategy. In contemplating approaches of this type, a plan must be formulated by matching requirements for regiocontrol with the availability of series of complexes in which the η values will vary as a consequence of each alkylation step.[300] Since activation for bond formation arises chiefly from the metal, substituent groups lose their normal central role in the design of the synthesis. As a consequence, synthetic approaches that are complementary to normal methods are often possible.

7.8 Anionic transition metal π complexes

Anionic π complexes have not received the same degree of attention from synthetic chemists as that enjoyed by their cationic analogues. Examples, however, have already been encountered in the intermediates formed from anion addition to organoiron (Section 7.3) and organochromium (Section 7.4) complexes. The organochromium cases are so far the more extensively applied.

A good many different types of anionic π complexes have been described over the years. In this short section we give a few examples of the types of reactions that are typical of these substances. They cannot, as yet, be recommended as general synthetic procedures for C-alkylation.

Allyl anions have been generated next to an arene Cr(CO)₃ complex.[301] An interesting chromium example is provided by the pentadienyl complex (307) which was formed by isomerisation of (306).[302] A similar isomerisation was used to form (308) which was alkylated by methyl iodide to form the *endo* product (309).[303] Isomerisation of the position of binding of an iron complex during the alkylation of an anionic intermediate has also been described, though in this case the alkylation proceeded in very much lower yield.[304]

The cycloheptatrienyl anion complex of Fe(CO)₃ (310) has been examined as a nucleophilic reagent. Deuteration[305] and reaction with chloroformates[306] both, in this case, give *exo* products. In a further example, (311) was obtained from a reaction that combines the use of organometallic nucleophiles and electrophiles.[305] Similar alkylations that use a σ bound nucleophile with a π bound electrophile are discussed in the next chapter.

An isoprene complex has been deprotonated with LDA to produce a cross-conjugated anion which was used successfully in a number of alkylation reactions. Good yields were obtained using potent electrophiles such as (312) but reaction with aldehydes, for example (313) which was used in a synthesis of a pheromone of a bark beetle, was less satisfactory. The 36% yield of (314) was improved by addition of an equivalent of zinc

bromide before introduction of the aldehyde. Removal of the metal by oxidation produced the pheromone (315).[307] A number of useful alkylations have employed anions stabilised by both an electron-withdrawing group and the metal π complex.[261, 308] An example of the use of this procedure with optically active complexes was discussed in Section 7.7. Deprotonation of dithianes next to Fe(CO)₃ complexes has been used to prepare anions with a variety of geometries. Linear, cross-conjugated, and trimethylene-methane derivatives have been obtained in this way.[309]

Many anionic organometallic complexes react with electrophiles by addition at the metal. Reactions of the cyclopentadienyl complex $CpFe(CO)_2^-$ to form σ bonded alkyl complexes have already been described in Chapter 5. Iron examples of this type[310, 311, 312] and related complexes of molybdenum,[310, 311, 312] vanadium,[313] tungsten,[310, 312] and tetracarbonyl complexes of titanium and zirconium[314] have been examined quite extensively. More recently, attention has turned to arene complexes[315] including dianions,[316] culminating in the characterisation of a simple benzene dianion complex of $Cr(CO)_2$ (316).[317] As might be expected, addition of Ph_3SnCl to (316) produced a metal-bound tin adduct.

(316)   (317)

An Fe(CO)₃ allyl anion (317), prepared by electrochemical reduction,[318] reacted with allyl bromide to produce a dicarbonyl bis-allyl complex.[319] Here, too, reaction has occurred at the metal. Some recent work on bimetallic dianions,[320, 321] effected C-alkylation and has led to a synthesis of O-methyljoubertiamine.[321]

In many cases, organometallic π complexes can serve as nucleophiles without being, themselves, anions. We have seen already in this chapter a good many examples of the protonation of neutral π complexes. This is a reaction that has been examined in considerable detail,[322] and, in cases where agostic products (see Chapter 1) are formed, is far from straightforward. Carbon–carbon bond formation can also be performed by the use of acylation reactions.[323] Care must be taken, however, since the use of strong Lewis acids such as $AlCl_3$ with metal carbonyl complexes can promote carbonyl insertion reactions (see Chapter 9). None the less, acylation has been used successfully on a number of occasions, and has been employed to give access to acyclic compounds[324] of potential importance in routes leading towards HETEs, discussed in Section 7.3, and in an organometallic approach to the synthesis of β-thujaplicin and β-dolabrin.[325] An alternative to Lewis acid catalysis employs a metal-stabilised anion in the reaction with acid chlorides.[326]

There are many questions concerning the stabilisation and control of reactions of anionic π complexes, however, that must be settled before they can be considered for general use as nucleophilic reagents in organic synthesis. In contrast, σ bound nucleophiles are very extensively used and generally follow simple and predictable reaction paths. These reagents are discussed in the next chapter.

8

σ Bound complexes as nucleophiles

Of the many types of transition metal alkyl complexes that are useful as reagents for nucleophilic displacements and additions, alkyl copper species are by far the most familiar to organic chemists. This chemistry has been developed to the point where many applications are becoming routine, and, when conjugate addition and halide coupling reactions are considered for use in synthesis, cuprates or copper catalysed Grignard reagents are usually the first methods that are tried. There are, however, a wide variety of metal complexes that show similar properties, and on occasions, the use of zirconium, nickel, titanium or manganese reagents may prove advantageous. In this chapter the main uses of organocopper reagents are described, and comparisons are made with alternatives that employ other metals.

8.1 Introduction to alkyl copper reagents

Most synthetic chemists will have encountered the use of solvated or complexed alkyl copper (RCu) reagents and the ether soluble cuprates (R_2CuLi) that have found extensive application for the introduction of alkyl, aryl, vinyl, and alkynyl groups in organic synthesis.[1] The main value of these reagents arises from an ability to couple with halides and to react selectively with enones.[1,2] This latter property is a consequence of the relatively low reactivity of RCu and R_2CuLi species in addition to ketone carbonyl groups. Consequently, in addition to enones, only 1,4- or 'conjugate' addition is observed. These typical types of reactions are discussed in detail in Sections 8.2 and 8.3.

$$R^1\text{—}R \xleftarrow{\quad R^1\text{—}X \quad} R_2CuLi \xrightarrow{\qquad\qquad}$$

Alkyl copper (1) and cuprate reagents (2) are simply prepared:

$$RLi \ + \ CuX \longrightarrow RCu \qquad (1)$$

$$RCu \ + \ RLi \longrightarrow R_2CuLi \qquad (2)$$

In their use in synthesis, however, only one of the alkyl groups in the cuprate, R_2CuLi, is normally utilised. Where the R group is of a simple, accessible kind, this is

234

not a serious disadvantage. When valuable alkyl groups are used, however, a second, non-transferable ligand must be included in the cuprate reagent. It has been found that the reactivity of groups attached to copper increases in the order alkynyl < aryl < alkenyl < alkyl. In a mixed cuprate, $R'(RC{\equiv}C)CuLi$ where R' is an alkyl group, the R' group therefore reacts selectively, an observation which has been developed[3] as a means to make more effective use of cuprates containing less readily available R' groups.

$$C_3H_7C{\equiv}CCu \ + \ R'Li \ \xrightarrow{\text{HMPA}} \ [C_3H_7C{\equiv}CCuR'] \ Li$$

In this procedure the copper ethynyl derivative is solubilised as the hexamethyl-phosphoramide complex. There are now many variants of this strategy, and a selection is discussed below.

As an example of the use of this mixed type of reagent in addition to an enone we quote[3] from the original work of Corey and Beames in which cyclohexenone (3) was converted into (4) in good yield. Selective alkyl group transfer has also been achieved

$$[C_3H_7C{\equiv}CCuR']Li \ \ + \ \ \text{(3)}$$

(3)

$$\longrightarrow \text{(4)}$$

$R' = {}^nBu, {}^tBu$

$-CH{=}CH_2$

(4)

by use of reagents in which a heteroatom ligand is attached to the copper. Effective procedures that employ $PhS(R)CuLi$[4,5] and $MeS(R)CuLi$,[6] and the use of a nitrile in $R_2Cu(CN)Li_2$[7] in the place of an acetylide, are finding widespread use. The dimethyl sulphoxide anion has also been used as a non-transferable ligand.[8] A selection of these procedures has been evaluated under comparable reaction conditions. Two salts, CuCN and CuBr.SMe$_2$, were found to be superior to the simple iodide, bromide or chloride salts for the formation of synthetically useful mixed cuprates. In some cases, considerable differences in ratios of 1,4- and 1,2-addition products were observed.[9]

Despite their relative lack of reactivity, copper derivatives can themselves be used to form heteroatom to carbon bonds; an example is found in the synthesis[10] of the penem skeleton from the amide (5). The Z-isomer of the E, Z-mixture of bromo derivatives (5), prepared as shown, was treated with Li^iPr_2N followed by $CuI(PBu_3)$ to yield (6).

(5)

(6)

R = p–nitrobenzyl

In a more recent example, the thiaarachidonic acid derivative (9), a potent irreversible inhibitor of leukotriene biosynthesis, was prepared[11] using a similar coupling of (7) with (8).

Even now, over fifty years since the first studies by Gilman and Straley[12] and following much subsequent development,[13] the mechanistic details of organocopper reactions remain unclear. Both direct nucleophile addition, and pathways that involve an initial electron transfer step, must be considered. Further complications arise when the exact nature of the reactive species is examined. Despite their selectivity, even simple cuprates are now known to exist in solution in several structural forms. 'Dimethylcuprate', for example, in solution in THF, has proved[14] to be in equilibrium with a higher order cuprate species:

$$2Me_2CuLi \underset{}{\overset{THF}{\rightleftharpoons}} MeLi + Me_3Cu_2Li$$

In this case the disproportionation can be prevented by addition of lithium iodide or use of ether as the solvent.

Higher order cuprates[15] can be prepared deliberately and may offer advantages in some cases. In particular, the presence of an alkyllithium species in solutions of organocuprates can be disadvantageous when selective 1,4-addition is required. This problem can be avoided by employing[16] higher order cuprates. The use of $R_2Cu(CN)Li_2$ in a vinyl addition to cyclopentenone (10), for example, is typical of methods now available.

A stable organocopper species (11) has been used as a precursor to higher order mixed organocuprates. The thienyl reagent can be stored in solution in THF for extended periods and can be converted, by addition of an organolithium reagent, into reactive higher order cuprate reagents which have been successfully used in conjugate addition and epoxide opening reactions. Coupling with the alkyl iodide (12) was also successful, proceeding almost quantitatively.[17]

Higher order mixed cuprates of the types discussed above exhibit many of the reactions expected of conventional cuprate reagents and are likely to find greater use in synthesis in situations where selectivity or controlled reactivity is of particular importance.[18] Clearly, in solution, there are a wide variety of stoichiometries available for cuprate reagents. Recent studies along these lines have sought different reaction characteristics through the use of mixed sodium/lithium counterions or copper/zinc cuprate analogues.[19]

Interest in the structure of cuprate reagents has also prompted extensive and difficult X-ray crystallographic examinations. Various cuprate species have been shown[20] to have ionic structures in the solid state, but in other cases[21] there is direct interaction with lithium counterions. This structural variety, both in the solid state and in solution, is at least in part responsible for the rich and varied chemistry of cuprate reagents and the relatively slight degree of fundamental understanding of a synthetic reaction of such great power. Of the many reagents available, it is currently the simple dialkyl cuprates and copper catalysed Grignard reagents which are the most popular in routine applications.

8.2 Cuprate additions to enones

Dialkylcuprates and related reagents have proved especially valuable for β-alkylation of conjugated enones. A simple example was noted for (3) above. The transformation of (13) into the isomers (14) 24 %, (15) 48 %, (16) 11 % and (17) 17 %,[22] illustrates the lack of stereocontrol operating in an addition process such as this where the directing chiral centre is too remote from the site of reaction.

Conjugate addition reactions have been used in a great many natural product syntheses and a few representative examples serve to indicate the generality of the approach.

In an interesting case, Me$_2$CuLi was used for β-alkylation of an α-chloro-α,β-enone system. This reaction was used in the conversion of (18) into (19) during a synthesis[23] of ($\pm$)-muscone (20) by Stork and Macdonald.

A simpler example is found in the methylation of (21), an initial step in a synthesis of the boll weevil sex attractant.[24]

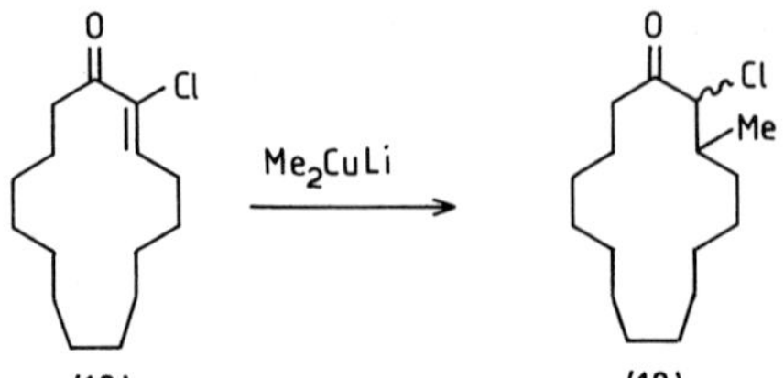

(18)   (19)

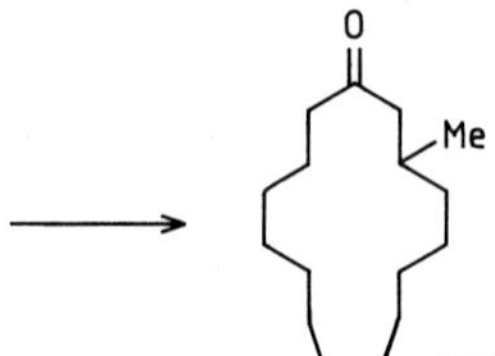

(20)

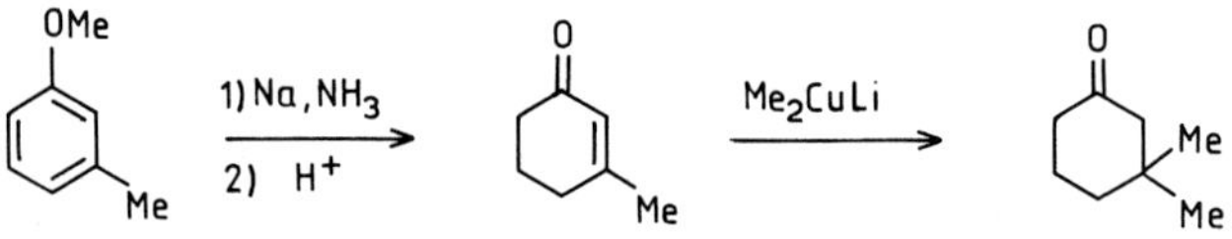

(21)

Chloroenones can undergo two cuprate additions. Wender has demonstrated a spiroannelation of chloroenones employing a new type of reagent, organobis-(heterocuprates).[25] The use of heterocuprate methodology enables two copper species to be included in the same reagent, as in examples (22) and (23).

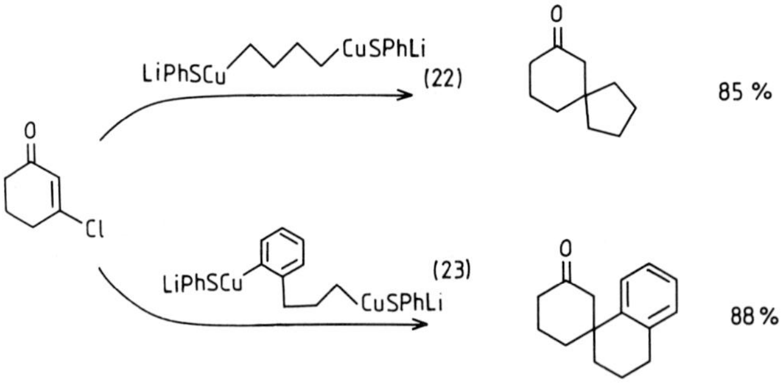

85%

88%

A synthesis[26] of (±)-β-vetivone (25), and its epimer (26), made use of Me$_2$CuLi to alkylate an α-formyl enone system (24). This produced (25) and (26) in a ratio of 2:1

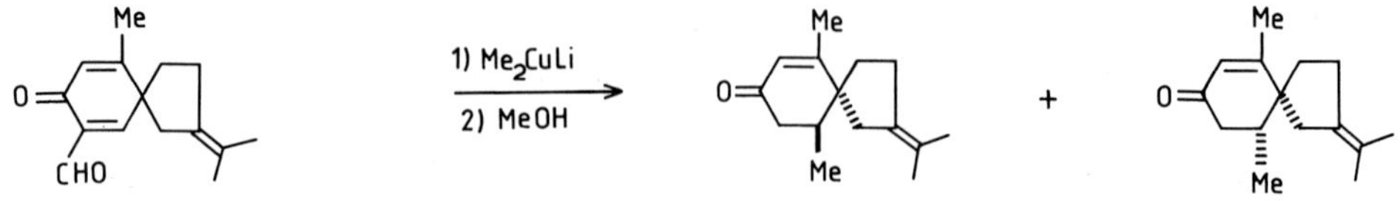

(24)   (25)   (26)

after removal of the formyl group. Similarly, a synthesis[27] of ($\pm$)-nootkatone made use of alkylation of an α-carbomethoxy enone system in the formation of (27).

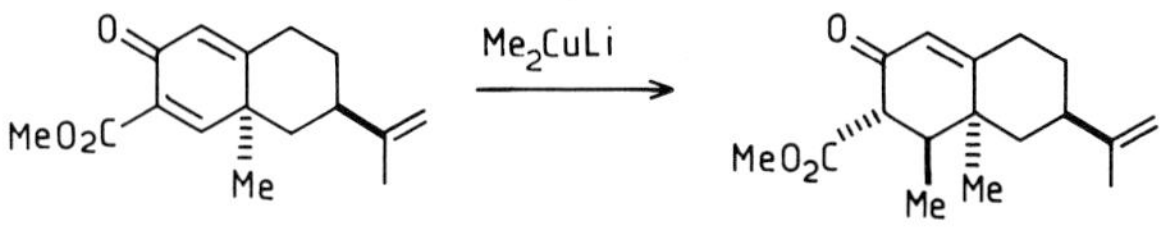

(27)

In the course of a synthesis[28] of hinesol, a similar use was made of addition of Me$_2$CuLi. In this sequence, addition to (28) occurred at the less substituted enone terminus, with preferential addition to the less hindered side of the molecule, (29) and (30) being formed in the ratio 1:4.

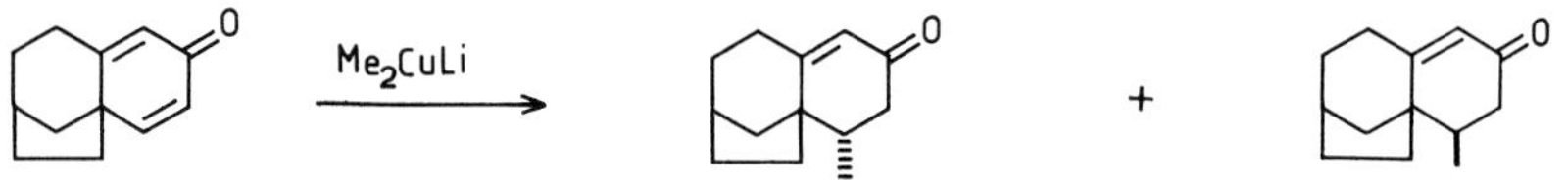

(28)   (29)   +   (30)

A class of cuprate reagents which have advantages, notably in alkylations with secondary or tertiary alkyl groups, is exemplified by the thiophenylcuprate (31), prepared as indicated.

$$PhSH + \tfrac{1}{2}Cu_2O \longrightarrow PhSCu \xrightarrow{RLi} [PhSCuR]Li \quad (31)$$

Applications of this type of reagent have been discussed by Posner and the example[29] illustrates the use of a cuprate reagent in reaction with an oxazolinone (32) to afford (33).

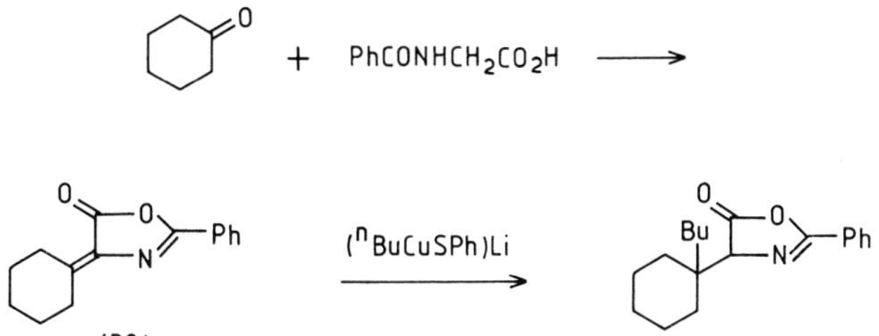

$$+ \quad PhCONHCH_2CO_2H \longrightarrow$$

(32)   (^{n}BuCuSPh)Li   (33)

In other cases, Lewis acids have been employed in combination with organocopper reagents to increase reactivity with enones. Examples of the types of reagents suited to this approach include MeCu.BF$_3$, Me$_2$CuLi.BF$_3$ and higher order cuprates of the types based on cyano or thienyl groups.[30]

Attachment of a chiral auxiliary to the enone can be used to effect an asymmetric induction during the addition reaction, as in the example from the work of Oppolzer *et al.*[31] in which alkylcopper reagents are used (Scheme 8·1).

Asymmetric inductions have also been examined using dimethyl cuprate reagents, the corresponding magnesium bromide reagents, and mixed and higher order cuprates. In

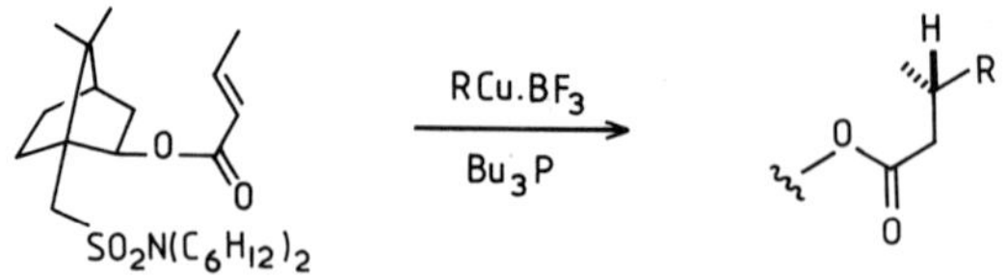

Scheme 8.1

these examples, non-transferable chiral ligands derived from (S)-proline were used. Optical yields varied considerably depending on the conditions employed, the best examples giving enantiomeric excesses in the range 75–88 %.[32]

Organocopper reagents can be formed from elaborate organic halides, often in the presence of other functional groups. The highly functionalised copper reagent (34) was used in an initial step in a synthesis of 22-hydroxy steroids:[33]

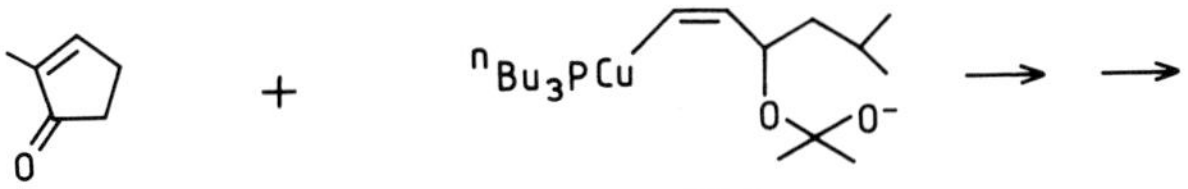

(34)

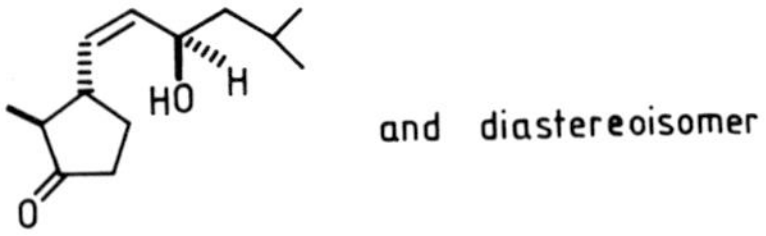

and diastereoisomer

Copper catalysed Grignard reagents are also valuable for conjugate addition reactions. The sequence (35) to (36), in the course of synthesis of isocomene,[34] employed a Grignard reagent in combination with $CuBr . Me_2S$ in THF as a means to effect the β-alkylation of an enone.

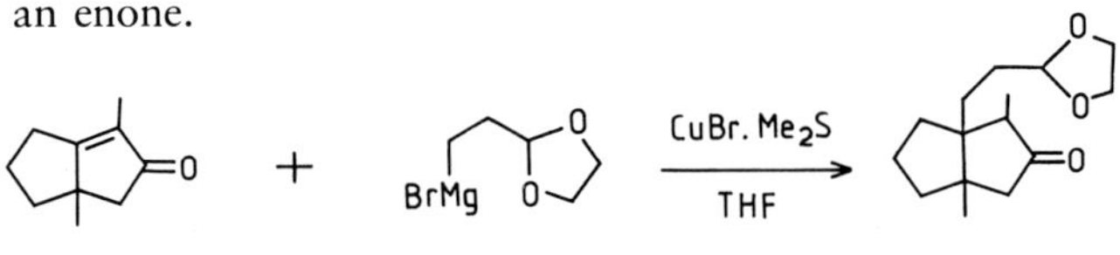

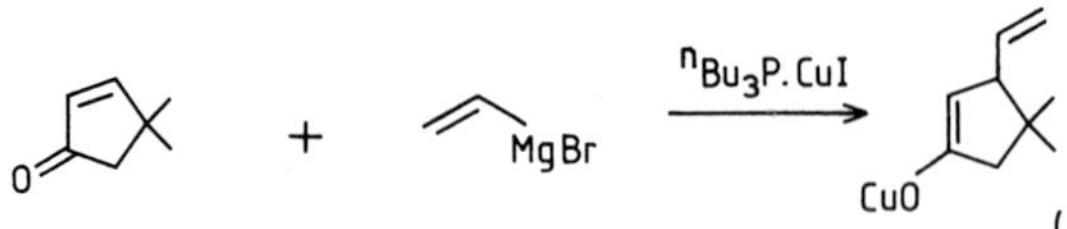

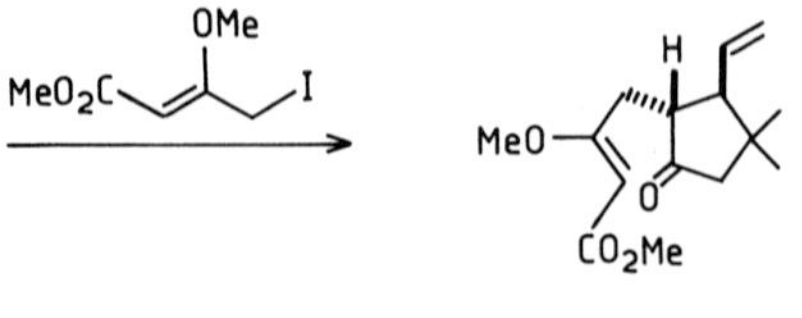

An interesting double alkylation that makes use of the intermediate copper enolate (37) is found as a step in the synthesis of quadrone.[35] Many examples of this type of process are described in Section 8.3. Cuprate reagents and copper catalysed Grignard reactions are both widely used in this conjugate addition – enolate trapping strategy, though, of course, simple organic nucleophiles such as ketone enolates can also be employed.[2]

8.3 Conjugate addition/enolate trapping

In a cuprate, RR'CuLi, the alkyl groups are part of an anionic metal complex. An alkyl group is therefore transferred effectively as an alkyl anion. In addition to an enone system this leads to the formation of a copper enolate intermediate which may be used in further reactions.[2] The cyclisations of (38) and (39) illustrate the reaction of the enolate in initiating an internal aldol reaction.[36]

A further example of enolate interception is found in the entry into the synthesis of substances in the quilenin series using (40).[36]

An instance[37] in which the copper enolate intermediate is used to displace a halogen is found in a synthesis of valerane (41). Copper enolate intermediates, formed following the conjugate addition step, have been shown to react successfully with a variety of alkyl halides,[38] acyl halides[39] and alkyl chloroformates[40] as well as aldehydes[41] and ketones.

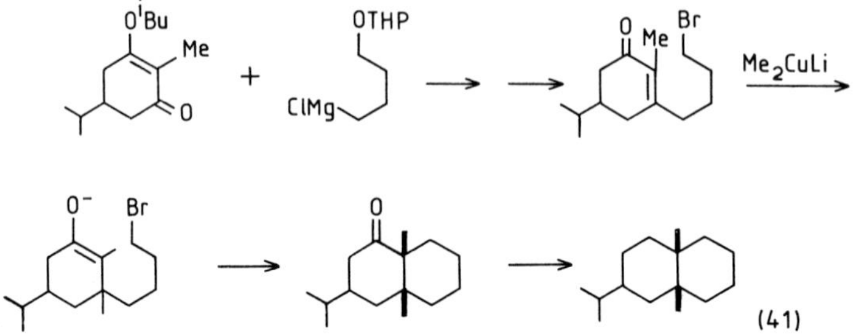

(41)

A considerable degree of stereocontrol can be obtained, and a *trans* relationship between the two new C–C bonds is the typical outcome. This result can be obtained even in cases such as (42) where several other chiral centres are present in the molecule.[43]

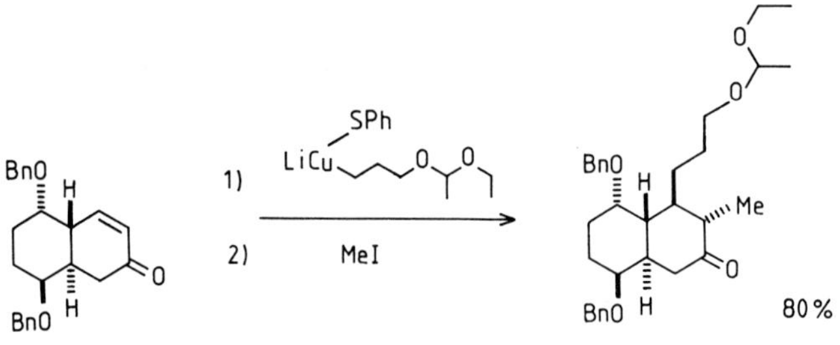

(42)

In some instances it can be advantageous to intercept the enolate initially as a silylenol ether, as in the example (43) which is drawn from a polyquinane synthesis.[44] This device can provide additional opportunity for control. In a remarkable example,[45]

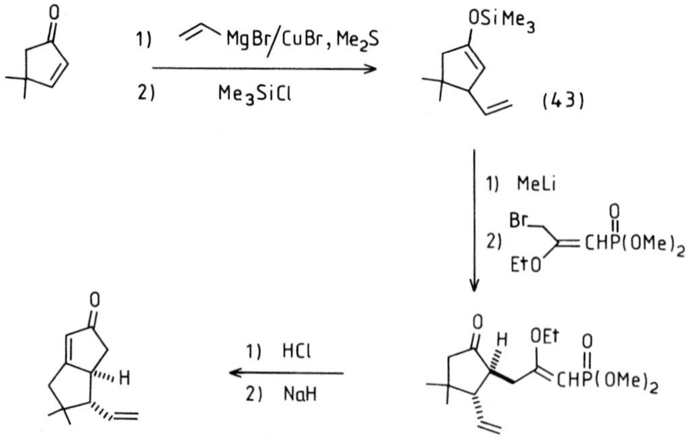

(43)

37% overall

it is possible to introduce a methylene group selectively at either side of the ketone group in (44).

The combination of dimethylaminopyridine and trimethylchlorosilane has been used to promote 1,4-addition in circumstances where otherwise this would not be possible.[46] The alkylations of (45) and (46) are typical.

Enolate trapping in copper catalysed Grignard additions has been promoted in a similar way by the use of hexamethylphosphoramide.[47] Tetramethylenediamine has also recently been used to bring about conjugate addition/enolate trapping with both alkyl copper and copper catalysed Grignard additions.[48]

The high reactivity of acid chlorides allows them to be alkylated selectively in the presence of other electrophilic centres,[49] a convenient property that has been put to good use in a synthesis of pinguisone (47). The formation of the furan ring is straightforward since the α-chloroketone is intercepted by the carbonyl group of the original enone which has participated in the formation of three key bonds in a single step.

In alkylation it may be important that the rate of entry of the alkyl group is fast relative to any proton transfer process that could lead to an isomeric enolate.[38] Thus

(48) is alkylated in high yield, but in a similar sequence, (49) gives 23 % of (50) and 47 % of (51). In a similar way, (52) gives (53) as major product via isomerisation of enolate (54) to form (55).

Enolate equilibration of this type has been a notable problem in attempts to use conjugate addition in the synthesis of prostaglandins that bear an 11-alkoxy group. The very direct approach[50] indicated in Scheme 8.2 proved initially to be difficult to achieve,

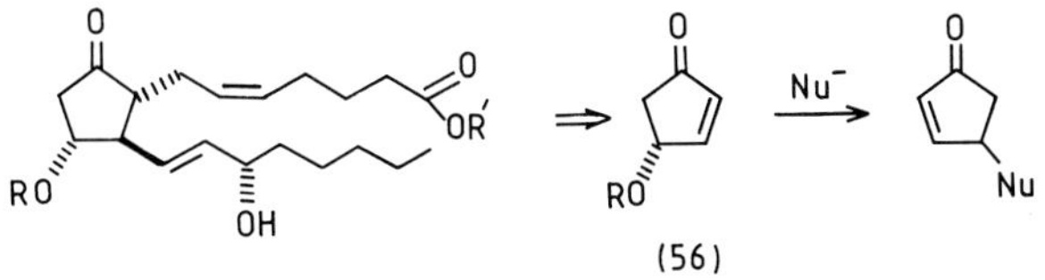

Scheme 8.2

since elimination of the OR group from (56) was rapid, following enolate equilibration. The use of highly reactive electrophiles to trap the enolate can overcome this problem. An example of the reaction with an acyl halide is found in the prostaglandin synthesis of Tanaka *et al.*[39] in which the disubstituted cyclopentanone enonate (57) was trapped to form (58).

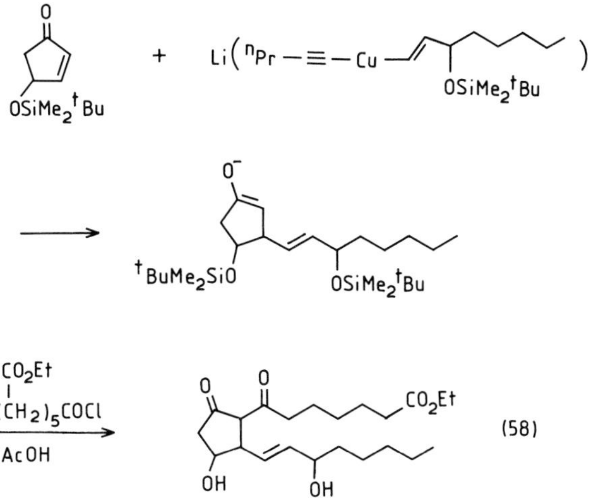

Recently the electrophile (59) was used in the formation of the allyl carbonate intermediate (60). This was easily decarboxylated using a palladium catalyst (see

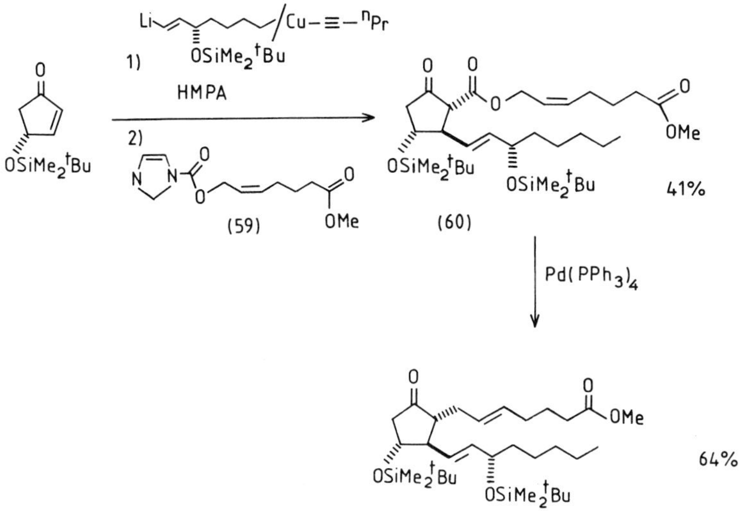

Chapter 6) to complete the carbon chain in a reaction that also converted the Z alkene stereochemistry of (59) into an E linkage in the product.[51]

A great attraction of this approach to prostaglandins lies in the ability of the OR group to control the entry of the cuprate completely. Similar stereocontrol has been achieved in the alkylation of the bicyclic enone (61) in a route to prostacyclin analogues.[52]

(61)

An alkyl cuprate reagent may also be used to alkylate the $-C\equiv CCO_2Et$ grouping by β-addition, as in the formation of (62) in the synthesis of the codling moth sex pheromone.[53] The steric course of addition is to be noted.[54]

(62)

Further examples of additions to alkynes are discussed in the next section.

8.4 Additions to alkynes

The reaction of alkyl cuprates with alkynes provides[55] a simple route to vinylcopper species that are of considerable value in synthesis. In many reactions, both vinyl groups may be put to use. Detailed work, mainly by the Normant group, over the last few years has led to a high degree of empirical understanding of these reactions.[56]

When substituted alkynes are employed, lithium based cuprate reagents gave only moderate yields, and, although there are examples[57] of successful additions, the use of magnesium salts was more satisfactory. Even this reaction is slow and requires the use of a large excess of dimethylsulphide. However, since the vinyl species that is produced is the only organometallic species present in solution, subsequent utilisation is straightforward, a crucial advantage when this method is compared with others[58] that require an excess of a Grignard reagent. A high degree of control is apparent in these reactions and the branched chain, *cis*-addition product obtained[59] from (63) represents a typical outcome.

$$C_6H_{13}-C \equiv C-H \xrightarrow{MeMgBr/ CuBr/ Me_2S}$$

(63)

C₆H₁₃ H ... Me Cu.MgBr₂.Me₂S

90%

When appropriate functional groups are present in substituents on the alkyne, possibilities of chelation arise that complicate regiocontrol. While a branched isomer is the major product from the ether (64), the acetal (65) shows good selectivity for the linear product.[60]

OMe OMe

BuCu.MgBr₂
67%

Cu Bu   +   Bu Cu

(64)

91 : 9

BuCu . MgBr₂
64%

Cu Bu   +   Bu Cu

(65)

13 : 87

The use[60] of a coordinating solvent such as THF reduces the influence of chelating substituents and tends to restore the normal (branched product) mode of addition. Bulky groups, as in the trimethylsilyl ether (66) used for the synthesis[56] of myrcen-8-ol, also assist the formation of branched products.

OSiMe₃

1) MgBr₂.Cu—
2) H⁺/H₂O

OH

81%

(66)

Propargyl ethers can give predominantly linear products. For this purpose THF must be avoided since chelation control is required. The inclusion of two equivalents of dimethyl sulphide in the example shown, led to the formation of almost pure linear product (67) (97:3) in 80% yield.[56] Similar high selectivity was obtained with thiol modified cuprates.[61]

OMe

1) BuCu . MgBr₂.2Me₂S, ether
2) H⁺/ H₂O

OMe

Bu

(67)

Assistance gained from chelation also made possible addition to the disubstituted alkyne (68) in a synthesis[62] of the geranial derivative (69). For this purpose, lithium based reagents must be used. Once again, THF is not a suitable solvent.

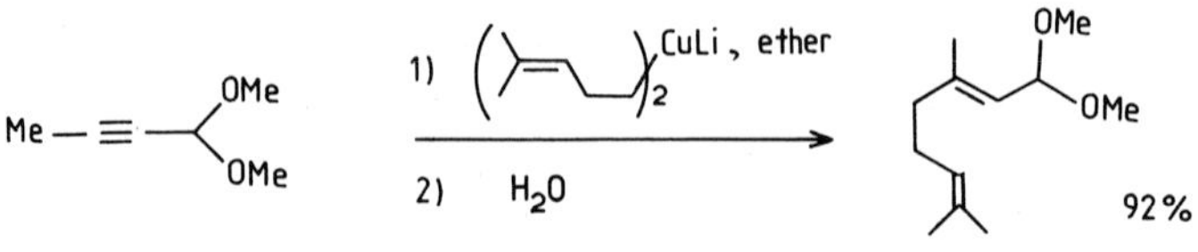

(68)   (69)

Formation of allenes can occur when propargyl acetates are employed.[63] Similar displacements are also possible with copper catalysed Grignard reagents. The stereochemistry of processes of this type depends on the nature of the halogen and of the Grignard derivative. In all these displacement reactions, intermediate vinylcopper species are formed. These subsequently undergo either *syn* or *anti* β-eliminations to produce the allene.[64]

The formation of the vinyl allenes (70) and (71) illustrates the way that vinyl cuprates

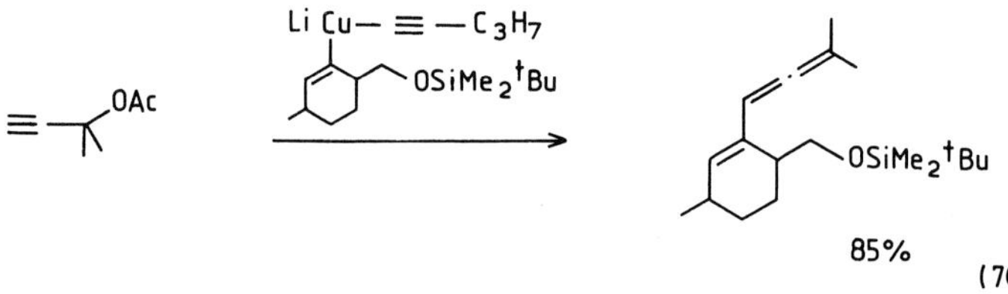

85%
(70)

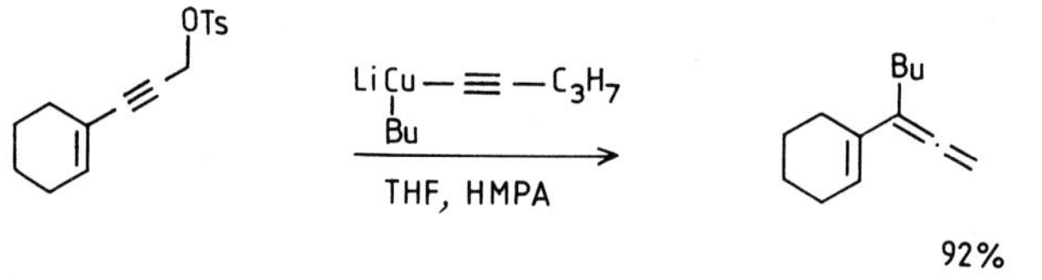

92%
(71)

can themselves be used in such addition reactions. The process fails, however, for primary propargyl derivatives, and in such cases tosylate leaving groups are required.[65]

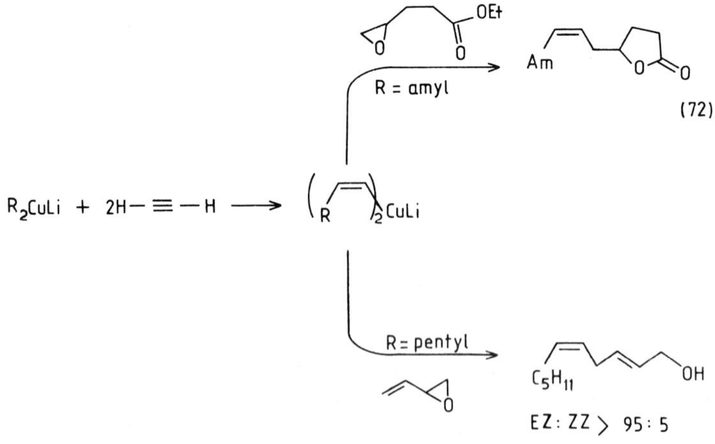

(72)

$R_2CuLi + 2H—\equiv—H \longrightarrow \left(\underset{R}{\diagup}\hspace{-2mm}=\hspace{-2mm}\diagdown \right)_2 CuLi$

(73)

EZ : ZZ > 95 : 5

Once formed, vinyl cuprates derived from alkynes can be utilised[66] in the normal way, the whole reaction sequence constituting a valuable synthetic process in which two C–C bonds are formed in sequence. Epoxide openings (see Section 8.6), for example, have been used in syntheses of the naturally occurring lactone (72),[66] a scent of the black-tailed deer, and of the sex pheromone of the potato tuberworm moth (73).[67]

As indicated above, further reaction between vinyl cuprates and alkynes is possible. Control of this process leads to an elegant route[68] to *cis,cis*-dienes that can be concluded by a normal cuprate addition reaction to an α,β-unsaturated carbonyl compound or an alkyl halide. The method has provided a very direct route to the Navel orangeworm pheromone (74).

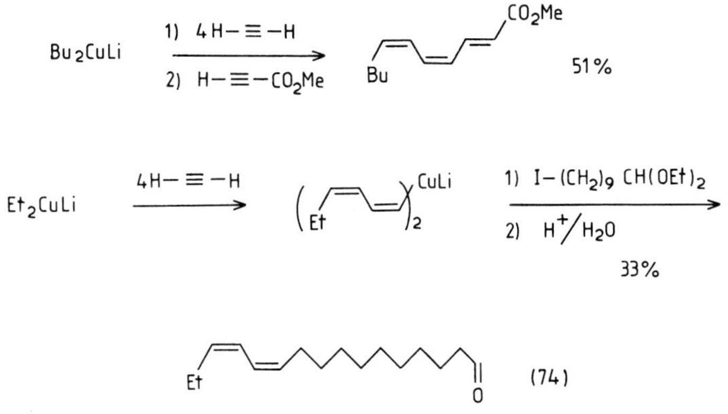

Highly substituted 1,3-dienes can be obtained by coupling the vinyl product with a vinyl halide (Scheme 8.3). The reaction requires an intriguing series of metal exchanges,[69] culminating in a palladium catalysed coupling reaction of a type discussed in detail in Chapter 10. Metal exchanges of this type, which are valuable in the formation of specific reagents, indicate the degree of sophistication now achieved in these methods.

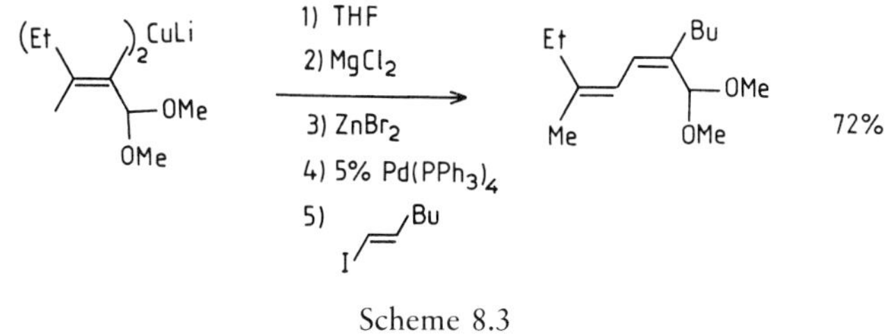

Scheme 8.3

8.5 Displacement of halides

The examples of the coupling of an organic halide with a vinyl cuprate that arose in Section 8.4 reflect a general process that constitutes a major application of cuprate chemistry (related coupling reactions using other metals are discussed in Chapter 10). The reaction is by no means restricted to vinyl cuprates, but in cases where they are used, direct preparation from a vinyllithium reagent offers an alternative to the

Normant route. The reaction appears to proceed with retention of configuration at the vinyl centre, and must be performed at low temperature. Such copper derivatives are stable under these conditions (e.g. at $-30\,^\circ$C), but at ambient temperature may dimerise,[70] as in the formation of (75).

An example of the direct use of a vinyl cuprate is encountered in the synthesis[71] by Büchi and Carlson of fulvoplumierin (78). The chlorofulven derivative (76) was treated with lithium di(*trans*-1-propenyl)cuprate (77) which was itself obtained from the propenyl halide.[72] Retention of stereochemistry in the propenyl moiety, and in the displacement reactions, is to be noted, as well as the lack of attack on the ester and lactone substituents.

Replacement of vinyl halogen by methyl is found in the synthesis[73] of sirenin at the step leading to (79). Again there is stereospecific displacement of halogen, and the adjacent alcohol function does not interfere.

An example in which a dihalovinyl system was the substrate in the reaction is provided[74] by the synthesis of selinadiene (80) by Posner and co-workers.

The halogen in structures containing 1,1-dibromocyclopropanes (e.g. 81) may also be displaced by Me₂CuLi. This is seen in the synthesis of ($\pm$)-globulol[75] (82). The reaction time with Me₂CuLi was, however, prolonged. The replacement of the halogens in a *gem* dihalocyclopropane is sequential. This permits a second step using a different alkylating agent, R′X, to effect the conversion of (83) into (84), as exemplified in the reaction sequence leading to (85).

This procedure has been applied by Hiyama *et al.*[76] to achieve a key step in synthetic work in the sesquicarene–sirenin series, in which (86), by sequential reaction with lithium bis(4-methylpent-3-enyl)copper and methyl iodide, gave the useful intermediate (87).

The lack of reactivity of copper alkyls towards carbonyl centres makes it possible to use cuprates containing such functional groups. The example[77] (88) illustrates the use

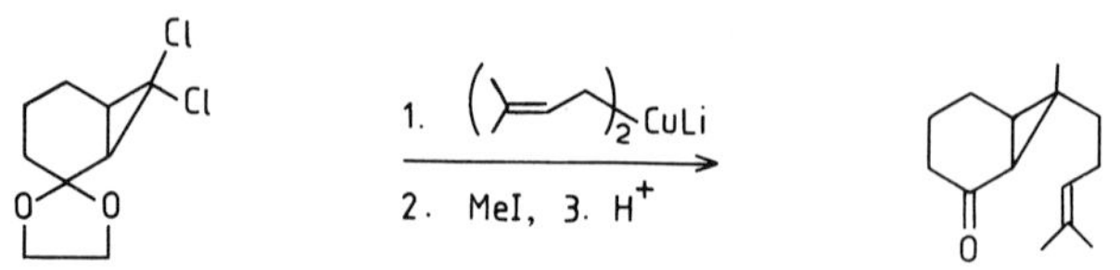

(86)   (87)

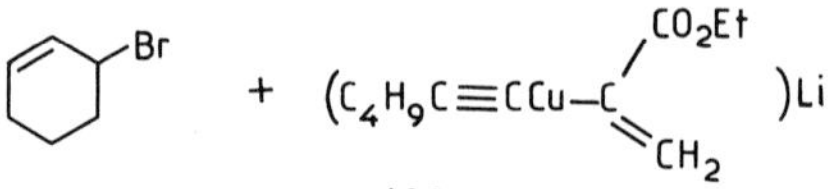

(88)

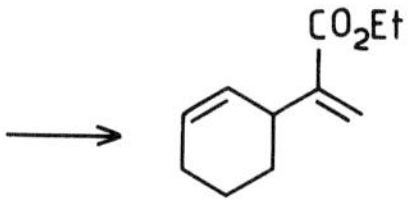

of such a derivative containing a carboethoxy group. An alkynyl ligand is included as a second, non-transferred group.

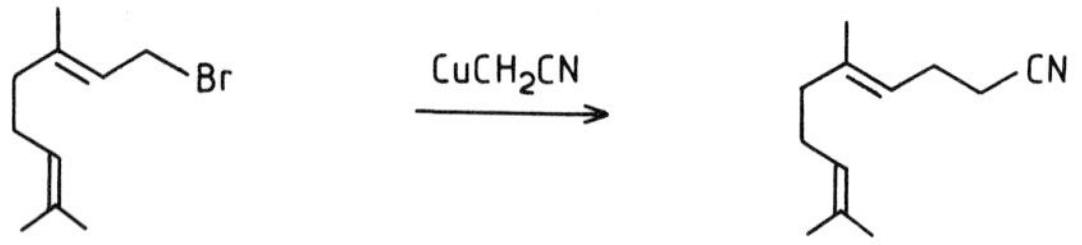

(89)   (90)

A related example[78] uses $CuCH_2CN$ as reagent in the transformation of (89) into (90). The copper reagent, $CuCH_2CN$, which is obtained by lithiation of CH_3CN to form $LiCH_2CN$ followed by transmetallation with CuI in THF, appears to react well only with allylic halides. Aryl halides also undergo alkylation with suitable organocopper reagents, even those bearing alkynyl groups. Examples[79] include the formation of (91) where R may be $CH(OEt)_2$, CH_2OTHP, or COPh.

$$PhI + CuC\equiv C-R \longrightarrow Ph-C\equiv C-R \qquad (91)$$

Use of a copper aryl may be illustrated by the synthesis of freelingyne[80] starting from 3-iodo furan (92) and the iodo alkyne (93). The product (94) could be elaborated to give isomers of freelingyne (95) and (96).

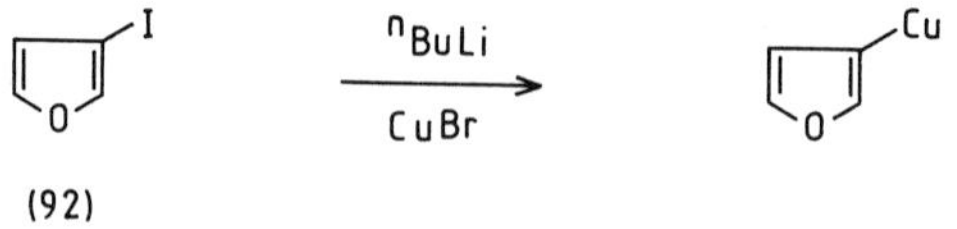

(92)

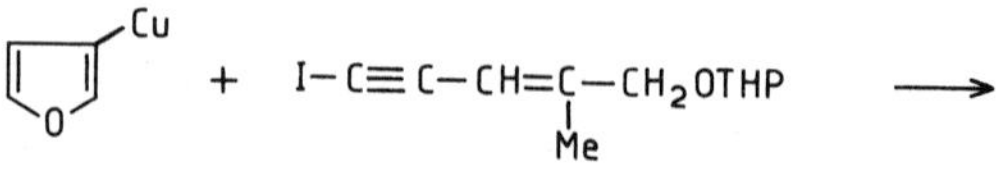

(93)

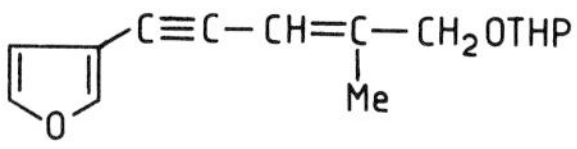

(94)

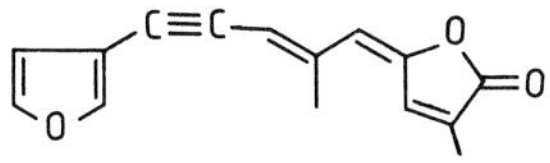

(95)

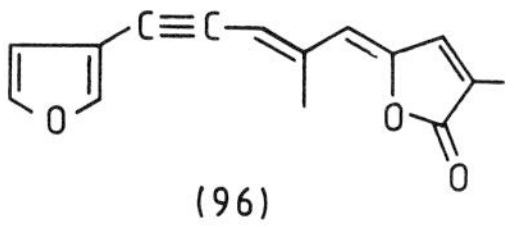

(96)

An alternative vinylalkyne formation is used in a synthesis[81] of 5,15-diHETE (98) in which a palladium catalyst (see Chapter 10) couples a Cu(I) derivative of the alkyne (97) with an appropriate vinyl bromide.

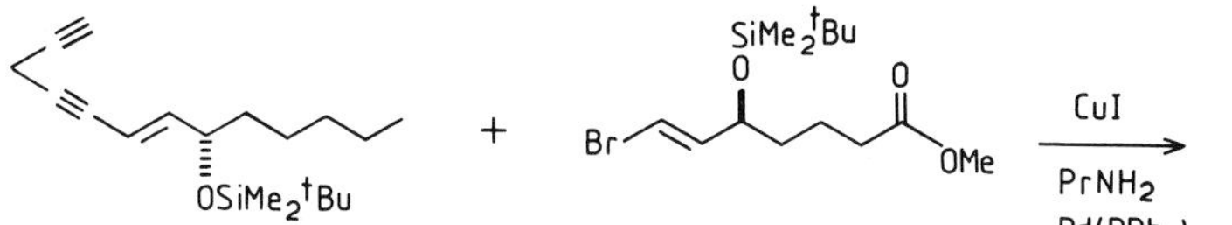

(97)

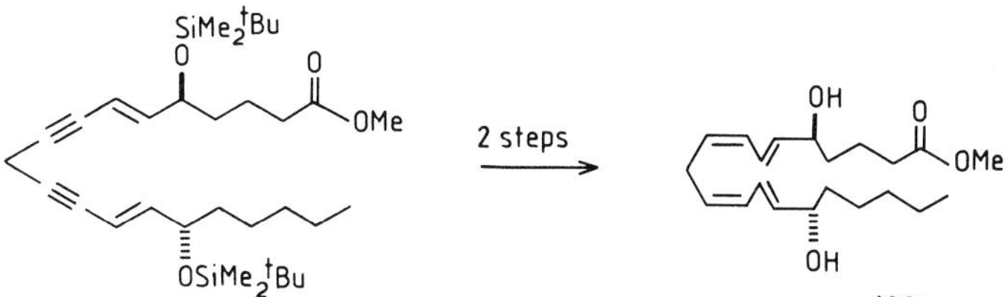

(98)

Reactions with bromoallenes have also been examined. Substituted alkynes are produced in a reaction that showed very high *anti* selectivity. When optically active allenes are used, an efficient transfer of chirality was obtained.[82]

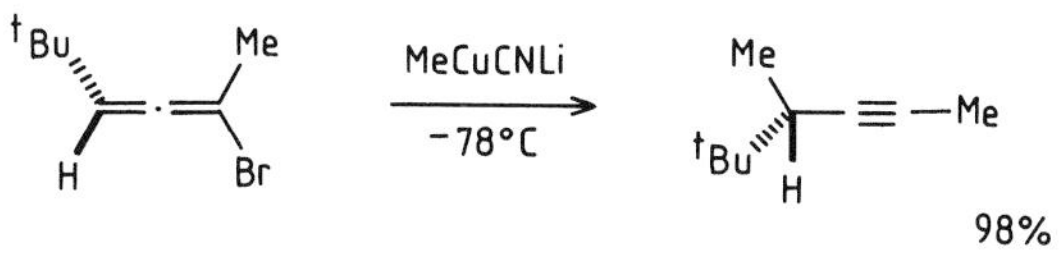

98%

8.6 Opening of epoxides, aziridines and lactones

Cuprates are effective reagents for the nucleophilic opening of epoxides; some examples of applications of this type of reaction have already been described (pp. 33–4, 69). A good indication of relative reactivities can be gained from some studies[83] in prostaglandin synthesis in which, despite the low reactivity of cuprates towards ketones, protection of the ketone (99) was needed to obtain reaction at the epoxide. Regiocontrol of the ring opening is also an important consideration. In this case, the product was obtained as a 4:1 mixture of regioisomers.

Vinyl epoxides can be opened by direct addition. The transformation of (100) into (101), for example, was used as a step in the synthesis of Prelog–Djerassi lactone.[84]

Vinyl epoxide systems may also undergo 1,4-addition.[85] Reaction of (102) with Me$_2$CuLi in ether gave (103) and (104) in 94% yield by conjugate addition. Only 6% of the product (105) was formed by direct reaction with the epoxide. Products (103) and (104) were obtained in a ratio of 4:1.

Competition between 1,2- and 1,4-addition to cyclopentene epoxides has been the subject of a systematic investigation. Alkyl cyanocuprates were found to react with complete regio- and stereocontrol, providing *trans*-4-alkylcyclopentenols in excellent yields. In contrast, vinyl, allyl, and phenylcyanocuprates produced mixtures of 1,2- and 1,4-adducts.[86]

Alcohol substituents adjacent to vinyl epoxides can be used to promote stereocontrolled addition relative to the orientation of the epoxide. Good yield of the *anti* addition product was obtained with (106), the E-isomer of the substrate. Selectivity was reversed with the Z-isomer (107).[87]

Epoxides derived from vinyl silanes had originally been found to be unreactive towards opening by cuprate reagents. More recently, the use[88] of Lewis acids in combination with cuprates has increased reactivity sufficiently to overcome this difficulty. Addition to the epoxide occurs at the site substituted by the silyl group, allowing subsequent elimination to produce an alkene. When this technique was combined with the stereocontrolled formation of vinyl cuprates from alkynes (see Section 8.4), an efficient stereocontrolled synthesis of E,Z-dienes such as (108) became available.[89]

The growing importance of the elaboration of epoxides derived from allyl alcohols has been discussed in Chapter 3. Two recent examples employing higher order cyanocuprates are to be found in the work of Chong[90] and Kurth[91] and their co-workers. With epoxy alcohols themselves, moderate selectivity for alkylation adjacent to the OH group was observed under most conditions, although this was reversed by

(108)

addition of a Lewis acid at low temperature. The higher order cuprate (109) showed better regiocontrol than the simple dimethylcuprate.[90]

	67	:	33
+ DMEU	84	:	16
+ BF$_3$ · OEt$_2$	21	:	79

Use of a mesylate derivative allowed a sequence of two alkylations to be performed. The product contained an alcohol substituent flanked by the two alkyl groups introduced via the cuprate additions. In this way, (110) has been converted into the secondary alcohol (111).[91]

Other strained rings can also afford alkylation products. Normally the opening of aziridines by cuprate reagents proceeds in low yield, but a considerable improvement can be obtained if the reaction is performed at low temperature in the presence of boron trifluoride etherate.[92]

Strained lactones shun reaction at the carbonyl group, forming carboxylic acid derivatives instead upon reaction[93] with copper catalysed Grignard reagents. In a similar way, an allene was obtained from the alkyne (112). The product was used in a synthesis of pelliterine (113).

Ring strain is not an essential factor in all organocopper mediated ring-opening reactions. Acetals can be opened by RCu by means of Lewis acid catalysis. When chiral acetals are used, reasonable asymmetric induction can be achieved. Allylic acetals undergo a conjugate form of addition to afford enol ethers[94] (see Chapter 14).

8.7 Alkylation of allylic acetates

Displacement of acetate from allylic positions, often with rearrangement of the alkene bond, can be conveniently effected by cuprate reagents.[95] The reaction provides an interesting comparison to palladium catalysed displacements discussed in Chapter 6. A typical example of the alkylation using cuprates is provided by the conversion[96] of (114) into (115). The R group may be Me, ^{n}Bu, or Ph.

This reaction has been applied[97] to a synthesis of the insect juvenile hormone (117) by reaction of the bis allylic acetate (116) with Me_2CuLi. The *trans-cis-trans* isomer

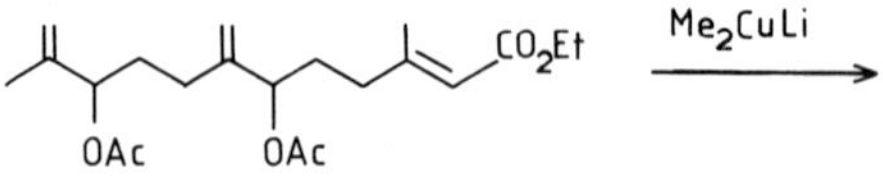

(116)

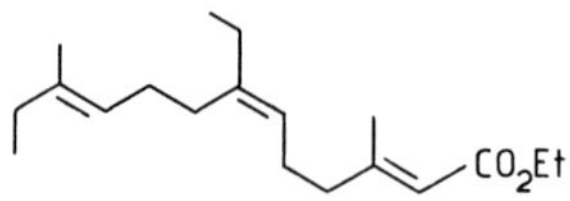

(117)

(117) was the major product, formed as mixture with about 14% of the all-*trans* and 8% of the *trans-cis-cis* isomers. Although reaction occurs by preferential γ attack, the side of entry of the alkylating agent is inevitably dependent on conformational preference and hence on the nearby substitution, as indicated in examples (118) and (119).[98]

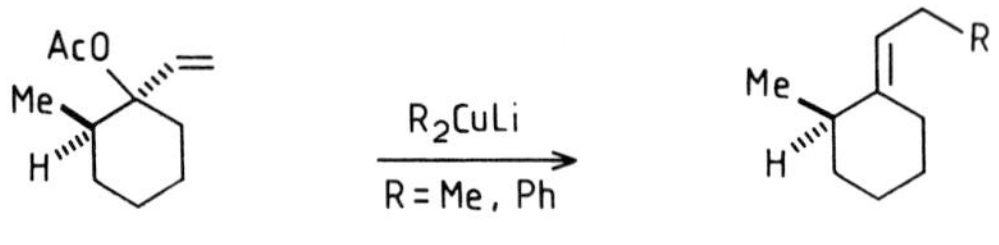

(118)

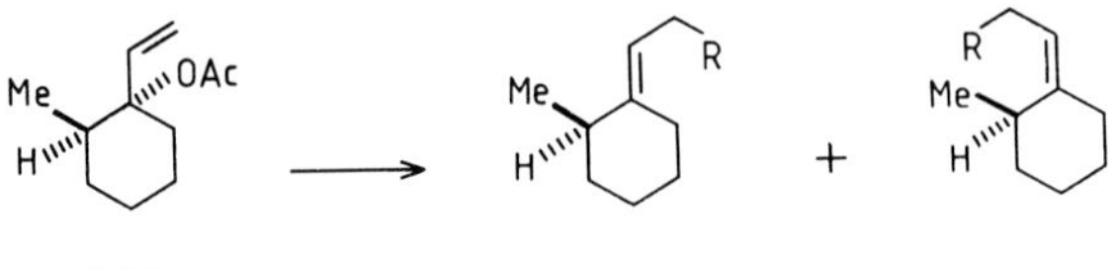

(119)

In these examples the entry of the alkyl group at a position γ to the leaving group is no doubt assisted by the location of the acetoxy group at a tertiary centre. Other examples indicate that the selectivity of entry may also depend on the solvent, and on the nature of the copper alkyl reagent. Both geranyl (120) and neryl acetate (121) with Me_2CuLi yielded[99] respectively (122) and (123). In contrast, linalyl acetate (124) gave[99]

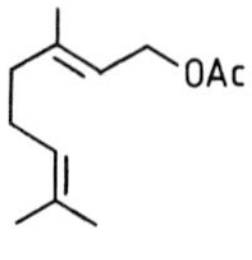

(120)

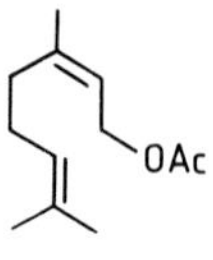

(121)

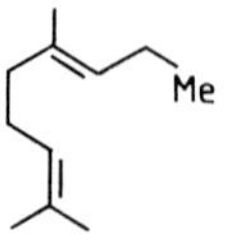

(122)

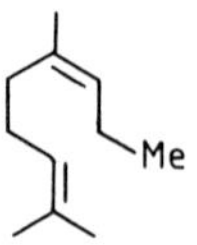

(123)

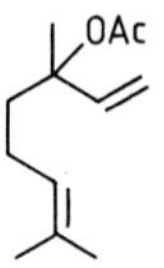

(124)

a mixture of (122) and (123). With MeCuCNLi as the reagent, however, (120) and (121) gave (125), and the linalyl ester (124) produced a mixture of (122) and (123), as before. This somewhat complex outcome is interpreted in terms of the involvement of a Cu-π-allyl intermediate.[99]

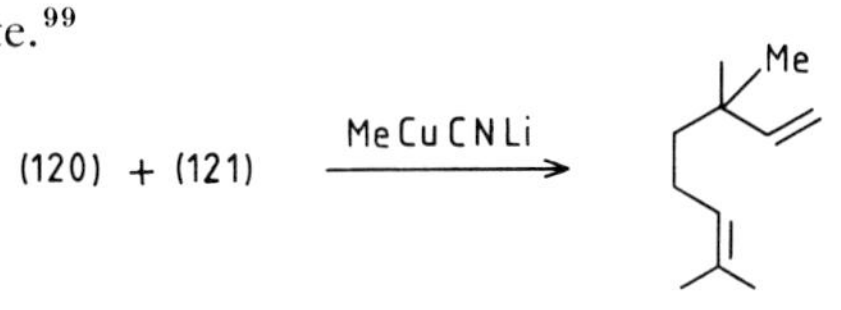
(120) + (121)
MeCuCNLi
Me

(125)

The examples (126) and (127)[100] illustrate the analogy between this alkylation process and the nucleophilic displacement of allylic acetates in the presence of $Pd(PPh_3)_4$ referred to above (see Chapter 6).

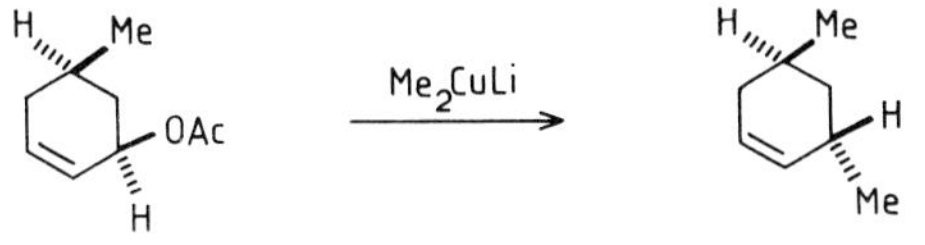
H
Me
OAc
H
Me2CuLi
H
Me
H
Me

(126)

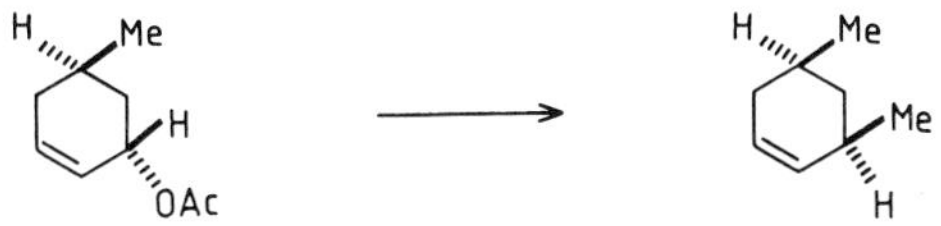
H
Me
H
OAc
H
Me
Me
H

(127)

These examples, which illustrate the displacement of OAc with inversion, may be understood[100] in terms of displacement[101] by the anionic Me_2Cu^- with transfer of Me within an intermediate Cu complex such as (128).

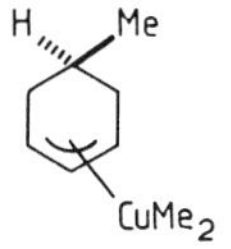
H
Me
CuMe2

(128)

An interesting application of this highly stereoselective displacement is found as an element in prostaglandin synthesis[102] in the transformation of (129) into (130).

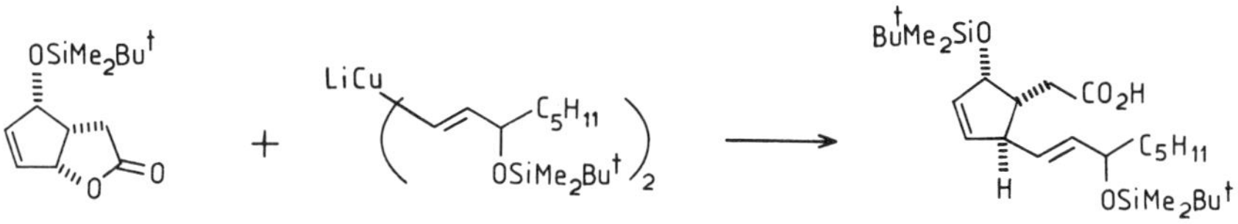
OSiMe2But
O
O
+
LiCu
C5H11
OSiMe2But
2
ButMe2SiO
CO2H
C5H11
H
OSiMe2But

(129) (130)

In this case we see displacement with inversion, with the entry of the alkyl group directed to the side remote from the silylether substituent. Not all reactions show such a high degree of control, as might be expected in view of the evidence above for a Cu-π-allyl intermediate. The examples[97] (131) and (132) which, with $^nBu_2CuLi$, gave the products (133) and (134) in the same ratio 82:18, are typical of many displacements of this type.

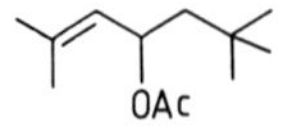

(131)

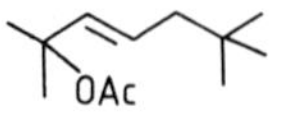

(132)

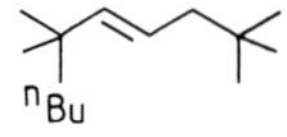

(133)

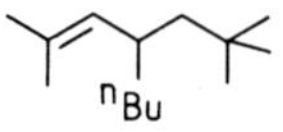

(134)

It was also found that (135), for example, gave (136) in greater than 90% yield, with only minor amounts of the Z-isomer (137) and the direct displacement product (138), when R was Me or ^nBu. However, when R was a phenyl group, more (20%) of the Z-isomer (137) was produced. Changing solvent from ether to THF, similarly, gave increased amounts of (137) and (138) when the reagent used was Me_2CuLi.

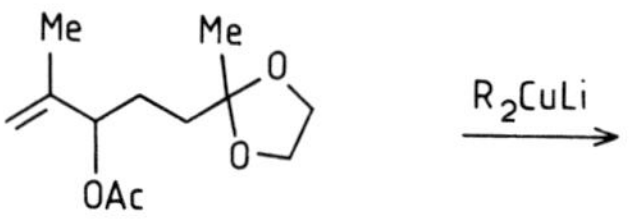

(135)

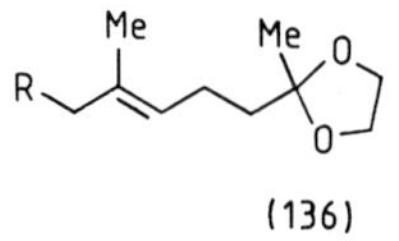

(136) 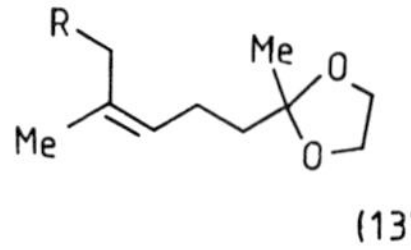   +   (137)

+

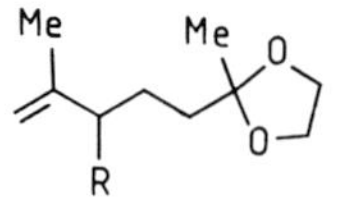

(138)

It is clear from the above that a Cu-π-allyl intermediate may react at either terminus in proportions determined by solvation as well as local steric hindrance.

In the example (139) to (140), alkyl displacement of an alkynyl acetate was shown to lead to the allene, a stage in the synthesis of the bean beetle sex attractant.[103]

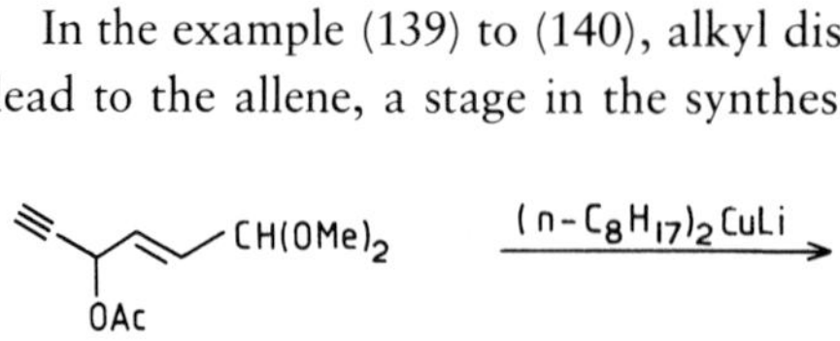

(139)

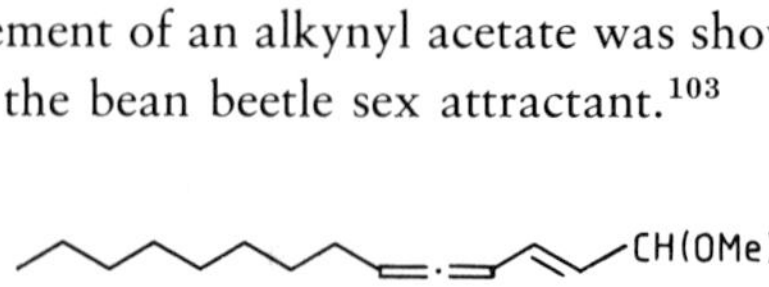

(140)

8.8 Reactions of allylic alcohols

Perhaps surprisingly, allylic alcohols can be alkylated directly by organocopper reagents despite the presence of a relatively acidic hydrogen. Comparison of the reactions[104] of (141) and (142) with $^{n}BuCuBF_3$ emphasises that regiocontrol is determined by the structure of the substrate.

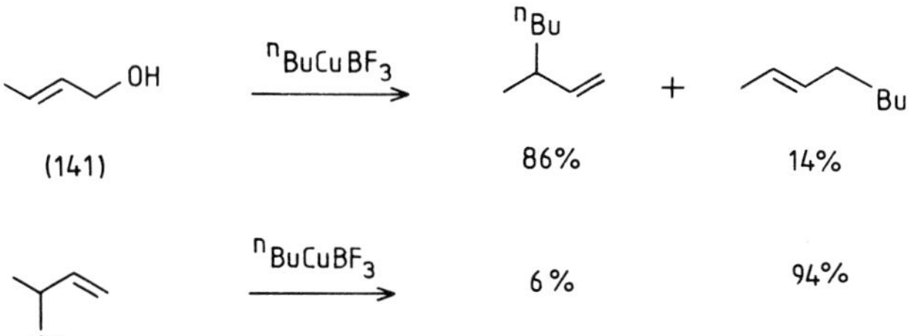

(141)　　　　　　　　　　86%　　　　14%

(142)　　　　　　　　　　6%　　　　94%

A related procedure[105] involves the reaction of an intermediate alkoxycuprate (143), formed from the alcohol, CuI and an alkyllithium, with the phosphonium salt (144).

$$R^1OCuR_3^2Li_3 \; + \; MePhN \cdot \overset{+}{P}Ph_3 \, I^- \; \longrightarrow$$

$$(143) \qquad\qquad (144)$$

$$R^1 - R^2 \; + \; Ph_3PO \; + \; R_2^2 Cu\,NMePhLi_2$$

Alkylation occurs preferentially at the site of the hydroxyl group, and the geometry of the alkene bond appears to be retained. Examples (145), (146) and (147) illustrate these features. The case of (147) illustrates that displacement of hydroxyl by alkyl occurs with inversion.

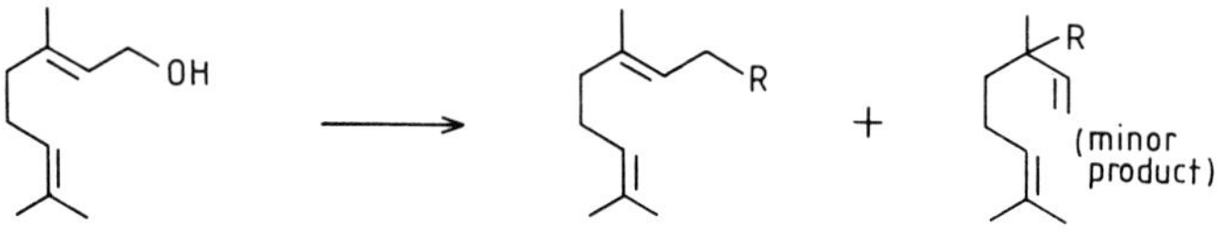

(145)　　　　　　　　　　　　　　　(minor product)

$$R = Me, \, ^{n}Bu, \, ^{s}Bu, \, C_4H_9C \equiv C-$$

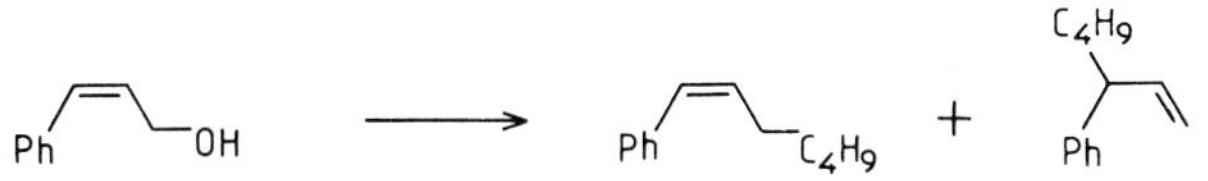

(146)　　　　　　　　　96%　　　　4%

An interesting comparison can be found in a synthesis[106] of ($\pm$)-shionone in which the vinyl halide system of (148) reacted normally with Me_2CuLi to yield (149). A similar example in which normal halide displacement was achieved in the presence of an allylic alcohol was encountered in Section 8.5.

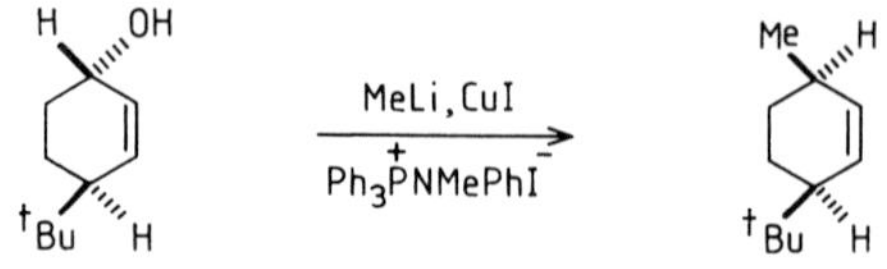

(147)

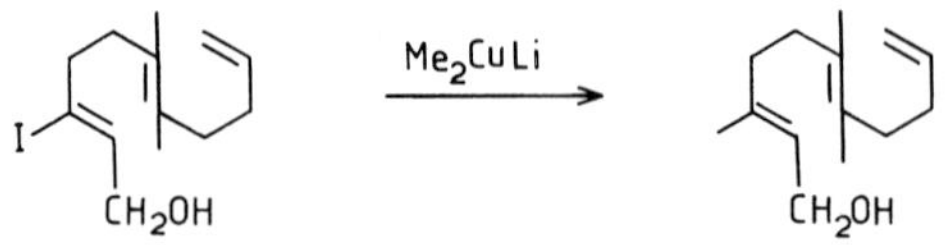

(148)   (149)

In many cases other metal systems can offer alternatives to the use of copper reagents for a particular purpose. Several examples will be given in the next section. The reactions of allylic alcohols provide an opportunity to introduce this topic since Grignard additions in this case are best catalysed by nickel compounds rather than the more generally used copper species. A typical example is the reaction with Grignard reagent in the presence of $(Ph_3P)_2NiCl_2$. In this way either (150) or (153) afford (151) and (152) as a 3:1 mixture.[107] This reaction has been exploited[108] in the course of a synthesis

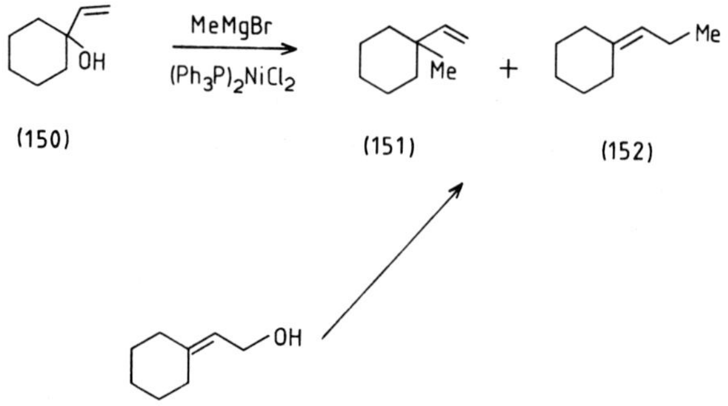

(150)   (151)   (152)

(153)

of diterpenes. The products from (154) were the mixture of isomers, (155), (156) and (157) (72:4:24). This, like examples (150) and (153), indicates preferential entry of the

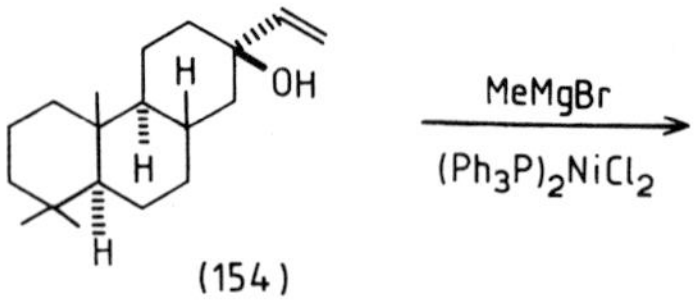

(154)

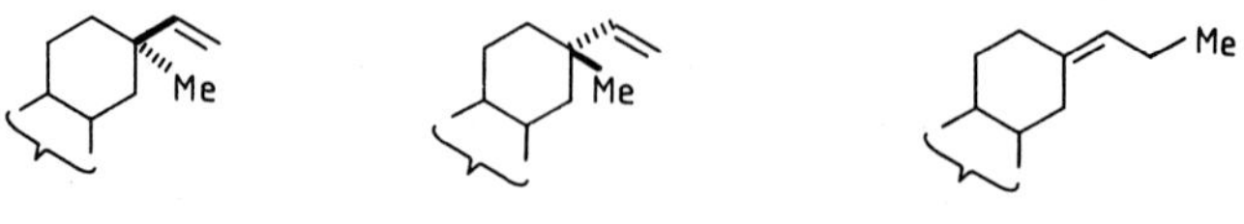

(155)   (156)   (157)

(158)

alkyl group at the more substituted carbon centre. The course of reaction is presumably via a Ni-allyl complex (158) which reacts further with $(Ph_3P)_2Ni(R) . MgBr$. The stereochemistry for the displacement of the hydroxyl group has been examined.[109]

8.9 σ bound complexes of other metals

While organocopper reagents offer a wide range of useful synthetic reactions, it is perhaps not surprising that the use of other metals for specific purposes can sometimes be advantageous. Among the potential benefits are changes in basicity, of steric requirements, and improvements in the range of temperatures to which the reagents may be exposed, a troublesome aspect of the use of some of the more sensitive cuprate reagents which may require inconveniently low temperatures.

Considerable effort has been devoted to the study of organozirconium reagents, once these had become generally available following the development of the hydrozirconation reaction.[110] Particularly notable in this field is the work of Schwartz's group.[111] Related C–C bond formation by addition to alkynes (carbozirconation) is limited in scope, but in some cases zirconium catalysed additions to substituted alkynes can provide an attractive reaction that does not suffer from the long reaction times that afflict the corresponding cuprate additions (see Section 8.5). In this case, of course, the product is a vinylalane (159).[112] Organoaluminium species of this type are also valuable synthetic reagents.

A wide range of methods for the preparation of alkyl and vinyl zirconium reagents are now available.[113] These species are far less reactive towards carbon electrophiles than their copper or magnesium equivalents and require catalysis, often using nickel complexes, if they are to be used in alkylation reactions. A nickel catalyst, prepared by treating $Ni(acac)_2$ with $^iBu_2AlH$, promotes efficient conjugate addition to (160) and (161). These reactions are similar to the analogous reactions of cuprates (Sections 8.2 and 8.3), except that less basic zirconium enolates are produced.[114] These may be

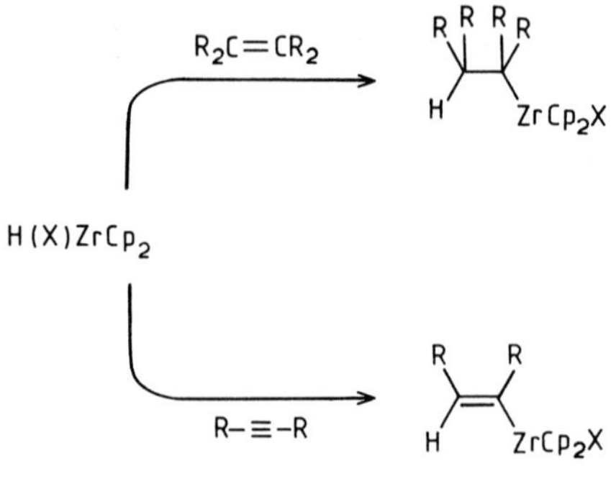

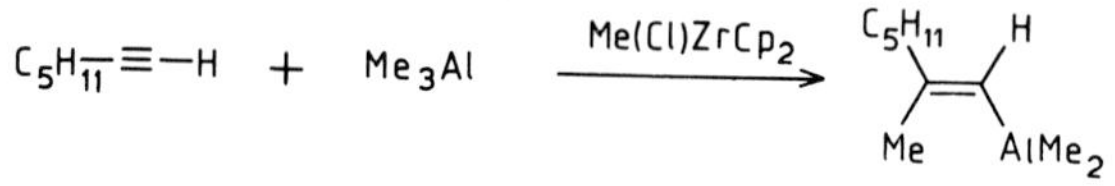

$$C_5H_{11}-\!\!\equiv\!\!-H \;+\; Me_3Al \xrightarrow{\;Me(Cl)ZrCp_2\;} \underset{Me\quad AlMe_2}{\overset{C_5H_{11}\quad H}{\diagdown\!\diagup}}$$

(159)

intercepted by electrophiles in the normal way.[115] Cobalt and palladium acetoacetonate complexes can also be used as catalysts.[114]

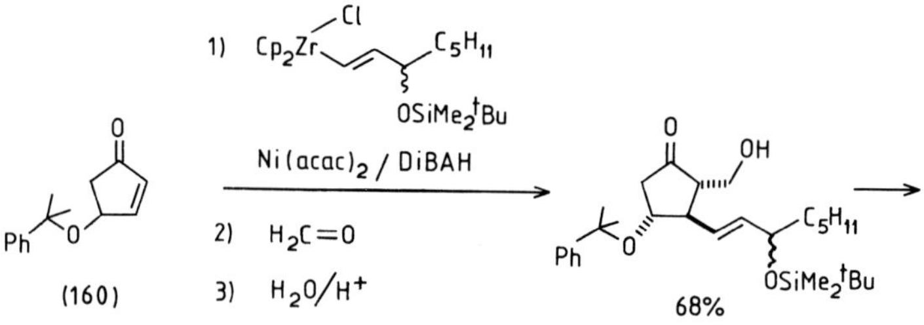

(160)

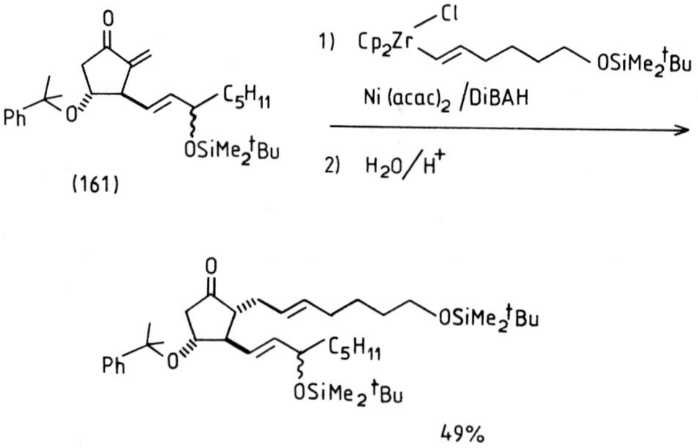

68%

(161)

49%

The nickel catalyst has been used[116] with organoaluminium reagents in a reaction that allows conjugate addition to be performed with alkynyl nucleophiles such as (162). Another use of nickel, this time Ni(acac)$_2$ itself, catalyses conjugate addition reactions that would not be possible with cuprate reagents. Organozinc reagents have been induced in this way to alkylate the highly hindered termini of the enones (163) and (164) to form β-cuparenone.[117] Nickel and palladium catalysts are also notable[113] in promoting cross-coupling reactions between organozirconium reagents and organic halides. Examples of this type of reaction are described in Chapter 10.

The highly electrophilic character of some metal complexes allows direct alkylation by weakly nucleophilic, σ bound metal complexes. Examples of addition of zirconium-based nucleophiles to palladium π complexes have been encountered in Chapters 5 and 6. These resemble the coupling reactions with organic halides, in that ligand transfer from zirconium to palladium is thought to precede C–C bond formation, since the product results from *endo* addition to the metal π complex. In contrast, the very weakly nucleophilic Fp complex (165) has been used for direct (*exo*) alkylation of the highly reactive cationic metal complexes discussed in Chapter 7.[118]

Like organocopper reagents, which are unsuitable for direct addition to carbonyl groups, most zirconocene (Cp_2Zr) derivatives fail to react with ketones and aldehydes. Allyl complexes are a notable exception. Reaction with aldehydes results in an efficient alkylation that has the advantage of giving predominantly *threo* products such as

(166).[119] Zirconocene adducts of dienes resemble allyl derivatives and react with aldehydes, ketones and even esters and nitriles.[120]

(166)

90-95%

Zirconium alkoxides are much more nucleophilic than zirconocene derivatives, and consequently can transfer a far wider range of substituents. The main advantage of these reagents is their very low basicity.[121] Related titanium species can also be useful as nucleophiles[121] and in both cases there are considerable opportunities for stereocontrol and for the introduction of chiral auxiliaries.[122]

(167)

Metal alkyl complexes of tungsten, molybdenum and rhenium, such as (167), can be regarded as carbon-bound enolates. Reaction with benzaldehyde under photolytic conditions effects an efficient alkylation.[123] Other organometallic enolates such as the iron acyl systems developed by Davies and Leibeskind have now become important general reagents. These are discussed separately in Chapter 9.

The manganese equivalents of Grignard reagents have been claimed[124] to have some advantages over alkyl copper and other metal reagents in the alkylation of acid chlorides to give ketones. Formation from a Grignard reagent and MnI_2 leads[124] to alkyl manganous complexes with a range of alkyl groups, for example, Me, ^iPr, ^nBu and ^tBu, $^nC_7H_{15}$.

$$RMgBr \ + \ MnI_2 \ \longrightarrow \ RMnI \ + \ MgBrI$$
(168)

Reaction of the organomanganese species (168) with an acid chloride gave good yields of ketones. This is exemplified by the transformation of bilianic acids into (169) and (170) via a reaction of their acid chlorides.[125] Yields were good and the steroid residue could still contain OAc, OCHO, and CO groups without serious side reactions. The manganese reagent was also found to be compatible with the use of dichloromethane as co-solvent.

In summary, the availability of a wide variety of σ bound organometallic nucleophiles

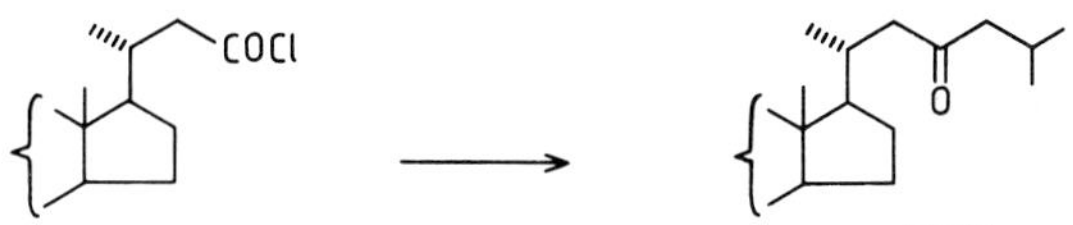

(169)

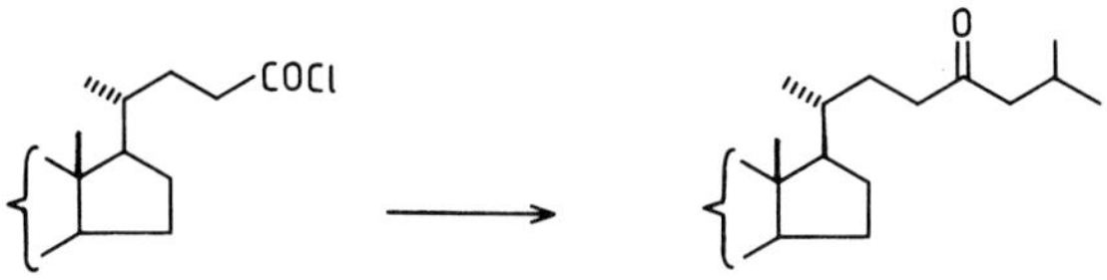

(170)

illustrated by examples given in this chapter, has provided valuable opportunities for selectivity in alkylation reactions. Careful planning can put the subtle differences between reagents to good use. In routine applications, however, Grignard, organo-lithium, and cuprate reagents are still the most suitable candidates for initial investigations.

9
Carbonyl insertion

The migration of a σ bound alkyl group to a π bound ligand (see Chapter 1) is one of the most important reactions that occur within the coordination sphere of a transition metal centre. When migration to a carbonyl ligand takes place, an acyl complex is produced. Upon removal of the metal this can form a variety[1] of organic compounds that arise from the overall insertion of carbon monoxide into the metal–carbon σ bond. An example (Scheme 9.1), drawn from the organozirconium chemistry just discussed in Chapter 8, illustrates the formation of aldehydes, esters, and carboxylic acids. Acyl halides can also be formed in this way.[2]

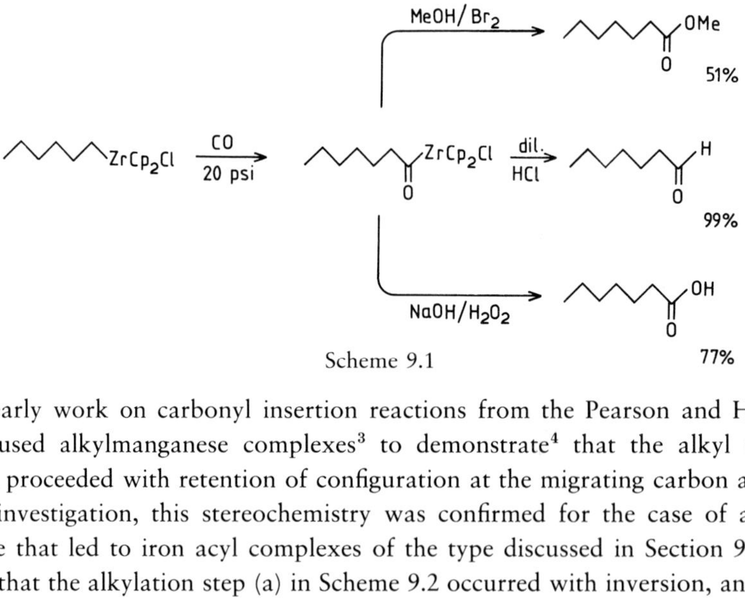

Scheme 9.1

The early work on carbonyl insertion reactions from the Pearson and Haszeldine groups used alkylmanganese complexes[3] to demonstrate[4] that the alkyl migration reaction proceeded with retention of configuration at the migrating carbon atom. In a similar investigation, this stereochemistry was confirmed for the case of a reaction sequence that led to iron acyl complexes of the type discussed in Section 9.7. It was shown[5] that the alkylation step (a) in Scheme 9.2 occurred with inversion, and that the carbonyl insertion step (b) occurred with retention of stereochemistry. For the reaction M = Fe, L = PPh$_3$ and R—X = D-(+)-sec-butylbromide, the oxidation step (c) in Scheme 9.2, effected by Cl$_2$/H$_2$O, led to L-(−)-2-methylbutanoic acid. More recent work[6] has examined the effect on the stereochemistry of the metal centre. For tetrahedral complexes, two distinct outcomes are possible. The acyl group in the

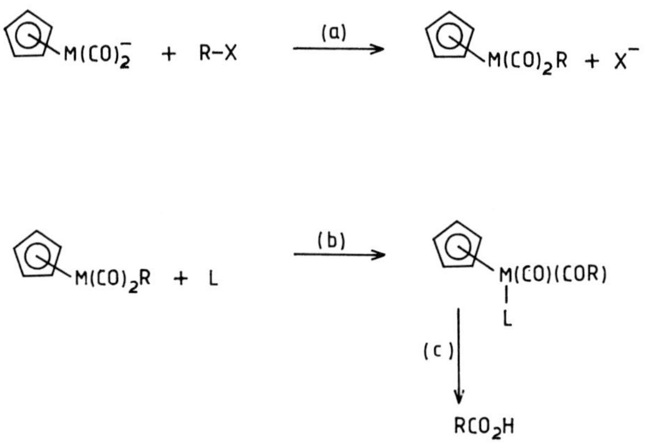

Scheme 9.2

product may be located either at the site originally occupied by the alkyl group (inversion) or by the carbonyl group (retention). Stereochemical results have been found[7] to be highly sensitive to changes of solvent, proceeding with substantial retention of configuration in HMPA, but with almost complete inversion in nitromethane, in the example of the ethyl complex (1) shown in Scheme 9.3.

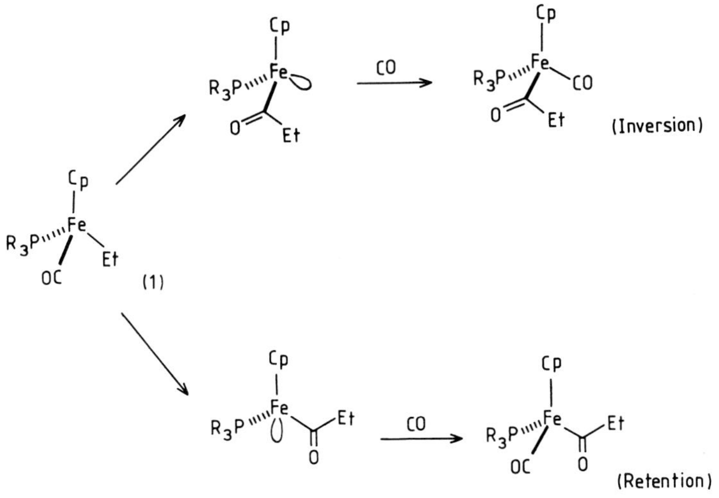

Scheme 9.3

Carbonyl insertion is of widespread importance in chemistry, with applications ranging from large-scale industrial catalysis to sophisticated stereocontrolled organic synthesis. In many cases, carbonyl insertion is one of a series of reaction steps in which several new bonds are formed in a single catalytic process.

9.1 Hydroformylation using cobalt and rhodium

The metal catalysed addition of the elements of formaldehyde across an alkene bond, hydroformylation, is a widely studied organometallic reaction.[8] Hydroformylation came to prominence following Roelen's pioneering studies on the cobalt carbonyl catalysed reaction of alkenes with synthesis gas[9] (CO/H_2) which culminated in the development of the OXO process.

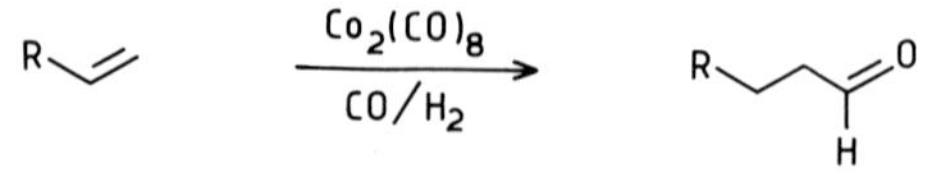

$$R \diagdown\diagup \xrightarrow[\text{CO}/\text{H}_2]{\text{Co}_2(\text{CO})_8} R \diagdown\diagdown\diagup\!\!\!\!\!\!^{O}_{H}$$

Today this process is operated around the world for the large-scale production of a variety of aldehydes and alcohols, although the use of the more selective rhodium catalysts is becoming increasingly important (see p. 273). Alcohols are often more important products and are formed either *in situ* using the same catalyst, or in a separate reduction unit. Examples of typical OXO process feedstocks and products are listed in Table 9.1.

Table 9.1

Feedstock	Product	Uses
Ethene	propionaldehyde	
Propene	butyraldehyde	dimerisation to 2-ethyl-hexanol → plasticisers
C_7–C_9 alkenes	C_8–C_{10} alcohols	plasticisers (usually as diphthalate esters)
C_{11}–C_{15} alkenes	C_{12}–C_{16} alcohols	surfactants

A widely accepted reaction mechanism due to Heck and Breslow[10] involves initial coordination of the alkene by the coordinately unsaturated species $HCo(CO)_3$ (2).

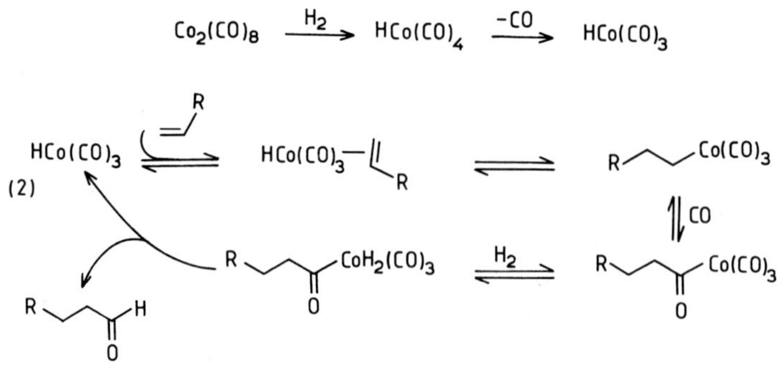

Scheme 9.4

A feature of the hydroformylation reaction for higher alkenes is the formation of branched chain aldehydes. These may arise from addition of $HCo(CO)_3$ in the opposite sense to that shown in Scheme 9.4 and/or via isomerisation (see Chapter 2) of the α-alkene bond with subsequent hydroformylation of internal alkenes as indicated in Scheme 9.5.

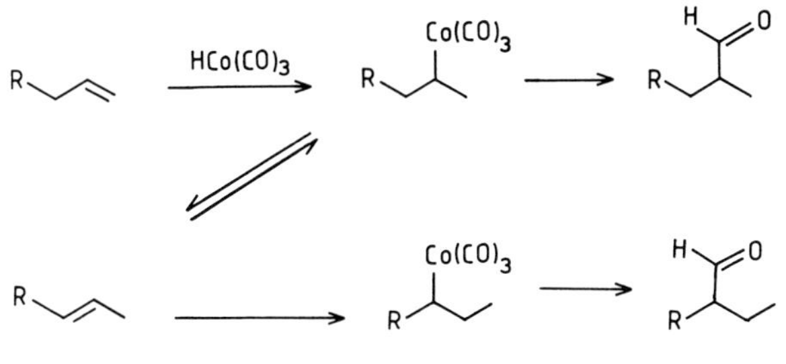

Scheme 9.5

By using optically active (S)-3-methylpent-1-ene (3), Pino *et al.*[11] demonstrated that the degree of isomerisation under normal 'OXO' reaction conditions was slight. The products were 4-methylhexanal (4), 3-ethylpentanal (5), and 2,3-dimethylvaleraldehyde (6), formed in the ratio 93:4:3. The product (4) was racemised to the extent of only 1·8%. At higher temperatures the racemisation of (4) increased to 32% at 145 °C and 94% at 180 °C, whilst the yield of (4) at 180 °C dropped to 61%.

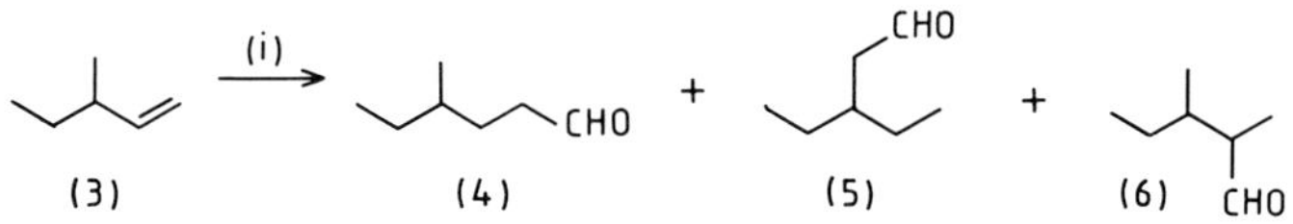

(i) $Co_2(CO)_8$, H_2, CO in dioxan, 100°C, 110 atm.

Decreasing the pressure of carbon monoxide was also found to increase the degree of racemisation. These results indicate that under conventional 'OXO' conditions, (4) is mainly formed by a mechanism which does not change the stereochemistry at C-3, whilst the degree of racemisation can be explained by the reversible formation of a pent-1-ene complex.

The double bond isomerisation encountered in the formation of (5) can be a common side reaction in hydroformylation. Further typical examples are found in extensive studies of the hydroformylation of cycloocta-1,5-diene (7), where product distributions can vary considerably, depending on the conditions employed.[12] A second common side reaction in hydroformylation is hydrogenation (see Chapter 10). In the case of (7), use of $Co_2(CO)_8$ gave the completely reduced hydroformylation product (8). With rhodium catalysis and high CO pressure, however, double hydroformylation is favoured, and mixtures of (9) and (10) were obtained. Reduction of the aldehyde can be controlled with a phosphine-modified rhodium catalyst. Only (10) was produced and under milder conditions, alkene hydrogenation was also prevented. A 1:1 mixture of cyclooctene aldehydes such as (11) and (12) was then formed, while, in more recent work, the

200 atm., Cat. = $Co_2(CO)_8$		72%	–
200 atm., Cat. = Rh_2O_3		43%	19%
1000 atm., Cat. = Rh_2O_3		11%	70%

catalyst $Rh(OAc)CO(PPh_3)_2$ has been used for the selective production of (12) in 75% yield. This aldehyde was converted into azelaic acid (13). Further carbonylation reactions of cyclooctene are discussed in Section 9.2.

From this selection of examples,[12] it is clear that subtle changes can exert a profound influence of the course of hydroformylation.[8] Competing hydrogenation reactions in particular, can often be controlled by adjusting reaction conditions. Two examples which illustrate this further, and also indicate the *cis* relative stereochemistry of hydroformylation, are provided by the conversion of (14) into (15),[13] and of (16) into (17) and (18).[14]

Formation of (15) demonstrates *cis*-addition at the less substituted terminus of the alkene, and from the less hindered side of the steroid. The products (17) and (18) similarly arise from *cis*-addition.

High selectivities for formation of terminal aldehydes have been achieved by use of organophosphine ligands as exemplified by the complex $Co_2(CO)_6[P(^nC_4H_9)_3]_2$.[15] Presumably the steric bulk of the phosphine ligand favours addition to the terminal alkene carbon. Hydroformylation reactions using cobalt phosphine complexes have been developed by the Shell company, and have been used industrially since the mid sixties.[8,16]

Rhodium complexes effectively catalyse the hydroformylation reaction at low pressures, with the added advantage of a high n:iso product ratio. Particular reference is made to the elegant studies of Wilkinson and co-workers who demonstrated the usefulness of hydridocarbonyltris(triphenylphosphine)rhodium (19) as a catalyst for

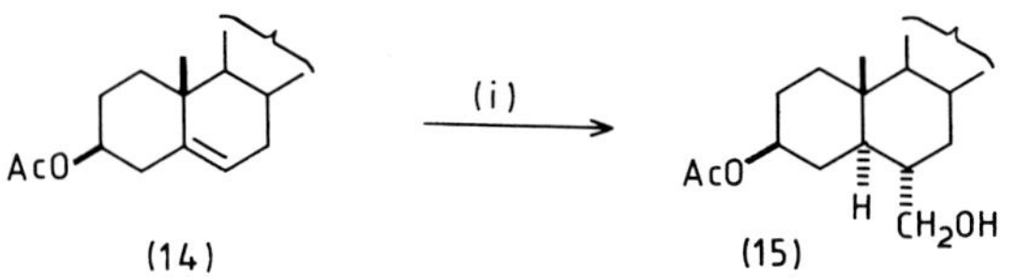

(14) → (15)

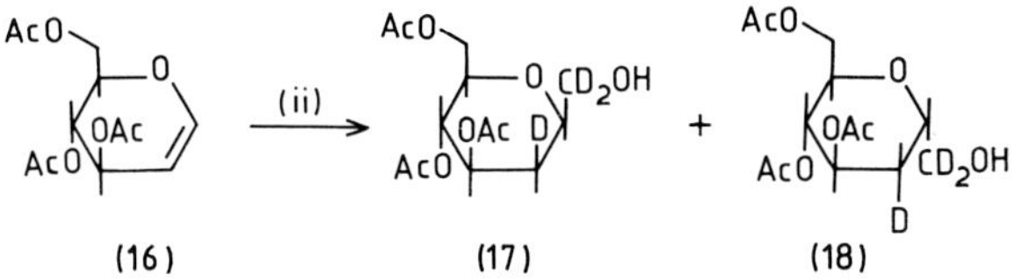

(16) → (17) + (18)

(i) $Co_2(CO)_8$, $CO + H_2$, toluene, 180°C 88 Kg. cm^{-2}.

(ii) $Co_2(CO)_8$, 125°C, D_2.

the hydroformylation of alkenes.[17] The reaction sequence is essentially the same as for the hydridocobaltcarbonyl reaction discussed earlier.

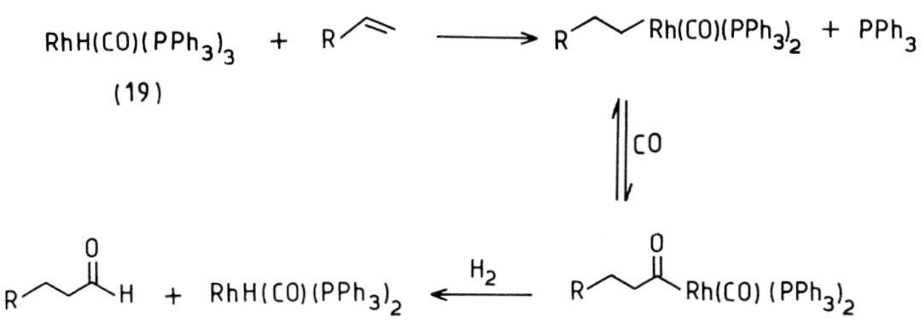

$$RhH(CO)(PPh_3)_3 + R\diagdown\diagup \longrightarrow R\diagup\diagdown Rh(CO)(PPh_3)_2 + PPh_3$$

(19)

Further development in this area resulted in the Johnson Matthey/Union Carbide/Davey McKee low pressure phosphine-modified rhodium process for the hydroformylation of propylene.[18] The advantages of the rhodium catalysed process are highlighted in Table 9.2.

Table 9.2. *Comparison of Propylene OXO Processes*

Feature	Cobalt carbonyl	Rhodium-phosphine complex
Pressure atm	200–300	< 20
Temperature °C	140–180	100
Linear–branched aldehyde ratio	3–4:1	8–16:1

An example of use of the complex (19) with higher alkenes is seen in the reaction[19] of 3-methylpent-2-ene (20) which gave (21) in 85 % yield, accompanied by a little of (23)

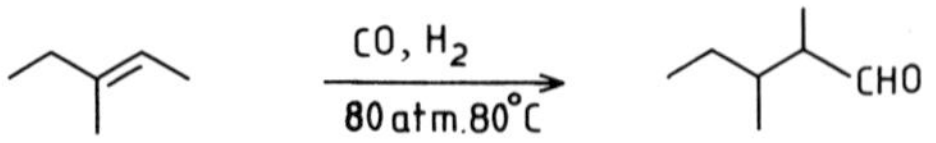

(20) $\xrightarrow[\text{80 atm.80°C}]{\text{CO, H}_2}$ (21)

(22)   (23)

arising from alkene isomerisation yielding the terminal alkene (22) which reacts more rapidly. Similarly RhH(CO)(PPh₃)₃ in benzene has been used in an effective preparation of homocitronellal (25), a useful intermediate in the synthesis of perfumery additives.[20] Aldehyde (25) was obtained from the diene (24) in 80% yield, accompanied by (26) and its *cis*-isomer. Under the very mild conditions of this reaction no other carbonylation

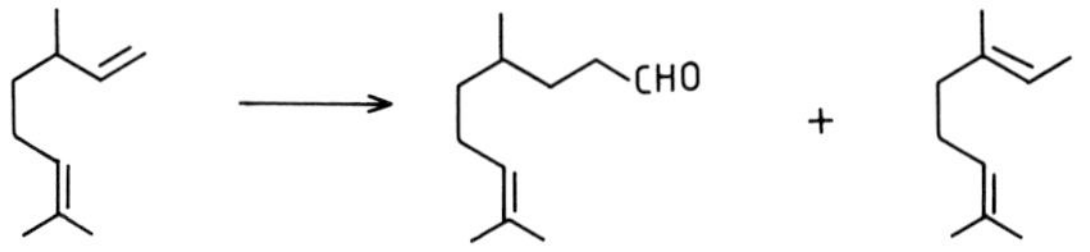

(24)   (25)   +   (26)

(i) RhH(CO)(PPh₃)₃ , benzene, CO + H₂ , 1 atm, 70°C.

product was detected. Hydroformylation reactions may be carried out in the presence of other functional groups. The allylic alcohol (27) gave the aldehyde (28) which, upon hydrogenation, yielded the diol (29), a source of an alkyldihydrofuran. Carbonylation of (30) was effected in a similar manner.[21]

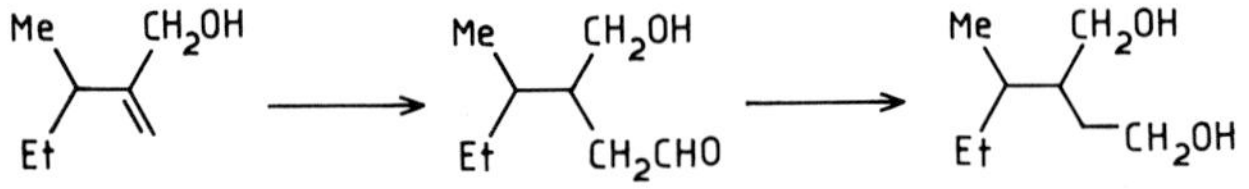

(27)   (28)   (29)

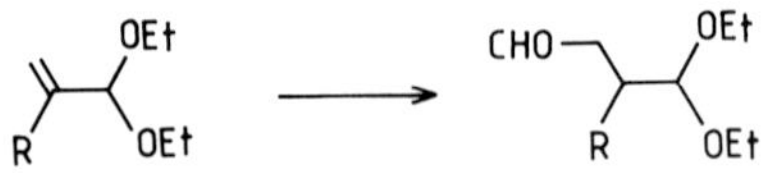

(30)

R = Me, CH(Me)(Et)

An interesting synthesis[22] in the guaiazulene group of sesquiterpenes made use of a chlororhodium cyclooctadienyl complex to effect the carbonylation step in a sequence starting from dehydrolinalool (31).

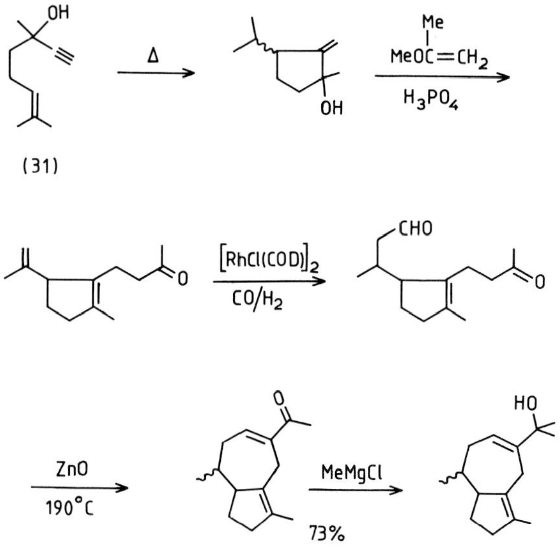

(31)

Ketone synthesis

Hydroformylation using RhH(CO)(PPh$_3$)$_3$ may be modified to effect ketone synthesis[23] by replacing carbon monoxide by an acid chloride as the inserting reagent. The rhodium complex (19) was used again in the example given in Scheme 9.6.

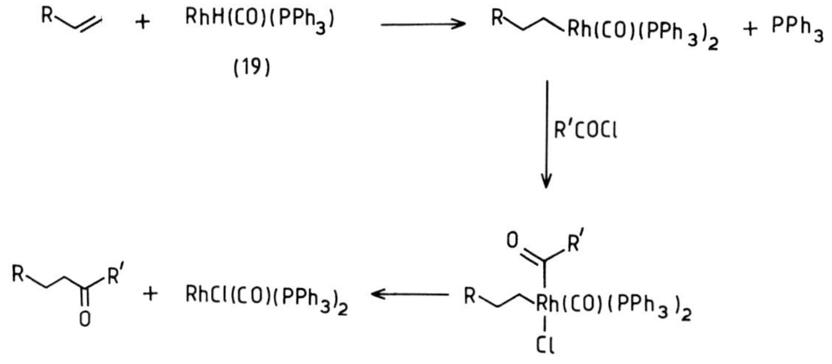

Scheme 9.6

Lactone synthesis

Hydroformylation can produce lactones by intramolecular interception of the metal acyl intermediate, when a hydroxy group is present elsewhere in the molecule. Lactones such as (32) and the Prelog–Djerassi lactone have been prepared[24] in this way.

The literature on cobalt/rhodium hydroformylation is extensive, and further discussion of examples of its application is beyond the scope of this book; the interested reader is referred to a number of reviews,[25, 26] in particular the review by Cornil.[25]

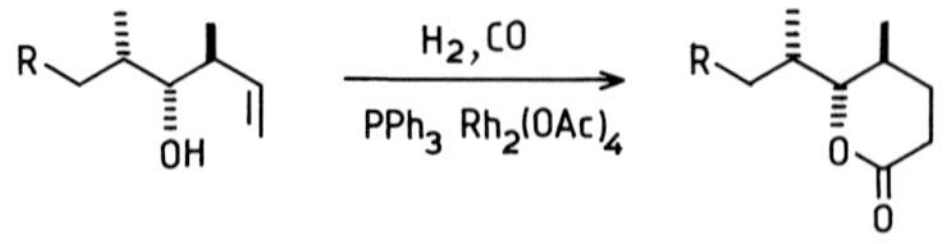

(32)

However, before closing this section, mention should be made of the use of chiral ligands to induce optical activity in hydroformylation products.[27] For example, the catalyst RhH(CO)(PPh$_3$)$_3$ in combination with (−)-DIOP a chiral phosphine ligand, is effective for inducing optical activity as in the formation of (33), (34) and (35).

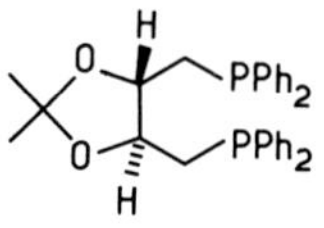

(−)-DIOP

Olefin	Chiral Product	optical purity	absolute config.	Ref.
	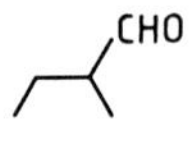 (33)	27%	S	28
	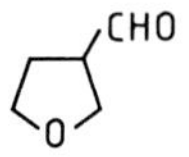 (34)	7%	R	28
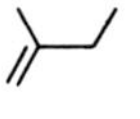	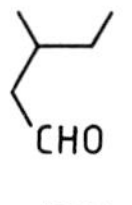 (35)	32%	S	29

However, as seen in these examples, the asymmetric induction is less efficient than may be achieved in hydrogenation or epoxidation using a chiral catalyst. Asymmetric induction in the transition metal catalysed hydroformylation reaction has been reviewed by Consiglio and Pino.[30]

9.2 Carbonylation using palladium

Alkenes and dienes

As we have already noted (Chapter 5), palladium(II) salts offer generally effective reagents to bring about nucleophile addition to alkenes. Since the organometallic intermediates formed after nucleophile addition contain σ bonds, they are susceptible to carbon monoxide insertion. It is possible to effect the synthesis of β-substituted carboxylic acid derivatives (36) in this way.

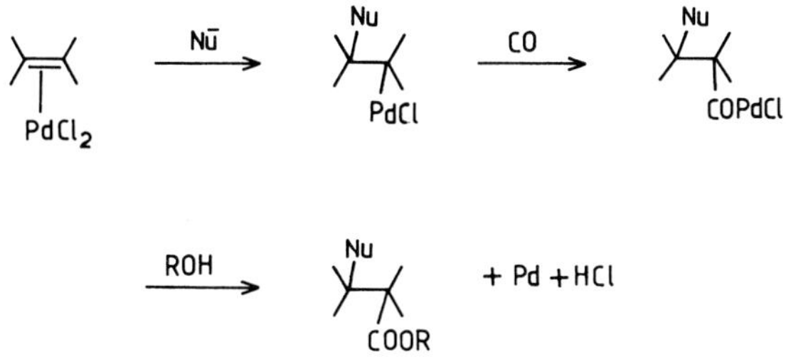

(36)

The carbonylation of cycloocta-1,5-diene[31] illustrates a case where water is the nucleophile. It will be noted that the lactone (37) is formed stereospecifically; addition to the alkene bond occurs by *trans* addition of water and insertion of carbon monoxide occurs with retention. In contrast, under catalytic conditions with $PdI_2(PBu_3)_2$ in the absence of nucleophiles, CO can be induced to span the eight-membered ring to form an unsaturated bicyclic ketone in 40% yield.[31]

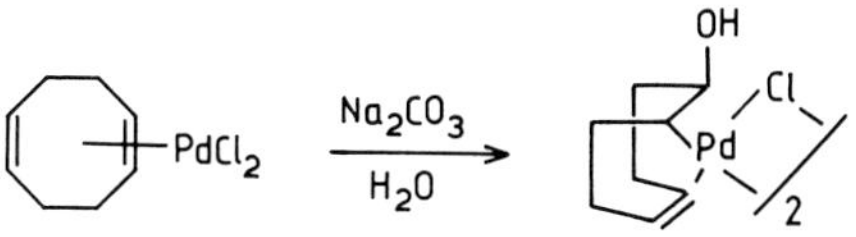

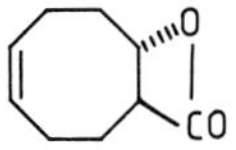

(37)

The reaction[32] of *cis* (38) and *trans*-but-2-ene (40) in the presence of $PdCl_2$ with $CuCl_2$ as re-oxidant for Pd, in MeOH at 28 °C with carbon monoxide (2 atm.) demonstrates the same stereospecificity. The alkene (38) gave the *threo*, and (40), the *erythro* methoxy esters. Examples of this type are related to the Wacker oxidation discussed in Chapter 4 (see also below).

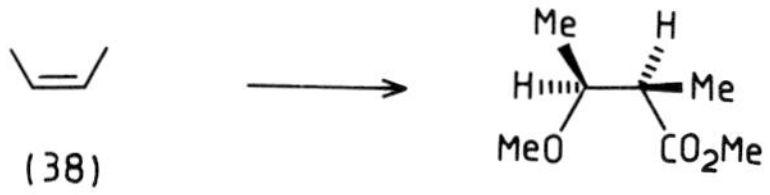

(38)

(39)

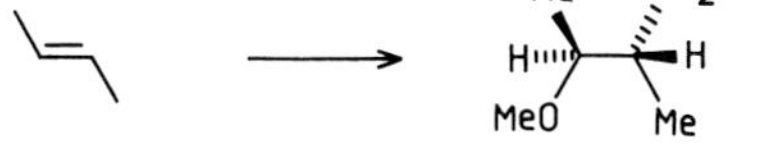

threo

(40)

(41)

erythro

The products (39) and (41) are, however, accompanied respectively by *meso-* and *rac-*dimethylbutane-2,3-dicarboxylate, indicating that the CO_2Me group may become the addend in the initial addition to the alkene. By carrying out the reaction in the presence of sodium acetate the dicarbomethoxylation can be made the dominant reaction. The process can be rationalised by a sequence shown in Scheme 9.7 in which CO_2Me is required as the σ bound ligand that is transferred from palladium.

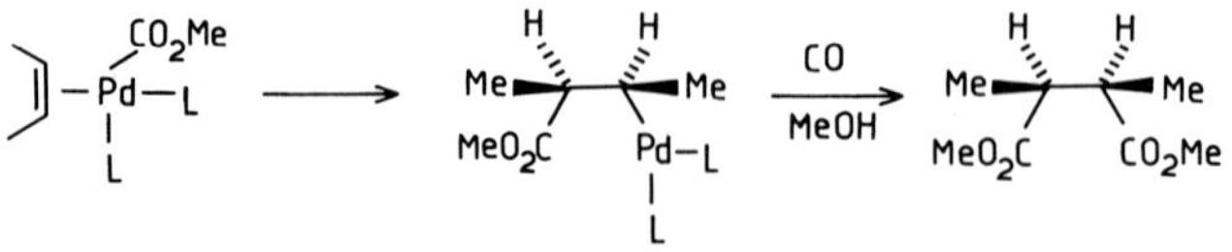

Scheme 9.7

Some other examples where the dicarboxylation process has been applied to dienes are shown in Scheme 9.8.[33]

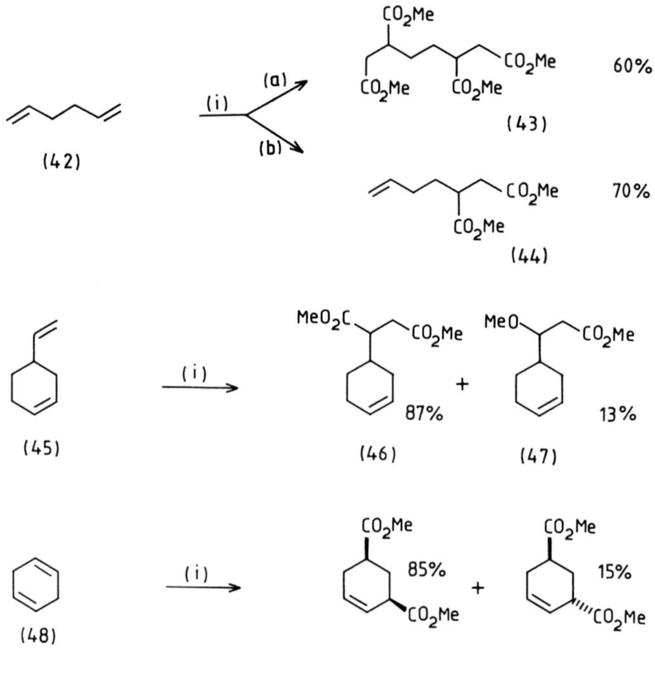

(i) $PdCl_2$, $CuCl_2$, CO, MeOH. (a) CO 1–3 atm. (b) CO >5 atm.

Scheme 9.8

These examples also indicate some points of general interest. In the conversion of (42) into (43) or (44), preferential coordination of CO rather than an alkene bond occurs when the carbon monoxide pressure is raised. For (45), the preferential reaction is that of the less substituted alkene, and with (48), rearrangement is possible by elimination and re-addition of a palladium hydride (Scheme 9.9). Example (45) also indicates some

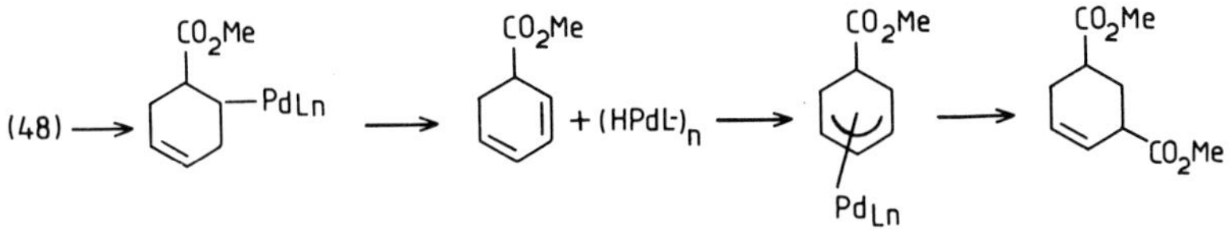

Scheme 9.9

competition from the addition of methanol to the alkene rather than to a carbonyl ligand. This accounts for the formation of both (46) and (47). Scheme 9.10[33] illustrates the use of the dicarboxylation process in the presence of other functional groups, in this case ketones.

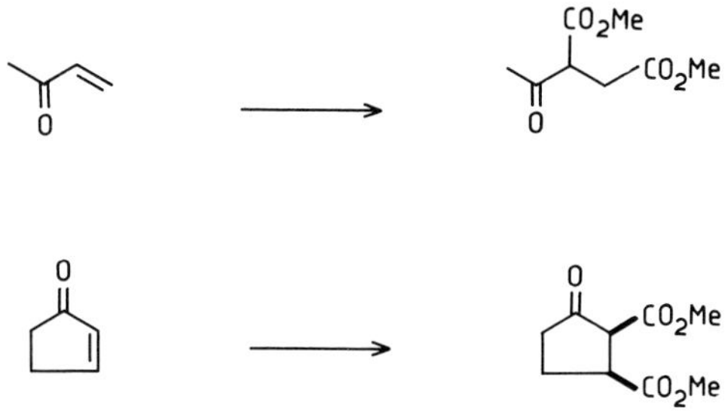

Scheme 9.10

Terminal alkynes can also be dicarboxylated to yield maleic esters.[34] The process is admirably illustrated by Buchanan's synthesis[35] of showdomycin (49), an antitumour and antibiotic substance. Mercuric chloride is used in this synthesis as a source of $ClHgCO_2Me$.

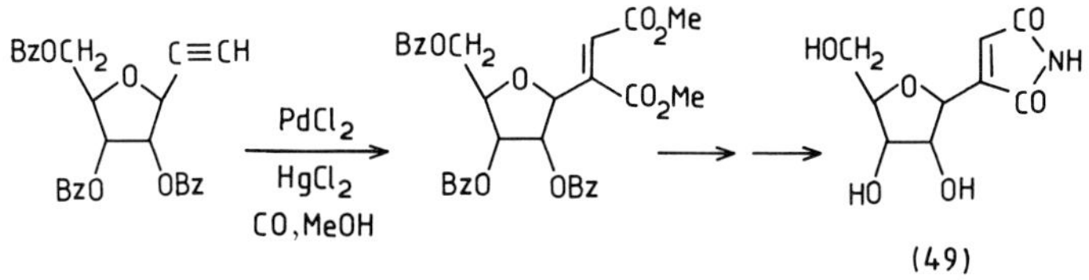

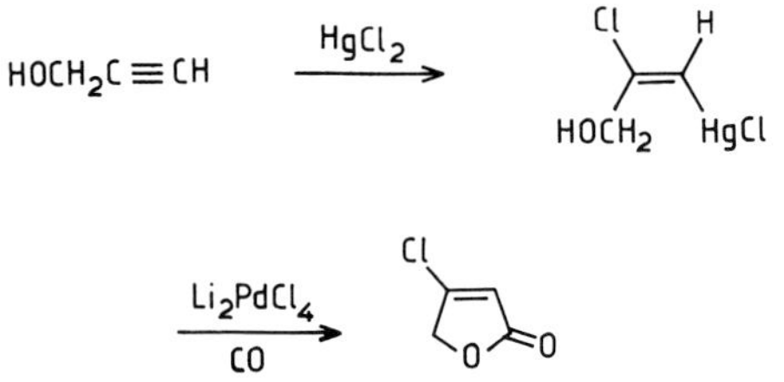

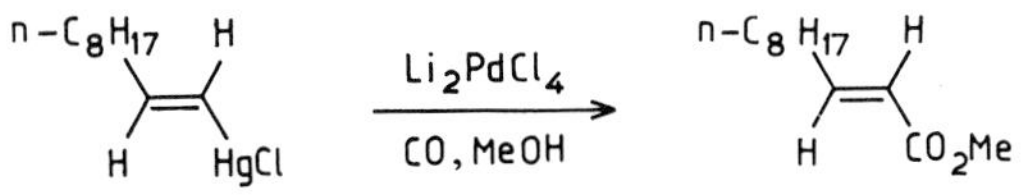

A similar application of this method is illustrated by the formation of (50).[36] The mercury salt here initiates the reaction by addition to the alkyne bond. This is followed by a metal exchange reaction, and carbon monoxide insertion into the newly formed carbon–palladium bond.

We noted above the insertion of carbon monoxide into a palladium complex to yield a β-lactone (37). A related process has been used[37] in the synthesis of β-lactams shown in the sequences (51) to (52) to (53), and (54) to (55), and (56) to (57) (see also Chapter 15).

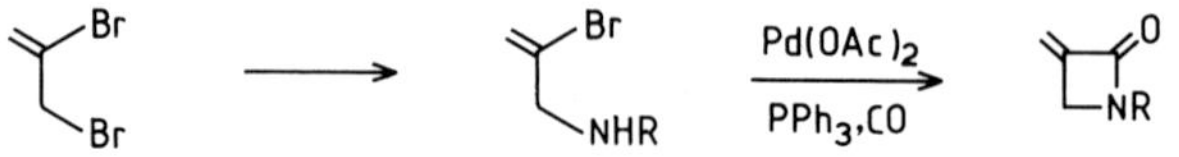

(51)   (52)   (53)

$$R = CH_2Ph, \quad CH_2(CH_2)_2 \, OTHP, \quad CH_2CH_2CO_2Me$$

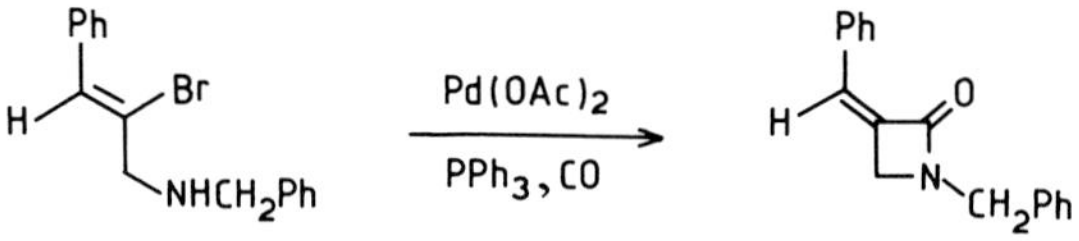

(54)   (55)

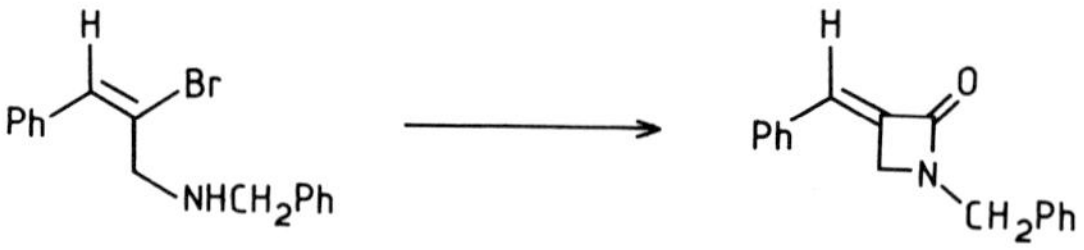

(56)   (57)

Another palladium catalysed reaction that commences with insertion into a carbon–bromine bond has been used to convert allenyl bromides into the corresponding amides and esters. This reaction, however, required the use of moderate (1–20 bar) carbon monoxide pressure.[38]

Alkynes

As seen above, alkynes also exhibit carbon monoxide insertion under catalysis by, for example, $PdCl_2$. The course of reaction is indicated by the carbonylation of *cis*-2-ethynylcyclohexanol (58) to the lactone[39] (59) in good yield. Comparison of the

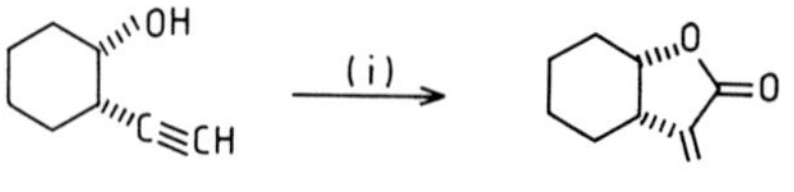

(58)   (59)

reactions of the *cis* and *trans* stereoisomers (60) and (61) has shown[40] that ring strain in the product does not affect the rate of consumption of the starting material, an observation consistent with the initial irreversible formation of a carboalkoxy intermediate of type (62).

observation consistent with the initial irreversible formation of a carboalkoxy intermediate of type (62).

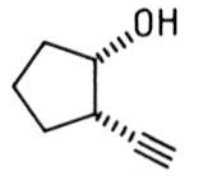

(60) 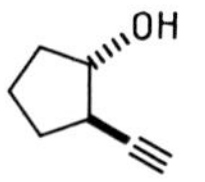 (61) 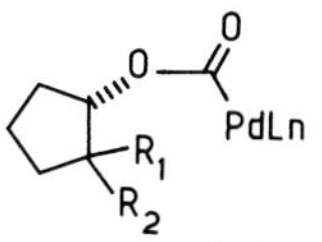 (62)

Alkyl and aryl halides

Hydroxy substituted organic halides can also be converted to lactones. Some examples are shown in Scheme 9.11. In this case, migration of a metal–carbon, rather than a metal–oxygen, σ bond has been proposed.[41]

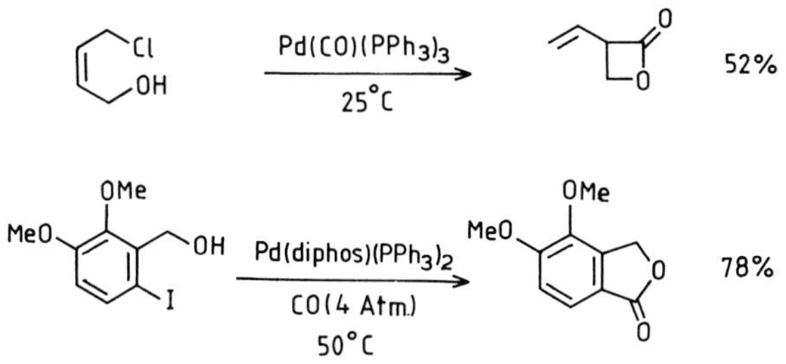

Scheme 9.11

Carbonylation of alcohols themselves can be used to produce carboxylic acids, esters, or lactones. Palladium chloride was used in this way to couple an aryl iodide with the alcohol (63) in a synthesis of the dimethyl ether of zearalenone (64).[42] Similar reactions

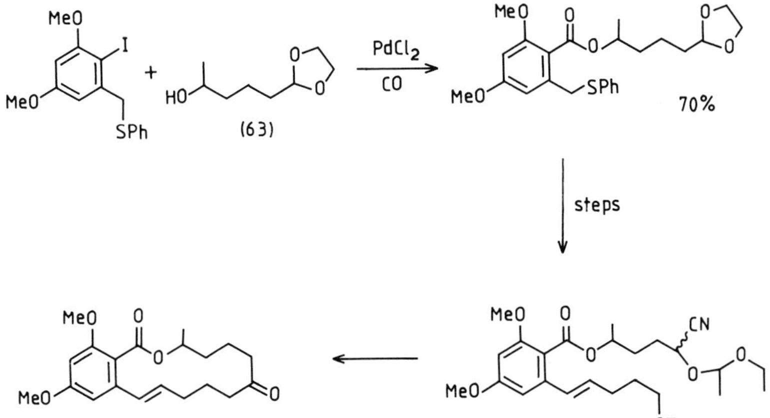

have been used to form quinones by cyclisation to an alkene, and in the closure of five-membered rings in reactions that culminate in a second carbonyl insertion step instead of β-elimination, in a fashion that resembles the modified Wacker reactions discussed below.[43]

The presence of a halogen substituent is not always necessary in palladium catalysed carbonylation reactions. Thiophene, furan and pyrrole have been converted into the corresponding 2-substituted ester derivatives by reaction with carbon monoxide and palladium chloride. Naphthalene has also been converted into β-naphthoic acid by carbonylation catalysed by palladium acetate. In this latter reaction, addition of oxygen increased the yield of the carboxylic acid.[44]

Alcohols

Carbonyl insertion into the carbon–oxygen bond of alcohols is of importance in the bulk preparation of simple chemical feedstocks. The Monsanto acetic acid process[45] provides an impressive example in which selectivities for acetate production are 99% based on methanol, and 90% based on carbon monoxide.[46] This direct route to acetic acid will gradually displace the Wacker process (Chapter 4), where the acetaldehyde produced was oxidised to acetic acid. The process involves a reaction of methanol with carbon monoxide catalysed by a rhodium complex and promoted by iodide.

Modified Wacker oxidations and other routes to lactones

A further palladium catalysed carbonylation involves the cyclisation of β-hydroxy-alkenes such as (65) in a reaction[47] that is also related to the Wacker oxidation (see

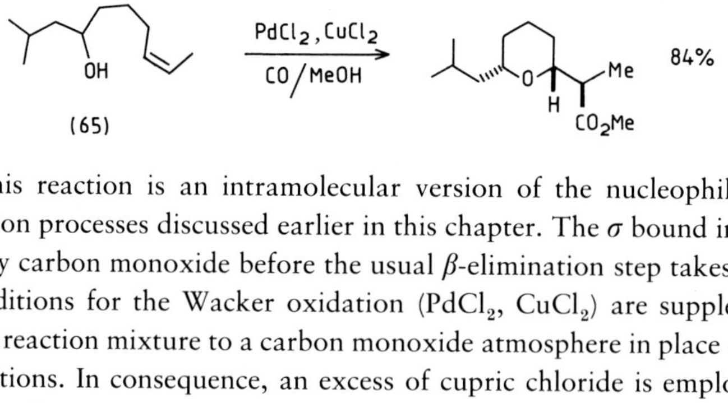

(65)

Chapter 4). This reaction is an intramolecular version of the nucleophile addition, carbonyl insertion processes discussed earlier in this chapter. The σ bound intermediate is intercepted by carbon monoxide before the usual β-elimination step takes place. The customary conditions for the Wacker oxidation (PdCl$_2$, CuCl$_2$) are supplemented by exposure of the reaction mixture to a carbon monoxide atmosphere in place of the usual oxidising conditions. In consequence, an excess of cupric chloride is employed.

An intramolecular version of the final carbon–oxygen bond formation is possible when an hydroxy group is present. A variety of pyran lactones have been synthesised in this way.[48,49] When used in combination with the directed lithiation techniques of Seebach,[50] the method provided an attractive approach to the elaboration of aromatic rings (Scheme 9.12) for natural product synthesis.[47]

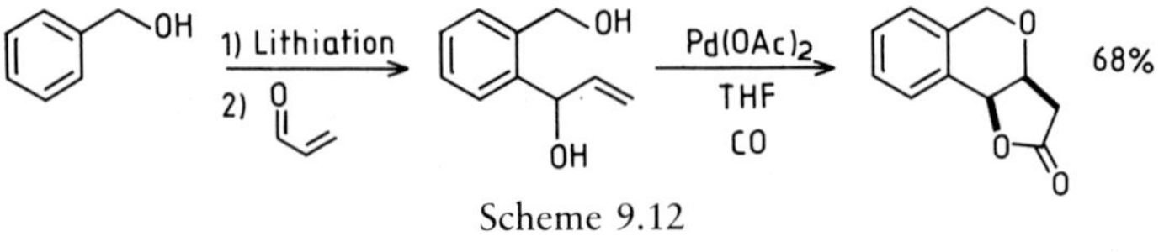

Scheme 9.12

Cyclisations involving the formation of a nitrogen–carbon bond can be promoted in a similar way. The formation of secondary amines from allenes such as (66) provides an interesting example of this type of process.[51]

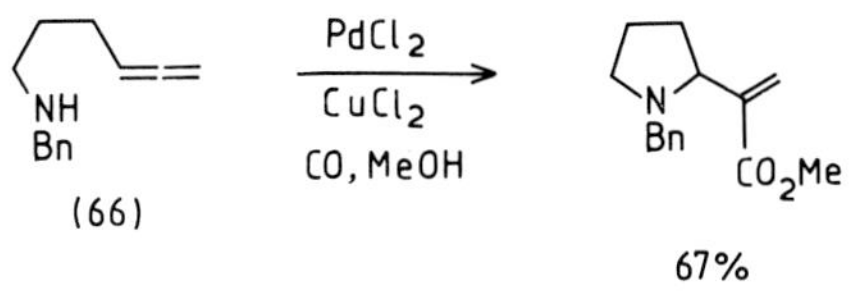

(66)

67%

Apparently similar reaction conditions can combine lactonisation and ester formation into a single process in which both carbonyl groups are obtained from carbon monoxide. It is reasonable to suppose that this reaction initially follows a path involving an oxygen-bound palladium species resembling intermediates in the formation of α-methylene lactones from alkoxy alkynes discussed above. In this example, however, the resulting alkylpalladium intermediate is intercepted by a second carbonylation reaction.

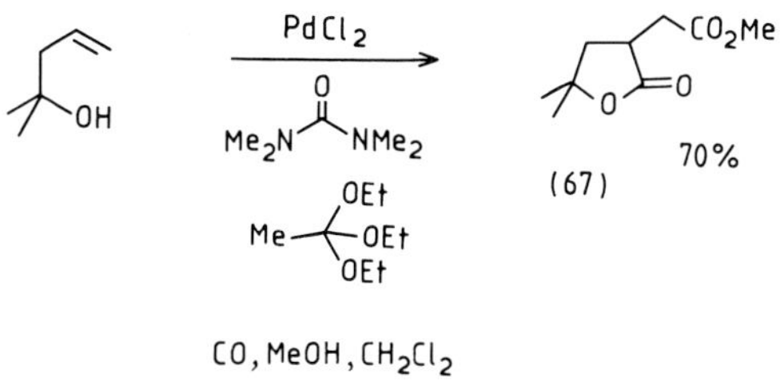

(67) 70%

$CO, MeOH, CH_2Cl_2$

The formation of (67) is typical of reactions of this type, which can also be used to form C–N bonds as seen in the aminocarbonylation of ureas (Scheme 9.13).[52]

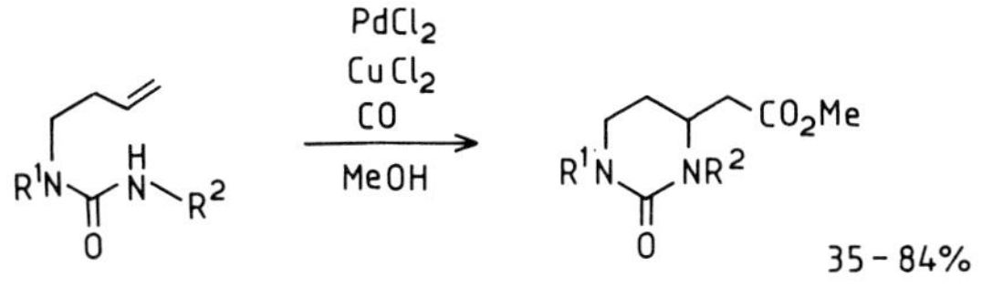

Scheme 9.13

35–84%

Alper[53] has developed a palladium catalysed lactonisation that employs acidic conditions and presumably proceeds by a quite different mechanism.

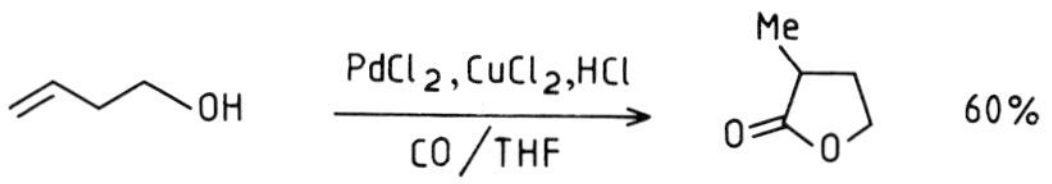

60%

A similar reaction has been reported by Samsel and Norton.[54] Allyl complexes can also undergo efficient carbonyl insertion reactions.[55]

9.3 Carbonylation using iron carbonyls

There are many examples where iron carbonyls react to yield carbonyl insertion products. These usually arise through the intermediacy of an iron σ,π-allyl complex, although the reaction of diiron nonacarbonyl with benzyl halides to yield the symmetrical ketone (68) has also been reported,[56] and requires a different mechanism.

Ph—Cl $\xrightarrow[C_6H_6]{Fe_2(CO)_9}$ Ph—C(O)—Ph   56%

(68)

Reaction of the cycloheptatriene tricarbonyliron complex (69) with acid and then with NaBH$_4$, followed by carbon monoxide insertion, gave the bridged ring ketone (70) in good yield.[57]

(CO)$_3$Fe [structure] $\xrightarrow{\text{1) H}^+ \text{ 2) NaBH}_4}$ [structure] Fe(CO)$_3$ $\xrightarrow{CO}$

(69)

[structure] (CO)$_3$Fe—CO $\longrightarrow$ [structure] O

(70)

Treatment of the cyclooctatetrene Fe(CO)$_3$ complex (71) with AlCl$_3$ in the presence of carbon monoxide also effected carbonyl insertion and provided a good synthesis of barbaralone (73) via the complex (72).[58]

[structure] Fe(CO)$_3$ $\xrightarrow[C_6H_6]{CO/AlCl_3}$ [structure] Fe(CO)$_3$ $\xrightarrow[100 \text{ atm.}]{CO, 120°C}$ [structure] O

(71)   (72)   (73)

Another AlCl$_3$ induced carbonylation reaction is seen in a synthesis of 2-indanone (75) from the complex (74).[59]

[structure] Fe(CO)$_3$ $\xrightarrow{AlCl_3}$ [structure] =O

(74)   (75)

Simple tricarbonyliron complexes have been shown[60] to undergo a Lewis acid mediated ring expansion to form the insertion products (76) and (77). Although such transformations typically give mixtures of regioisomers and proceed in only moderate yield, a more recent examination[61] of the influence of substituents on the course of the reaction has provided a remarkable example in which (78) is formed selectively and in an unusually efficient way.

Far more consistent yields can be expected in cases where metal alkyl intermediates arise from the opening of strained rings. Vinyl cyclopropanes may be ring-expanded in this way to form cyclohexenones upon treatment with $Fe(CO)_5$ or $Fe_2(CO)_9$,[62] as shown in the synthesis of the enones (79). Similarly, vinyl epoxides may be converted into lactones[63] such as (80).

R = Me, Ph, p–chlorophenyl, p–anisyl, 2–thienyl.

The initial step of this reaction involves the opening of the epoxide to produce a π-allyl intermediate. The oxygen of the epoxide cyclises onto the metal, and carbonyl insertion follows to produce a ferralactone intermediate. These initial stages of the lactonisation process have been investigated in some detail.[64]

This lactonisation reaction has now been developed considerably, by the work of Ley's group,[65] to afford either four- or six-membered rings, depending on the conditions used. The method has been applied in a number of efficient natural product syntheses.[66] A typical example is the synthesis of malingolide, to be discussed in Chapter 14. A case in which the size of the lactone ring can be readily controlled[64,66] is to be found in studies leading to the synthesis of Massoia lactone (81).

The combination of methods of this type with the Sharpless methodology for the preparation of resolved epoxides discussed in Chapter 3, provides a powerful route for the formation of optically active ester derivatives. In an example related to synthetic work on the ionophore antibiotic CP 61405, an initial Sharpless epoxidation kinetic resolution is followed by oxidation of the alcohol control group to an aldehyde to allow the introduction of a vinyl group adjacent to the epoxide by a Wittig reaction. Complexation with $Fe_2(CO)_9$ produced the terralactone intermediate which, upon heating with carbon monoxide, was converted into the optically active intermediate (82).[67]

Use of tetracarbonylferrate

Collman and co-workers have demonstrated the effectiveness of disodium tetra-carbonylferrate[68,69] for a wide range of carbonylation reactions[69] indicating the considerable selectivity that can be obtained. The reagent is easily prepared.

$$Fe(CO)_5 + Na \xrightarrow[C_6H_5COC_6H_5]{dioxane, 100°C} Na_2Fe(CO)_4 \ 1.5 \ dioxane$$

Some typical applications are shown in Scheme 9.14. These reactions proceed via an anionic tetracarbonyliron acyl intermediate which then reacts with a second alkyl halide. In this way, unsymmetrical ketones such as (83) can be produced. By employing other electrophiles, a variety of products can be obtained.

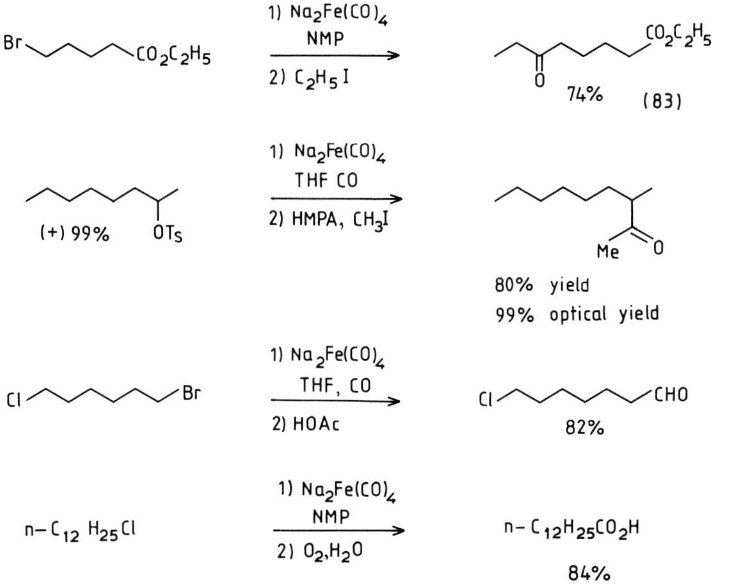

Scheme 9.14

Displacement of a tosylate leaving group from the alkene (84) can be used to produce a cyclic ketone.[69,70] This type of reaction has been employed[71] in the later stages of a

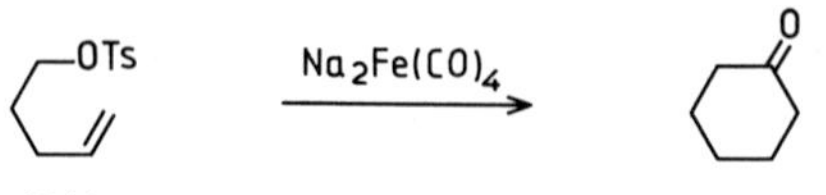

(84)

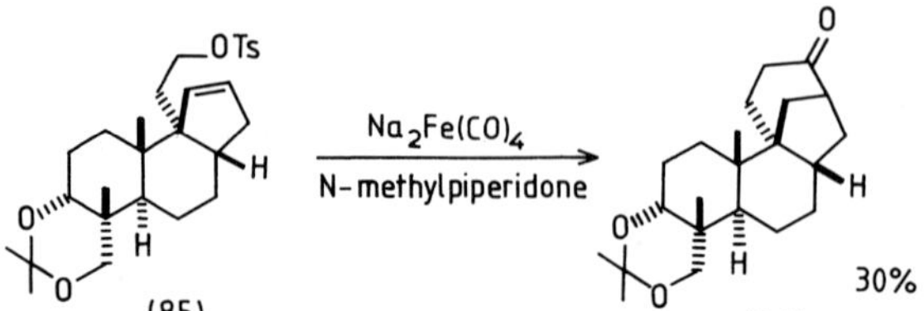

(85)   (86)   30%

synthesis of aphidicolin. The conversion of (85) into the ketone (86), however, proceeded in only 30 % yield. The advantages and limitations of $Na_2Fe(CO)_4$ have been reviewed, and the mechanism of the reaction has been discussed in detail.[69, 72]

The $[Fe(CO)_4]^{2-}$ anion may be generated *in situ* by mixing $Fe(CO)_5$ in an organic solvent by addition of aqueous NaOH and a phase transfer catalyst. By this means, alkyl iodides and benzyl bromides can be converted into the corresponding symmetrical ketones in high yield ($\approx 90 \%$) as shown in Scheme 9.15.[73]

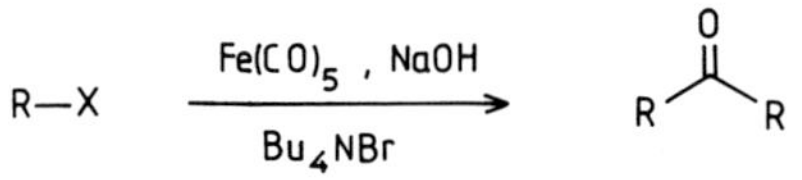

$$R—X \xrightarrow[\text{Bu}_4\text{NBr}]{\text{Fe(CO)}_5 \,,\ \text{NaOH}} \overset{O}{\underset{R \quad R}{\|}}$$

a: $R = C_6H_5CH_2$, $X = Br$,   c: $R = p\text{-}CH_3C_6H_4CH_2$, $X = Br$,   e: $R = n\text{-}C_8H_{17}, X = Br$

b: $R = p\text{-}ClC_6H_4CH_2$, $X = Br$,   d: $R = n\text{-}C_8H_{17}$, $X = I$.

Scheme 9.15

Since alkyl bromides were found to be less reactive, it was possible to use these for the synthesis of unsymmetrical ketones (Scheme 9.16).[73]

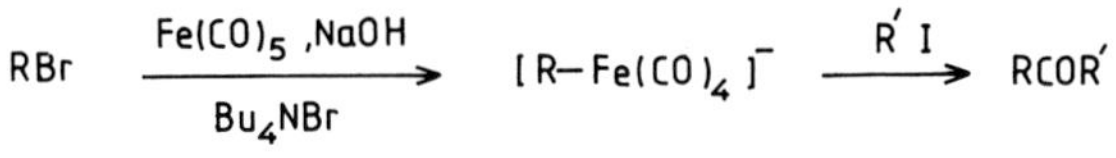

$$RBr \xrightarrow[\text{Bu}_4\text{NBr}]{\text{Fe(CO)}_5 \,,\text{NaOH}} [R—Fe(CO)_4]^- \xrightarrow{R'I} RCOR'$$

a: $R = n\text{-}C_8H_{17},$ $R' = C_2H_5$ 45%

b: $R = n\text{-}C_{16}H_{33},$ $R' = C_2H_5$ 60%

Scheme 9.16

A similar phase transfer procedure has been reported[74, 75] for the synthesis of carboxylic acids via a cobalt carbonyl catalysed carbonyl insertion reaction of, for example, benzyl halides (Scheme 9.17).

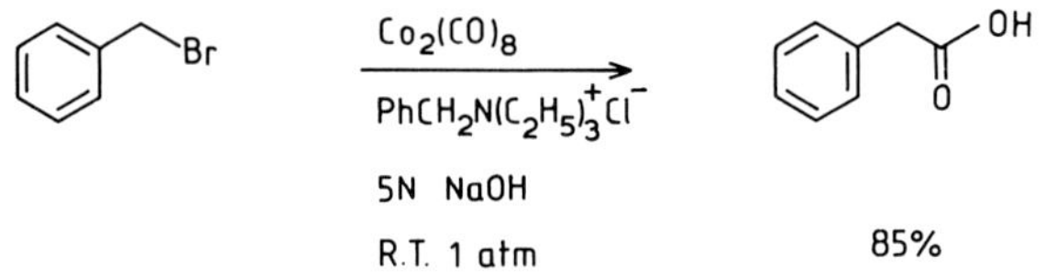

Scheme 9.17

9.4 Carbonylation using cobalt carbonyls: the Pauson–Khand reaction

In recent years, initially through the efforts of Pauson's group, a method has been developed to convert $Co_2(CO)_6$(alkyne) complexes into cyclopentenones. This reaction, referred to as the 'Pauson–Khand reaction', can show a high degree of stereo- and regiocontrol in the cyclisation of an alkene, an alkyne, and carbon monoxide. Typical procedures[76] (Scheme 9.18) involve heating an alkyne complex and the alkene, or the free alkene and alkyne in the presence of dicobaltoctacarbonyl, at about 110 °C in a sealed tube.

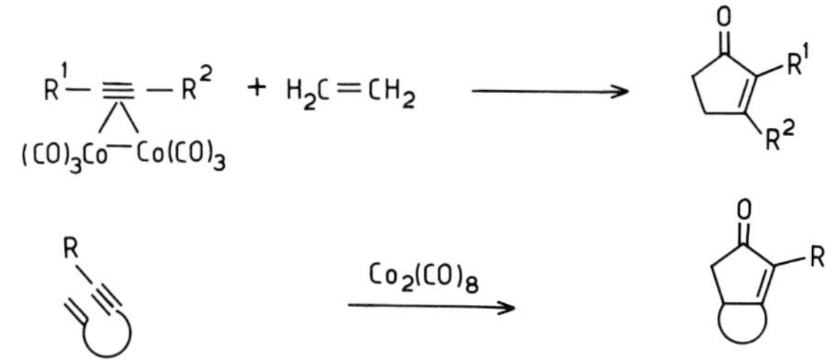

Scheme 9.18

This method has now found application by a number of groups for the synthesis of prostaglandins,[77] carbocyclines,[78] and polyquinanes.[76] Some examples are shown in Scheme 9.19.

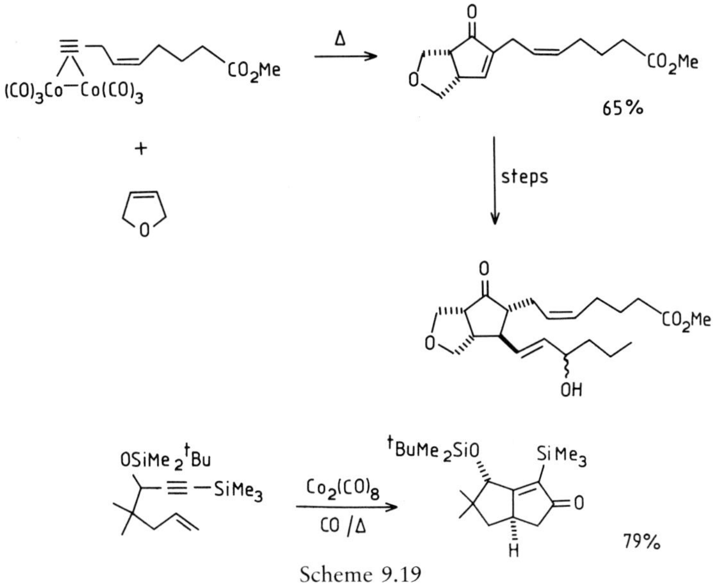

Scheme 9.19

Reaction via the alkyne complex most probably proceeds by attack of the alkene to form a bimetallic metallocycle of the type (87), so introducing the metal carbon σ bond

required for the alkyl migration step. Collapse of the metal acyl intermediate (89) to give the alkene complex (90) then leads to the formation of the cyclopentenone product after loss of $Co_2(CO)_6$. The most important aspect of this cyclisation method is the ability of $Co_2(CO)_8$ to effect the cyclisation of alkenes and alkynes that are present in the same substrate molecule, since intramolecular interactions in intermediates such as (87) and (88) can then efficiently determine the regiochemical and stereochemical outcome of the cyclisation process.[79]

A zirconium catalysed version of the Pauson–Khand reaction has recently been developed and offers an alternative entry to polyquinane systems.[80] The starting material (91) was prepared from a dialkyne using coupling reactions of the type discussed in Chapters 8 and 10.

This reaction has now been examined with a variety of enyne substrates.[81]

Organoiron reagents can also be used in the conversion of alkynes into cyclopentenones[82] and indenones.[83] In the first instance, $Fe_2(CO)_9$ combined an alkyne, carbon monoxide, and an allene to produce β-methylenecyclopentenones. In the second case, diphenylacetylene was combined with a $CpFe(CO)_2$-substituted arene. Although several examples have been described for both these processes, yields, in general, were low. In a similar process, either $Fe_2(CO)_9$ or $Fe(CO)_5$ was used to combine alkynes, allenes and CO in moderate yields.[84]

(92)

Pauson–Khand reactions can also offer synthetic chemists a remarkable route to more elaborate polycyclic systems.[85] The formation of (92) illustrates the high degree of stereocontrol possible in these reactions. In this case, chiral centres in the starting material establish the orientation of the ring junction between the two five-membered rings. Similar cyclisations in which an alkoxy group is placed adjacent to the alkyne portion of the starting material give similarly high stereocontrol arising from a 1,2-asymmetric induction.[86] A further example[87] places all the substituents in the cyclopentenone ring.

Another similar cyclisation of the substrate (93), this time without isolation of the cobalt alkyne complex, afforded a further tricyclic polyquinane product.[88]

(93)   30%

Until recently,[89] good regiocontrol in Pauson–Khand cyclisations has only been available by use of intramolecular reactions. The selection of suitable substituents on the alkyne ligand and on the alkene reactant has now been used to provide satisfactory regiocontrol in the intermolecular formation of cyclopentenones from dicobalthexacarbonyl alkyne complexes. Steric effects appear to be responsible for the regioncontrol in these reactions.[90]

When used in combination with the electrophilic properties of the propargyl cation complexes (see Chapter 5), the interesting possibility emerges that nucleophile addition can be followed by removal of the metal by the introduction of the cyclopentenone ring. The impact of this concept upon synthesis design is considerable, and is discussed in detail in Chapter 14. Two examples, however, will be given here.

The prostaglandin analogue (94) was prepared using this approach.[91] A similar strategy has been used by Schreiber to gain access to medium ring compounds. Reaction of the complexed propargyl acetal (95) with Lewis acid catalysis effects addition of the allylsilane to an intermediate cobalt stabilised carbenium ion, so forming the substituted cyclooctyne intermediate (96). A Pauson–Khand cyclisation was used in the next step to combine the alkyne with the remaining allyl ether substituent to form the cyclopentenone (97).[92] In this last example, the application of the intermolecular addition of allyl silanes to cobalt stabilised propargyl cations (discussed in Chapter 5) provides a good illustration of the high degree of sophistication now apparent in organometallic methods of this type. The control of relative stereochemistry between the substituents in the eight-membered ring, and subsequently at the ring junction between the two five-membered rings following the Pauson–Khand step, is notable in the reaction sequence. The use of the propargyl cation to form the eight-membered ring also constitutes an essential part of the strategy. Otherwise, an alkyne linkage could not be accommodated in intermediates containing a ring of this size. The cobalt complex is used to control four carbon–carbon bond formations and to establish three chiral centres, in addition to distorting[93] the alkyne sufficiently from its normal linear geometry to permit its presence in an eight-membered ring intermediate.

9.5 Carbonylation using nickel

Carbonylation reactions induced by nickel catalysts have been widely exploited in industry. They are commonly termed Reppe reactions, after the German chemist who explored and developed the field during the 1930s and 1940s. Perhaps the most important Reppe reaction is the carbonylation of ethyne to yield acrylic acid, which, in turn, is used in the production of acrylic resins.[94]

The reaction is written as a stoichiometric transformation; however, the process is operated catalytically using nickel salts with halide promoter as catalyst. In this form,

$$H-\equiv-H + Ni(CO)_4 \xrightarrow[\substack{H_2O \\ 40°C,\ 1\ atm.}]{HCl} \text{(acrylic acid)} + NiCl_2$$

higher temperatures and pressures (250 °C and 100 atm) are required.[94] Other examples are indicated by the formation of (98) and (99).[95]

$$R-\equiv-H \xrightarrow[EtOH]{Ni(CO)_4} \quad (98)$$

$$R = n-C_6H_{13}, C_6H_5$$

$$^nPr-C\equiv C-^nPr \xrightarrow[H_2O]{Ni(CO)_4} \quad (99)$$

An interesting application of this reaction to a somewhat more complex molecule is seen in the formation of (101), which was used in a synthesis of ($\pm$)-sirenin to introduce an alkoxycarbonyl group into a readily accessible alkyne.[96] Reaction occurs preferentially at the alkynyl bond of (100) rather than at the trisubstituted alkene site.

$$\text{(i) } ^nBuLi \quad \text{(ii) } (CH_2O)_n$$

$$\xrightarrow[EtOH,\ AcOH]{Ni(CO)_4,\ H_2O}$$

(100)   (101)

Further instances of the use of nickel carbonyl are discussed below under the carbonylation of allylic halides.

Carbonylation of allylic halides

Nickel or palladium π-allyl derivatives may be carbonylated in the presence of an alcohol to yield an ester (Scheme 9.20).[97] In a somewhat more sophisticated application, the reaction of $Ni(CO)_4$ and carbon monoxide with an allyl derivative (in this case the methane sulphonate) is found in a synthesis of the α-methylene lactone (102).[98] This reaction was incorporated into a successful route to frullanolide (103), demonstrating

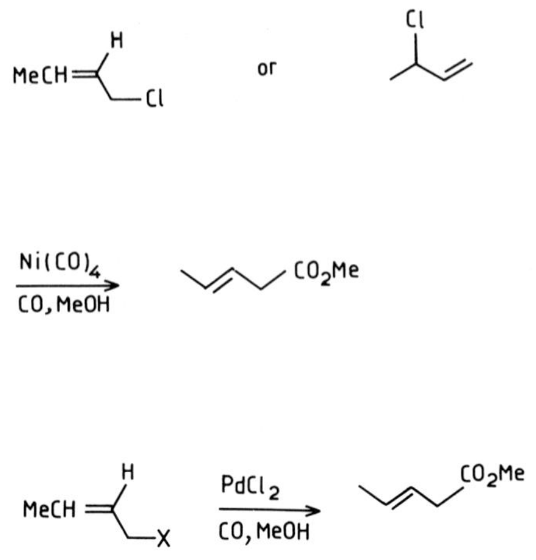

MeCH=CH—CH₂Cl   or   [structure]

$$\xrightarrow[\text{CO, MeOH}]{\text{Ni(CO)}_4}$$ [structure] CO₂Me

MeCH=CH—CH₂—X $\xrightarrow[\text{CO, MeOH}]{\text{PdCl}_2}$ [structure] CO₂Me

Scheme 9.20

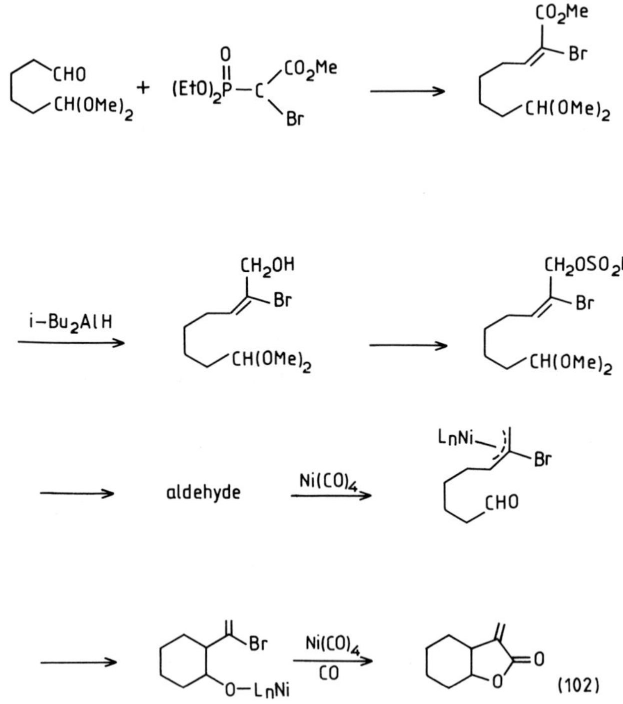

[structure: CHO / CH(OMe)₂] + (EtO)₂P(O)—C(Br)—CO₂Me ⟶ [structure]

i-Bu₂AlH ⟶ [structure] ⟶ [structure]

⟶ aldehyde $\xrightarrow{\text{Ni(CO)}_4}$ [structure]

⟶ [structure] $\xrightarrow[\text{CO}]{\text{Ni(CO)}_4}$ [structure] (102)

an alternative to the palladium catalysed approach to α-methylene lactones discussed earlier and in Chapter 15.

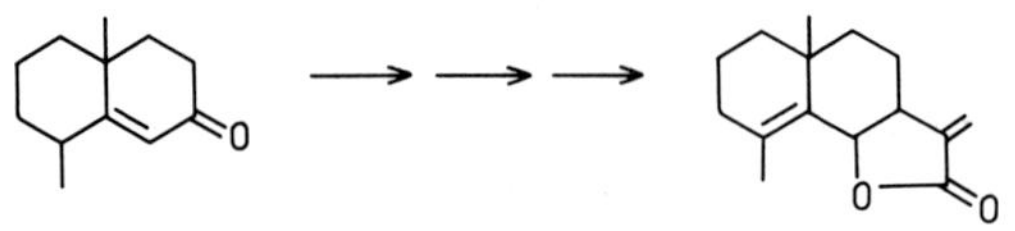

[structure] ⟶ ⟶ ⟶ [structure] (103)

9.6 Some unusual carbonylation reactions

In the Pauson–Khand reaction (Section 9.4) we encountered an example where two unsaturated hydrocarbons were linked together by a transition metal prior to carbonyl insertion. A number of telomerisation reactions show a similar overall outcome, but produce linear rather than cyclic products (see also Chapter 10). One example uses palladium catalysis to link two butadiene molecules to form the 2,7-dialkene (104) which was used as a substrate for a synthesis of endo-brevicomin[99] (see Chapter 14).

When other palladium catalysts were selected in place of the acetate, alkoxy-carbonylation occurs on the monomer to give mainly the 1,4-addition product (105).[100]

A curious combination of hydrosilation (see Chapter 10) and carbonylation takes place when $Co_2(CO)_8$ is used to add carbon monoxide to an alkene in the presence of $HSiMeEt_2$.[101] An enol ether (106) is produced. A similar carbonylation cleaves an ether to produce the aldehyde (107). Indeed, epoxides can also undergo much the same reaction.[102]

n = 3 to 6

Palladium catalysis has been used to form an aldehyde by cleaving a carbon–oxygen bond within an ester. The reducing component in this carbonylation reaction was provided by the use of synthesis gas (CO/H_2).[103] This reaction has the advantage over the conventional production of ethanal from methanol and synthesis gas, that low pressures are adequate when methyl acetate is used as the substrate.

A combined hydrogenation/carbonylation has been used by Eilbracht's group[104] to form a trisubstituted cyclopentanone in a synthesis of α-cuparenone (108).

Carbonylation by the use of isocyanides

Use of an isocyanide in place of carbon monoxide may at times be a convenient alternative means to effect an overall carboxylation. This is illustrated by the synthesis[105] of (±)-muscone (110) which starts from the butadiene trimer–nickel complex (109) by condensation with allene followed by reaction with ᵗBuNC.

A version of the Pauson–Khand cyclisation (Section 9.4) has also been developed to use an isocyanide in the place of carbon monoxide. In this case, nickel catalysis is used and the conversions take place under more moderate reaction conditions.[106]

It has also been found possible[107] to transfer a carbon fragment (C—R′) from an amidine, derived from a 2-aminopyridine, to afford the ketone (111). In this case, $RhCl(PPh_3)_3$ was used as the catalyst.

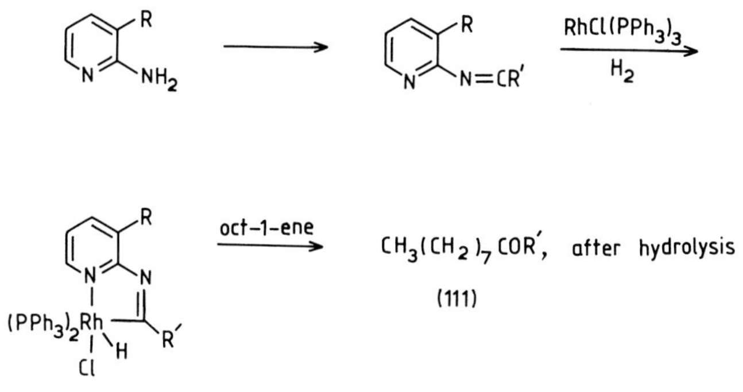

$$CH_3(CH_2)_7 COR', \quad \text{after hydrolysis}$$

(111)

$(R = Ph. \ p-ClC_6H_4 . \ p-MeC_6H_4)$

9.7 Uses of metal acyl complexes in synthesis

The wide range of organic products (described at the start of this chapter) that can be obtained from metal acyl complexes upon removal of the metal, make these compounds useful as versatile intermediates in synthesis. More profound advantage can be gained from this versatility if the reactivity properties of the acyl intermediates themselves can also be exploited. This is of particular note when the metal atom is a chiral centre, since extremely high diastereoselectivities have been obtained in such circumstances. Elaboration of acyl complexes (Scheme 9.21) has provided some of the best stereocontrolled enolates currently available.

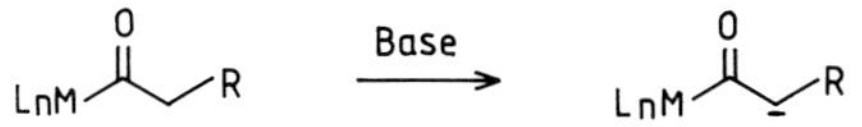

(112)

Scheme 9.21

The basic principle of the reaction is simple. Deprotonation of metal acyl complex (112) produces an 'enolate' (the bonding is rather different to organic enolates, see below). In cases in which the metal atom carries a sufficient number of different ligands, for example the cyclopentadienyl and phosphine ligands in (113, L = PPh$_3$), this enolate is in the unsymmetrical environment of the Cp and L ligands on the metal. This results

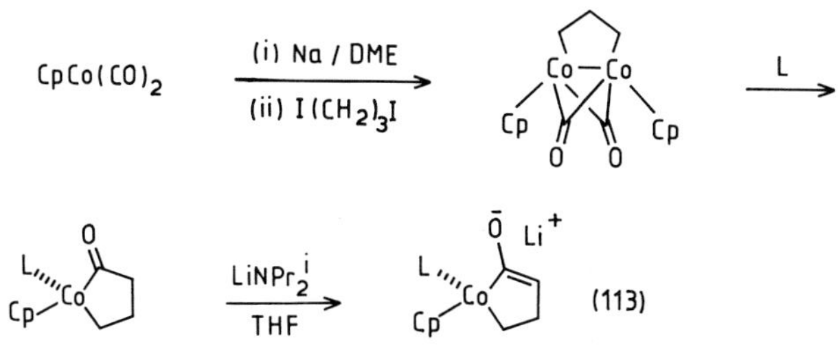

(113)

in a degree of stereoselectivity in the reaction, and raises the possibility of an asymmetric induction in the enolate/electrophile reaction to produce optically active compounds.

These findings may be illustrated by some examples[108] in which the ratio of stereoisomers could be estimated by n.m.r. (Scheme 9.22). The results indicate a considerable steric preference for entry of the electrophile *cis* to the Cp ligand.

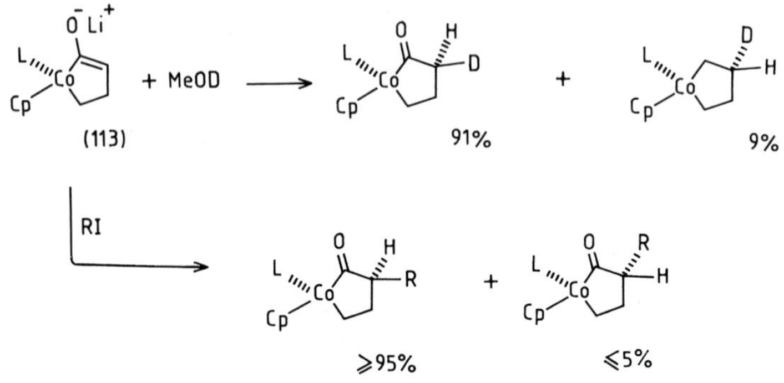

Scheme 9.22

The stereoselectivity of the reaction was confirmed[108] by the examples of aldol condensation of the enolate (113, Scheme 9.23).

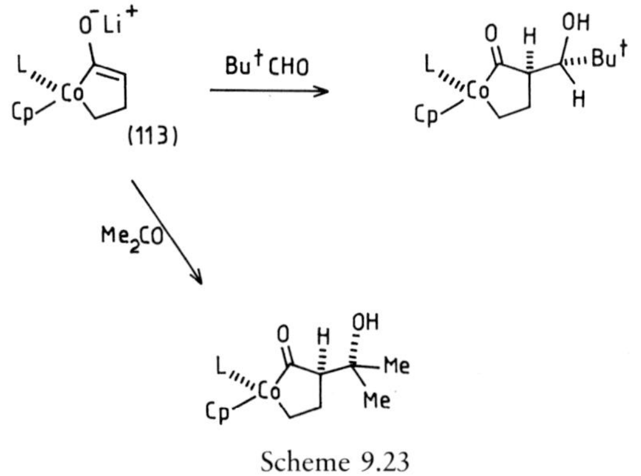

Scheme 9.23

Induced optical activity was observed as the result of the sequence shown in Scheme 9.24 using (+)-PhCH(Me)N(Me)PPh$_2$ as chiral phosphine ligand. The two isomers of (114) were formed in approximately equal amounts, but separation by crystallisation gave the (−)-isomer pure, which led to (115) with high stereoselectivity. The oxidation step, using FeCl$_3$, gave the cyclobutanone product (116) optically pure and in high yield.

These reactions using cobalt complexes have been parallelled in similar developments using the now far more extensively applied cyclopentadienyliron carbonyl complexes with which quite remarkable degrees of stereoselectivity have been obtained despite the possibility that rotation might occur in the absence of the metallocyclic ring. The

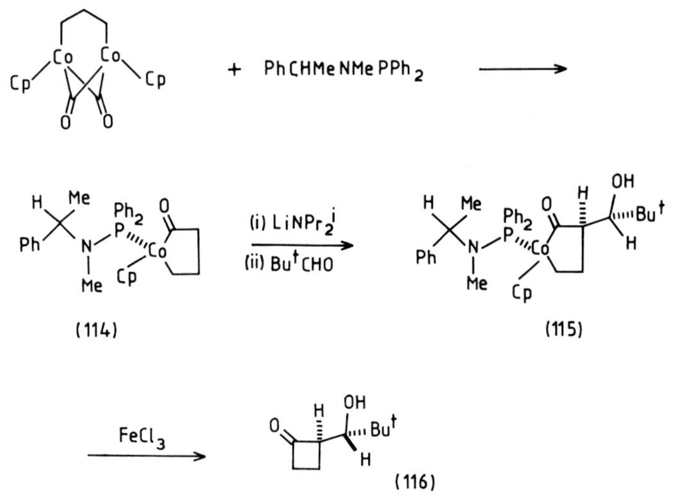

(114)   (115)

(116)

Scheme 9.24

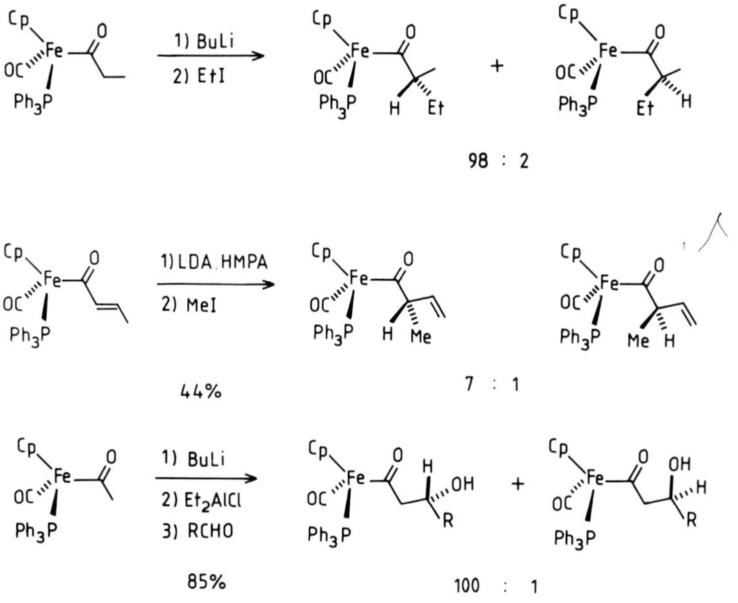

Scheme 9.25

examples in Scheme 9.25 indicate that the faces of the enolate are very efficiently differentiated and that a single conformation and enolate stereochemistry appear to dominate the reaction path. The simple nature of the metal complex intermediates makes this a most suitable reaction for synthetic application, particularly in cases where resolution of the metal centre leads to enantioselective reactions based on the metal mediated asymmetric induction process. Similar alkylations of metal carbene complexes are discussed in Chapter 12. The deprotonation step requires a strong base such as ^{n}BuLi or LiN(iso-Pr)$_2$. Deprotonation of (acyl)Fe(CO)$_2$Cp complexes initially proved less straightforward[109] than the reaction of phosphine substituted analogues,[110] although good results have been obtained using lithium hexamethyldisilazide.

Over the last few years, the procedures for iron acyl complexes have been further developed to the point where stereoselectivities better than 200:1 are commonplace in many reactions.[111] By the use of high field n.m.r., the Davies group has been able to make accurate measurements of product ratios of this order. The most important development, however, in this area, has been the large scale resolution by Davies of the acetyl complex (117) which is now commercially available as either enantiomer in optically pure form. The resolution procedure has not been disclosed but details of the determination of the absolute configuration of the (+) and (−) isomers of (117) have been described.[112] X-ray crystallography of menthyl derivatives (118) provided the absolute configuration information. N.m.r. spectroscopy at 500 MHz demonstrated the optical purity of both the (+) and the (−) isomers of (118) and so, also, of the two enantiomers of (117).

(R)-(−)- (117)   (−)- (118)

By simple elaboration of (117) through the reaction of its enolate with alkyl halides, a range of substituted fully resolved acyl starting materials are available. Several examples have now appeared[111] of the use of these compounds in enantioselective synthesis.

(R)- (119)

(SS)-(−)-(120)   100 %   83%

The ethyl ketone (119) has been converted into the antihypertensive agent captopril, an inhibitor of the angiotensin-converting enzyme (ACE). In this synthesis, the stereocontrolled enolate methodology was used to introduce the sulphur containing side-chain and the metal was removed to produce the required amide by introduction of the t-butyl ester of L-proline. The product (120), which had identical melting point and optical rotation to an authentic sample, was obtained in 59% overall yield from (119).[113]

In another example, homochiral iron acyl complexes have been employed as chiral succinate equivalents. The intermediate (121), obtained from (S)-(+)-(117) by reaction with t-butyl bromoacetate, can be deprotonated selectively next to the ester. With the site of reaction moved further from the chirality of the iron acyl control group, more moderate stereocontrol was observed in reactions of the resulting enolate. In a typical example, reaction with iso-butyl iodide gave (122) as a 15:1 mixture of diastereoisomers. Oxidation of (122) with NBS, followed by removal of the ester in acid, afforded the succinic acid derivative (123) in 86% yield. Alternatively, oxidation in the presence of 1-phenylethylamine formed the amide (124) as a single diastereoisomer in 90% yield. Similar reactions with (117) have been examined using t-butyl 2-bromopropionate.[114]

Studies of this type by the groups of Davies and Liebeskind have demonstrated the generality of these reactions for the control of stereochemistry adjacent to the metal centre, in both simple[115] and extended[116,117] enolates and in aldol reactions.[118] This latter reaction is noteworthy since stereocontrol better than $100:1$ was obtained at the β-position without the requirement for substitution at the α-carbon. Reaction of enolates with epoxides in the presence of either Et_2AlCl or of $BF_3 . OEt_2$ also proceeds with a high degree of stereocontrol.[119]

Used in sequence, such reactions provide a well controlled route to *erythro* products such as (125). The stereocontrol of direct aldol alkylations was found to be sensitive to

(125)

the nature of the counterion,[120] an observation that has been put to good use in the development[121] of direct alkylation routes to both *erythro* and *threo* aldol condensation products.

The high degree of stereocontrol exhibited in these reactions can be best explained[122] on the basis of a 'pseudooctahedral' geometry about the metal atom. The carbonyl, phosphine and acyl ligands are thus aligned at approximately 90° to one another, compared to the much wider angle encountered about a tetrahedral centre. This difference accentuates the blocking effect of the phosphine ligand which effectively precludes electrophile addition to the lower face of the enolate (126).

(126)

The α,β-unsaturated carbonyl group in (127) can also be used in conjugate addition/enolate trapping reactions (see Chapter 8). Alkyl lithium reagents, rather than

(127)

the usual cuprate or copper catalysed Grignard reagents, were used for the conjugate addition. The enolate formed in this way was then alkylated by methyl iodide in the usual fashion. Because of the blocking effect of the phenyl group (see above) both nucleophilic and electrophilic reagents approach the acyl ligand from the top face.[123] A similar *cis*-addition of nucleophile and electrophile, arising for a different reason, was emphasised in the discussion of the stereodirecting effects of $Cr(CO_3)$ π complexes in Chapter 7. Stereocontrolled cyclopropanation has also been performed.[124]

In an interesting variation on the aldol theme, alkylation of the enolate (126, R,R′ = H, M = Fe, L = CO) by imines, has been used in routes to β-lactams (128) which were formed upon removal of the metal.[125]

Conjugate addition has also been employed to form β-lactams using resolved iron acyl complexes.[126] A more detailed discussion of these and other organometallic methods for the formation of β-lactams is given in Chapter 15.

In most of the examples of chiral iron acyl chemistry discussed so far in this chapter, the metal centre, apart from its unusual pseudooctahedral geometry, has served simply, though extraordinarily effectively, as a chiral auxiliary blocking one face of a reactive pendent group attached to the metal. Other ligands on the metal[127] or the metal atom itself[128] can, however, become more intimately involved in the reactions in a fashion that can promote special stereocontrol effects. For example, treatment of the aldol addition product (129) with two equivalents of butyllithium produced a cyclised enolate

intermediate with an anionic centre. This reacted with complete stereocontrol with methyl iodide to produce (130) and ultimately (131) after oxidation with aqueous bromine.[127]

The metal centre intervenes in rearrangements of α-alkoxy iron acyl complexes to form esters such as (133). The stereochemistry of these interconversions reveals the role of the metal. The reaction proceeds stereospecifically with inversion of configuration of carbon. A mechanism has been proposed that involves a metal assisted displacement of the alkoxy substituent to form a cationic ketene π complex (132). This intermediate then

reacts with benzyl alcohol at the complexed carbon atom of the ketene carbonyl group, so producing an ester.[128] Involvement of the metal in this case produces an intermediate with planar chirality in addition to the metal centred chirality of the starting material. This type of CpFe(CO)L alkene complex is well known and has been discussed in Chapter 5, but the generation of a ketene complex with planar chirality, presumably with complete asymmetric induction, is an unusual stereochemical feature.

Not only does the chemistry of chiral iron acyl complexes offer synthetic organic chemists one of the most potent stereocontrol groups currently available, but the availability in intermediates of other reactive ligands or of the wider-ranging reaction properties of transition metals in their complexes, if properly controlled, demonstrates further prospects for stereocontrol effects. This work is related to broader studies of organometallic carbene and enolate chemistry (see Chapter 12). The distinction between σ bound auxiliary and planar chirality in π complexes, for example, is further illustrated by the work of Bergman and Heathcock *et al.* with π bound η^3 heteroallyl enolate ligands that can interconvert with O, or C bound σ bonded formulations.[129]

9.8 Double carbonyl insertions

A reaction of metal acyl complexes which exploits the continued presence of a metal carbon σ bond, involves a second carbonyl insertion step. Overall, this constitutes double carbonyl insertion. Catalytic systems, typically employing palladium and cobalt complexes,[130] can effect, sequentially, two carbonyl insertions in a single reaction. Palladium catalysis can give access to α-keto acids,[131] esters[132] and amides,[133] although often other products such as acids and aldehydes are also formed. At its present state of development, the reaction is highly sensitive to both the solvent and the nature of reactants. The overall transformation is shown in Scheme 9.26.

Some insight into the control of these reactions can be gained by consideration of likely mechanisms. The overall process, two carbonyl insertions followed by nucleophile (e.g. ROH, R_2NH) addition, can follow the two distinct routes indicated in Scheme

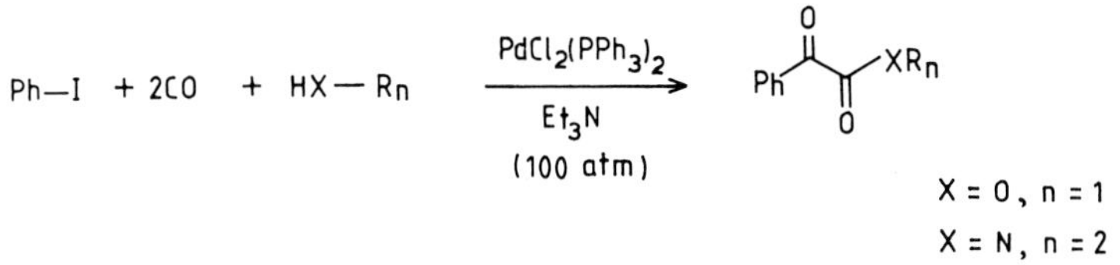

Scheme 9.26

9.27. In the case of palladium catalysed formation of $PhCOCONEt_2$, at least, the involvement of path B has been established.[134]

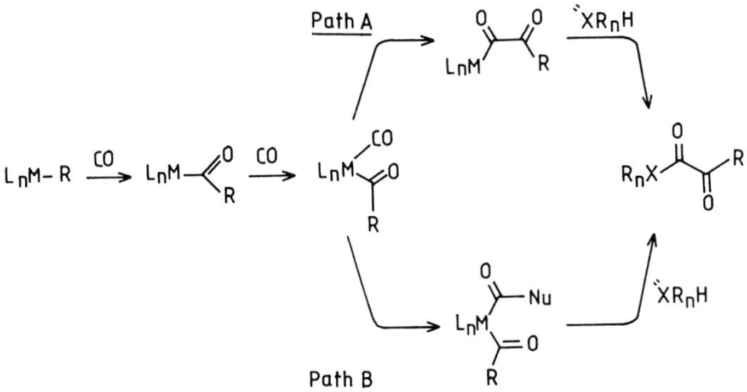

Scheme 9.27

Detailed kinetic studies have led to the proposal by Yamamoto and co-workers that nucleophile addition to the acyl palladium species and its carbon monoxide adduct determines the rate of the reaction.[135]

An intramolecular example of the nucleophile addition step is found in a cobalt catalysed reaction that results in the formation of a lactone (134) from styrene oxide.

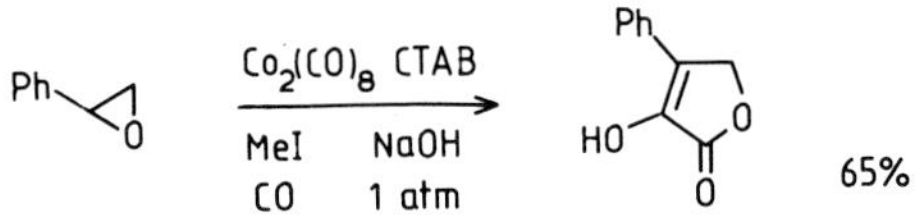

The involvement of methyl iodide was proposed to lead to an acetyl cobalt complex, the active catalytic species, which followed path A after opening the epoxide to form a metal alkyl intermediate. Phase transfer conditions were used to obtain this result.[136]

A similar double carbonylation cyclisation has been applied to the synthesis of oxamates by Murahashi's group.[137] Yields varied from 67 to 86%. The bicyclic product (135) was obtained in 70% yield.

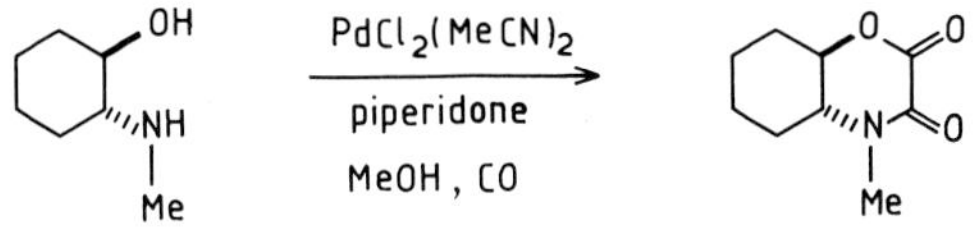

A double insertion process that inserts first carbon monoxide and then an alkyne into an alkylmanganese bond, has been described by DeShong *et al.*[138] An analogous reaction was applied in the synthesis of a number of ketoesters by the use of alkynes as the second ligand in a sequence of insertion steps.[139]

Access to bulk chemical feedstocks by double carbonylation is of growing importance. In particular, processes involving synthesis gas (CO/H_2) have received much attention over the past 20 years, a result of the oil/energy 'crisis' of the 1970s. Projections then were that synthesis gas derived from coal would, by the end of the century, be a more economical feedstock than naphtha from oil. An interesting reaction born of these concerns is the Union Carbide high pressure ethandiol synthesis which is shown in Scheme 9.28.[140]

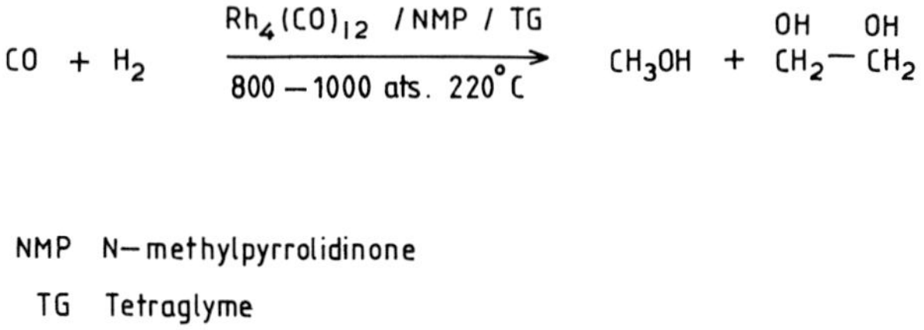

$$CO \ + \ H_2 \quad \xrightarrow[\text{800 - 1000 ats. 220°C}]{\text{Rh}_4(CO)_{12} \ / \text{NMP} \ / \ \text{TG}} \quad CH_3OH \ + \ \overset{OH \quad OH}{CH_2 - CH_2}$$

NMP   N—methylpyrrolidinone

TG   Tetraglyme

Scheme 9.28

In an analogous ruthenium catalysed reaction, use of radiolabelled materials has shown that methanol and the diol are both primary products. No ^{14}C-ethandiol was produced when the reaction was conducted in the presence of ^{14}C-methanol.[141]

9.9 Decarbonylation

Decarbonylation of aldehydes has been successfully effected with rhodium and ruthenium complexes. Perhaps the most widely used reagent is Wilkinson's complex $RhCl(PPh_3)_3$.[142]

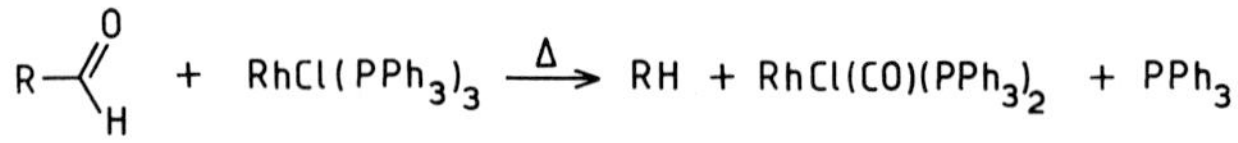

$$R-\overset{O}{\underset{H}{\overset{\|}{C}}} \ + \ RhCl(PPh_3)_3 \ \xrightarrow{\Delta} \ RH \ + \ RhCl(CO)(PPh_3)_2 \ + \ PPh_3$$

The reagent $RhCl(PPh_3)_3$ is, however, converted into the non-catalytic carbonyl complex $RhCl(CO)(PPh_3)_2$ during the process of decarbonylation.[143]

Using an aldehyde with a chiral centre, the process was found to be stereospecific[144] with retention. An example is the conversion of (R)-(−)-(136) into (S)-(+)-(137).

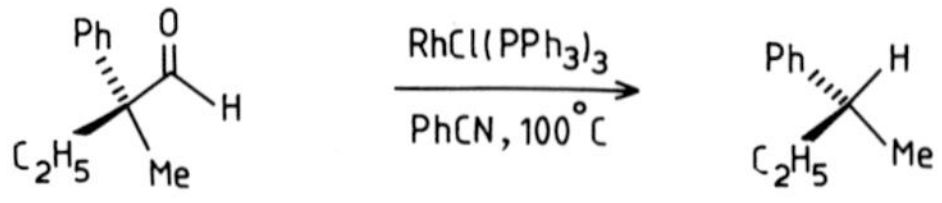

(136)   (137)

The R-group may be aromatic (e.g. Ph, o-OHC$_6$H$_4$, PhCH=CH) or aliphatic, although in the latter case a minor amount of the alkene derived from RH may be formed.[145]

Applications indicate that the procedure is compatible with a range of structures and functional groups in the aldehyde that is decarbonylated. A number of examples are given. The conversion of (138) into (139) required heating at 160°C.[145] The steroid derivative (140) could be decarbonylated using RhCl(PPh₃)₃ in benzene.[146] Similarly, the unsaturated[147] steroid aldehyde (141) gave (142). In the decarbonylation of (143) there was no indication of methyl group migration.

A much more significant and complicated example due to Heusler,[148] is the step (144) to (145) in the course of a transformation of a penicillin into a cephalosporine. The loss of the tertiary *O*-acetyl group should also be noted in this case. Another example in which a rather complex type of molecule successfully decarbonylated using RhCl(PPh₃)₃ is offered by the formation of (146).[149]

(R = tetracetylfuranose)

Decarbonylation using $RhCl(PPh_3)_3$ was an essential step in an ingenious synthesis by Hobbs and Magnus[150] of the boll weevil pheromone agent, grandisol (149, R = H). The cyclobutane aldehyde (148), generated by photolysis of (147), gave grandisol as the acetate by refluxing with $RhCl(PPh_3)_3$ in CH_2Cl_2.

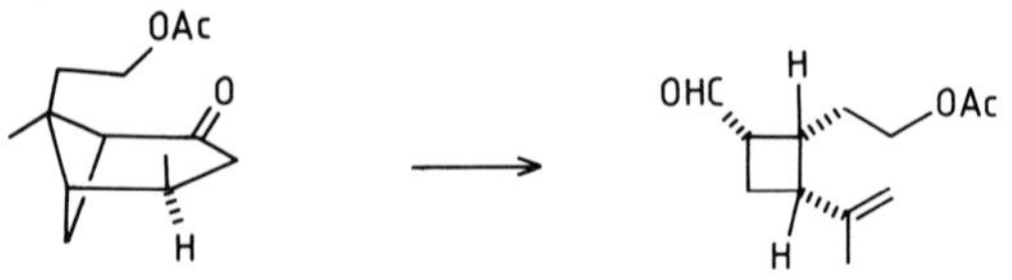

(147)   (148)

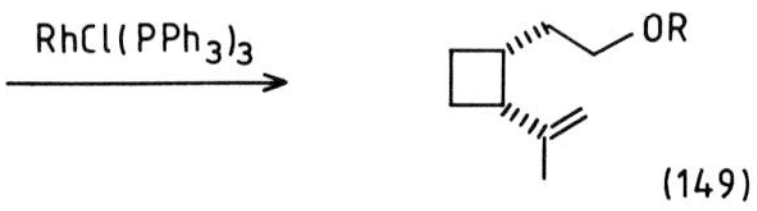

(149)

The decarbonylation procedure may also be applied to effect methylation. The sequence (150) to (151) represents such a case,[151] and (152) to (153) provides a further example.[152]

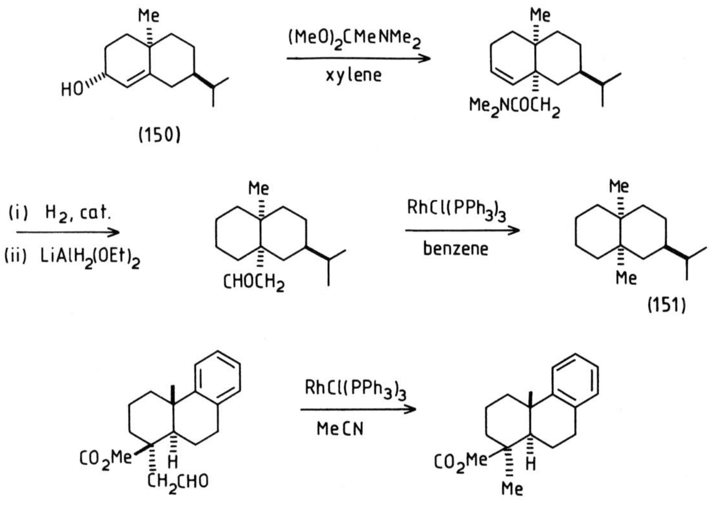

(150)

(151)

(152)   (153)

It is notable, however, that in suitably constituted molecules in which formyl and alkene groups can become close to one another, the action of $RhCl(PPh_3)_3$ may lead to cyclisation. This reaction was observed with (+)-citronellal (154) which gave a mixture of (−)-isopulegol (155) and (+)-neoisopulegol (156) upon treatment[153] with $RhCl(PPh_3)_3$.

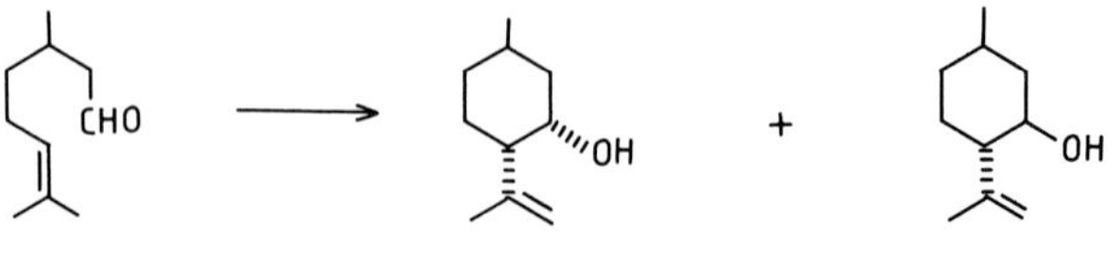

(154)   (155)   +   (156)

Acyl halides have been successfully decarbonylated[154] by warming with $RhCl(PPh_3)_3$ in benzene or CH_2Cl_2. The method has been applied to a range of aromatic acid halides, but aliphatic acid chlorides yield alkenes.

Certain anhydrides of aromatic dicarboxylic acids suffer decarbonylation on heating with $Co_2(CO)_8$ in the presence of H_2 and CO at 200 °C.[155] Examples include (157) and yields are good.

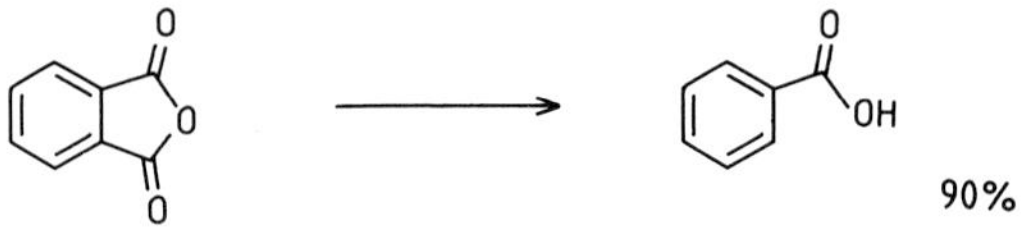

(157)   90%

However, in an aliphatic example (158)[156] with $Ni(CO)_2(PPh_3)_2$ as reagent, both CO and CO_2 are eliminated.

(158)

Similar reactions have been employed in the formation of alkenes such as (159).[157] A reasonable mechanism for this transformation involves initial insertion of palladium into the carbon-chlorine bond of the acid chloride, followed by decarbonylation and β-elimination.

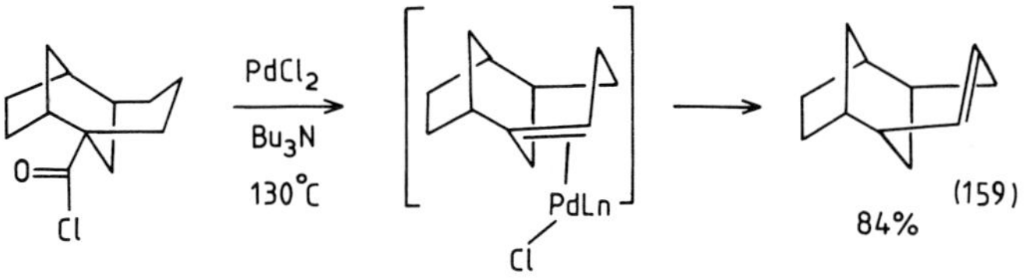

(159)   84%

The subsequent isomerisation of the position of the alkene is a common process of a type described in Chapter 2.

10
Further insertion and coupling reactions

10.1 Insertion of carbon dioxide

The insertion of carbon dioxide into a carbon–metal σ bond parallels the insertion of carbon monoxide discussed at length in the previous chapter. The formation of the iridium complex (1), for example, may be followed under an atmosphere of carbon dioxide by formation of the cyanoacetate complex (2) from which methyl cyanoacetate may be isolated.[1] Similar reactions have been found using an iron complex.[2]

$$[Ir(dpe)_2]Cl \;+\; MeCN \longrightarrow \underset{(1)}{\overset{NC}{\underset{H}{\diagdown}}Ir(dpe)_2}$$

$$\xrightarrow{CO_2} \underset{(2)}{\overset{NC}{\diagdown}\overset{O}{\diagup}\!\!\overset{O}{\diagdown}\underset{H}{Ir(dppe)_2}} \xrightarrow[BF_3]{MeOH} \underset{(3)}{\overset{NC}{\diagdown}\overset{O}{\diagup}\!\!\overset{}{OMe}}$$

$$dppe \;=\; Ph_2P\frown PPh_2$$

Cyanoacetate has itself been used as a source of carbon dioxide by decarboxylation in reactions catalysed by copper (I) salts. The transformation of (3) into propene carbonate (4) illustrates one such example,[3] and more complex cases have been

$$\underset{(3)}{\overset{}{\bigtriangleup\!\!O}} \;+\; NC\diagdown CO_2Cu \xrightarrow{130^\circ C} \underset{(4)}{\overset{}{\text{carbonate}}} \quad (83\%)$$

described.[4] The stereochemistry of carbon dioxide insertion has been investigated using a tungsten complex. Like CO insertion discussed in Chapter 9, migration of a σ bound ligand to carbon dioxide occurs with retention of configuration at the α-carbon of the migrating ligand.[5]

310

More significant examples[6] of the use of carbon dioxide insertion as a synthetic method combine the carbon dioxide insertion step with the dimerisation of butadiene catalysed by $Pd(OAc)_2$ in the presence of a phosphine, a reaction discussed in Section 10.8. The effective catalyst is a Pd(0) phosphine complex which may be introduced as $Pd(PR_3)_4$ or as $[(\eta^3$-2-methylallyl)PdOAc]$_2$, from which the active complex is formed *in situ*. Under 80–100 atm pressure of CO_2 at 70 °C the lactone (6), nonatrienoic acid and esters (7) and (8) are formed together with a small amount of octa-1,3,7-triene.

(5)

(6)

(7)   +   (8)

The reaction can be understood in terms of an interconversion of η^3- and η^1-allyl complexes to make available intermediates with σ bonds, such as (5), to participate in the carbonyl insertion step.

Use of phosphines PL_3, $L = C_6H_{11}$, or ^iPr, gave the lactone (6) in 27–28% yield. The esters were major products (44%) when L was CH_2Ph.

In related work using $Pd(Ph_2PCH_2CH_2PPh_2)_2$, a lactone product (9) was isolated

(9)

from reaction of butadiene in the presence of carbon dioxide, but in only small yield, along with octa-1,3,7- and -1,3,6-trienes as major products. More recently, a rather more efficient form of this type of reaction has been developed using the bimetallic intermediate (10), which is obtained from butadiene and palladium bis(hexafluoroacetoacetonate). Treatment with tri-isopropyl phosphine and carbon dioxide gave an intermediate which was converted into (11) with HCl. Catalytic hydrogenation afforded (12) in 44% yield from (10).[7]

In another synthesis of carboxylic acids by carbon dioxide insertion into Pd π-allyl phosphine complexes,[8] the 2-methyl allyl-PdCl complex with bisdiphenylphosphino-

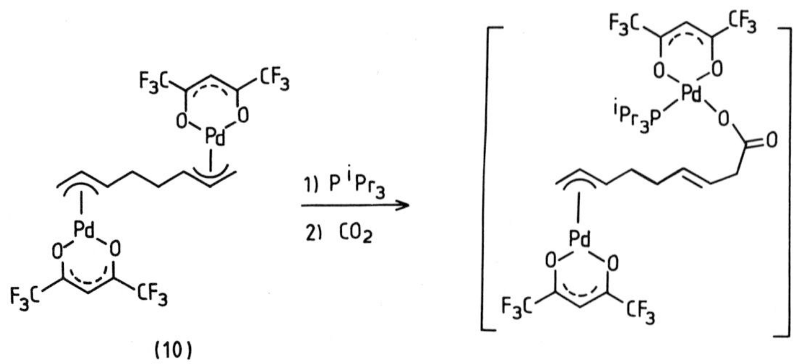

(10)

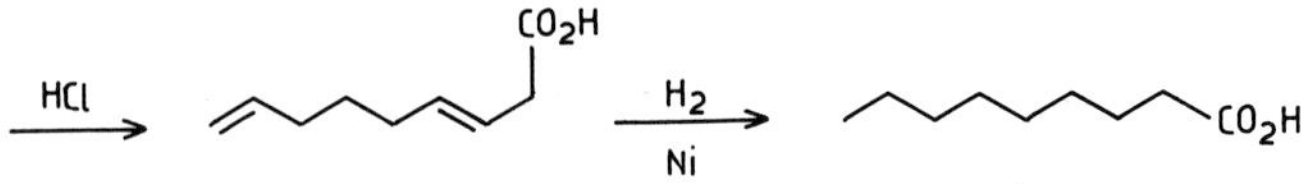

(11)   (12)

ethane was shown to react with carbon dioxide at 50 atm in acetonitrile solution at 80 °C. The product of reaction was found to comprise the esters (13) and (14) which could be hydrolysed to form (15), (16) and (17). In the presence of excess ligand $(Ph_2PCH_2CH_2PPh_2)$, however, carbon dioxide insertion led to complex (18) which on acid hydrolysis gave (15) and (16).

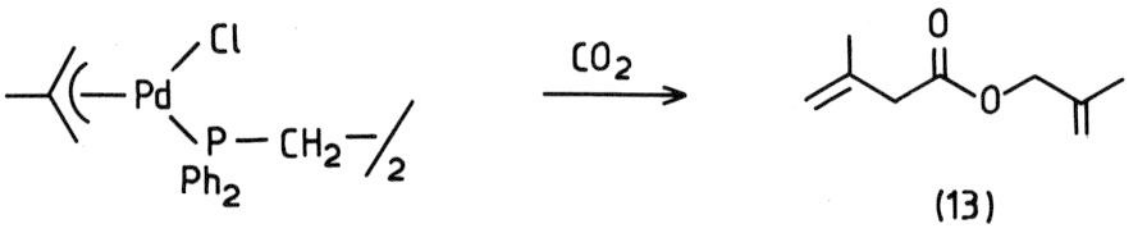

(13)

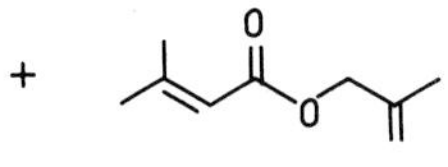

(14)

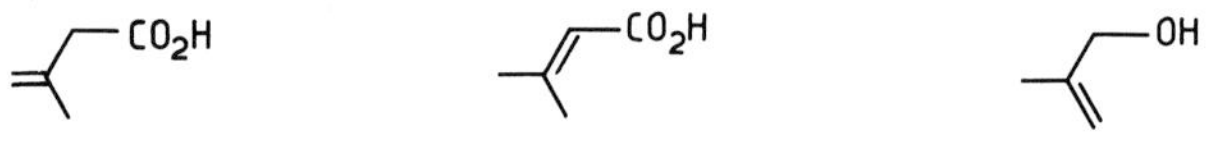

(15)   (16)   (17)

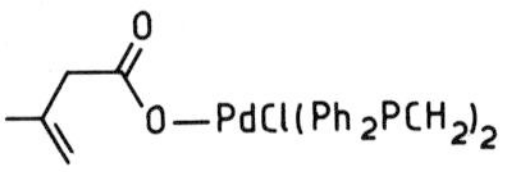

(18)

In a further example that shows obvious parallels with palladium catalysed alkylation of epoxides and presumably involves similar palladium allyl intermediates, carbon dioxide is inserted into a vinyl epoxide, to form a cyclic carbonate (19), without the need for high temperatures and pressures.[9]

A nickel complex has been used to combine diynes with carbon dioxide to form bicyclic α-pyrones.[10,11] Coupling of the two alkyne linkages in a fashion discussed later

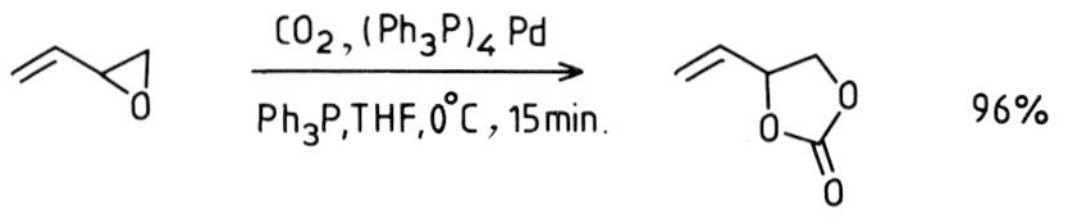

(19)

in this chapter is followed by carbon dioxide insertion in the final step forming the pyrone ring. Yields for a variety of examples range from 11 to 62%.[10] Carbon dioxide insertion combined in this way with a coupling of alkynes resembles the insertion step of Pauson–Khand cyclisations discussed in Chapter 9. Other alkyne coupling processes are discussed in Section 10.4.

10.2 Hydrogenation, hydrocyanation and hydrosilation

The use of transition metal complexes as soluble catalysts for alkene hydrogenation has been extensively investigated in the expectation that a high degree of selectivity could be achieved. Together with hydrocyanation and hydrosilation, (and also hydro-formylation, discussed in Chapter 9), these reactions provide the means to add the elements of HX (where X may be H, CN, SiR_3, or CHO) to an alkene. A typical example (Scheme 10.1) is a catalytic cycle using rhodium. This involves an initial reaction sequence in which oxidative addition of HX to a coordinatively unsaturated complex is followed by a ligand exchange reaction that brings the alkene into the coordination sphere of the metal.

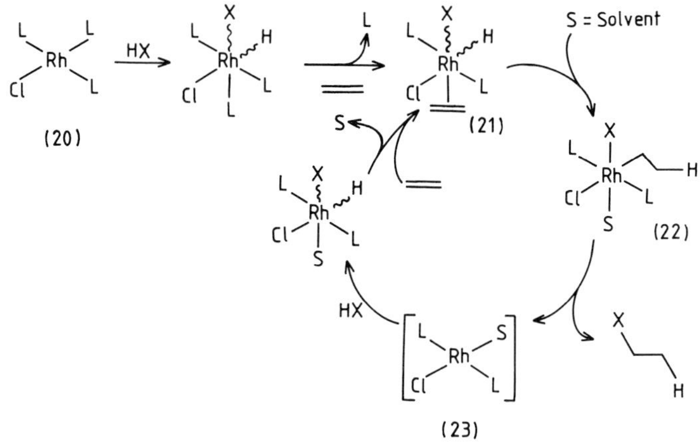

Scheme 10.1

The catalytic cycle begins with a slow migration of H to the alkene ligand. This is followed by a reductive elimination step that liberates the organic product to form an intermediate (23). This intermediate is related to the starting complex (20), but has one ligand replaced by solvent. The remaining steps shadow exactly the initiation sequence of oxidative addition with subsequent ligand replacement. In this final step the solvent molecule is lost to re-form (21). The normal *cis* addition of HX is a natural consequence of this mechanism since the C–H and C–M bonds are formed at the same face of the alkene.

The above description provides a simplified picture which draws out the main features of the catalytic cycle applicable to complexes of this type. There are many variations possible on this theme. For example, a second substrate molecule could be included in place of the solvent (S) in the formation of (22). This considerably complicates the possibilities available to effect the overall conversion. Coupling of the alkenes could be envisaged, as is seen in oligomerisation reactions of alkenes in Section 10.8. Another mechanism with both alkyl and alkene units bound to the same metal will be encountered when rhodium catalysed hydrosilation processes are discussed at the end of this section. In the hydrosilation case, coupling of the alkyl and alkene portions does not occur, and reductive elimination dominates.

In all these reactions, typical sequences involving combinations of ligand association, oxidative addition, alkyl migration, and reductive elimination steps bring the substrates through to the products in cycles that usually require variation of the electron count and oxidation state of the metal centre. It is apparent that appropriate metals will be those that are stable both as 16 and 18 electron complexes, and that are able to undergo ligand dissociation without bringing about complete decomposition. The intermediates involved in these cycles are related to those implicated in metal catalysed isomerisation reactions discussed in Chapter 2, and indeed, many pertinent aspects of this chemistry have been introduced there.

Hydrogenation

A typical hydrogenation catalyst that meets these requirements is 'Wilkinson's catalyst', $RhCl(PPh_3)_3$.[24] This reagent, first described as a homogeneous catalyst for alkene hydrogenation in three independent reports[12] in 1965, has become the best known and most popular of the many reagents now available. Examples of hydrogenations using Wilkinson's catalyst are shown in Scheme 10.2 to give an indication of the scope and selectivity of this method.

The formation[14] of (25) and (26) demonstrates the potential for selective reduction of the less substituted alkenes in polyenes, an effect employed in the reduction[15] of the guaiadiene (27). Acyclic alkenes can be reduced selectively in the presence of enones, as seen in the reactions of nootkatone[17] and eremophilone[18] forming (28) and (29). Good selectivity is achieved despite the fact that α,β-unsaturated aldehydes and ketones can themselves be reduced by $RhCl(PPh_3)_3$ and related catalysts.[20] α-Methylenelactones in confertiflorin and psilostachyine have also been reduced to saturated lactones but a similar reaction for damsin resulted in isomerisation.[21]

Enol ethers in the 1,4-dienes (30) and (31) survived[14] the reduction conditions. In the case of thebane (32), the 1,3-diene was selectively reduced to leave the enol ether linkage.[19]

Wilkinson's catalyst can also reduce alkenes in the presence of cyclopropanes, to form products such as (33), (34),[22] and the terpene derivative (35).[23] In the case of (33) and (34), complications might be anticipated through opening of the cyclopropane by

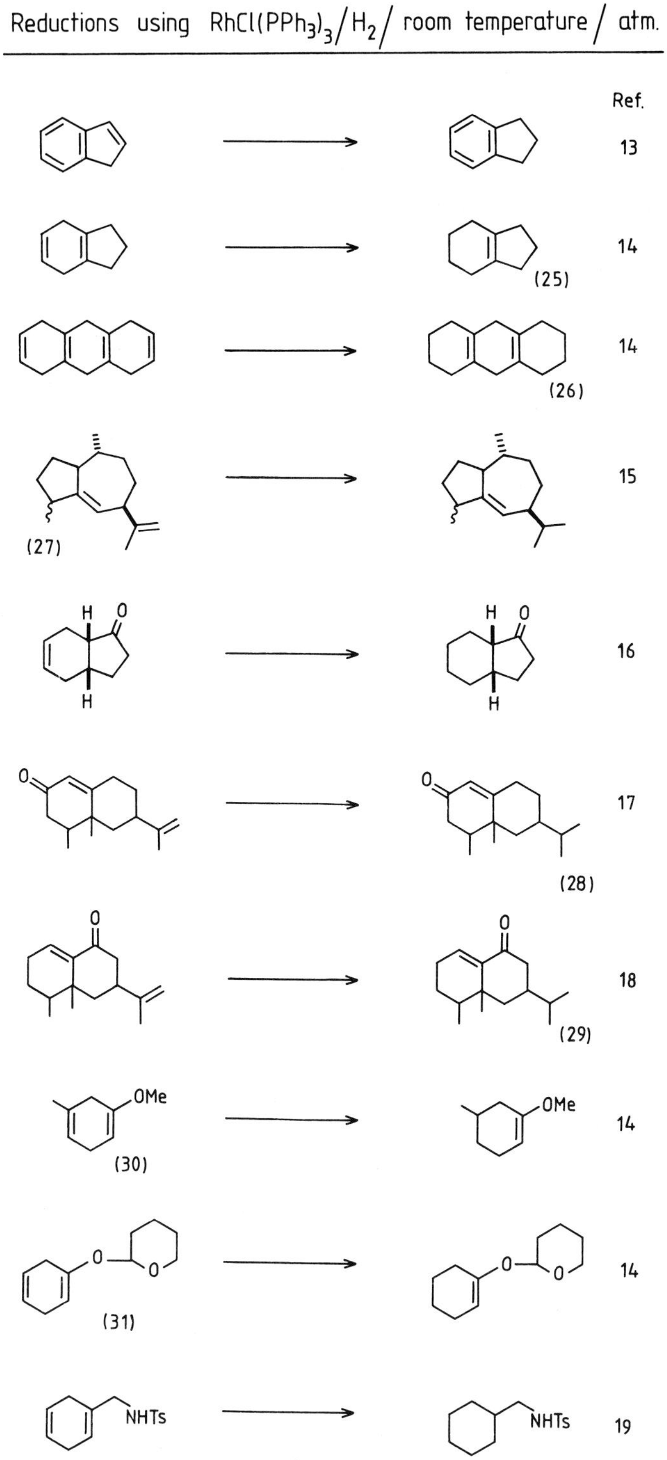
Reductions using RhCl(PPh3)3/H2/ room temperature / atm.
Ref.
13
14
(25)
14
(26)
15
(27)
16
17
(28)
18
(29)
OMe
14
(30)
14
(31)
NHTs
19

Scheme 10.2

(32)

rhodium in intermediate complexes (see Chapter 2), and in other examples, mixtures of cyclopropanes and alkanes are produced.[22]

(33)   93%

(34)   99%

(35)

Interesting selectivity is observed in the reduction of quinones to enediones. The formation of (36)[14, 24] further illustrates the relatively low reactivity of alkoxy-substituted alkenes. Similar reduction has been performed with juglone (37),[24] and diospryrin (38)[25] which contains two naphthoquinone groups. In the latter case, it is, once again, the less substituted alkene linkage that is reduced.

In reactions employing Wilkinson's catalyst, three preliminary steps are crucial to the final outcome: ligand dissociation to provide a free coordination site, addition of hydrogen to form the metal hydride intermediate, and association of the alkene to bring it into the coordination sphere of the metal prior to hydrogen transfer. The reagent $RhCl(PPh_3)_3$ is thus really a catalyst precursor which must lose PPh_3 and react with hydrogen to form the true catalytically active species, usually considered to be $RhCl(PPh_3)_2H_2$.

In other cases, the catalyst precursor may itself contain hydrogen and the details of the catalytic cycle will be different, although the key step of hydrogen transfer to the π bound alkene will remain the same. An example of this is provided by $MeC(CH_2PPh_2)_3RhH(C_2H_4)$, which contains a triply chelating phosphine ligand. This complex catalysed the quantitative reduction of hexene to hexane.[26]

In the main, hydrogenation mechanisms differ most importantly in the manner in which H_2 is taken up by the catalyst. In the mechanism described above (Scheme 10.1) this was achieved by oxidative addition, but two other types of addition process, heterolytic addition and homolytic addition, should also be taken into account. These three possibilities are outlined below:

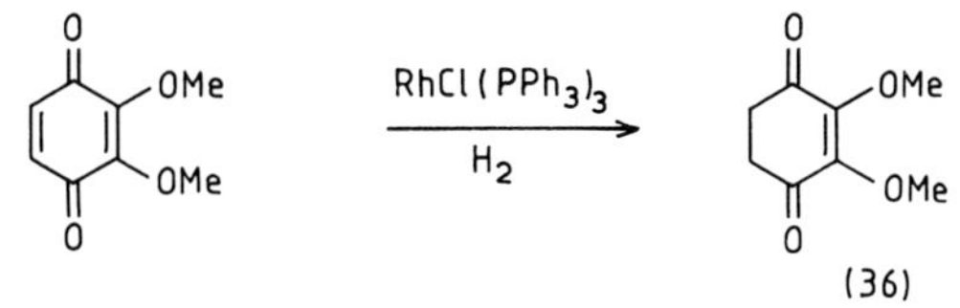

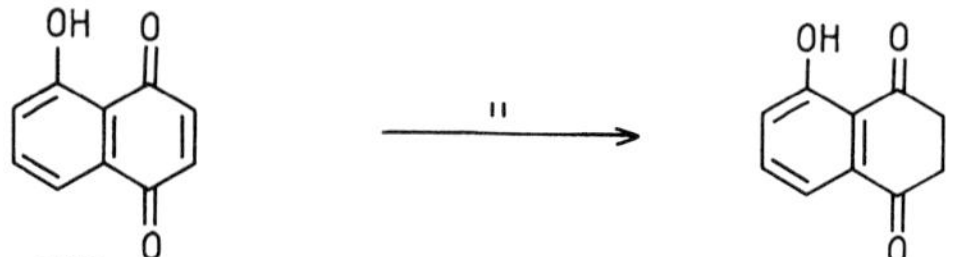

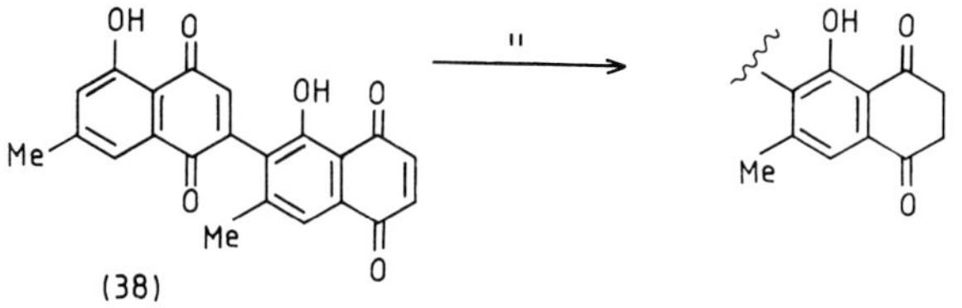

oxidative addition

$$M + H_2 \longrightarrow M \big\langle {}^H_H$$

heterolytic addition

$$M{-}X + H_2 \longrightarrow M{-}H + H{-}X$$

homolytic addition

$$M + \tfrac{1}{2}H_2 \longrightarrow M{-}H$$

While the use of Wilkinson's catalyst is typical of reactions beginning by oxidative addition, several important catalytic systems involve either heterolytic or homolytic additions, and in this respect homogeneous hydrogenation is more complicated mechanistically than many other types of insertion reaction. A typical example of a heterolytic addition is provided by $RuCl_2(PPh_3)_3$, while $[Co(CN)_5]^{3-}$ follows a homolytic path. Examples of hydrogenations using ruthenium and cobalt catalysts of this type are given in Schemes 10.3 and 10.4.

A further possibility for the interaction of H_2 with a transition metal centre is the formation of a molecular hydrogen complex (see Chapter 1), in which H_2 is bound side-on ($M{-}(H_2)$) to the metal.[27] In a sense, this can be regarded as an initial stage in the oxidative addition process, and in some cases the two adducts (39) and (40) are in equilibrium.[28] The $M{-}(H_2)$ form (39) can often be stable, as in $Fe(H_2)H_2(PEtPh_2)_3$,[28]

$$M + H_2 \longrightarrow \quad (39) \quad \rightleftharpoons \quad (40)$$

$Ru(H_2)H_2(PPh_3)_2,[28]$ $Ir(H_2)_2H_2(PCy_3)_2^+,[29]$ and $Re(H_2)H_5(PPh_3)_2.[27]$ An interesting case related to chromium hydrogenation catalysts discussed later in this chapter is the formation of $(H_2)Cr(CO)_5$ from H_2 and the fragment $Cr(CO)_5$. In other chromium complexes, however, the oxidative addition process may well be complete, as it is in the iron analogue $H_2Fe(CO)_4.[27]$ While spectroscopic methods[27] can be brought to bear to determine the bonding mode of hydrogen in isolated complexes, the binding of hydrogen in intermediates in catalytic cycles cannot be determined with certainty. Molecular hydrogen complexes are, however, in some cases, known to be active hydrogenation catalysts.[27, 30]

A more significant role for molecular hydrogen complexes may arise in scrambling processes which often complicate the introduction of deuterium by homogeneous catalysts. The formation of an intermediate such as (41) leaves the oxidation state of the

(41)

metal unchanged, requiring only a free coordination site. Complete oxidative addition increases the oxidation state of the metal by two, a change that may not always be reasonable. On the basis of kinetic studies, an unusually rapid isotopic scrambling observed in a hydrogenation reaction of (42) using $[Ir(cod)(PCy_3)py]BF_4$ has been ascribed to the involvement of molecular hydrogen complexes.[31]

$$\xrightarrow[H_2]{Ir(COD)(PCy_3)py^+}$$

(42)

In the examples in Schemes 10.3 and 10.4 reactions using heterolytic and homolytic processes for hydrogen addition to the metal centre are illustrated for a range of substrates.

While simple cycloalkenes such as (43) are reduced efficiently to saturated products by $RuHCl(PPh_3)_3,[32, 33, 34]$ partial reduction[34] of 1,5-cyclooctadiene (46) has been observed with a similar ruthenium carbonyl catalyst. Cyclooctene was also obtained from (46) by use of $Co_2(CO)_8$ as the catalyst precursor.[36] In contrast, a reagent derived from nickelocene and $LiAlH_4$ gave complete reduction to cyclooctane.[37] Partial reduction of 4-vinylcyclohexene (44) with $RuHCl(PPh_3)_3$ shows[34] once again selectivity similar[38] to Wilkinson's catalyst, with reaction occurring preferentially at the less substituted alkene.

Reductions using $RuHCl(PPh_3)_3$ / H_2 / room temperature / 1 atm.

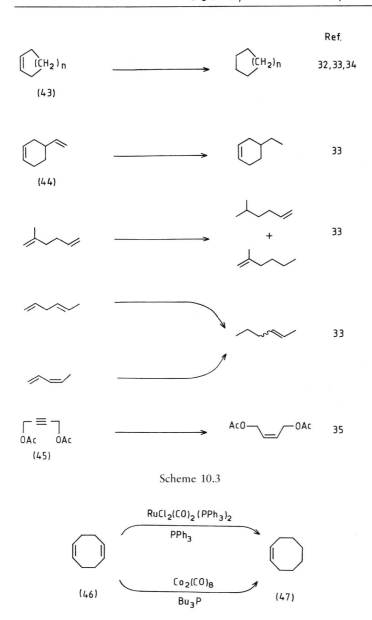

Scheme 10.3

Reduction of norbornadiene (48) and norbornene (49) provides an opportunity to compare the reducing power of some of these reagents. The relatively mild action of $RuCl_2(CO)_2(PPh_3)_3$ allows the selctive formation of (49) from (48) while $RuHCl(PPh_3)_3$,[34] $RhNO(PPh_3)_3$,[40] and $RhCl(CO)(PPh_3)_2$[41] are all capable of reducing norbornene to norbornane. $RuCl_2(PPh_3)_3$ has been used for the complete reduction of norbornadiene.[42]

(48)

(49)

The example of the partial reduction of the alkyne (45)[35] seen in Scheme 10.3 provides an interesting comparison with results obtained for the terminal alkyne (50). Here, complete reduction to hexane was observed.[43] Similarly, diphenylethyne (51) was partially reduced to *cis*-stilbene (52) with $RuCl_3/PPh_3$,[42] but use of $RhCl(PPh_3)_3$[44] or $Co_2(CO)_8$[45] afforded 1,2-diphenylethane.

(50)

(51)

(52)

Unlike the reagents discussed above, $[Co(CN)_5]^{3-}$ is normally used in water and many examples concern the reduction of carboxylic acids. Dienes can be partially reduced to alkenes, but activated alkenes such as styrenes and α,β-unsaturated acids and aldehydes can also be reduced, as seen in the examples in Scheme 10.4.

Phase-transfer catalysis has been successfully used with many substrates, for example cyclohexa-1,3-diene, which was partially reduced to cyclohexene.[50] Besides $[Co(CN)_5]^{3-}$, many coordination complexes of cobalt, including vitamin B_{12} itself,[51] have been used as hydrogenation catalysts. Cobaloximes, for example $Co(DH)_2$ (54),[52] and phthalocyanins, for example CoPc (53),[53] are commonly used in synthesis as substitutes for vitamin B_{12}. These have been found to be active as reduction catalysts. Although $Co(DH)_2$ is used with hydrogen gas, vitamin B_{12} and CoPc use Zn/HOAc and $NaBH_4$, respectively, as their sources of hydrogen. Reduction of the terpene carvone (Scheme 10.5) provides a comparison of these and other reagents. Either alkene linkage can be selectively reduced.

Phase-transfer catalysis can also be used successfully with other metal catalysts. In two recent examples, $RhCl_3$ has been used to reduce styrenes in two-phase reaction

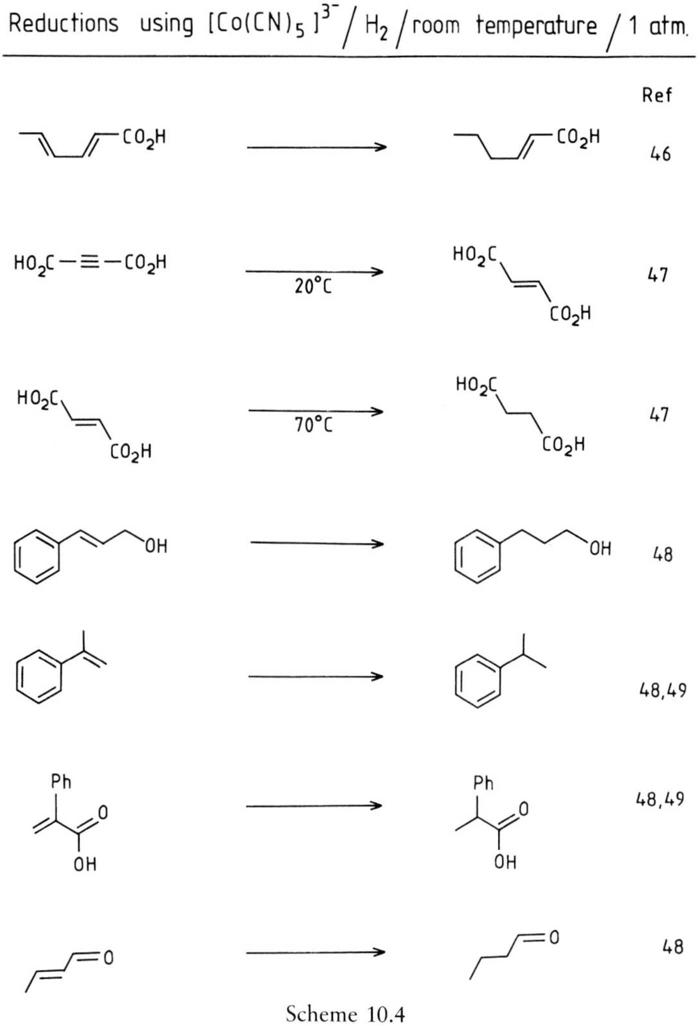

Reductions using $[Co(CN)_5]^{3-}$ / H_2 / room temperature / 1 atm.

Scheme 10.4

mixtures. In one case, Aliquat 336 was used as the phase-transfer catalyst.[54] The other employed a water soluble phosphine to promote the reaction.[55]

The examples employing the three types of catalyst $RhCl(PPh_3)_3$, $RuCl_2(PPh_3)_3$, and $[Co(CN)_5]^{3-}$, given above (Schemes 10.2, 10.3 and 10.4), provide an illustration of uses of three of the most common hydrogenation systems which employ oxidative addition, heterolytic addition, and homolytic addition, respectively, as the key step in which hydrogen is introduced into the catalytic cycle. Despite this mechanistic distinction, the overall transformations achieved are analogous, and to a large extent, selectivity is determined by other factors such as steric or electronic effects influencing the approach of the alkene to bind to the metal. An understanding of the different mechanisms for hydrogen addition, however, is important for the design of new catalyst systems, and has been discussed in greater detail elsewhere.[56] Here, our objective is to emphasise the similarities of the insertion steps in which the hydrogen is transferred to the substrate.

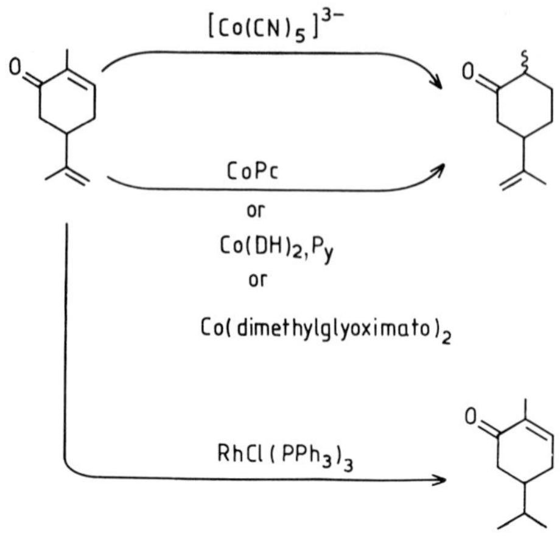

Scheme 10.5

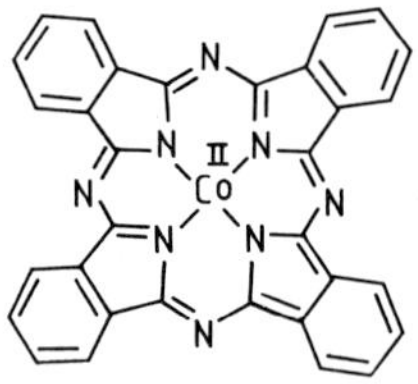

The relationship between this and other insertion reactions is discussed in this chapter. In particular, the *cis* stereochemistry of hydrogen addition is common to almost all systems in which metal hydride species serve as intermediates in the catalytic cycle.[57]

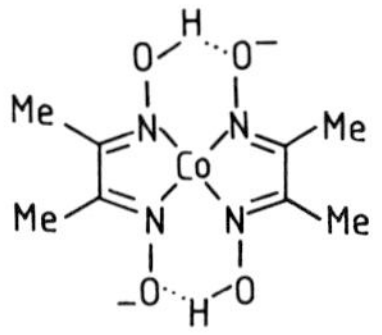

In this sense (and only this sense), the principles of homogeneous catalytic hydrogenation are simple. However, for the reasons indicated above, the details of mechanisms, and practical differences in reaction conditions required by different catalysts, make this a complicated and difficult subject. While a more detailed treatment of the topic is beyond the scope of this book (there are specialised books and reviews[58, 59, 60] that concentrate solely on this topic) it is important here to provide an indication of the factors that underlie the selectivity that can be achieved.

Throughout the development of this field, and particularly in recent work,[61] selectivity, the original goal of research in this area, has been achieved in good measure. Examples given earlier in this chapter have emphasised differences available by the use of alternative reagents. These distinctions are possible because of the intimate involvement of the metal centre in the reduction process. Hydrogen is transferred to the alkene within the coordination sphere of the metal, and interactions between ligands, or restrictions due to steric effects or availability of free coordination sites, must be looked to for the sources of differentiation between competing reaction paths. The reduction of 1,3-dienes provides a good example to illustrate this. As already seen, a number of catalysts will partially reduce dienes to alkenes. A well studied case involves the use of $[Co(CN)_5]^{3-}$ (see Scheme 10.4) or Cp_2MoH_2 to effect a 1,2-addition of hydrogen[62], while chromium carbonyl catalysts act in a 1,4-fashion.[63] Caution is required, however, since analogies between these catalysts are not straightforward. When used with D_2, (arene)$Cr(CO)_3$ promotes[63,64] selective deuteration (Scheme 10.6),

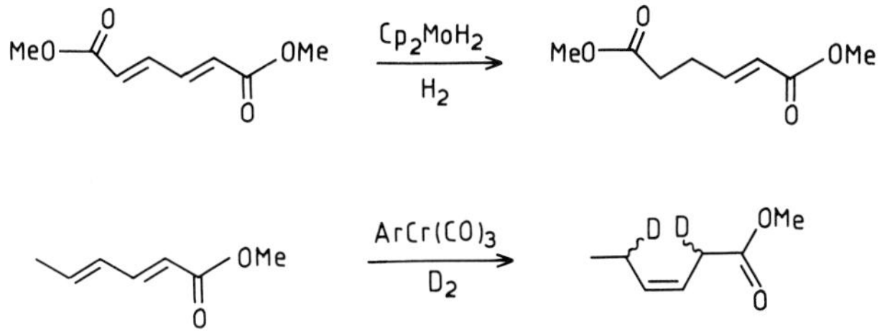

Scheme 10.6

but $[Co(CN)_5]^{3-}$ scrambles[65] the position of deuterium in the product. The differences between the molybdenum and chromium examples may be due to limitations on bonding to the diene. The molybdenum complex is itself coordinatively saturated, but might accommodate an incoming ligand by the revision of bonding to a cyclopentadienyl ligand. Initial binding of the alkene in an η^2 fashion can be promoted in this way. $ArCr(CO)_3$, on the other hand, must add both the diene and H_2. This can be achieved by replacement of the arene ligand to form an intermediate in which the diene binds as an η^4 ligand. Kinetic studies[66] of a naphthalene complex support this view.

An example of the use of chromium 1,4-reduction catalysts can be found in a prostacyclin synthesis[67] leading to a carbocyclic analogue of the anti-ulcer compound nileprost (Scheme 10.7). A further example of the use of this type of reduction can be found in Section 10.9.

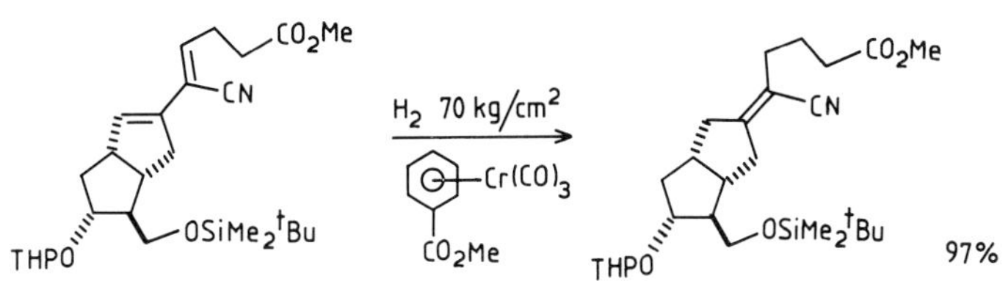

Scheme 10.7

Although examples so far have concentrated on reactions involving alkenes and alkynes, the applicability of homogeneous catalytic hydrogenation is more general. Carbonyl groups, imines, oximes, nitro groups, and nitriles can also be reduced by homogeneous catalysts.[58,60] Aromatic rings, too, can be hydrogenated under some conditions,[68] but, as seen in examples above, arenes are usually retained unchanged in products formed by selective reduction elsewhere.

Depending on the catalyst, carbonyl groups can be reduced to alcohols, or are completely removed to form alkanes. Reduction of benzophenone with a rhodium catalyst, for example, produced[69] the alcohol (55), but $Co_2(CO)_8$ effected[70] complete reduction forming (56). Similar removal of carbonyl groups with the same catalyst gave (57),[70] and a selectively reduced product (58) in which the furan group remained unchanged.[71]

Two more catalysts ($RuCl_2(PPh_3)_3$[72] and $Co(dimethylglyoximato)_2$[73]) encountered already in reductions of alkenes, also reduce carbonyl groups, as does H_2IrCl_6 which afforded[74,75,76] the *trans* product (59). This same catalyst has been employed[77] in the stereoselective reduction of (60) producing (61) in 76% yield. A small amount (8%) of the epimeric alcohol was also formed. Reduction of 5α-cholestan-3-one[74,76] and selective partial reduction of 5α-pregnane-3,11,20-trione[78] have also been effected with this reagent.

The $[Co(CN)_5]^{3-}$ reagent has proved useful for reductive amination of ketones using aqueous ammonia. The amino acid (62) was obtained[79] in this way (see below for other amino acid syntheses). The product (63)[80] provides an interesting comparison with (59) which has the opposite stereochemistry.

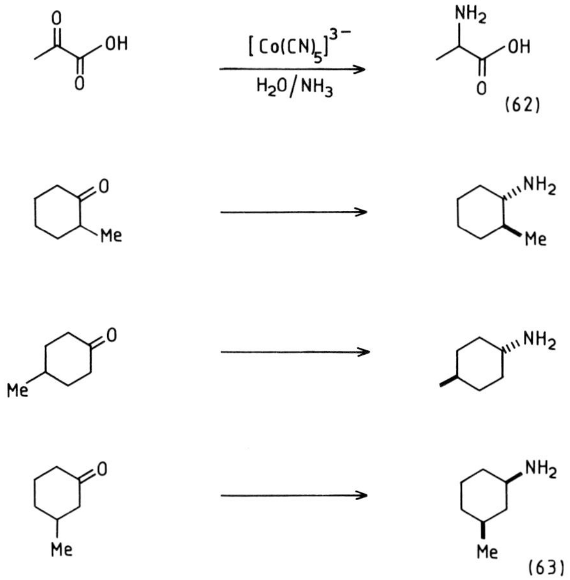

When piperidine or N-methyl aniline are used in place of ammonia, water is no longer required as the solvent and cobalt and rhodium catalysts have been used in benzene to effect reductive aminations (Scheme 10.8).[81]

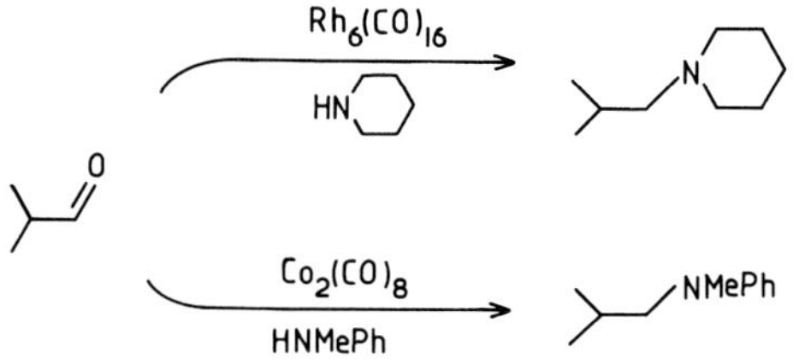

Scheme 10.8

Asymmetric hydrogenation

Asymmetric synthesis based on the modification of catalysts by optically active ligands has also received a great deal of attention.[82] The topic has recently been discussed in detail.[82,83] Three examples of asymmetric hydrogenation are given in Scheme 10.9.

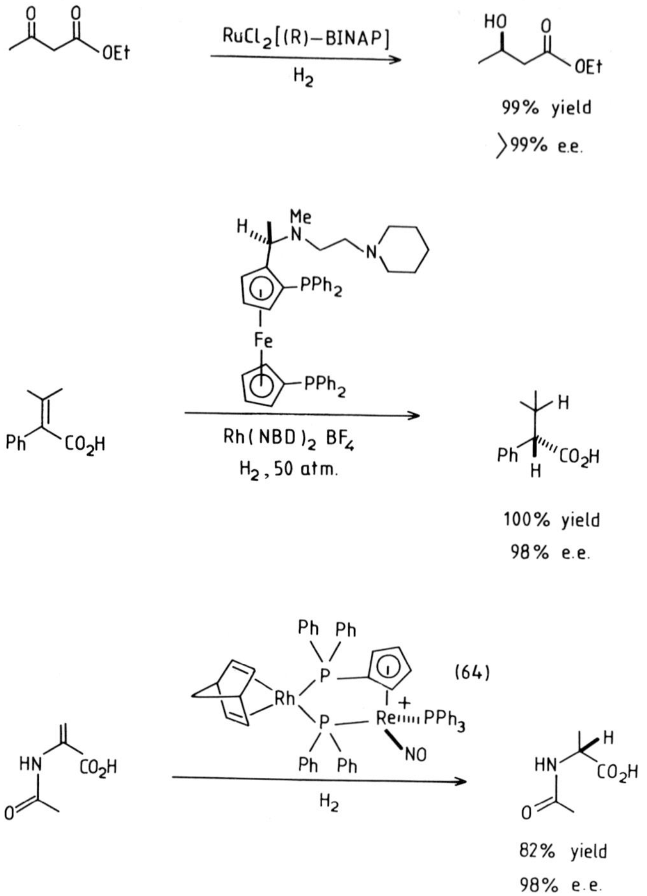

Scheme 10.9

These illustrate particular features of modern applications of homogeneous asymmetric hydrogenation. In the first,[84, 85] the conventional chiral phosphine ligand BINAP is used in hydrogenation reactions of functionalised ketones. Consistently high optical purity in the range 90–100% is typical of the degree of control now expected in successful applications of this type of reaction. A rather different type of chiral auxiliary, a bis-phosphine based on a chiral ferrocenyl unit provides an example of the degree of sophistication possible in the design of chiral auxiliaries. Here rhodium catalysts are used in the asymmetric hydrogenation of tri-substituted acrylic acids in a similarly efficient process.[86] The third example involves a new type of bimetallic catalyst combining a chiral rhenium centre with the catalytically active rhodium norbornadiene complex.[87] This catalyst system was successfully employed in the hydrogenation of enamide precursors to α-amino acids, an approach to amino acid synthesis first made popular by the successful Monsanto process for the synthesis of DOPA.[88] Bimetallic systems of the type (64) bring the directing chiral centre into very close proximity to the catalytically active hydrogenation site, a consideration that may lead to a new generation of powerful chiral catalysts.[87] Related bimetallic complexes combining rhodium and an asymmetric molybdenum atom have been investigated in Brunner's group.[89]

Another successful asymmetric hydrogenation process used a sulphoxide chiral auxiliary with a more conventional catalyst.[90] In a further case,[91] asymmetric hydrogenation has been used to prepare dipeptides (65) with diastereoisomer excess as high as 92%. Both cobalt[92] and ruthenium[93] catalysts have also been modified for

(65)

asymmetric hydrogenation. The example of asymmetric ruthenium catalysis shown in Scheme 10.10 uses the popular BINAP ligand.[94]

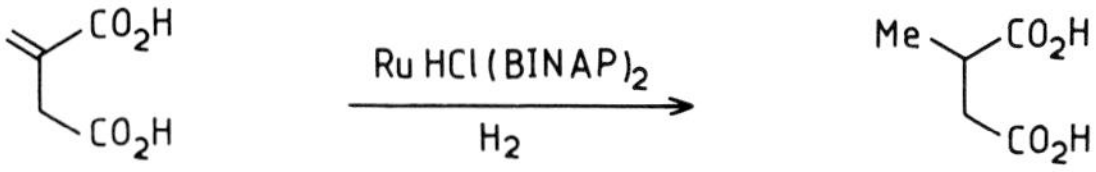

Scheme 10.10

As was encountered with the Sharpless epoxidation reaction (see Chapter 3), an efficient asymmetric catalyst can also be used to effect kinetic resolution of racemic substrates. An example[95] drawn from Brown's extensive work on asymmetric hydrogenation will illustrate the point (Scheme 10.11). In this reaction, as in the

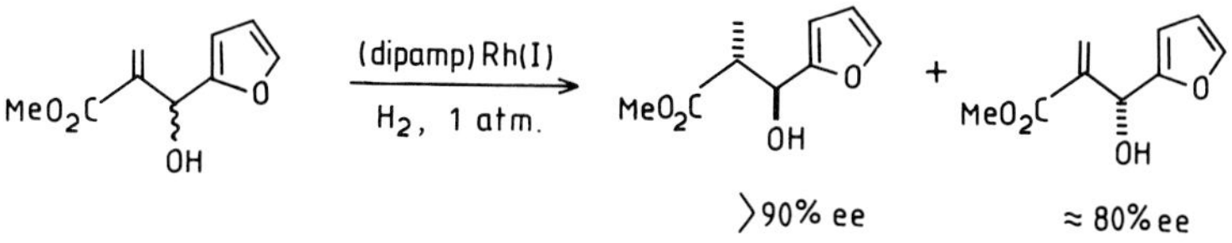

dipamp = (R,R)-1,2-bis[(2-methoxyphenyl)phenylphosphino]ethane

Scheme 10.11

Sharpless case, interaction with the OH group in the substrate can have an important role in the control process. This type of reaction has also been successfully performed in diastereoselective hydrogenation of homoallylic alcohols.[96]

The mechanism of asymmetric induction in homogeneous hydrogenation has been extensively investigated in efforts to determine which steric interactions between ligands are the most important. In catalytic cycles involving selectivity between competing reaction paths, it is the most reactive (i.e. the fastest to react) intermediates that often determine the stereochemical outcome. These intermediates need not be the least hindered, and, indeed, often are not. A detailed discussion of the stereochemistry of these processes appears in a recent monograph.[97]

C–H activation

The importance of hydrogenation catalysts has stimulated great efforts in the field of metal hydride chemistry. Over the years, a number of substances have been discovered that are capable of interacting with organic molecules by cleavage of unactivated C–H bonds, the reverse of the final step of the hydrogenation process. A typical example is the dehydrogenation of alkanes with iridium[98] or rhenium[99] catalysts (Scheme 10.12). In these reactions, t-butylethene is used as a hydrogen acceptor.

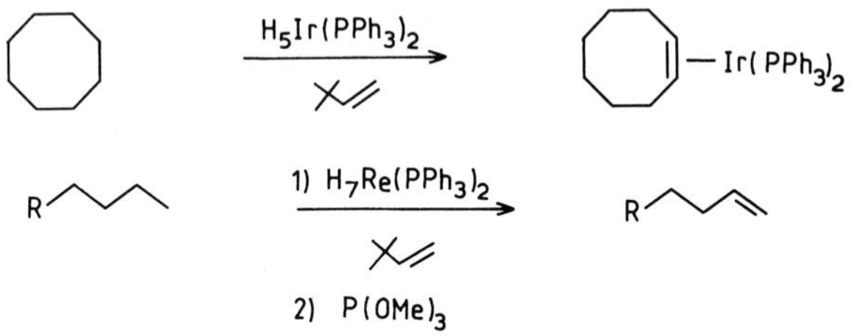

Scheme 10.12

The types of addition process discussed earlier in this section for H–H addition to a transition metal centre, also offer reasonable mechanisms for C–H activation. In the same way, direct interaction of the H–X σ bond with the metal to form a three centre bonding system could constitute an important stage in the process. The involvement of such 'agostic' interactions (see Chapter 1) would suggest a stepwise mechanism for C–H activation.

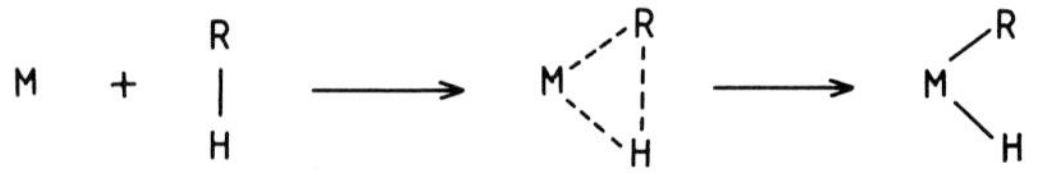

By the analysis of structural information for a series of agostic complexes, Crabtree's group have proposed[100] a more detailed depiction of this process which takes the initial approach of the metal through a sequence of stages to an eventual oxidative addition product. This view of the reaction is supported by reaction-coordinate calculations by Hoffmann.[101]

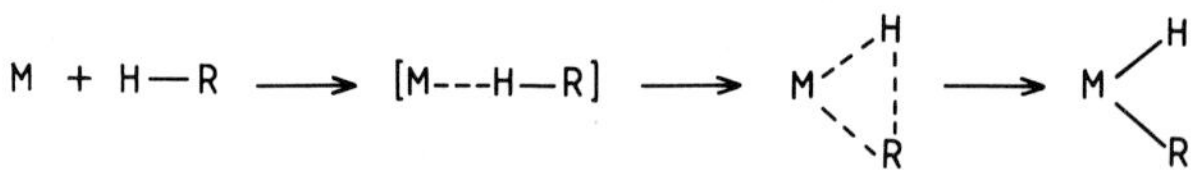

The agostic interaction of C–H bonds with metals is a well established phenomenon. A series of cobalt complexes (66) provide a typical recent example,[102] although instances[102] have been known since the early days of organometallic chemistry.

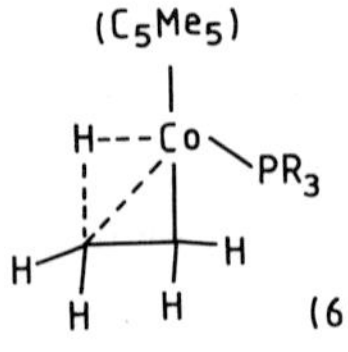

As was the case for H–H complexes, C–H interactions can be inferred from spectroscopic data, but X-ray or neutron diffraction studies give the most definitive evidence. While it is a large step to conclude from this that such species are involved in intermediates in catalytic cycles effecting C–H activation, matrix isolation evidence for alkane–metal interactions[104] and results from isotope labelling studies are available and lend weight to such an assertion. An example of isotope exchange that involves an iridium catalyst[105] related to that seen in the dehydrogenation of cyclooctane (Scheme 10.12) is of particular interest in this respect. The intermediate (67) requires only β-elimination to follow its cyclooctene homologue down the path to dehydrogenation.

Although most studies of C–H activation have focused on simple alkanes[106] (see, for example, Watson's use[107] of lutetium complexes with methane), organometallic methods of this type are destined to grow in importance and may offer alternatives to selective remote activation using free radical methods. Already, hydrogen transfer from alkanes, alkenes, and arenes in metallation reactions, particularly *o*-metallation, have found many applications. These processes are discussed elsewhere in this book, where other types of insertion reaction are surveyed.

Hydrosilation and hydrocyanation

We now turn to some examples of hydrosilation and hydrocyanation, reactions that are rather less fully developed than the hydrogenation processes that occupy most of this section. The use, once again, of examples employing Wilkinson's catalyst makes clear the similarity between these processes and hydrogenation and isomerisation reactions. Much faster reaction, however, can be obtained when the catalyst is H_2PtCl_6. Cobalt complexes can also be used, although the reactions in this case are relatively slow.[108] It is noteworthy that when employed with acrylate substrates, substitution, rather than addition, occurs.[109] Alkene isomerisation (see Chapter 2) is often a complication in hydrosilylation reactions, as seen in the example of 2-pentene (68).[110]

Depending on the conditions used, the alkene functionality can also be retained in the product by loss of the metal in a β-elimination reaction. Ruthenium catalysis with $HSiEt_3$ has been used to prepare allylsilanes in this way.[111]

The mechanism of hydrosilation using the rhodium catalyst (69) has recently been studied in detail.[112] Investigation of deuterium exchange has provided information which distinguishes between a number of otherwise plausible mechanisms. The two cycles shown are consistent with all the information currently available. The first involves transfer of a silyl group to a bound ethene in the presence of an ethyl group in (70). The alternative of reductive elimination of SiEt$_4$ from (70) has been ruled out. The reductive elimination step, instead, awaits the formation of the Rh(V) bis-alkyl complex (71), but must selectively lose the ethyl group in preference to the CH$_2$CH$_2$SiEt$_3$ group.

A second mechanism interconverts Rh(I) and Rh(III) species. The requirement for selective reductive elimination from (71) is replaced by a selective β-elimination from (72).

A comparison of these two cycles with the general (and simplified) cycle at the start of this section reveals some interesting differences that arise in the hydrosilation case. The similarities between these three cycles and those given elsewhere in this chapter are also notable. As indicated earlier, the same basic steps are used in each case. What is at issue when these mechanisms are studied in detail is the order of the steps and the exact constitution of each intermediate species.

Hydrocyanation has also been examined using a variety of catalysts. Nickel catalysts can be used for the hydrocyanation of both dienes[113] and alkynes[114] to produce allyl and vinyl nitriles, respectively. *Cis* addition is apparent in both cases (Scheme 10.13).

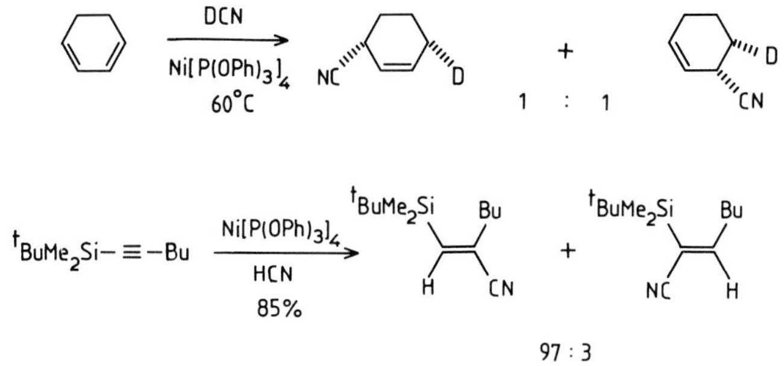

Scheme 10.13

Titanium catalysts have been used for an asymmetric hydrocyanation of aldehydes to produce optically active cyanohydrins.[115] Palladium catalysts can also be used in this type of process, effecting a *cis* hydrostannation of alkynes which closely resembles the hydrocyanation reaction shown in Scheme 10.13.[116]

Asymmetric hydrosilation can also be achieved, although the optical purity of products from these reactions is currently relatively low.[117] A typical example is the hydrosilation of styrene with $HSiCl_3$ shown in Scheme 10.14. The product was converted to 1-phenylethanol which was obtained in 52% e.e.[118]

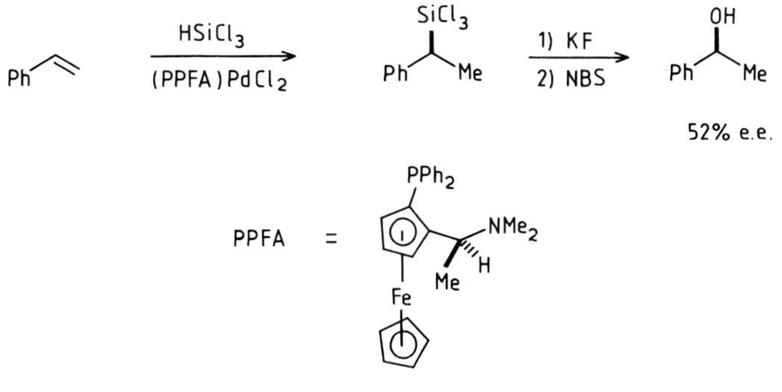

Scheme 10.14

A more important reaction is the hydrosilation of ketones which can give products in higher enantiomeric excess. The example shown in Scheme 10.15 employs the popular DIOP chiral phosphine ligand and two bulky substituents on the silicon.[119] The use of the dihydride R_2SiH_2 is important; $PhMe_2SiH$ has been found to be less satisfactory.[120] While bulky groups on silicon may also assist efficient asymmetric induction, this effect is variable. Hydrosilation of $PhCO^iPr$, for example, was more

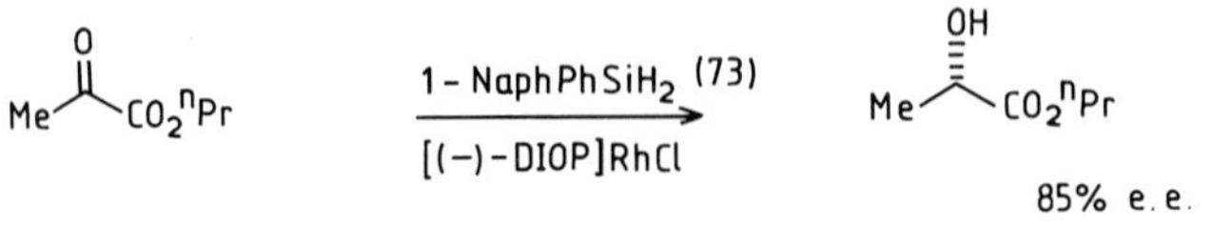

$$Me-CO-CO_2{}^nPr \quad \xrightarrow[{[(-)-DIOP]RhCl}]{1-NaphPhSiH_2 \ (73)} \quad Me-CH(OH)-CO_2{}^nPr$$

85% e.e.

Scheme 10.15

selective with Ph_2SiH_2 than with the napthyl reagent (73).[121] Optically active amines such as (74) can also be obtained by asymmetric hydrosilation, but in lower optical yield.[122]

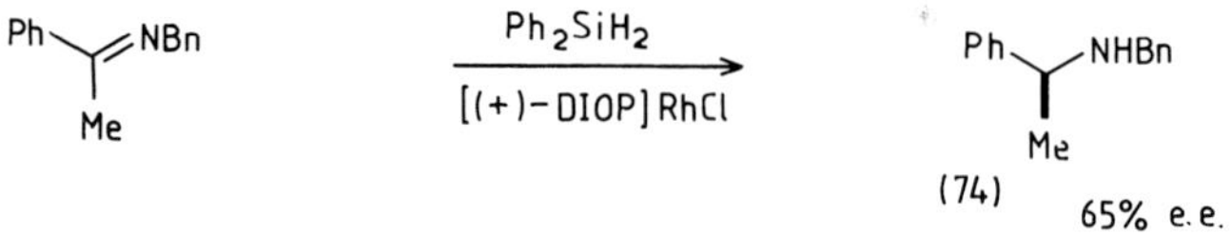

$$\underset{Me}{Ph-C(=NBn)} \quad \xrightarrow[{[(+)-DIOP]RhCl}]{Ph_2SiH_2} \quad \underset{Me}{Ph-CH(NHBn)}$$

(74)   65% e.e.

10.3 Nature of metal mediated coupling reactions

The coupling of two ligands bound to a transition metal centre provides the basis for many valuable synthetic reactions. A number of examples of C–C bond formations, effected in this way, have been encountered earlier in this book as a part of proposed mechanisms for organometallic reactions, without attracting detailed discussion at that stage. Chapter 9 in particular, where the migratory insertion reactions of an alkyl ligand and carbon monoxide were discussed, has examined in detail a special case of what is in reality a general type of reaction in which σ and π bound ligands are combined so forming a new σ bond (e.g. an η^2 ligand in the starting material becomes an η^1 ligand in the product). A similar reaction can convert an η^n π ligand to an $\eta^{(n-1)}$ π ligand. In

$$L_xM \cdots \xrightarrow{\quad\quad} L_xM \cdots$$

both cases, the result is the formation of a product or reaction intermediate which still contains the metal. The overall effect of the process can either be to change both the oxidation state of the metal and the electron count of its valence shell by two (e.g. A in Scheme 10.16), or to maintain the same oxidation state and reduce only the electron count (e.g. B in Scheme 10.16). General examples shown here to illustrate this are relevant to reactions discussed below and subsequent sections of this chapter.

Not all reactions in which a new bond is formed between two ligands at a metal

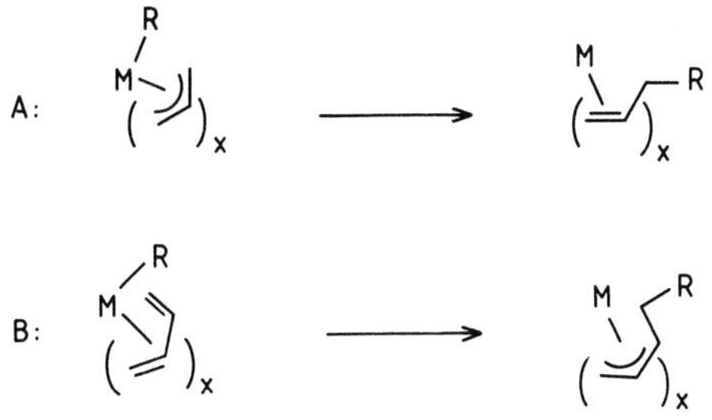

Scheme 10.16

centre need follow this pattern. An alternative is for the metal to undergo the formal reduction, with loss of the ligands as an organic reaction product (a reductive elimination, see Chapter 1). A simple representation of this process is shown:

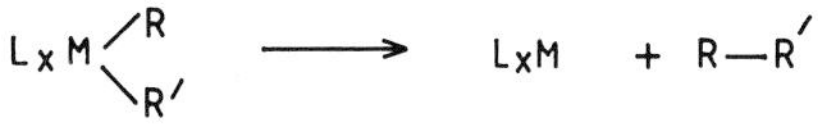

$$L_x M \overset{R}{\underset{R'}{<}} \longrightarrow L_x M \ + \ R—R'$$

$$(75) \hspace{5cm} (76)$$

When R and R′ are alkyl, vinyl or aryl groups, then (76) contains a new carbon–carbon bond, formed at the expense of two metal–carbon bonds in (75). Examples of this type of process are found in mechanisms offered for many catalytic cycles, although generally the involvement of a final reductive elimination step has not been confirmed experimentally. The same equation could, with R = H, for example, equally well describe the final step of an hydrogenation reaction, an observation that emphasises the relationship between processes of these types.

When this chemistry is employed to couple two carbon ligands, a number of methods are important for the initial formation of the key intermediate (75). An obvious approach commences with an oxidative addition to an alkyl complex, the reverse of the required coupling reaction. In other cases (75) can be formed by orthometallation or by

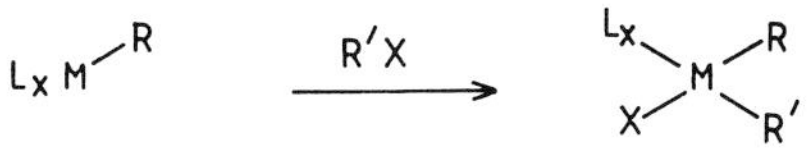

$$L_x M \overset{R}{—} \xrightarrow{R'X} \overset{L_x}{\underset{X}{>}} M \overset{R}{\underset{R'}{<}}$$

nucleophilic addition or substitution at the metal centre (Scheme 10.17). While changes

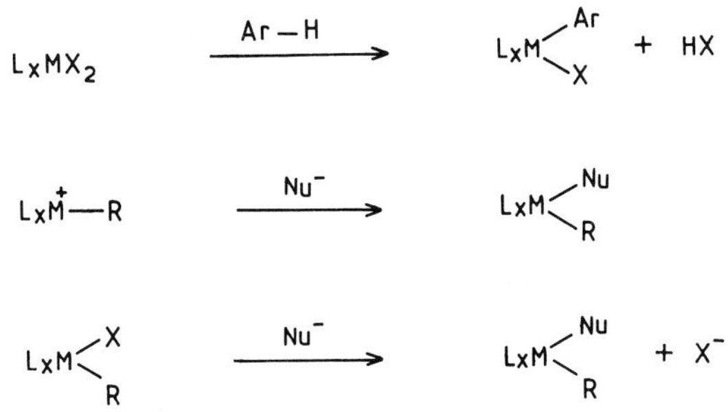

Scheme 10.17

in oxidation state and electron count of complexes in these reactions leading to (75) will vary depending on the route involved, the final step to form (76) in each case changes both the oxidation state and the electron count by two.

The overall reaction will thus normally proceed in one of two ways. Complexes with a full valence shell may either give rise to intermediates with a free coordination site, or may undergo ligand exchange or nucleophile addition reactions of types that do not change the electron count. Alternatively, metals such as palladium, which form complexes that are stable with a 16 electron valence shell, can increase their electron count to 18 on formation of (75), returning to a 16 electron product with the completion of the reaction. This provides a particularly simple route to the key insertion step, and accounts to a large extent for the success and popularity of catalyst systems based on palladium and rhodium chemistry.

Both alkyne and alkene insertion into palladium–carbon σ bonds have been studied in detail by kinetic and n.m.r. investigations[123] and by an analysis by Thorn and Hoffmann of molecular orbital interactions during the insertion process.[124] This palladium catalysed coupling process has now been employed in a synthesis of β-necrodol,[125] and an unusual approach to carbapenem.[126]

Although most commonly employed, palladium is not the only metal that can be selected for catalysis of this type. Examples of a nickel/chromium catalyst,[127] and the organocobalt mediated Pauson–Khand reaction and its zirconium equivalent discussed in Chapter 9, show some of the alternatives that are available. In fact, the range of metals that can be used for this type of reaction is wide. Further examples will be encountered in the coupling of alkynes discussed in the next section.

10.4 Coupling reactions of arenes, alkenes and alkynes

Coupling of arenes

Palladium salts such as $Pd(OAc)_2$, or $PdCl_2$ with NaOAc, effect a coupling[128] of aromatics in which the initial step is metallation of the aromatic ring. Reaction of benzene with $PdCl_2$, NaCl, and NaOAc in HOAc at 90 °C, for example, gives biphenyl (77). $Pd(OAc)_2$ in TFA is found[129] to be effective at lower temperature, forming (78) and

(77)

(79). Yields are, however, only modest, and in *p*-xylene, or mesitylene, the side chain Me groups are subject to oxidation, although this occurs more slowly than the coupling

(78)   (79)

reaction.[130] By control of conditions it may therefore be possible to combine both processes, as indicated by the reaction of mesitylene (80).

(80)

In their original studies at Shell, van Helden and Verberg found that inclusion of acetate was essential to promote a successful reaction, and that coupling failed if a bromide or iodide were used in the place of $PdCl_2$.[128] Subsequent studies by Bryant's group have indicated that the coupling of toluene to form bitolyls was critically sensitive to the ratio of acetate to palladium.[131]

It has also been shown[132] that arylmercury salts may be coupled to an arene in good yield by copper metal in the presence of a catalytic quantity of $PdCl_2$, e.g.

$$Ar-HgX \; + \; PdCl_2 \longrightarrow \; Ar-PdX$$

$$Ar-PdX \; + \; ArH \longrightarrow \; Ar-Ar \; + \; HX \; + \; Pd$$

Instances include the cases where Ar = C_6H_5, 1-chloronaphthyl, 2-chlorofuryl, 2-chlorothiophenyl, 2- and 4-$MeOC_6H_4$, 4-$NH_2C_6H_4$, 4-$AcNHC_6H_4$, or 4-ClC_6H_4.

Biaryls have indeed most commonly been obtained using organocopper chemistry. In a development[133] of the Ullmann coupling, the arylcopper derivative (81) was used in

(81)

a reaction with CF_3SO_3Cu. This reaction may be extended[134] for the preparation of unsymmetrical coupling products. For example, an arylcopper (82), prepared as shown, reacted with an aryl iodide (83) to form (84). It is important to note that the group X was chosen to coordinate to copper.

(83)

(82)   (84)

eg X = NMe_2,

This reaction, which occurs at room temperature in THF solution, has been applied by Ziegler et al.[134] to a synthesis of the anti-leukemic lactone, steganacin (85), using the coupling process illustrated. A more recent version of an Ullmann type of coupling has been described for aryl triflates. The reaction conditions employed ultrasound irradiation of zinc, nickel chloride, triphenylphosphine and sodium iodide in DMF. Yields of coupled biphenyls typically varied from 67 to 95%.[135]

X = successively
Br , Li , Cu

(85)

Coupling of arenes with alkenes

Carbopalladated intermediates encountered previously can be stabilised[136] by intramolecular chelation. Cyclopalladated complexes of this type offer high yields in coupling reactions with alkenes, particularly enones, in the presence of triethylamine.[137] As seen, for example, in (86), the product retains the enone group. Although the metal can be recovered, reactions of this type suffer from the disadvantage that stoichiometric

96%   (86)   92%

conditions are required. None the less, the procedure is versatile, and satisfactory for a wide range of substrates. Indeed, cyclopalladated arenes have been employed successfully as intermediates[138] in a wide variety of organometallic reactions including carbonylation, halogenation and additions to acid chlorides (see Section 10.8).

The presence of coordinating groups to direct carbopalladation is not essential. Benzene and substituted arenes have been coupled with a range of furan and pyrrole heterocycles using palladium catalysis. Regiocontrol, however, in these reactions was poor.[139]

In Trost's route to ibogamine (87)[140] and catharanthine[141] (see also Chapter 14), intramolecular addition to an alkene provided a key in the reaction sequence. Subsequently, a second palladium catalysed coupling step was required to effect a key cyclisation. This step was followed by a reductive work-up using borohydride to replace the palladium by hydrogen, an illustration of an alternative to the more normal concluding reaction, a β-elimination. β-Elimination, however, is encountered in an intermolecular version of the same reaction (Scheme 10.18), in which $PdCl_2$ and

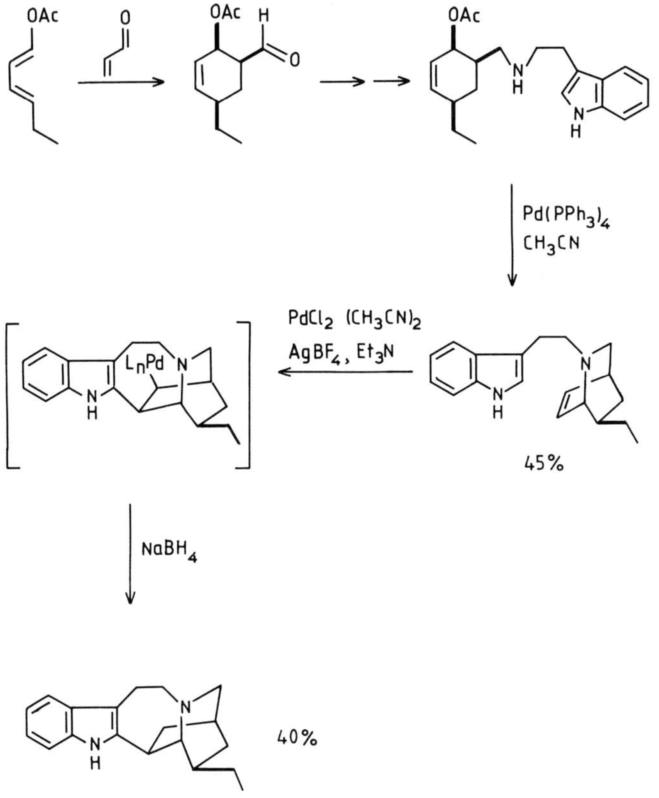

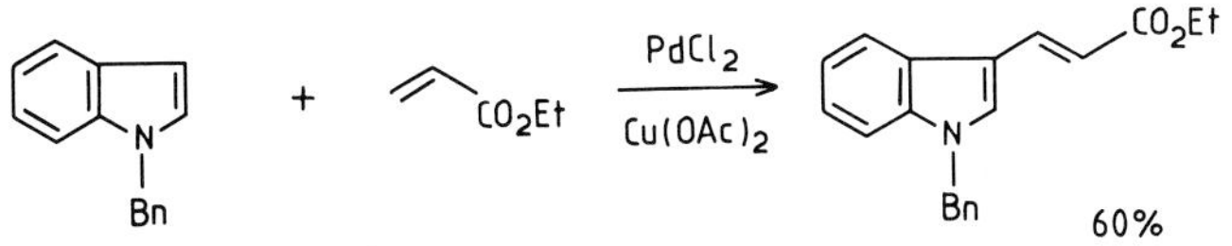

Scheme 10.18

$Cu(OAc)_2$ were used to produce vinyl indoles.[142] Trost's choice of route to (87) is also of interest since the Diels-Alder cycloaddition is followed by an intramolecular palladium catalysed addition[143] of the nitrogen to the allylic acetate, a type of reaction discussed in Chapter 6. This cyclisation had been considerably more efficient (56–67%) in less sterically demanding model studies.[143]

Coupling alkenes

Rhodium catalysis has been used[144] to cyclise the diene (88) to a methylene cyclopentane, in a reaction that is restricted to primary alkenes. Palladium acetate effects a similar cyclisation with internal alkenes but produces a cyclopentene.[145]

Coupling alkenes and alkynes

Trost's group have demonstrated an analogous reaction between an alkene and an alkyne which produces a 1,3-diene that was employed in a Diels–Alder reaction in a synthesis[146] of sterepolide (91). Catalysis using $(Ph_3P)_2Pd(OAc)_2$ effected the formation of (90) without interfering with the allylic substituent in (89).

1,4-Dienes can also result from cyclisations of this type. When used in combination with the displacement reactions of allylic acetates, the process provides direct routes to ring systems such as (92) and (93).[147] Because of the β-elimination step at the concluding

stage of coupling reactions of this type, two alkene linkages are left in the product. The first remains following partial reduction of the alkyne, while the second, derived from the alkene linkage of the substrate, is introduced at the β-elimination stage. A recent modification of the cyclisation process by inclusion of silanes in the reaction mixture has provided an alternative process. This creates an alkene rather than a diene in the product. The mechanism is similar to the conventional reaction but concludes by formation of a palladium hydride intermediate which undergoes reductive elimination rather than β-elimination.[148] In this way the reduced coupling product (94) was obtained. The conventional cyclisation reaction formed (95) which could not be hydrogenated selectively to afford (94). The high degree of stereocontrol in both these reactions is notable. Wilkinson's catalyst, $RhCl(PPh_3)_3$, has also been used to promote cyclisations of this type.[149]

Coupling alkynes

Exocyclic dienes can also be produced from diynes by treatment with a titanium reagent obtained by reduction of Cp_2TiCl_2 in the presence of sodium amalgam and $MePPh_2$ (Scheme 10.19). An aqueous work-up provides the required additional hydrogens by

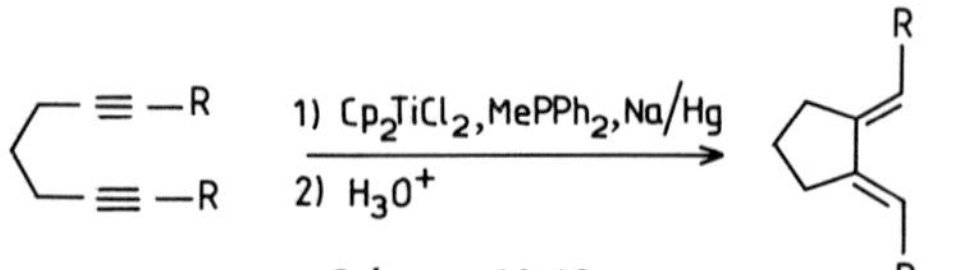

Scheme 10.19

decompositions of a titanocyclic intermediate. The reaction is unsatisfactory with terminal and trimethylsilylalkynes.[150] An unusual version of this reaction combines a nitrile triple bond with a benzyne metal complex, generated *in situ* from a diaryl zirconocene. A metallocyclic intermediate can be isolated from this reaction and hydrolysed to produce aryl ketones.[151] The formation of (96) in this way illustrates the selectivity for insertion into the nitrile rather than the alkene linkage.

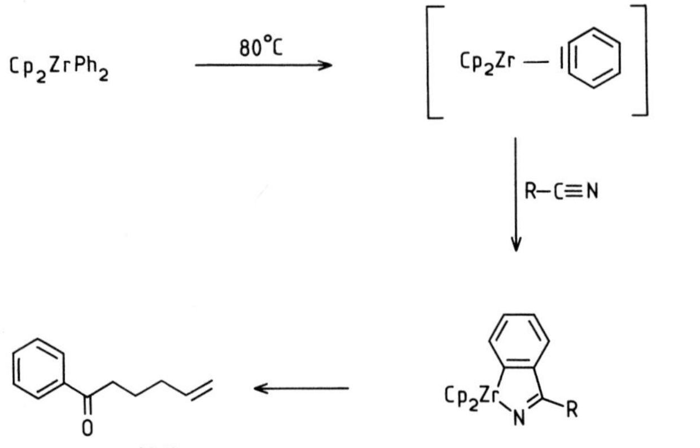

The dimerisation of ethyne by hydrogen transfer (Straus coupling) to give vinylalkynes such as (97) has been known for some time.[152] Terminal alkynes are also dimerised in

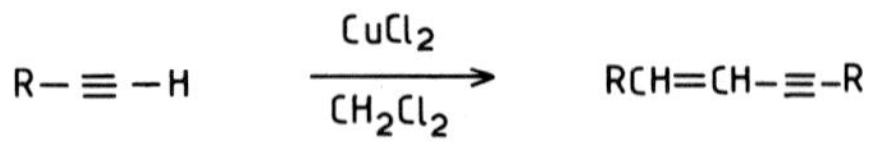

the presence of $RhCl(PPh_3)_3$, as in the formation of (99) and (100) from oct-1-yne (98). The branched isomer (100) was the major product.[153]

The well known tetramerisation of ethyne, catalysed by a Ni(II) catalyst such as $Ni(CN)_2$ or $Ni(acac)_2$,[154] leads to cyclooctatetraene (101) in good yield. However, this tetramerisation depends on ligand displacement at nickel, and addition of strongly

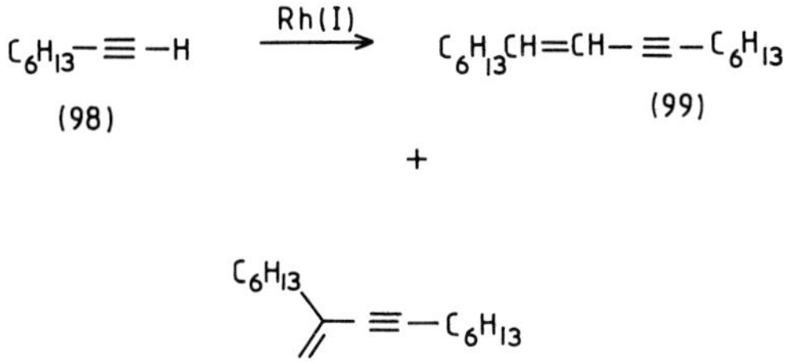

$$C_6H_{13}-\!\!\equiv\!\!-H \xrightarrow{\text{Rh(I)}} C_6H_{13}CH=CH-\!\!\equiv\!\!-C_6H_{13}$$

(98)   (99)

+

(100)

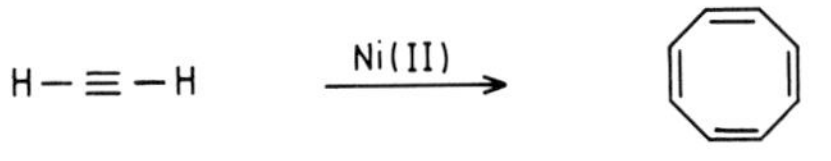

$$H-\!\!\equiv\!\!-H \xrightarrow{\text{Ni(II)}}$$

(101)

coordinating ligands such as a phosphine alters the course of reactions so as to lead to trimers, i.e. benzene or the derivatives (102) and (103).

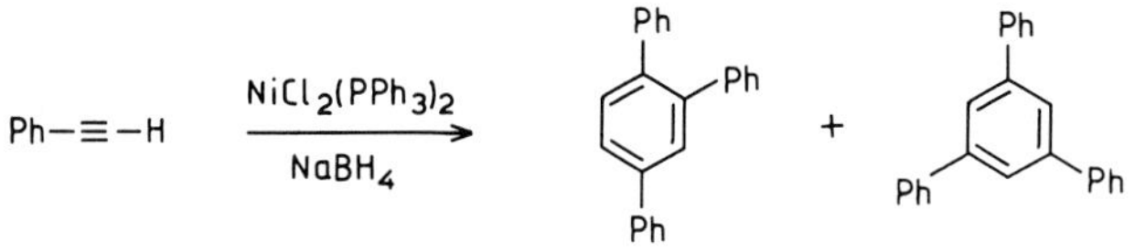

$$Ph-\!\!\equiv\!\!-H \xrightarrow[\text{NaBH}_4]{\text{NiCl}_2(\text{PPh}_3)_2}$$

(102)   (103)

Alkynes also react with $PdCl_2(PhCN)_2$ in CH_2Cl_2 and other solvents[155] to give benzenoid trimers. The reaction lacks regiocontrol but gives highly substituted products such as (104), (105) and (106).

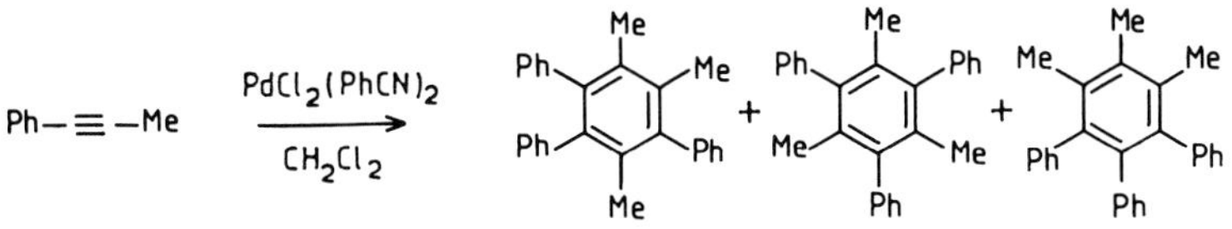

$$Ph-\!\!\equiv\!\!-Me \xrightarrow[\text{CH}_2\text{Cl}_2]{\text{PdCl}_2(\text{PhCN})_2}$$

(104)   (105)   (106)

With relatively more bulky substituents such as ^{t}Bu, reaction stops at the dimer stage with formation of a complex (107), which leads to the cyclobutadiene derivative (108).[156] The alkyne ^{t}Bu-C≡C-Bu^t proved too hindered to participate in the reaction at all, failing even to dimerise.[157]

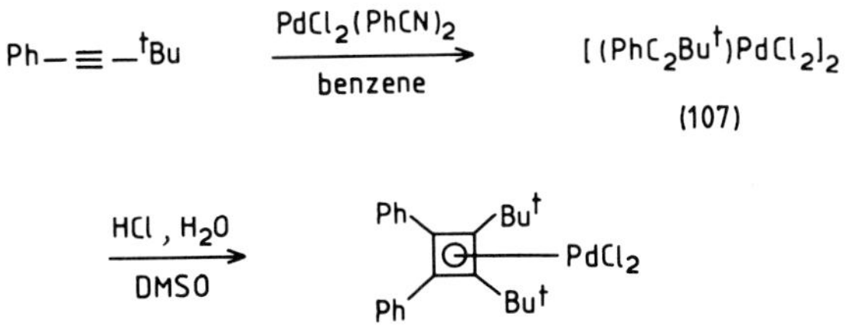

$$Ph-\!\!\equiv\!\!-^tBu \xrightarrow[\text{benzene}]{\text{PdCl}_2(\text{PhCN})_2} [(PhC_2Bu^t)PdCl_2]_2$$

(107)

$$\xrightarrow[\text{DMSO}]{\text{HCl, H}_2\text{O}}$$

(108)

In an alcohol solvent, RC≡CR with $PdCl_2$ gave an intermediate (109) arising from alkoxy addition. This could also be converted into the cyclobutadiene complex.[158]

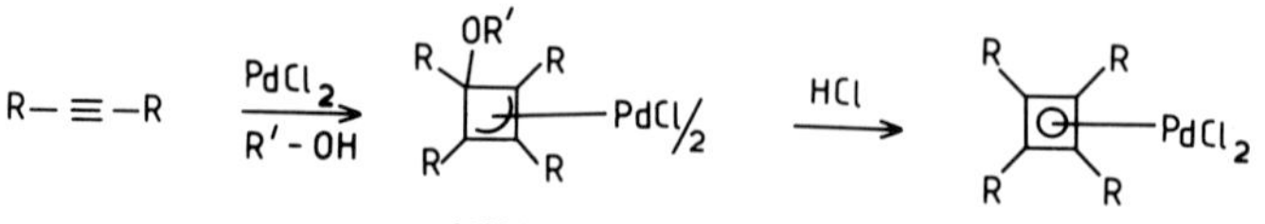

(109)

$$R = Ph, \; p\text{-}Cl\,C_6H_4, \; p\text{-}Me\,C_6H_4; \quad R' = Et$$

A quite different coupling of two alkyne linkages spanning the two sides of a medium-sized ring, has been achieved using diiron nonacarbonyl. In this case, unusual rearrangements occurred to produce trimethylenemethane[159] and methylenecyclo-propene[160] complexes with quite different geometrical arrangements of the carbon atom originating in the alkynes. These results provide a timely warning that the metal-promoted combination of two alkynes can follow a great variety of reaction paths depending on the metal and the conditions.

An intermediate (110) in the Pd-catalysed trimerisation reaction, can be isolated and may be reduced[161] to substituted cyclopentadiene derivatives (111) or (112) and (113).

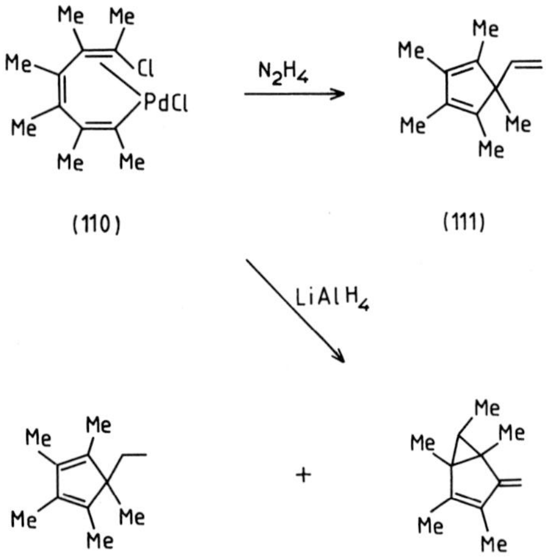

(110)   (111)

(112)   +   (113)

Despite the possible complexities, the oligomerisation of alkynes has been developed into an efficient general synthetic route by Vollhardt and co-workers. This may be

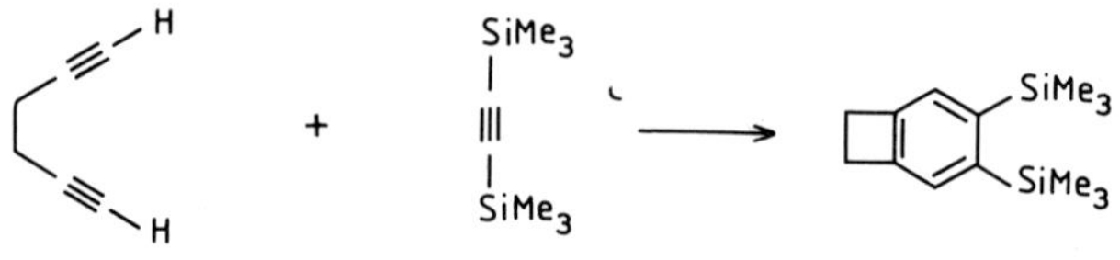

(114)   (115)   (116)

illustrated by the simple case of (114) and (115) yielding (116) by reaction in the presence of a catalytic amount of CpCo(CO)$_2$ in a hydrocarbon solvent.[162]

Since the benzocyclobutane formed in this way may undergo further cycloaddition, it has been possible to elaborate[163, 164] this reaction sequence to provide a synthesis of estrone. Estrone derivatives have also been obtained by a similar method. Although MeOC≡CSiMe$_3$ can be used in the reaction with (117) and CpCo(CO)$_2$, the regioisomers (118) and (119) are formed in a 2:1 ratio. The most practical route[164] to estrone made use of a selective desilylation of (120), which is obtained by reaction of (115) with (117). The intermediate (117) is obtained[164] via an organocopper addition (see Chapter 8).

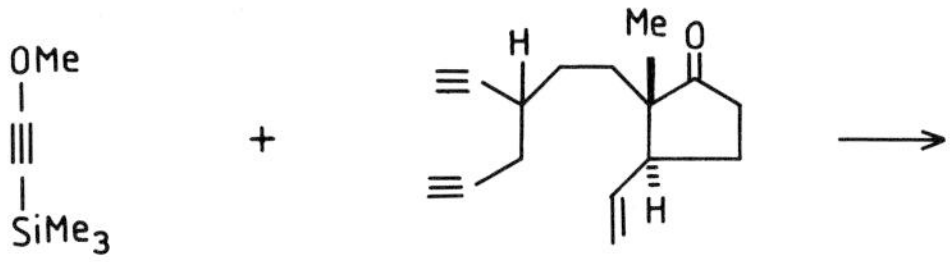

(117)

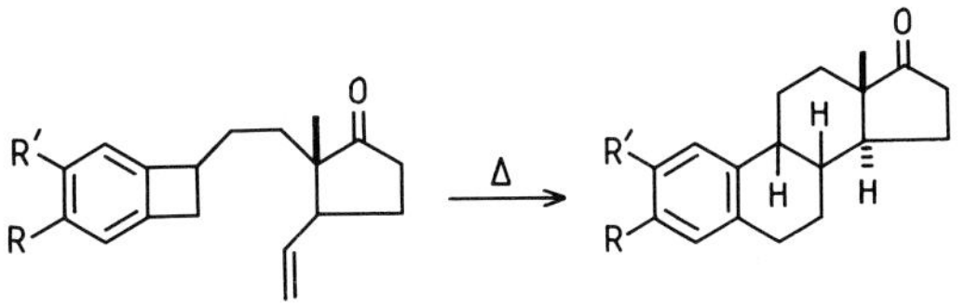

(118): R = OMe, R′ = SiMe$_3$.
(119): R = SiMe$_3$, R′ = OMe.
(120): R = R′ = SiMe$_3$.

In a related entry to the steroid ring system, cobalt catalysed cyclisation was used to form the B, C, and D rings in (121).[165] The use of cobalt mediated alkyne cyclisation in steroid synthesis has recently been reviewed.[166]

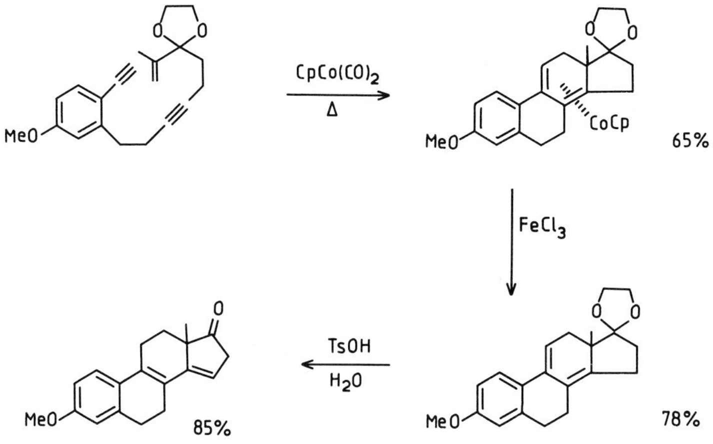

(121)

The cyclisation process has now been applied in a great many situations,[167] as illustrated by its use in the synthesis of protoberberine alkaloids[168] and some strained arenes[169, 170] shown in Scheme 10.20. By replacement of an alkene by a nitrile, the same procedure constitutes a novel pyridine synthesis. The scope of this modification is

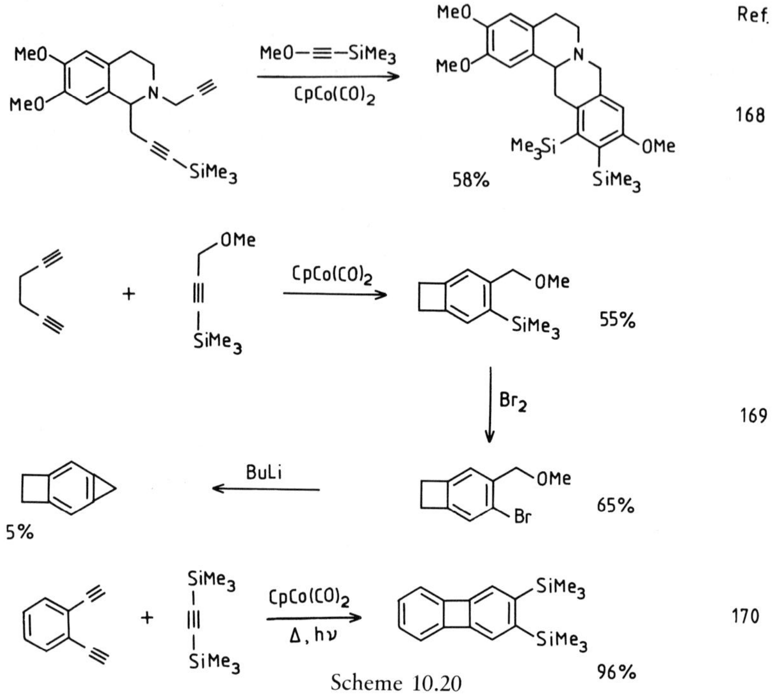

Ref.

168

169

170

Scheme 10.20

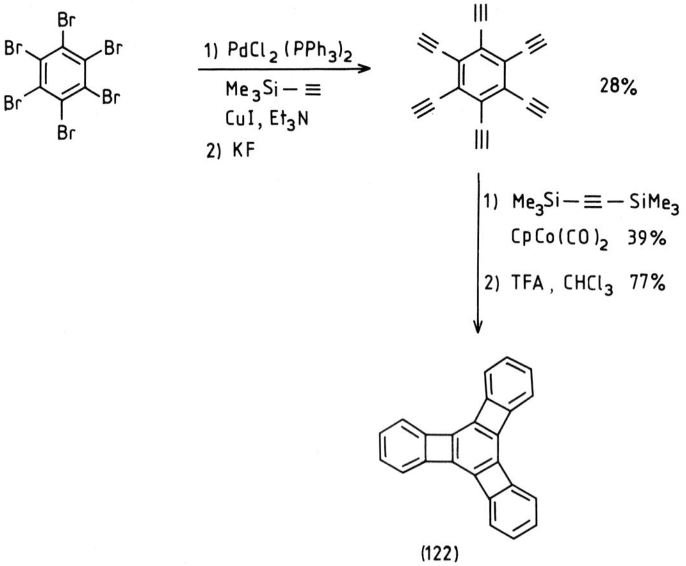

(122)

discussed in Chapter 15. A remarkable example of the synthesis of strained polyaromatics converts hexabromobenzene to the heptacyclic compound (122). In addition to the cyclisation step in which three bis(trimethylsilyl)ethyne sub-units are combined with a central hexa-substituted arene ring, a palladium catalysed coupling reaction was also used in this synthesis.[171] Couplings of alkynes with allyl halides of this type are discussed later in this chapter. Although the yields of the organometallic steps in this reaction sequence are low, the process does bear witness to the useful ability of transition metal catalysed reactions to occur in crowded or highly strained situations.

Whilst it is by far the most extensively investigated example, CpCo(CO)$_2$ is not the only catalyst capable of cyclisation reactions of this type. Wilkinson's catalyst has also proved effective for the formation of products such as (123) and (124).[172]

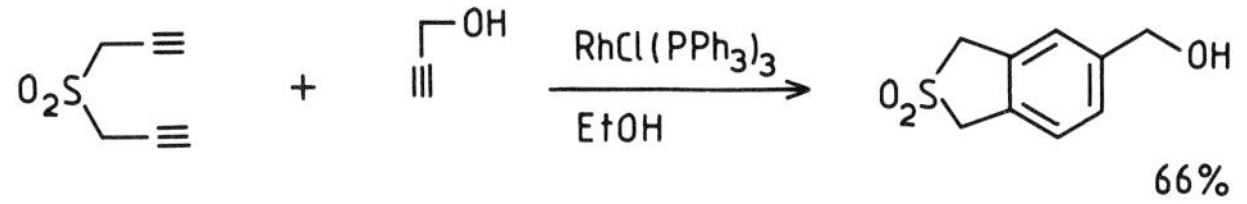

(123)

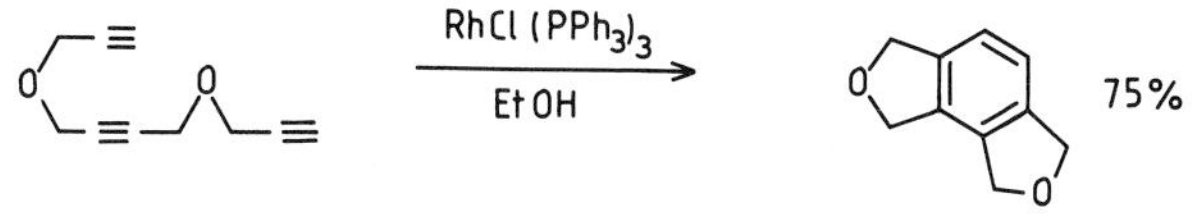

66%

(124)

75%

10.5 Coupling reactions of alkyl, vinyl and aryl halides

The discovery that organopalladium complexes could be generated from alkyl and aryl halides and palladium phosphine complexes, led to the development[173] of coupling reactions with alkenes that are similar to those described for carbopalladated arenes in Section 10.4. The example shown in Scheme 10.21, the Heck reaction, proceeds in a stereocontrolled manner that is typical of these reactions. The outcome is consistent with *syn* addition of the organopalladium intermediate R-PdL$_n$ to the alkene, followed by *syn* elimination of palladium hydride to re-form the alkene linkage. In order to obtain the correct alignment for the β-elimination, a change in conformation is required following the completion of the insertion of the alkene.

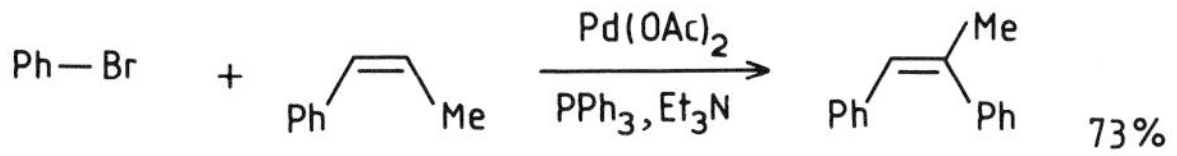

Scheme 10.21

Coupling alkyl halides with alkenes and arenes

In an intramolecular process which is related to the reaction shown in Scheme 10.21, the predominant product (125) arose[174] from a reductive elimination that produced an alkyl halide. Small amounts of alkenes from competing β-elimination were, however, also formed in some cases.

The coupling of arenes and alkyl halides can be effected by employing cyclopalladated intermediates such as (126). The reaction may proceed by either direct addition of the alkyl halide to the ring or by oxidative addition at the metal followed by reductive elimination of the substituted arene.[175] A version of this coupling process has been used to introduce radio-labelled methyl iodide into the aromatic amide (127). The product (128) was converted in two steps to ^{14}C-ropivacaine (129), a local anaesthetic agent with low heart toxicity. This product was shown to have enantiomeric purity greater than 99.7%.[176]

Coupling vinyl halides with alkenes

In an extension of their studies of alkene cyclisations, Grigg's group has developed a further route to 1,3-diene products using the vinyl bromide (130). The ratio of cisoid and transoid products depends of the conditions, and whether rhodium or palladium catalysis is used.[177]

$$M = Pd(PPh_3)_4 \quad : \quad 1 : 10 \quad (74\%)$$
$$M = RhCl(PPh_3)_3 \quad 5 : 1 \quad (63\%)$$

An analogous reaction completed the 16-membered macrocyclic ring of (131) in a model study for a synthesis of the antibiotic carbonolide B.[178] Similar palladium chloride catalysts couple vinyl triflates with alkenes and alkynes.[179]

(131)

Halogen substituted quinones also undergo coupling reactions of this type. The reaction has been applied[180] to the synthesis of the mitomicin analogue (133). Palladium acetate was required for this purpose. $PdCl_2(PhCN)_2$ produced the *o*-quinone (132).

An interesting example[181] of a coupling reaction of a vinyl halide from Heck's group, is provided by the following sequence in which an amine reacts as a nucleophile with a Pd π-allyl complex. The *N*-dimethylheptadienyl morpholine product (134) could be

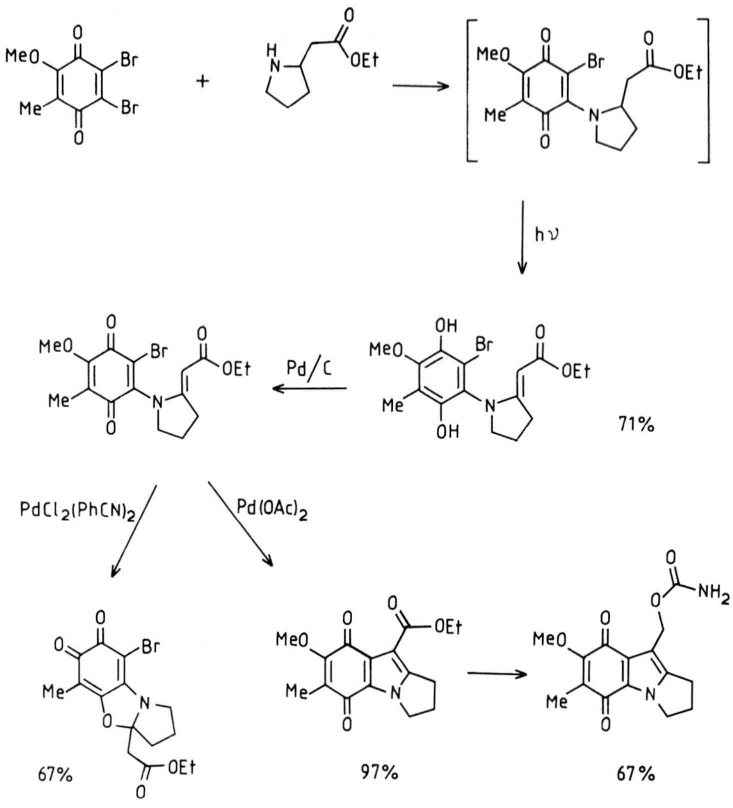

(132)   (133)

converted into ethyl geranoate (135). The alkyl cyclic ether (136) was produced in a similar way.[182]

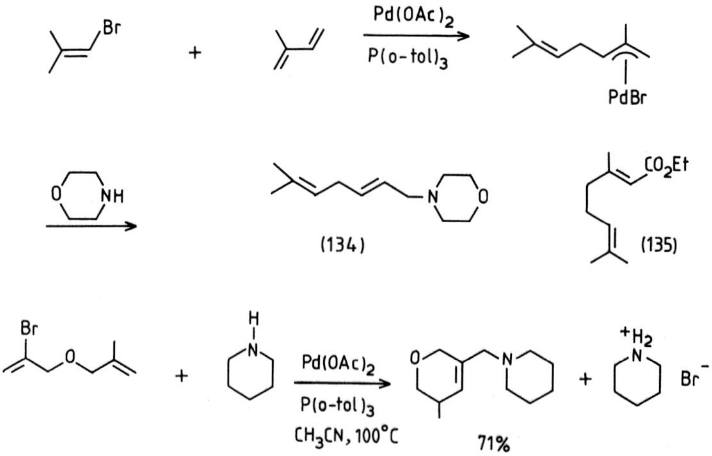

(134)   (135)

(136)

Coupling aryl halides with alkenes

The Heck coupling of vinyl halides is also successful when aryl halides are used as substrates. Together these reactions constitute a general and well tried process that is commonly referred to as the Heck reaction.[183] Examples[173,184] comparing the aryl and vinyl halide versions are shown below:

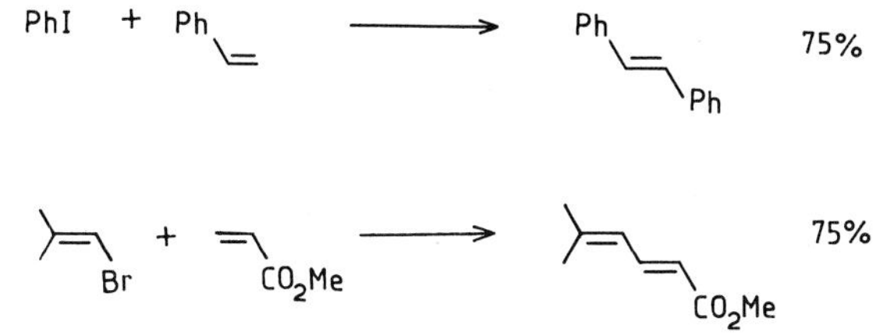

PhI + Ph⟍═ ⟶ Ph⟍═⟍Ph   75%

⟩═⟍Br + ═⟍CO$_2$Me ⟶ ⟩═⟍═⟍CO$_2$Me   75%

These reactions could be effected in good yield using a catalytic amount of Pd(OAc)$_2$ in the presence of (o-tolyl)$_3$P to stabilise what is probably a Pd(0) complex. In some cases an amine is added. An alternative procedure, employing a 9-BBN derivative of the alkene, combines coupling with reduction of the double bond.[185]

Coupling to cycloalkenes is an attractive arylation procedure because it is well controlled. The *syn* addition of the arylpalladium species to the double bond ensures that the β-elimination step can proceed in only one direction. This is possible because a suitably placed hydrogen in the intermediate is available only at one side of the carbon bearing the metal. The formation of (137) in this way is typical.[186]

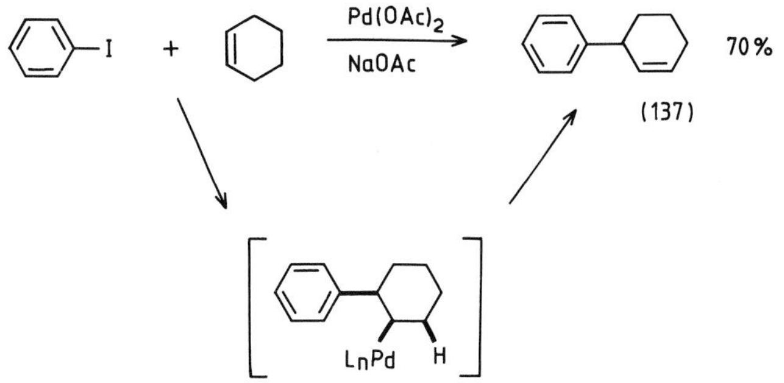

⟨⟩—I + ⟨⟩ $\xrightarrow[\text{NaOAc}]{\text{Pd(OAc)}_2}$ ⟨⟩—⟨⟩   70%

(137)

[⟨⟩—⟨⟩]
LnPd   H

The lack of a suitable adjacent hydrogen does not always block the formation of an alkene. This can be seen in the conversion of (138) into the tetracyclic (139) which was taken on in four steps to munduserone (140).[187]

Vinyl silanes[188] and vinyl stannanes[189] can be coupled with aryl iodides to form styrenes. Normally loss of silicon or tin replaces the β-elimination step in such cases, but by addition of silver nitrite to the reaction mixture, desilylation can be suppressed.[188] Palladium catalysts were used, and under conditions that avoid the presence of phosphine ligands, very rapid reaction occurred (Scheme 10.22).[189] The involvement of an arylpalladium intermediate is supported by the observation that pre-formed phenylpalladium acetate undergoes the same reaction with vinyl silanes.[190]

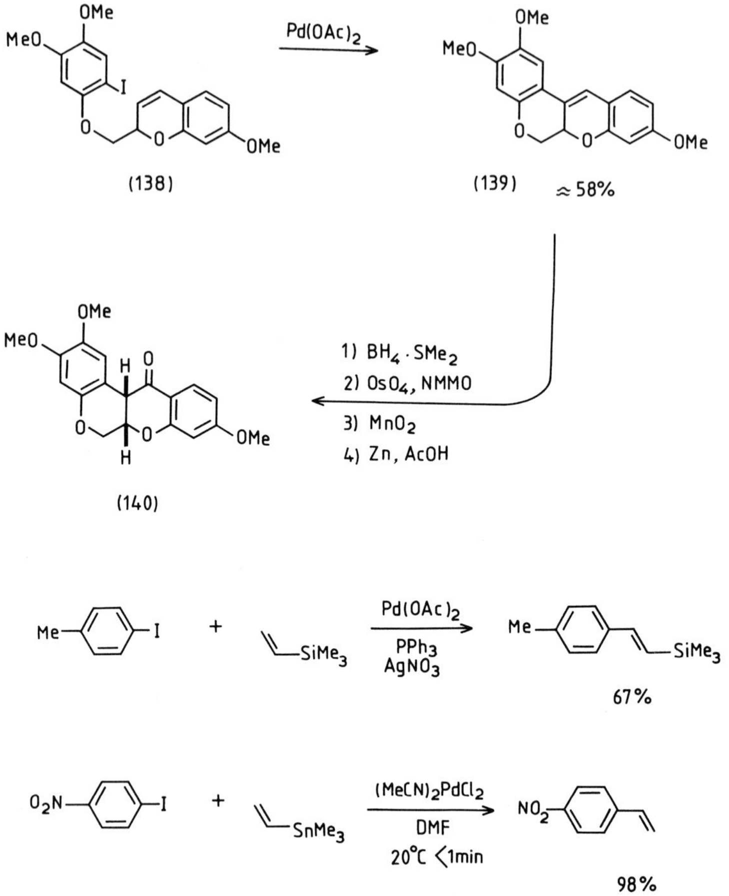
OMe
MeO
I
O
O
OMe
(138)
Pd(OAc)2
OMe
MeO
O
O
OMe
(139)
≈ 58%
1) BH4 · SMe2
2) OsO4, NMMO
3) MnO2
4) Zn, AcOH
OMe
MeO
H
O
O
O
H
OMe
(140)
Me
I
+
SiMe3
Pd(OAc)2
PPh3
AgNO3
Me
SiMe3
67%
O2N
I
+
SnMe3
(MeCN)2PdCl2
DMF
20°C <1min
NO2
98%

Scheme 10.22

Addition of the arylpalladium species to an allyl alcohol[191] is an interesting case since elimination of a palladium hydride led to an enol, and so to an aldehyde (141). This is

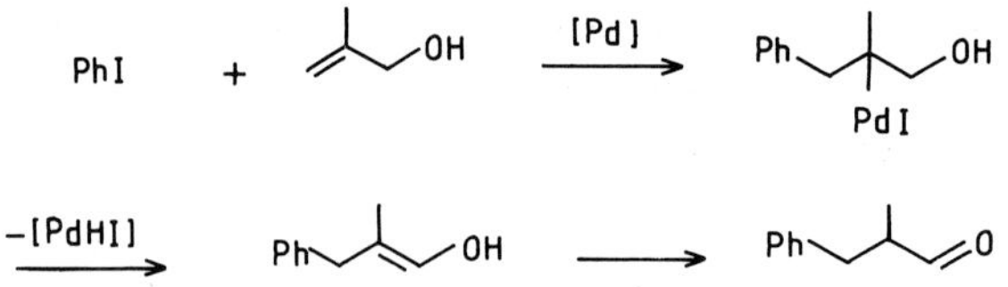
PhI
+
OH
[Pd]
Ph
OH
PdI
–[PdHI]
Ph
OH
Ph
O

(141)

a further example of a useful type of addition to the alkene bond of an allyl alcohol that was discussed in Chapter 6. Scheme 10.23 illustrates some instances in which the reaction is used to elaborate thiophenes.[192] The isomers (142) and (143) were formed in only minor amounts.

This reaction has been applied by Yoshida in a synthesis[192] of honeybee queen substance (144).

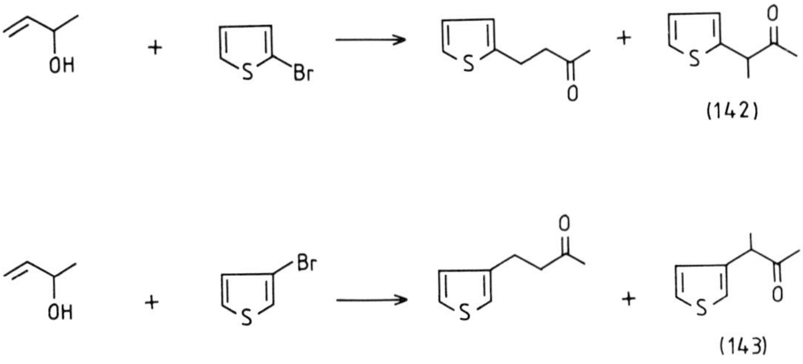

(142)

(143)

Scheme 10.23

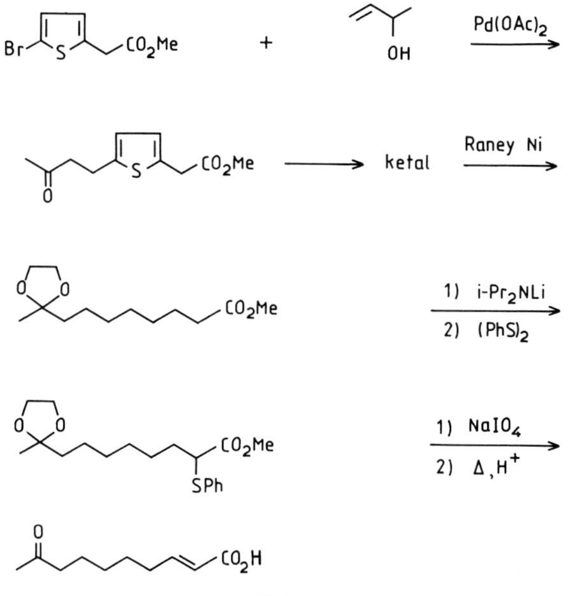

(144)

The procedure is paralleled in similar reactions with 3-bromoquinoline, N-acetyl-3-bromoindole, and in the example depicted, between 2-bromothiopene (145) and 4-vinylpyridine (146) to give the product (147).[193]

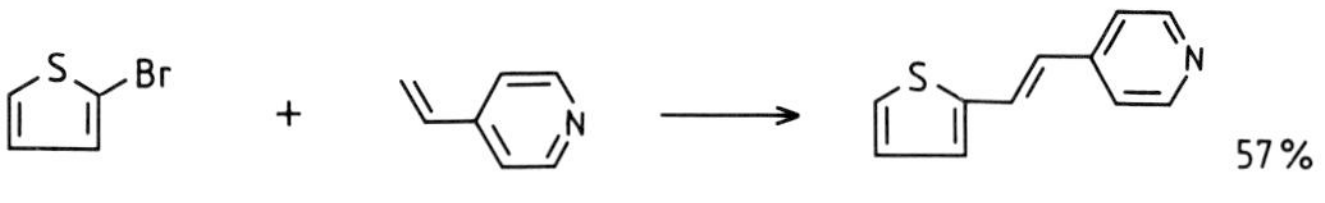

| (145) | (146) | (147) | 57% |

Mercury aryls are also excellent substrates for the coupling reactions, as seen in an example used in a synthesis[194] of pterocarpin (148).

Michael additions can be effected using palladium catalysis, but, if followed by a β-elimination, the transformation retains the alkene linkage in the product. Furthermore,

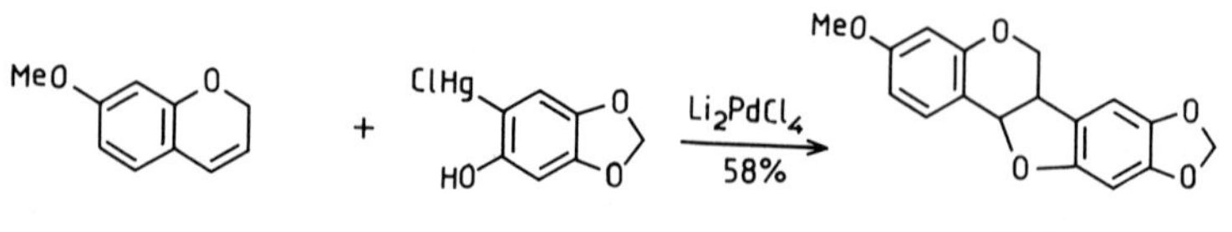

(148)

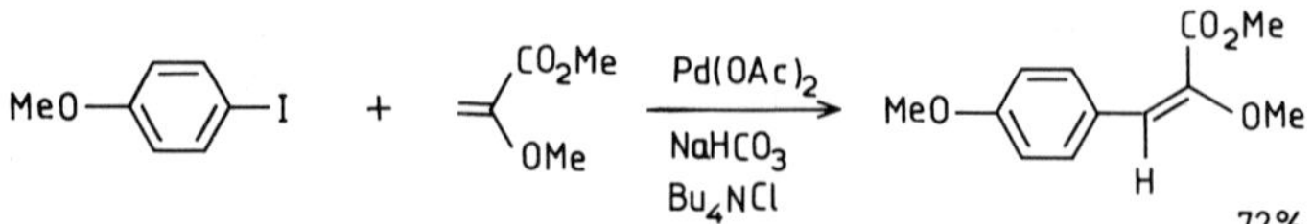

(149)   72%

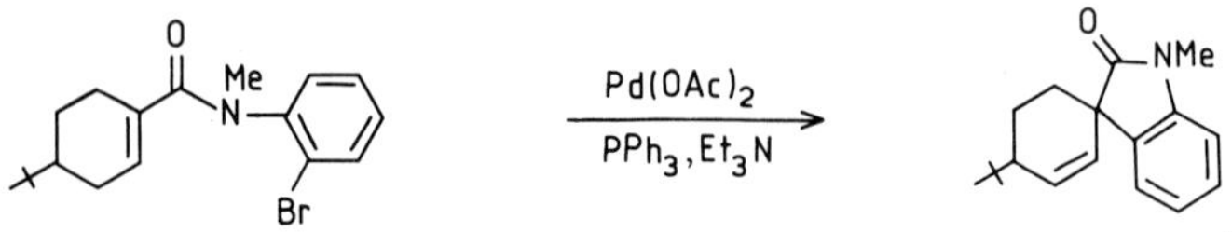

(150)   75%

the carbon–carbon forming reaction need not occur at the β-position. The formation of (149)[195] and (150)[196] illustrates these two points. When β-elimination is avoided, conventional Michael addition products are obtained, as seen in the preparation of (151), a compactin analogue, by combination of an aryl iodide and the α,β-unsaturated lactone. This reaction is also notable for its selectivity in preference of the aryl iodide over the aryl chloride substituents in the lactone side-chain.[197]

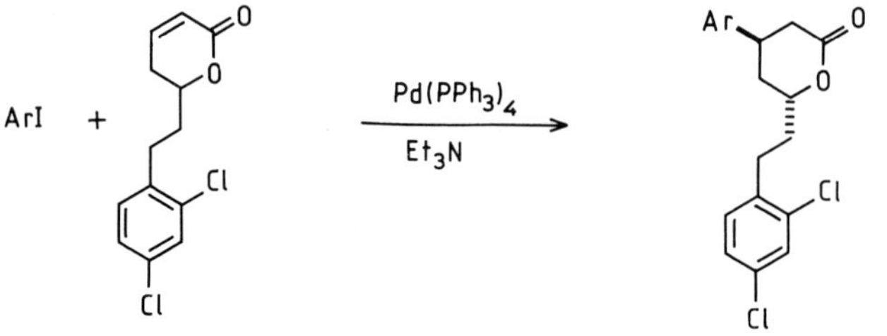

(151)

A further application of coupling reactions using unsaturated carbonyl compounds is found in the elaboration of bromochromones explored by the Davies group.[198]

Palladium catalysed coupling reactions can be run in tandem since the intermediate produced by the first coupling step contains a new palladium–carbon σ bond which is available to carry the reaction on further. A tandem cyclisation of this type has been examined using the substrate (152) which produced a mixture of tetracyclic products in 90% combined yield.[199]

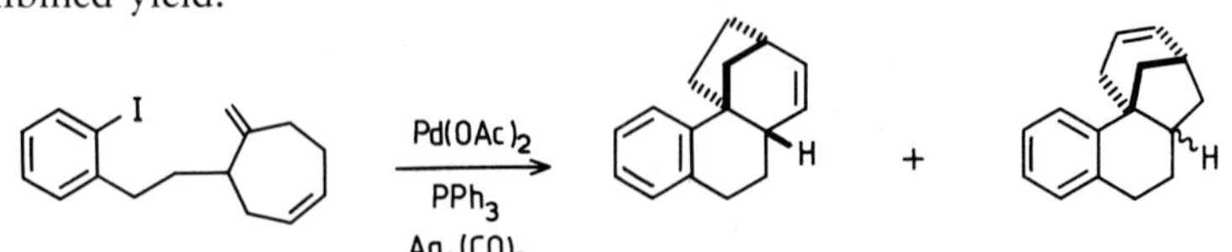

(152)   4   :   3

Aryl and vinyl halides with other aryl and vinyl derivatives

Aromatic and heteroaromatic tin and boron derivatives can be coupled with aryl halides in a versatile reaction. Examples include the formation of (154) using the tin derivative (153),[200] and the combination of bromopyridines and phenylboronic acid to give products such as (155).[201] The selective formation of (154) shows another example of the relatively unreactive character of aryl chlorides. This lack of reactivity, however, can be overcome by complexation of the aromatic ring with $Cr(CO)_3$.[202] The coupling of a chromium complex of a chloroarene with a tin substituted benzofuran has been successfully used in a series of studies leading to the synthesis of moracin M.[203]

Further examples of tin[204] and boron[205] reagents have been used in the introduction of masked aldehydes to aromatic rings and in the preparation of stereocontrolled alkenes.

Vinyl derivatives can be coupled together as seen in the dimerisation of (156) which produced an (E, E)-diene.[206] The t-butylhydroperoxide was added to this reaction to re-

oxidise the palladium after the coupling step, and so permit a catalytic cycle. A similar reaction[207] employs borane reagents under basic conditions, again forming stereocontrolled products (157). Since the stereochemistry of the product corresponds to that

of the starting materials, the reaction provides an efficient method for the stereocontrolled production of dienes.[208] Reactions of this type have now been used quite extensively (see Scheme 10.24).

A simple example is found in a reaction sequence that provided a short route to the red bollworm moth pheromone (158).[209] These products were obtained in good yield and high stereochemical purity.

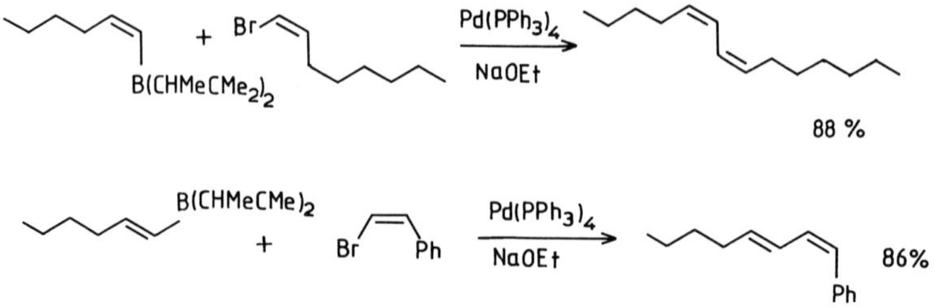

Scheme 10.24

The procedure for coupling with a borane has been extended by Rossi *et al.* to permit reaction with an alkynyl bromide.[209] The method has also been applied to the synthesis of a grapevine moth pheromone (160) using the borane (159).

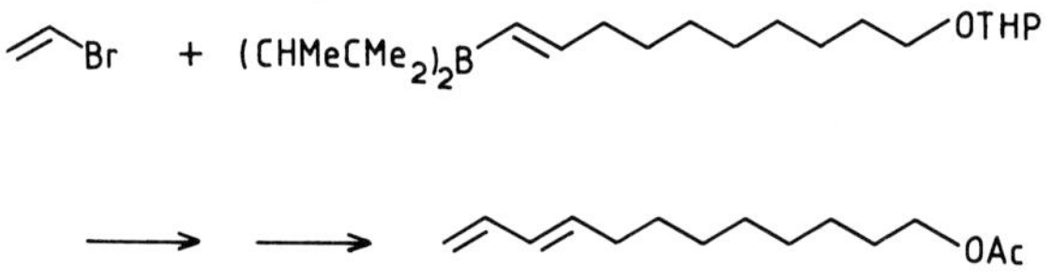

(158)

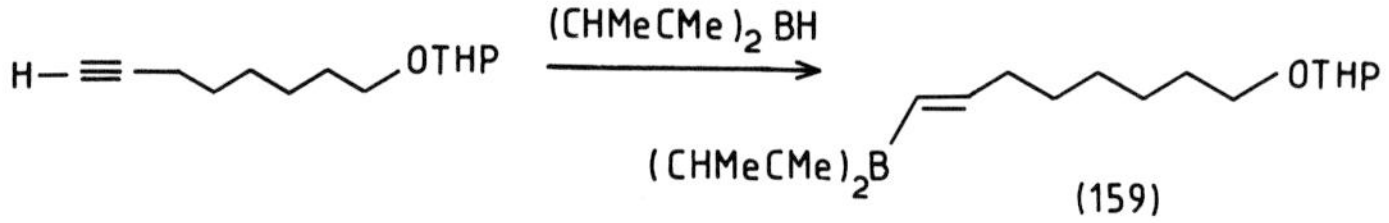

(159)

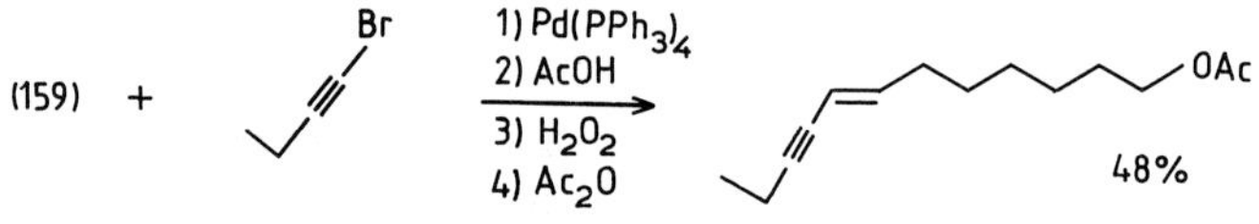

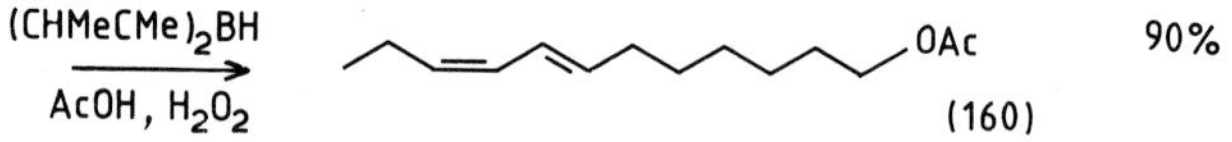

(160)

There are also many examples of the use of other vinyl derivatives in synthetic applications based on these types of coupling reactions. The use of vinyl triflates in coupling reactions[210] is typical of procedures of this type. An intramolecular coupling of vinyltin and vinyl triflate groups has been used by Stille and Tanaka to form the macrocycle (161) in about 56% yield.[211]

Even in intermolecular coupling reactions, unsymmetrical products can be produced. A nice example is the preparation of the pheromone (162) by combination of a vinyl stanane and a vinyl iodide. The (E, Z) product was obtained in 73% yield in a reaction that did not require protection of the alcohol function.[212] Vinyl iodides have been coupled with alkynyltin reagents in a similar way.[213]

Boronate esters can be used in place of the vinyltin reagent. The (E, Z) diene bombykol (163) has been synthesised in this way in a reaction that again occurred in high yield (82%) in the presence of the OH group.[214]

1,3-Dienes bearing a boron substituent are easily obtained from vinyl alkynes. Organoboranes of this type can also be used in couplings with vinyl iodides, as seen in a synthesis of leukotriene B$_4$ (164). The coupling step proceeded in 70% yield.[215] Aryl boronates can also be used in reactions with vinyl halides.[216]

Coupling of alkyl, allyl, and aryl halides

Nickel π-allyl derivatives[217] such as (165) have found application[218] as nucleophilic reagents in synthesis and for coupling reactions by displacement of halogen.

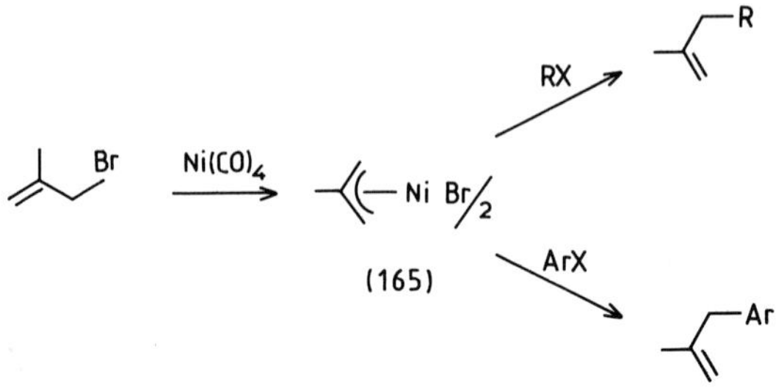

(165)

While the nucleophilic character of nickel complexes is clear from reactions[218] with other electrophiles such as ketones and epoxides, displacement of halogens can occur from situations where direct nucleophilic substitution cannot proceed, as in the example of the aryl halides. In some cases, the mechanism of this coupling process is probably related to coupling reactions discussed earlier in this chapter, whilst other examples may proceed via a bis allyl complex of the type (166).

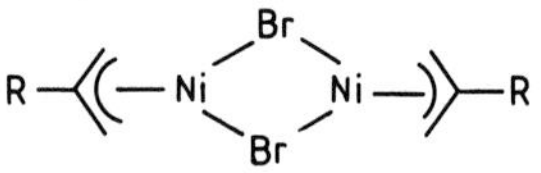

(166)

The synthesis of dictyolene[219] (169) is a good illustration. Combination of the iodide (167) and a nickel π-allyl derivative of isoprene, (168), was followed by LiAlH$_4$ reduction to give (169).

(167)   (168)

(169)

It is important to note that this reaction is mechanistically different from examples of displacement of palladium from a Pd π-allyl complex by a carbanion. Palladium is displaced as Pd(0), whereas in the above case, nickel retains its Ni(II) state:

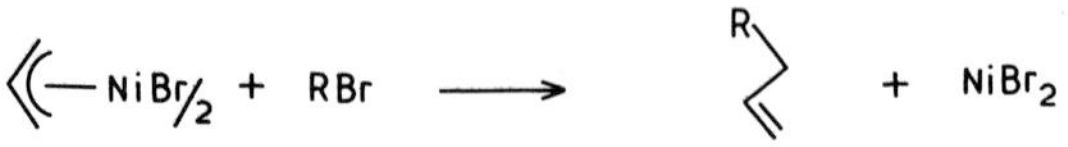

An aryl halide may also be displaced, as exemplified by the reaction[220] of (170) with (171).

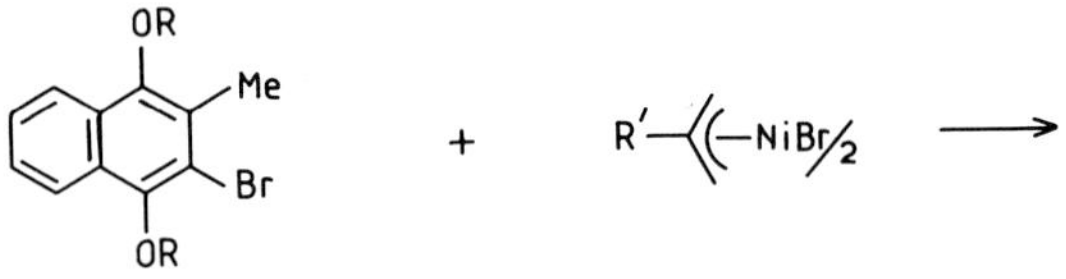

(170)   (171)

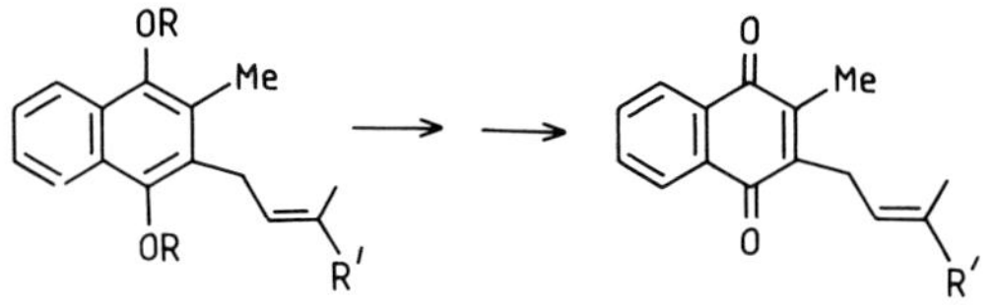

(172)

The protecting group R = Ac or CH_2OMe can be hydrolysed by alkali or by acid, respectively. When R' is $(CH_2CH_2CHMeCH_2)_3H$, the product (172) after ferric chloride oxidation to the quinone, is Vitamin K. It is usual to employ an *N*-alkyl amide solvent in these reactions. The choice of solvent has been found to influence the *cis/trans* ratio of the product.

The nickel π-allyl complex required for this synthesis was obtained by reaction of phytyl bromide with $Ni(CO)_4$. Inoue *et al.* employed a similar sequence starting from solanesyl bromide (173) to obtain Vitamin K_2. Furthermore, starting from the hydroquinone (174), it has been possible to synthesise members of the coenzyme Q series (175).[221]

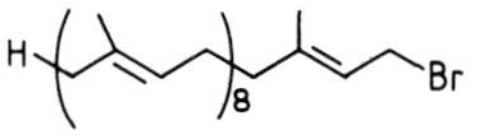

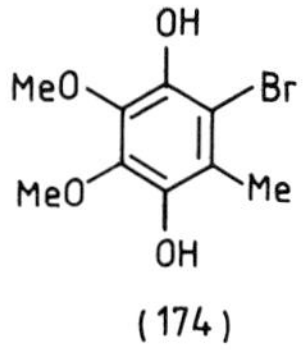

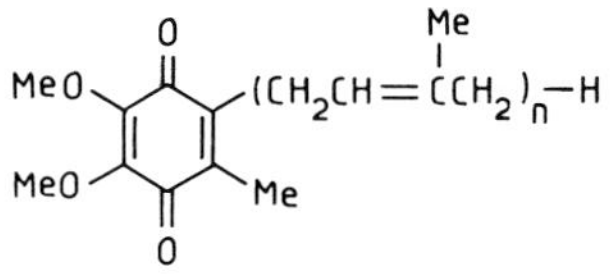

(173)   (174)   (175)

In related work[222] nickel π-allyl alkylation of (176) gave access to naturally occurring 8-methyltocol (177, R = $(CH_2CH_2CHMeCH_2)_3H$).

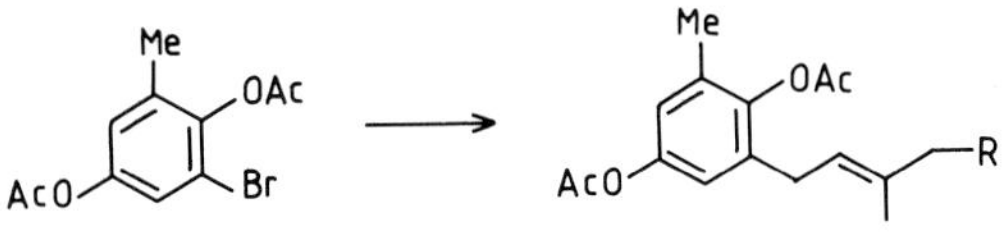

(176)

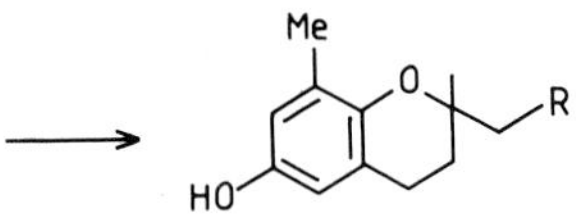

(177)

Bis-allylic dihalides have been cyclised[223] by reaction with $Ni(CO)_4$, no doubt through initial nickel π-allyl formation by one halide. *Trans–trans* cycloalkadienes can be obtained in good yields for $n = 6$, 8, or 10, by cyclisation of (178).

(178)

The kinetic problem of cyclisation of these large rings is, however, illustrated in a synthesis of humulene (181) by this method. It was found that with $Ni(CO)_4$, the dibromide (179), which contained the transoid alkene bond required for (181), did not cyclise effectively. The synthesis was, however, completed starting from the cisoid isomer (180) which underwent a satisfactory cyclisation. Humulene (181) was obtained by stereomutation ($h\nu$ and PhSSPh) of the product.[223, 224]

(179)

(180)

(181)

Another application of the coupling of dibromides with nickel tetracarbonyl can be found in a synthesis of the macrocyclic diterpene ($-$)-casbene by Crombie, Kneen, Pattenden, and Whybrow.[225] Again, nickel carbonyl was used to close a large ring by creating a bond between two allyl units. This procedure successfully closed a 14-membered ring (182), although in rather low yield.

(182)

A lactone synthesis by Semmelhack depended on a cyclisation reaction between a nickel π-allyl and an aldehyde function, followed by carbonylation of a second nickel π-allyl complex. The process was applied to a synthesis of frullanolide (185).[226] The allylic methane sulphonate (184), elaborated from (183), was converted by reaction with nickel carbonyl via the sequence shown, into a bicyclic product which underwent carbon monoxide insertion.

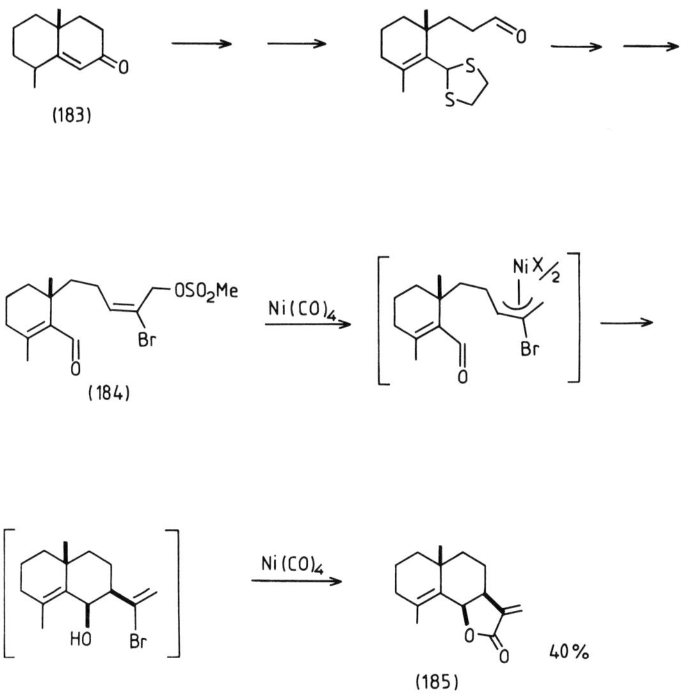

(183)

(184)   Ni(CO)₄

(185)   Ni(CO)₄   40%

10.6 Coupling of vinyl, aryl, and alkyl halides with nucleophiles

Vinyl halides

Grignard reagents can be coupled[227] to vinyl halides of the type (186) in the presence of Pd(PPh₃)₄. A similar reaction has been performed using a nickel catalyst. By exploiting

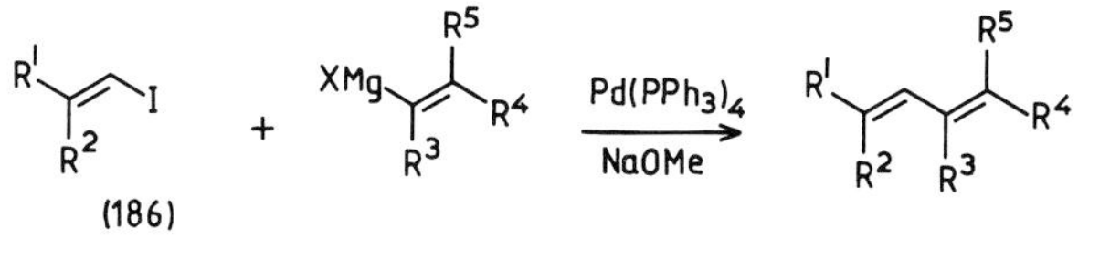

(186)   Pd(PPh₃)₄ / NaOMe

$R^1 = R^2 = H$, C_6H_{13} R^4, $R^5 = H$ or H, Me, $R^3 = H$ or Me.

the difference in reactivity between allyl and vinyl bromides, a controlled sequence of two coupling reactions can be performed.[228] In a similar process, vinyl triflates and a manganese catalyst, Li₂MnCl₄, have been used in couplings with Grignard reagents. Alternatively, trialkylmanganese derivatives can be coupled with enol phosphates or triflates using palladium catalysis.[229]

Nickel catalysed coupling with Grignard reagents has been applied by Kociénski in the preparation of a fragment for the polyether natural product premonensin B. In this case, the vinyl oxygen linkage in dihydrofurans and dihydropyrans was cleaved during the introduction of a methyl group from methylmagnesium bromide. The result was a highly stereoselective preparation of trisubstituted alkenes.[230] When applied to premonensin B, a sequence involving introduction of the vinyl group as a lithiated enol ether, followed by reaction with the Grignard reagent, was used twice. Both the alkene products (187) and (188) were formed with complete stereocontrol.[231]

A vinylzirconium reagent has been used by Crombie in a coupling reaction with the vinyl iodide (189).[232] Allylzinc reagents have also been used as the nucleophilic component in cross coupling reactions with vinyl iodides.[233] A similar reaction has been described with enol triflates.[234] In a good many cases, either Grignard reagents themselves, or the derived organozinc reagents, can both be used to effect palladium catalysed versions of these coupling reactions.[235] In a final example, a vinylaluminium reagent was combined with vinyl bromide to produce (190).[236]

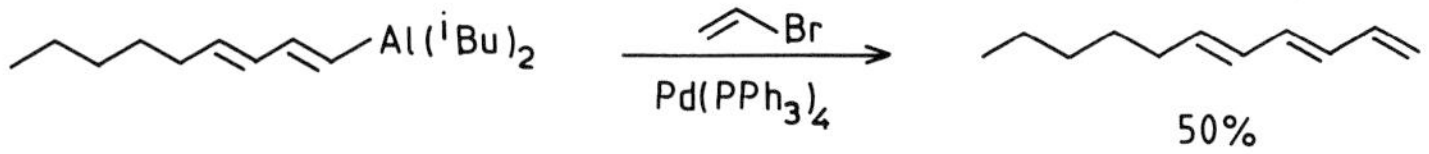

(190)

A widely used type of reaction is illustrated by the coupling[237] of a vinyl bromide and an alkyne using $(Ph_3P)_4Pd$ and CuI under phase transfer conditions. In this case, the conditions employed used $PhCH_2NEt_3Cl$ as the phase transfer agent and NaOH as base, and benzene was used as the organic solvent. Zinc derivatives of alkynes can also be used to couple with vinyl halides. Double coupling with 1,2-dibromoethene provides an interesting example.[238] The $(Ph_3P)_4Pd/CuI$ procedure has been applied to a synthesis of the pheromone (191).

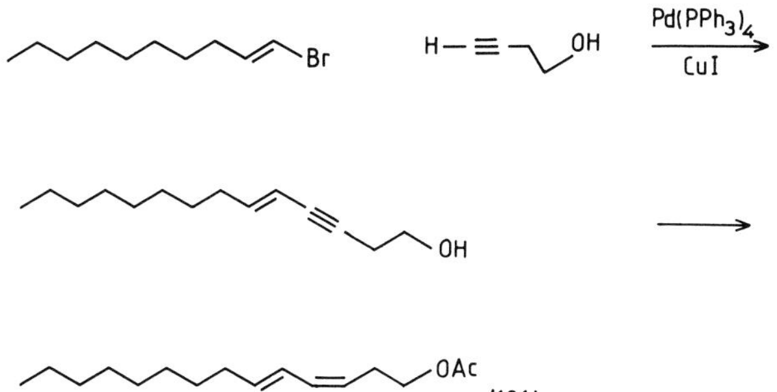

(191)

Coupling reactions of this type are suitable for use in the presence of other substituents, as is evident from an example taken from the synthesis of the arachidonic acid metabolic (192).[239] This type of coupling has also been used with a vinyl triflate

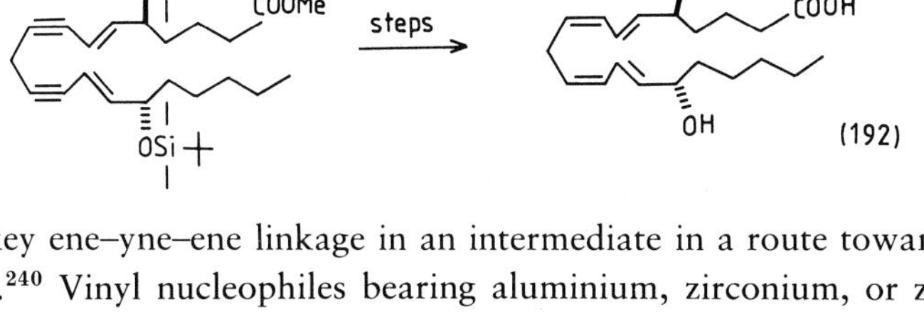

(192)

to form a key ene–yne–ene linkage in an intermediate in a route towards a dihydroxy vitamin D_3.[240] Vinyl nucleophiles bearing aluminium, zirconium, or zinc, introduced by an initial addition to an alkyne (see Chapter 8), can be combined with a variety of organic halides, including the vinyl halide (193). Both nickel and palladium catalysts

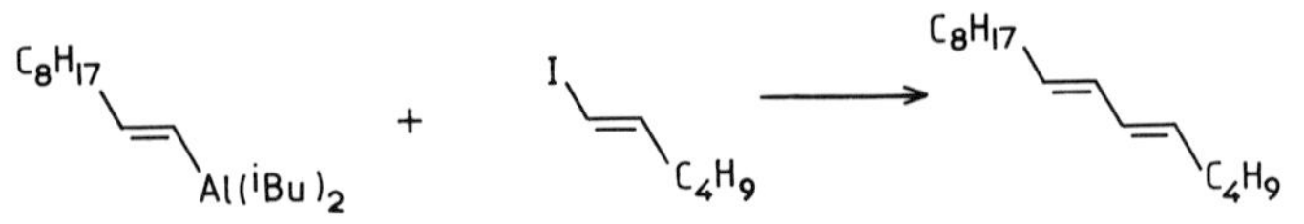

(193)

have been examined.[241] This process makes an interesting comparison with methods using dialkylboranes discussed earlier. While both constitute attractive synthetic procedures for linear dienes, the organoaluminium and zirconium methods are also applicable to the formation of trisubstituted alkenes.

Aryl halides

Nucleophilic displacement of halide from aryl halides was encountered in the case of chromium arene complexes in Chapter 7. A similar outcome can be achieved by a totally different mechanism, through catalytic cross coupling reactions. Examples in Scheme 10.25 demonstrate the use of stabilised enolates[242] and thiolate anions.[243]

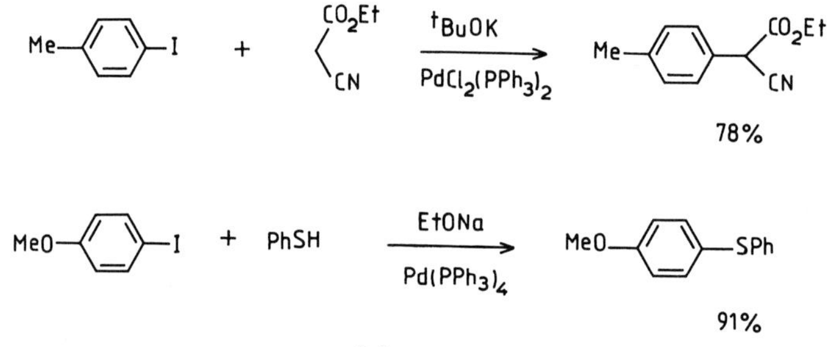

Scheme 10.25

Similar coupling to a heteroatom provides the final step of a synthesis of the CDE ring system of the antitumour antibiotic lavendamycin. For this purpose Pd(PPh₃)₄ was used in excess to effect the cyclisation of (194).[244]

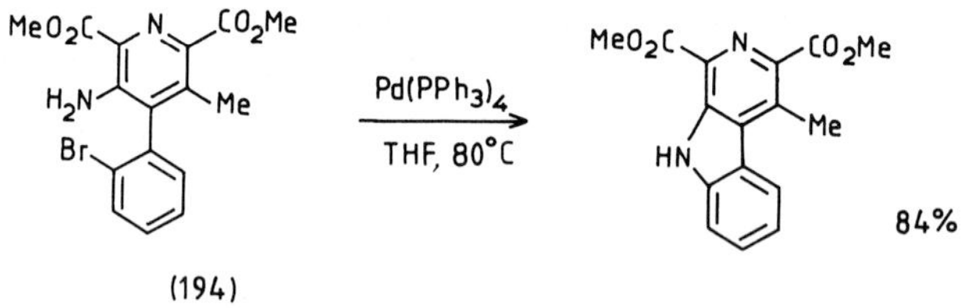

(194)

In another example, taken from work from the Pfizer research laboratories at Sandwich, lithiated imidazoles have been used to couple with 2-bromopyridine in the presence of (Ph₃P)₄P and ZnCl₂.[245] Alkynes have also been coupled to a pyridine ring. This reaction employed CuI in a manner similar to that discussed earlier in this chapter (pp. 340, 344 and 361) for other coupling reactions of alkynes.[246]

Phosphorous has also been employed as the nucleophilic centre in a reaction that converted tyrosine-containing peptides into arylphosphonates in a search for inhibitors of tyrosine protein kinases.[247] This interconversion involved a palladium catalysed replacement of OTf in the triflate derivatives (195).

(195)

A synthetic application of the stabilised enolate coupling to aryl iodides (described above) has been explored by Ciufolini. The spirocyclic product (196) was obtained in model studies for a route to fredericamycin A.[248]

Quite a variety of nucleophilic reagents have been examined in coupling reactions with aryl halides. The procedures used are similar to those described for vinyl iodides. For example, alkynes have been coupled with aryl[249] and heteroaryl[250] halides by reaction with copper iodide, base, and a palladium catalyst. A similar example involving dibromobenzene and propargyl alcohol has been performed using *n*-propylamine without the need for a copper salt.[257]

Organoaluminium reagents derived from alkynes, similarly, have been coupled with aryl halides, in this case in a synthesis of tamoxifen isomers.[252] These examples point to an impressive variety of nucleophiles compatible with the coupling procedures. Indeed the range is extensive, spanning unusual reagents such as $Me_3SiF_2^-$,[253] to relatively mild alkylating agents such as allyl stannanes.[254] In this latter case, the formation of (197), in which alkylation has occurred by reaction at the alkene end of

(197)

the allyl unit, is consistent with a nucleophilic addition of the allyl stannane. Other mechanistic possibilities, however, cannot be ruled out since this coupling process could proceed by oxidative addition of the aryl halide to a π-allyl palladium complex.

Reactions of vinyl silicon, tin, and boron derivatives discussed earlier most probably follow this path, rather than direct nucleophile addition. It is not realistic to draw a definitive general dividing line between these two processes when comparing reactions of neutral, weakly nucleophilic species.

A similar coupling of allyl acetates in the presence of a tin nucleophile and a palladium catalyst has been used in a series of cyclisation reactions.[255] Cross coupling of allyl stannanes with aryl bromides, and of aryl stannanes with allyl acetates, have both been described.[256] In the case of the formation of (198), two features are notable. The first

(198)   76%

is that the reaction is stereocontrolled. The second concerns the regiochemistry of the cyclisation. It is tempting to compare this with the intermolecular reaction leading to (197). In fact, however, the cyclisation is dominated by the preferred formation of a five-membered ring and the position of the acetoxy group and the alkene are immaterial. Similar examples using acetate derivatives of vinyl substituted tertiary alcohols give rise to the same type of products.[255]

Alkyl halides with nucleophiles

While alkyl halides, and particularly allyl and benzyl halides, react well with nucleophiles themselves, there can, none the less, be advantages gained from the use of palladium catalysed cross coupling.[257] Reactions of this type, employing both aryl and alkyl Grignard reagents with several alkyliodides, have been examined. Organolithium and zinc nucleophiles have also been used.[258] With allyl phosphonates and halides, there is the possibility of the formation of π-allyl intermediates. A variety of coupling reactions with Grignard reagents,[259] organotin reagents,[260] stabilised enolates,[260] have all been found to proceed with satisfactory yields.

10.7 Coupling of alkenes with nucleophiles

Heterocyclic ring closures similar to the preceding example can be effected by palladium catalysed nitrogen addition to an alkene. The reaction requires the Z isomer (199). When the E isomer (200) was used, β-elimination was blocked by the methyl group. The corresponding palladium complex formed from (199) could not be isolated, but both complexes could be converted to esters by carbonyl insertion, a reaction that provides confirmation for the stereochemistry of the intermediates.[261] These results indicate that both nucleophile and metal add to the same face of the alkene, perhaps by coordination of the alkene, and nitrogen transfer to the ligand via the metal.

A cyclisation that formed a carbon–carbon bond results from palladium catalysed

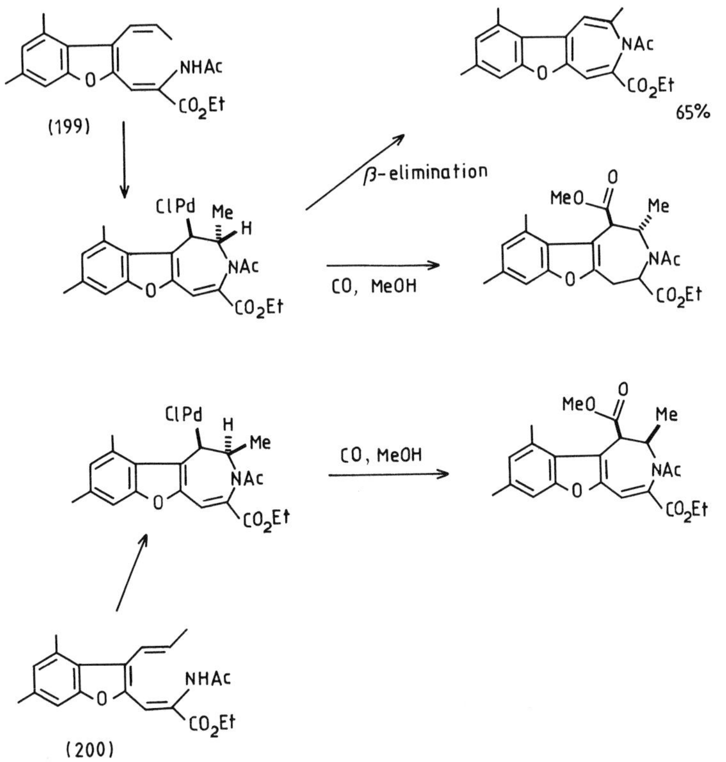

(199)

β-elimination

65%

ClPd

CO, MeOH

ClPd

CO, MeOH

(200)

addition of a silylenol ether to an alkene that has been used in a formal synthesis of quadrone (204).[262] The intermediate (203)[263] was obtained from the cyclisation product (201) by an aldol reaction of the ketone (202). A further palladium catalysed coupling step[264] was used in the preparation of (202). In another case, uracils have been coupled with alkenes using palladium catalysis.[265]

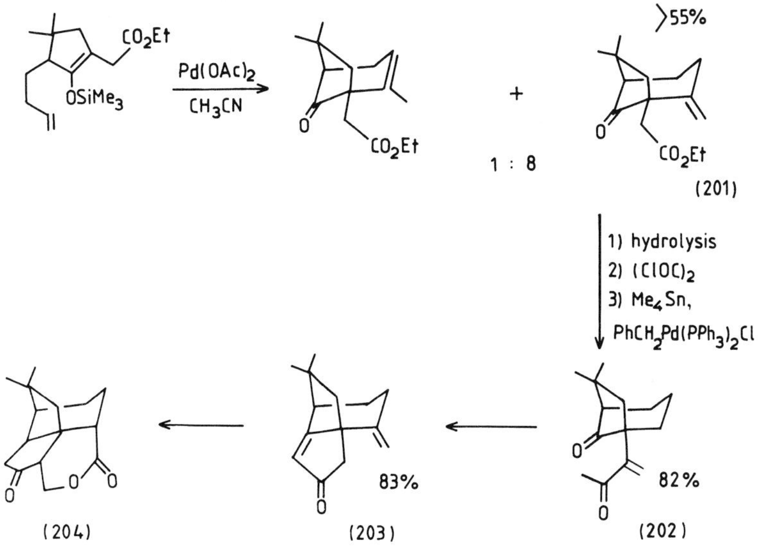

CO_2Et

$OSiMe_3$

$\dfrac{Pd(OAc)_2}{CH_3CN}$

CO_2Et

+

>55%

CO_2Et

1 : 8

(201)

1) hydrolysis
2) $(ClOC)_2$
3) Me_4Sn,
   $PhCH_2Pd(PPh_3)_2Cl$

(204)

(203) 83%

(202) 82%

All these palladium catalysed additions to alkenes are essentially further examples of the nucleophile addition to alkene π-complexes discussed in Chapter 5. Since these examples culminate with a β-elimination step to reform an alkene linkage, they also resemble other coupling processes discussed earlier in this chapter. Often it is difficult to identify accurately the mechanism of reactions of this type. This problem is apparent in the following examples. A palladium catalysed Claisen rearrangement[266] could be viewed as involving either an allyl ether or an enol ether. In fact, in common with palladium catalysed Cope and hetero Cope rearrangements[267] of allylic esters[268] (see Chapter 2), and of allyl imidates[269] and S-allyl thioimidates,[270] these processes may all proceed by comparable reactions involving coupling of the nucleophilic centre with a metal bound alkene.[267, 268]

10.8 Coupling with acid chlorides

In the quadrone synthesis discussed above, tetramethyltin was coupled with an acid chloride to produce (202). Similar couplings using tetraalkyl lead reagents and simple acid chlorides have also been described.[271] With mixed organotin reagents, at least for vinyltrimethylstannanes, selective transfer of the vinyl group was observed. This can be seen in the preparation of (205) from the corresponding acid chloride.[272]

(205)   70%

Organotin and lead reagents have no special part to play in these processes, serving simply as a convenient source of an alkyl group in the catalytic cycle. The insertion of acid chlorides into metal–carbon σ bonds is likely to be a fairly general process, but has not been extensively used in synthesis, perhaps because there are alternative methods for the transformations. Examples of the reaction with isolated palladated intermediates in stoichiometric reactions can be seen in the formation of (206)[273] and (207).[274]

(206)   81%

(207)

10.9 Oligomerisation reactions of dienes

The well known di- and trimerisation of buta-1,3-diene,[154] brought to prominence by Wilke and his co-workers, has formed the basis for a synthesis of large ring compounds. The best known trimerisation, which is promoted by a wide range of catalysts, yields 1,5,9-cyclododecatriene (208). The reaction is diverted to yield dimers (209), (210), and (211) in the presence of strongly coordinating ligands, such a phosphines.[275]

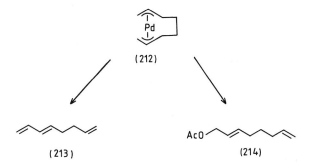

Dimerisation has also been brought about by Pd(II).[276] This forms the basis of a number of syntheses of natural products. The intermediate complex (212) in dimerisation with palladium, may undergo internal hydrogen transfer to yield an octatriene (213), or, in the presence of a suitable nucleophile such as HOAc, an octadienyl derivative is formed. In this instance the product is octadienyl acetate (214).

The relative acidity of hydrogens in the ligands can exert an influence to control the reaction. An interesting example is the selective formation of (215) by palladium catalysis. A different result was obtained when a nickel catalyst was employed.[277]

A good example in which this type of reaction is put to use is provided by a synthesis of 12-acetoxydodeca-1,3-diene (222)[278] by Tsuji and co-workers. Buta-1,3-diene, with $PdCl_2(PPh_3)_2$ and phenol/sodium phenoxide in benzene, gave the phenoxyoctadiene (216) which, with 9-BBN and hydrogen peroxide, yielded (217, X = HO). This product was converted to the allyl iodide via a *p*-toluenesulphonate. Tetrahydrofuran was the source of (218) and hence, of the derivative (219). Coupling (219) with the iodide (217, X = I) in presence of CuI and α,α-bipyridyl gave (220) which, after hydrolysis of the

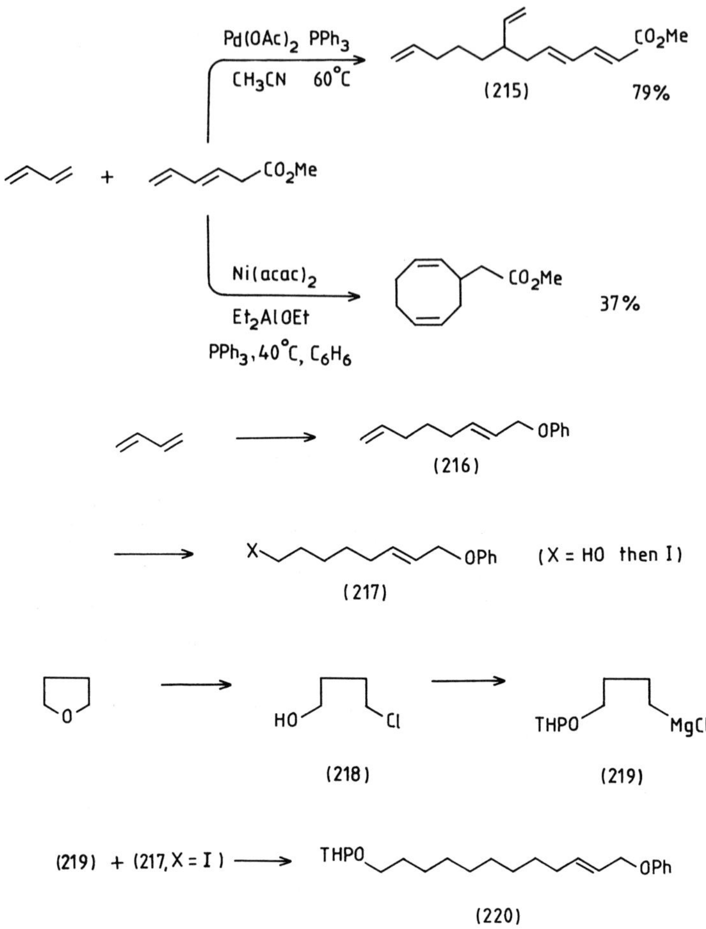

$$\text{Pd(OAc)}_2 \quad \text{PPh}_3 \xrightarrow{\quad} \qquad (215) \qquad 79\%$$

$$\text{CH}_3\text{CN} \quad 60°C$$

$$\text{(215)}$$

$$\begin{array}{c} \xrightarrow{\quad \text{Ni(acac)}_2 \quad} \\ \text{Et}_2\text{AlOEt} \\ \text{PPh}_3, 40°C, \text{C}_6\text{H}_6 \end{array}$$

37%

$$\xrightarrow{\quad\quad} \qquad \qquad \text{OPh}$$

(216)

$$\xrightarrow{\quad\quad} \text{X} \qquad \qquad \text{OPh} \qquad (X = HO \text{ then } I)$$

(217)

$$\xrightarrow{\quad\quad} \text{HO} \qquad \text{Cl} \qquad \xrightarrow{\quad\quad} \text{THPO} \qquad \text{MgCl}$$

(218)   (219)

$$(219) \ + \ (217, X = I) \ \longrightarrow \ \text{THPO} \qquad\qquad\qquad \text{OPh}$$

(220)

tetrahydropyranyl ether, was converted to the acetate (221). The required diene (222) was obtained by heating at 160 °C under reduced pressure in the presence of Pd(OAc)_2 and PPh_3, an application of the elimination reaction discussed in Chapter 6. The product was shown to be a 1:1 mixture of (222) and the *cis*-isomer (223).

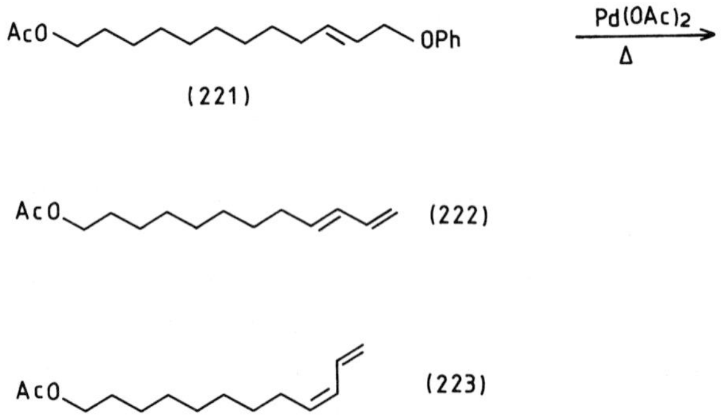

$$\text{AcO} \qquad\qquad\qquad \text{OPh} \xrightarrow[\Delta]{\text{Pd(OAc)}_2}$$

(221)

$$\text{AcO} \qquad\qquad\qquad\qquad (222)$$

$$\text{AcO} \qquad\qquad\qquad (223)$$

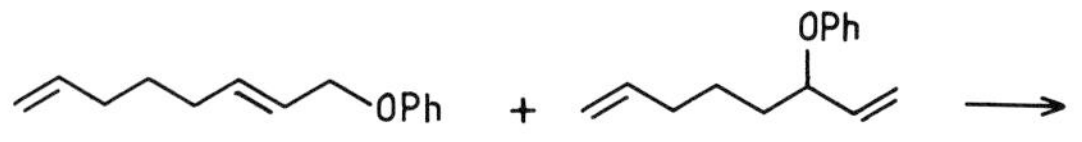

The structures (224), (225), (226), (227), (228), and (229) are shown as chemical diagrams.

(224) + (225) →

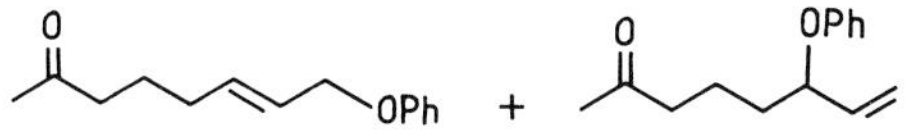

(226) + (227)

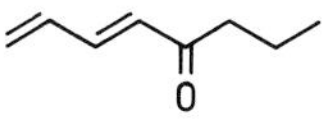

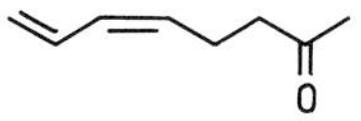

(228)   (229)

A somewhat similar sequence of reactions formed the basis of a synthesis of (*E*)- and (*Z*)-pyrethrolones (230) and (231).[279] The 1:1 mixture of phenoxyoctadienes (224) and (225), produced from a catalysed dimerisation of butadiene with addition of phenol, was oxidised to the ketones (226) and (227) by means of the Wacker conditions PdCl$_2$, CuCl, and air, in aqueous DMF (see Chapter 4). Refluxing these products in dioxane in the presence of Pd(OAc)$_2$ and PPh$_3$ brought about elimination of phenol to produce (228) and (229) in the ratio 7:3.

The synthesis of the pyrethrolones was completed by already established procedures as shown in Scheme 10.26.[280]

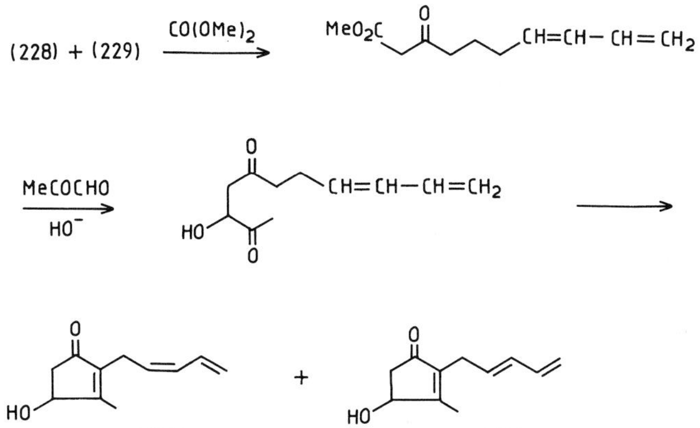

(228) + (229) $\xrightarrow{\text{CO(OMe)}_2}$ MeO$_2$C—...—CH=CH—CH=CH$_2$

$\xrightarrow[\text{HO}^-]{\text{MeCOCHO}}$ ...—CH=CH—CH=CH$_2$ →

(230)   +   (231)

Scheme 10.26

Another example of heteroatom addition to the butadiene during dimerisation is provided by combination of butadiene and aminopyridines promoted by Pd(acac)$_2$, PPh$_3$, and Et$_3$Al. Both mono and bis adducts at the amino group were obtained.[281]

A range of active methylene compounds may be incorporated in the final step in palladium catalysed butadiene dimerisation.[282] Malonic esters, β-ketoesters, nitro-alkanes, and 1,3-diones are all suitable. An example of the use of diethyl malonate relates to the synthesis of the insecticidal compound, pellitorine (235, R = NH^{i}Bu).[283] The first product (232) was partially hydrogenated (see Section 10.2) at the terminal alkene centre using RuCl$_2$(PPh$_3$)$_3$ catalysis in benzene/ethanol. The resultant diethyl nonenoate was converted to the mono potassium salt and treated with diphenyl-diselenide. The phenylseleno product (233) could be oxidised by periodate to (234), and thence to the carboxylic acid (235 R = H) by elimination of PhSeOH and hydrolysis.

(232)

(233)

(234)   (235)

A parallel reaction in which butadiene was dimerised in the presence of a nitroalkane[284] is the basis of a synthesis of *cis*-civetone (238).[285] The initial reaction, which is brought about by PdCl$_2$(PPh$_3$)$_2$ in the presence of NaOPh in ^{i}PrOH at 25 °C led to (236) with the mono-octadienyl derivative, $n = 1$, as the major product.[284]

(236)

Starting from the bis-octadienyl substituted nitromethane (237), which is available in this way, the synthesis of civetone[285] proceeded by the steps shown in Scheme 10.27.

This approach to macrocycles has been further explored by Tsuji's group in other examples using the mono-octadienyl substituted nitroethane (239) available from palladium catalysed condensation of butadiene with nitroethane.[286] The sequence of reactions shown led from (239) to recifeiolide (240). An alternative route starting from (239)[287] has also been described.

A method for introducing the hydroxyl function which does not make use of the nitroalkane coupling is outlined in Scheme 10.28.[286] The octadienyl acetate (241), obtained from butadiene with (Ph$_3$P)$_2$Pd(maleic anhydride) in acetic acid,[288] was oxidised at the terminal alkene. The carbonyl and acetoxy groups were modified to give (242). The elaboration of (242) to (243) afforded a product related to diplodialide.

An octadienyl acetate (244), also available from butadiene, provided a route for the

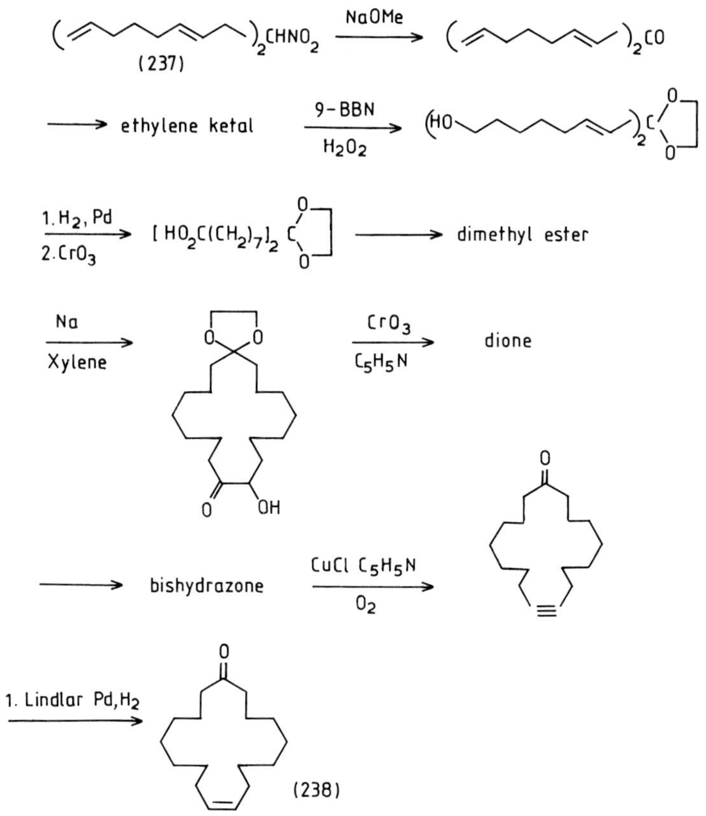

Scheme 10.27

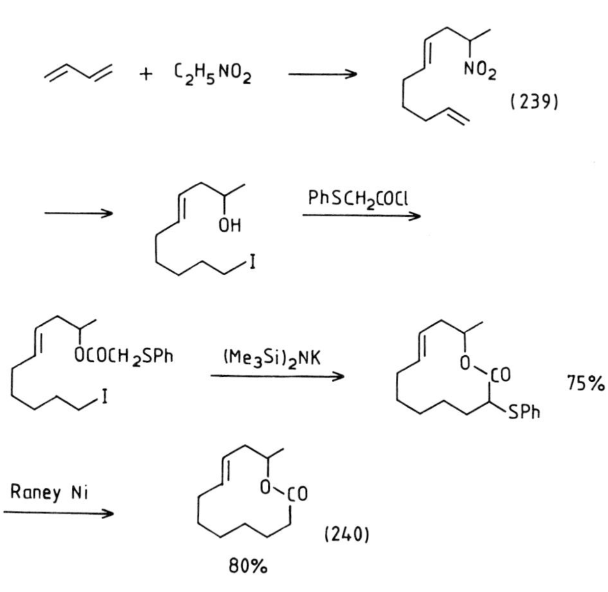

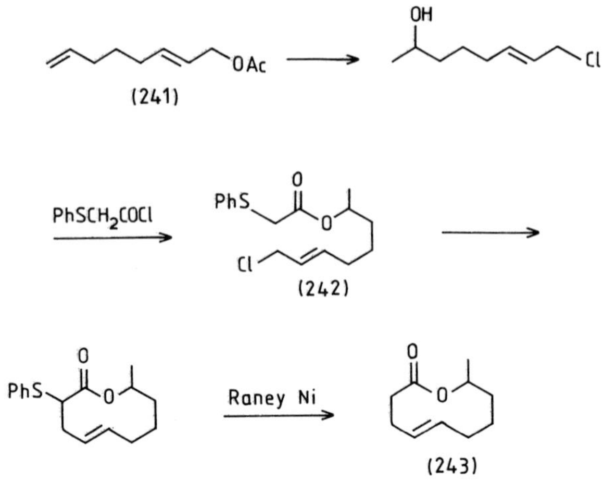

(241)

(242)

(243)

Scheme 10.28

synthesis of (±)-zearalenone.[289] The acetate (244) was converted via the alcohol to the ketone (245) which, by addition of diethyl malonate, gave (246). This product was converted into the mono ester, and then into the toluene *p*-sulphonate (247) in which the carbonyl group was protected as an acetal. Oxidation of the vinyl group of (247) by means of PdCl$_2$, CuCl in aqueous DMF (see Chapter 4) yielded a methyl ketone and thence (248), which was condensed with (249) to yield (250) after cyclisation. The PhS group was eliminated by NaIO$_4$ oxidation to give (±)-zearalenone (251).

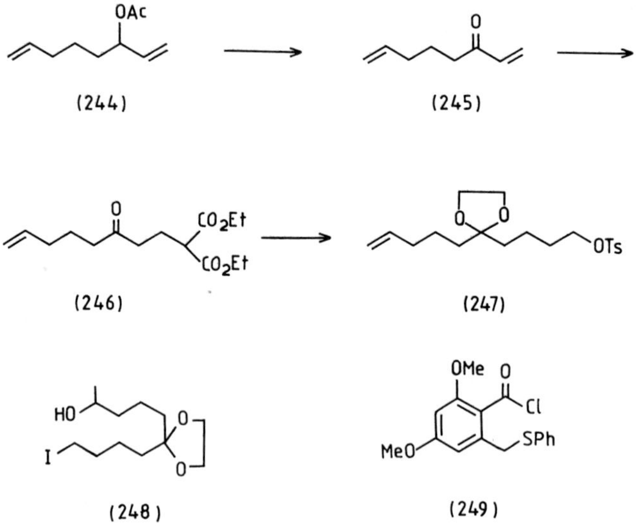

(244)

(245)

(246)

(247)

(248)

(249)

The octadienone (245) has also provided a useful building block in a route[290] to steroid-related structures such as 19-nortestosterone (253), starting from (252). Once again, use was made of the conversion of a vinyl group into a methylketone to form

(i) $PdCl_2$, $CuCl$, H_2O;

(ii) Pd, H_2;

(iii) HCl , $MeOH$.

(253). The telomer (232) gave access[291] to a convenient synthesis of queen substance (254), the honey bee pheromone, in a route that included another Wacker oxidation step.

The dimerisation of 1,3-dienes has been applied by Hidai *et al.* to a synthesis of citronellol (258).[292] Isoprene (255), when reacted with PdCl$_2$ and (nBu)$_3$P in methanol containing NaOMe, gave the head-to-tail dimer (256). If the reaction was carried out in the presence of neomenthyldiphenylphosphine in place of (nBu)$_3$P, an optically active product was obtained. 1,4-Addition of hydrogen at the diene centre by hydrogenation catalysed by the [Cr(CO)$_3$(PhCO$_2$Me)] complex provides a further example of the application of the method which was discussed in Section 10.2. Diene products such as (257) can be obtained directly by the use of a catalyst derived from (2-methylallyl)Pd(cyclooctadiene)$^+$ salts and phosphines.[293]

The dimerisation of butadiene has been explored by Chanvin *et al.*[294] as a route to some α-amino acids. Thus, reaction of butadiene with diethyl acetamidomalonate (259) in the presence of Pd(OAc)$_2$/PPh$_3$/PhONa produced the diethyl 1-octa-2,7-dienyl-acetamidomalonate (260) in 91% yield. Hydrogenation and hydrolysis yielded the α-amino acid (261).

Carbonyl insertion can also provide a concluding step to butadiene oligomerisation. A good example is the formation of esters by addition[295] of carbon monoxide and an alcohol in the manner discussed in Chapter 9. In common with the reactions described above, the products, 3,8-nonadienoate esters such as (262), are formed with excellent control of double bond position and stereochemistry. This suggests that a similar bis-allyl intermediate may be involved in oligomerisation reactions of this type.

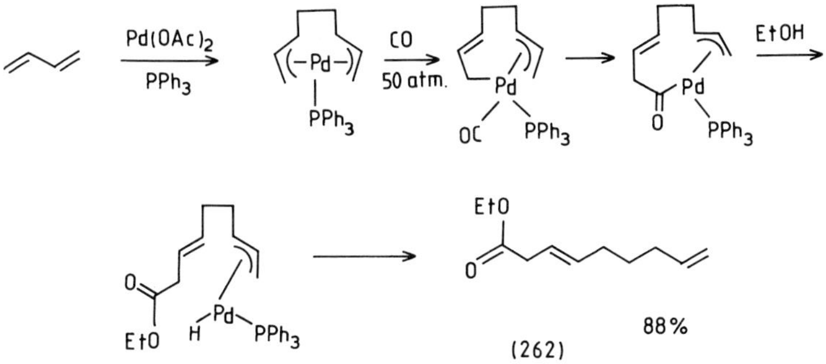

Carbon dioxide undergoes a similar insertion (see Section 10.1) to form a δ-lactone, but in substantially lower yield.[6] This example is of note, however, since the independent preparation[7] of a stoichiometric bis-palladium complex, which produced acyclic insertion products with CO_2, provides an interesting comparison with the monometallic intermediates implicated in the catalytic oligomerisation reaction.

10.10 Reductive coupling of metal π complexes

Organometallic π complexes can also couple with one another. Reductive conditions, typically using zinc, as shown in Scheme 10.29, are generally employed,[296,297,298,299] although photolytic[300] and electrochemical[299,301] methods have also been used.

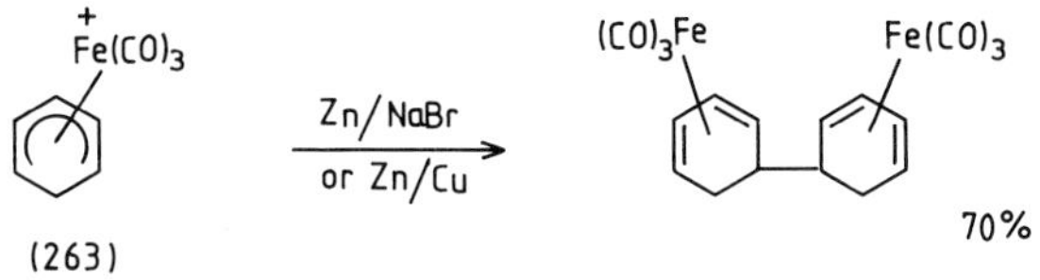

Scheme 10.29

Recently, similar reductive couplings with cationic cobalt and iron cyclopentadienyl complexes with sodium amalgam have been described.[302] Basic reagents can also effect reductive dimerisation of cationic π complexes, a result often accompanying an unsuccessful nucleophile addition. The dimer (265) was formed, for example, by reaction of (264) with 2,6-dimethylpyridine.[303] In other cases, reaction with alkali metal hydroxides in acetone[296] or with trialkylborohydride reagents[304] has resulted in coupling reactions. In these reactions electron transfer, rather than nucleophile addition has occurred.

The reductive dimerisation of tricarbonyliron complexes with zinc to form (270) provides an example that has been used in medium ring synthesis.[305] The bimetallic system required for this purpose was prepared by Friedel-Crafts acylation of the butadiene complex (267) with an appropriate bis acid chloride, and treatment with

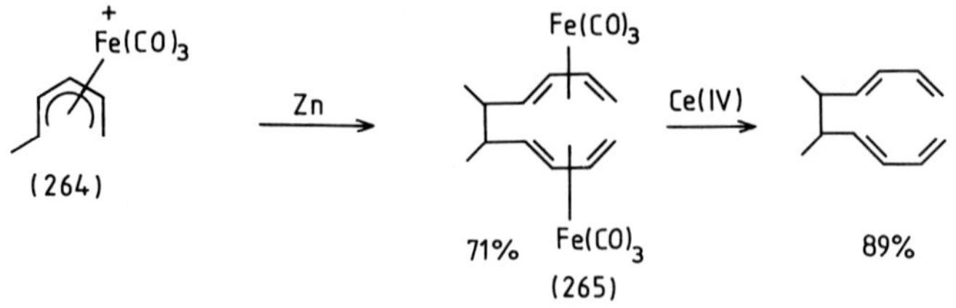

(264)   Zn   71% (265)   Ce(IV)   89%

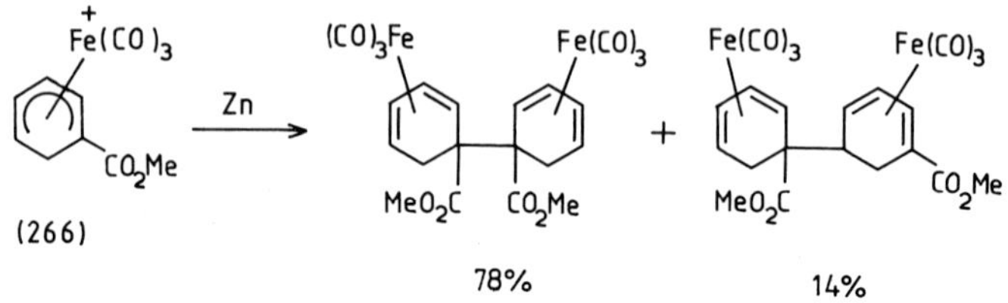

(266)   Zn   78%   +   14%

Ph⤳OAc   Pd(PPh₄), Zn, THF   85%

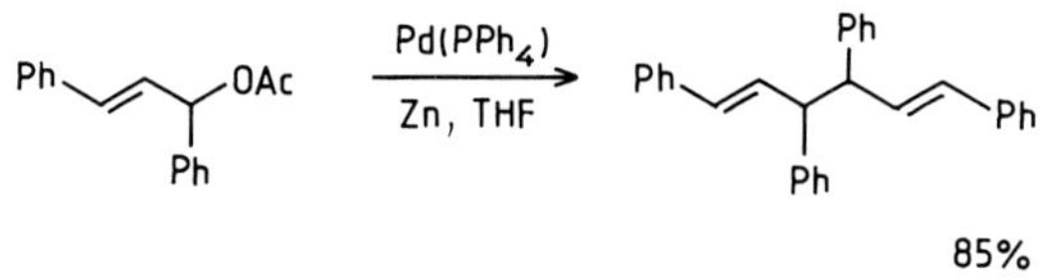

1) (267), AlCl₃, Fe(CO)₃
2) NaOMe

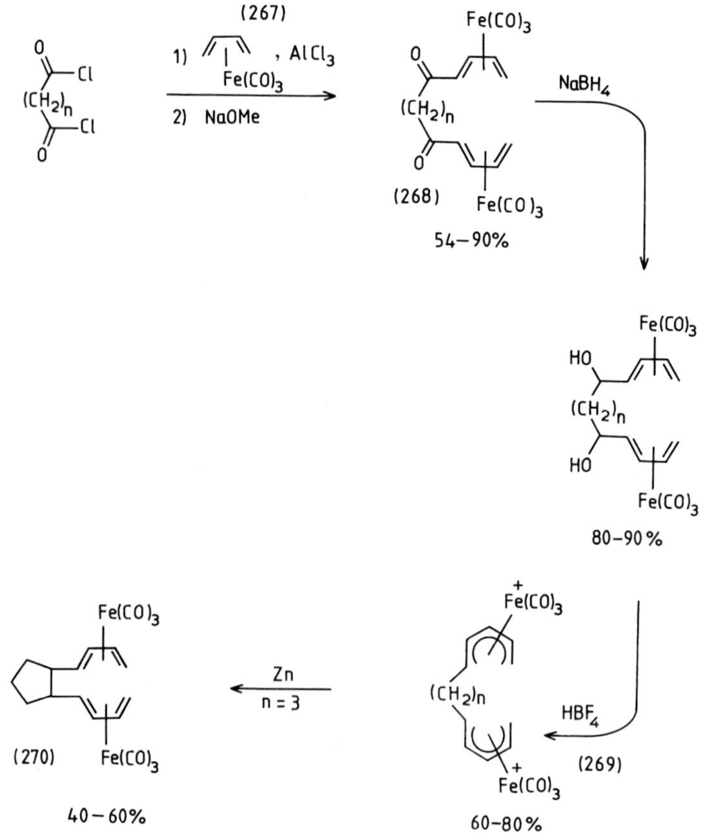

(268) 54–90%   NaBH₄

80–90%

(270) 40–60%   Zn, n = 3   (269) 60–80%   HBF₄

sodium methoxide to form the *trans* isomer (268). Borohydride reduction and reaction with HBF$_4$ produced the required dication (269) which was coupled with zinc. In addition to the example shown, six- and ten-membered rings have been formed in this way. The metal can be removed from the products by the normal oxidation methods (see Chapter 7). In contrast to (264), which afforded a mixture of stereoisomers,[296] the intramolecular case of (269) gave only (270) which was characterised by X-ray crystallography. The dimer from (263) is also described[297] as a single isomer, but the products from (265) were both 1:1 mixtures of diastereoisomers.[303]

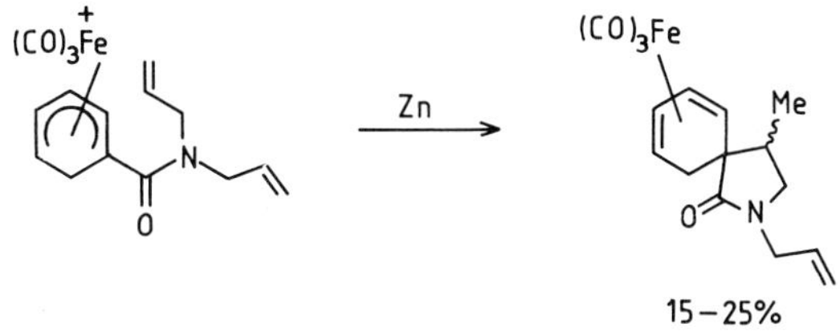

Scheme 10.30

An important development of this type of coupling process employs a conventional organic group to intercept the organometallic species formed by reduction of the cationic π complex (Scheme 10.30). Although this cyclisation reaction proceeded in low yield, it represents a reaction of particular significance since moderate (3:1) stereoselectivity was obtained at the centre adjacent to the ring junction.[299]

11

Organometallic cycloaddition reactions

The cycloaddition reactions of free alkenes and polyenes have been extensively studied, and utilised as a key method for organic synthesis through the development of uses of the Diels–Alder reaction, its variants, and other cycloaddition processes. Far less attention has been paid to related reactions when one or both partners in the reaction are bound as ligands within metal complexes. Involvement of the metal commonly leads to the formation of products that are different to those expected, on the basis of the Woodward–Hoffmann rules, from simple symmetry-controlled cycloaddition reactions of the free ligands. In the organometallic version of the cycloaddition reaction, and also in the case of some pericyclic reactions related to the alkene isomerisation processes discussed in Chapter 2, no such general systematic understanding of the reaction has emerged, although the topic has been examined[1] in detail from both the theoretical and experimental point of view. Some recent progress in the understanding of sigma-haptotropic rearrangements has drawn attention to the importance of analysing the structural changes that occur during the reaction within the metal coordination sphere.[2,3] For synthetic chemists, however, such deliberations are not enlightening since it is not always possible to distinguish with confidence between concerted and stepwise reaction mechanisms, the most important point at present being the empirical observation that many disfavoured conventional reactions may be promoted by the use of metal complexes as substrates.

11.1 Cycloaddition reactions to metal bound alkenes and polyenes

When a coordinatively saturated metal π complex is required to react with an alkene, three types of process can be envisaged. The alkene might approach the coordinated π system from the opposite face to that attached to the metal, as is shown in path A of Scheme 11.1, or may approach on the bound face, either by displacement of a ligand from the metal (path B) or reduction of the extent of binding of the organic ligand involved in the reaction (path C). The first possibility is not entirely satisfactory, since path A will inevitably lead to the formation of a coordinatively unsaturated intermediate. In practice, most organometallic cycloaddition reactions are brought

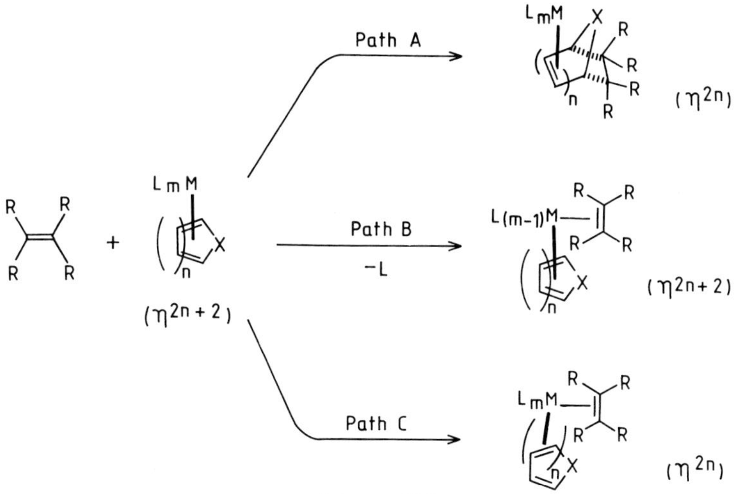

Scheme 11.1

about under photochemical conditions in which the reaction is initiated by displacement of a ligand promoting addition by path B or path C.

A typical example is the addition[4] of alkynes to (η^4-dimethylcyclobutadiene)Fe(CO)$_3$ to form unstable Dewar benzene complexes (1) and ultimately arenes. The use of an alkyne leads to coordinatively saturated intermediates. A typical alkene dienophile, maleic anhydride, on the other hand, gives rise to an insoluble Fe(CO)$_2$ complex (2)

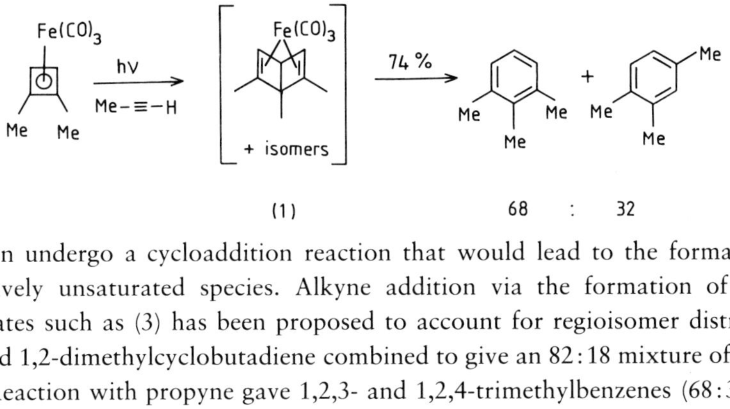

rather than undergo a cycloaddition reaction that would lead to the formation of a coordinatively unsaturated species. Alkyne addition via the formation of σ,π-allyl intermediates such as (3) has been proposed to account for regioisomer distributions. Ethyne and 1,2-dimethylcyclobutadiene combined to give an 82:18 mixture of *o*- and *p*-xylenes. Reaction with propyne gave 1,2,3- and 1,2,4-trimethylbenzenes (68:32) whilst 2-butyne, which proved far more regioselective, afforded 1,2,3,4- and 1,2,4,5-tetramethylbenzenes (97:3).

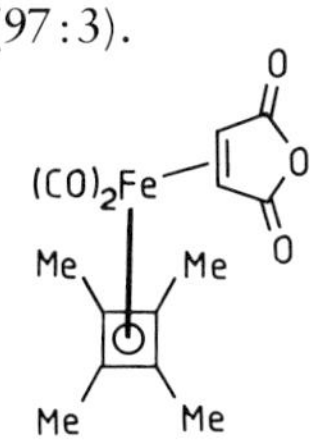

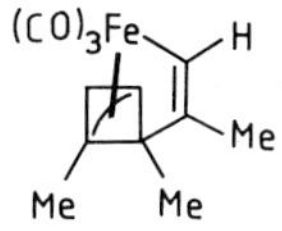

Production of (3) by direct oxidative addition to the metal complex has been suggested,[4] but reasonable alternative routes to (3) follow mechanisms involving initial formation of π complexed intermediates. Path C in Scheme 11.1 clearly retains the $Fe(CO)_3$ unit in the molecule, but for path B, too, this is also possible if carbon monoxide is scavenged by the intermediate. This latter route has been proposed[5] for additions to the related cyclohexadiene complex which resulted in the formation of the double adduct (4), rather than a cycloaddition product.

Participation of η^2 alkene complexes in cycloaddition reactions with dienes is not common, as is to be expected since the alkene fragment would no longer contain a ligand site in the cycloaddition product. Binding to the metal during the approach to the transition state would inevitably be interrupted. The reaction can be brought about, however, when the required additional ligand is incorporated in the dienophile.[6]

The first example[7] of a formal [4+2] cycloaddition in which the dienophile was present as a metal complex is of particular interest since the product (5), obtained in

moderate yield by removal of the metal with trimethylamine-N-oxide, is a potential intermediate for anthracyclinone synthesis. In this reaction there is an additional driving force available from aromatisation by loss of the oxygen bridge. The mechanistic details of the reaction are unclear and reactions of this type are far from straightforward. Direct treatment of the organic ligand with $Fe_2(CO)_9$, produced an *ortho*-quinodimethane complex (6) among the products, raising the possibility that deoxygenation may precede the cycloaddition step. The reaction is also sensitive to the nature of the metal complex used to promote the addition. Reaction with $Ru_3(CO)_{12}$ produced the furan (7) in addition to (5), indicating that loss of ethyne can compete with deoxygenation. In this case it seems probable that the cycloaddition step occurs before the fragmentation. A photochemical addition to an η^2 complex resulted in the observation of the unstable $Fe(CO)_2$ complex (8), which underwent carbonylation above $-30\,^{\circ}C$ to form (9). The

same product could be obtained by direct combination of the ligands and $Fe(CO)_5$ under photolytic conditions.[8] Alkyne metal complexes serve more readily as dienophiles, and a number of examples have been reported.[9]

In a case related to the incomplete cycloaddition described above, an intermediate similar to (9) was easily converted by heating to the [4+2] adduct (10).[10]

An example of a [6+2] cycloaddition can be found in the reaction[11] of (η^6-cycloheptatriene)$Cr(CO)_3$ with the spirocyclic diene (11). Here again, an additional ligand site allows the formation of a stable metal complex, the η^2, η^4 product (12).

Reactions of this type are sensitive to the nature of the substrate, a clear disadvantage when judged as candidates for synthetic applications in new situations. This can be seen by comparison of the product (12), obtained from a cyclopentadiene, and (15), obtained from cyclohexadiene. Reaction with cyclohexa-1,3-diene followed a [4+2] path, despite the probable involvement of coordinatively unsaturated intermediates (13) and (14). Rearrangement of the site of one of the alkenes in the seven-membered ring has taken place to form the new cyclohexadiene unit in (15). Isotopic labelling has been used to demonstrate hydrogen transfer from the cyclohexadiene ring to the seven-membered ring of the product.[11]

A number of butadienes have been found[12,13] to undergo [6+4] cycloadditions under similar conditions. The example given in Scheme 11.2 demonstrates the liberation of the organic product by displacement by trimethylphosphine.[13] The metal can be removed from the complexes (9) and (12) by the same method.[11]

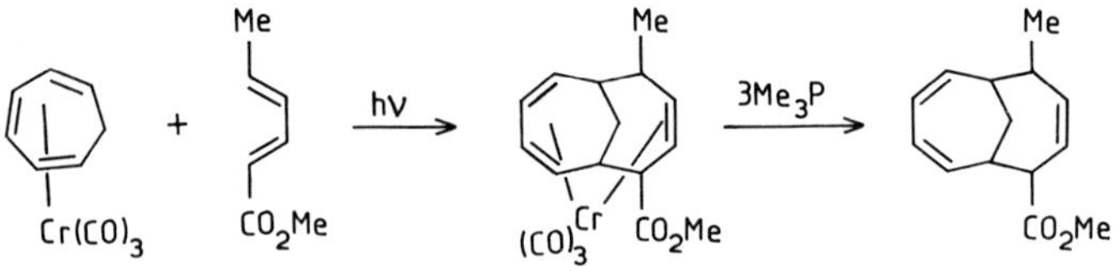

Scheme 11.2

As with the iron complexes discussed earlier, the addition process may occur by a stepwise mechanism. This possibility was suggested[14] by Kreiter, who has been responsible for much of the development of this field, to account for the formation of the η^5,η^3-product (17) from the (dimethylheptafulvene)$Cr(CO)_3$ complex (16).[14] This reaction is similar to the alkyne additions leading to the formation of (4), but in this case loss of carbon monoxide following the oxidative addition step resulted in the attachment of the butadiene fragment as a π ligand. Carbonylation can be used to force the cycloaddition to completion.[14]

With other butadienes, the heptafulvene ligand demonstrates[15] reactivity analogous to that of cycloheptatriene, and cycloadducts are formed directly. While a stepwise route to these products seems reasonable, the isolation of (17) cannot prove the point conclusively since other processes that would lead to products such as (18) may be blocked in this case. The [4+2] reaction between cyclohexadiene and cycloheptatriene was not repeated when the heptafulvene ligand is used. The more common [6+4] product (19) is obtained instead.[15] Butadiene cycloaddition with a bis [$Cr(CO)_3$] complex of heptafulvalene has also been examined and proceeds in a similar way at each of the η^6 chromium complexes in turn.[16]

Two butadienes have been combined in a [4+4] manner to form the cycloocta-1,5-diene ring in (20) in a reaction catalysed by $Ni(COD)_2/2PPh_3$ at 60 °C. This intramolecular reaction produces a *trans* ring junction between the six- and eight-membered rings and was strongly influenced by the orientation of the ester group. The equatorial product (20) was obtained in a 70:1 ratio in a mixture with the diastereoisomer in which the ester was axial. Not all substituents, however, provide such powerful control. A methyl group in the same place gave a ratio of 20:1 and the corresponding nitrile was still less controlled producing a 3:2 mixture.[17]

The examples given comprise $[2+2]$, $[4+2]$, $[4+4]$, $[6+2]$ and $[6+4]$ cycloaddition reactions, mostly promoted by a photochemical activation step. This permits the entry of the free polyene into the coordination sphere of the metal via initial η^2 binding in the metal complex. Subsequent steps typically (but by no means always) result in addition of the polyene across the termini of the π ligand. If a stepwise mechanism is followed, then steric interactions and effects of ring size must be considered when examining the outcome of the reaction. Reactions of this type show considerable selectivity but a very wide variety of results. This complicates their application in synthesis.

An example of a thermal cycloaddition reaction that led to a product that corresponds to a thermally 'forbidden' reaction of the free ligands shows the potential of metal complexes to circumvent the restrictions of the Woodward–Hoffmann rules. Again, as is shown in Scheme 11.3, reaction occurs at the termini of the bound π system.

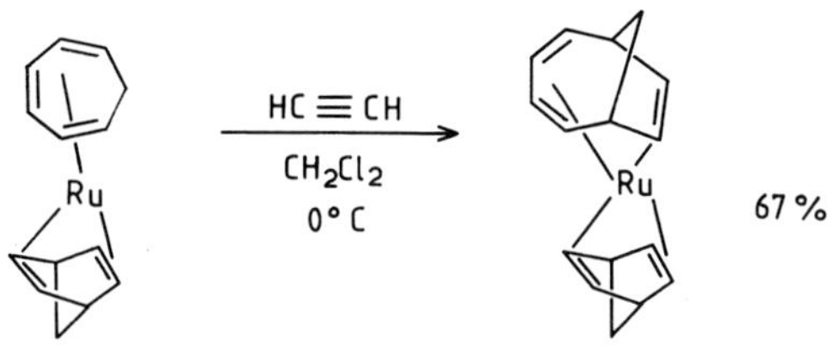

Scheme 11.3

This facile $[6+2]$ cycloaddition with ethyne also demonstrates that selectivity is possible between ligands within the same metal complex, since a $[2+2+2]$ reaction of the norbornadiene ligand is unfavoured.[18] Similar results have been obtained with an η^6-cyclooctatriene complex.[19]

Another thermal reaction is shown in Scheme 11.4. In this case it is an alkyne complex that undergoes reaction with a free polyene. The reaction has also been examined with

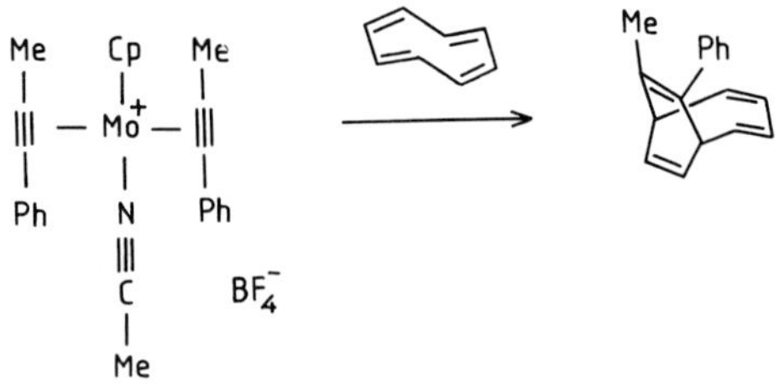

Scheme 11.4

cycloheptatriene, and in both cases the complex can also serve as a catalyst for the $[6+2]$ combination of phenylalkynes and polyenes.[20] It is notable that in this example the metal complex is charged. This raises the possibility that an ionic mechanism may operate. Some examples of addition reactions proceeding by a stepwise ionic mechanism are to be found in Sections 11.3 and 11.4.

A cationic metal catalyst, the rhodium complex (21), has been used to combine nonactivated alkynes with 1,3-dienes. Monosubstituted alkynes, and 2-substituted dienes mostly produced 1,4-disubstituted 1,3-dienes such as (22) as the major product, often as mixtures with other 1,3 and 1,4-dienes which were present in minor (10–20%) amounts. In all cases, the 1,4 relationship between the two substituents was retained. Yields were mostly in the range 58–85%, but with (22), the only example not produced as a mixture,

the chemical yield was lower.[21] These reactions presumably proceed via a regio-controlled cycloaddition, followed by alkene isomerisation steps of the type discussed in Chapter 2.

11.2 Cycloaddition to free alkenes within metal π complexes

In many cases, as expected from the 18 electron rule, polyenes in metal π complexes bind at only some of the available ligand sites. In these circumstances, unbound double bonds may participate independently in cycloaddition reactions, or may react in conjunction with the metal complex. Typical examples that have been studied extensively, are additions of highly activated dienophiles such as tetracyanoethene (tcne).

Following the first report[22] of an addition of this type to the η^4-cycloheptatriene complex (23), it was soon realised that the normal mode of reaction was a 1,3-addition

of tcne. The examples mentioned above are taken from Green's synthetic studies, which showed this to be the case with a variety of triene ligands in complexes of both iron and ruthenium.[23] In some circumstances products can isomerise following addition, as in the formation of a 1,5 product from (24)[23] or a 1,6 product from (25).[24] A similar isomerisation with a 4-phenyltriazoline-3,5-dione adduct of (η^4-tropone)Fe(CO)$_3$ has also recently been reported.[25]

In a frontier orbital analysis,[24] McArdle *et al.* have confirmed that a 1,3 approach should predominate, although both 1,3 and 1,6 pathways are topologically related to Woodward–Hoffman [4s + 2s] and [8s + 2s] cycloadditions.[26] The frontier orbitals represented in Figure 11.1 show that good overlap is possible between the HOMO of the complex and the LUMO of the diene.

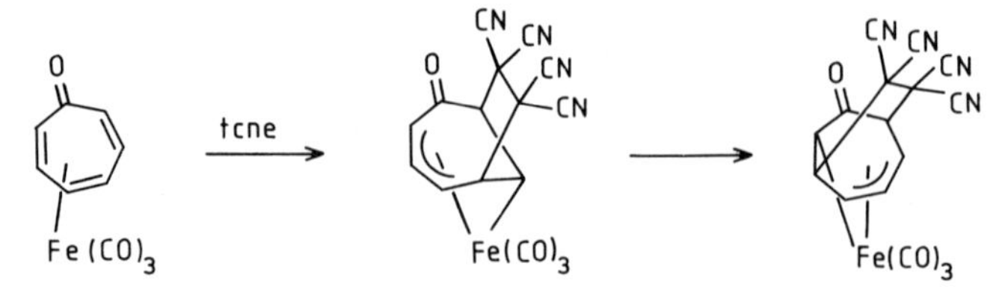

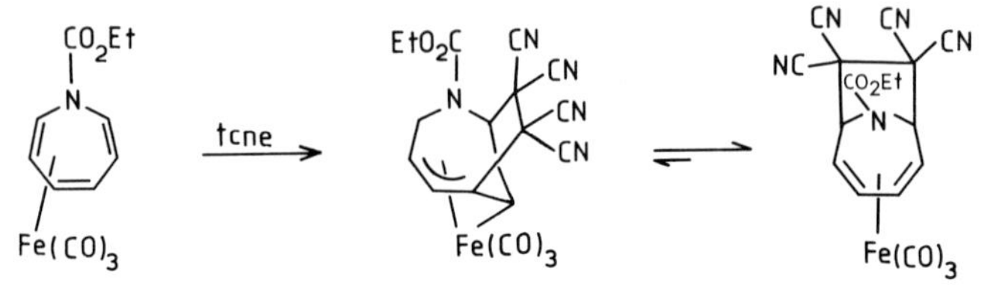

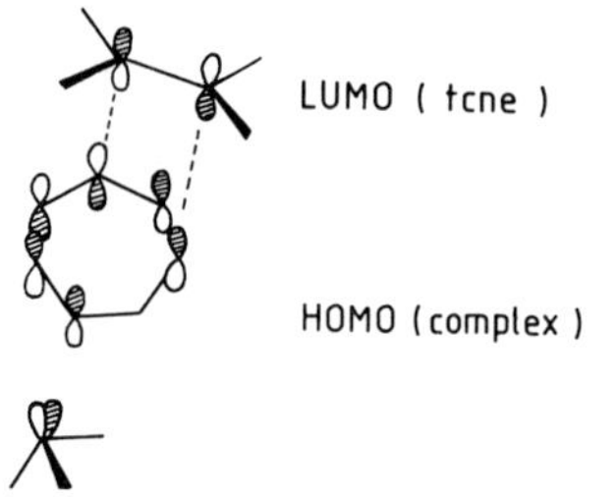

Figure 11.1

Cycloaddition of tcne involves the free alkene linkage of the η^4-triene complex and occurs from the face of the ligand opposite to that attached to the metal.[22, 27] Other highly activated dienophiles have given similar results.[28] In some cases, chemical yields can be rather low, but synthetically useful examples have been applied to give access to novel heterocycles (see Chapter 15). The presence in the dienophile of groups that are well able to stabilise negative charge led to consideration of stepwise ionic mechanisms for the reaction. A kinetic examination[29] of the 1,3 addition reaction of (23), however, ruled out both ionic and free radical mechanisms, leaving a concerted reaction as by far the most probable alternative. In an analysis of the reaction of the trimethyl substituted complex (26), from which both 1,3 and 1,6 tcne adducts were obtained, Goldschmidt and Antebi concluded[30] that the formation of a 'tight' transient ion pair could account for the 1,6 addition. From Figure 11.1, it is clear that direct 1,6 approach is symmetry forbidden.

A ketene addition to (23) provides a contrast to these results, since reaction occurs

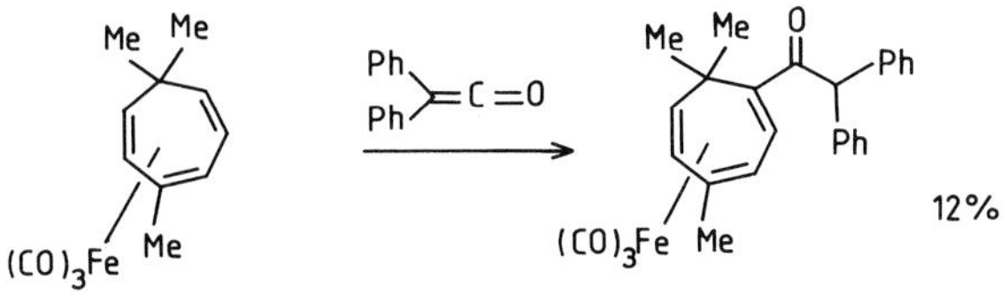

(26) → (12%)

in a [2+2] fashion at the free alkene. The reaction proceeded in moderate yield but with good stereo- and regiocontrol, which is interpreted as evidence for a concerted mechanism. An X-ray analysis of the product (27) showed the phenyl substituent oriented over the metal complex.[31]

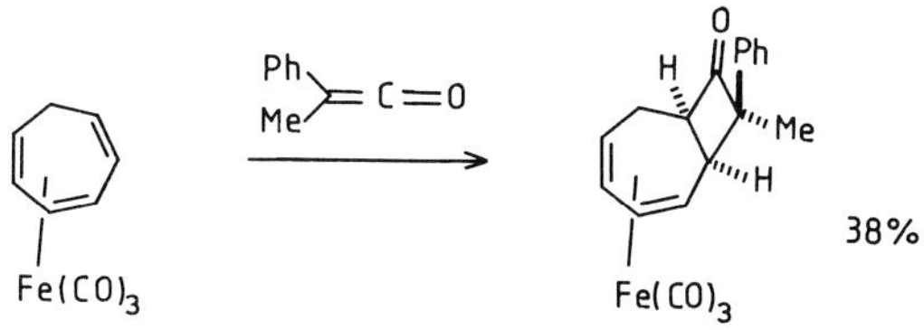

(23) → (27) 38%

Diphenylketene also exhibits [2+2] additions with η^4 complexes of both cycloheptatriene and cyclooctatetraene. The adduct (28) was found to rearrange to a 1,3 product upon heating, but (27) was inert under the same conditions.[31] A diphenylketene adduct obtained from (26) also failed to rearrange to a 1,3 product, following in this case a hydride shift mechanism to form an acyl complex.[30]

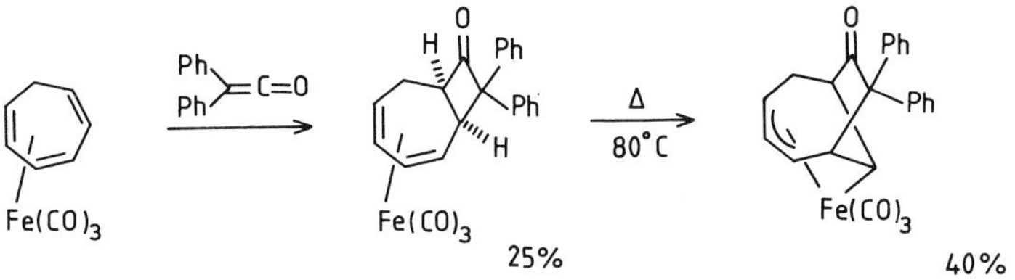

(23) → (28) 25% → Δ 80°C → 40%

A kinetic study of the rearrangement of (28) has suggested a concerted mechanism, although a moderate solvent effect points to a somewhat polar transition state. The unsymmetrically substituted ketene adduct (29) showed retention of configuration in the rearrangement process, giving further support to the view that the reaction is concerted.[2]

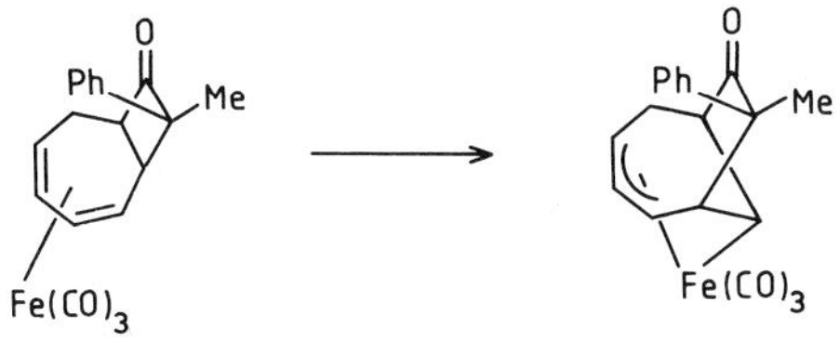

(29) →

Removal of the metal after the cycloaddition step can produce interesting polycyclic products. A tcne adduct has been used as a starting material in studies relating to dodecahedrane synthesis. Removal of the metal from the σ,π-allyl intermediate (30) formed a new carbon–carbon bond producing (31) which contained three fused five-

membered rings, one formed in the original cycloaddition, and two by bridging across the remains of the cyclooctatetraene ring. The four nitrile groups were converted to the more useful functionality in (32) by use of rather vigorous reaction conditions, conc. HCl at 130 °C. Deletion of the nitrile functionality is, however, expensive in terms of time. Six further steps are required to reach the ketone (33).[32]

Investigation of the dimerisation of cyclooctadiene by ruthenium complexes has led to the isolation of a wide variety of complexes.[33] An interesting product (34) arises from a formal Diels–Alder reaction between the two ligands, occurring within the coordination sphere of the metal.

In another Diels–Alder type of addition, reaction between a number of dienophiles and the σ bound cyclopentadienyl $CpFe(CO)_2$ complex (35) resulted in the stereocontrolled formation of (36) from which the carbomethoxy substituted organic product can be obtained by oxidation in methanol, a decomplexation process first encountered in Chapter 6. In this way (35) serves as a useful equivalent of 5-carbomethoxycyclopentadiene, itself too unstable to be used directly with all but the most reactive dienophiles.[34] Although formally a Diels–Alder addition, a tempting alternative might be to regard the reaction as a special case of the dipolar type of addition, common for allyl Fp complexes, which is discussed in the next section. Recent results, however, have shown[35] that this is not the case. From a study that revealed

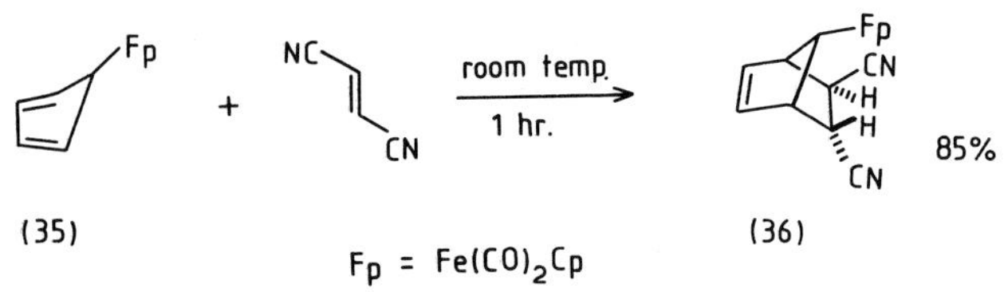

(35) Fp = Fe(CO)$_2$Cp (36) 85%

similar relative rates of reaction in solvents of very different polarity, Glass has concluded that the reaction most probably proceeds by a concerted [4+2] cycloaddition mechanism.

11.3 η^1-Allyl complexes in dipolar cycloaddition reactions

The nucleophilic character of σ bound metal allyl complexes (see Chapter 8) leads to an interesting [3+2] cycloaddition that proceeds in a stepwise fashion,[36] first by electrophilic attack on the metal complex by an activated alkene, and then by a nucleophilic ring closure of the dipolar intermediate (37).

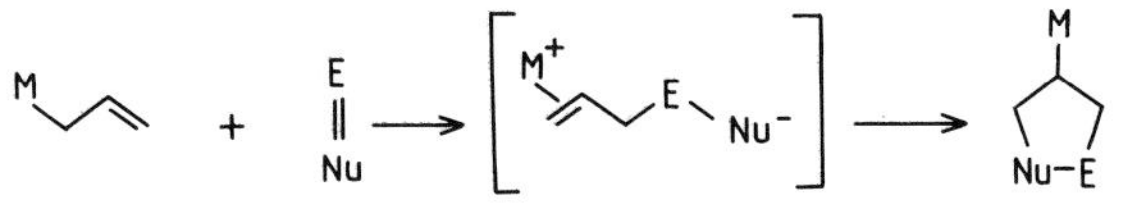

(37)

Once again, tcne found application in early examples[37] of this reaction, which are related to SO$_2$ addition reactions studied in the preceding decade.[38] Although dichlorodicyanoquinone served as a substrate (see Scheme 11.5), neither p-quinone nor the tetrachloro-derivative could be used.

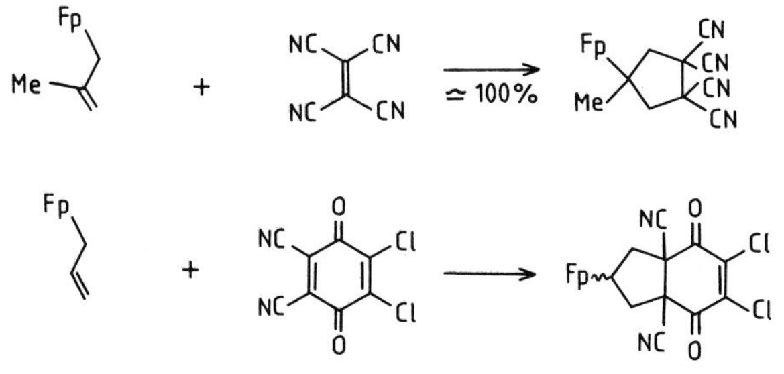

Scheme 11.5

Reactions with other dipolar reagents afforded heterocyclic products such as bicyclic γ-lactams[39] and [3+3] addition products such as (38).[40] This latter reaction, observed for iron, molybdenum, and manganese allyl complexes, is likely to proceed by initial [3+2] addition to form (39) which can fragment and re-cyclise to form the six-membered ring. A [4+2] version of the alkene addition, which used the cyclopropane (40), presumably follows a similar mechanism.[37]

It is, however, as a [3+2] annelation route to form cyclopentanoid derivatives, that

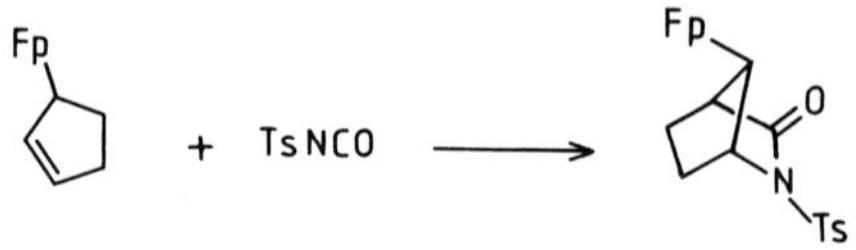

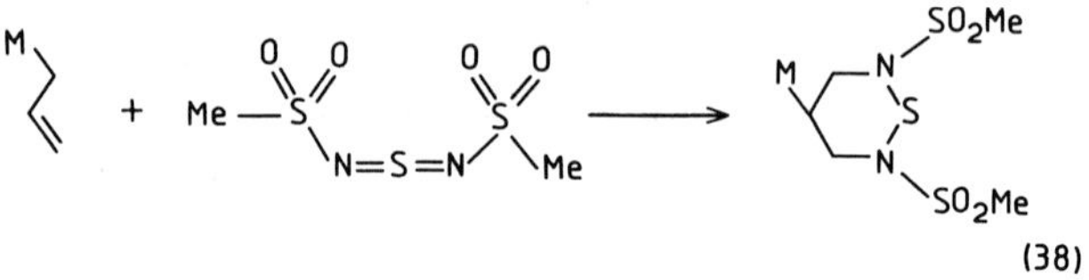

$$M = Fp \ (75\%) \ , \ CpMo(CO)_3 \ (63\%), \ Mn(CO)_5 \ (60\%)$$

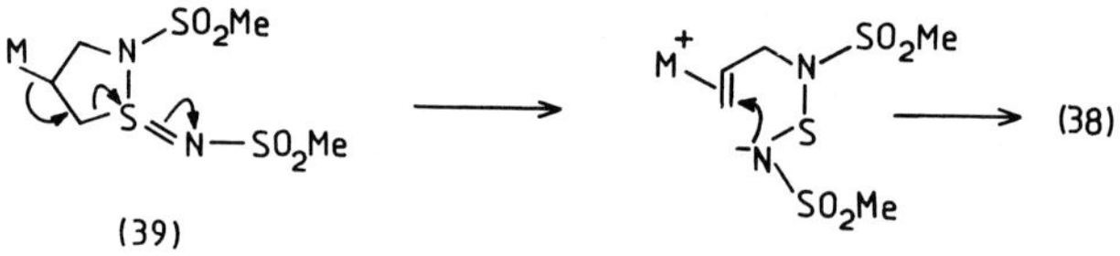

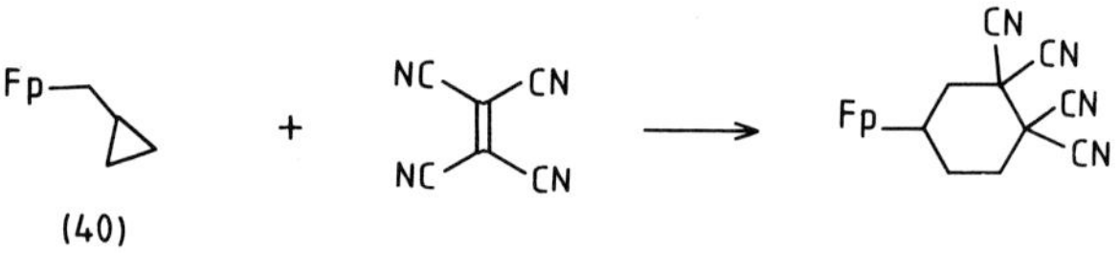

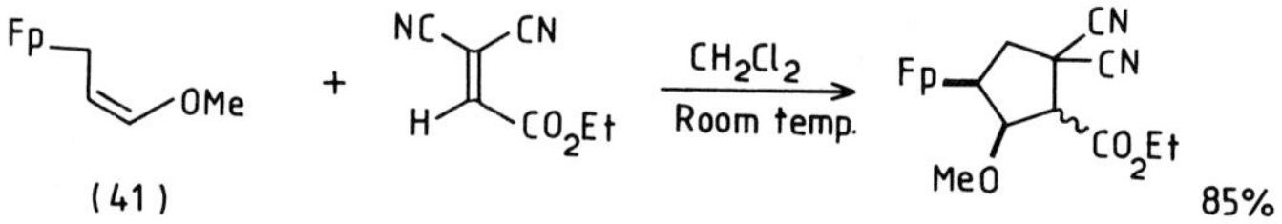

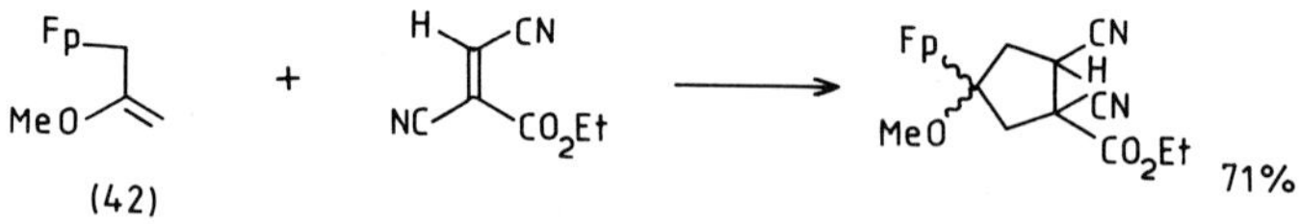

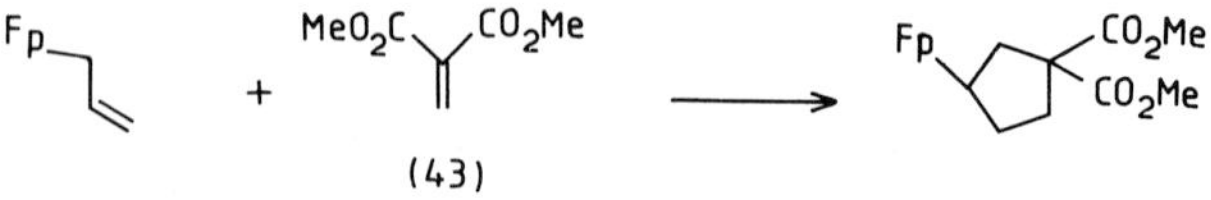

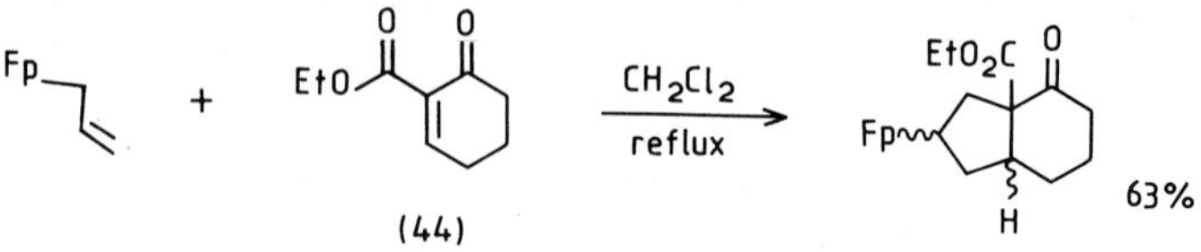

(38)

(39) → (38)

(40)

(41)   85%

(42)   71%

(43)

(44)   63%

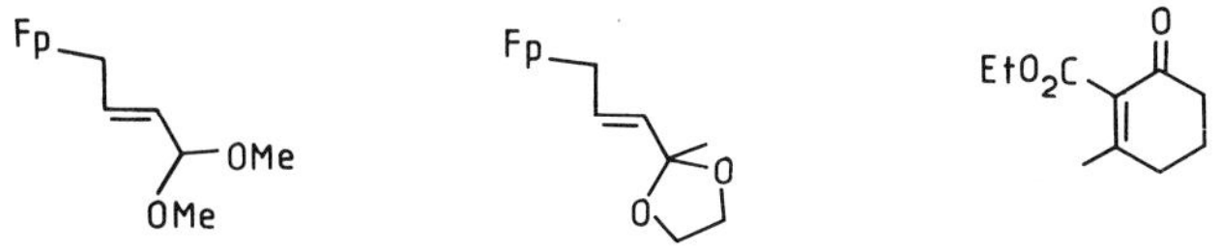

(45)   (46)   (47)

reactions of this type are gaining most attention. Work by both the Rosenblum and Baker groups has extended the range of substrates available for synthetic application. The enol ethers (41)[41] and (42)[42] provide useful three carbon fragments, and some less substituted alkenes, such as the methylenemalonate ester (43)[43] and the keto ester (44)[44] have been successfully employed. It is notable that the reaction is sensitive to steric effects.[44] Neither the terminally substituted allyl complexes (45) or (46), nor the tetrasubstituted alkene (47), will undergo cycloaddition, and in other cases, the initial addition is not followed by ring closure, and acyclic products result. More effective cyclisation can be achieved using metal complexes in which carbon monoxide has been replaced by a phosphite ligand,[45] a further example of a common tendency (see Chapter 9) for ligand substitutions of this type to alter the electrophilicity of metal π complexes.

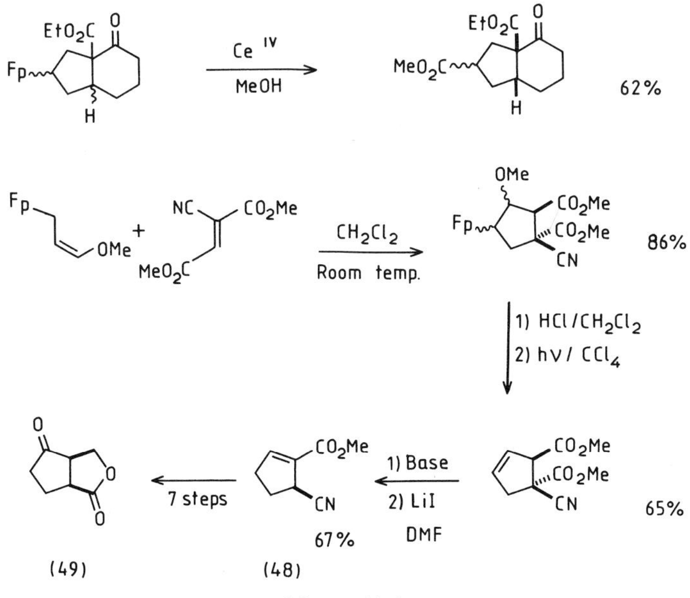

Scheme 11.6

Following cycloaddition, the Fp complex can be removed[46] to introduce the range of organic substituents described earlier, in Chapter 5 (Scheme 11.6). Removal following formation of an alkene complex was used to provide the intermediates (48) and (49) for a synthesis of sarkomycin. Three electron withdrawing substituents on the alkene ensured an effective cycloaddition; one was subsequently removed by decarbomethoxylation, but the remaining ester and nitrile groups were used to provide the methylene and carboxylic acid groups in the product.[47]

A further development of this cyclisation reaction uses cationic metal complexes as the electrophilic species. The high reactivity gained in this way is evident in the reaction of the hindered complex (50) in which the *trans* stereochemistry of the alkene was retained in the products. In less difficult circumstances better yields can be obtained and the resulting dienyl complexes can be further elaborated in the usual way.[48] Removal

of the metal afforded the hydroazulenes (51)[48] and the 10-methyl derivative (52) that was obtained[49] from a tropone metal complex.

11.4 Ionic allyl complexes in dipolar cycloaddition reactions

Another organometallic reaction that has found considerable use in the synthesis of cyclopentanoids is the palladium catalysed cycloaddition of silyl substituted allyl acetates, e.g. (53), to enones.[50] The method originates from observations by Trost and Chan[51] that although the acetate substituent could be displaced by nucleophiles in a normal palladium catalysed reaction of the type described in Chapter 7, a reaction could also be effected with electrophilic alkenes. Other allylic leaving groups can similarly be used to initiate the reaction,[52] further demonstrating the analogy with palladium allyl

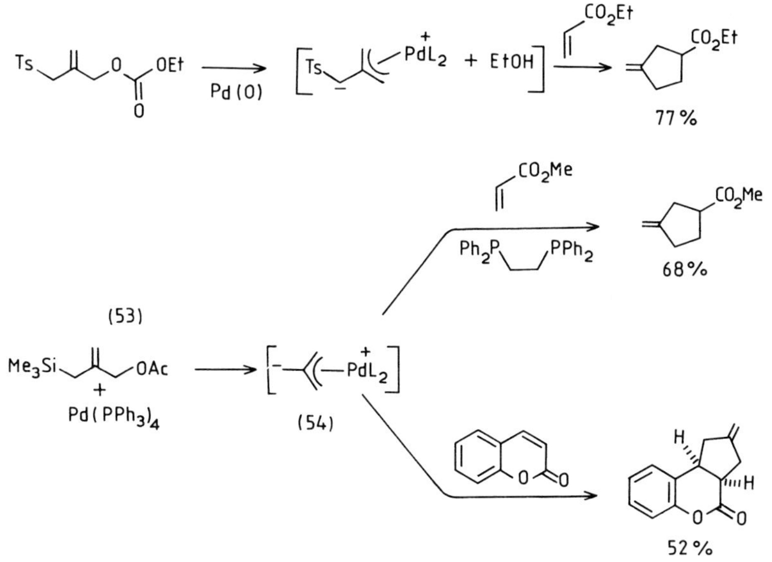

Scheme 11.7

chemistry. Some typical examples of annulation reactions of this type are given in Scheme 11.7.

The reaction has now been extensively studied and developed, mainly by Trost's group.[50] A remarkable feature to emerge from these efforts is the unusual regioselectivity arising when the allyl precursor bears further substituents. Both (55) and (57) were

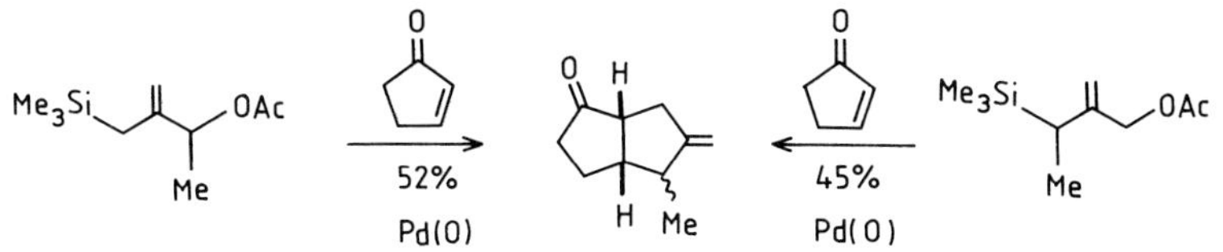

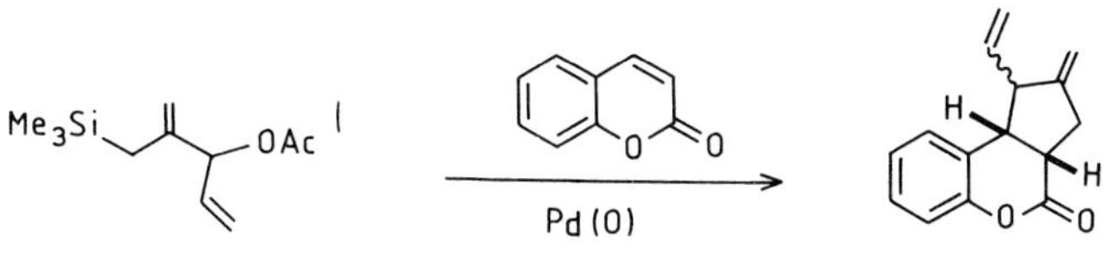

found[53] to give the same product (56) upon reaction with cyclopentenone, despite the fact that trapping with dimethylmalonate indicated the initial formation of different π-allyl intermediates. In general, all substituents, regardless of whether they are electron donating or electron withdrawing, become located in the products at the position bonded to the β carbon of the acceptor molecule.[50] A particularly interesting example[54] is provided by the vinyl substituted complex derived from (58), in which there is considerable potential for isomerisation. None the less, the product (59) was obtained as a single regioisomer in 75 % yield using a phosphite modified catalyst. The choice of catalyst can be crucial in such reactions. Use of triphenylphosphine, rather than triisopropyl phosphite, in this case reduced the yield to only 6 %.

Triethylphosphite and $Pd(OAc)_2$ have been used to effect a similar cycloaddition involving an alkene activated by both a ketone and an ester group.[55] In another case, alkene (53) is combined with tropone in a [6+3] addition process, which, in this instance, made use of the complete π system of the complex.[56]

Studies probing the mechanism of the reaction point to the involvement of an organometallic ylide such as (54) as the most likely explanation for results obtained from deuterium labelling experiments. There is an interesting comparison here with other organometallic trimethylenemethane complexes, which in some cases may offer quite different selectivity[57] for the introduction of a three carbon fragment by a cycloaddition reaction. Recently, attention has turned to whether the reaction is stepwise or concerted, with the balance of evidence currently favouring a highly ordered, but stepwise process.[58] The intermediates obtained by palladium catalysis appear to be unsymmetrically bound η^3 complexes,[59] a conclusion consistent with results obtained from molecular orbital calculations.

The stereochemistry of addition offers further support for this view of the mechanism. Unlike many organometallic addition reactions in which initial co-ordination at the metal brings the reactants together on the face of the ligand that is bound to the metal, palladium catalysed addition of (60) to dimethylbenzylidene-malonate (61) produced a 4:1 mixture of stereoisomers. Both products arose by

addition opposite to the metal, resulting in an overall retention of configuration with respect to the original carbonate substituent.[60] Also, as one would expect from the mechanism, the products are racemic despite the use of optically active starting materials derived from carveol.

Another example showing considerable stereocontrol used the carbohydrate-derived enone (62). A single stereoisomer (63) was obtained when one equivalent of the precursor to trimethylenemethane was used. When the reagent was used in excess, a double adduct was formed, again as a single stereoisomer which was tentatively assigned as (64).[62] This is an interesting product since it shows that carbonyl groups, as well as alkenes, can participate in the cycloaddition process.

Cycloaddition approaches to five-membered rings can open up synthetic strategies that were not previously available. This has proved particularly to be the case with the palladium catalysed reactions described above, in which the reactants are not

(63) 70%

(62)

(64) 54%

encumbered with excessive numbers of activating substituents. The reaction leaves, in the product, a methylene substituent at the central position derived from the C_3 fragment. Often, as in syntheses of albene,[60] loganin,[61] and brefeldin A,[63] a carbon atom is not required at this position and the alkene is converted at the appropriate stage into an oxygen substituent by, for example, ozonolysis. In a synthesis of hirsutene,[50] on the other hand, the alkene in (65) was elaborated to a *gem*-dimethyl group by reaction with CH_2I_2/Et_2Zn.

The synthesis[61] of loganin provides an example of the application of the regiocontrol effects discussed above. The cycloaddition product (66) was converted in four steps into the ester (67), a key intermediate in the synthetic strategy which rested on the opening of the original five-membered ring to form the heterocyclic ring of the loganin aglycon (68).

The same intermediate (66) has also been used in a synthesis of chrysomelidial (69), which, in this case, involved the opening of the other five-membered ring after formation of the ketone (70). In this synthesis, the oxygen functionality, needed originally to activate the alkene for the cycloaddition reaction, was used in an elimination reaction to position the alkene in (69).[50]

Functionalisation of the enone component has been used by Baker and Keen[64] to provide an approach to pentalenene by the preparation of the tricyclic compound (71), which should be convertible into the known intermediate (72).

The observation that alkene stereochemistry can be retained in the cycloaddition

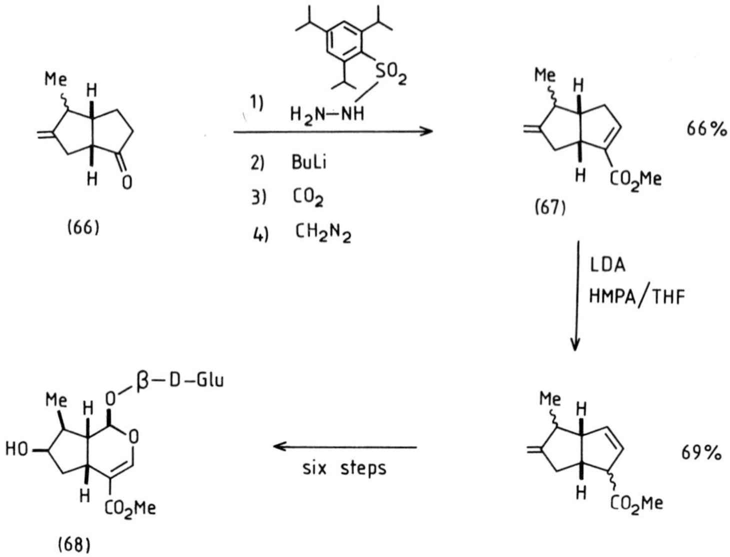

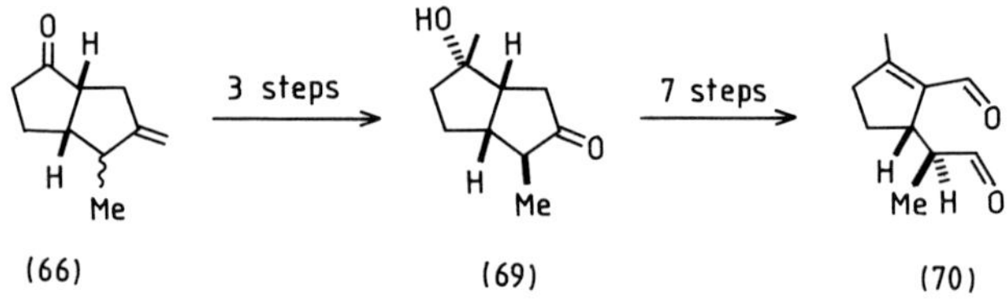

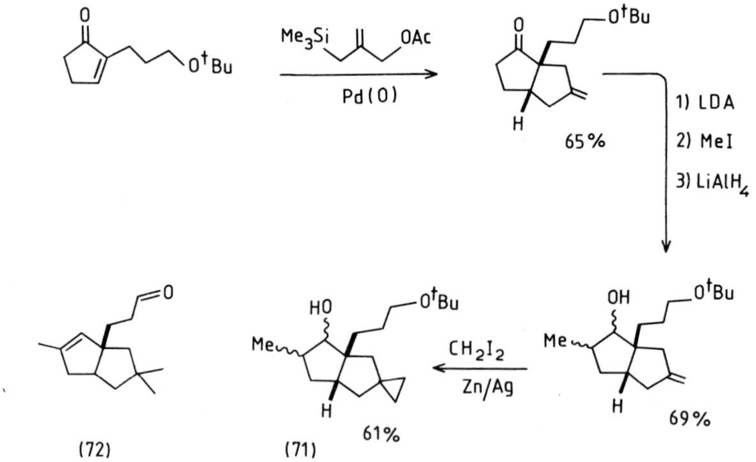

(66)   (67) 66%

(68)   six steps   69%

(66)   3 steps   (69)   7 steps   (70)

Me₃Si ⟍ OAc
Pd(0)   65%   1) LDA
2) MeI
3) LiAlH₄

(72)   (71) 61%   69%

CH₂I₂
Zn/Ag

product,[65] has found application in a cycloaddition of brefeldin A (74) which began with a reaction of the *trans* substituted alkene (73), to establish the *trans* stereochemistry required at the ring junction. With diastereoselectivity of 66% in the cyclisation step, an enantioselective route to (+)-brefeldin A was developed in which chirality was controlled by the allylic oxygen substituent in (73).[63]

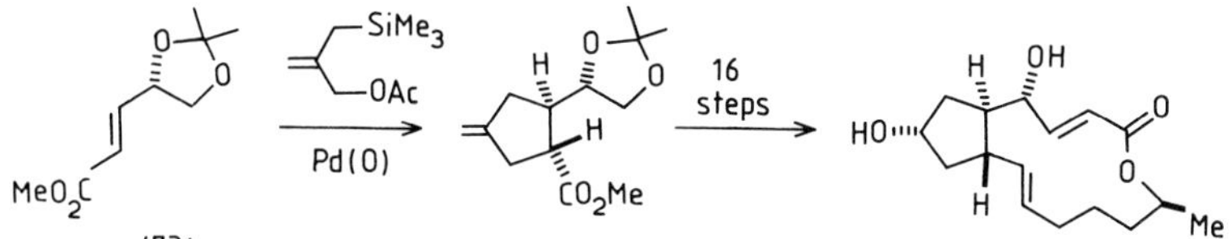

(73) (74)

Although attempts to employ the silyl reagent (53) with carbonyl groups to form heterocyclic products have been unsuccessful, the tin analogue has proved useful for this purpose. Examples, which include the formation of (75) in a model study for ionophore synthesis, show that the reaction has very good diastereoselectivity.[50] Use of Lewis acids to initiate a two step reaction sequence has allowed the addition to be used with imines to afford 3-methylenepyrrolidines.[66] An alternative stepwise procedure using allylzinc nucleophiles has been demonstrated in [3+2] additions to aldehydes, ketones, and imines.[67]

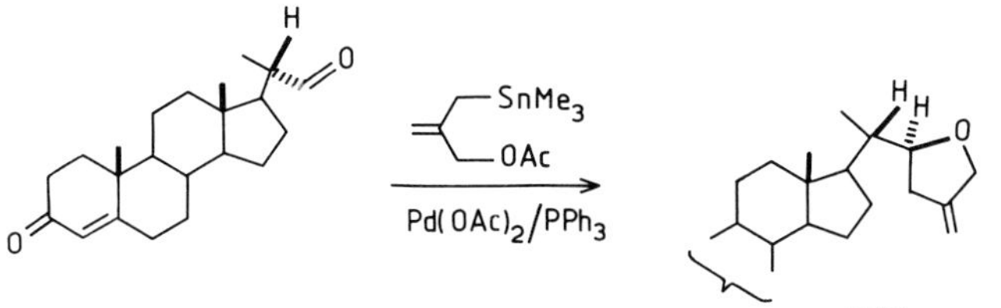

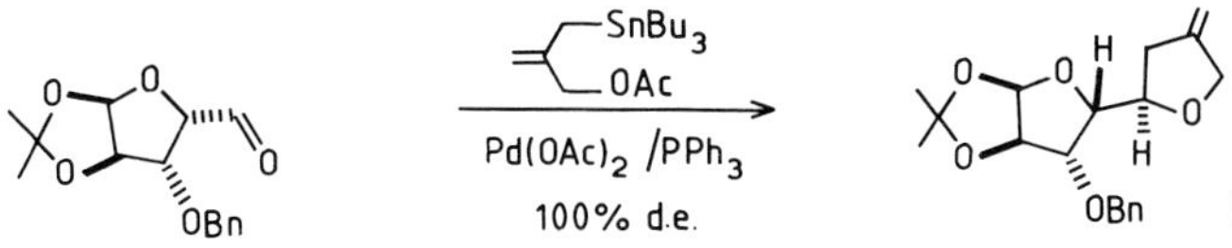

(75)

An intramolecular cyclisation can be effected by the correct choice of catalyst. The reaction shown in Scheme 11.8 forms two five-membered rings in the [3+2] addition step, but hydroindane systems can also be obtained by lengthening the chain between the reactive centres.[68]

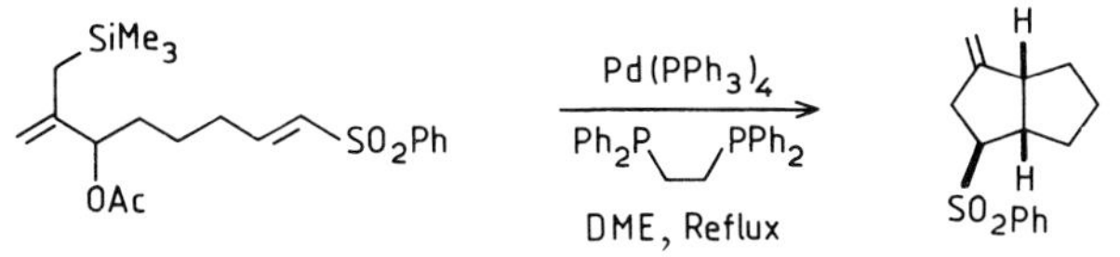

Scheme 11.8

As well as [3+2] addition reactions, formation of larger rings is possible when the π system is more extensive. Early examples[69] using dienes revealed a competition between the formation of five- and seven-membered rings. By placing substituents at C2 and C3 of the diene, the balance can be shifted in favour of the larger ring. Suitable dienes for the [3+4] reaction are available from palladium catalysed enyne couplings discussed in Chapter 10. The formation of (76) is typical of these reactions.

With pyrones, a [3+4] cycloaddition occurs with addition proceeding across the ends of the π system.[70] In a similar way, [3+6] addition has been performed using tropone

86%   88%   (76)

or tropone derivatives such as (77).[71] From this it can be seen that, in common with examples of cycloadditions discussed in Sections 11.1 and 11.2, a range of different sizes of π systems can be used in conjunction with the trimethylenemethane reagent. By far

(77)   74%

the most extensively applied combination, and also that which shows the most reliable control, is the [3+2] reaction. Examples based on the silyl compound (53) have combined in turn both nucleophilic and electrophilic properties during the cyclisation. With a diacetate starting material,[72] or by means of nickel η^3 allyl complexes,[73] alternative 1,3 dication or dianion equivalents can be achieved. Further versions of this type of cyclisation reaction have employed Cl and SO_2Ph groups with $Pd_2(dba)_3$ and phase transfer catalysis,[74] and OCO_2Et and SO_2Ph groups in an asymmetric modification in which a ferrocene derived chiral auxiliary introduced the asymmetry.[75]

A final example of a palladium catalysed [3+2] cycloaddition is really more akin to an enolate addition to palladium allyls. Opening of the vinylcyclopropane forms an enolate in the manner described in Chapter 6. Michael addition of the anionic centre of (78) to cyclopentenone, affords a new enolate species that cyclises by reaction at the allyl complex to form a five-membered ring.[76]

87%

(78)

Stoichiometric metal complexes that participate as three carbon fragments in cycloaddition reactions have been developed by Hoffmann *et al.* and Noyori using oxyallyl cations obtained by reaction of α,α'-dibromoketones with $Fe_2(CO)_9$.[77] The metal stabilised cations proved far more satisfactory than the simple organic species used initially when oxyallyl cation cycloaddition reactions were first discovered by Hoffmann[78] in 1972. Synthetic applications of both [3+4] and [3+2] reactions were soon examined. The adduct obtained from tetrabromoacetone and furan was debrominated to afford the ketone (79) in 63% overall yield, providing a useful intermediate for C-nucleoside syntheses using a Baeyer–Villiger reaction to open the seven-membered ring formed in the cycloaddition step.[79]

An adduct obtained in a similar way from 3-isopropylfuran underwent elimination and dehydrogenation in a synthesis of nezukone (80).[80] Both α- and γ-thujaplicin have also been prepared using this approach.[81]

[3 + 2] Addition with alkenes has been used in another terpene synthesis that provides an unusual solution to the problem of construction of the adjacent quaternary centres in α-cuparenone (81).[82] The use of the unsymmetrical dibromoketone demonstrates the high degree of regiocontrol possible in these reactions.

Intramolecular versions of both [3 + 2] and [3 + 4] reactions have been examined.[83] Lack of stereocontrol was evident in the cyclisation of (82) to form a 2:1 mixture of campherenone (83) and epicampherenone (84). The [3+4] reaction with the furan (85) showed better control, affording (86) in moderate yield.

The [3+4] addition with furans has been used to provide alkene precursors for Pauson–Khand cyclisations (see Chapter 9). In this way polycyclic compounds with ether bridges related to algal and fungal metabolites have been prepared.[84]

Regioisomer distributions have been examined to gain information about the mechanism of the addition process. Product mixtures from [3+4] additions are consistent with frontier orbital control, suggesting that this reaction is concerted. The [3+2] reaction, however, has been considered as a stepwise process.[85] The oxyallyl cations are thought to adopt a W-conformation in the transition state.[86] Alkene partners in this reaction typically bear substituents that will assist the stabilisation of positive charge. One proposal[85] to account for the regiocontrol leading to products like (81) is that the first step proceeds at the nucleophilic end of the alkene in the way that will leave the most stabilised enolate moiety in the intermediate (87).

The [3+2] cycloaddition reactions of nucleophilic alkenes described above provide

(88)

complementary alternatives to the methods employing electrophilic alkenes discussed earlier in this chapter. Although most examples use $Fe_2(CO)_9$ to form the Fe(II) intermediate, the precise nature of the reactive species is not clear. Similar cycloadditions have been observed in studies of a simple Fe(II) bromide adduct obtained from the reaction of dibromoketones and an iron–graphite reagent.[87]

Oxyallyl cation additions have also been reported in reactions where complexed alkenes were used as the substrate, in a manner similar to reactions described in Section 11.2. (Cyclohexatriene)Fe(CO)$_3$ gave a 1,3 adduct that underwent carbonyl insertion upon removal of the metal.[88] Reaction with the free ligand resulted in the formation of the ketone adduct. (88).[89]

12

Carbene complexes

Carbene chemistry impinges on organic synthesis in several distinct ways. Transition metals can be used as catalysts to facilitate normal carbene addition processes, or may be used to form stable carbene complexes that are valuable as reagents. In other cases, carbene complexes occur as intermediates in processes in which their formation is not self-evident from the transformation involved; an important example is alkene metathesis which is dealt with in Section 12.3. The first section of this chapter describes insertion and addition reactions involving carbene complexes. Although this is a fairly large topic, only a few representative examples have been chosen, since in these processes the metal, while important to ensure a reasonable yield, does little that is out of the ordinary.

12.1 Insertion and addition reactions of carbene complexes

Insertion reactions play an important role in the chemistry of diazoketones and diazoesters. From a synthetic viewpoint, many of the most valuable reactions are intramolecular, and frequently involve copper catalysis, a subject that has been extensively reviewed.[1] Recently it has become clear that transition metal catalysts can produce highly reactive species that undergo useful insertions with hydrocarbons[2] and ethers[3] which would normally be considered relatively inert to chemical attack. The use of rhodium catalysts is particularly of note. The normal addition and insertion reactions of carbenes have parallels in rhodium chemistry, as is shown in the examples[4] given in Scheme 12.1.

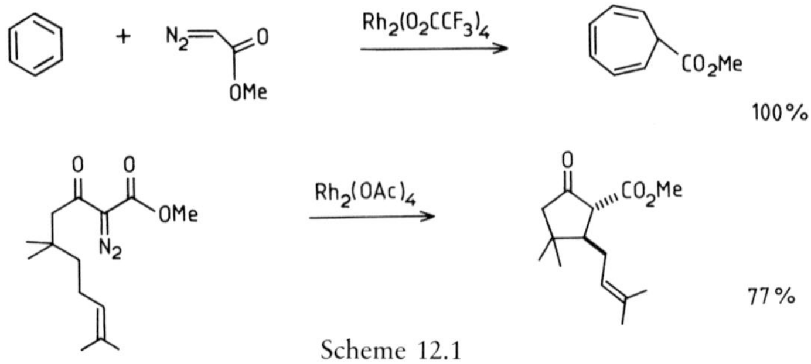

Scheme 12.1

402

67%   77%

In a synthesis of (+)-α-cuparenone, the cyclisation of (1) was found to proceed with retention of configuration.[5] Here, too, there is some similarity with copper mediated reactions.[6] This type of cyclisation has also been used by Taber to prepare the di-substituted cyclopentanone (2) which was used in a synthesis of (+)-estrone methyl ether.[7] In this case, a chiral auxiliary was included as the group, R*, in the diazo ester. The formation of (2) proceeded with reasonable diastereoselectivity, giving products in the ratio 92:8.

Rhodium catalysis of this type is now finding important applications. Chemists at SmithKline Beecham Pharmaceuticals used[8] the method to close the β-lactam ring in the formation of (3), which was converted into (4), an intermediate in the synthesis of 1-

63%   50%
(3)   (4)

methylcarbapenams in a synthetic approach that is discussed in more detail in Chapter 15. It is interesting, however, to compare this approach with that of Habich at Bayer AG who included the β-lactam intact but used a rhodium catalysed carbene insertion to form the second ring. The product was elaborated to the carboxylate derivative (5).[9] In this case, insertion occurred into an N–H bond.[10] This process, forming a five-membered ring, completely dominates reaction with the OH group which would form a seven-membered ring, or indeed, other possibilities for C–H insertion reactions at more substituted sites. The formation of seven, or even eight-membered rings in this way is, however, possible,[11] and has recently been discussed in competition with alternative C–H insertions.[12] Further examples of the uses of carbene complexes in β-lactam synthesis are discussed in Chapter 15.

57%

steps

(5)

The examples given above have all involved diazoketones bearing an additional ester group. Sulphone,[13] phosphonate, and phosphine oxide groups[14] have also been used at this position. Indeed, a second electron withdrawing group is not essential to allow these reactions to proceed. This can be seen in a study by Stork and Nakatani[15] which also indicated that electron withdrawing groups elsewhere in the molecule can have a profound effect on the regiocontrol of the cyclisation process. The formation of (6) and (7) both in high yield, demonstrates the power of the ester group in disfavouring cyclisation both at the α and β carbon atoms.

(6)   83%

(7)

81%

In another reaction that shows possibilities for regiocontrol, the 3,4-disubstituted phenol derivative (8) cyclised at C-6 in high yield. A similar result was obtained with a derivative of β-naphthol which again preferred cyclisation at a position remote from the second ring in the starting material.[16] It is interesting to compare this reaction with the intermolecular carbene addition shown at the beginning of this chapter which proceeded by a cyclopropanation mode of addition. In this intramolecular case, insertion into a C–H bond was preferred.

Attention has already been drawn to the possibility of insertion into heteroatom–H bonds. When the heteroatom bears alkyl groups, a reaction with the rhodium carbene complex is still possible by interaction with a lone pair at the heteroatom. This produces an ylide as, for example, in the case of (9). Subsequent rearrangement formed the bicyclic ether (10).[17] In a similar process using phenylthioethers, sulphur ylides were

isolated in reasonable yields.[18] In another case, the rhodium carbene complex was generated in the presence of a dienophile. The adduct (11) was obtained, a result that was interpreted in terms of an initial reaction of the carbene at a lone pair of the lactam carbonyl group.[18] This process appears to be preferred over the alternative insertion into a C–H bond as seen for the substrate (8) above. A double carbonyl insertion, using first a rhodium catalysed addition to an amide, followed by trapping of the resulting 1,3-dipole in an intramolecular cycloaddition reaction, shows promise as an entry to synthetically useful polycyclic systems.[19]

Organometallic carbene complexes can interact with alkenes to bring about cyclopropanation reactions. Organocopper and organorhodium reagents are popular for this type of process, as will be seen in the examples below. Other instances include the use of iron,[20] chromium,[21] tungsten,[22] and palladium[23] complexes. Copper complexes, however, have been most extensively used as catalysts. Salomon and Kochi[24] found copper(I) triflate to be a highly active catalyst for the cyclopropanation of alkenes with diazo compounds, a reaction shown in Scheme 12.2.

$$R = CH_3(CH_2)_3 ; \quad CH_3(CH_2)_5$$

Scheme 12.2

Using intra- and intermolecular competition reactions it was shown that copper triflate promotes cyclopropanation of the least alkylated alkene in contrast to other copper catalysts which favour the more nucleophilic highly substituted alkene. The former observation is attributed to initial alkene coordination by copper(I) species.

Copper(I) triflate was successfully employed as a catalyst in the vinyl cyclo-propanation of alkenes with vinyl diazomethane (12).[25]

(12)

In the course of a comprehensive study of transition metal catalysts for the cyclopropanation of alkenes with alkyl diazo esters, Hubert and co-workers found that palladium complexes, like copper triflates, were selective for reaction of mono-substituted alkenes. Rhodium(II) carboxylates were shown to be efficient for reaction at more highly substituted double bonds.[26] Thus, for example, in the reaction of *cis*-2-octene (13), with methyl diazoacetate catalysed by Cu, Pd, and Rh complexes, the yields of the cyclopropane (14) shown in Table 12.1 were obtained.

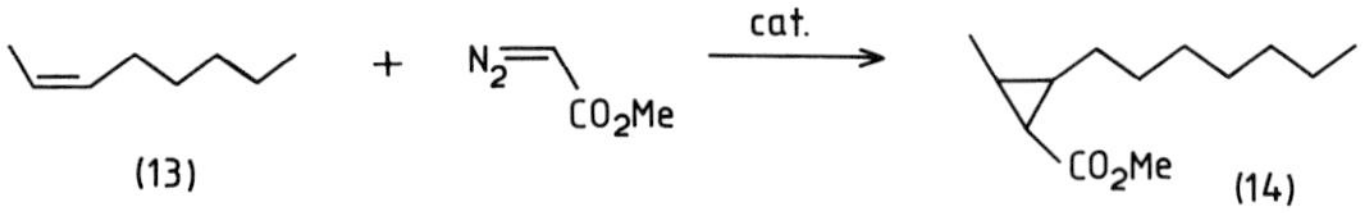

(13)   (14)

Table 12.1

Catalyst	Yield of (14)
Cu(II) triflate	40%
Pd(II) acetate	5%
Rh(OCO^{t}Bu)$_2$	89%

It was proposed that rhodium essentially promoted a carbenoid mechanism involving an electrophilic attack of uncomplexed alkenes, whereas initial alkene complexation occurred with palladium.[27] The former will be more reactive towards highly substituted double bonds whilst the latter would favour the sterically unhindered monosubstituted alkenes.

Most copper derivatives (e.g. Cu $(acac)_2$) behave as carbenoid catalysts unless associated with weak ligands as is the case with copper triflate.[27]

Rhodium(II) carboxylates also efficiently catalyse the reaction of alkynes with diazo esters[28] e.g.

With acetylenic alcohols, competition with insertion of the carbene into the C–O bond is seen. Strained polycyclic products can also be formed. Recent examples include

the simple formation of the tricyclic ether (15) which was converted by opening of the cyclopropane ring with a dimethylcuprate reagent (see Chapter 8) to produce an intermediate which has been converted into the terpene eucalyptol.[29]

In another case,[30] the cyclopropanation product (17) was converted by treatment with acid to the dihydrobenzofuranone (18), a product which would have been formed directly had the reaction proceeded by insertion of the carbene complex into a C–H bond of (16). This emphasises the general observation that trapping of the carbene complex by an alkene linkage completely dominates other possibilities such as insertion into a σ bond. A further case which relates to this point is the overall cycloaddition of vinyl carbene complexes to furans and dienes.[31] This has been shown[31,32] to proceed by initial cyclopropanation followed by a Cope rearrangement, rather than a concerted [3+4] cycloaddition reaction. A typical example is the formation of (19) from cyclopentadiene.

Transition metal catalysed addition of diazoester has been successfully applied to the preparation of permethric acid derivatives (21) from the diene (20). The product (21) is

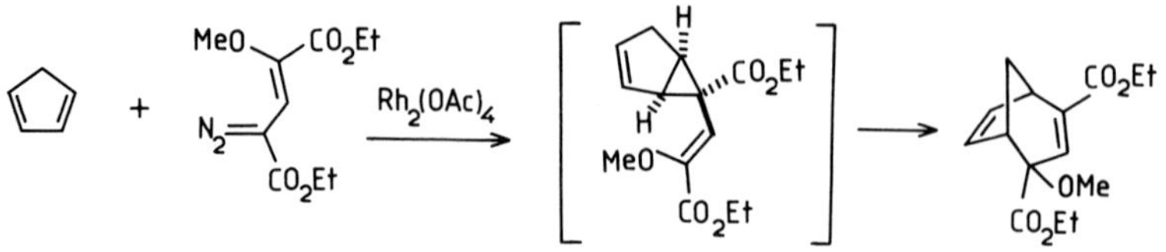
MeO
CO_2Et
N_2
CO_2Et
$Rh_2(OAc)_4$
H
H
CO_2Et
MeO
CO_2Et
CO_2Et
OMe
CO_2Et

(19) 89%

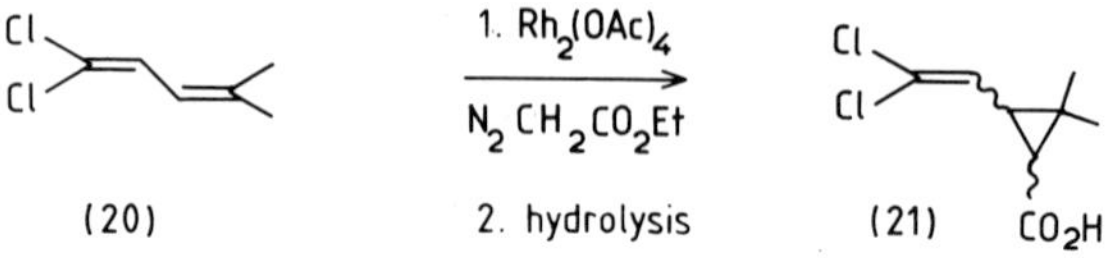
Cl
Cl
1. $Rh_2(OAc)_4$
$N_2CH_2CO_2Et$
2. hydrolysis
Cl
Cl
CO_2H

(20) (21)

an important intermediate in the synthesis of insecticidal pyrethroids.[33] Both rhodium(II) carboxylates[34] and copper complexes are effective catalysts.[35]

It would be misleading to give the impression that copper and rhodium catalysts invariably give the same results. In reactions with ketene acetals, copper catalysed carbene addition fails through the formation of an ester acetal.[36] By use of cyclic ketene acetals and rhodium catalysis, cyclopropane products have been satisfactorily obtained. These could be converted to further products by reduction or alkylation in the presence of the protected ketone,[37] as shown in Scheme 12.3.

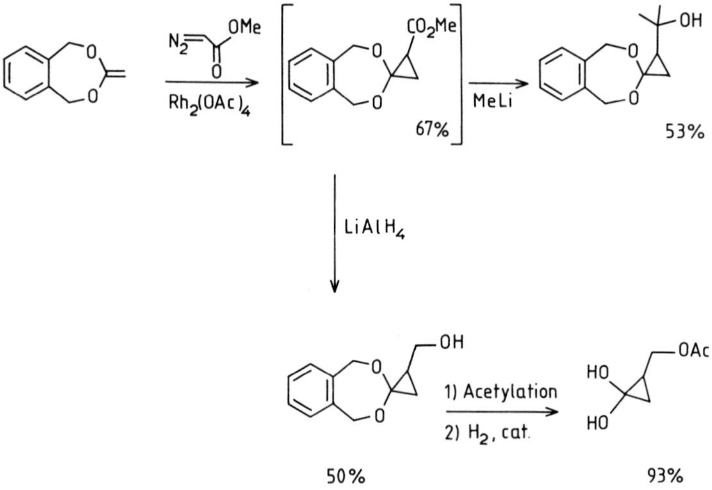
OMe
N_2
O
$Rh_2(OAc)_4$
CO_2Me
67%
MeLi
OH
53%
$LiAlH_4$
OH
1) Acetylation
2) H_2, cat.
HO
HO
OAc
50%
93%

Scheme 12.3

There have been many examples in this book of reactions in which the presence of a metal complex has provided a convenient site for the introduction of a chiral auxiliary, leading to the formation of optically active products by an asymmetric induction. In the case of copper carbene complexes, the use[35] of chiral ligands, e.g. (22) successfully enhanced selectivities for the insecticidal *cis*- and *trans*-(*R*)-isomers over the respective inactive (*S*)-isomers. In this way an exceptionally high enantiomeric excess of 99% has been obtained.[38] The tridentate binding of the chiral auxiliary may be important in promoting a high optical yield. A tetracoordinated copper complex with styrene, butadiene, and 1-heptene has also been found to effect similar cyclopropanation reactions in high yield and enantiomeric excess. In common with the iron case discussed below, however, these reactions lack diastereoselectivity.[39] A similar approach[40] used chelating ligands derived from camphor in conjunction with cobalt catalysis to obtain optical yields up to 88%.

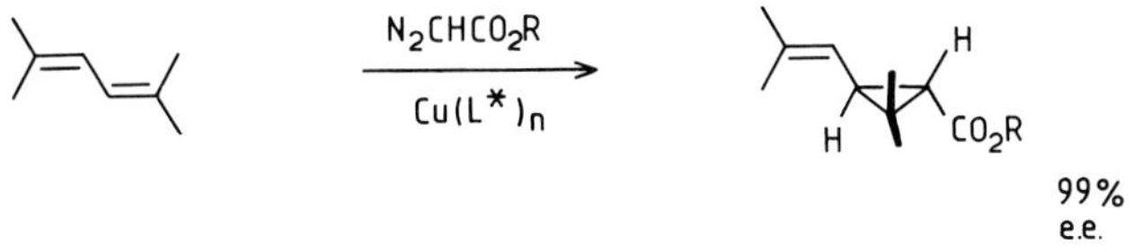

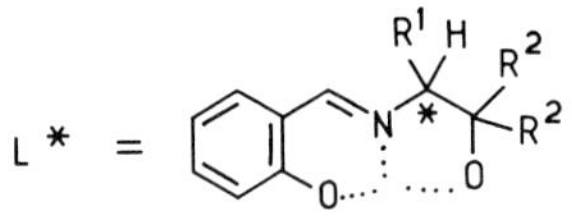

The stoichiometric nature of iron carbene complexes can become a virtue when asymmetric cyclopropanation is required. By correct substitution, the metal atom itself can become the chiral centre responsible for the asymmetric induction (Scheme 12.4). This ensures that the controlling asymmetry is very close to the reaction centre, and again, very respectable optical yields are possible.[41] The products, however, were mixtures of *cis* and *trans*-isomers. The selectivity of this class of reaction has been studied in detail.[42]

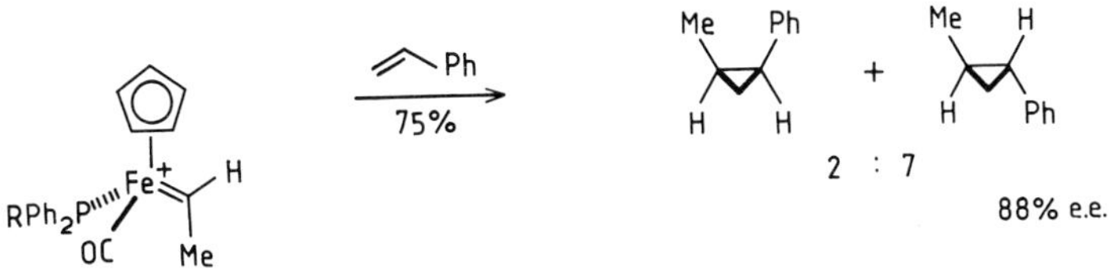

Scheme 12.4

Chiral iron reagents of this type are related stereochemically to the iron acyl complexes discussed in Chapter 9. They can perform similar aldol chemistry, which will be described at the end of this chapter in Section 12.5.

12.2 Carbene complexes in the synthesis of alkenes

Carbene complexes of titanium, zirconium and tantalum offer alternatives to the use of stabilised anions such as Wittig and Wittig–Horner reagents for the conversion of carbonyl groups into alkenes. Organometallic methods offer the advantage that non-basic conditions can be employed and are able to extend the range of carbonyl containing substrates. Early in the development of this subject, Schrock's study of tantalum complexes indicated the importance of a free coordination site to allow association of the carbonyl compound at the metal. The highly electron deficient carbene complex (23) proved effective for the olefination not only of ketones and aldehydes, but also of esters and amides.[43]

90%

$(Me_3CCH_2)_3Ta=$

(23)

E/Z = 50 : 50

77%

E/Z = 95 : 5

Practical applications of related titanium carbene complexes blossomed through the discovery of the aluminium stabilised complex (24)[44] which has now found extensive use as a methylenation reagent and has become known as the 'Tebbe reagent'. Under basic conditions, esters are converted[45] efficiently into enol ethers, a reaction that is not normally possible by Wittig methods.

$$Cp_2TiCl_2 \; + \; 2AlMe_3 \quad \xrightarrow[\substack{-CH_4 \\ -Me_2AlCl}]{} \quad Cp_2Ti\underset{Cl}{\overset{H_2 \atop C}{<}}AlMe_2$$

(24)

(24)

85%

(24)

2 equiv.

The requirement for basic conditions can be avoided by formation[46,47] of the carbene intermediate *in situ* from a metallocyclobutane which can be prepared in a separate step from the Tebbe reagent by reaction with an alkene and base. Reactions of this type are discussed in more detail in Section 12.3. Titanocycles such as (25) are remarkably stable,

$$(24) \quad \xrightarrow[base]{} \quad Cp_2Ti \overset{}{\underset{}{\diagup}} \quad \rightleftharpoons \quad Cp_2Ti=CH_2$$

(25)   (26)

(27)

(25)

60–70%

and can be stored as a source of the carbene reagent and, unlike (24), can even be handled as a solid in air for brief periods.[48] Mild conditions suffice[49] to liberate the carbene intermediate (26). The most widely used complex of this type, (25), can be employed in methylation reactions at 0 °C. Under typical reaction conditions, ketones do not appear to enolise, and so can be used in resolved form without fear of racemisation. This property has been exploited by Ireland's group[50] in the methylenation of (27) in an approach to lasalocid A. Use of $Ph_3P=CH_2$ gave a much lower yield and resulted in extensive epimerisation α to the carbonyl group.

Titanocycles are also the reagent of choice when acidic conditions must be avoided. They are suitable for use with acid-sensitive esters and even cyclic carbonates.[49] Aqueous work-up conditions can also be avoided, opening up promising routes to enamines from amides.[49] Another example of a reaction that exceeds the scope for the application of Wittig reagents, arises when acid chlorides are used[51] as substrates. The products are not vinyl chlorides, but rather enolates, arising by displacement of the halide leaving group. The enolate species can be used directly in aldol reactions.[49] Examples with carbonates and acid chlorides as substituents are shown in Scheme 12.5.

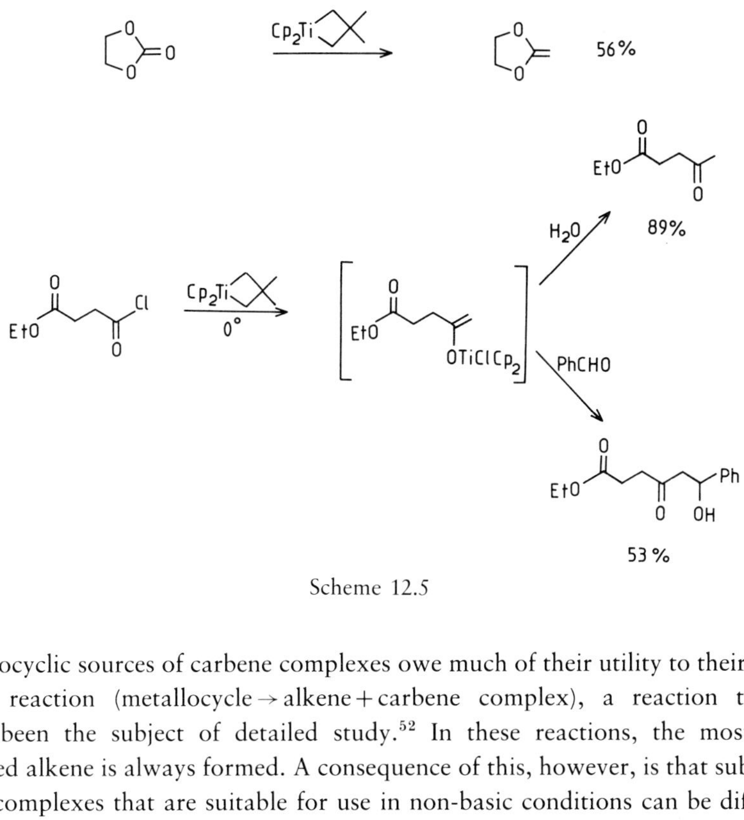

Scheme 12.5

Metallocyclic sources of carbene complexes owe much of their utility to their reliable cleavage reaction (metallocycle → alkene + carbene complex), a reaction that has recently been the subject of detailed study.[52] In these reactions, the most highly substituted alkene is always formed. A consequence of this, however, is that substituted carbene complexes that are suitable for use in non-basic conditions can be difficult to obtain, and, originally, special techniques were required in these cases leading to the development of reactions used for the preparation of substituted allenes.[53] Zirconium chemistry now offers a more general method that is well suited to applications where the alkylidene substituent is bulky. High yields can be obtained with ketones and lactones.[54] Alkenation of imines shows an increasing selectivity for *E*-stereochemistry in the products as the substituent on nitrogen is increased in size. By contrast, cyclic imino-ethers such as (28) show a preference for the Z-enol product (29).[55] Another approach has employed an α,α-dibromo substrate with titanium tetrachloride and zinc to form an olefination reagent *in situ*.[56]

The Tebbe reagent is now finding applications in natural product synthesis. A good example occurs in Paquette's synthesis of precapnelladiene (31). Methylenation of the lactone (30), followed by a Claisen rearrangement, is used to establish the required eight-membered ring.[57] A further example is found in the work of Evans and Shih who subsequently adjusted the alkene position prior to an ozonolysis step which left a ketone and an acetate group in the product (32).[49]

12.3 Alkene metathesis

The discovery by Banks and Bailey (Phillips Petroleum) of transition metal catalysed alkene disproportionation reactions, as exemplified for propene,[58] stimulated much research both in industry and academia. In this flurry of activity, both the search for potential uses of the reaction, and investigations to establish the reaction mechanism, became important goals. Initial work concentrated on heterogeneous systems such as molybdenum hexacarbonyl, tungsten hexacarbonyl and molybdenum oxide supported on alumina.

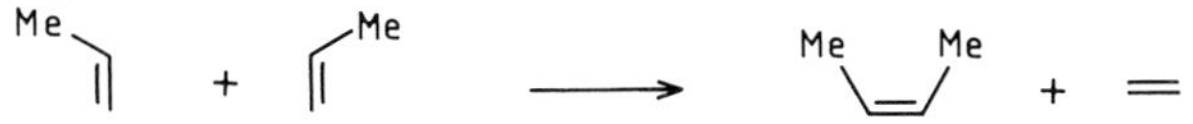

Later, the term metathesis was introduced by Calderon, Chen and Scott of Goodyear Tyres, when they reported the first homogeneous transition metal catalysed examples of this reaction.[59] The catalyst system here was comprised of tungsten hexachloride, ethanol and ethylaluminium dichloride. In this system, 2-pentene was found to undergo an interchange reaction which at equilibrium gave 50:25:25 mole% of 2-pentene, 2-butene and 3-hexene respectively.

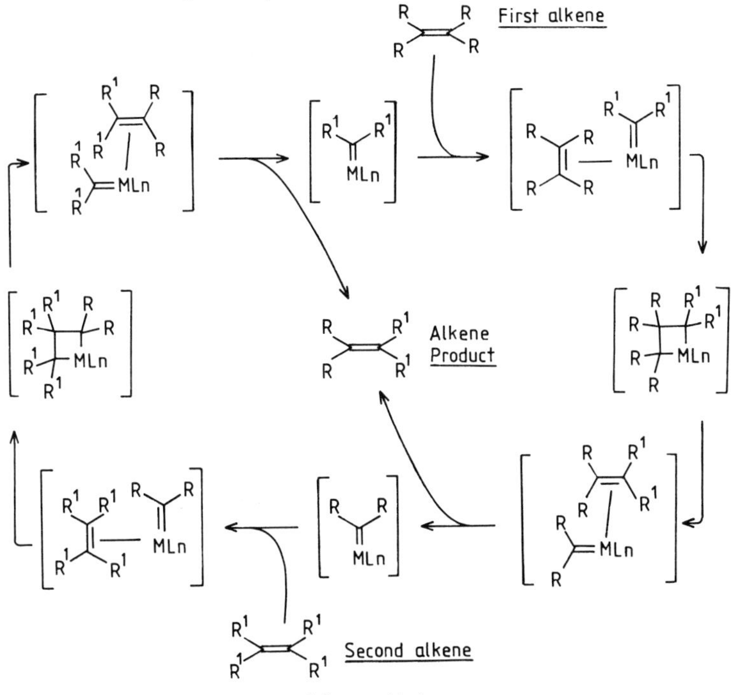

Scheme 12.6

Despite the formal implication of the formation of a cyclobutane intermediate from the two alkenes and subsequent scission, the true mechanism involves a carbene metal intermediate in a chain sequence. A detailed discussion of the reaction mechanism is beyond the scope of this book, but excellent reviews covering the topic are available.[60]

A representation of the carbene mechanism is shown in Scheme 12.6. A number of features are amenable to direct study. Examples drawn from Schrock's carbene synthesis and tungsten chemistry give information concerning the structure of (carbene) (η^2-alkene) complexes,[61] and of the reaction of carbene complexes with alkenes.[62] The alignment of the carbene and the alkene is believed to be important since only certain conformations will result in metathesis. Insight into the relative orientation of groups in the transition state is available from molecular orbital calculations by Hoffmann *et al.*[63]

Catalysts may be derived from the oxides or carbonyls of, for example, Mo or W, or, in a homogeneous system in inert solvents, from halides such as WCl_6 in combination with a metal alkyl (LiBu or Et_2AlCl). Systems using WCl_6 may be made more active by addition of protic solvents such as ethanol.[64]

A number of industrial applications for metathesis have been developed.[65] The most significant of these is found in the Shell higher olefins process.[66] This is incorporated into an ethene growth process to up-grade low value 'by-products'. Products required are α-alkenes in the range C_{12}–C_{18}. α-Alkenes $< C_{12}$ or $> C_{18}$ are isomerised to internal alkenes and then subjected to metathesis which yields a further 10–20 % of α-alkenes in the desired C_{12}–C_{18} range. Some recycling is necessary since metathesis is an equilibrium process (Scheme 12.7).

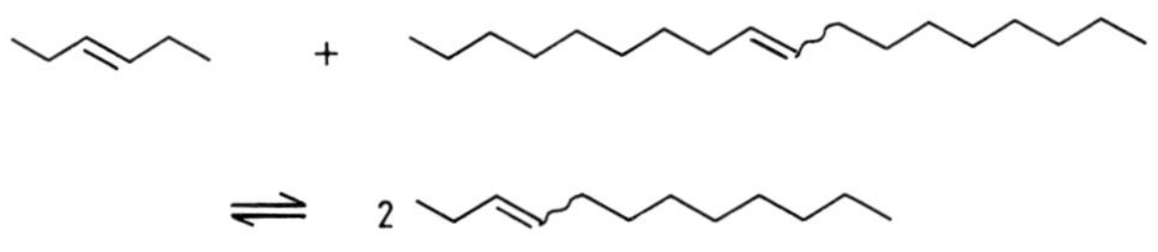

Scheme 12.7

An interesting route to styrene is available from toluene via a cross metathesis reaction of stilbene (33) with ethene.[67]

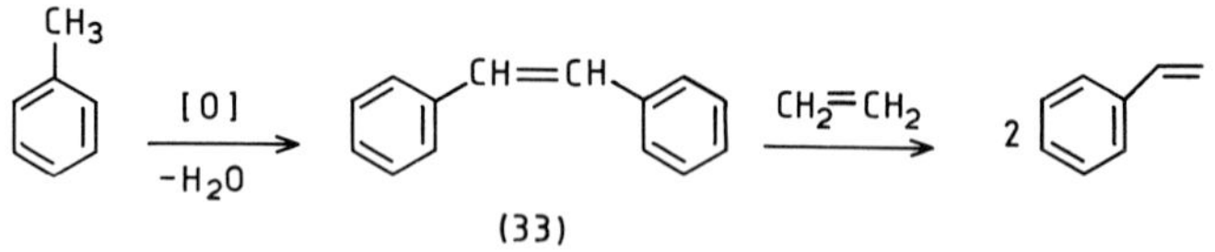

A study[68] of the metathesis of a series of aryl alkenes (34) has shown that reaction is most rapid when $n = 2$, attributed to coordination of the aryl residue to the metal carbene intermediate (35). Alkyl substitution retards, and especially where $R^2 = R^3 =$ Me, inhibits the reaction. The reaction is slower where X = Me, 'Bu, F, Cl, Br than for X = H, and reaction is inhibited when X = OMe. These studies used as catalysts, $W(CO)_5PPh_3$–$EtAlCl_2$, or $EtAlCl_2$–$Mo(NO)_2Cl_2(PPh_3)_2$, which is the more effective combination.

(34)   (35)

A number of useful synthetic applications of metathesis have also been reported. 1,9-Cyclohexadecadiene (36), obtained via dimerisation of cyclooctene, was oxidised to afford the ketone (37).[69] When the alkene bears strongly coordinating substituents such

(36)   (37)

as amines, the metathesis reaction is generally inhibited. This limitation can be overcome by the use of quaternary ammonium derivatives, such as (38), in which the nitrogen centre is no longer a coordinating group.[70] Further examples of applications of metathesis in synthesis may be found in a review by Hughes.[71]

(38)

An intramolecular alkene metathesis can also be employed to form large rings, although in rather low yield. The example shown uses a catalyst derived from $WOCl_4$ and Cp_2TiMe_2 with an α,ω-diene (39) obtained from commercially available 10-undecenoic acid.[72] Metathesis of functionally substituted alkenes would extend the

(39)   12%

synthetic utility of the reaction. Although many metathesis catalysts are deactivated in the presence of substituted alkenes, a number of interesting examples have been reported. Geranyl acetate (40) and 1-methylcyclobut-1-ene reacted with WCl_6 and $SnMe_4$ as catalyst[73] to give farnesol acetate (41). The yield of (41) was, however, very low. As we have already noted, metathesis is essentially an equilibrium process.

The method using $WCl_6 + SnMe_4$ has been applied[74] by Boelhouwer's group to alkenic esters with terminal unsaturation. In this way, a *cis/trans* mixture of products (43) was obtained from (42). The acetates (44) behaved similarly.

(40)   (41)

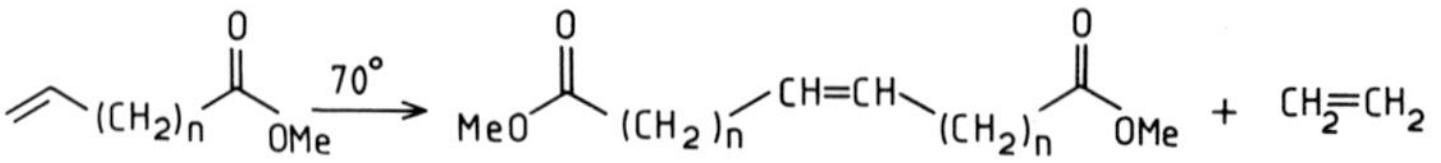

(42)   (43)

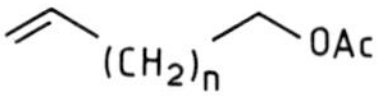

(44)

Internal alkenic esters also underwent metathesis[74, 75] as exemplified by the formation of (46) and (43, $n = 7$) from the octadecenoic ester (45). Conversion of (43, $n = 7$) to the corresponding diacid (47) provided starting material for the useful perfume additive civetone (48).[75]

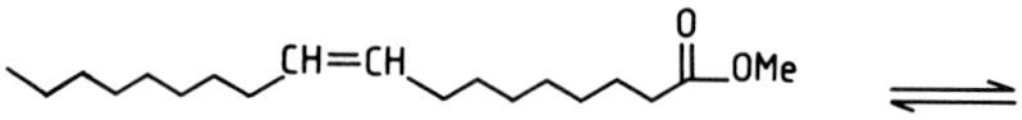

(45)

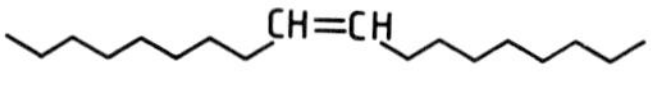

(46)   +   (43, n = 7)

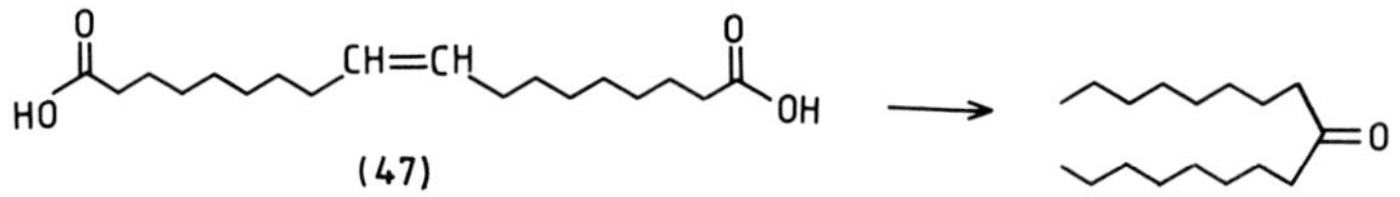

(47)   (48)

Use has been made of the metathesis process to elaborate terminal alkenes to produce long chain alkenes such as (49), required as insect pheromones.[76]

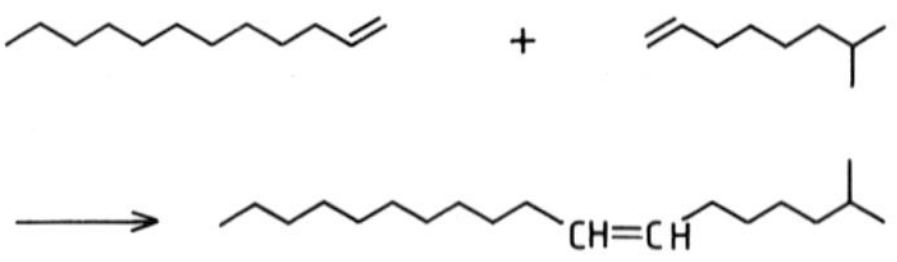

(49)

$WCl_6/SnMe_4$ (see p. 415) is a popular metathesis reagent system. Another example of its use can be found in an unusual, if rather brutal, entry to prostaglandin synthesis.[77] This employed a starting material (50) that is itself a metathesis product.[72]

Relatively low yields are a common problem in conversions that rely on metathesis, since equilibrium mixtures of products are obtained. A cunning reaction that uses a metallocyclic intermediate related to metathesis intermediates exploits the facile formation of titanocycles, discussed in Section 12.2, in a very efficient way. The 'metathesis' intermediate (52), obtained by addition of the Tebbe reagent to the alkene (51), is trapped by an intramolecular reaction with the ester substituent. This irreversible formation of (53) ensures an efficient conversion in this reaction, used by Stille and Grubbs in a synthesis of capnellene (54).[78] The relationship between the titanocycles of this type and alkene metathesis intermediates has been examined by Grubbs' group.[47,79]

Alkene metathesis is also implicitly in competition with cyclopropanation reactions. This aspect of the chemistry of carbene complexes has been explored by Casey in studies that use stable Fischer type carbene complexes with pendent double bonds. This can be illustrated by the example of the tungsten complex (55) which, in coordinating solvents, gave a carbene insertion complex (56), but in non-coordinating solvents formed (57) by alkene metathesis.[80]

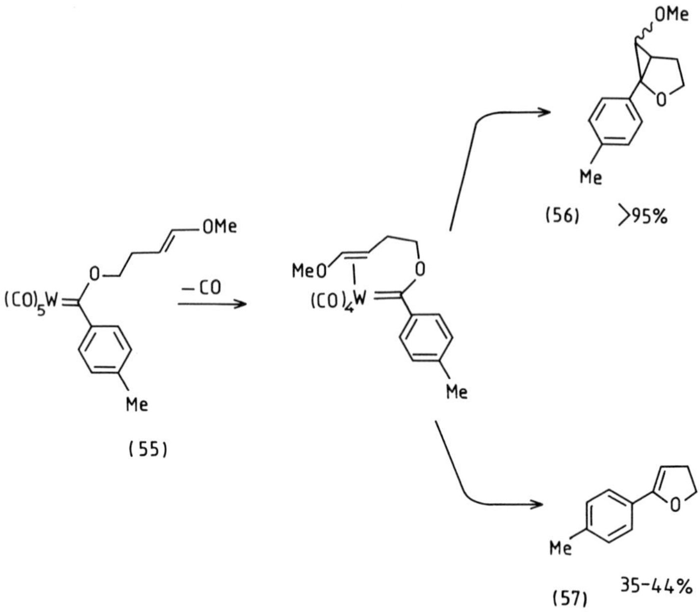

(56) >95%

(55)

(57) 35-44%

12.4 Carbene complexes in the synthesis of arenes

When transition metal carbene complexes bear heteroatom substituents, a typical example being a methoxy group, profound stabilisation results. Widely used chromium carbene complexes such as (58) are stable[81] to dilute acids and bases and to relatively high temperatures in excess of 100 °C, and can be handled in air. These are remarkable properties for a class of compounds unknown before 1964 when Fischer and Maasböl described[82] the formation of the first example of a transition metal carbene complex. Their synthesis of carbene complexes by a nucleophile addition, anion-trapping reaction of $Cr(CO)_6$, demonstrated that stoichiometric complexes of this class need not be inherently unstable. Stabilised carbene complexes of this type, now often referred to as Fischer carbenes, opened the way for the rapid development of stoichiometric applications of carbene complexes.

A particularly important example is the formation of arenes by insertion of alkynes and carbon monoxide into vinyl and aryl carbene complexes. This versatile synthetic reaction has been employed in many situations, principally through the efforts of the research groups of Dötz[83] and Wulff,[81] and is now beginning to find wider application.

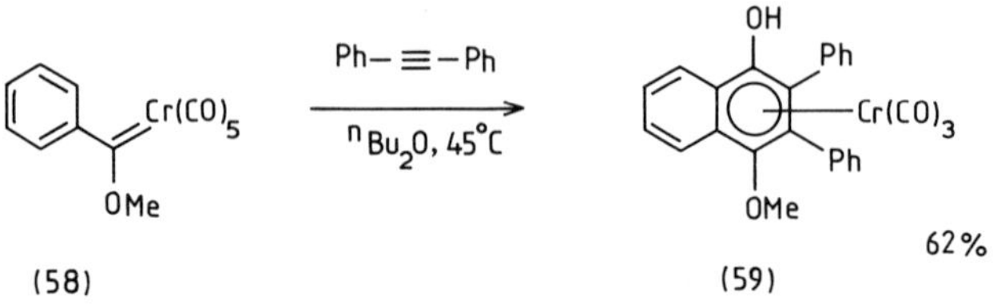

(58)

(59) 62%

The first example of this reaction was due to Dötz, who described[84] the formation of the naphthalene complex (59) by the reaction of diphenylethyne with (58). The free ligand can be easily obtained by the removal of the metal by carbonylation,[84] or now more commonly, by oxidative methods.[85] Air is often used, but more vigorous oxidation can afford quinone derivatives in one step, as seen in the formation of (60).

Vinyl carbene complexes were soon found to undergo analogous cyclisation reactions. These can show considerable selectivity, as is apparent in the formation[86] of (62) from the substituted alkyne (61). In this case, air was used for the oxidative decomplexation, and the product was converted to the benzofuran (63), confirming the regioselective nature of the reaction.

Regiocontrol is a typical feature of these reactions. Many considerations must be taken into account in the discussion of the origin of this regiocontrol effect: the orientation of approach of unsymmetrically substituted alkynes, selectivity between competing arene substituents on the carbene complex, and the geometry of the polycyclic aromatic products, are all important aspects where control is required.

Alkyne approach is particularly well controlled when alkyl substituted alkynes are employed.[83] The formation of (64) shows remarkable ability to distinguish between butyl and propyl groups. In competition between substituents on the carbene, there is a strong driving force for annellation of a phenyl group, although the formation of (65) and (66) show that both naphthalene and furan substituents are able to act as substrates in other situations. The phenanthrene complex (65) was produced exclusively. None of the alternative anthracene isomer was observed.[87] Further development of reactions forming benzofuran complexes such as (66) has led to syntheses[88] of sphondin, heratomin, and angelicin (see p. 424).

The regiocontrol of reactions with unsymmetrical alkynes has been examined. Electron withdrawing groups on the alkyne take up positions next to the OH group of the naphthol product. The reverse is true for electron donating substituents such as ethers. In other situations, such as when phenyl or alkyl groups are placed in competition with esters, mixtures of products result.[89]

Detailed mechanisms for reactions of this type have been proposed.[81,84] Such speculation is of particular value if a better understanding of the selectivity of the reactions results. The mechanism shown in Scheme 12.8 emphasises the two key steps in which bonds to the alkyne and carbon monoxide moieties are formed. The many 16 electron intermediates shown may in reality be stabilised by interactions with donor solvents such as ether or THF which are commonly required for these reactions. The regiocontrol of alkyne approach is established at the formation of the metallocycle (67), while selectivity of aromatic substitution is determined at a later stage following rearrangement of the carbene intermediate (68). Thus the formation of (64), for example, is consistent with a steric effect positioning of the smaller alkyne substituent adjacent to the hindered benzyl position in (67), but selectivity following carbonyl insertion, it seems, may be more dependent on orbital or electronic considerations.

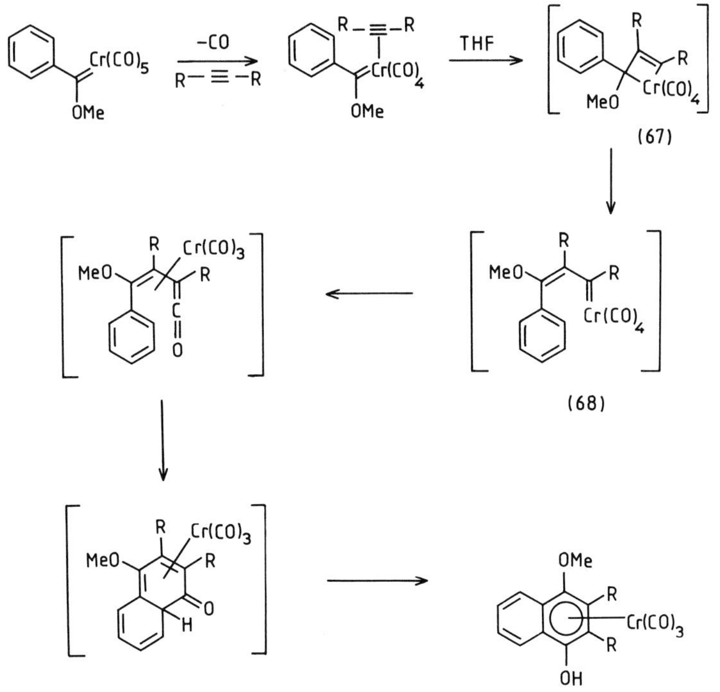

Scheme 12.8

Although polysubstituted naphthol derivatives are the usual product from these reactions, other compounds such as indenes,[81,90] furans,[91] and cyclobutanone[91,92] have all been observed as minor products. Recently, using morpholeno- or pyrrolidino-chromium carbene complexes it has been possible to shift the course of the reaction entirely towards the closure of five-membered rings.[93] Routes to anthraclinones based on chromium carbene chemistry have now also been extensively investigated.[94]

Another synthetic application arises in the work of Yamashita at the Upjohn Company.[95] The tendency of 1-alkoxyalkynes to undergo regiocontrolled reactions with chromium carbene complexes has been employed to produce the intermediate (69) for a synthesis of khellin.

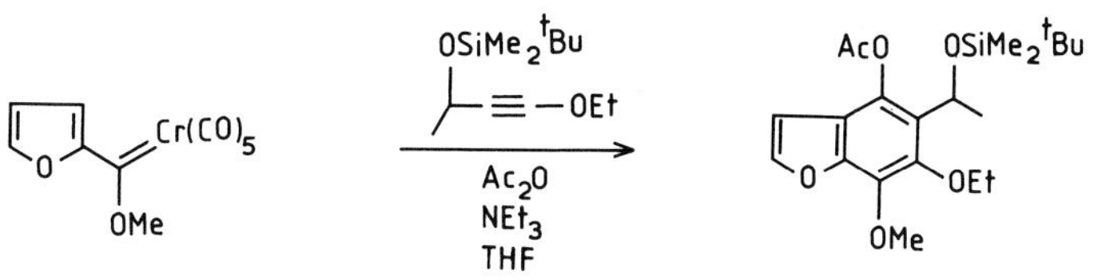

(69)   43%

Many synthetic sequences using chromium carbene complexes begin with a version of Fischer's original entry to methoxycarbene complexes, which has proved over the years to be an extremely reliable and generally applicable method. In this way, the vinyl carbene complex (70), required for a synthesis of vitamin E (73), was prepared from 2-

bromobut-2-ene by lithiation using t-butyllithium.[96] Reaction with the tetra-methylnonadecen-2-yne (71) was not completely regiocontrolled, but produced the correctly substituted arenes as a 7:3 mixture of regioisomers. The ligands were liberated under carbon monoxide pressure, and converted to (72) without the need to separate isomers.[83]

In other situations, a simple method to overcome difficulties with regiocontrol is to convert the reaction into an intramolecular process. This approach was used with (74) to bring the more hindered end of the alkyne adjacent to the OR group in the product.[97]

Efforts to apply this chemistry to the synthesis of anthracycline derivatives have met with considerable success. Approaches to the synthesis of daunomycinones can employ either aryl or vinyl carbene complexes. The arylcarbene (75), prepared from the corresponding bromide, was found to undergo loss of carbon monoxide above 50 °C to form the tetracarbonyl complex (76) which was characterised by X-ray crystallography. The complex (76) proved a very good substrate for the cyclisation reaction, forming a 4-demethoxydaunomycin precursor (78) following addition of the alkyne (77) and treatment with carbon monoxide.[98] An approach to fredericamycin A has also been based on chromium carbene methods.[99]

The vinyl carbene approach again began using methods based on the work of Fischer.[82] Formation of the required vinyllithium reagent from the hydrazone (79) initiated the preparation of the $Cr(CO)_5$ complex (80) which underwent a regio-controlled reaction with the alkyne (81) to produce (82).[100] A conventional cyclisation was then used to form the C-ring, affording (83) which was converted to the known daunomycinone intermediate (84). A similar vinyl carbene formation has been used by Quayle in a synthesis of 12-O-methyl royleanone.[101]

The synthetic applications of carbene complexes described so far in this section have two features in common, they use stable Fischer-type carbene complexes, and they produce hydroxyl substituted aromatic rings. Although reactions of this type have clearly now been established as a versatile synthetic methodology, these two features also impose limitations upon the situations in which the method can be applied. In recent work,[102] *in situ* generation of unstabilised carbene complexes has been used to extend the applicability of the reaction. In this case an intramolecular addition converts a Fischer carbene into a new vinyl carbene derivative (85) which was then trapped intramolecularly to form (86).[88]

A convenient method for reduction of the OH group typical in these products has recently been examined. By means of a palladium catalysed reduction of the derived triflate with formic acid, (86) has been converted into (87) in a synthesis of angelicin (88).[88, 103]

Related addition/carbonyl insertion reactions of chromium and molybdenum complexes have been employed in β-lactam synthesis, a topic covered in Chapter 15.

Two alkynes can be induced to undergo the cyclisation reaction by initial *in situ* formation of the vinyl carbene complex. In this case an alkylcarbene and carbon monoxide each contribute one carbon atom to the aromatic ring, which is formed by loss of OMe from the cyclisation product. In the formation of (90) from (89), chromium complexes gave good results, but in other cases, better yields were obtained by the use of tungsten.[104] These reactions required heating or photolysis and relatively long reaction times.

The carbyne complex (91) has been shown[105] to be a more reactive reagent for the formation of phenols from diynes. Reactions can be performed at −10 to −20 °C in a period of only a few minutes. The example leading to the formation of (92) emphasises the similarity between cyclisations of this type, and the coupling of dialkynes to form dienes that was described in Chapter 10. This parallel is further advanced by formation of (93) and (94) in a similar reaction using an ene–yne starting material. The mixture of E/Z isomers was converted to the diketone (95) in 32% overall yield.[106]

When aromatisation is blocked, dienone products can be isolated in good yield. Reaction between the vinylcarbene complex (96) and trimethylsilylethyne, gave predominantly the *trans* product (97) as a 92:8 mixture of stereoisomers.[107]

In a similar process,[108] iron carbene complexes and two alkynes are combined to form (pyrone)Fe(CO)$_3$ complexes. The structure of the product (98), which requires an apparent migration of the ethoxy substituent following insertion of two molecules of carbon monoxide, was determined by X-ray crystallography.

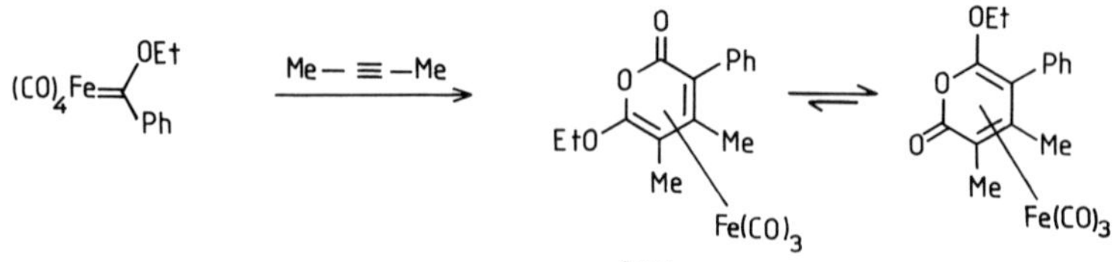

12.5 Carbene complexes as ketone mimics in aldol reactions

Carbene complexes can be deprotonated by butyllithium to produce anionic reagents that show a strong resemblance to a ketone enolate. Early attempts to perform alkylation reactions by addition of such reagents to aldehydes and ketones were thwarted by the low nucleophilicity of the metal stabilised anion. By the use of non-enolisable aldehydes as electrophiles, α,β-unsaturated complexes were obtained, but in disappointingly low yields.[109]

Enol ethers have also been used in the place of aldehydes in these reactions, in a procedure in which initial deprotonation is followed by addition of the enol ether and trifluoroacetic acid (see Scheme 12.9). Once again, rather low yields were encountered.[110]

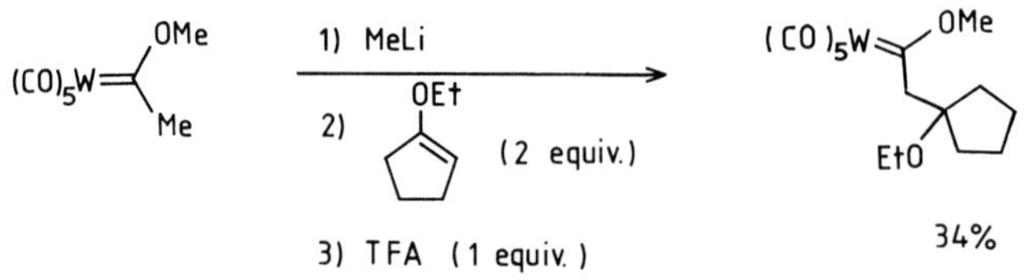

Scheme 12.9

The situation has recently been much improved by the use of Lewis acids to promote reaction with aldehydes and ketones.[111] Boron trifluoride etherate has been recommended for use with ketones, while aldehydes gave better results with titanium tetrachloride. In this way, the acetone adduct (99, M = Cr) was isolated in 70% yield.

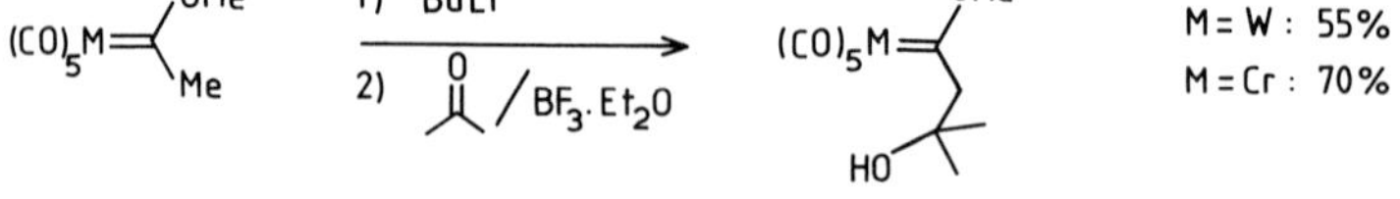

The reaction shows a strong similarity to processes involving metal acyl enolates described in Chapter 9, but has not yet been developed to give the same remarkable degree of stereocontrol. The diastereoselectivity of the reaction with aldehydes has, however, been examined in preliminary studies.[111] Lack of control of enolate stereochemistry, rather than lack of selectivity of the alkylation reaction itself, may account for the relatively modest degree of stereocontrol obtained. Reaction of (100) with benzaldehyde gave a 6:1 mixture in which the *anti* isomer (101) predominated. Alkylation of the methyl carbene complex (102) with the chiral aldehyde (103) gave a similar degree of selectivity, affording an 8:1 mixture of products. In this case, however, the *syn* product predominated.

Products from the 'aldol' reactions of carbene complexes are potentially useful in synthesis. Removal of the metal from (101) under oxidative conditions gave the corresponding ester, while reaction with diazomethane afforded the enol ether (104).[111]

Dehydration of the 'aldol' products can be used to provide vinyl carbene complexes for use in cyclisation reactions of the type discussed in Section 12.4. In this way the methyl carbene complex (102) has been converted[111] into the spirocyclic dienone (105).

13

Organometallic methods for protection, stabilisation and masking in organic synthesis

The properties of transition metal complexes are almost invariably different from those of their free organic ligands. This modification of reactivity can be employed to protect complexed functional groups while reactions are performed at positions elsewhere in the molecule. Indeed, the effects of complexation can be so pronounced that some ligands which have only a transient existence in the free state can be easily handled as metal complexes. Stoichiometric metal complexes can often be retained within the molecule through a series of synthetic steps, so that, at a later stage, sensitive or highly reactive functionality can be revealed by decomplexation. Masking functionality in this way can be very useful in synthesis design. These three related topics of protection, stabilisation and masking will be dealt with in this chapter.

Protection, in its simplest form, is the differentiation of similar functional groups (or functional groups with a common susceptibility to reaction with a particular reagent). Organometallic chemistry can be used to this end by the formation of a metal complex of one of the groups to be distinguished. To be useful, protecting groups must be inert to a range of reaction conditions, and yet be efficiently removed in a specific deprotection process. A variety of protecting groups that can be removed selectively, using different conditions, are of considerable value in the manipulation of functional group reactivity during a synthetic sequence. Here a potential advantage from transition metal chemistry becomes apparent; novel types of protecting group offer the opportunity to extend the range of stability and deprotection properties, and so increase flexibility and provide new effects.

The concept of masked functionality, or 'latent' functional groups, is a development that takes the principle of protection beyond the simple differentiation of similar groups into the strategic level of synthesis design. The use of protecting groups, though often necessary, is inherently inefficient since additional steps are required for their attachment and removal. Masked functional groups, on the other hand, are introduced in a 'natural' way at an appropriate point in a synthesis in a chemical form that, while having very different properties to the functional group that will be required, can be easily transformed into that group at a later stage. In this sense, the best masking groups are those which have different synthetic reactions to those of the functionality being masked, and so are able to be gainfully employed during the synthetic scheme, before the point of unmasking. Of course, properties such as chemical stability and the generality of the unmasking reaction are also essential requirements for success. By the stabilisation of highly reactive structures, transition metal chemistry can offer some unusual and spectacular masking effects, in terms of the contrast between the masked and unmasked states. Some examples will be given in Section 13.3.

13.1 Transition metal complexes as protecting groups

Protection of alkenes against hydrogenation has been effected by the formation[1,2] of $Fe(CO)_2Cp^+$ and $Re(CO)_2Cp$ complexes. This approach has been used to shift the site of reduction of the tricyclic diene (1) to its less reactive double bond, by hydrogenation of the metal complex (2).[1] The ligand can be subsequently released in the normal way (see Chapter 5) by treatment with sodium iodide. The method has since been extended to include a number of other dienes.[3]

Cationic complexes such as (3) and (4) can protect an alkene from attack by electrophiles.[1] This has been used to permit selective reaction at other alkenes or even arenes within the molecule.

Alkynes can also be protected against electrophilic attack, so allowing Friedel–Crafts acylation of diphenylethyne to be performed, by reaction of the $Co_2(CO)_6$ complex (5).[5] The chemistry of complexes of this type has been discussed in Chapters 5 and 9. Simple recovery of the alkyne by oxidative decomplexation makes these complexes convenient for use as protecting groups.

(5)

Reactions can be performed at alkenes in the presence of 1,3-dienes elsewhere in the molecule if the diene is first converted into a metal complex. Tricarbonyliron complexes have been most extensively examined for this purpose, but even so, relatively few cases have been reported. An early example from Givaudan[6] in the patent literature concerns transformations of terpenes for use in perfumes. The (myrcene)Fe(CO)$_3$ (6) complex

was oxidised at its free double bond by osmium tetroxide in pyridine to afford the corresponding diol which was liberated with ceric ammonium nitrate.[6] More recently, hydroboration,[7] reduction with anthracene-9,10-diimine,[7] and reactions with dichlorocarbene[8] and under Friedel–Crafts conditions[9] at the free double bond have also been described. The diimine reduction, however, proceeded in low yield, and attempts to use Adams' catalyst, or hydrazine hydrate were unsuccessful.

Interest in the synthesis of vitamin D metabolites by modification of the side-chain alkene of the steroid ergosterol in the presence of its B-ring diene, stimulated a re-examination of methods of this type by several groups. Formation of the steroid complex itself proved far from straightforward[10] and the main distinctions between the different approaches lie in the details of the initial stages. The Lewis group examined[11, 12] complexation of ergosteryl acetate, while Barton *et al.* worked[13] with the benzoate. Birch *et al.* examined[14] the complexation of the enol acetate (ergosta-3,5,7,22-tetraen-3-ol-acetate), which underwent an efficient direct complexation with Fe$_3$(CO)$_{12}$ in 83% yield.

The acetate (7, X = COMe) was an efficient substrate for a side-chain oxidation with osmium tetroxide to afford the diol complex in 82% yield. The metal was removed in 99% yield with ferric chloride.[11] Hydrogenation of (7) and removal of the metal from the product has also been described.[12] Barton examined hydrogenation of the benzoate (7, X = COPh) with Adams' catalyst, which, unlike the myrcene example, was successful in this case, giving (8) in 94% yield. Removal of the metal also removed the benzoyl group. Hydroboration of the acetate with diborane in THF proved to be more efficient (83% yield) than use of BF$_3$.Et$_2$O/NaBH$_4$ (45% yield), but both reactions gave a mixture of alcohol products (Scheme 13.1).[11] Hydroboration of the benzoate was also

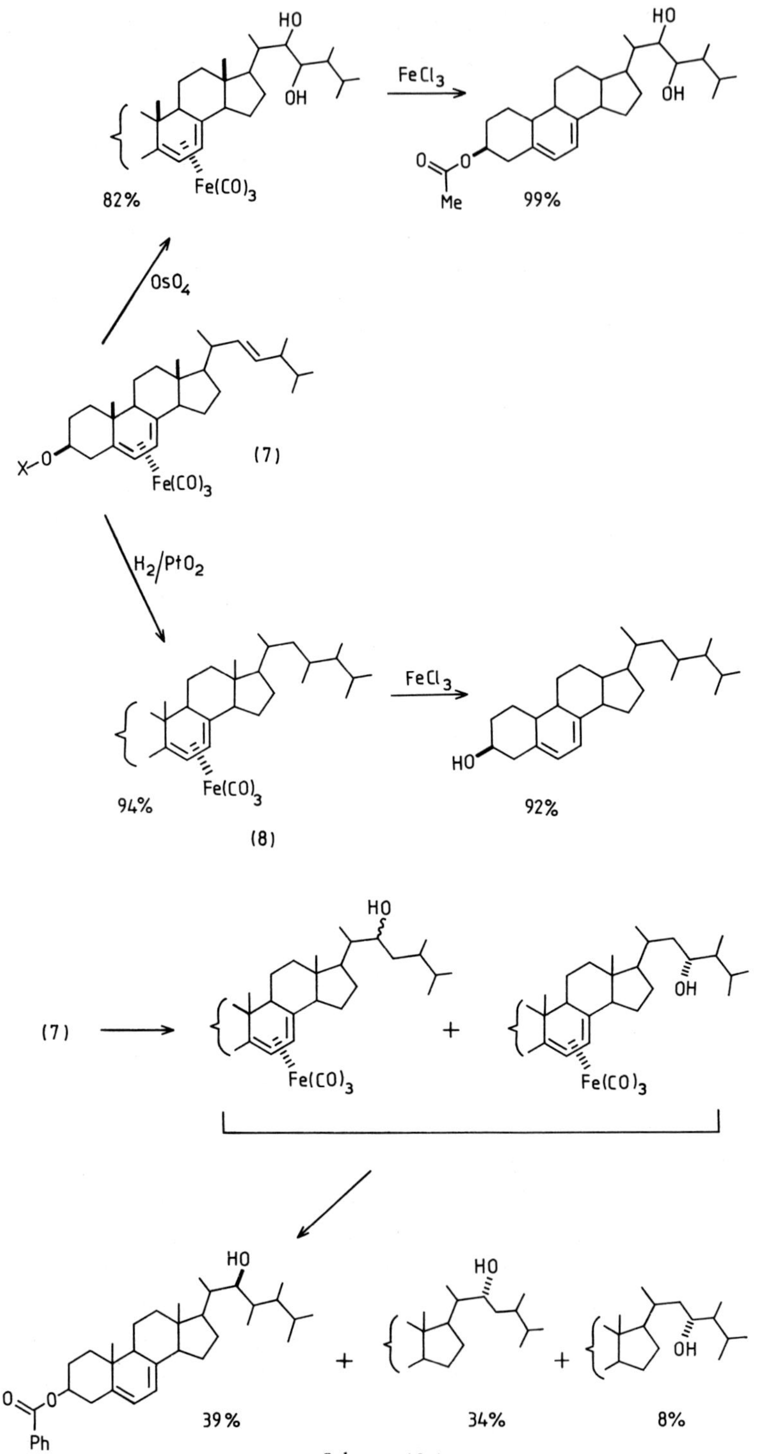

Scheme 13.1

performed using diborane and formed the first stage of a series of interconversions of side-chain oxygenation patterns following removal of the metal.[10]

Reduction and hydroboration of alkenes has also been achieved in the presence of alkynes, protected as $Co_2(CO)_6$ complexes.[15]

Use of a tricarbonyliron protecting group has allowed the conversion of thebaine to northebaine (12). Demethylation with BrCN cannot be performed using thebaine itself without skeletal rearrangement, but Birch *et al.* have shown[16] that the reagent can be employed successfully with the diene complex (9). The reaction replaces the N-methyl group by a nitrile that can then be hydrolysed in 91% yield to form (10). Demethylation of (9) with 2,2,2-trichloroethylchloroformate/1,2-epoxypropane gave (11). Decomplexation in 61% yield was followed by reduction of the carbonate to afford northebaine (12) in 64% yield. Other reactions of thebaine complexes are discussed in Chapter 7.

Controlled carbene additions to polyenes have also been examined in the presence of metal complexes. Comparison of reactions of η^4 Fe(CO)$_3$ and CoCp complexes with the Simmons–Smith reagent (CH$_2$I$_2$, Zn/Cu) shows the different effects that can be promoted by making use of the different bonding preferences of isoelectronic metal complexes. The addition of three CH$_2$ groups to (13) is unexpected and the mechanistic details of this reaction have not been examined.[17] The cyclooctatetrane complex (13) does not react with dichlorocarbene, but the free ligand (14) forms the mono-adduct (15).[8] Dichlorocarbene additions to a variety of partially complexed conjugated trienes have also been studied.[8, 18]

The inert nature of complexed dienes when exposed to carbenes has been used to promote carbene insertion into a C–H bond in the presence of a protected diene. Phase-transfer conditions were employed to generate the carbene reagent. Yields in these reactions, however, were rather low,[19] as can be seen from the examples given in Scheme 13.2.

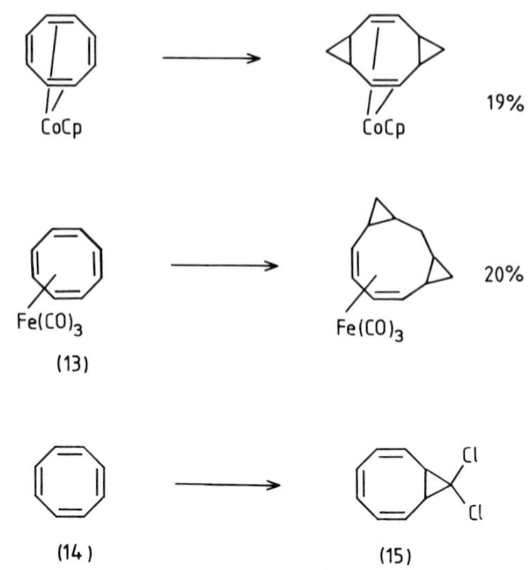

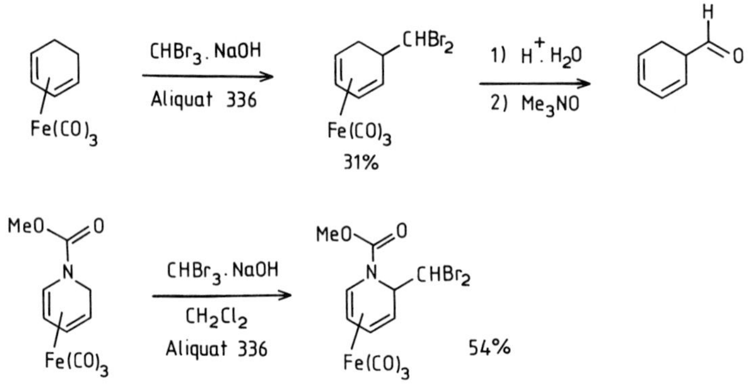

Scheme 13.2

Electrophilic substitution reactions of partially complexed polyenes have also been examined. Processes of this type are related to dipolar cycloaddition reactions discussed in Chapter 11 in that initial electrophilic attack takes place at a free double bond. Both Friedel–Crafts and Vilsmeyer conditions have been used.[20]

In a synthesis of β-thujaplicin (18) and β-dolabrin (19), Friedel–Crafts acylation of tropone was required to introduce a directing group to control the addition of an isopropyl group by the action of 2-diazopropane. (For an example of the alternative mode of addition, see Chapter 14.) Although tropone itself does not undergo Friedel–Crafts substitution,[21] reaction using the tricarbonyliron derivative (16), followed by acid catalysed equilibration of the resulting regioisomers, proceeded in good yield.[22] Diazoalkane addition afforded (17) which was converted to the two required synthetic intermediates by removal of the metal with Me_3NO.

Another example of Friedel–Crafts acylation, in which the metal complex also helps to ensure the correct double bond stereochemistry, can be found in Knox and Thom's synthesis[23] of a 1,4-disubstituted butadiene pheromone (20) of the codling moth *Laspeyresia (Carpocapsa) pomonella*. A complementary Friedel-Crafts entry to 2-substituted butadiene complexes is discussed in Section 13.3.

13.2 Transition metal complexes as masked functional groups

The use of methoxydiene complexes as precursors to enones, described in Chapter 7, provides an excellent example of masked functionality. Unlike the free diene ligand, methoxydiene complexes are not hydrolysed by aqueous acids, nor do they undergo reactions with many reagents commonly used with ketones and enones. Because of their excellent chemical stability, diene complexes of $Fe(CO)_3$ are well suited for use as masked functional groups. An example taken from Pearson's synthesis of tricothecanes illustrates all the essential elements of this approach. The metal complex is introduced in the course of a key C–C bond formation by the alkylation[24] of the corresponding dienyl cation (see Chapter 7). Two reduction steps (NaBH$_4$ and DIBAL) and epoxidation (VO(acac)$_2$/^{t}BuO$_2$H) (see Chapter 3) were accomplished before the unmasking of the 4,4-disubstituted cyclohexenone prior to cyclisation to form (22).[25] A series of standard interconversions produced (23) in which the ketone was correctly placed for conversion to the simple tricothecane (24). The synthetic route has since been developed[26] to give more general access to tricothecanes such as the more highly oxygenated analogue (25). After epoxidation and esterification, the functionality around the five-membered ring was further manipulated before cyclisation and decomplexation. A key step involved another example of an osmium tetroxide oxidation in the presence of the iron complex. A further modification employs a tin enolate to build the quaternary centre at the junction of the two rings, and a silyl group to introduce functionality in the C ring.[26] In this way the synthetic route has been made far more efficient.

A fairly broad range of reagents has been employed with molecules containing tricarbonyliron complexes. The use of a number of reducing agents has been examined. Many are illustrated by examples given earlier in this chapter. $NaBH_4$,[20,25] $LiAlH_4$,[27,28], $LiAlH_4/AlCl_3$,[23] DIBAL,[25,29,30,31] $LiAl(^tBuO)_3H$,[27] diborane,[11,15,32] $H_2B.SMe_2$ and 9-BBN,[33] Zn/HCl,[34] catalytic hydrogenation,[13,36] and diimine reducing agents[7] have all been successfully applied in particular situations. DIBAL appears to be one of the best for general use.

Despite the common removal of tricarbonyliron complexes by oxidation, several oxidising agents are compatible with masking groups based on this chemistry. Collins' reagent,[37] N-chlorosuccinimide/Me_2S,[27] DMSO/$(COCl)_2$/Et_3N,[28] and MnO_2[37] (with alcohols) and VO(acac)$_2$/$^tBuO_2H$,[25] OsO_4/pyridine[6,11] and hydroboration/H_2O_2[7,11] (with alkenes) have all been used, although manganese or lead dioxide may sometimes induce[38] neighbouring group interactions with the complexed diene (see Chapter 7). Collins' reagent has also been shown to be capable of inducing decomplexation.[39]

Many functional groups exhibit their normal reactivity in the presence of the metal complex. Ketones can form oximes[27] and acetals[40] and can be used in olefination reactions,[41,42,43] and with Grignard reagents.[42] Nitriles have been alkylated with Grignard[44] and alkyllithium reagents,[45] and have been reduced to aldehydes.[29,31] Nitro groups have been reduced to amines.[34] Alcohols have been esterified,[35] protected [(ClCH$_2$OMe/$^iPr_2$NEt),[37] ($^tBuMe_2SiI/CH_3CN$),[26] (PhCOCl/pyridine)[26]], converted to ethers,[25] and to tosylates[30,46] and mesylates[37] and hence to nitriles,[30] amines[46] and malononitrile adducts.[46] Alcohols[25,42,47] and alkyl halides[26] have been eliminated to form alkenes. Esters have been hydrolysed,[32] and β-ketoesters[41] and malonate derivatives[30,48] have been decarboxylated.

This extensive range of conditions under which tricarbonyliron complexes can survive shows the general prospects for synthetic operations in the presence of masked enones and other protected dienes. Several illustrative examples from the above list are given in Scheme 13.3. The calciferol complex (26) was used to overcome the problems usually encountered with isomerisation of alkenes during conventional functional group interconversions. Similar oxidation and reduction methods have converted the ergosterol complex (7, X = H) into epiergosterol.[27] The complex (27), obtained by alkylation of (21), was used in a synthesis of steroid analogues.[41]

In an example in which a metal complex itself serves as a masked functional group, Fe(CO)$_2$Cp is used in place of a CO$_2$Me substituent in cycloaddition reactions of 5-substituted cyclopentadienes. The complex (28) undergoes[49] useful cycloaddition reactions which cannot be performed directly with 5-carbomethoxycyclopentadiene which is thermally unstable.[50] Finally, oxidation in methanol converts the metal complex into a methyl ester.

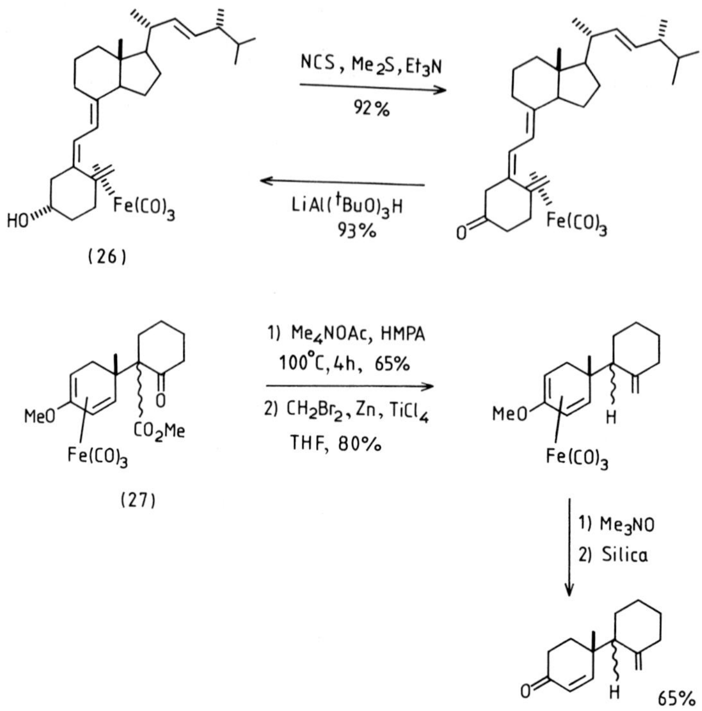

Scheme 13.3

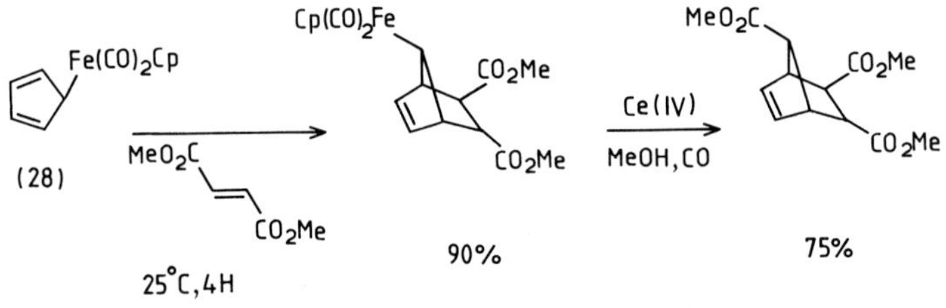

The specialised reactions of transition metal complexes can sometimes provide a convenient selective procedure by which to unmask functionality by the interconversion of conventional functional groups. Examples of this were encountered in the discussion of the Wacker oxidation in Chapter 4, where the conversion of a terminal alkene to a methyl ketone was used as a key step in several syntheses. In these reactions the alkene is used as a masked ketone, as shown in Scheme 13.4.

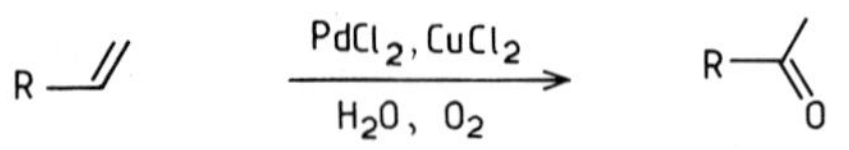

Scheme 13.4

A metal catalysed isomerisation reaction (see Chapter 2) has been used by Tsuji's group,[51] to reveal a vinyl ether for use in a Claisen reaction in a synthesis of an intermediate for dihydrojasmone. This reaction sequence employs a number of techniques discussed in earlier chapters. Oligomerisation of butadiene and homogeneous catalytic hydrogenation (see Chapter 10) provided convenient access to (30) which isomerised to the vinyl ether (31) and so produced (32) by a Claisen reaction. The alkene

formed in this step was converted into a methyl ketone. In this case the allyl alcohol in (29) was employed as the masked methyl ketone, a further example of the methods discussed above.

13.3 Stabilisation of highly reactive species

The effect of complexation on the properties of alkenes can result in the stabilisation of unstable tautomers that would have only a fleeting existence in the free state. The complexes of the enol tautomer of ethanal (33),[52] and the ketonic tautomer of phenol, (34)[53] clearly demonstrate this effect.

In another example[54] aromatisation of the triene (35) is prevented by the use of tricarbonyliron complexes. Though often somewhat unstable, triene complexes of this type are reasonable synthetic intermediates when prepared for immediate use.

(33)

Fe(CO)$_3$

(34)

Fe(CO)$_3$

(35)

By far the most widely known case where coordination confers synthetically valuable stability is found in the chemistry of cyclobutadiene. Both η^2-(36)[55] and η^4-(37)[56] complexes are known. Examples of synthetic application of these materials are given later in this section.

$^+$ Fe(CO)$_2$Cp

(36)

Fe(CO)$_3$

(37)

Similar stabilisation of a number of polyenes has been reported. Norcaradiene,[57] heptafulvene,[58] cyclopentadieone,[59] and *o*-quinodimethane[60] have all been shown to be highly unstable structures. The corresponding metal complexes (38),[61] (39, R = Me),[62] (40),[63] (41)[60] and the related tropoquinodimethane complexes such as (42),[64] however, are all relatively stable substances.

Fe(CO)$_3$

(38)

Fe(CO)$_3$

(39)

Fe(CO)$_3$

(40)

—Fe(CO)$_3$

(41)

—Fe(CO)$_3$

(42)

Of all these examples, work by Pettit on the Fe(CO)$_3$ complexes of cyclobutadiene has proved to have the most extensive application in organic synthesis, not just because of its initiation during the early stages of the development of transition metal organometallic chemistry, but also because of the remarkable stability of the complex produced and the simplicity of cyclobutadiene as a synthetic intermediate. This has stimulated applications by a number of workers in widely differing syntheses.

Attempts to synthesise cyclobutadiene using normal double bond forming reactions, e.g. dehydrohalogenation of 1,2-dichlorocyclobutane, all resulted in failure. However, following the predictions of Longuet-Higgins and Orgel,[65] Pettit isolated the cyclobutadiene tricarbonyliron complex (37) from reaction of dihalocyclobutenes with

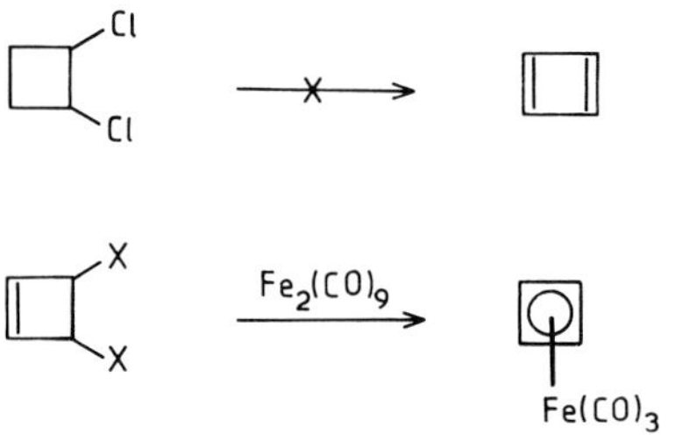

Fe$_2$(CO)$_9$.[56] The complex (37) is susceptible to electrophilic attack, as exemplified by the acetoxymercuration reaction[66] shown in Scheme 13.5. With excess mercuric acetate all

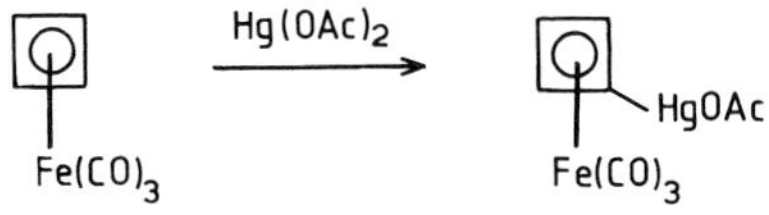

Scheme 13.5

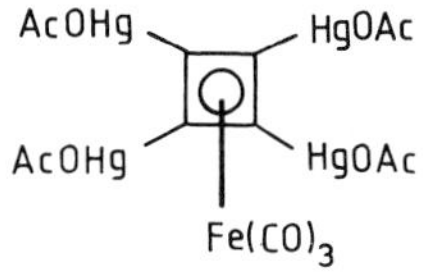

possible acetoxymercury derivatives including the tetrasubstituted complex (43) were obtained. Free cyclobutadiene may be released from (37) by oxidative removal of the Fe(CO)$_3$ moiety, and can be isolated as an adduct in the presence of a suitable trapping agent, i.e. a dienophile. For example, treatment of (37) with Ce^{4+} in the presence of alkynes leads to formation of Dewar benzene derivatives (44).[67]

In another example, liberation of the cyclobutadiene was followed by trapping with quinone to produce the adduct (45) which was used in a synthesis of cubane (47).[68] Photolysis produced the strained diketone (46) which was decarbonylated to give the required product.

The trapping of cyclobutadiene in reactions of this type has been shown[69] to be a reaction of the free ligand, rather than an electrocyclic reaction in the coordination sphere of the metal of the type described in Chapter 11. By the formation of η^2 complexes, however, cycloaddition reactions can be performed at the free alkene, which now reacts in the normal way as a two electron component. Subsequent decomplexation, in this case, produces the free organic product.[55]

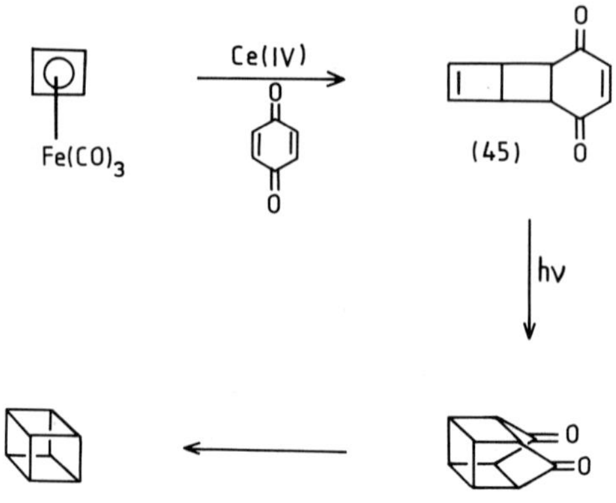

The stabilisation of cyclobutadiene complexes is reminiscent of aromatic stabilisation, and several aspects of the chemistry of the complexed cyclobutadiene ring resemble the reactions of arenes. Various substituted derivatives can be obtained from the parent complex by reaction with electrophiles. Mercuration has already been discussed. The examples shown in Scheme 13.6 indicate a variety of C–C bond formations that can be achieved in this way.[70]

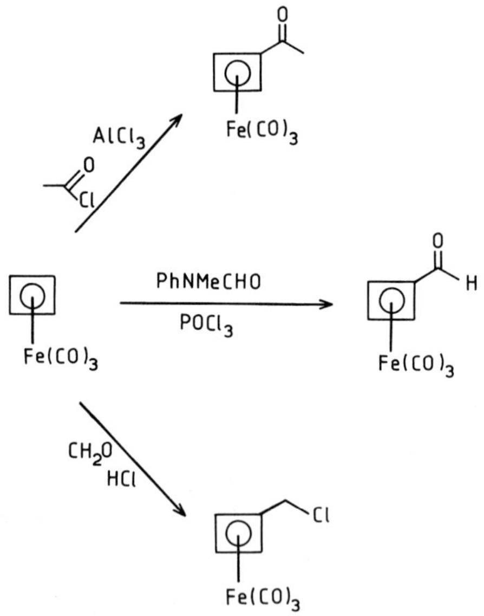

Scheme 13.6

A further method to functionalise cyclobutadiene complexes is by lithiation. The formation of (48) shows that this reaction can be performed selectively at the four-membered ring. Acylation of the corresponding rhodium complex was also regioselective in this way.[71]

The tricarbonyliron complex corresponding to (48) has been prepared from the pyrone (49).[72] Complexation of bicyclic intermediates of the type (50), obtained by the photolysis of pyrones, provides a useful general route to cyclobutadiene complexes.[73]

Cyclobutadiene methodology has been used to produce the strained tetracyclic 9,10-Dewar-anthracene (53) from (51). Photolysis with 1,2-dichloroethene, followed by decarboxylation afforded the cyclobutadiene precursor that was complexed with $Fe_2(CO)_9$ in the normal way. Oxidation with ceric ammonium nitrate in the presence of a second equivalent of (51) produced the dihydro product (52) which was aromatised by treatment with DDQ.[74]

An unusual organometallic route to another ligand system that was not available in the free state has been used to produce 2-acylbutadienes for use in Diels–Alder reactions. Again, the first step is a dehalogenation reaction,[75] which has now been further developed as a route to 2-acyl complexes such as (54). Attempts to isolate the butadiene ligand (57) produced only the dimer (56), but decomplexation in the presence of a diene gave the expected cycloaddition products, formed by reaction at the more activated of the two alkene linkages in (57).[76] The products were converted to the bicyclic ketone (55).

Unstable reaction intermediates can sometimes be intercepted and stabilised by complexation. Photolysis of ergosterol with $Fe(CO)_5$ led to the formation of two complexes (58) and (59) that afforded the products tachysterol$_2$ and precalciferol$_2$ respectively, upon oxidation.[35]

In other cases, unstable species in electrocyclic reactions have been intercepted and

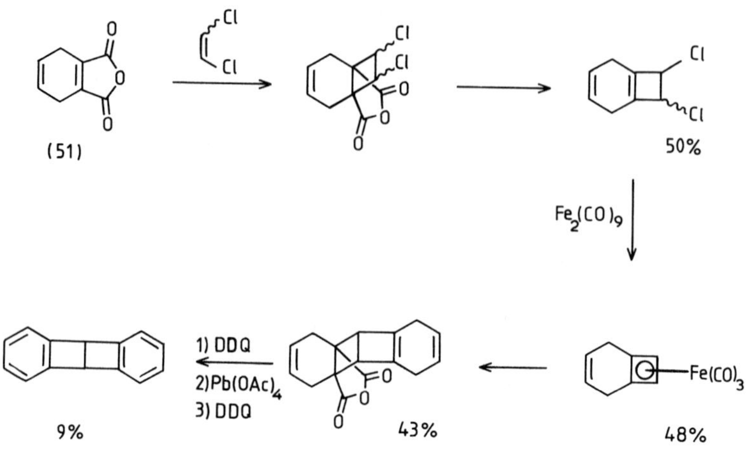

(51)

50%

Fe$_2$(CO)$_9$

9%
(53)

1) DDQ
2) Pb(OAc)$_4$
3) DDQ

43%
(52)

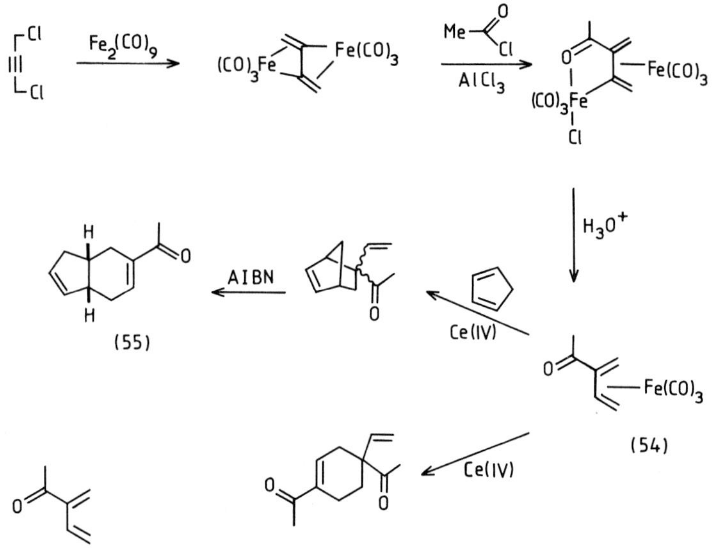

48%

Fe(CO)$_3$

Fe$_2$(CO)$_9$

(CO)$_3$Fe
Fe(CO)$_3$

Me—C=O / Cl

AlCl$_3$

(CO)$_3$Fe
Cl
Fe(CO)$_3$

H$_3$O$^+$

(55)

AIBN

Ce(IV)

Fe(CO)$_3$
(54)

Ce(IV)

(57)

(56)

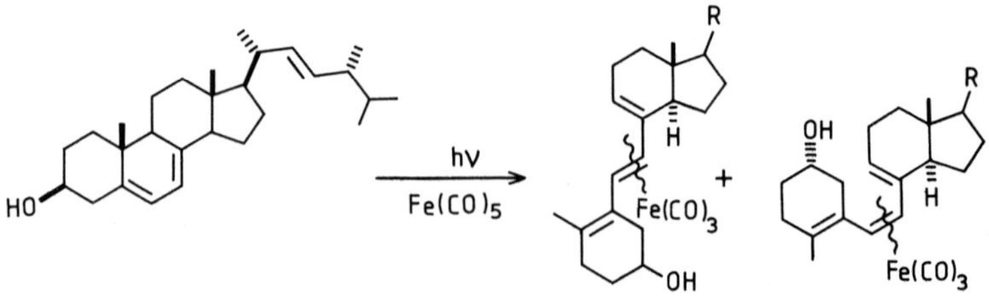

HO

hν
Fe(CO)$_5$

R

Fe(CO)$_3$

OH

(58)

+

R

OH

Fe(CO)$_3$

(59)

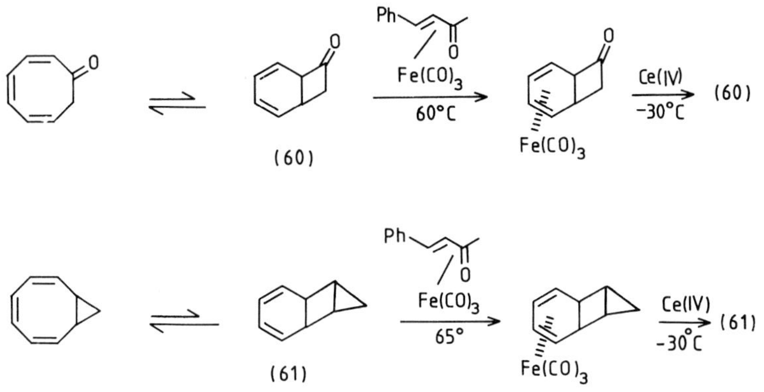

Scheme 13.7

then liberated at low temperature by ceric oxidation.[77] Some typical examples, the formation of (60) and (61), are indicated in Scheme 13.7.

13.4 Organometallic methods for the removal of protecting groups

The isomerisation of alkenes effected by transition metals (see Chapter 2) can offer a selective reaction to labilise protecting groups under exceptionally mild conditions. Studies by Corey and Suggs[78, 79] have shown that both alcohols and ketones can be deprotected in this way. Thus allyl ethers are converted to enol ethers (62) which can be easily cleaved by hydrolysis.

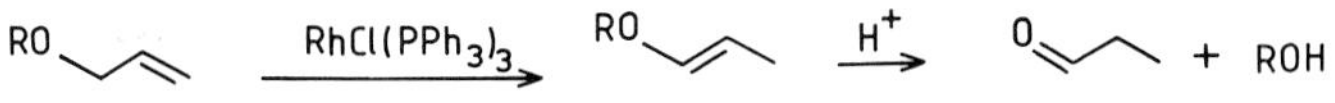

(62)

Treatment of the allyl acetal (63) with a catalytic quantity of RhCl(PPh₃)₃, Wilkinson's catalyst, afforded an enol ether which could then be removed under very mild conditions using $HgCl_2$ to give the free ketone in 80 % overall yield.

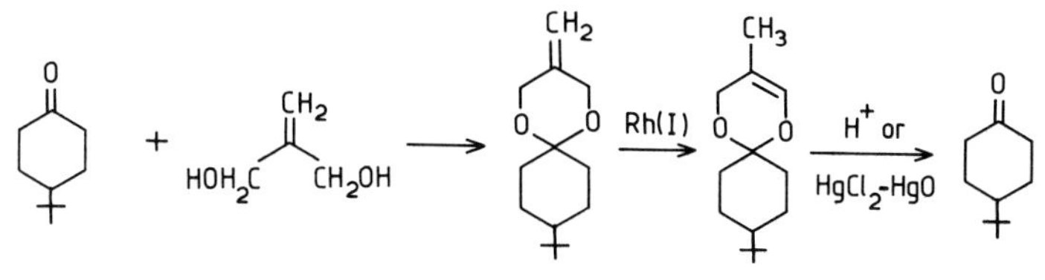

(63)

Wilkinson's catalyst has been successfully applied by Gent and Gigg to some instances of the allyl to propenyl isomerisation of carbohydrate ethers.[80] They found the relative rates of isomerisation to be allyl ether > 2-methylallyl ether > but-2-enyl ether.

Using RhCl(PPh₃)₃ added to the allyl ether dissolved in refluxing ethanol/benzene/water (7:3:1), isomerisation of 6-O-allyl-1,2:3,4-di-O-isopropylidene-α-D-galactopyranose (64) for example, was very rapid. This produced the 6-O-propenyl

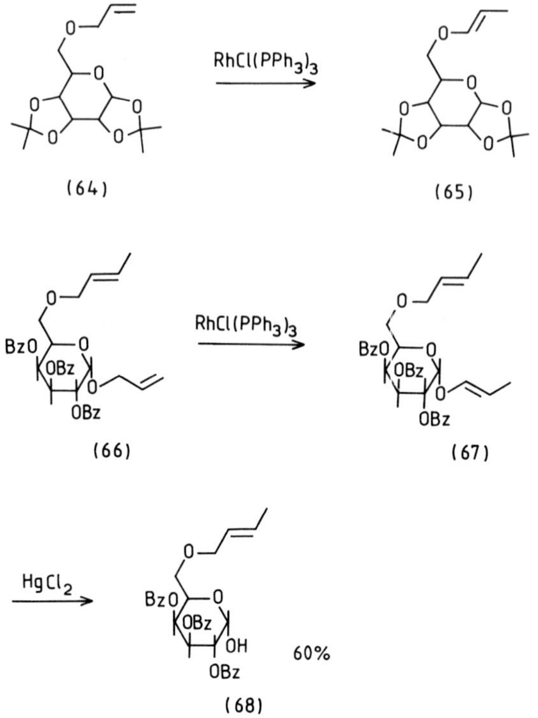

(64)   (65)

(66)   (67)

(68)

ether (65). The but-2-enyl ether (66) from galactopyranose gave, as major product of isomerisation with RhCl(PPh$_3$)$_3$, the propenyl butenyl ether (67) and thence (68) by hydrolysis. Although aldehydes can be decarbonylated by RhCl(PPh$_3$)$_3$ (see Chapter 9), such reactions appear not to interfere in the case of sugar derivatives such as (68).

Allyl protecting groups can also be removed from amines by conversion to the corresponding enamine using RhCl(PPh$_3$)$_3$, RhCl$_3$, or PtCl$_2$(PhCN)$_2$ catalysts (Scheme 13.8).[81]

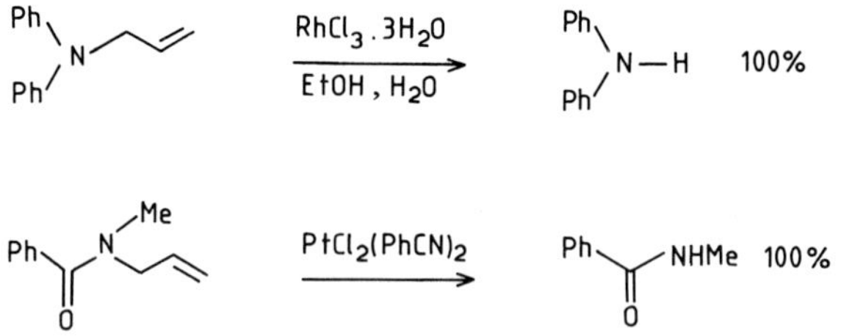

Scheme 13.8

The method has been employed in a synthesis of the amino acid antibiotic (+)-anticapsin (70). Removal of the allyl protecting groups in the presence of the epoxide was achieved by treatment with RhCl(PPh$_3$)$_3$ which resulted in the formation of propanal. After hydrolysis, the product (70) was obtained by cellulose chromatography. Partial racemisation had occurred during the reaction sequence, although epimerisation of the α-aminoester (69) during the de-allylation was ruled out.[82]

Attempts to use a similar deprotection procedure in a synthesis of biotin were less successful, giving rise to partial isomerisation, or a mixture including a product arising from partial reduction. The enamines (71) and (72) formed in these isomerisations, however, were easily hydrolysed.[81]

Allyl carbamates have been used[83] in a similar way to protect a number of primary and secondary amines, ranging from simple examples such as morpholine to several amino acids. In the case of the amino acids, it was found that after a cycle of protection and deprotection, optical purity was essentially unchanged. In a further development of this approach, an isopropylallyloxycarbonyl group has been used for amine protection in a peptide synthesis. Deprotection with palladium involves removal of the isopropylallyl group and CO_2.[84]

An allyl protecting group has been removed from oximes under mild conditions by reaction with Pd(OAc)$_2$. This worked well, for example, in the presence of an acetal as can be seen in the formation of (73).[85]

Allyl protecting groups can be removed from esters by the direct intervention of the transition metal to form a π-allyl complex (see Chapter 6). This reaction offers a selective method to cleave allyl esters under mild conditions[86] (PdCl$_2$(PPh$_3$)$_2$/HCO$_2$NH$_4$) and has been applied by workers at Merck Sharp & Dohme in a synthesis of alkylthio-substituted penems.[87] At the conclusion of the synthesis of (74), the removal of three protecting groups was required. Following treatment with fluoride to effect desilylation in the usual way, the allyl ester was cleaved with Pd(PPh$_3$)$_4$ in the presence of the *p*-nitrobenzyloxycarbonyl. Finally, hydrogenation afforded thiathienamycin (75) which was converted to the penem analogue (MK-787).

Since π-allyl complexes can recombine with carboxylic acids, the deprotection reactions discussed above must rely on favourable equilibria. In a further development of the method, a silyl group was included in the allyl moiety to drive the reaction to completion by irreversible formation of a silyl ester.[88] (Butadiene will react with palladium complexes in the presence of suitable nucleophiles, see Chapters 9 and 10.) After work-up, carboxylic acids, or their salts were obtained. The method has also been used in the synthesis of peptides[89] and β-lactams.[88] Scheme 13.9 illustrates the synthesis of β-lactams.

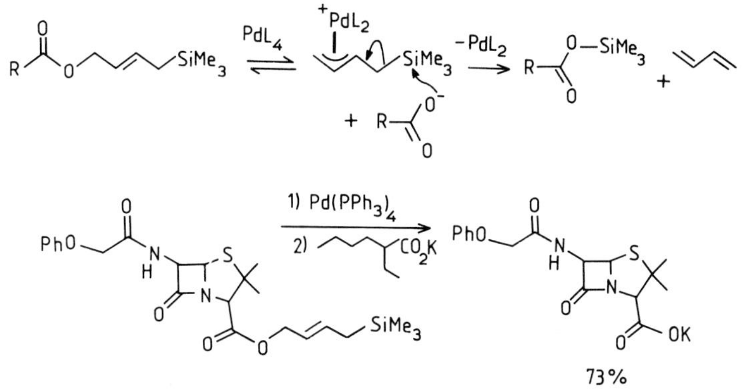

Scheme 13.9

Deprotection of phosphate groups provides another example of selectivity obtained by the use of transition metal chemistry. An allyl group has been removed from (76) in the presence of a propargyl group.[90] The propargyl group can itself be removed by palladium catalysed hydrostannolysis.[91]

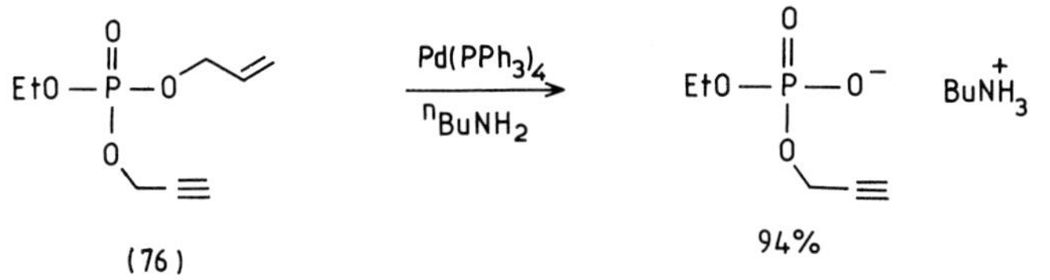

Palladium catalysed de-allylation has been used in the synthesis of nucleoside 3'- and 5'-monophosphates.[92]

14

Organometallic strategies in natural product synthesis

In many cases the availability of unusual synthetic reactions, developed through the exploration of transition metal organometallic chemistry, can alter the way synthetic chemists approach their work, and, as these methods become more generally used, an increased awareness of their potential will exert a broader influence on the way organic synthesis is planned. In addition to a conventional analysis of functional group position in target molecules, new types of disconnections can now be considered when retrosynthetic analysis is performed, so that organometallic methods, which are often complementary in their scope to more normal synthetic reactions, can be taken fully into account. This chapter examines a number of natural product syntheses to draw attention to the lessons that can be learnt from their design, and provides a commentary on the execution of syntheses that rely on key organometallic steps.

The following chapter examines heterocyclic synthesis in a similar way. However, although they are strictly speaking heterocycles, a number of syntheses of lactone natural products are discussed in this chapter. Several other heterocyclic natural products are also included here, but an examination of the extensive use of organometallic methods in β-lactam synthesis will be reserved for Chapter 15.

14.1 Grigg's synthesis of endo-brevicomin by means of the Wacker reaction

A variety of cyclic acetal bark beetle pheromones, a class of compounds that includes the pheromone of the Dutch elm beetle, have provided synthetic chemists with a selection of relatively simple target molecules with a pleasing combination of requirements for functional groups and stereochemical control. Endo-brevicomin (1) has been particularly popular with chemists wishing to demonstrate the power of new synthetic methods, and a great many syntheses of this molecule have appeared in recent years.

Grigg's organopalladium synthesis[1] was performed fairly early in the development of this area. When examined from an organometallic viewpoint, the formation of the acetal from the ketone (2) in the final step of a synthesis suggested an approach based on the use of the Wacker oxidation to produce the required methyl ketone from a terminal alkene, a reaction that will ensure the correct position of the carbonyl group

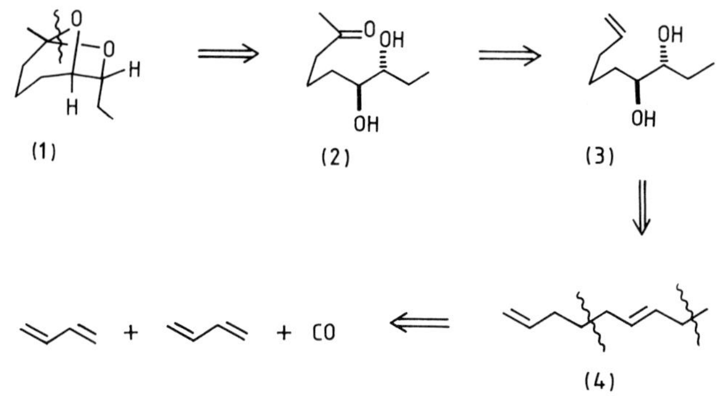

Scheme 14.1

(see Chapter 4). With the diol portion, one is again led to an alkene precursor. In this way, the problem is reduced to the control of alkene position and stereochemistry in the 9-carbon diene (4). This 1,6 arrangement of alkenes can be recognised as that formed in telomerisation reactions of butadiene (see Chapter 10), suggesting the formation of (4) from 2 equivalents of butadiene and one of carbon monoxide as shown in Scheme 14.1.

The synthetic sequence began with a reaction developed by Tsuji, who formed the ethyl nonadienoate (5) from butadiene and carbon monoxide in ethanol by palladium catalysis.[2] Reduction, tosylation and a second reduction gave the intermediate (4) in the usual way in 64% yield. Epoxidation and hydrolysis afforded (3), which was converted directly into endo-brevicomin under the Wacker conditions by use of an aprotic solvent.

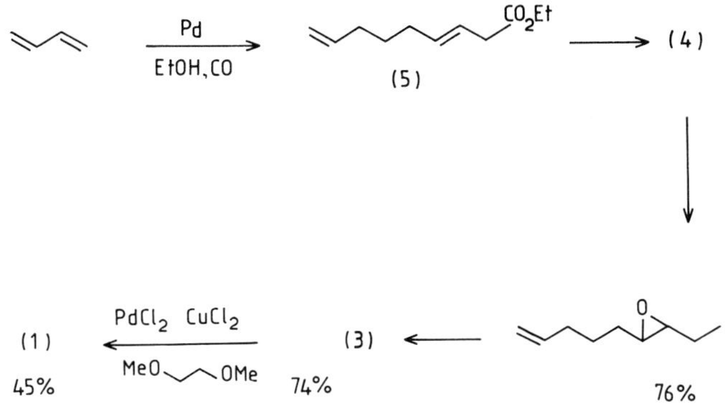

14.2 Sharpless epoxidation routes to (+)- and (−)-endo-brevicomin: syntheses by Oehlschlager, Mori, and Takano

In a recent approach to the same target, Oehlschlager and Johnston also chose an epoxide intermediate (6). Because of the now generally accepted importance of the development of enantioselective synthetic routes, this synthesis opted for a Sharpless epoxidation to introduce asymmetry and so used a disconnection for (6) which allowed the use of an allylic alcohol. In common with several examples discussed in Chapter 3, the asymmetric induction is paid for with extra synthetic steps to replace the alcohol by a methyl group. Powerful though the Sharpless epoxidation is, there is still a place for a reaction of similar generality that does not need an alcohol directing group. The allylic alcohol in (7) can be introduced in a simple way using propargyl alcohol dianion and a functionalised alkyl bromide.

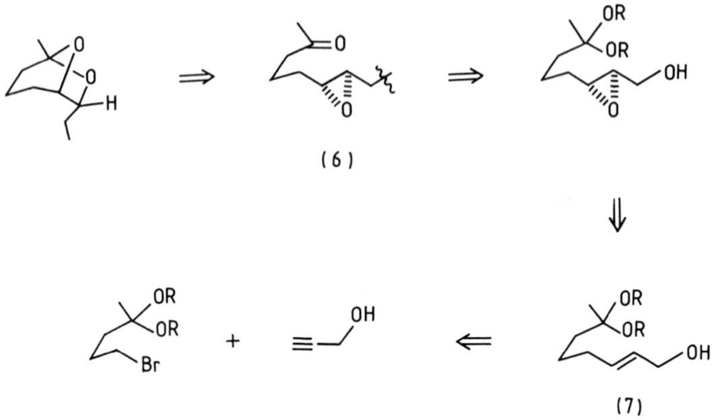

(6)

(7)

Combination of propargyl alcohol and an alkyl bromide bearing a suitably placed acetal, followed by reduction with sodium in liquid ammonia, produced the *trans* alkene (8). Epoxidation under the normal conditions with either (+)- or (−)-DIPT gave access to the two enantiomers of (9), each in better than 90% e.e. Conversion to the tosylate and displacement with bromide gave the intermediate (10) which was converted to (11) using dimethylcuprate (see Chapter 8). Finally cyclisation in pentane with dilute HCl produced endo-brevicomin in 69% yield.[3]

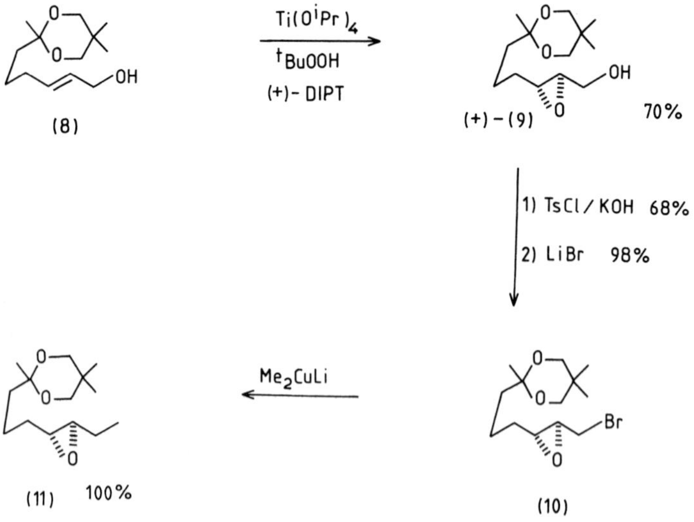

Another Sharpless epoxidation approach has used a kinetic resolution in place of the asymmetric induction. This route by Mori and Seu combines the Sharpless and Wacker oxidation (see Section 14.1) approaches, and is able to use the epoxide in a more direct way, shown below, by means of copper catalysed Grignard addition (see Chapter 8). The overall yield from (12), however, is somewhat lower than the Oehlschlager route (starting from 5-bromopentan-2-one), even after allowance is made for the 50% maximum yield in the kinetic resolution. To effect the opening of the epoxide with the organocopper reagent, protection and deprotection steps were needed.

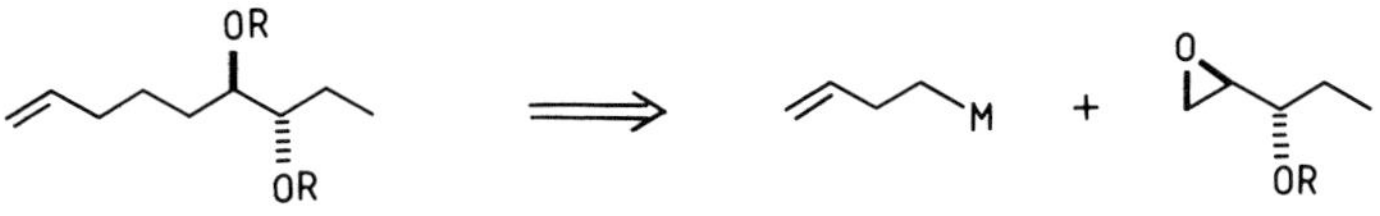

Sharpless epoxidation of (12), followed by protection of the alcohol, gave the epoxide (13). Opening with the Grignard reagent derived from 1-bromo-3-butene and catalytic amounts of copper (I) bromide, introduced the alkene needed later for the Wacker step. Removal of the protecting group with dilute acid gave a diol which was recrystallised to afford (14) in 96% e.e. and 54% chemical yield. The Wacker oxidation with concomitant cyclisation, as seen in the previous section, produced (+)-(1).[4]

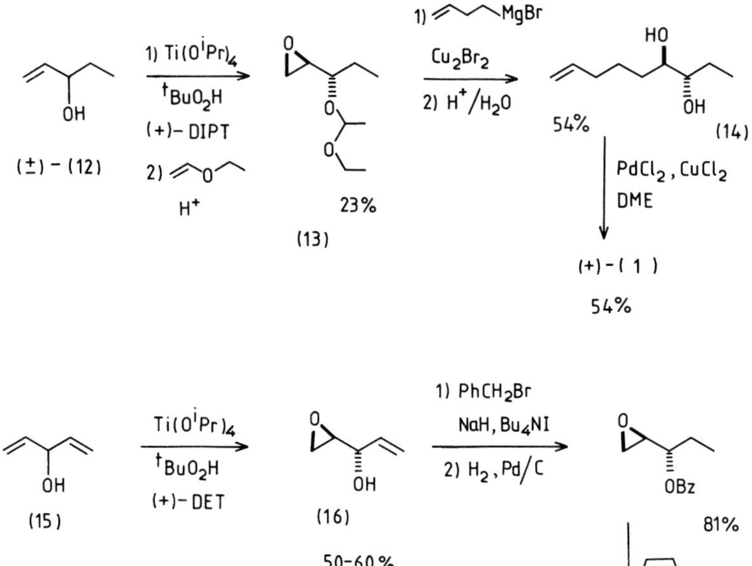

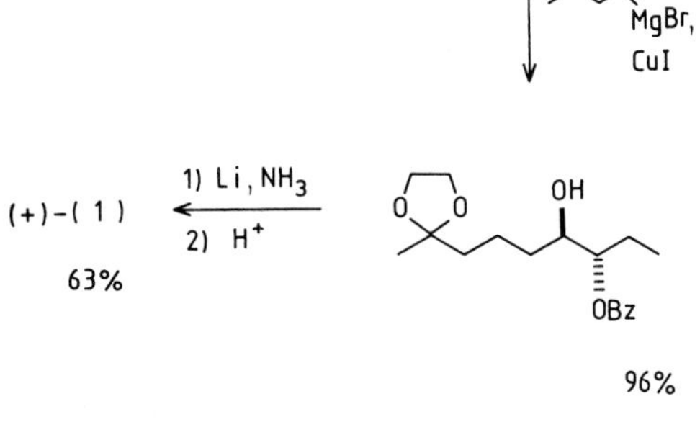

The epoxidation of hydroxydienes such as 3-hydroxy-1,4-pentadiene (15) is a special case among kinetic resolutions. This type of reaction, in which diastereofaces and enantiotopic groups are distinguished in the same process, can be relied upon to provide efficient enantioselective transformations (see Chapter 3). Takano's route[5] to (+)-endo-brevicomin commenced with the formation of (16) which was further elaborated by benzylation and catalytic hydrogenation. Once again, a copper catalysed Grignard was used to open the epoxide. Finally, removal of the benzyl protecting group and cyclisation in acid provided the product in almost 50% overall yield.

Despite the need to manipulate protecting groups this synthesis is the most efficient of the four discussed in Sections 14.1 and 14.2. The optical purity of the product (78.5% e.e.), however, was rather low. Unless racemisation has occurred in the conversion of (16) into (1), this route by Takano, none the less, offers good prospects for both efficiency and high optical purity, if the efficiency of the kinetic resolution can be improved along the lines described by Schrieber (discussed in Chapter 3).

14.3 The Normant–Alexakis organocopper approach to the California red scale pheromone

Two organocopper nucleophiles contribute to the preparation of a key intermediate (18) which was used by Alexakis, Mangeney and Normant in an examination of routes to the polyene (17), a pheromone of Californian red scale. Choice of a conventional olefination approach to introduce one of the alkene links puts the aldehyde in (18) two

carbons away from the chiral centre in (17). Functional group interconversion to the enol ether suggests the possibility of using an asymmetric induction by the BF_3/RCu opening of chiral acetals, described in Chapter 8. In this case, reaction of the acetal (19) with a vinylcopper nucleophile would move the alkene to the required position, after detachment of the chiral auxiliary by hydrolysis.

A second reaction discussed in Chapter 8 is used at the start of this synthesis. Cuprate addition to the alkyne (20) provides convenient access to the 1,5-diene (21) which was converted into the chiral acetal (19) by transacetalisation. The asymmetric induction step afforded (18) in 85% e.e. and 50% overall yield.[6]

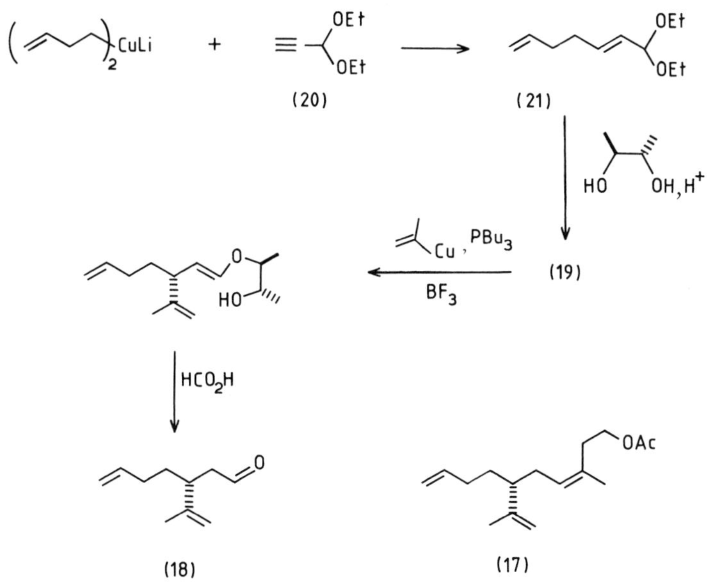

(20)   (21)   (19)

(18)   (17)

14.4 Tsuji's approach to the monarch butterfly pheromone via palladium allyl complexes

Another insect pheromone (22), a compound produced[7] by the monarch butterfly, provides a similar problem in terms of alkene stereocontrol but with a different spacing between the double bonds, and so was approached by Tsuji in a quite distinct fashion.[8] Synthesis of the intermediate (23)[9] can be based on nucleophile additions using the ester

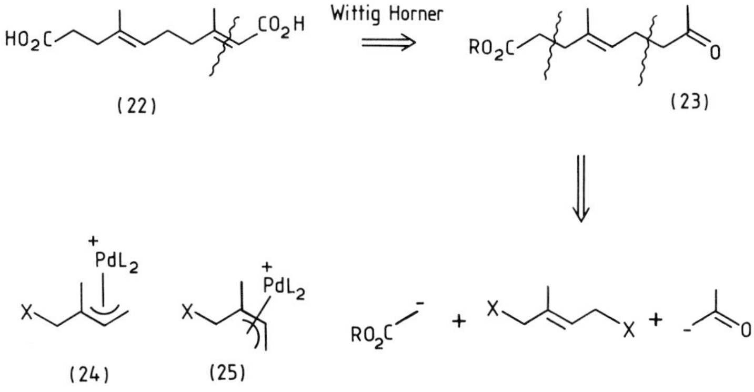

(22)   (23)

(24)   (25)

and ketone end groups in their natural polarities in two allylic displacement reactions. Control of both the stereochemistry and the sequence of alkylation reactions around the alkene linkage in the central 5-carbon fragment is thus of key importance in the development of an efficient synthetic approach. Palladium allyl complexes of the type (24), formed from an isoprene derivative, can ensure the stepwise execution of the two alkylations.

For this purpose, the isoprene unit required two leaving groups, one each for the preparations of the two palladium allyl intermediates. The first alkylation was performed using the isoprene epoxide (26) in a reaction in which the allylic C–O bond of the epoxide is cleaved, presumably forming the intermediates (24 and 25, X = O$^-$). Although the alkylation products from these intermediates lacked stereocontrol, control in the reaction sequence as a whole was not doomed, since the interconversion of π-allyl stereoisomers[10] makes possible the subsequent equilibration of the mixture. In fact, in this reaction sequence, it is only the second palladium catalysed step that must be examined when considering the stereochemistry of the products.

After ester hydrolysis and decarboxylation of the resulting β-keto acid, the alcohol group formed in the epoxide-opening reaction was acetylated to provide a suitable leaving group for the second alkylation step. Treatment of the E/Z mixture (27) with Pd(PPh$_3$)$_4$ in the presence of an acetoacetate nucleophile produced only the E isomer (28), which was converted into the required intermediate (23, R = Me) by removal of the acetyl group with sodium methoxide. The formation of a single alkene stereoisomer in the second palladium catalysed alkylation step indicates either a rapid equilibration of the *syn* and *anti* complexes (29) and (30) prior to combination with the nucleophile, or kinetic control by a rapid alkylation of (29) in the presence of (30), which then may be converted to (29) in a relatively slow reaction. Such a difference in rates of reaction for (29) and (30) seems unreasonable, and the well established thermodynamic preference for *syn* substitution is consistent with the first explanation, in which stereoconvergence depends on thermodynamic control.

14.5 Trost's organopalladium approach to isoquinuclidines

The isoquinuclidine ring system is common to a variety of alkaloids, including catharanthine, ibogamine and cannivonines.[11] Catharanthines, in particular, are an important target for synthesis, since they can be converted biomimetically[12] and enzymatically[13] into the structurally more complex vinblastine alkaloids that are valuable as antileukemic agents.

Following the development of a palladium catalysed cyclisation to form ibogamine, discussed in Chapter 10, Trost and Romero[14] have examined a palladium catalysed method for the formation of the azabicyclo[2.2.2]octane ring system in (31) which bears functionality on either side of the nitrogen bridge. The prospect that the allylic C–N bond could be formed by amine addition to a π-allyl complex again suggested a palladium catalysed epoxide opening reaction, since this would provide the alcohol group at the required position. It is interesting to note that this approach also ensures the *cis* relative stereochemistry between the C–N and C–O bonds, though this is not essential for a successful entry to *iboga* alkaloids. Opening of the epoxide (33) was expected to occur from the lower face to produce the π-allyl complex (32). This has the correct stereochemistry for nitrogen addition, which must occur *trans* to the metal.

The synthetic route started with a series of manipulations to obtain (35) from a suitable optically active starting material, D-(−)quinic acid (34). Formation of the acetonide and lactonisation were performed in the same step. Reduction of the lactone produced a *cis*-diol which was converted regioselectively to the alkene (35). After acetylation, hydroboration produced the primary alcohol as the major product. This reaction was not completely stereocontrolled, giving a 2:1 mixture of diastereoisomers. The acetate group was removed during the hydroboration reaction.

Having established in this way the important chiral centres around the six-membered ring, the next stage in the synthesis involved a series of interconversions needed to introduce the indole portion of (37). Conversion of the primary alcohol to a tosylate (92% yield) and then to an azide (91% yield), followed by silylation (77% yield) and reduction (97% yield) produced an amine intermediate which was combined with indole-3-acetic acid to give (36) in 64% yield.

Reduction of the amide with DIBAL (95% yield), protection of the amine (90% yield) and exchange of the silyl group for the MeSO$_2$ group (Bu$_4$NF, THF/H$_2$O, 25 °C, 82%; MeSO$_2$Cl, Et$_3$N, 91%), completed preparations for the formation of the epoxide. The acetonide was hydrolysed to form (37) (98% yield) which was treated with sodium hydroxide to effect the cyclisation by displacement of the mesylate. Oxidation of the remaining alcohol provided a ketone which was converted to an alkene by a Wittig reaction producing the protected amine intermediate (38). Removal of the final protecting group involved a second desilylation, this time performed under more vigorous conditions at 60 °C to form (33) which was cyclised in the expected fashion by treatment with Pd(PPh$_3$)$_4$.

This long synthetic sequence can be envisaged as a progression through a series of objectives: first, the entry from an optically active starting material; second, the arrangement of chiral centres to ensure the correct relative stereochemistry between the epoxide and the CH$_2$NHR unit; and third, the introduction of the heterocyclic component. In the final stage, the key elements for the palladium catalysed cyclisation, the epoxide and the exocyclic alkene, are introduced in a sequence that culminates in the long awaited organometallic step.

An important lesson, from the point of view of synthesis design, can be drawn from this second example of the use of an allylic epoxide bearing a substituent on the alkene. Unlike the case encountered in Section 14.4, reaction of the Z isomer (39) formed a Z-product (40), presumably via the *anti*-methyl complex (41) which was not converted to

(39)　　　　　　(40)　91%

the more stable *syn*-methyl isomer (42). It is clear from this that reaction via the *syn* isomer cannot always be guaranteed, and careful thought must be given to the mechanism of the equilibration reaction, which is believed to proceed via η^1-allyl intermediates (see Chapters 6 and 7 for details of similar processes), when a synthesis relies on stereoconvergent effects of this type. In the case of (39), an explanation for the Z-product is easily found. Inclusion of the allyl unit into a ring reduces the opportunities for rotation in the η^1 intermediates. Only attachment of the metal to the CHMe terminus can lead to rotation to form the *syn* geometry, and this will produce (43) in which the metal is on the wrong face of the ligand to undergo direct nitrogen addition. Thus, even if the *syn* isomer were formed, the reaction could proceed solely via the *anti* isomer (41).

(41)　　　　　　(42)　　　　　　(43)

14.6 Holton's carbopalladation route to prostaglandin intermediates

In the design of a stereocontrolled route to the prostaglandin intermediate (44) using palladium complexes, Holton[15] combined nucleophile addition, which occurs *trans* to the metal (see Chapter 5), and carbopalladation, which proceeds *cis* to the metal (see Chapter 10), to obtain the correct relative stereochemistry. Formation of the lactone ring with inversion by displacement of X in (45), offered the attractive possibility that X might be used as a control group for the introduction of the other three chiral centres around the five-membered ring. By use of the ketone oxidation state for the prostaglandin 'lower chain' [—CH=CH—CO—(CH$_2$)$_4$—CH$_3$] it is possible to attach the vinyl group to the ring by carbopalladation. This suggested the σ bound organometallic intermediate (46) which could be formed by nucleophilic addition to η^2 metal alkene complexes of the type (47) or (48) (see Chapter 5). The choice of the dimethylamino group as X was based on a considerable number of earlier examples[16] where binding to –NMe$_2$ had served to control organopalladium reactions. The expectation that quaternisation would provide a suitable means to form the lactone ring was confirmed in advance by model studies.[15]

The synthetic sequences began by the addition of HCl to cyclopentadiene to form the

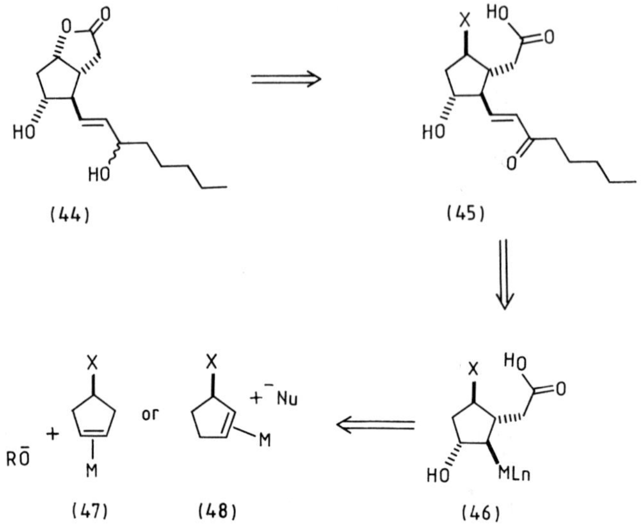

(44) (45)

(47) (48) (46)

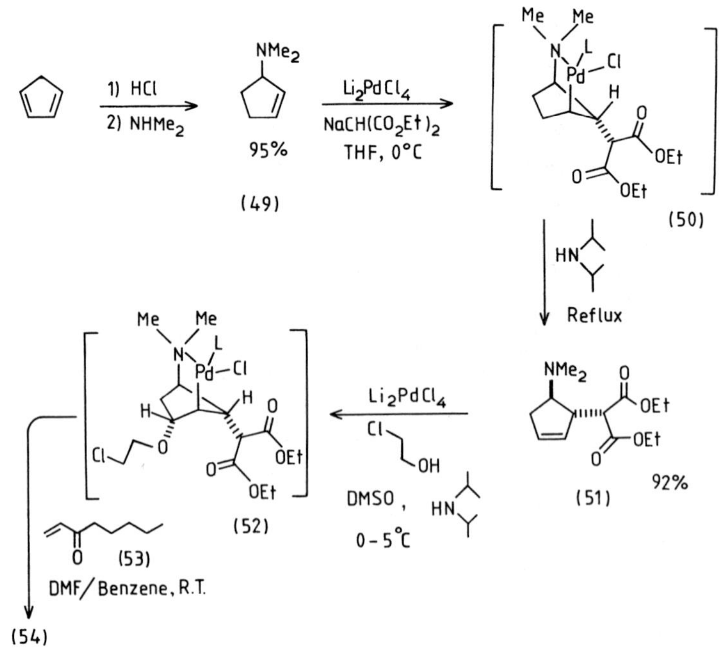

highly reactive 3-chlorocyclopentene[17] which was converted directly to the amine (49). Reaction with tetrachloropalladate and the malonate nucleophile produced the η^1-cyclopentane intermediate (50) which underwent β-elimination upon heating with diisopropylamine to form the alkene (51). A similar organopalladium mediated nucleophile addition was then used to form the corresponding η^1-cyclopentane complex (52). In this case, the metal was removed by its participation in the insertion of the vinylketone (53), which led to the direct formation of the required product (54) in 50% yield, together with a small amount (20%) of an impurity arising through a competing β-elimination reaction. After separation by chromatography, (54) was converted into (55) by reduction of the ketone, displacement of the chloride and quaternisation of the amine to provide the leaving group for lactonisation. Reaction with KOH both formed the lactone, and effected decarboxylation and deprotection, to afford (44), an intermediate which has been converted[18] by Corey to PGF$_{2\alpha}$ in two steps in 80% yield.

The intermediate (44), was obtained as a mixture of diastereoisomers at the carbon destined to be C–15 of PGF$_{2\alpha}$. When this synthetic work was performed, control of stereochemistry at this position was an unsolved problem, but subsequently reducing agents capable of the task have been developed.[19]

14.7 Pearson's organoiron route to *Aspidosperma* alkaloids

Limaspermine (56), a member of the *Aspidosperma* class of alkaloids,[20] bears a relatively unusual functionalised angular side chain at C–20.[21] The efficient introduction of this side chain (CH$_2$—CH$_2$OH in the case of limaspermine) is an important requirement in syntheses of these compounds. In a route[22] developed by Pearson using organoiron chemistry, this is achieved by the inclusion of a methoxyethyl substituent on a tricarbonyl(η^5-2-alkoxycyclohexadienyl)iron(1+) complex. Compounds of this type have a general utility as sources of substituted cyclohexenones (see Chapter 7). The synthesis of limaspermine in this way by Pearson's group can be regarded as a significant milestone in the development of applications of organoiron chemistry, since its completion established the precedent for the application of functionalised tri-complexity. A further, and more recent, example, Pearson's revised route to tricothecanes, has already been discussed in Chapter 13 (p. 437).

Retrosynthetic analysis for the *Aspidosperma* alkaloids began in a conventional fashion leading to an intermediate (57) of a type available from a Michael addition to cyclohexenones.[23] In this case, the intermediate (58) appeared a promising candidate for synthesis by alkylation of a tricarbonyliron complex (59) (for an alternative approach, see below). The alkoxyethyl substituent was easily obtained by selection of the starting material (60), 4-hydroxyphenylacetic acid, which contains a suitable two carbon chain. The isopropyl ether in (59) was chosen on the basis of earlier studies[24] that had shown that this gave superior regiocontrol during alkylation.

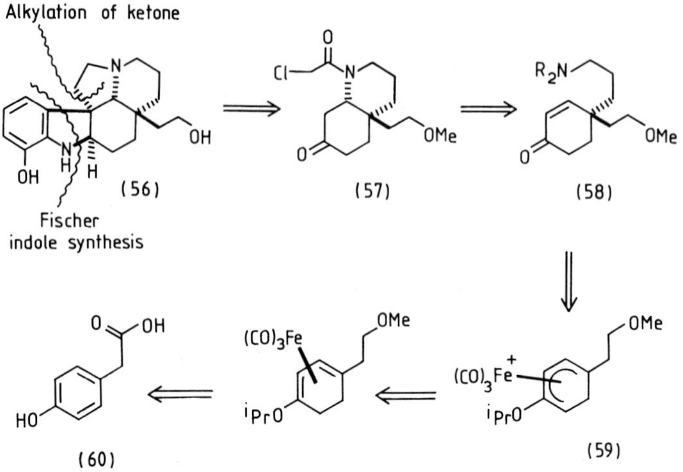

Alkylation of ketone

(56)

Fischer
indole synthesis

(57)

(58)

(60)

(59)

(60) $\xrightarrow{\begin{array}{l}1) \text{Br}, \text{OH, NaOH}\\ 2) B_2H_4\\ 3) \text{NaH, MeI}\\ 4) \text{Li, NH}_3\end{array}}$ (61) $\underset{\text{TsOH}}{\rightleftharpoons}$ $\xrightarrow{\text{Fe}_2(\text{CO})_9}$ (62) 50%

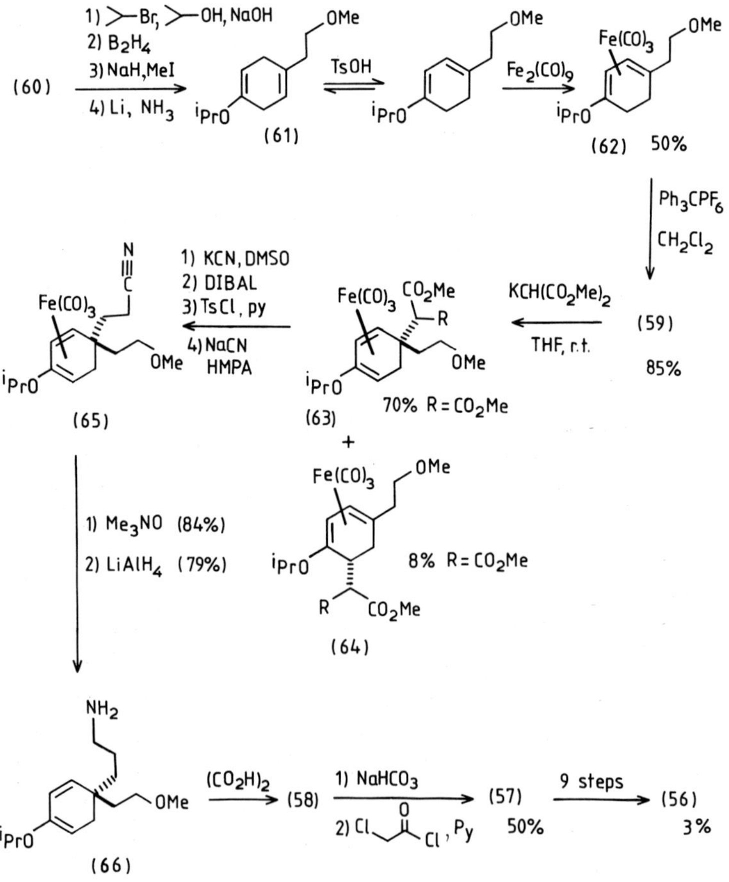

$\Big|\begin{array}{l}\text{Ph}_3\text{CPF}_6\\ \text{CH}_2\text{Cl}_2\end{array}$

(65) $\xleftarrow{\begin{array}{l}1) \text{KCN, DMSO}\\ 2) \text{DIBAL}\\ 3) \text{TsCl, py}\\ 4) \text{NaCN HMPA}\end{array}}$ (63) 70% R = CO_2Me $\xleftarrow[\text{THF, r.t.}]{\text{KCH}(CO_2Me)_2}$ (59) 85%

+

(64) 8% R = CO_2Me

$\Big|\begin{array}{l}1) \text{Me}_3\text{NO (84%)}\\ 2) \text{LiAlH}_4\ (79\%)\end{array}$

(66) $\xrightarrow{(CO_2H)_2}$ (58) $\xrightarrow[\substack{2) \text{Cl}\ \overset{O}{\underset{}{\text{C}}}\ \text{Cl, Py}\\ 50\%}]{1) \text{NaHCO}_3}$ (57) $\xrightarrow[3\%]{9 \text{ steps}}$ (56)

The first few steps of the synthetic sequence are relatively straightforward. Alkylation of the phenol (62 % yield), reduction of the acid (100 % yield), and methylation (92 % yield), followed by Birch reduction (93 % yield), gave convenient access to the 1,4-diene (61) which was conjugated prior to complexation with $Fe_2(CO)_9$. In this way the diene complex (62) was obtained as a single regioisomer. Many 1,4-disubstituted diene complexes have been prepared by complexations of this type. Two points, however, are worth noting. First, when entry to functionalised alkoxydienes is required, if it can reasonably be avoided it is better not to have side-chain acids or alcohols present in the molecule during the reduction step. The need for careful neutralisation during the work-up can then be avoided. Secondly, the decision to conjugate the diene prior to complexation was taken to avoid the formation of a mixture of isomeric products. Either $Fe_2(CO)_9$ or $Fe(CO)_5$ gave good regiocontrol in this case. Conversion to (59) was also regiocontrolled.

Alkylation of the dienyl cation would ideally require a three carbon nucleophile bearing terminal functionality. However, no suitable reagent was available, so a route using a homologation reaction was followed. As expected[24, 25] alkylation with dimethyl malonate gave the best regiocontrol [9:1 (63):(64)] when performed using a potassium counterion. Subsequent decarbomethoxylation and reduction produced a hydroxyethyl group. The regiocontrol of the alkylation of (59) is an important consideration, and surprising results were obtained using S,S'-diethyl thiomalonate anion which gave solely the wrong type of alkylation product (64). This was unfortunate, since direct conversion of this sidechain to the required hydroxyethyl group with Raney nickel[26] was anticipated. Tosylation (96 % yield) and substitution by cyanide (94 % yield) completed the route to (65) in which the correct two and three carbon units were in place.

An alternative to the homologation sequence has also been examined. If the three carbon unit was present originally as a substituent in the dienyl cation, then a two carbon nucleophile would be required, which could then be used in its natural polarity if a carbonyl group were placed at the far end. Once again, malonate addition (requiring subsequent decarbomethylation) was used.[27] The choice of stabilised enolates such as malonate as the nucleophiles in these reactions, and in similar examples encountered in Chapters 7 and 13, arises because of their excellent ability to alkylate a substituted terminus of the dienyl cation, a position that can be severely hindered and somewhat base-sensitive in these circumstances.

With the nitrile (65) in hand, the remaining steps in the synthesis were fairly standard and make no further use of the metal. Decomplexation and reduction of the nitrile gave the enol ether (66) which was hydrolysed, cyclised by a Michael addition to the enone (58, R = H) and treated with $ClCH_2COCl$ in pyridine to form the amide (57). This compound has been converted[28] to limaspermine (56) in nine steps.

A significant advantage of this approach (and in examples of uses of stoichiometric allyl and arene π complexes discussed in Sections 14.8 and 14.9) is that the metal complexes that are employed as intermediates are chiral, and so, in principle, can provide an enantioselective entry to the synthetic route (see Chapter 7).

14.8 Ley's synthesis of malyngolide by an iron mediated lactonisation

The antibiotic malyngolide (67), obtained from the blue-green alga *Lyngbya majuscula*,[29] offers an interesting synthetic challenge, since the lactone is flanked by chiral centres. Ley's approach[30] to the synthesis of (67) was based on an organometallic reaction producing an α,β-unsaturated lactone by a carbonyl insertion procedure developed in his group (see Chapter 9). By selecting the intermediate (68) Ley sets the stage for the introduction of the lactone carbonyl group from carbon monoxide by an application of organoiron chemistry, through the opening of the epoxide (69). The presence of the CH_2OH group in (69) suggested the epoxidation of (70) (Scheme 14.2), which should be easy to control, using reagents that are selective for allylic alcohols (see Chapter 3).

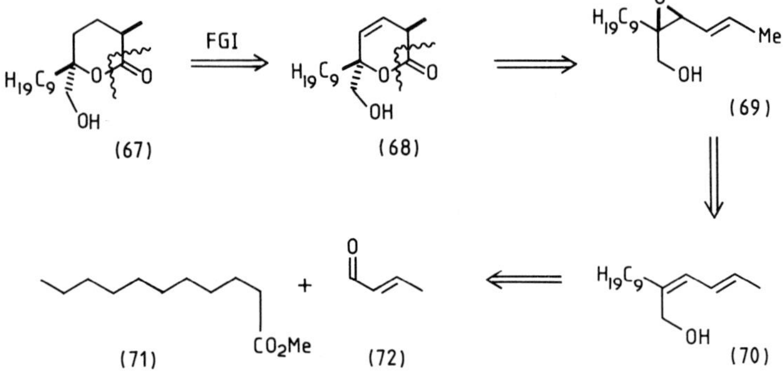

Scheme 14.2

Condensation of (71) and (72) gave the alcohol (73). The conversion of (73) to (70) (Scheme 14.3) required a one-pot procedure in which the elimination and reduction

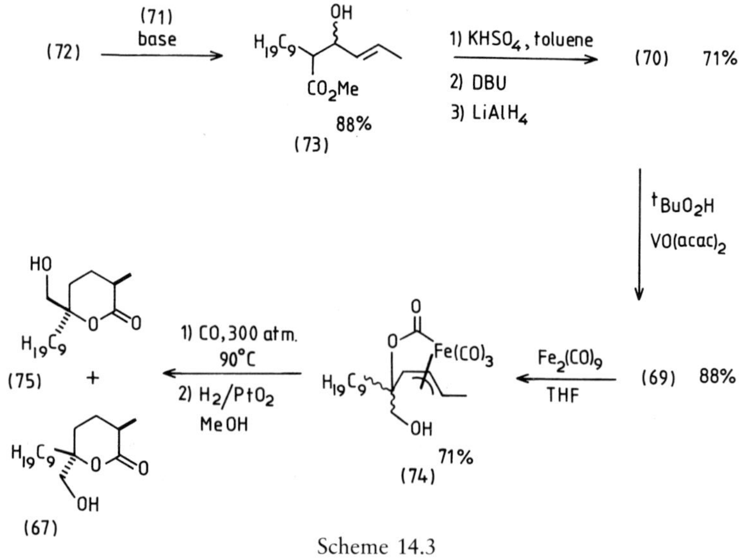

Scheme 14.3

steps were performed without the isolation of intermediates. Epoxidation of (70) with t-butylperoxide and a vanadium catalyst produced the epoxide (69) which reacted with $Fe_2(CO)_9$ in the expected fashion to form a mixture of diastereomeric iron complexes (74). Epimerisation at the carbon adjacent to the π-allyl complex is commonly encountered during the carbonylation step, which requires fairly severe conditions, and this mixture was taken on in this synthesis without separation. Carbonyl insertion gave the required product (68) and its diastereoisomer. Malyngolide (67) was obtained in 74% yield by catalytic hydrogenation of (68). The diastereoisomers ($\pm$)-(67) and ($\pm$)-(75) can be equilibrated by epimerisation adjacent to the lactone carbonyl group to give a 9:4 mixture with LDA at $-78\,°C$ followed by an acid work-up.[31]

14.9 Semmelhack's organochromium syntheses of frenolicin and deoxyfrenolicin

Frenolicin (76)[32] and its chemical reduction product deoxyfrenolicin (77) both show antibiotic and antifungal activity.[32,33] Since (76) can be obtained from (77) by direct epoxidation,[34] a route to (77) was the main synthetic requirement. As in the previous example, there are two important chiral centres in the target molecule, separated in this case by an ether linkage. The retrosynthetic analysis can be approached once again by considering carbon monoxide as a source of the carboxylic acid carbonyl group. This requires an insertion into a metal–carbon σ bond, and hence the intermediate (78). Since this intermediate could be produced by nucleophilic addition to a metal π complex such as (79), a synthesis based on the key alkene intermediate (80) looked attractive.

Two routes to (80) have been examined, both by Semmelhack's group, and, although they both employ organochromium intermediates, they could scarcely be more different. The first approach used the chromium carbene cyclisation reactions of Dötz and Wulff (see Chapter 12) to construct the quinone ring. The second employed two reactions of chromium arene complexes, their efficient *ortho* lithiation, and their regiocontrolled reactions with nucleophiles (see Chapter 7). Here, then, we have an interesting opportunity to compare two related syntheses, both based on organometallic methods.

The chromium carbene route

When considered in its hydroquinone oxidation state (81), the ring system of (80) becomes accessible from an arylcarbene complex, an alkyne and carbon monoxide, which provides the phenolic C–OH group in the carbonyl insertion step. Efficient addition to the alkyne requires regiocontrol, a problem which Semmelhack solved by linking the two components with an ethanediol strap.[35]

The complex (82) was prepared from 2-bromoanisole (83), which was lithiated to effect addition[36] to $Cr(CO)_6$ in the usual way. Reaction with the alcohol (84) produced (82) which cyclised to afford (85) after removal of $Cr(CO)_3$ and oxidation with DDQ. The $HOCH_2CH_2O-$ group was removed with dilute acid to give a ketone (86) arising from an internal redox reaction. The resulting hydroquinone was converted to (80, R = Me) by reduction of the ketone, and oxidation of the hydroquinone ring with DDQ. The final organometallic step employed a modification of the Wacker process (see Chapters 4 and 9) in which intramolecular nucleophilic attack was followed by carbonyl insertion. This was achieved by performing the reaction under a carbon monoxide atmosphere to trap the Wacker intermediate before the usual β-elimination step. Although this reaction gave a 3:1 mixture of *cis* and *trans* stereoisomers, these were equilibrated in the next step during cleavage of the methyl ether to give a single product with the required stereochemistry. Finally, saponification of the ester afforded deoxyfrenolicin (77).

The chromium arene approach

A second disconnection[37] for the naphthoquinone ring system suggested the epoxide (87) and hence to alkene (88). Attachment of the two side-chains with the correct

regiocontrol can be approached using chromium arene complexes. To introduce the acyl chain by the nucleophilic addition of an acyl anion equivalent to an anisole complex, an attractive possibility because such additions occur *meta* to the methoxy group, requires an additional directing group to force reaction at the more hindered of the *meta* positions. The silyl substituted complex (89) is an appropriate choice for this purpose. The allyl chain offered a convenient electrophilic precursor, the allylbromide (91), which could react with nucleophiles obtained by *ortho* lithiation of the arene ligand of a $Cr(CO)_3$ complex. In this approach, the arene ring is required to act both as a nucleophile and as an electrophile, an effect that is readily obtained by complexation to chromium (see Chapter 7).

As in the previous route, 2-bromoanisole (83) provided the first aromatic ring. After silylation,[38] complexation with Cr(CO)$_6$ in dioxane produced (92) in 94% yield. Lithiation occurred in the correct position *ortho* to the OMe group, but (90) could not be induced to react with (91). The problem was solved by formation of an arylcopper reagent by reaction with CuI. In this way, a most satisfactory alkylation was achieved, providing an efficient synthesis of the intermediate (89). Addition of the nucleophile (93), prepared from 5-hexenenitrile and LDA, followed by protodesilylation with dilute HCl, gave the trisubstituted aromatic (94) which was converted to (88) in a one-pot procedure by Selikson and Watt's method.[39] Epoxidation of the more substituted alkene gave (87), which was cyclised to (95) via silylenol ether intermediates. This mixture of stereoisomers was further cyclised and then carbonylated in a similar fashion to the previous example to give (96) as an 81:19 mixture of isomers. Aromatisation and oxidation of the resulting naphthol gave (97), which was converted to the required *trans* isomer during ether cleavage and saponification, as described above.

14.10 Polyquinanes via the Pauson–Khand reaction: Magnus' coriolin synthesis

In recent years there has been a growing interest in the development of methods for the control of stereochemistry around five-membered rings. Following the description of an intramolecular version of the Pauson–Khand reaction by Croudace and Schore[40] in 1981 (see Chapter 9), there have been several applications of this process for the synthesis of polyquinanes, a class of natural products with highly functionalised fused five-membered rings that offer a challenging proving-ground for new methods of stereocontrol.

Coriolin (98),[41] an antitumour sesquiterpene, provides a good example[42] of the use of the Pauson–Khand reaction in this type of synthesis, combining both conventional and organometallic methods to form cyclopentenones. Simplification to the tricyclic intermediate (99) suggested the construction of the enone by an aldol condensation. Introduction of the $-CH_2COCH_2-$ fragment by using allylbromide as a masked ketone (see Chapters 4, 8 and 13) would make (100) a suitable precursor. Formation of this enone by use of the Pauson–Khand reaction required the propargyl derivative (101) which should be conveniently available by acetylide addition to the aldehyde (102).

In fact, the silyl ether of (101) was prepared by addition of lithio(trimethylsilyl)ethyne to (102) (a reaction that had already been performed in a series of preliminary experiments), followed by desilylation and methylation. Reaction with $Co_2(CO)_8$ and carbon monoxide in heptane at 110 °C in a sealed tube, gave the enone (103) (50 % yield) which was separated from a small amount of a minor diastereoisomer (15 % yield) by chromatography. Catalytic reduction, and alkylation with allyl bromide gave (104) which was converted to the methyl ketone by a Wacker oxidation (Chapter 4). Cyclisation and isomerisation of the resulting double bond gave (105) which was converted to (99),[42] Trost's intermediate[43] for coriolin.

The stabilisation of propargyl cations by $Co_2(CO)_6$ complexes (see Chapter 5) offers a powerful alternative for the construction of precursors for the Pauson–Khand reaction. A retrosynthetic analysis for the bicyclic enone (106), based on this approach, arose from a collaboration between Smit's Moscow team and Caple at Minnesota, and required the intermediate (107) in which the OMe group could be introduced via a propargyl cation generated by electrophile addition to an enyne complex.[44] Since the $Co_2(CO)_6$ group will be required for the Pauson–Khand step, it is an attractive ploy also to employ the same complex in the preceding steps.[45]

Reaction of (108) with *E*-crotonyl tetrafluoroborate formed the cationic intermediate (109) which combined with methanol to give the adduct (110). This complex did not cyclise upon heating, but conversion to (111) and (112) afforded satisfactory substrates for the Pauson–Khand step. Treatment of (111) at 60 °C on a silica support[46] completed the construction of (106) in a synthesis[45] in which all five new bonds were formed using organocobalt chemistry.

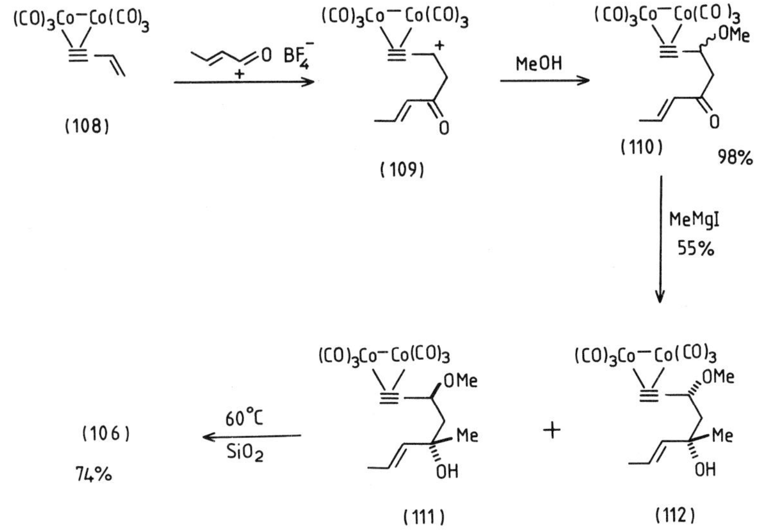

(108)   (109)   (110)   98%

MeMgI
55%

(106)   60°C   (111)   +   (112)
74%   SiO₂

14.11 Nicholas' organocobalt synthesis of cyclocolorenone

Cyclocolorenone (113), a guiane sesquiterpene, is another target molecule that contains a cyclopentenone ring. In his synthesis[47] of cyclocolorenone, Nicholas chose to form the cyclopentenone by a conventional aldol approach in which a cobalt stabilised propargyl cation could be employed in a C–C bond formation. This synthetic methodology has

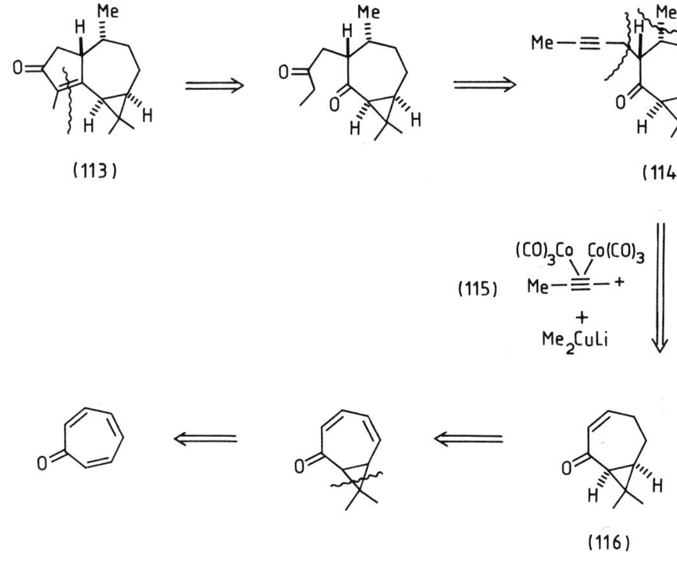

(113)   (114)

(115)   Me—≡—⁺
         +
         Me₂CuLi

(116)

been extensively developed by the Nicholas group (see Chapter 5). Construction of the key intermediate (114) could be based on a conjugate addition/enolate trapping sequence (see Chapter 8) in which the electrophile was the cation (115) prepared from 1-hydroxy-2-butyne. Formation of the cycloheptenone intermediate (116) from tropone required the use of an $Fe(CO)_3$ complex as a protecting group (see Chapter 13).

(Tropone)$Fe(CO)_3$ reacted with 2-diazopropane to give the cycloadduct (117). It is interesting to compare the regiochemistry of this reaction with the case discussed on p. 435 where an acyl directing group was present to reverse regiocontrol. Thermal displacement of N_2 was followed by an application of an unusual procedure[48] for decomplexation which results in a concomitant partial reduction of the diene to an alkene. The β,γ-enone (118), was produced in this way in about 40% crude yield from tropone. The intermediate (118) was isomerised to (116) in base. Conjugate addition using dimethylcuprate and trapping of the resulting enolate with trimethylchlorosilane gave (119) which was alkylated by reaction with (115). Decomplexation by oxidation with ceric ammonium nitrate gave (114) which was converted to the ethyl ketone by oxymercuration, and cyclised in an aldol reaction to complete the synthesis of (113).

14.12 Applications of organopalladium and organocobalt complexes in routes towards esperamicins by Magnus and by Schreiber

Since the reports of esperamicins[49] and calichemicin γ_1[50] in 1987, the unusual enediyne unit within a ten-membered ring has caught the eye of many synthetic chemists. Speculation that this strained sub-unit might be involved in the biological action of compounds of this type, which are potent antitumour agents, further encouraged attempts to synthesise the natural products and analogues needed for mechanistic studies. A particular problem in work towards this goal is the closure of the 10-membered ring, which also carries two exocyclic alkenes, as seen in the structure of esperamicin A_2 (120). The bending of alkynes by complexation by $Co_2(CO)_6$ (discussed in Chapter 9) raises the prospect that this might ease the ring closure step by bringing the reaction centres closer together. This possibility prompted the Magnus group to examine access to the bicyclic enediyne (121) by means of an alkylation at a cobalt stabilised propargyl cation (see Chapter 5). The intermediate (122) has the required functionality with an enol ether present to provide the nucleophilic component in the reaction.

Palladium catalysed coupling reactions (Chapter 10) were used twice in the construction of (122). An alkyne, attached to the six-membered ring by a simple 1,2-addition to the enone (123), was first coupled to 1,2-dichloroethene by reaction in the presence of Pd(PPh$_3$)$_4$ and CuI/^{n}BuNH$_2$. This reaction, like examples seen in Chapter 10, was performed without protection of the alcohol, and gave (124) in 80% yield, clearly indicating good control of competing double coupling. A protecting group was now attached, and a second coupling under the same conditions, but employing the methyl ether of propargyl alcohol, afforded (125). The MEM ether was replaced by a silyl ether at this stage, and the product was complexed by reaction with Co$_2$(CO)$_8$ in heptane to afford (122) in 90% yield. Ring closure was effected by Lewis acid catalysis to give a complex which was converted into (121) by oxidation of the metals with iodine. The product (121) has unsaturation in the correct place for the formation of the exocyclic alkene at the bridging position. The reaction sequence has now been modified to leave an alcohol group at the required position at the site of ring closure.[51]

Similar organopalladium methods have been used by Schreiber and Kiessling in work which correctly positions the bridgehead alkene by use of a cycloaddition approach for the simultaneous formation of the six- and 10-membered rings.[52] This synthesis starts from the 1,2-dichloroethene attaching first (127) and then (128) in palladium catalysed steps. The product (129) was deprotected and alkylated with a lithiated reagent (130) derived from the corresponding bromodiene. The expected reactions at the aldehyde gave the alcohol (131) which was desilylated with t-butylammonium fluoride. A further palladium coupling reaction using the vinyl iodide (132) followed. Through the introduction of first (130) and then (132), the product from this last coupling step now contains the diene and alkene fragments needed for the cycloaddition step. Silylation of (133), followed by heating at 80 °C in the presence of a radical inhibitor, successfully completed the bicyclic core of the target (120).

14.13 Davies' iron acyl enolate route to homochiral pyrrolizidine alkaloids

Enantioselective access to 1-hydroxypyrrolizidine alkaloids such as (134) can be based on a chiral enolate strategy in which double stereodifferentiation is used to reinforce stereocontrol in cases where the chiral centres in the prolinal derivative (135) and in the chiral auxiliary X*, employed in the enolate portion, work in concert in the aldol step. In the disconnection shown in Scheme 14.4, the enolate portion contributes two carbons, and the amide bond is formed during detachment of X*. Relative stereochemistry between the two chiral centres in (134) is established in the aldol step.

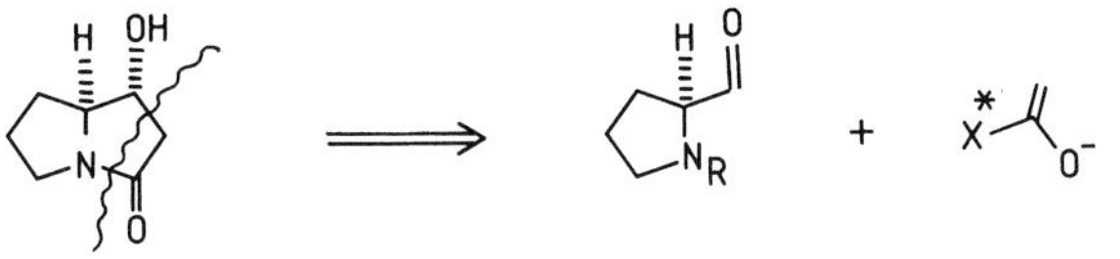

Scheme 14.4

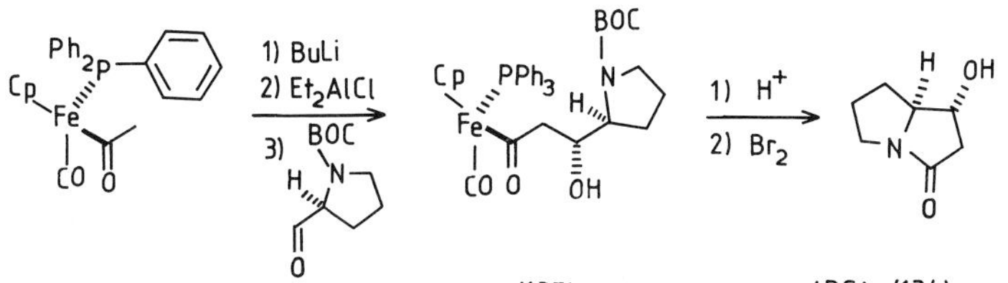

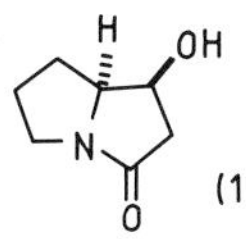

By means of the aluminium enolate modification discussed in Chapter 9, Beckett and Davies have successfully employed the iron acyl enolate approach to the stereocontrolled construction of (RS)-(134).[53] Stereocontrolled alkylation of a protected prolinal derivative produced a single diastereoisomer of (137) in 75% yield, by addition of the enolate derived from the homochiral iron acyl complex (S)-(136). The product was shown to have a diastereoisomer ratio greater than 300:1. Deprotection of the nitrogen and oxidation with bromine produced the expected amide (RS)-(134) in 57% yield. When the (R) isomer of (136) was used, the alkylation was less well controlled but still proceeded with 35:1 selectivity. The product (138) was obtained in a similar way and was taken to diastereomeric purity by a single crystallisation.

The considerable stereocontrol available in the aldol routes to both diastereoisomers emphasises the dominance of the iron acyl enolate in the control of the reaction, overcoming the disadvantage anticipated with double stereodifferentiation in situations where the controlling centres are mismatched. Aldol addition to (135, R = BOC) naturally favours (134) by a factor of 4:1,[54] but in the formation of (138), this effect is overcome by the selectivity imposed by the iron acyl enolate approach.

14.14 Design of transition metal mediated organic synthesis

The examples discussed in this chapter have been selected to illustrate the high degree of selectivity and stereocontrol that can be expected in transition metal mediated reactions. Catalytic systems, particularly asymmetric catalysts, have great promise in the design of efficient synthetic routes. Since ligands on the metal can often be varied, subtle changes in the steric and electronic effects at the metal centre can be manipulated in search of high chemical and optical yields. When considered as control centres for enantioselective synthesis, it is clear that catalysts can only control chirality at centres formed in a single step. Subsequently, conventional diastereoselective processes or further, perhaps different, metal catalysed reactions must be employed as the synthesis progresses.

Stoichiometric complexes can serve as control centres at many stages of the execution of a synthetic plan, since the same metal centre can be retained in the molecule from one step to the next. Indeed, if they are in general to rival catalytic methods, it is essential that stoichiometric complexes are used in this way.

Uses of transition metals in the synthesis of heterocycles

Transition metal complexes have been used to prepare a great variety of heterocyclic ring systems but scarcely any of these reactions have been developed to the point where they can be regarded as routine, general syntheses. A detailed review by Davidson and Preston[1] provides a broad coverage of literature up to 1979. Our intention in this chapter is not to duplicate this effort, but rather to draw together examples of reactions of types discussed elsewhere in this book to bring attention to their use in heterocyclic synthesis, and to discuss the way in which such applications can be planned. Consequently, the emphasis here will be on reaction type, not synthetic target; readers seeking to discover whether the synthesis of a particular heterocyclic ring system has been achieved using transition metals will find the organisation of material in the Davidson and Preston review of more assistance.

One area of heterocyclic chemistry in particular has become popular as a proving ground for organometallic processes. This topic, β-lactam synthesis, is of such importance that many types of organometallic reactions have been employed for this purpose, providing an excellent opportunity for the comparison of different approaches. In common with other examples of the applications of these processes covered in subsequent sections, the organometallic routes to β-lactams cannot yet properly be considered to be general methods. The use, however, of well precedented and well understood types of organometallic reactions for this purpose should encourage the further development of these methods, since the fact that they have been used reliably in other situations provides a reasonable basis to assess whether a heterocyclic application has good prospects for success.

15.1 Organometallic routes to β-lactams

Organometallic reactions can be used in many different ways[2] to construct the β-lactam ring system and any of the four bonds in the ring may be formed using transition metal reagents. This provides considerable flexibility in synthesis design, and complements conventional synthetic methods. Scheme 15.1 shows a variety of disconnections that are possible in the retrosynthetic analysis of these target molecules. In many cases, however, the synthetic reactions require specific substitution patterns, and so are useful only in

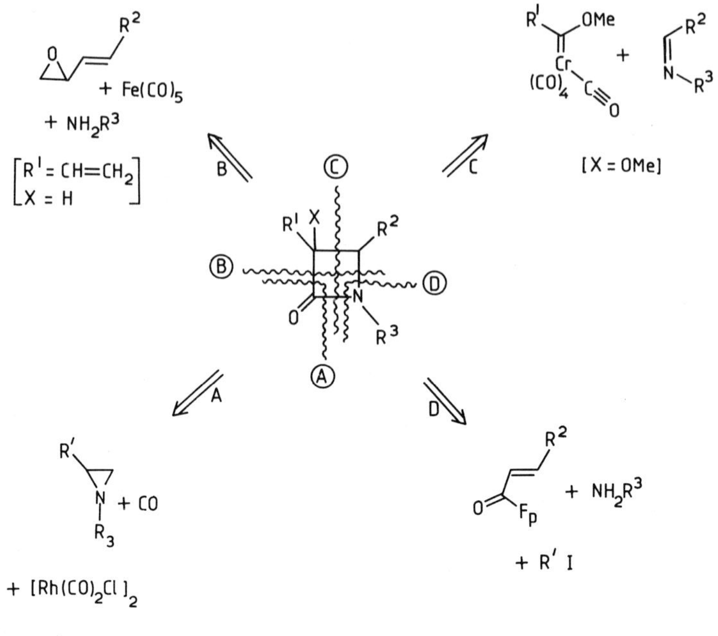

R² ... + Fe(CO)₅

+ NH₂R³

$\begin{bmatrix} R^1 = CH=CH_2 \\ X = H \end{bmatrix}$

R¹ ... OMe / Cr / (CO)₄ C≡O   +   R² ... N–R³

[X = OMe]

B   Ⓒ   C

R¹ X ... R²

Ⓑ ... Ⓓ

O ... N

Ⓐ

R₃

A   Ⓐ   D

R' ... N + CO / R₃

R² ... Fp + NH₂R³

+ R' I

+ [Rh(CO)₂Cl]₂

[R² = H]   Scheme 15.1

particular cases. Synthetic routes developed by Alper[3] (A), Ley[4] (B), Hegedus[5] (C), and Davies[6] and Liebeskind[7] (D) and their co-workers, summarised in Scheme 15.1, are typical of the different approaches to the problem.

Carbonyl insertion (see Chapter 9) provides a common organometallic method to introduce a carbonyl group into a molecule. It is natural that several groups have employed this procedure for the formation of β-lactams. The reaction requires an intermediate with a σ bond, which in reactions leading to the formation of an amide may be either a metal–carbon or metal–nitrogen bond. The former suggests a synthesis that begins by nucleophilic addition to a metal η^2-alkene complex to produce an intermediate of type (1). (A similar argument led to the strategy behind the concluding stages of the syntheses of frenolicins discussed in Chapter 14.) The alternative of an insertion into a carbon–nitrogen bond leads to a requirement for a metalocycle (2), though here insertion into either C–N σ bond of the aziridine could be expected to afford the required product. Both types of mechanism have been proposed in different examples. These routes, portrayed in Scheme 15.2, together with several related approaches, provide realistic methods for syntheses based on disconnection (A). Indeed, several of the other processes shown in Scheme 15.1 ultimately rely on a carbonyl insertion step, but are perhaps more easily discussed later when considered in terms of the overall transformation effected.

Rosenblum's extensive development of uses of Fe(CO)₂Cp⁺ alkene complexes as electrophiles (see Chapter 5) led to his exploration[8] of the nitrogen addition approach indicated in Scheme 15.2. The heterocycle is formed by oxidation of an intermediate

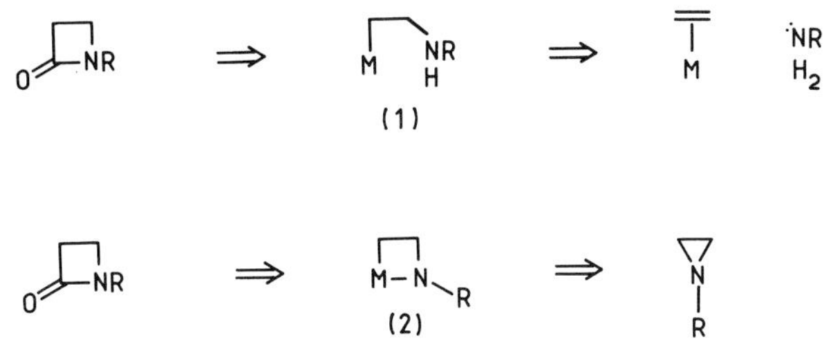

Scheme 15.2

metal complex following carbonyl insertion. Initially chlorine was used but milder variants[8,9] using lead or silver oxides are preferable. Examples are shown in Scheme 15.3.

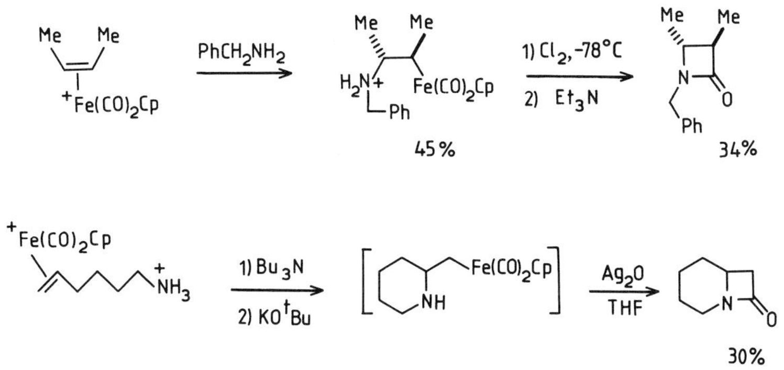

Scheme 15.3

The method has been applied[10] in a synthesis of 3-carbomethoxycarbapenem. Transfer of the metal from isobutene led to the formation of the required cationic alkene complex which reacted with ammonia to produce the cyclic product (3) in 96% yield. Reduction with sodium borohydride gave a 2:1 mixture of stereoisomers. The major product was converted into the metallocycle (4) in an insertion reaction driven by coordination of the nitrogen to the metal. Finally oxidation served to form the amide during removal of the metal from the complex, to afford (5) in rather disappointing yield.

Metal halogen exchange has also been used (Scheme 15.4) to prepare the way for a carbonyl insertion step. 2-Bromo-3-aminopropene derivatives have been carbonylated by use of a catalytic quantity of palladium acetate to afford products such as (6), presumably by reductive elimination of the metal from a metallocyclic metal acyl complex.[11] This approach has been developed further at Upjohn in search of access to convenient precursors to a new class of monocyclic β-lactam antibiotics.[12]

The opening of aziridines by [Rh(CO)$_2$Cl]$_2$ affords β-lactam products.[3] The process (Scheme 15.5) begins by oxidative addition to the nitrogen–carbon bond. About 20 atmospheres of carbon monoxide pressure is used to effect the insertion. A similar transformation has been described using nickel tetracarbonyl.[13]

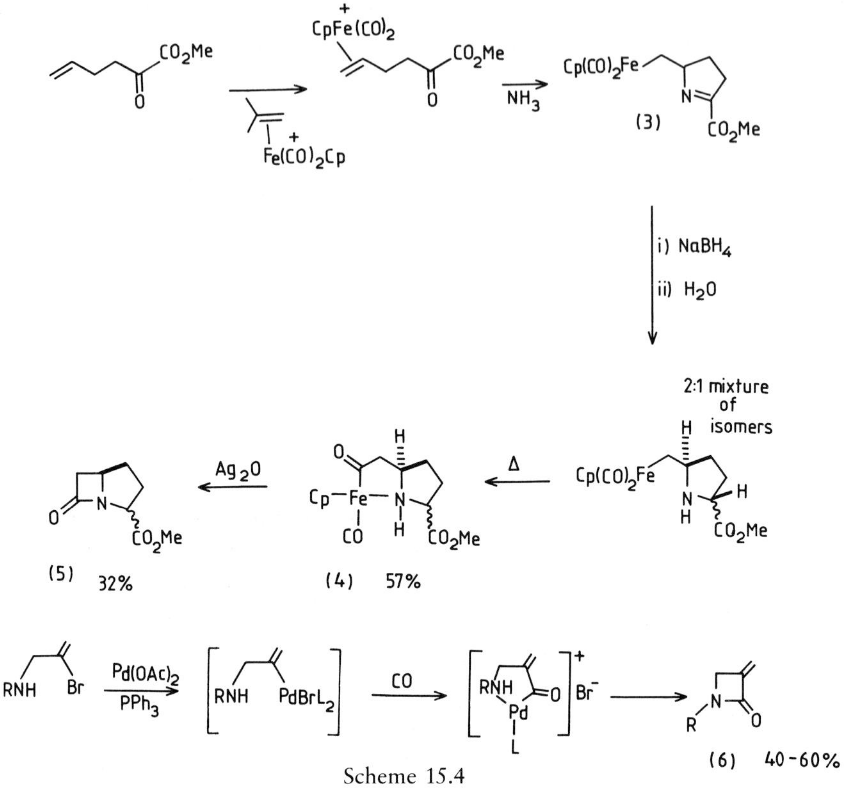

Scheme 15.4

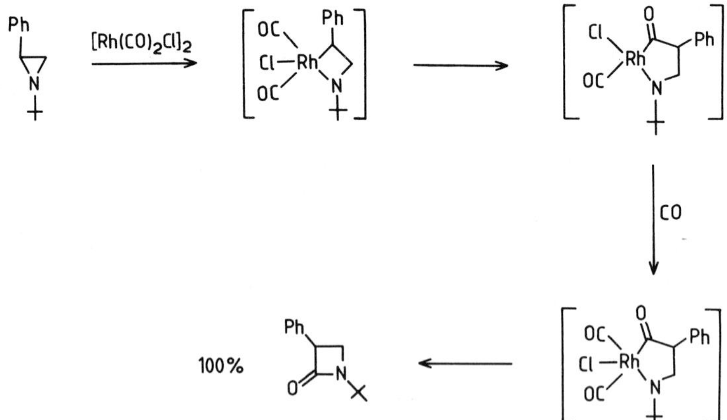

Scheme 15.5

Azirines can produce a variety of heterocyclic products when opened by transition metal reagents. In one case, however, through the use of Pd(PPh$_3$)$_4$, carbonyl insertion was effected at atmospheric pressure to produce the bicyclic β-lactam (7).[14] In this case carbon monoxide insertion into a metal–nitrogen bond was proposed. This reaction is remarkably selective. Use of a different catalyst, (dba)$_2$Pd, resulted in the exclusive formation of a vinyl isocyanate, an observation that suggested the involvement of a common intermediate (8) in both reactions.

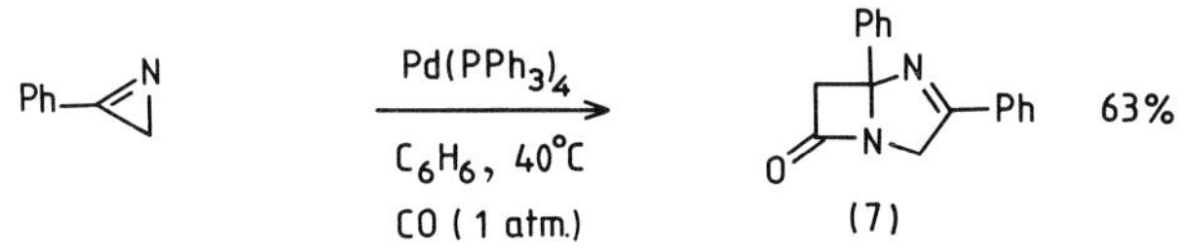

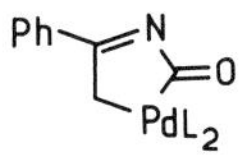

Another insertion[15] involving a metal–nitrogen bond led to the formation of (10, R = Me) in a reaction that proceeded by ring opening followed by decarboxylation to produce a metal η^3-allyl complex. For (9, R = H), the alkenyl substituted β-lactam (11) is produced in very poor yield.[16] While clearly not of use synthetically, the process follows an interesting mechanism and shows parallels with far more efficient reactions of allyl intermediates described below.

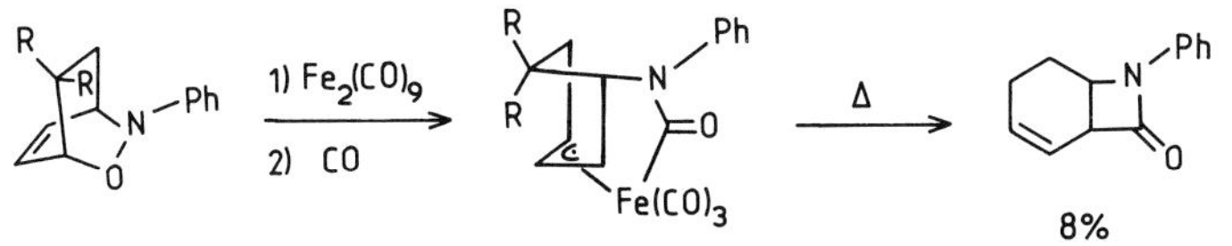

Disconnection (B) requires the introduction of both nitrogen and carbonyl centres to a common precursor. The Rosenblum approach discussed above can be regarded as a step-wise example of such a process. The development of the chemistry of π-allyltricarbonyliron lactone complexes by Ley's group has provided an efficient means to elaborate vinyl epoxides to vinyl substituted β-lactams (Scheme 15.6) in which the

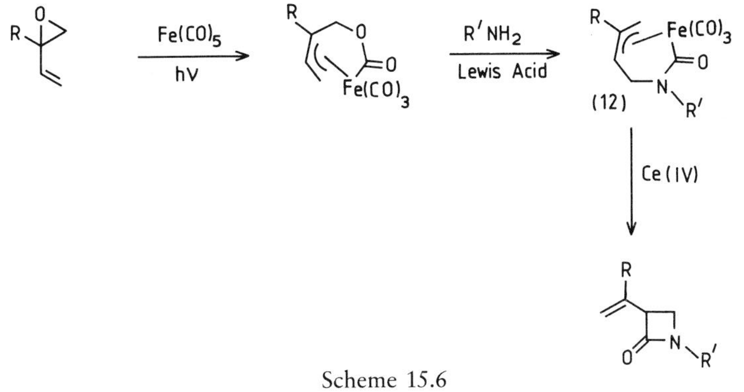

Scheme 15.6

amine and carbonyl components are introduced in the same reaction.[17] A key step in this process is the S_N2' addition of the amine to form the ferrilactam intermediate (12), a reaction that requires the presence of a Lewis acid. The best results were obtained using zinc chloride/TMEDA. Oxidative removal of the metal reliably produced the β-lactam in preference to the δ-lactam alternative in most cases.

The presence of the vinyl substituent in the product, a legacy of the η^3-allyl intermediate, proved convenient in the use[4] of this approach in a synthesis of the antibiotic $(+)$-thienamycin (20). Methylenation of the enone (13) by dimethyl-

sulphonium methylide gave the racemic epoxide (14) which was converted to a racemic mixture of π-allyltricarbonyliron lactone complexes by the action of $Fe_2(CO)_9$ under reflux in THF (84%) or by photolysis with $Fe(CO)_5$ (90%). Addition of $(-)$-1-phenylethylamine provided an entry to optically active complexes by separation of diastereomeric ferrilactam complexes (15) and (16) which were formed in roughly equal amounts. Oxidation of (15) gave the β-lactam (17) in 87% yield, together with a small quantity of the δ-isomer. The *cis* stereochemistry of (17) is consistent with that shown for (15)/(16), which arises from addition of the amine to the allyl ligand with *trans* stereochemistry relative to the metal and rotation to align the nitrogen for addition to the metal-bound carbonyl group. Ozonolysis and equilibration gave the more stable isomer (18), which was reduced and deprotected to give the thienamycin intermediate[18] (19), which was taken on to (20) in several steps.

The disconnection (C) can be regarded as dividing the ring in two in the opposite sense, suggesting the cycloaddition of a two-carbon fragment to an imine. An organometallic approach that uses imines as the source of the C–N unit employs a Fischer carbene complex (see Chapter 12) in the place of the other component. Both chromium and molybdenum complexes have been used. Because of the requirement that the carbene complex be stabilised by an electron donating group, the method is only appropriate for the synthesis of alkoxy substituted β-lactams.

These methods have initially been developed through the work of the Hegedus group using photolysis. A variety of imines undergo stereocontrolled addition to give a single diastereoisomer of products such as (22).[19]

Thiazolines proved to be reactive substrates, giving rise to stereocontrolled products containing the penam ring system shown in Scheme 15.4. The case of (23), an optically active starting material, is noteworthy since a single resolved product (24) was obtained from photolysis with the carbene complex (21). The relative stereochemistry of the product was proved by X-ray crystallography, but the optical purity was not confirmed.[19]

An attempt to extend the scope of these reactions to include oxapenams and oxacephams by photolysis of chromium complex (21) with oxazines and oxazolines was unsuccessful, resulting only in the slow decomposition of the organometallic reagent. Use of the molybdenum complex (25), however, proved more satisfactory, affording adducts such as (26) in moderate yield. When the same reaction was attempted using oxazoline substrates, the process was less selective giving much lower yields of the corresponding products (27) together with similar amounts of (28) which might arise from the presence of free methoxymethyl ketene in the reaction mixture.[5]

The same method has been applied to reactions of aminocarbene complexes to form β-lactams which bear an amino substituent adjacent to the carbonyl group of the lactam. The required carbene complex (29) was prepared by reduction of chromium hexacarbonyl with sodium naphthalenide and reaction of the resulting dianion with a Vilsmeier salt. Reactions of (29) with imines and oxazolines proceeded in a similar fashion to those discussed above.[20]

Reaction[21] between the carbene anion (30) and the imine (31) produced azetidinylidene complex (32), which might arise by a formal [2+2] cycloaddition with a chromium vinylidine[22] intermediate. Cationic iron vinylidene complexes[23] have been used[24] in much the same way to afford rather unstable products such as (33). Oxidation of (32) or (33) removed the metal producing a β-lactam (34).

The iron complex contains a chiral centre at the metal (see Chapter 9) which raises the interesting possibility that an asymmetric induction might occur during the cycloaddition. Use of the dimethyl substituted complex (35) afforded stable adducts, for example (36), which were amenable to study. Rather disappointingly, they proved[24] to be mixtures of diastereoisomers, formed in similar amounts (4:3 – 8:5 mixtures). Conversion into β-lactams, in this case, was, at first, rather inefficient, but subsequent procedures using phosphite complexes have improved the yields into the range 51–82%.[25]

Rhodium catalysed carbene insertion (see Chapter 12) into a C–H bond was used by chemists at SmithKline Beecham Pharmaceuticals to complete the β-lactam ring in (38) in a stereocontrolled route[26] to optically active 1-methyl carbapenems. Resolution of 3-amino-2-methylpropan-1-ol and elaboration[27] to the α-diazoketone (37), followed by cyclisation with $Rh_2(OAc)_4$ gave the acyl product (38). Reduction with K-selectride, protection of the alcohol and cleavage of the tetrahydro-1,3-oxazine ring produced the intermediate (39) which was converted to the bicyclic carbapenem (40) by well established methods.

The final disconnection (D) in Scheme 15.1 requires the addition of a nitrogen containing moiety to a three carbon unit, coupled with a lactonisation step. Iron acyl complexes provide the means to effect the required lactonisation. An example of this process from the Davies group has already been encountered in Chapter 9. Conjugate addition of an amine to an α,β-unsaturated iron acyl complex, which serves as the three carbon electrophile, has also been examined by Liebeskind and Welker.[7] Both nucleophile addition and enolate trapping reactions proved to be well stereocontrolled, giving products such as (41) in 15:1 to 30:1 diastereomeric excess. Oxidation afforded the β-lactam as expected. The reaction has been examined in some detail and gave good yields of β-lactams in several cases (R = Me, Et, CH_2Ph), although (R = $CH_2CH{=}CH_2$) was not satisfactory.[28] The conjugate addition step was successful even when the site of alkylation was disubstituted. The β-lactam (42) has been prepared in this way.[29]

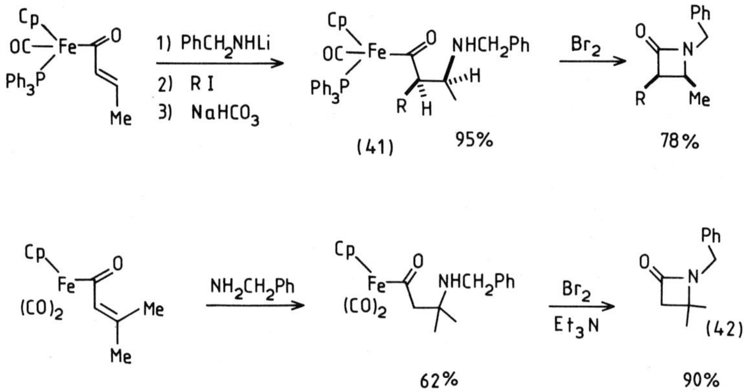

There are several examples where organometallic chemistry has been used to elaborate β-lactam systems. The examples in Scheme 15.7 will give an impression of the

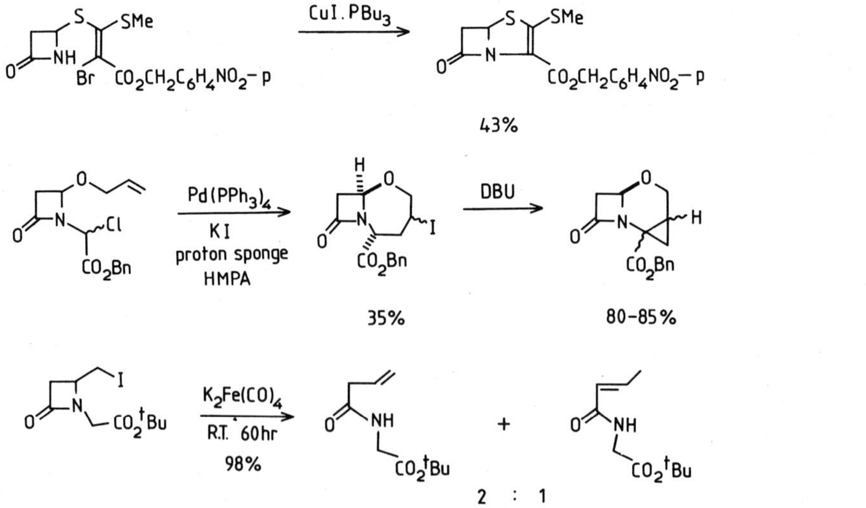

Scheme 15.7

types of reactions that can be performed in the presence of this sensitive functional group. Both organocopper[16] (see Chapter 8) and organopalladium[30] (see Chapter 10) methods have been used to close a second ring to the β-lactam system in moderate yield. In another case[31] an attempt to convert an iodomethyl side-chain to a dialkyl ketone by use of Collman's reagent (see Chapter 9) was examined, but only ring-opened products were obtained. Thus, while some organometallic species are easily tolerated by the β-lactam ring, the use of highly charged reagents may lead to severe difficulties.

15.2 Nucleophile addition to metal complexes

The use of complexation by transition metals to activate organic ligands for nucleophile addition has been a major topic in this book, discussed in detail in Chapters 5, 6, and 7. An example in which palladium catalysis is used to effect the cyclisation of an allyl substituted aniline has already been encountered (see Chapter 5). This reaction is typical of methods in which the heterocyclic ring is closed by use of the heteroatom as a nucleophile in an addition reaction with an organometallic complex.

Further examples of this type of reaction are shown for (43), (44) and (45).[32] These last two indicate a preference for addition to the butenyl residue so as to form a six-membered ring, and also show the intervention of a dehydrogenation step. *Ortho*-allyl benzylamines such as (46) may be cyclised[32] in the same manner.

(43) $\xrightarrow{\text{PdCl}_2}$

(44) $\longrightarrow$

(45) $\longrightarrow$

(46) $\longrightarrow$

Synthesis by addition of an amino function to an alkene bond may be combined with carbon monoxide insertion into the palladium intermediate.[33] The formation of the ester (47) provides a typical example.

(47)

A more extensive cyclisation using $PdCl_2(MeCN)_2$, was reported[33] by Hegedus *et al.* for the acylated *ortho* allyl aniline (48). Here the σ bound organometallic intermediate arising from formation of the C–N bond undergoes an insertion reaction with the second alkene to produce a further ring.

(48)   38%   54%

In another reaction which involves an amide nitrogen, the butadiene derivatives (49) are cyclised to form pyridones by the action of lithium chloropallidate. The reaction most probably proceeds by the familiar sequence of nucleophilic addition to the coordinated alkene, followed by β-elimination, although questions concerning alkene stereochemistry arising in such a mechanism were not addressed in this report.[34]

(49)   62%

The examples above, in which amides act as the nucleophilic group, emphasize the point that relatively modest nucleophilicity is sufficient to promote a useful reaction. A further instance[35] is the addition of nitriles to the N-allylindole (50). Despite the nucleophilic character of the indole, initial nucleophile addition did not occur from this position, but instead involved a relatively inert nucleophile, the nitrile solvent, to produce an *ortho*-palladated intermediate (52), presumably by addition of the indole to the nitrile adduct (51). Access to heterocyclic systems by use of orthometallated reagents will be discussed in more detail in Section 15.3. In the case of (52), reaction with benzylamine resulted in the formation of (53). Carbonyl insertion produced the ester (54), in a reaction that is similar to that leading to (47), discussed above. In another palladium catalysed cyclisation, a carbon–oxygen bond was formed in the reaction of a ketone with an alkyne to produce furans.[36] Substrates of either type (55) or (56) can be used. The former allows substituents to be placed at C-3 of the furan.

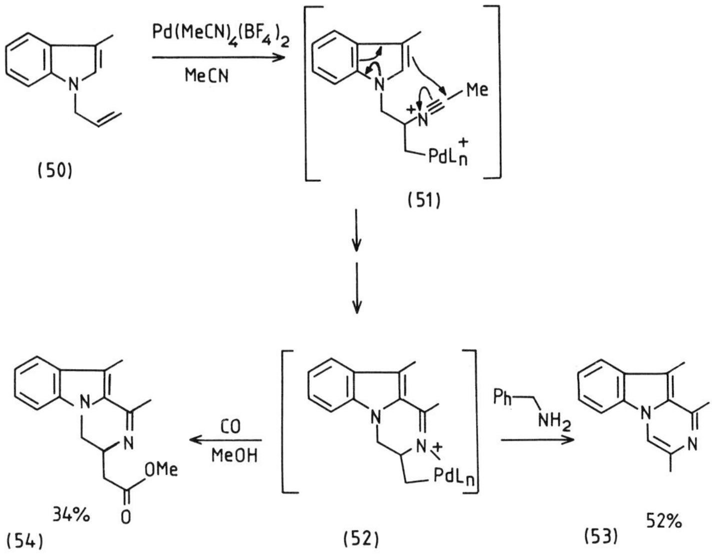

(50) (51)

(54) 34% (52) (53) 52%

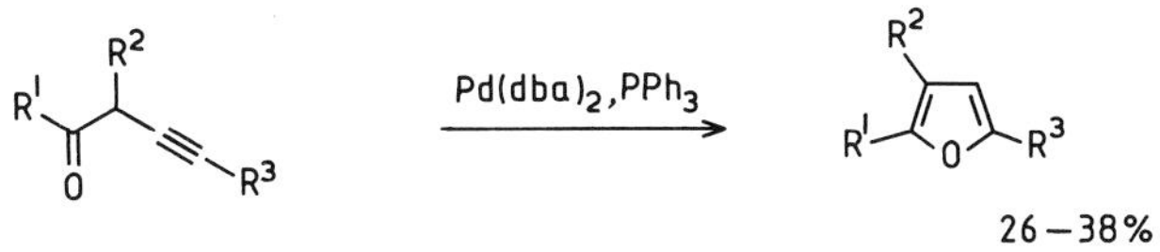

(55) 26–38%

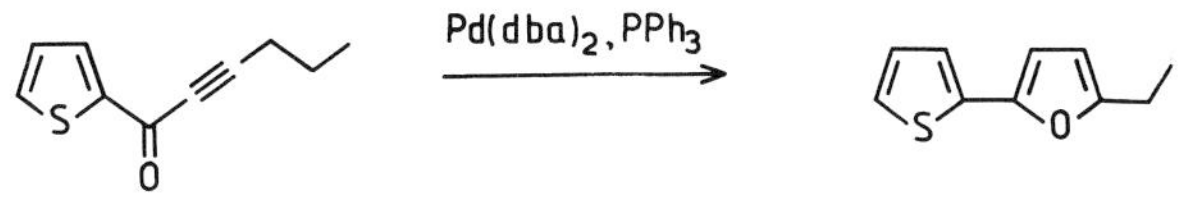

(56) 41%

Trost's work on the synthesis of *iboga* alkaloids has already received some comment in Chapter 14. In his synthesis of ibogamine and catharamine, reactions employing nitrogen addition to palladium allyl and alkene complexes play a major part.[37] In a simple example, the sequence shown led to 6-benzyl-6-azabicyclo[3,2,1]oct-2-ene (57).

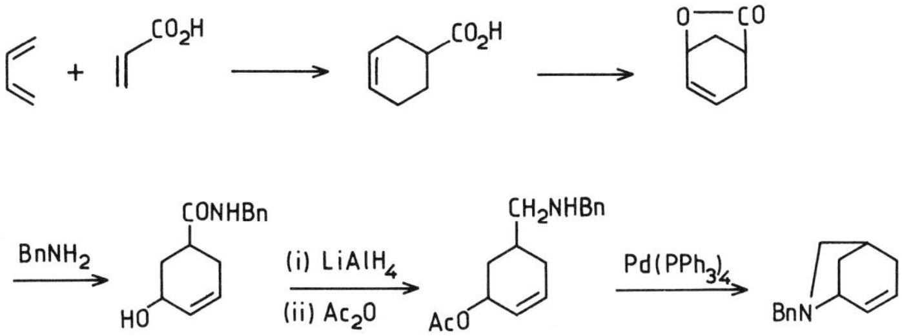

(57)

A somewhat similar sequence led to a synthesis of the hexahydroindole (58). By incorporating tryptamine it was possible to achieve the transformation of (59) into (60), and, by cyclisation, to form desethylibogamine (61).

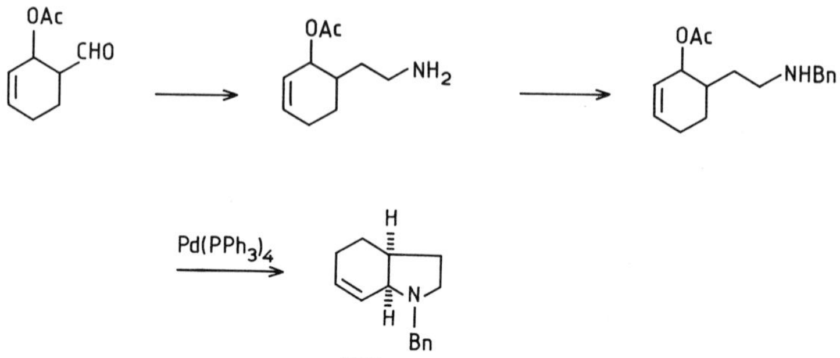

(58)

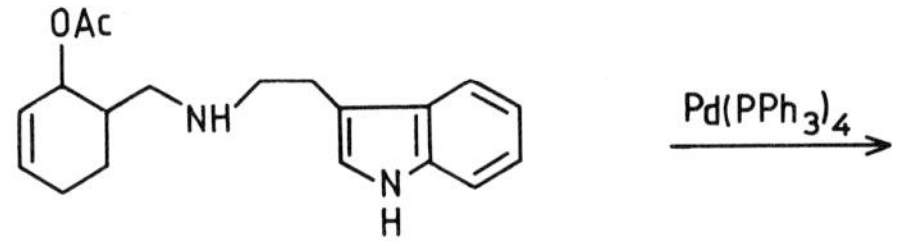

(59)

(60)   (61)

A similar route starting from (62) led to a synthesis of ($\pm$)-catharanine[38] (63).

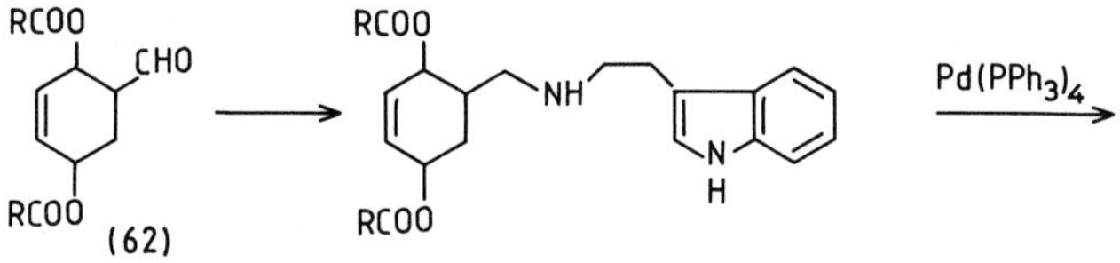

(62)

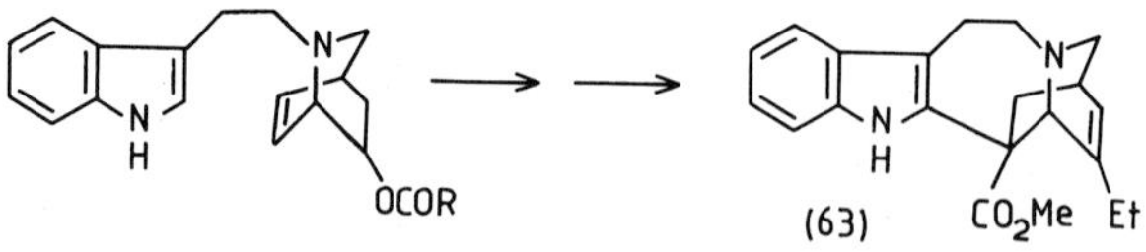

(63)   CO₂Me Et

The synthesis of ibogamine[39] (65) itself by this method, incorporated an additional feature by starting from compound (64), which contains a chiral group R = PhCH(OMe). The final step of cyclisation on the indole ring is to be noted as a further application of organometallic methods. Reaction may depend on catalysed addition to the alkene bond with the indole *o*-position acting as nucleophile, or by an *o*-metallation of the indole and subsequent insertion of the alkene to form a metal–carbon bond, the

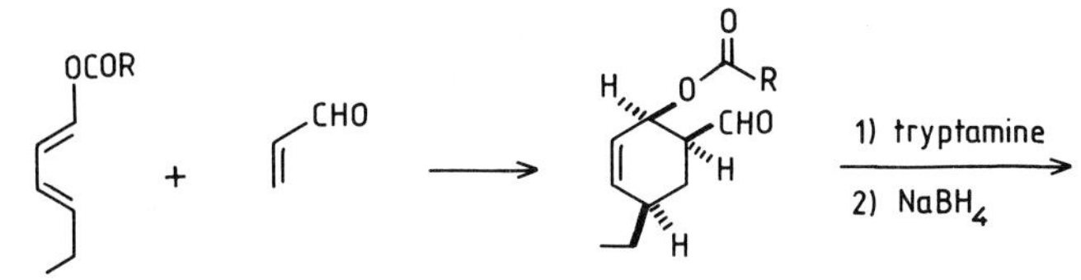

(64)

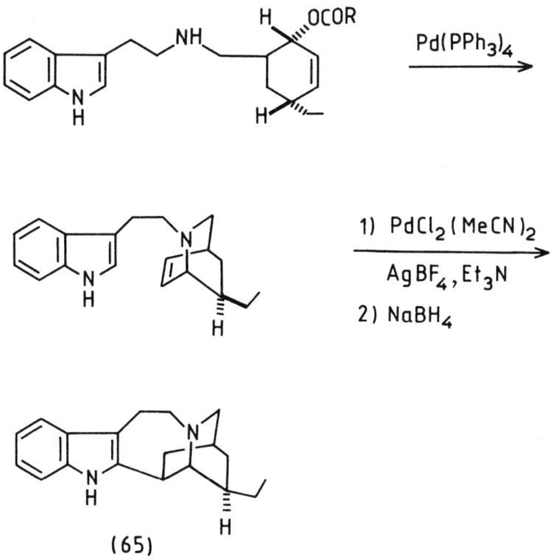

(65)

more probable route since reduction with NaBD$_4$ gave a ^{2}H-ibogamine. The added AgBF$_4$ presumably increases the electrophilic character of a palladium salt adduct. Other examples of cyclisation onto the indole ring are noted above.[38]

Palladium catalysed addition of bifunctional vinyl or aryl mercury reagents to dienes can produce an intermediate palladium allyl complex that undergoes a cyclisation reaction to the second nucleophilic centre. Products such as (66), (67), and the chroman

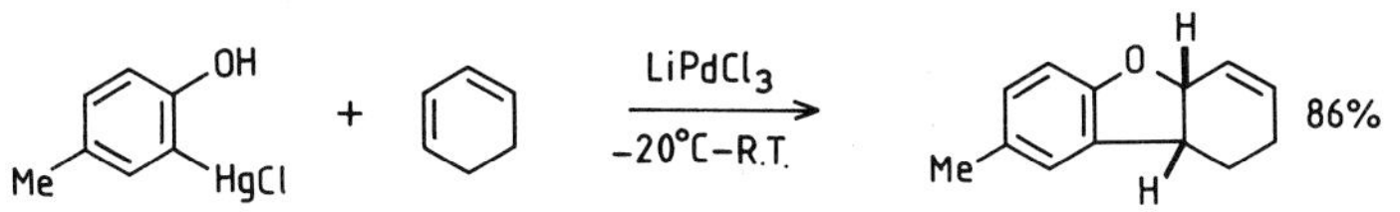

(66) 86%

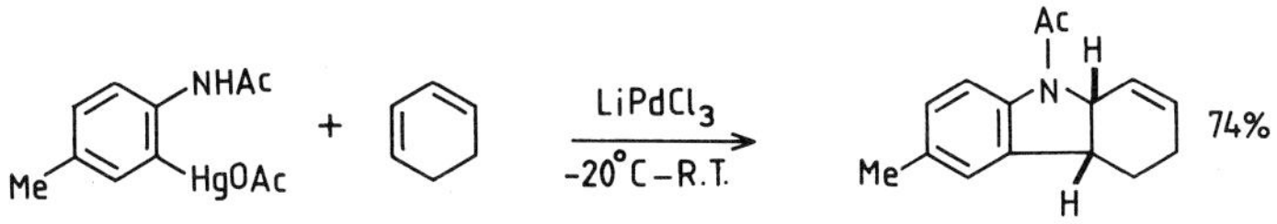

(67) 74%

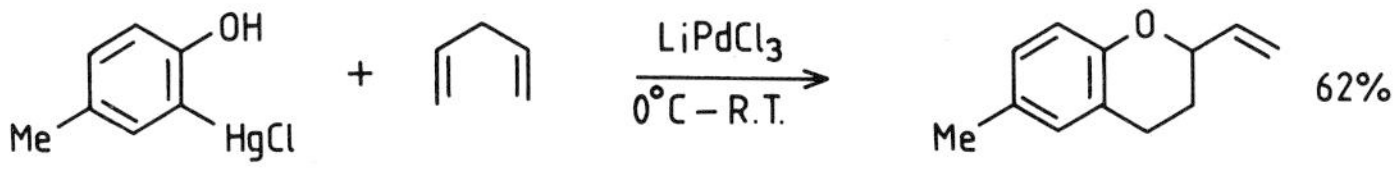

(68) 62%

(68) were obtained in reasonable yield in this way.[40] These cyclisation methods by Larock's group provide an interesting comparison with other organopalladium cyclisation procedures, for example, those of Hayashi, Yamamoto and Ito,[41] and Bäckvall et al.[42]

Another chroman synthesis employed the nucleophilic displacement of fluoride from 3-(2-fluorophenyl)propanols. In Chapter 7 we encountered several examples in which complexation by tricarbonylchromium was used to promote nucleophilic aromatic substitution. In this case, the need for stoichiometric complexes is dispensed with by the use of the catalytic system shown in Scheme 15.8 which is based on dicationic pentaalkylcyclopentadienyl rhodium(III) complexes.[43]

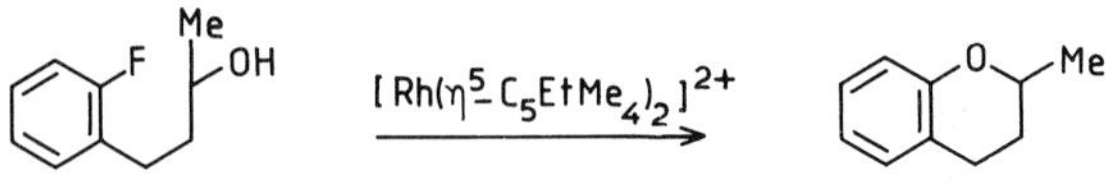

Scheme 15.8

Nucleophile addition to the tricarbonyl chromium complexes of N-protected indoles afforded substituted products after oxidation of the anionic intermediates (Scheme 15.9). Regiocontrol in these reactions depended on the nature of the nucleophile employed. While addition of LiCMe$_2$CN proceeded predominantly (97:3) at C-4, use of lithiated dithiane as the nucleophile substantially reversed this selectivity giving a 14:86 product ratio in favour of C-7 attack. When used with an indoline complex, exclusive C-4 addition (61% yield) was obtained. The product was dehydrogenated to produce the corresponding indole.[44] In an earlier study, the same dithiane adduct was obtained in 41% yield after chromatography. This product was converted to the aldehyde (69). The same paper[45] describes a C-7 alkylation production from addition of LiCMe$_2$CN, a result in conflict with the later investigation. Nucleophilic displacement of nitro groups from indoles has also been achieved using organometallic intermediates. Stoichiometric cationic RuCp complexes were employed in this work.[46]

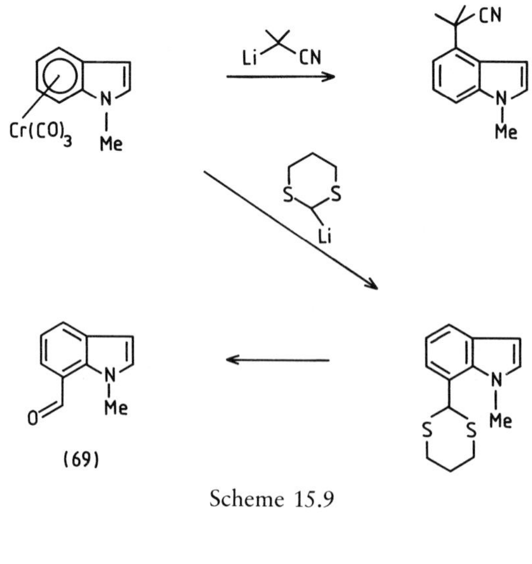

Scheme 15.9

Regiocontrolled additions to dihydropyridines have been achieved using organo-chromium chemistry.[47]

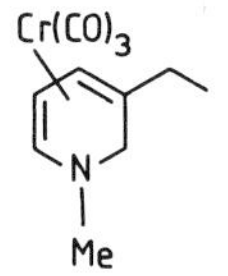

(70)

Attempts to alter the regiocontrol of these reactions by hydride abstraction from (70) to form a cationic η^6-pyridinium complex (71), or by production of the corresponding anion (72) (by the action of butyllithium on (70)) were unsuccessful, but treatment with methyllithium produced the coupled products (73) as a mixture of diastereoisomers, perhaps via an initial lithiation step.[48] A smaller amount of the regioisomer (74) was also formed. Lithiation of the N-methylindole complex (75) with butyllithium in

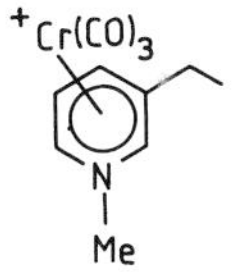

(71)

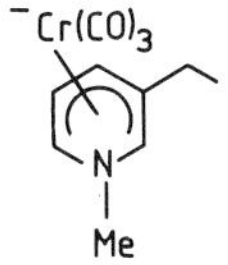

(72)

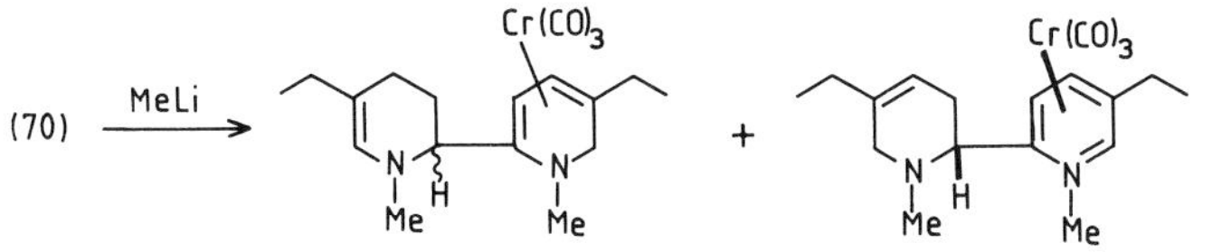

(73) 59%

(74) 25%

TMEDA/THF occurred selectively at C-2 allowing the production of (76) and (77) upon addition of trimethylchlorosilane or ethyl chloroformate. A second lithiation of (77) was

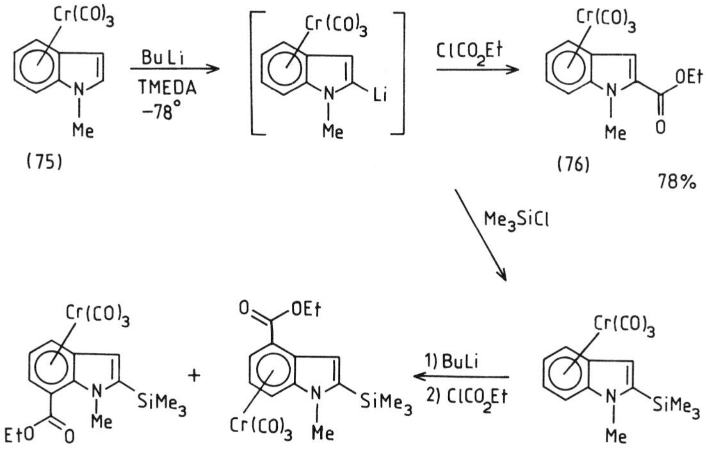

(75)

(76) 78%

(78)

(79)

(77) 60%

less well controlled, producing a 4:1 mixture of (78) and (79) after addition of ethyl chloroformate. The major product from this reaction was isolated in 67% yield by chromatography and converted to the free indole in 78% yield by photolytic decomplexation.[49] With a silyl protecting group on the nitrogen, lithiation has similarly been directed to C-4 in approaches towards tremorgenic mycotoxins such as ergotamine.[50]

15.3 Synthesis via orthometallation and cyclisation reactions

The elaboration of orthometallated species encountered in the preceding section is an example of a fairly common process for the production of heterocyclic rings. Simple cyclopalladation reactions are frequently followed by carbonyl insertion and reductive elimination.

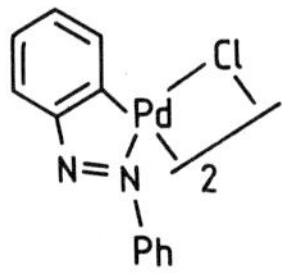

(80) 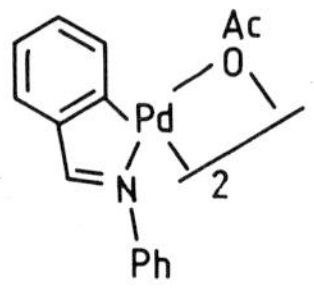   (81)

Azobenzene with $PdCl_2$ yielded the ring metallated derivative (80), and benzylidene aniline afforded the analogous derivative[51] (81). Carbon monoxide insertion[52] into the carbon–metal bond then leads to synthesis such as in Scheme 15.10 (see also Chapter 9).

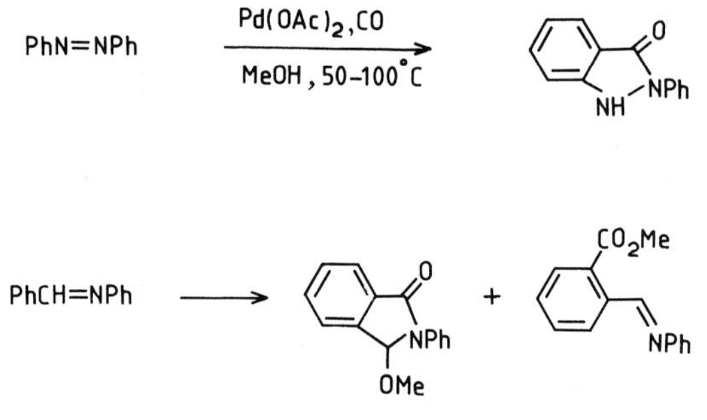

Scheme 15.10

Benzylamines also give orthometallation derivatives which may similarly undergo carbon monoxide insertion to give amides such as (82). One should note, however, the loss of an alkyl group in this example. 2-Phenylpyridine and similar heterocycles can also exhibit metallation.[53]

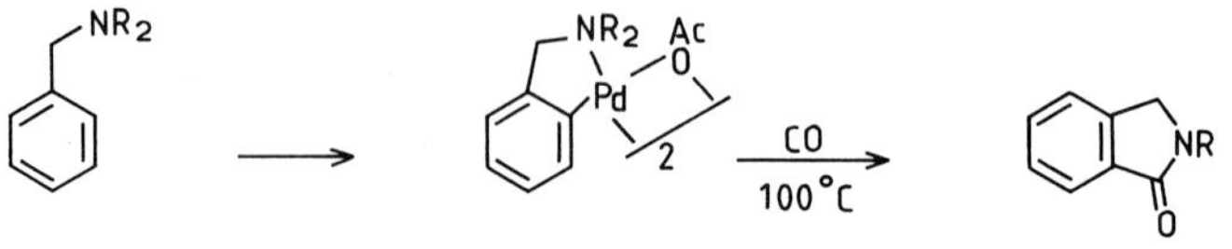

(82)

A quite extensive range of heterocyclic products has been obtained by the step-wise addition of alkynes to *ortho*-palladated benzyl amines and the corresponding α-aminonaphthalene derivatives. Loss of a methyl group was again observed in the formation of (83), but with the cationic starting material (84) a tetrafluoroborate salt, (85), was obtained. The formation of (86) from a cationic complex formed from benzylamine demonstrates a step-wise mode of addition.[54] A step-wise insertion of

ethene has also been suggested to account for a reaction forming 2-methylquinoline from aniline in a process catalysed by rhodium trichloride.[55]

A number of substituted benzylidene anilines have been converted to phthalimidines (87) by reaction with $Co_2(CO)_8$,[56] and similar reactions have been reported for azobenzene. Insertion of isocyanides (see Chapter 9) to form (88) proceeded in an analogous way.[57]

An interesting catalytic cycle using a ruthenium catalyst in a sealed tube converted the aryl isonitrile (89) into 7-methylindole.[58] Although yields are good, the reactions

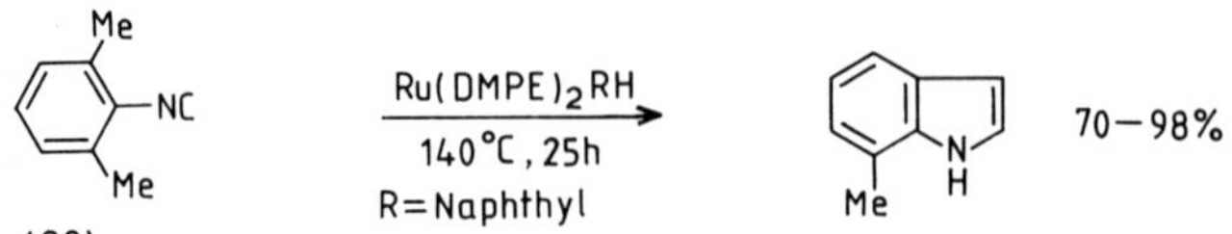

reported were performed on a very small scale, and when low catalyst:isocyanide ratios were employed, problems were encountered when a second isocyanide insertion produced a dimeric product. The mechanism proposed for the reaction is illustrated in Scheme 15.11. The cycle involves a series of steps, metallation, insertion, and reductive elimination, discussed in earlier chapters.

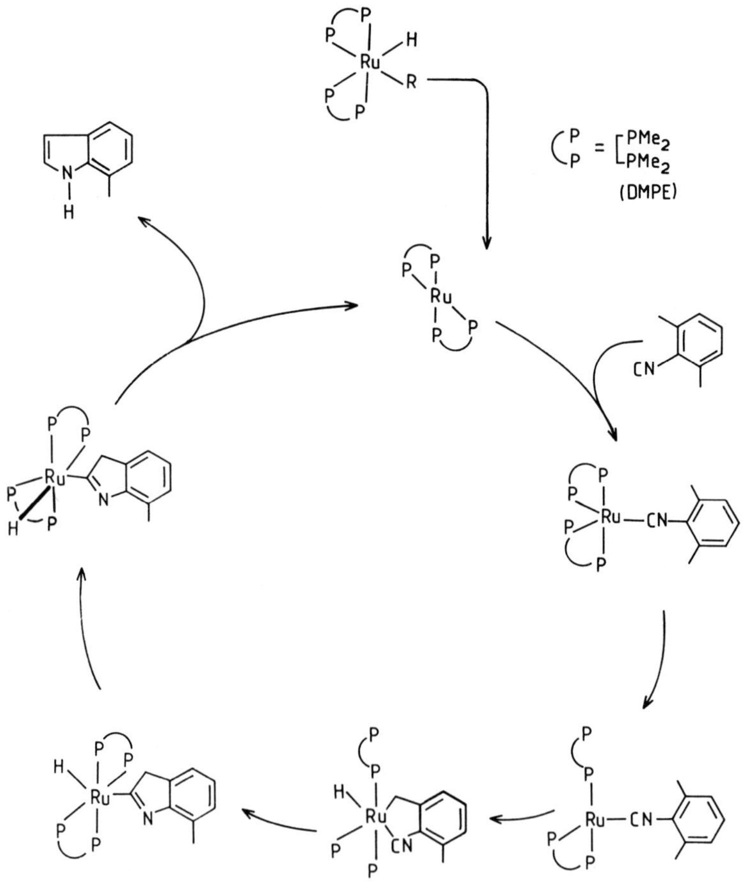

Scheme 15.11

Another indole synthesis is based on more conventional lines, and used a $PdCl_2(MeCN)_2$/benzoquinone system to form vinyl 2-bromoanilines which were then cyclised by palladium(0) catalysed orthometallation, which in this case was used to produce a carbon–carbon bond in (90).[59]

Acylindoles such as (91) have been prepared from N-phenylhydroxamic acids and vinyl acetate by the action of lithium tetrachloropalladate.[60]

Bromo- and fluoro-substituted arenes have a somewhat central role in transition metal mediated routes to heterocycles and have already figured in examples given in earlier sections of this chapter. Nickel catalysts can be used to couple an alkene to the site of halogen substitution in a reaction that provided an approach[61] to the synthesis of indoles and hydroquinolines. o-Halo-N-alkenyl anilines (92), (93) and (94), are cyclised by NiCl$_2$(PPh$_3$)$_2$ and MeMgBr. Yields were moderate to good.

An alternative organometallic indole synthesis (Scheme 15.12) by palladium catalysed cyclisation of o-allylanilines has been discussed in Chapter 5. The allyl anilines employed in these syntheses could be obtained[32] by allylation of o-bromoanilines using a nickel π-allyl complex. The order of the bond-forming reactions is reversed. These two cyclisation processes provide an interesting comparison since they have in common the ability to elaborate o-haloanilines.

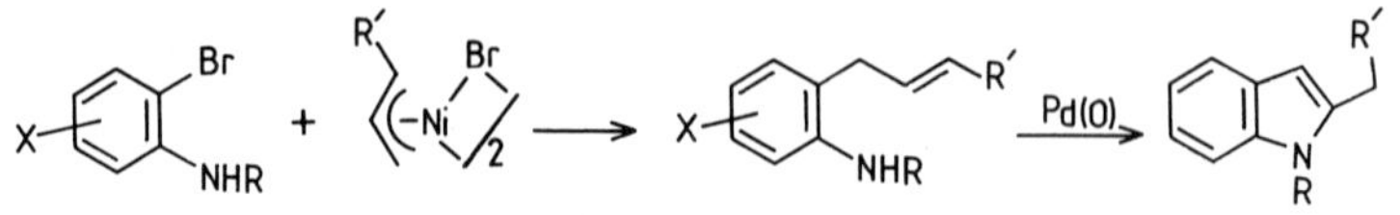

Scheme 15.12

Activation of aryl halogen via formation of a nickel derivative was also involved in a synthesis of cephalotaxinone[62] in which the conversion of (95) into (96) was effected by treatment with nickel(cyclooctadiene)$_2$, which reacts with the iodide, and Ph$_3$CLi which forms the enolate of the $-CH_2-CO$ residue.

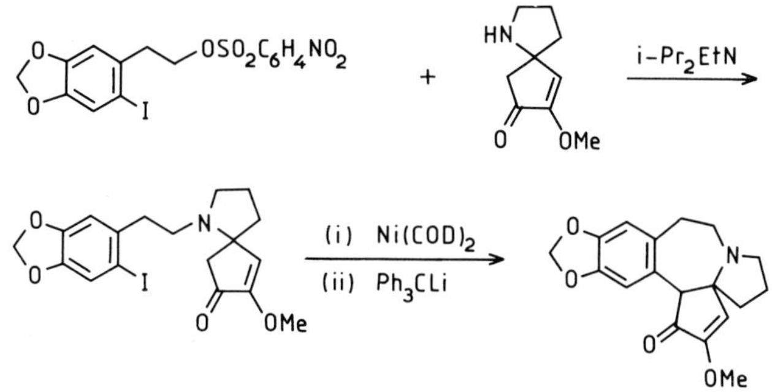

(95)   (96)

A range of simpler examples illustrate the addition of an aryl group to an alkene brought about in the presence of Pd(OAc)$_2$ and (o-tolyl)$_3$P. A route to nornicotine (97) is indicated.[63] The amino mercuration cyclisation step should be noted.

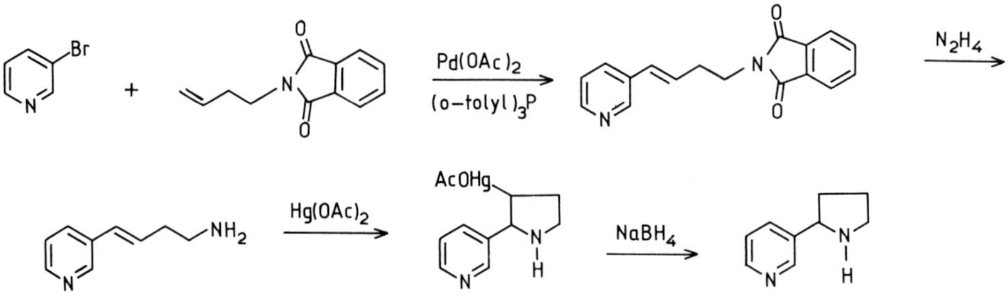

(97)

In some cases when stoichiometric π complexes are used, cyclisation is effected upon removal of the metal. An example drawn from the extensive chemistry of tricarbonyliron complexes (see Chapter 7) is provided by the cyclisation of (98) to form a carbazole upon oxidation with iodine in pyridine. The precursor (98) was easily obtained by alkylation of a cyclohexadienyl complex with *p*-toluidine.[64] This route to carbazoles has recently been improved by Knölker and applied to the synthesis of koenoline and mukonine. Activated manganese dioxide is used in this case to remove the metal and close the central ring.[65] Organoiron complexes have also been used as precursors to tetrahydrocarbazolones.[64]

A quite different process, removing tricarbonyliron from (99), introduced a two carbon fragment and effected cyclisation to form the pyrrole (100). This reaction presumably begins by addition of methyllithium to a coordinated carbon monoxide ligand, as seen in analogous reaction of α,β-unsaturated ketone complexes discussed in Chapter 9. Thus one of the two additional carbons comes from methyllithium while the other originated as a carbonyl ligand in (99).[66]

Cyclisation of (101), following addition of a one carbon fragment from a Vilsmeier reagent, produced the tetracyclic product (102). Dehydrogenation of the crude reaction mixture produced the cation (103) in 60% yield. This product is an intermediate in the synthesis of clivacine and guatambuine. In these reactions, the chromium tricarbonyl group in (101) is serving to protect the dihydropyridine ring during the initial addition of the Vilsmeier reagent to the indole (see Chapter 13).[67]

In another reaction involving insertion of a two carbon unit derived from methyllithium and carbon monoxide, the cobalt complex (104) was converted into the pyrone (105).[68] In this case, a more extensive rearrangement of the ligand is required, since the four-membered ring must open during the reaction.

15.4 Metal promoted rearrangements

Organometallic interconversions of azirines, which exploit the highly reactive nature of these strained rings, have already been encountered in Section 15.1 in the context of β-lactam synthesis. This unusual dimerisation reaction, however, does not reflect the normal outcome following the opening of an azirine ring by a transition metal catalyst. Rearrangement or dimerisation to form a larger ring is far more typical. In cases where there is an unsaturated substituent, cleavage of the C—N bond and rearrangement of the resulting metallocycle can account for the products (106), (107), and (108).[69]

Although chromium, molybdenum or tungsten catalysts are typical of those used in these reactions, there are clear possibilities for other catalyst systems. Isomura *et al.*[70] found that $PdCl_2(PhCN)_2$ induced rearrangement of the azirine (109) into the indole (110). In this reaction, the ring expansion requires the substitution of one of the adjacent phenyl rings.

A related rearrangement to form the six-membered ring of a tetrasubstituted pyridine occurred on treatment of (111) with $Fe_2(CO)_9$.[71] Low yields and poor selectivity make this version of the reaction unsuitable for use as a synthetic procedure.

When a suitable unsaturated substituent is absent from the sp^3 position of the ring, dimerisation by the action of $Mo(CO)_6$ can occur[72] to give diarylpyrazines and the

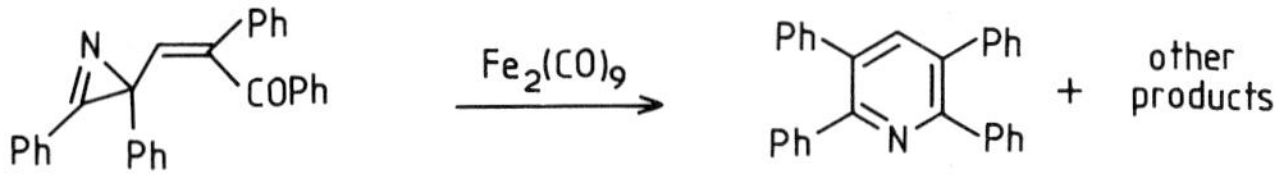

(111)   (8-10 %)

isomeric dihydropyrazine products shown in Scheme 15.13. With another catalyst,[73] dimerisation took a different course to give a symmetrically substituted pyrrole in low yield, a comparison that emphasises the sensitivity of these processes to reagents and conditions. Aziridines undergo a related reaction with dicobalt octacarbonyl to produce dihydrooxazoles.[74]

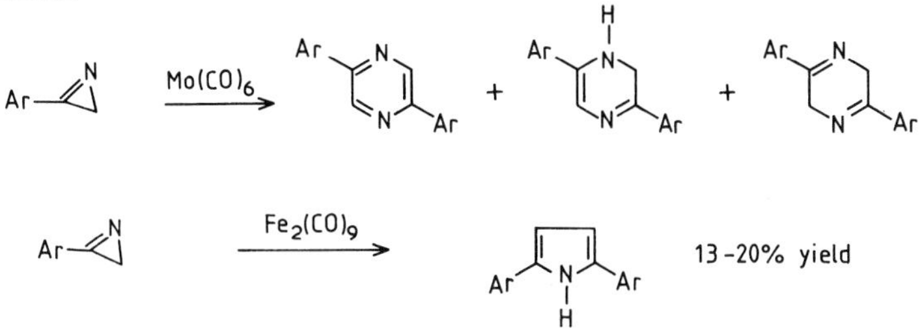

Scheme 15.13

An interesting comparison to the low yielding preparation of pyridines from (111), referred to above, is provided by the rearrangement of phenyl substituted 2-oxa-3-azabicyclo[3.2.0]hepta-3,6-dienes by treatment with either molybdenum[75] or iron[76] carbonyls. Phenyl substituted isoquinolines have also been prepared in the same way. Reactions of this type are of particular interest from a mechanistic, rather than from a synthetic, viewpoint. In the presence of water, open chain products were formed. These products could arise from reaction of an intermediate (112) with water, and so offer

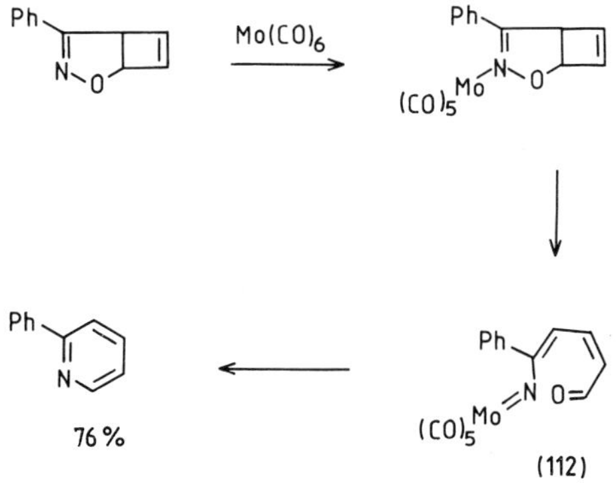

76 %   (112)

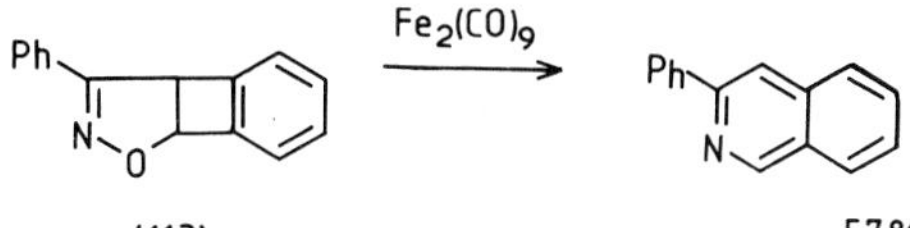

(113)   57%

(114)   91%

(115)   74%

some support for the mechanism shown. When the N–O linkage was placed in a six-membered ring, as in (114) and (115), pyrroles and imidazoles were obtained. In these cases, triiron dodecacarbonyl was used.[77]

Starting materials, such as (113), of the type required for the rearrangement process, can themselves by prepared by an organometallic method which applies the use of $Fe(CO)_3$ complexation to stabilise cyclobutadienes in the way discussed in Chapter 13. Cycloaddition of arylnitrile oxides to benzocyclobutene, which was liberated from its complex (116) by oxidation, afforded (113) in 65% yield.[78]

(116)

Although this reaction is limited in application since other 1,3-dipolar reagents fail to react in the same way, several cycloaddition reactions involving different substrates have found use in heterocyclic synthesis. These reactions are discussed in the next section.

15.5 Organometallic cycloadditions

While the example of a cycloaddition process that concludes Section 15.4 shows a reaction that occurs by addition to the free ligand, there are several interesting cases where addition occurs in the coordination sphere of the metal and so becomes susceptible to its control influence, as discussed in Chapter 11. In a reaction that is similar to the one encountered above, in that it uses a rearrangement involving cleavage of an N–O bond, intermolecular trapping of the intermediate by cycloaddition with alkynes such as (117) produced a substituted pyridine in low yield.[79]

(117)   16%

Cycloaddition between a diene (118) and a nitrile, catalysed by a cobalt complex, offers a rather more efficient process for the construction of a pyridine ring.[80]

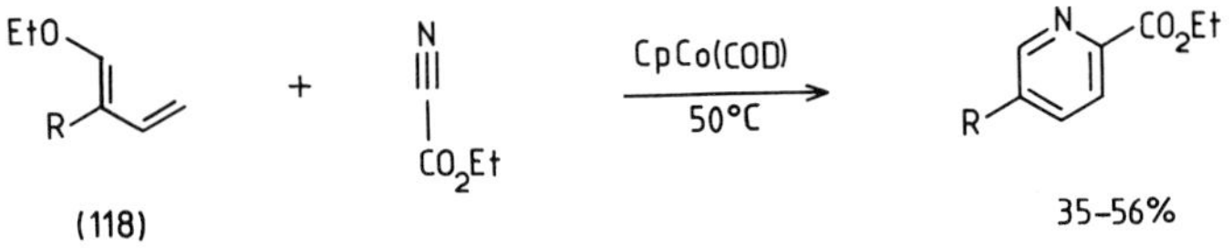

(118)   35–56%

An extensively studied method for the incorporation of nitriles into pyridine is related to the cobalt catalysed alkyne trimerisation discussed in Chapter 10. Some early studies by Bönnemann and Brinkmann showed that the CpCo(cyclooctadiene) catalyst encountered above was effective for the production of dipyridyls from cyanopyridines and ethyne under pressure. Examination of 2-, 3-, and 4-substituted pyridines as substrates showed that the corresponding linkage in the dipyridyl made all three isomers accessible in high yield. The 2-cyanopyridine example given in Scheme 15.14 is typical of results obtained.[81] When methyl or phenyl alkynes were used, relatively little regiocontrol was observed.

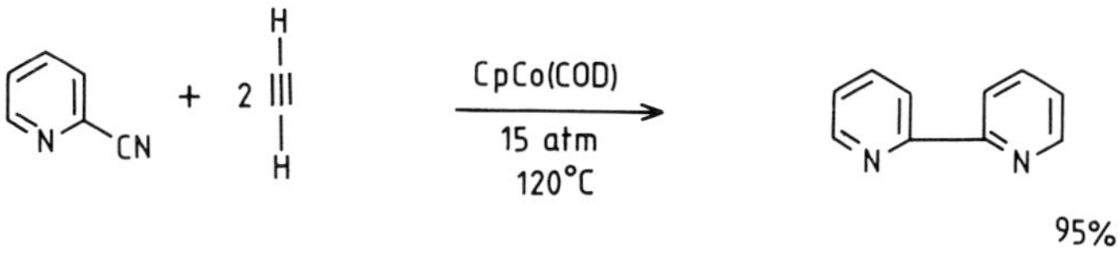

Scheme 15.14

A catalyst used in the highly successful trimerisation of alkynes, Cp_2Co, can itself be used to effect pyridine synthesis. Two commercially important pyridines have been obtained in this way. Ethyne combines with acetonitrile under pressure in the presence of Cp_2Co to afford 2-methylpyridine (119) in 60% yield. The yield of vinylpyridine obtained in this way was somewhat lower.[82]

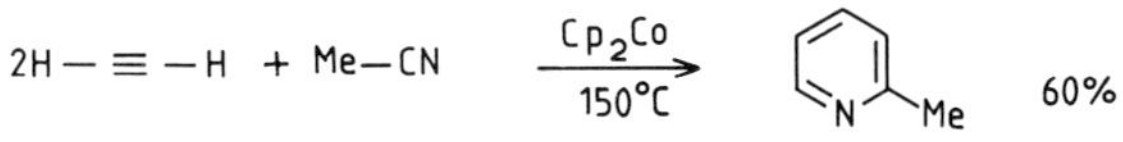

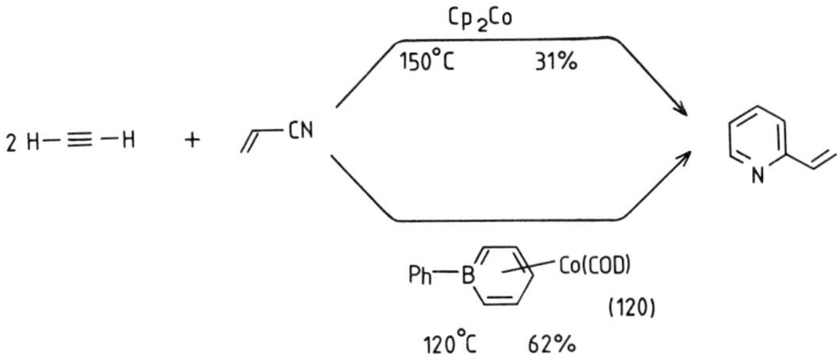

Reactions of this type have been extensively developed by the Bönnemann group with the intention of achieving bulk conversions. Their strategy for optimisation by variation of ligand substitution has proved highly successful, offering a fine case-study illustrating methods of catalyst development.[83] Parameters against which catalysts were assessed included the yield and selectivity of the reaction, and the turnover number of the catalyst. The key to the production of vinylpyridine, for example, was the search for a catalyst capable of promoting useful rates of reaction at relatively low temperatures. The use[84] of the borinato cobalt catalyst (120)[85] with ethyne and acrylonitrile at 120 °C and 50 bar raised the yield of 2-vinylpyridine to 62% in a reaction that is described as reaching productivity ranges near 4 tonnes of product per kg of cobalt.[83]

Vollhardt's development of the cyclisation of α,ω-diynes has been extended to methods for pyridine synthesis. Both the reaction with nitriles,[86] and the use[87] of the corresponding cyanoalkynes in reactions with alkynes have been examined (Scheme 15.15).

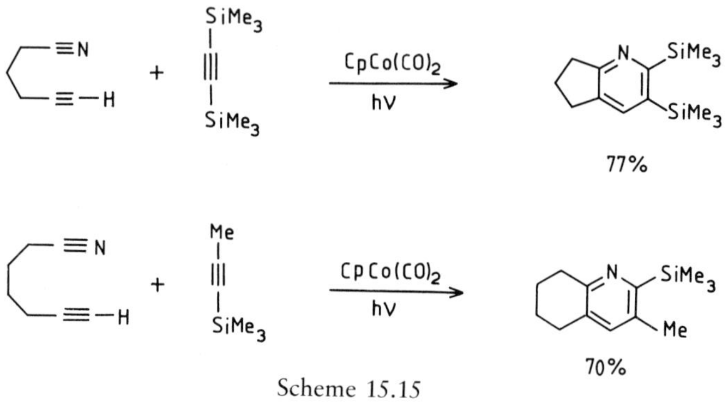

Scheme 15.15

An excellent example of the application of these methods is to be found in the synthesis[88] of pyridoxine (vitamin B$_6$, 121) by chemists at the Hoffmann La Roche laboratories at Basel. Here the 2-methylpyridine sub-unit suggested a cyclisation approach using acetonitrile, based on the methods discussed above. The two adjacent hydroxymethyl substituents could be introduced in the correct position by use of an

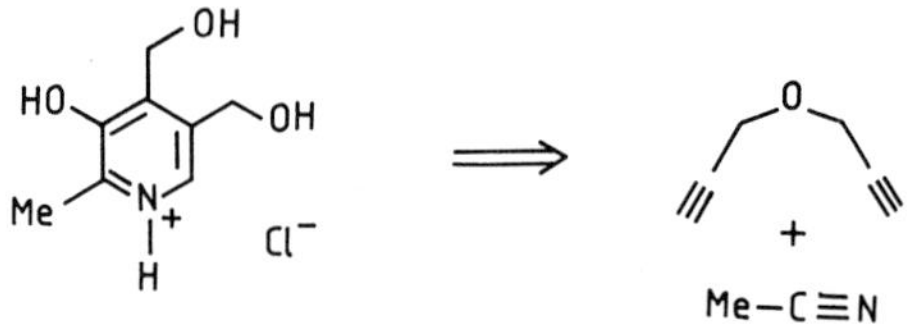

(121)

ether strap to link the two alkyne components. In common with Vollhardt's route to estrone, discussed in Chapter 10, a symmetrical alkyne substitution pattern, combined with a subsequent selective protodesilylation, was used to circumvent potential problems with regiocontrol in the cyclisation step. The reaction sequence has been performed on a large scale, beginning with the reaction of 1 kg of the diyne (122) with acetonitrile using 40 g of cobaltocene as the catalyst precursor. The pentasubstituted pyridine (123) was produced in 73% yield. Removal of the 6-silyl substituent (73% yield) and conversion to the N-oxide (69% yield) gave (124). Introduction of the 3-OH group was less well controlled, forming the required product (125) as a mixture with

(126). Cleavage of the ether with HBr and then acetylation converted (125) into a triacetate from which pyridoxine was obtained by hydrolysis with HCl.

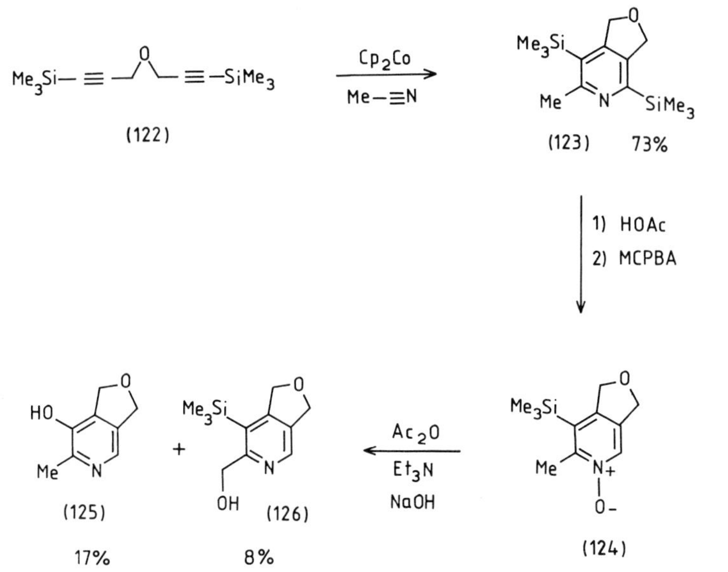

Cyclisation of unsymmetrically substituted diynes was also studied. The ester (127) produced a 15:1 mixture of regioisomers, with the major product as that shown. Other heterocyclic systems can be obtained by modifications of the cyclisation reaction. Thus replacement of the nitrile with an isocyanate group led to the formation of pyridones.[89]

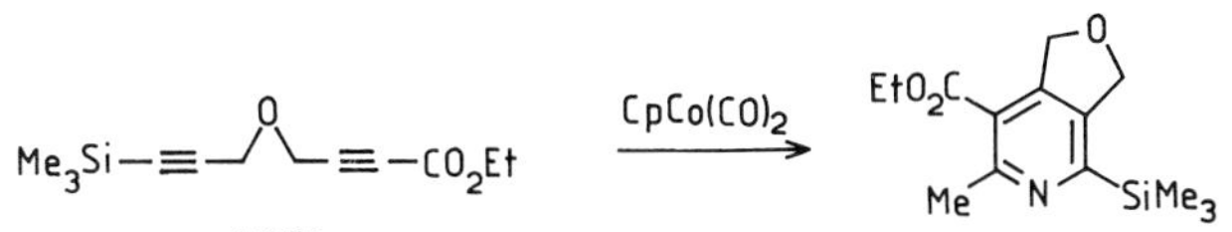

Direct combination of alkynes with sulphur to form thiophenes (Scheme 15.16) has also been described.[83]

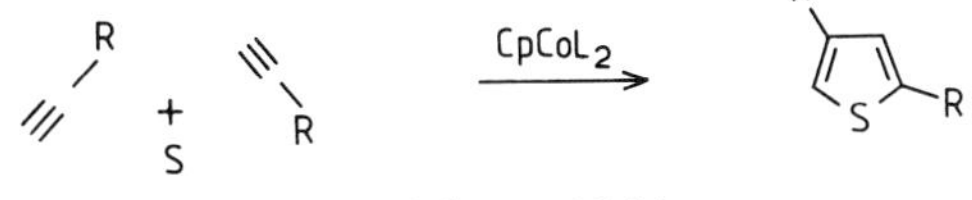

Scheme 15.16

Work from Vollhardt's group has moved these methods into several new areas. Cyclisation of three nitrile groups by an iron catalyst has been used to form 1,2,4-triazines such as (128).[90]

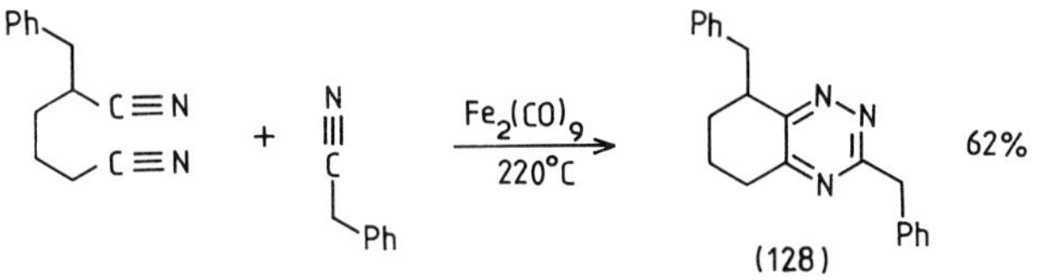

Another development has been the replacement of the nitrile by a heterocyclic ring. Thus Grotjahn and Vollhardt describe[91] the combination of the indole (129) with (130) by means of the usual $CpCo(CO)_2$ catalyst and irradiation by a projector lamp to produce the tetracyclic product (131). Small amounts of the cyclobutadiene complex (132) were also formed under these conditions, but this could be completely avoided by the use of the $CpCo(CH_2CH_2)_2$ catalyst system, a ploy that can give dramatic improvements in some examples of cyclisation reactions of this type.[92]

(129)   (130)   (131)   64%

(132)

Of all the reactions discussed in this chapter, it is the cobalt mediated cyclisation reactions described above that have been developed most fully as a general synthetic method. A clear sign of the maturity of methods of this type appears when they are taken up by synthetic chemists who are not transition metal specialists. Workers contemplating an attempt at a cobalt mediated cyclisation for the first time should keep in mind that the catalysts used, especially some of the more unusual ones, are relatively air-sensitive, and so need to be handled with considerable care.

A number of other types of cyclisation process also lead to the formation of heterocyclic products. The stoichiometric iron complex (133) contains a metallocyclic ring which is reminiscent of metallocyclic intermediates in cobalt catalysed alkyne oligomerisations (see Chapter 10). In (133), however, the complex is stabilised as an η^4

(133)

ligand to a second tricarbonyliron group.[92] Reaction of (173) with nitrosoarenes or nitrosylchloride afforded substituted pyrroles. Direct cycloaddition of t-butylisocyanide with alkynes, effected by nickel catalysis, also resulted in the formation of pyrroles, such as the rather surprising product (134).[93]

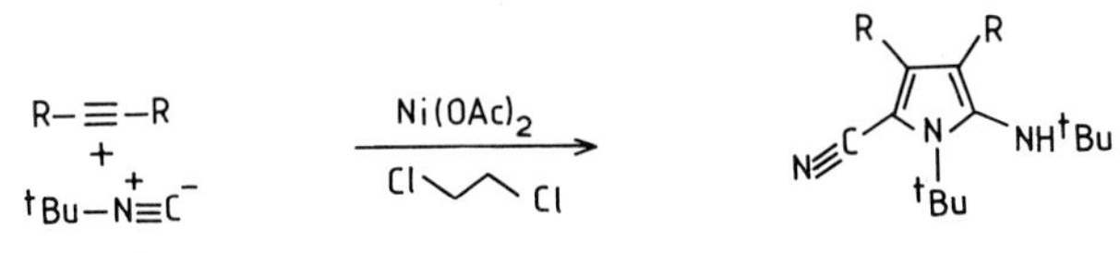

(134)

Addition of an isocyanate to a phenylethyne is catalysed by iron pentacarbonyl. Good yields of the adduct (135) were obtained[94] but the reaction was sensitive to the nature of the substrate. With diphenylbutadiyne and diarylcarbodiimides, a similar adduct has been obtained by reaction with $Fe(CO)_5$, although the product was formed in lower yield.[95]

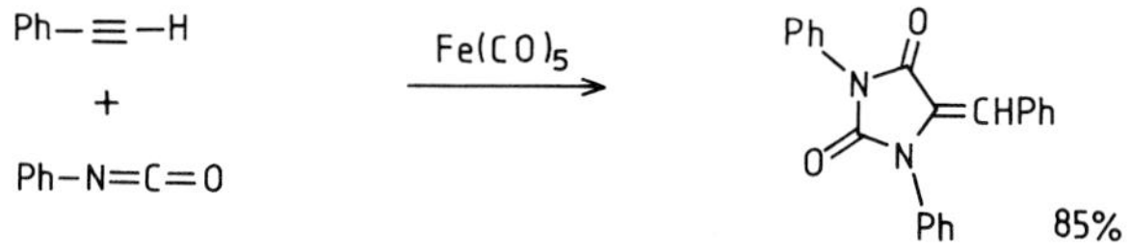

(135)

Another cyclisation of diphenylethyne and phenylisocyanate has been effected by nickel catalysis. In this case, substituted 2-pyridones were produced.[96] Comparison of the results obtained with iron and nickel catalysts shows the potential for a complete differentiation between alternative reaction paths in cyclisations of this type.

In Chapter 11, emphasis was given to the fact that two types of cycloaddition process can occur. Fully complexed polyenes react by binding of the reactant at the metal, a process that often required initial photolytic displacement of a ligand. When only part of the polyene is bound, as is the case with complexes (136) and (141), then initial reaction can occur through the involvement of this free alkene position. Reaction of (136) with the heterocycle (137) in a [4+2] cycloaddition process, followed by loss of

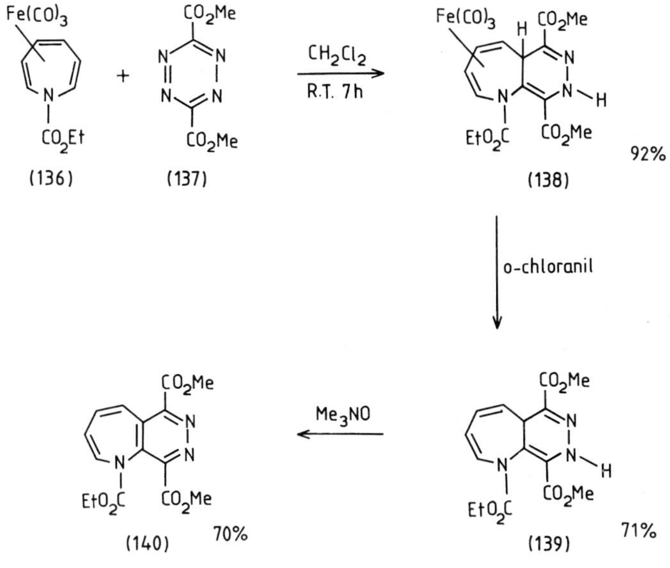

N_2, produced the adduct (138) in a very efficient reaction. Removal of the metal with chloranil afforded the free ligand (139) which could be aromatised by treatment with trimethylamine N-oxide. Direct formation of (140) from (138) by decomplexation with the N-oxide was less efficient, proceeding in only 32% yield.[97]

A furan derivative has been obtained in rather low yield by heating (141) and the diazoketone (142) at 45 °C. Although one feasible mechanism for the reaction might involve a cyclopropane intermediate, Goldschmidt's group discounts this possibility and offers evidence to support a 1,3-dipolar addition mechanism which makes an interesting parallel with the reaction leading to (138).[98]

15.6 Carbene additions as routes to heterocycles

Rhodium catalysed carbene insertions (see Chapter 12) have been applied to heterocyclic synthesis. Following initial work using copper catalysis to produce dihydropyrrolizines and indolizines from diazoketones,[99] Jefford's group developed the most efficient rhodium methods to give access to products of the type (144) in a synthesis[100] of ipalbidine (146). The required precursor (143) was obtained by fairly standard methods. Regiocontrol of the cyclisation step depended on the arene substituent. With a nitro group (X = NO$_2$), only the required mode of addition occurred, but with methoxy substitution (X = OMe) much less control was available and a 7:2 mixture of products was obtained. The required oxygenated substituent, however, could be introduced with acceptable regiocontrol in 82% yield by use of the acetate (143, X = OAc). Reduction of the pyrrole ring was performed by two routes. Hydrogenation with Pd/C gave an enone intermediate which was further reduced in a second step. Alternatively, complete reduction with platinum oxide required a subsequent oxidation step to produce the ketone (145). Reaction with methyllithium, addition of acetic anhydride and elimination/hydrolysis in aqueous HBr completed the synthesis of (146).

An intermolecular reaction between dimethyl diazomalonate and a variety of aryl and alkyl nitriles has been used to form trisubstituted oxazoles such as (147). Once again, rhodium acetate catalysis was used to promote carbene insertion.[101] Addition of stoichiometric carbene complexes to isocyanides has also been used to form heterocycles.[102, 103]

Another type of carbene reaction that received much attention in Chapter 12 is the combination of Fischer-type complexes with alkynes. Wulff[104] has developed an interesting application of this process in which an initial dipolar addition introduced a heterocyclic substituent in the place of the normal vinyl or aryl group. This reaction is then followed by annellation via addition of an alkyne and a carbonyl group in the usual way.

Both alkenyl and alkynyl substituted transition metal carbene complexes have proved to be highly reactive dienophiles.[105] By employing the tungsten carbene complex (148) in a reaction with a 1,3-dipole, a method for the production of pyrazole substituted carbene complexes was discovered. A pyrazole tungsten complex was also formed, as had been anticipated from the much earlier results of Fischer's group, who studied the

M = Cr   76%

M = W   87%

reaction of carbene complexes with diazomethane.[106] A much more efficient reaction occurred when trimethylsilyldiazomethane was used with either chromium or tungsten dienophiles. Oxidation yielded the pyrazole ester.[107] These reactions offer much better regiocontrol [> 300:1, (149):(150)] than the direct reaction between a propargyl ester and diazomethane [4:1, (149):(150)] or trimethylsilyldiazomethane [1:2, (149):(150)].

The organometallic intermediate (149) can be made to undergo further reaction by addition of an alkyne. The carbonyl insertion product was oxidised to pyrazolepyridine quinones such as (151) in moderate yield. In this reaction sequence the carbene complex has been used twice, providing a high degree of control at each step of the construction of (151), which is obtained as a single regioisomer despite its origin from four separate components.[107] Thus by means of Fischer carbene intermediates, the two different alkynes, diazomethane, and carbon monoxide can be assembled in the proper order in just two steps.[107]

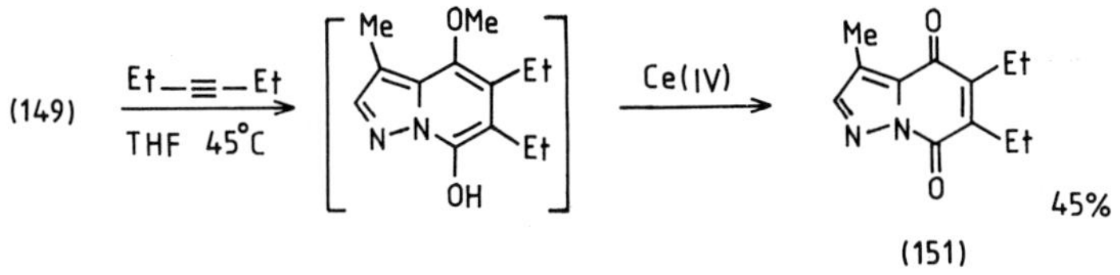

A further variation on this general theme, again developed by the Wulff group, is the use of cobalt carbene complexes to form tetrasubstituted furans.[108] Yields for reactions of this type are rather variable, spanning a range from 93% to 39%, with the best results obtained with an aryl substituted carbene. Reaction of (152) with but-2-yne

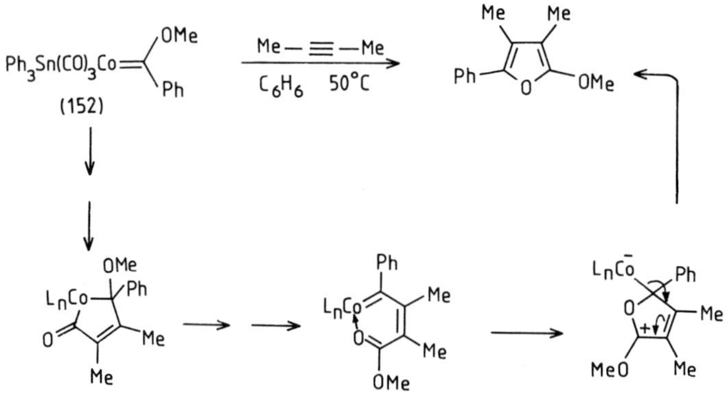

Scheme 15.17

provides a good example of the process. The key steps shown in Scheme 15.17 are taken from a mechanism proposed to account for the migration of the OMe group.

This synthetic method has been applied by Wulff[108] to the synthesis of the natural product bovolide (155) via a furan intermediate which was not isolated. In this synthesis, the convenient access to the carbene complex (see Chapter 12) makes the overall route a most attractive proposition. Lithium/tin exchange converted (153) into the corresponding alkyllithium reagent which reacted with a carbonyl group in the cobalt complex (154) to produce the usual type of anionic intermediate which was then alkylated with Me$_3$O$^+$. Heating this complex with an excess of but-2-yne gave the required intermediate which was converted to the butenolide (155) by the action of trimethylsilyliodide.

15.7 Further carbonyl insertion routes to heterocycles

Earlier in this chapter, several examples of carbonyl insertion have been encountered as part of more extensive reactions. At the start of this section, reactions are discussed in which the introduction of an oxygenated carbon atom by carbonyl insertion is the main purpose of the procedure. The cyclisation of (156)[109] forming (157) by cobalt catalysis provides a typical example. Such reactions, however, require high pressures and are sensitive to the conditions. At higher temperature, for example, (157) undergoes a second carbonyl insertion to afford (158).

When an oxime is used as the substrate, the product (159) is similar to that obtained from (156), but the reaction requires hydrogen and proceeds with loss of the oxygen atom of the oxime.[110]

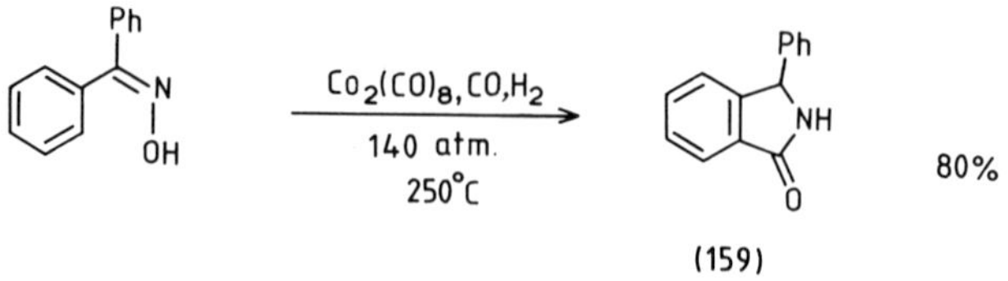

(159)   80%

Substrates for the carbonyl insertion step often arise as reaction intermediates. Thus the reaction of 1,1-dibromo-2-phenylcyclopropane and phenylisocyanate with nickel tetracarbonyl produced (160) in which a C–O unit in the ether linkage originated as carbon monoxide.[111]

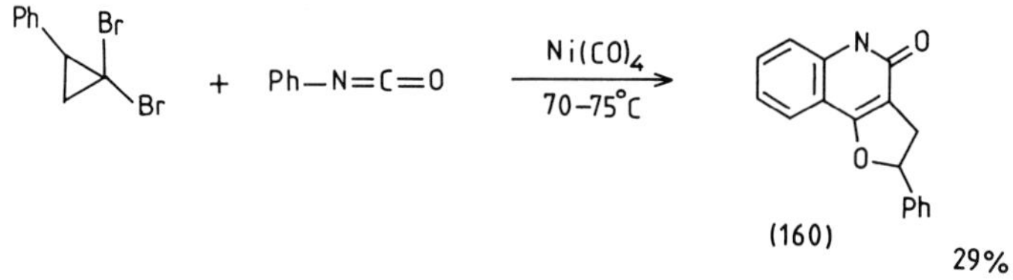

(160)
29%

1,2-Diazepin-3-ones have been produced from cyclic diazenes and alkynes.[112] In this case, the fate of the CO molecule as the carbonyl group of (163) is more apparent. Photolysis of (161) with $Fe(CO)_5$ produced a bimetallic complex[113] which took up two alkynes in a separate reaction to form the intermediate (162). Oxidative removal of the metal with bromine completed the carbonyl insertion process.

Just as we saw at the start of Chapter 10, carbon dioxide can also participate in

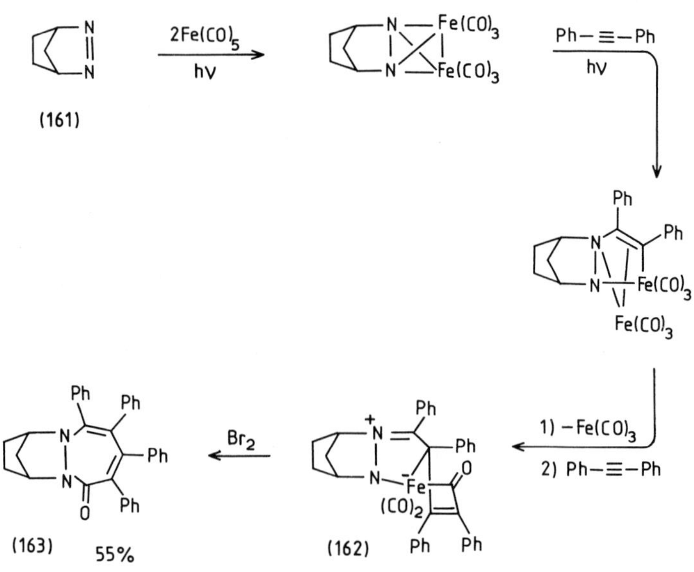

(161)

(163)   55%   (162)

insertion reactions. A recent example forms an α-pyrone from a diyne (164) using a nickel catalyst and triphenylphosphine. Quite high temperatures were used and the reaction proceeded in rather low yield.[114]

(164) CO_2 / $Ni(COD)_2$ / $2PPh_3$ / 130°C   34%

Insertions of alkenes, also seen in Chapter 10, have been used in transformations that are related to reactions commencing with orthometallation discussed in Section 15.3. 3-Substituted indoles, for example, have been obtained from N-allyl derivatives of 2-iodoaniline in good yields ranging from 73 to 97 %.[115,116] By choosing nickel in place of palladium, chloro substrates can also be used.[117] These reactions involve insertion of the

$Pd(OAc)_2$ / Bu_4NCl 92%

(165)

$Pd(OAc)_2$ / Bu_4NCl 39%

(166)

transition metal into the carbon–halogen bond, and so are not dependent on orthometallation. This provides additional flexibility as can be seen in the formations of the quinoline (165) and the isoquinoline (166) from appropriate aniline and benzylamine derivatives.[116]

A number of lactone syntheses have already been discussed in Chapter 14. At the conclusion of this chapter, however, we will deal with the construction of α-methylene lactones by methods that introduce the carbonyl group by an insertion of carbon monoxide. Since it is a common functional group occurring in many classes of natural products, the α-methylene lactone moiety has received considerable attention from synthetic chemists and in several cases organometallic methods have been used for its construction. Thus, as was the case for β-lactams considered in Section 15.1, it is possible to compare a number of alternative methods which share a common target.

An approach in which the carbonyl group was already present in the starting material as an ester has been described in Chapter 5. This is summarised by the disconnection shown in Scheme 15.18 which makes use of the ability of the OMe group to serve to re-form the alkene following addition of the nucleophile.

A carbonylation alternative for the same target molecule is shown in Scheme 15.19. The alkyne intermediate is readily accessible in the *trans* stereochemistry by opening an

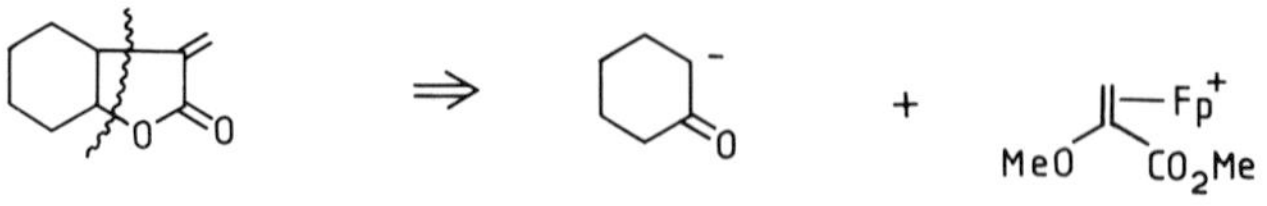

Scheme 15.18

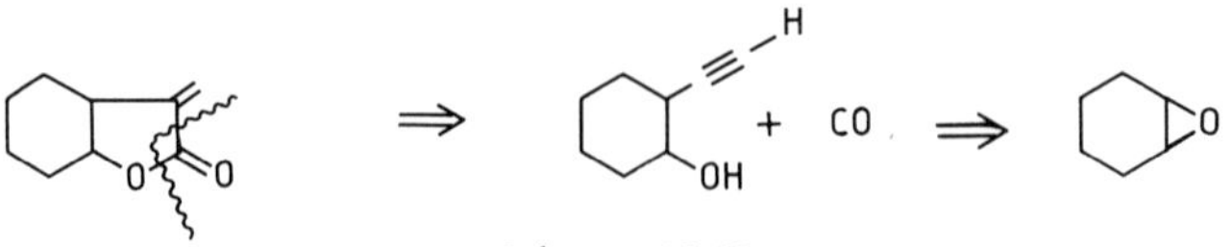

Scheme 15.19

epoxide by the ethyne anion. The reaction is thought to proceed (Scheme 15.20) by
initial attack of the alcohol on a coordinated CO group.[118]

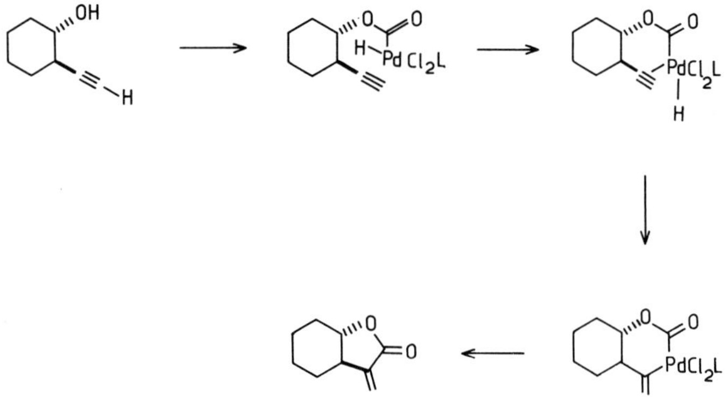

Scheme 15.20

A further method (Scheme 15.21)[119] which is capable of forming both rings in the
same reaction, employs a nickel allyl complex (167) as a nucleophile in an initial
addition to an aldehyde. The subsequent closure of the alkoxide to the vinylbromide has
also been used to form six-membered ring products such as (168).[120]

As was also observed in Chapter 14, a clear feature of the closing examples in this
chapter, and of the β-lactam syntheses discussed earlier, is the observation that where
a rational synthesis design can be attempted, the methods used to plan a synthetic route
using an organometallic approach are quite distinct from those required for a
conventional retrosynthetic analysis. Functional groups, which normally are pressed
into service to drive bond-forming reactions, can to some extent be freed from this role
when transition metals are present as activating groups. As the use of organometallic
methods in heterocyclic synthesis develops further, an increasing range of general and
well understood reactions should gradually replace the diverse selection of organometallic procedures currently available, a change that will have a significant impact in
opening up new ways of thinking in heterocyclic synthesis design. In the meantime,
organometallic methods that use well established types of reactions offer the best

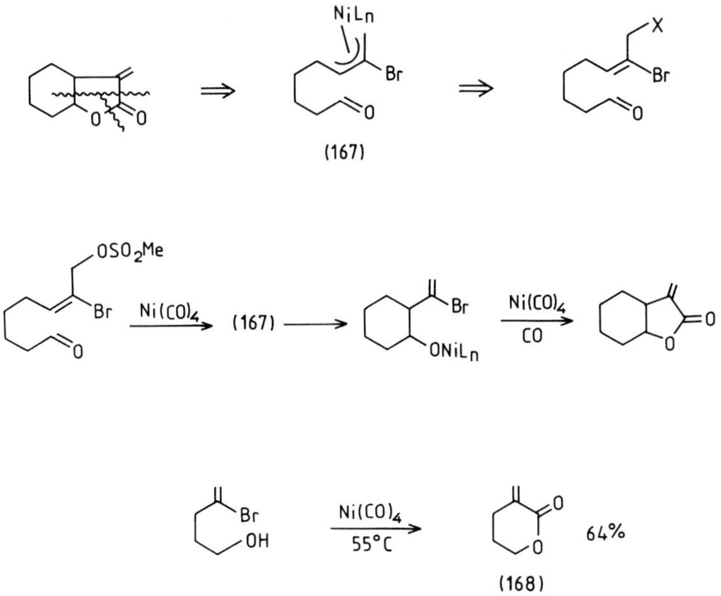

(167)

OSO₂Me

Ni(CO)₄   (167) ⟶

Ni(CO)₄
CO

Ni(CO)₄
55°C

64%

(168)

Scheme 15.21

prospects, and heterocyclic chemists would need to look beyond the bounds of their discipline, and this chapter, to properly gauge the chances of success. It is hoped that the provision in this chapter of cross referencing to other parts of the book will assist in the task.

References

Chapter 1

1 J. P. Collman, L. S. Hegedus, J. R. Norton, and R. G. Finke, *Principles and Applications of Organotransition Metal Chemistry*, University Science Books, 1987, Ch. 3, pp. 57–234.

2 W. C. Zeise, *Ann. Phys.*, 1827, **9**, 932.

3 A. Whitaker and J. W. Jefferey, *Acta Crystallogr.*, 1967, **23**, 977; A. Jost, B. Rees, and W. B. Yelon, *Acta Crystallogr.*, 1975, **B31**, 2649; D. E. Sherwood, Jr. and M. B. Hall, *Inorg. Chem.*, 1980, **19**, 1805.

4 F. A. Cotton, *J. Organomet. Chem.*, 1975, **100**, 29.

5 R. G. Hayes and J. L. Thomas, *J. Am. Chem. Soc.*, 1969, **91**, 6876; K. O. Hodgson and K. N. Raymond, *Inorg. Chem.*, 1972, **11**, 171; C. W. DeKock, S. R. Ely, T. H. Hopkins, and M. A. Brault, *Inorg. Chem.*, 1978, **17**, 625.

6 G. E. Coates, M. H. L. Green, and K. Wade, *Organometallic Chemistry*, Methuen, 1968, Vol. 2, pp. 2–6; F. A. Cotton and G. Wilkinson, *Advanced Inorganic Chemistry*, Edition 5, Wiley, 1988, pp. 36–7.

7 R. G. Pearson, *Chem. Rev.*, 1985, **85**, 41.

8 O. Eisenstein and Y. Jean, *J. Am. Chem. Soc.*, 1985, **107**, 1177; N. Koga, S. Obara, K. Kitaura, and K. Morokuma, *J. Am. Chem. Soc.*, 1985, **107**, 7109; M. Brookhart, M. L. H. Green, and R. B. A. Pardy, *J. Chem. Soc., Chem. Commun.*, 1983, 691.

9 M. Brookhart and M. L. H. Green, *J. Organomet. Chem.*, 1983, **250**, 395.

10 M. Brookhart, W. Lamanna, and M. B. Humphrey, *J. Am. Chem. Soc.*, 1982, **104**, 2117; M. Brookhardt and A. Lukacs, *J. Am. Chem. Soc.*, 1984, **106**, 4161.

11 R. B. Cracknell, A. G. Orpen, and J. L. Spencer, *J. Chem. Soc., Chem. Commun.*, 1984, 326.

12 M. L. H. Green and L.-L. Wong, *J. Chem. Soc., Chem. Commun.*, 1984, 1442.

13 S. J. La Placa and J. A. Ibers, *Inorg. Chem.*, 1965, **4**, 778.

14 J. Evans, *Adv. Organomet. Chem.*, 1977, **16**, 319; E. Ban and E. L. Muetterties, *Chem. Rev.*, 1978, **78**, 639.

15 F. A. Cotton, *Prog. Inorg. Chem.*, 1976, **21**, 1; C. P. Horwitz and D. F. Shriver, *Adv. Organomet. Chem.*, 1984, **23**, 219.

16 C. A. Rusik, M. A. Collins, A. S. Gamble, T. L. Tonker, and J. L. Templeton, *J. Am. Chem. Soc.*, 1989, **111**, 2550.

17 A. J. Deeming, *Adv. Organomet. Chem.*, 1986, **26**, 1.

18 R. B. King and C. A. Harmon, *J. Organomet. Chem.*, 1975, **88**, 93.

19 E. N. Jacobsen and R. G. Bergman, *J. Am. Chem. Soc.*, 1985, **107**, 2023; D. L. Davies, B. P. Gracey, V. Guerchais, S. A. R. Knox, and A. G. Orpen, *J. Chem. Soc., Chem. Commun.*, 1984, 841; S. R. Allen, R. G. Beevor, M. Green, N. C. Norman, A. G. Orpen, and I. D. Williams, *J. Chem. Soc., Dalton Trans.*, 1985, 435; F. J. Feher, M. Green, and R. A. Rodigues, *J. Chem. Soc., Chem. Commun.*, 1987, 1207.

20 C. P. Casey, M. S. Konings, R. E. Palermo, and R. E. Colborn, *J. Am. Chem. Soc.*, 1985, **107**, 5296; C. P. Casey, M. W. Meszaros, P. J. Fagan, R. K. Bly, and R. E. Colborn, *J. Am. Chem. Soc.*, 1986, **108**, 4053; C. P. Casey, M. W. Meszaros, S. R. Marder, R. K. Bly, and P. J. Fagan, *Organometallics*, 1986, **5**, 1873; C. P. Casey, L. K. Woo, P. J. Fagan, R. E. Palermo, and B. R. Adams, *Organometallics*, 1987, **6**, 447.

21 J. W. Faller, *Adv. Organomet. Chem.*, 1977, **16**, 211.

22 B. F. G. Johnson, *J. Chem. Soc., Chem. Commun.*, 1986, 27; B. F. G. Johnson and R. E. Benfield, *Transition Metal Clusters*, ed. B. F. G. Johnson, Wiley, 1980.

23 G. Wilkinson and T. S. Piper, *J. Inorg. Nucl. Chem.*, 1956, **2**, 32; M. J. Bennett Jr., F. A. Cotton, A. Davison, J. W. Faller, S. J. Lippard, and S. M. Morehouse, *J. Am. Chem. Soc.*, 1966, **88**, 4371.

24 K. J. Karel, T. A. Albright, and M. Brookhart, *Organometallics*, 1982, **1**, 419; B. E. Mann, *J. Chem. Soc., Chem. Commun.*, 1977, 626; J. A. Gibson and B. E. Mann, *J. Chem. Soc., Dalton Trans.*, 1979, 1021; J. Browning, C. S. Cundy, M. Green, and F. G. A. Stone, *J. Chem. Soc. A*, 1971, 448; J. Browning and B. R. Penfold, *J. Cryst. Mol. Struct.*, 1974, **4**, 335; J. Browning, M. Green, J. L. Spencer, and F. G. A. Stone, *J. Chem. Soc., Dalton Trans.*, 1974, 97; A. Greco, M. Green, and F. G. A. Stone, *J. Chem. Soc. A*, 1971, 285; A. Westerhof and H. J. de Liefde Meijer, *J. Organomet. Chem.*, 1978, **149**, 321.

25 T. A. Albright, P. Hofmann, R. Hoffmann, C. P. Lillya, and P. A. Dobosh, *J. Am. Chem. Soc.*, 1983, **105**, 3396; B. E. Mann and S. D. Shaw, *J. Organomet. Chem.*, 1987, **326**, C13; P. Berno, A. Ceccon, F. Dapra, A. Gambaro, and A. Venzo, *J. Chem. Soc., Chem. Commun.*, 1986, 1518; N. Morita, T. Asao, A. Tajiri, H. Sotokawa, and M. Hatano, *Tetrahedron Lett.*, 1986, **27**, 3873; P. L. Timms and R. B. King, *J. Chem. Soc., Chem. Commun.*, 1978, 898; M. Y. Darensbourg and E. L. Muetterties, *J. Am. Chem. Soc.*, 1978, **100**, 7425; I. B. Benson, S. A. R. Knox, R. F. D. Stansfield, and P. Woodward, *J. Chem. Soc., Dalton Trans.*, 1981, 51.

26 N. T. Ahn, M. Elian, and R. Hoffmann, *J. Am. Chem. Soc.*, 1978, **100**, 110.

27 F. A. Cotton and D. L. Hunter, *J. Am. Chem. Soc.*, 1976, **98**, 1413.

28 D. Ciappenelli and M. Rosenblum, *J. Am. Chem. Soc.*, 1969, **91**, 6876; K. J. Karel and M. Brookhardt, *J. Am. Chem. Soc.*, 1978, **100**, 1619.

29 H. W. Whitlock, Jr. and Y. N. Chuah, *Inorg. Chem.*, 1965, **4**, 424; H. W. Whitlock, Jr. and Y. N. Chuah, *J. Am. Chem. Soc.*, 1965, **87**, 3605; H. W. Whitlock, Jr. and R. L. Markezich, *J. Am. Chem. Soc.*, 1971, **93**, 5290; R. C. Kerber in *The Organic Chemistry of Iron*, Vol. 2, ed. E. A. Koerner Von Gustorf, F.-W. Grevels, and I. Fischer, Academic Press, 1981, pp. 16–17.

30 F. A. Cotton, D. L. Hunter, and P. Lahuerta, *J. Am. Chem. Soc.*, 1975, **97**, 1046.

31 R. Aumann, *Chem. Ber.*, 1975, **108**, 1974.

32 M. Tsutsui and A. Courtney, *Adv. Organomet. Chem.*, 1977, **16**, 241; K. Tatsumi and M. Tsutsui, *J. Mol. Catal.*, 1981, **13**, 117.

33 J. S. Merola, R. T. Kacmarcik, and D. Van Engen, *J. Am. Chem. Soc.*, 1986, **108**, 329.

34 J. K. Kochi, *Organometallic Mechanisms and Catalysis*, Academic Press, 1978.

35 R. Hoffmann, *Acc. Chem. Res.*, 1971, **4**, 1.

36 M. S. J. Dewar, *Bull. Soc. Chim. Fr.*, 1951, **2**, C71; J. Chatt and L. A. Duncanson, *J. Chem. Soc.*, 1953, 2939.

37 F. A. Cotton and R. M. Wing, *Inorg. Chem.*, 1965, **4**, 314.

38 F. A. Cotton, *Helv. Chim. Acta*, 1967, *Fasc. Extraordinarius*, (Alfred Werner Commemoration Volume, Verlag) p. 117.

39 T. A. Albright, *Tetrahedron*, 1982, **38**, 1339.

40 M. Elian and R. Hoffmann, *Inorg. Chem.*, 1975, **14**, 1058.

41 T. A. Albright and R. Hoffmann, *Chem. Ber.*, 1978, **111**, 1578.

42 J. W. Chin, Jr. and M. B. Hall, *Organometallics*, 1984, **3**, 284; A. B. Anderson, and G. Fitzgerald, *Inorg. Chem.*, 1981, **20**, 3288; S.-Y. Chu and R. Hoffmann, *J. Phys. Chem.*, 1982, **86**, 1289.

43 T. A. Albright, R. Hoffmann, and P. Hofmann, *Chem. Ber.*, 1978, **111**, 1591.

44 R. Hoffmann, T. A. Albright, and D. L. Thorn, *Pure. Appl. Chem.*, 1978, **50**, 1.

45 J. W. Lauher and R. Hoffmann, *J. Am. Chem. Soc.*, 1976, **98**, 1729; R. Hoffmann and P. Hofmann, *J. Am. Chem. Soc.*, 1976, **98**, 91; J. W. Lauher, M. Elian, R. H. Summerville, and R. Hoffmann, *J. Am. Chem. Soc.*, 1978, **98**, 3219; R. H. Summerville and R. Hoffmann, *J. Am. Chem. Soc.*, 1979, **101**, 3821; R. Hoffmann, M.-L. Chen, and D. L. Thorn, *Inorg. Chem.*, 1977, **16**, 503; T. A. Albright, P. Hofmann, and R. Hoffmann, *J. Am. Chem. Soc.*, 1977, **99**, 7546; D. L. Thorn and R. Hoffmann, *Inorg. Chem.*, 1978, **17**, 126; N. T. Anh, M. Elian, and R. Hoffmann, *J. Am. Chem. Soc.*, 1978, **100**, 110; R. Hoffmann, T. A. Albright, and D. L. Thorn, *Pure. Appl. Chem.*, 1978, **50**, 1; A. Dedieu and R. Hoffmann, *J. Am. Chem. Soc.*, 1978, **100**, 2074; B. E. R. Schilling, R. Hoffmann, and D. L. Lichtenberger, *J. Am. Chem. Soc.*, 1979, **101**, 585; B. E. R. Schilling and R. Hoffmann, *J. Am. Chem. Soc.*, 1979, **101**, 3456; T. A. Albright, R. Hoffmann, J. C. Thibeault, and D. L. Thorn, *J. Am. Chem. Soc.*, 1979, **101**, 3801; T. A. Albright, R. Hoffmann, Y.-C. Tse, and T. D'Ottavio, *J. Am. Chem. Soc.*, 1979, **101**, 3812; A. Dedieu, T. A. Albright, and R. Hoffmann, *J. Am. Chem. Soc.*, 1979, **101**, 3141; S. Shaik, R. Hoffmann, C. R. Fisel, and R. H. Summerville, *J. Am. Chem. Soc.*, 1980, **102**, 4555; R. J. Goddard, R. Hoffmann, and E. D. Jemmis, *J. Am. Chem. Soc.*, 1980, **102**, 7667; E. D. Jemmis, A. R. Pinhas, and R. Hoffmann, *J. Am. Chem. Soc.*, 1980, **102**, 2576; A. R. Pinhas, T. A. Albright, P. Hofmann, and R. Hoffmann, *Helv. Chim. Acta*, 1980, **63**, 29; R. Hoffmann, *Science*, 1981, **211**, 995; K. Tatsumi and R. Hoffmann, *J. Am. Chem. Soc.*, 1981, **103**, 3328; O. Eisenstein and R. Hoffmann, *J. Am. Chem. Soc.*, 1981, **103**, 4308.

46 E. Elian, M. M. L. Chen, D. M. P. Mingos, and R. Hoffmann, *Inorg. Chem.*, 1979, **15**, 1148.

47 B. J. Nicholson, *J. Am. Chem. Soc.*, 1966, **88**, 5156; D. W. Clack, M. Monshi, and L. A. P. Kane-Maguire, *J. Organomet. Chem.*, 1976, **107**, C40; D. W. Clack, M. Monshi, and L. A. P. Kane-Maguire, *J. Organomet. Chem.*, 1976, **120**, C25; D. M. P. Mingos, *J. Chem. Soc., Dalton Trans.*, 1977, 20, 26, and 31; D. W. Clack and K. D. Warren, *J. Organomet. Chem.*, 1978, **162**, 83; D. W. Clack and L. A. P. Kane-Maguire, *J. Organomet. Chem.*, 1979, **174**, 199; P. A. Dobosh, D. G. Gresham, C. P. Lillya, and E. S. Magyar, *Inorg. Chem.*, 1976, **15**, 2311; A. J. Pearson and P. R. Raithby, *J. Chem. Soc., Dalton Trans.*, 1981, 884; O. Eisenstein, W. M. Butler, and A. J. Pearson, *Organometallics*, 1984, **3**, 1150; M. F. Semmelhack, G. R. Clark, R. Farina, and

M. Saeman, *J. Am. Chem. Soc.*, 1979, **101**, 217; R. A. Kok and M. B. Hall, *J. Am. Chem. Soc.*, 1985, **107**, 2599; T. A. Albright and B. K. Carpenter, *Inorg. Chem.*, 1980, **19**, 3092; A. Solladié-Cavallo and G. Wipff, *Tetrahedron Lett.*, 1980, **21**, 3047; J. C. Boutonnet, L. Mordenti, E. Rose, O. le Martret, and G. Precigoux, *J. Organomet. Chem.*, 1981, **221**, 147.

48 K. Wade, *Inorg. Nucl. Chem. Lett.*, 1972, **8**, 559; D. M. P. Mingos, *Nature*, 1972, **236**, 99; D. M. P. Mingos, *J. Chem. Soc., Chem. Commun.*, 1983, 706; D. M. P. Mingos and D. G. Evans, *J. Organomet. Chem.*, 1983, **251**, C13; Y. L. Slovokhotov and Y. T. Struchkov, *J. Organomet. Chem.*, 1983, **258**, 47; B. K. Teo, *J. Chem. Soc., Chem. Commun.*, 1983, 1362; D. G. Evans and D. M. P. Mingos, *Organometallics*, 1983, **2**, 435; B. K. Teo, *Inorg. Chem.*, 1984, **23**, 1251; B. K. Teo, G. Longoni, and F. R. K. Chung, *Inorg. Chem.*, 1984, **23**, 1257.

49 P. S. Braterman, *Topics in Current Chemistry*, 1980, **92**, 149; A. Stockis and R. Hoffmann, *J. Am. Chem. Soc.*, 1980, **102**, 2952; R. J. McKinney, D. L. Thorn, R. Hoffmann, and A. Stockis, *J. Am. Chem. Soc.*, 1981, **103**, 2595.

50 R. Hoffmann, *Angew. Chem. Int. Ed. Engl.*, 1982, **21**, 711.

51 F. G. A. Stone, *Angew. Chem. Int. Ed. Engl.*, 1984, **23**, 89.

Chapter 2

1 M. J. D'Aniello, Jr. and E. K. Barefield, *J. Am. Chem. Soc.*, 1978, **100**, 1474.

2 B. Cruikshank and N. R. Davies, *Aust. J. Chem.*, 1966, **19**, 815.

3 H. Suzuki, Y. Koyama, Y. Moro-Oka, and T. Ikawa, *Tetrahedron Lett.*, 1979, 1415.

4 J. K. Nicholson and B. L. Shaw, *Tetrahedron Lett.*, 1965, 3533; R. E. Rinehart and J. S. Lasky, *J. Am. Chem. Soc.*, 1964, **86**, 2516.

5 T. A. Manuel, *J. Org. Chem.*, 1962, **27**, 3941.

6 R. Cramer, *J. Am. Chem. Soc.*, 1966, **88**, 2272.

7 H. Alper, P. C. LePort, and S. Wolfe, *J. Am. Chem. Soc.*, 1969, **91**, 7553.

8 F. G. Cowherd and J. L. von Rosenberg, *J. Am. Chem. Soc.*, 1969, **91**, 2157.

9 D. V. Banthorpe, H. Fitton, and J. Lewis, *J. Chem. Soc., Perkin Trans. 1*, 1973, 2051; K. J. Karel, M. Brookhart, and R. Aumann, *J. Am. Chem. Soc.*, 1981, **103**, 2695.

10 D. G. Bourner, L. Brammer, M. Green, G. Morgan, A. G. Orpen, C. Reeve, and C. J. Schaverien, *J. Chem. Soc., Chem. Commun.*, 1985, 1409.

11 C. P. Casey and C. R. Cyr, *J. Am. Chem. Soc.*, 1973, **95**, 2248.

12 J. F. Harrod and A. J. Chalk, *J. Am. Chem. Soc.*, 1964, **86**, 1776.

13 von E. Bertele and P. Schudel, *Helv. Chim. Acta*, 1967, **50**, 2445.

14 G. Brieger, T. J. Nestrick, and C. McKenna, *J. Org. Chem.*, 1969, **34**, 3789; F. J. McQuillin and D. G. Parker, *J. Chem. Soc., Perkin Trans. 1*, 1975, 2092.

15 J. Andrieux, D. H. R. Barton, and H. Patin, *J. Chem. Soc., Perkin Trans. 1*, 1977, 359.

16 A. B. Holmes, K. Russell, E. S. Stern, M. E. Stubbs, and N. K. Wellard, *Tetrahedron Lett.*, 1984, **25**, 4163.

17 P. A. Grieco and N. Marinovic, *Tetrahedron Lett.*, 1978, 2545.

18 D. Ma, Y. Lin, X. Lu, and Y. Yu, *Tetrahedron Lett.*, 1988, **29**, 1045.

19 R. J. McKinney, *Organometallics*, 1985, **4**, 1142.

20 A. J. Birch and G. S. R. Subba Rao, *Tetrahedron Lett.*, 1968, 3797.

21 A. Fischli, H. Mayer, W. Simon, and H.-J. Stoller, *Helv. Chim. Acta*, 1976, **59**, 397.

22 T. Hudlicky, T. M. Kutchan, S. R. Wilson, and David T. Mao, *J. Am. Chem. Soc.*, 1980, **102**, 6351.

23 G. Lesma, G. Palmisano, and S. Tollari, *J. Chem. Soc., Perkin Trans. 1*, 1984, 1593.

24 D. H. R. Barton, S. G. Davies, and W. B. Motherwell, *Synthesis*, 1979, 265.

25 P. A. Spanninger and J. L. von Rosenberg, *J. Org. Chem.*, 1969, **34**, 3658.

26 H. Alper and J. T. Edward, *J. Organomet. Chem.*, 1968, **14**, 411.

27 E. J. Corey and G. Moinet, *J. Am. Chem. Soc.*, 1973, **95**, 7185.

28 A. J. Birch, *Ann. N. Y. Acad. Sci.*, 1980, **333**, 107.

29 K. Tani, T. Yamagata, S. Otsuka, S. Akutagawa, H. Kumobayashi, T. Taketomi, H. Takaya, A. Miyashita, and R. Noyori, *J. Chem. Soc., Chem. Commun.*, 1982, 600;

30 R. Noyori, Royal Society Discussion Meeting, London, 24–25 Feb. 1988.

31 E. J. Corey and J. W. Suggs, *J. Org. Chem.*, 1973, **38**, 3224; *J. Am. Chem. Soc.*, 1984, **106**, 5208; S. Inoue, H. Takaya; K. Tani, S. Otsuka, T. Sato, and R. Noyori, *J. Am. Chem. Soc.*, 1990, **112**, 4897; W. Strohmeier and L. Weigelt, *J. Organomet. Chem.*, 1975, **86**, C17; Y. Sasson and G. L. Rempel, *Tetrahedron Lett.*, 1974, 4133.

32 M. Kitamura, K. Manabe, R. Noyori, and H. Takaya, *Tetrahedron Lett.*, 1987, **28**, 4719.

33 P. Golborn and F. Scheinmann, *J. Chem. Soc., Perkin Trans. 1*, 1973, 2870.

34 C. F. Lochow and R. G. Miller, *J. Org. Chem.*, 1976, **41**, 3020.

35 D. Baudry, M. Ephritikhine, and H. Felkin, *J. Chem. Soc., Chem. Commun.*, 1978, 694.

36 A. J. Hubert, A. Georis, R. Warin, and P. Teyssie, *J. Chem. Soc., Perkin Trans. 2*, 1972, 366.

37 J.-C. Fiaud and L. Aribi-Zouioueche, *J. Chem. Soc., Chem. Commun.*, 1986, 390.

38 J.-E. Bäckvall, E. E. Björkman, L. Pettersson, P. Siegbahn, and A. Strich, *J. Am. Chem. Soc.*, 1985, **107**, 7408.

39 M. R. A. Blomberg, P. E. M. Siegbahn, and J.-E. Bäckvall, *J. Am. Chem. Soc.*, 1987, **109**, 4450.

40 F. J. McQuillin and K. G. Powell, *J. Chem. Soc., Dalton Trans.*, 1972, 2123; R. J. Puddephatt, M. A. Quyser, and C. F. H. Tipper, *J. Chem. Soc., Chem. Commun.*, 1976, 626.

41 F. J. McQuillin and K. C. Powell, *J. Chem. Soc., Dalton Trans.*, 1972, 2129.

42 L. Cassar, P. E. Eaton, and J. Halpern, *J. Am. Chem. Soc.*, 1970, **92**, 3515.

43 P. G. Gassman and R. R. Reitz, *J. Am. Chem. Soc.*, 1973, **95**, 3057; L. A. Paquette, G. R. Allen, Jr., and R. P. Henzel, *J. Am. Chem. Soc.*, 1970, **92**, 7002.

44 P. G. Gassman and F. J. Williams, *J. Am. Chem. Soc.*, 1970, **92**, 7631.

45 H. W. Voigt and J. A. Roth, *J. Catal.*, 1974, **33**, 91.

46 P.-W. Chum and J. A. Roth, *J. Catal.*, 1975, **39**, 198.

47 R. G. Salomon, M. F. Salomon, and J. L. C. Kachinski, *J. Am. Chem. Soc.*, 1977, **99**, 1043.

48 A. de Meijere, *Tetrahedron Lett.*, 1974, **15**, 1845; A. de Meijere and L.-U. Meyer, *Tetrahedron Lett.*, 1974, **15**, 1849.

49 A. Stockis and E. Weissberger, *J. Am. Chem. Soc.*, 1975, **97**, 4288.

50 R. M. Giddings and D. Whittaker, *Tetrahedron Lett.*, 1978, **19**, 4077.

51 G. R. Clark and S. Thiensathit, *Tetrahedron Lett.*, 1985, **26**, 2503.

52 L. S. Liebeskind, D. Mitchell, and B. S. Foster, *J. Am. Chem. Soc.*, 1987, **109**, 7908.

53 V. Rautenstrauch, *J. Org. Chem.*, 1984, **49**, 950.

54 L. E. Overman and F. M. Knoll, *J. Am. Chem. Soc.*, 1980, **102**, 865.

55 N. Bluthe, M. Malacria, and J. Gore, *Tetrahedron Lett.*, 1983, **24**, 1157.

56 L. E. Overman and A. F. Renaldo, *Tetrahedron Lett.*, 1983, **24**, 3757.

57 L. E. Overman, *Angew. Chem., Int. Ed. Engl.*, 1984, **23**, 579.

58 L. E. Overman and A. F. Renaldo, *Tetrahedron Lett.*, 1983, **24**, 2235.

59 T. G. Schenck and B. Bosnich, *J. Am. Chem. Soc.*, 1985, **107**, 2058.

60 P. M. Henry, *J. Am. Chem. Soc.*, 1972, **94**, 5200.

61 L. E. Overman and F. M. Knoll, *Tetrahedron Lett.*, 1979, **20**, 321.

62 K. Dunne and F. J. McQuillin, *J. Chem. Soc. (C)*, 1970, 2196; K. Dunne and
   F. J. McQuillin, *J. Chem. Soc. (C)*, 1970, 2200.

63 B. M. Trost, T. R. Verhoeven, and J. M. Fortunak, *Tetrahedron Lett.*, 1979, **20**, 2301.

64 K. H. Meyer and K. Schuster, *Chem. Ber.*, 1922, **55**, 819; H. Rupe and E. Kambli, *Helv.*
   Chim. Acta, 1926, **9**, 672; A. Swaminathan and K. V. Narayanan, *Chem. Rev.*, 1971, **71**,
   429; A. W. Johnson, *The Chemistry of the Acetylenic Compounds*, Vol. I, Edward
   Arnold & Co., London, 1946.

65 G. Saucy, R. Marbet, H. Lindlar, and O. Isler, *Helv. Chim. Acta*, 1959, **42**, 1945.

66 P. Chabardes, E. Kuntz, and J. Varagnat, *Tetrahedron*, 1977, **33**, 1775.

67 G. L. Olson, H.-C. Cheung, K. D. Morgan, R. Borer, and G. Saucy, *Helv. Chim. Acta*,
   1976, **59**, 567.

68 S. Pürro, A. Pryde, J. Zsindely, and H. Schmid, *Helv. Chim. Acta*, 1978, **61**, 266.

69 S. Saito, S. Hamano, H. Moriyama, K. Okada, and T. Moriwake, *Tetrahedron Lett.*,
   1988, **29**, 1157.

70 R. W. Ashworth and G. A. Berchtold, *Tetrahedron Lett.*, 1977, **18**, 343.

71 R. Grigg, R. Hayes, and A. Sweeney, *J. Chem. Soc., Chem. Commun.*, 1971, 1248.

72 G. Adames, C. Bibby, and R. Grigg, *J. Chem. Soc., Chem. Commun.*, 1972, 491.

73 S. Achab, J.-P. Cosson, and B. C. Das, *J. Chem. Soc., Chem. Commun.*, 1984, 1040.

74 Y. D. Vankar, N. C. Chaudhuri, and S. P. Singh, *Synth. Comm.*, 1986, **16**, 1621.

75 A. F. Noels, J. J. Herman, and P. Teyssie, *J. Org. Chem.*, 1976, **41**, 2527.

76 W. M. Best, P. A. Collins, R. K. McCulloch, and D. Wege, *Aust. J. Chem.*, 1982, **35**, 843.

77 H. Glombik and W. Tochtermann, *Chem. Ber.*, 1984, **117**, 2422.

78 A. Bruggink and H. Hogeveen, *Tetrahedron Lett.*, 1972, **12**, 4961; H. Hogeveen and
   T. B. Middelkoop, *Tetrahedron Lett.*, 1973, **44**, 4325.

79 R. Roulet, J. Wenger, M. Hardy, and P. Vogel, *Tetrahedron Lett.*, 1974, **15**, 1479.

80 M. Suzuki, R. Noyori, and N. Hamanaka, *J. Am. Chem. Soc.*, 1981, **103**, 5606;
   M. Suzuki, R. Noyori, and N. Hamanaka, *J. Am. Chem. Soc.*, 1982, **104**, 2024.

81 J. H. Hoare, P. P. Policastro, and G. A. Berchtold, *J. Am. Chem. Soc.*, 1983, **105**, 6264.

82 J. L. Pawlak and G. A. Berchtold, *J. Org. Chem.*, 1987, **52**, 1765.

Chapter 3

1 E. G. E. Hawkins, *J. Chem. Soc.*, 1950, 2169.

2 M. I. Farberov, L. V. Mel'nik, B. N. Bobylev, and V. A. Podgornova, *Kinet. Catal. (Engl.*
  Transl.), 1971, **12**, 1018.

3 Y. M. Paushkin, S. A. Nizova, and B. T. Shcherbanenko, *Dokl. Chem. (Engl. Transl.)*,
  1971, **198**, 539.

4 R. Landau, World Petroleum Congress paper, in *Hydro. Process.*, 1967, **46**, 141.

5 H. Mimoun, *J. Mol. Catal.*, 1980, **7**, 1; N. N. Sheng and J. G. Zajacek, *Adv. Chem. Ser.*,
  1968, **76**, 418.

6 J. Sobczak and J. J. Zidkowski, *J. Mol. Catal.*, 1981, **13**, 11; R. A. Sheldon, *J. Mol. Catal.*, 1980, **7**, 107.

7 R. A. Sheldon and J. A. Van Doorn, *J. Catal.*, 1973, **31**, 427.

8 V. P. Rajan, S. N. Bannore, H. N. Subbarao, and S. Dev, *Tetrahedron*, 1984, **40**, 983.

9 M. Yamazaki, H. Endo, M. Tomoyama, and Y. Kurusu, *Bull. Chem. Soc. Jpn.*, 1983, **56**, 3523.

10 M. Pralus, J. C. Lecoq, and J. P. Shirmann, in *Fundamental Research in Homogeneous Catalysis*, Vol. 3, ed. M. Tsutsui, Plenum, New York, 1979, pp. 327–43.

11 Y. Matoba, H. Inoue, J. Akagi, T. Okabayashi, Y. Ishii, and M. Ogawa, *Synth. Commun.*, 1984, **14**, 865.

12 C. Venturello, E. Alneri, and M. Ricci, *J. Org. Chem.*, 1983, **48**, 3831.

13 M. A. Andrews, T. C.-T. Chang, C.-W. F. Cheng, and K. P. Kelly, *Organometallics*, 1984, **3**, 1777.

14 B. Meunier, E. Guilmet, M.-E. de Carvalho, and R. Poilblanc, *J. Am. Chem. Soc.*, 1984, **106**, 6668.

15 F. Chauvet, A. Heumann, and B. Waegell, *J. Org. Chem.*, 1987, **52**, 1916.

16 R. Sinigalia, R. A. Michelin, F. Pinna, and G. Strukul, *Organometallics*, 1987, **6**, 728.

17 H. Arakawa, Y. Moro-Oka, and A. Ozaki, *Bull. Chem. Soc. Jpn.*, 1974, **47**, 2958.

18 H. Mimoun, *Israel J. Chem.*, 1983, **23**, 451.

19 K. B. Sharpless and R. C. Michaelson, *J. Am. Chem. Soc.*, 1973, **95**, 6136.

20 W. C. Still, *J. Am. Chem. Soc.*, 1979, **101**, 2493.

21 S. Tanaka, H. Yamamoto, H. Nozaki, K. B. Sharpless, R. C. Michaelson, and J. D. Cutting, *J. Am. Chem. Soc.*, 1974, **96**, 5254; see also ref. 29.

22 K. C. Nicolaou, D. A. Claremon, and W. E. Barnette, *J. Am. Chem. Soc.*, 1980, **102**, 6611.

23 N. Indictor and W. F. Brill, *J. Org. Chem.*, 1985, **30**, 2074.

24 T. Itoh, K. Jitsukawa, K. Kaneda, and S. Teranishi, *J. Am. Chem. Soc.*, 1979, **10**, 159.

25 B. M. Trost and Y. Masuyama, *Israel J. Chem.*, 1984, **24**, 134.

26 O. Bortolini, S. Campestrini, F. Di Furia, and G. Modena, *J. Org. Chem.*, 1987, **52**, 5467.

27 R. K. Boeckmann, Jr. and E. W. Thomas, *J. Am. Chem. Soc.*, 1979, **101**, 987.

28 J. J. Partridge, V. Toome, and M. R. Uskoković, *J. Am. Chem. Soc.*, 1976, **98**, 3739.

29 B. E. Rossiter, T. R. Verhoeven, and K. B. Sharpless, *Tetrahedron Lett.*, 1979, 4733.

30 H. Tomioka, T. Suzuki, K. Oshima, and H. Nozaki, *Tetrahedron Lett.*, 1982, **23**, 3387.

31 A. S. Narula, *Tetrahedron Lett.*, 1982, **23**, 5579.

32 E. D. Mihelich, K. Daniels, and D. J. Eickhoff, *J. Am. Chem. Soc.*, 1981, **103**, 7690.

33 Y. Kobayashi, H. Uchiyama, H. Kanbara, and F. Sato, *J. Am. Chem. Soc.*, 1985, **107**, 5541.

34 T. Hiyama and M. Obayashi, *Tetrahedron Lett.*, 1983, **24**, 395.

35 M. Isobe, M. Kitamura, S. Mio, and T. Goto, *Tetrahedron Lett.*, 1982, **23**, 221.

36 R. D. Bach, G. J. Wolber, and B. A. Coddens, *J. Am. Chem. Soc.*, 1984, **106**, 6098.

37 P. Chaumette, H. Mimoun, L. Saussine, J. Fischer, and A. Mitschler, *J. Organomet. Chem.*, 1983, **250**, 291

38 A. O. Chong and K. B. Sharpless, *J. Org. Chem.*, 1977, **42**, 1587.

39 K. F. Purcell, *Organometallics*, 1985, **4**, 509.

40 R. A. Sheldon and J. A. Van Doorn, *J. Organomet. Chem.*, 1975, **94**, 115.

41 V. I. Garmonov and A. B. Nazarova, *J. Gen. Chem. USSR (Engl. Transl.)*, 1984, **54**, 1454.

42 H. Mimoun, L. Saussine, E. Daire, M. Postel, J. Fischer, and R. Weiss, *J. Am. Chem. Soc.*, 1983, **105**, 3101.

43 H. Mimoun, P. Chaumette, M. Mignard, L. Saussine, J. Fischer, and R. Weiss, *Nouv. J. Chim.*, 1983, **7**, 467.

44 K. B. Sharpless and T. R. Verhoeven, *Aldrichimica Acta*, 1979, **12**, 63.

45 K. Takai, K. Oshima, and H. Nozaki, *Tetrahedron Lett.*, 1980, **21**, 1657.

46 T. V. Lubben and P. T. Wolczanski, *J. Am. Chem. Soc.*, 1987, **109**, 424.

47 M. B. Groen and F. J. Zeelen, *Tetrahedron Lett.*, 1982, **23**, 3611.

48 P. A. Bartlett and W. S. Johnson, *J. Am. Chem. Soc.*, 1973, **95**, 7501.

49 M. Kobayashi, S. Kurozumi, T. Toru, and S. Ishimoto, *Chem. Lett.*, 1976, 1341.

50 M. Shibasaki, J. Ueda, and S. Ikegami, *Tetrahedron Lett.*, 1979, **20**, 433.

51 T. Fukuyama, B. Vranesic, D. P. Negri, and Y. Kishi, *Tetrahedron Lett.*, 1978, **19**, 2741.

52 T. Nakata and Y. Kishi, *Tetrahedron Lett.*, 1978, **19**, 2745.

53 M. Kitamura, M. Isobe, Y. Ichikawa, and T. Goto, *J. Am. Chem. Soc.*, 1984, **106**, 3252.

54 S. Masamune, Y. Hayase, W. Schilling, W. K. Chan, and G. S. Bates, *J. Am. Chem. Soc.*, 1977, **99**, 6756.

55 S. Yamada, T. Mashiko, and S. Terashima, *J. Am. Chem. Soc.*, 1977, **99**, 1988.

56 R. C. Michaelson, R. E. Palermo, and K. B. Sharpless, *J. Am. Chem. Soc.*, 1977, **99**, 1990.

57 T. Katsuki and K. B. Sharpless, *J. Am. Chem. Soc.*, 1980, **102**, 5974; K. B. Sharpless, *Chem. in Britain*, 1986, 38.

58 S. L. Schreiber, T. S. Schreiber, and D. B. Smith, *J. Am. Chem. Soc.*, 1987, **109**, 1525.

59 S. L. Schreiber and M. T. Goulet, *J. Am. Chem. Soc.*, 1987, **109**, 4718.

60 B. E. Rossiter and K. B. Sharpless, *J. Org. Chem.*, 1984, **49**, 3707.

61 S. Ikegami, T. Katsuki, and M. Yamaguchi, *Chem. Lett.*, 1987, 83.

62 R. Jackson and G. R. John, European Patent, 1983, 0070618 A1.

63 R. M. Hanson and K. B. Sharpless, *J. Org. Chem.*, 1986, **51**, 1922.

64 Z.-M. Wang, W.-S. Zhou, and G. Q. Lin, *Tetrahedron Lett.*, 1985, **26**, 6221.

65 Z.-M. Wang and W.-S. Zhou, *Tetrahedron*, 1987, **43**, 2935.

66 L. D.-L. Lu, R. A. Johnson, M. G. Finn, and K. B. Sharpless, *J. Org. Chem.*, 1984, **49**, 728.

67 V. S. Martin, S. S. Woodard, T. Katsuki, Y. Yamada, M. Ikeda, and K. B. Sharpless, *J. Am. Chem. Soc.*, 1981, **103**, 6237.

68 M. Kusakabe, H. Kato, and F. Sato, *Chem. Lett.*, 1987, 2163.

69 Y. Hanzawa, K. Kawagoe, M. Ito, and Y. Kobayashi, *Chem. Pharm. Bull.*, 1987, **35**, 1633.

70 K. B. Sharpless, S. S. Woodard, and M. G. Finn, *Pure Appl. Chem.*, 1983, **55**, 1823.

71 S. F. Pedersen, J. C. Dewan, R. R. Eckman, and K. B. Sharpless, *J. Am. Chem. Soc.*, 1987, **109**, 1279.

72 S. S. Woodward, M. G. Finn, and K. B. Sharpless, *J. Am. Chem. Soc.*, 1991, **113**, 106; K. B. Sharpless, Oral communication at an Organic Chemistry Colloquium at Cambridge University, January 1988.

73 K. A. Jorgensen, R. A. Wheeler, and R. Hoffmann, *J. Am. Chem. Soc.*, 1987, **109**, 3240.

74 J. M. Hawkins and K. B. Sharpless, *Tetrahedron Lett.*, 1987, **28**, 2825.

75 C. E. Adams, F. J. Walker, and K. B. Sharpless, *J. Org. Chem.*, 1985, **50**, 422.

76 A. C. Oehlschlager and E. Czyzewska, *Tetrahedron Lett.*, 1983, **24**, 5587.

77 R. Baker, C. J. Swain, and J. C. Head, *J. Chem. Soc., Chem. Commun.*, 1985, 309.

78 J. M. Klunder, S. Y. Ko, and K. B. Sharpless, *J. Org. Chem.*, 1986, **51**, 3710.

79 R. D. Tung and D. H. Rich, *Tetrahedron Lett.*, 1987, **28**, 1139.

80 J. M. Palazon and V. S. Martin, *Tetrahedron Lett.*, 1988, **29**, 681.

81 K. C. Nicolaou, R. A. Daines, J. Uenishi, W. S. Li, D. P. Papahatjis, and T. K. Chakraborty, *J. Am. Chem. Soc.*, 1987, **109**, 2205; K. C. Nicolaou, R. A. Daines, and T. K. Chakraborty, *J. Am. Chem. Soc.*, 1987, **109**, 2208; K. C. Nicolaou, T. K. Chakraborty, R. A. Daines, and Y. Ogawa, *J. Chem. Soc., Chem. Commun.*, 1987, 686.

82 B. H. Lipshutz, H. Kotsuki, and W. Lew, *Tetrahedron Lett.*, 1986, **27**, 4825.

83 S. T. Russell, J. A. Robinson, and D. J. Williams, *J. Chem. Soc., Chem. Commun.*, 1987, 351.

84 M. Aziz and F. Rouessac, *Tetrahedron Lett.*, 1987, **28**, 2579.

85 L. Pettersson, T. Frejd, and G. Magnusson, *Tetrahedron Lett.*, 1987, **28**, 2753.

86 K. S. Reddy, O.-H. Ko, D. Ho, P. E. Persons, and J. M. Cassady, *Tetrahedron Lett.*, 1987, **28**, 3075.

87 W. R. Roush and M. A. Adam, *J. Org. Chem.*, 1985, **50**, 3752.

88 G. P. Howe, S. Wang, and G. Procter, *Tetrahedron Lett.*, 1987, **28**, 2629.

89 H. Mimoun, I. Seree de Roch, and L. Sajus, *Bull. Soc. Chim. Fr.*, 1969, 1481; E. Vedejs, *J. Am. Chem. Soc.*, 1974, **96**, 5944; E. Vedejs, D. S. Engler, and J. E. Telschow, *J. Org. Chem.*, 1978, **43**, 188.

90 K. Yamada, Y. Kyotani, S. Manabe, and M. Suzuki, *Tetrahedron*, 1979, **35**, 293.

91 J. P. McCormick, W. Tomasik, and M. W. Johnson, *Tetrahedron Lett.*, 1981, 607.

Chapter 4

1 F. C. Phillips, *Amer. Chem. J.* 1894, **16**, 255.

2 J. Smidt, W. Hafner, R. Jira, J. Sedlmeier, R. Sieber, R. Ruttinger, and H. Kojer, *Angew. Chem.*, 1959, **71**, 176; J. Smidt, W. Hafner, R. Jira, R. Sieber, J. Sedlmeier, and A. Sabel, *Angew. Chem. Int. Ed. Engl.*, 1962, **1**, 80; J. Smidt, *Chem. & Ind.*, 1962, 54.

3 R. E. Kirk and D. F. Othmer, Ed., *Encyclopedia of Chem. Tech.*, 3rd edn., 1978, **1**, 97.

4 C. W. Bird, *Transition Metal Intermediates in Organic Synthesis*, Academic Press, 1966.

5 W. G. Lloyd and B. J. Luberoff, *J. Org. Chem.*, 1969, **34**, 3949.

6 N. T. Byrom, R. Grigg, and B. Kongkathip, *J. Chem. Soc., Chem. Commun.*, 1976, 216.

7 N. T. Byrom, R. Grigg, B. Kongkathip, G. Reiomer, and A. R. Wade, *J. Chem. Soc., Perkin Trans. 1*, 1984, 1643.

8 B. Kongkathip and N. Kongkathip, *Tetrahedron Lett.*, 1984, **25**, 2175.

9 J. Tsuji, *Synthesis*, 1984, 369.

10 J.-E. Bäckvall, B. Åkermark, and S. O. Ljunggren, *J. Am. Chem. Soc.*, 1979, **101**, 2411.

11 I. I. Moisseev, M. N. Vargaftik, and Y. K. Syrkin, *Dolk. Akad. Nauk. SSSR*, 1980, **130**, 820; W. H. Clement and C. M. Selwitz, *J. Org. Chem.*, 1964, **29**, 241; M. Kolb, E. Bratz, and K. Dialer, *J. Mol. Catal.*, 1977, **2**, 399.

12 H. Mimoun, R. Carpentier, A. Mitschler, J. Fischer, and R. Weiss, *J. Am. Chem. Soc.*, 1980, **102**, 1047.

13 J. Tsuji, H. Nagashima, and K. Hori, *Chem. Lett.*, 1980, 257.

14 J. Tsuji and M. Minato, *Tetrahedron Lett.*, 1987, **28**, 3683.

15 T. Hosokawa, T. Uno, S. Inui, and S.-I. Murahashi, *J. Am. Chem. Soc.*, 1981, **103**, 2318.

16 J. Muzart and J. P. Pete, *J. Mol. Catal.*, 1982, **15**, 373.

17 J.-E. Bäckvall and R. B. Hopkins, *Tetrahedron Lett.*, 1988, **29**, 2885.

18 I. Rico, F. Couderc, E. Perez, J. P. Laval, and A. Lattes, *J. Chem. Soc., Chem. Commun.*, 1987, 1205.

19 J. Tsuji, K. Mizutani, I. Shimizu, and K. Yamamoto, *Chem. Lett.*, 1976, 773.

20 C S. Subramaniam, P. J. Thomas, V. R. Mamdapur, and M. S. Chadha, *J. Chem. Soc., Perkin Trans. 1*, 1979, 2346.

21 T. Hosokawa, T. Ohta, and S.-I. Murahashi, *J. Chem. Soc., Chem. Commun.*, 1983, 848.

22 T. Hosokawa, T. Ohta, S. Kanayama, and S.-I. Murahashi, *J. Org. Chem.*, 1987, **52**, 1758.

23 H. Alper, K. Januszkiewicz, and D. J. H. Smith, *Tetrahedron Lett.*, 1985, **26**, 2263.

24 S. Takano, Y. Imamura, and K. Ogasawara, *Tetrahedron Lett.*, 1981, **22**, 4479.

25 H. Nagashima, K. Sakai, and J. Tsuji, *Chem. Lett.*, 1982, 859.

26 J. Tsuji, H. Nagashima, and K. Hori, *Tetrahedron Lett.*, 1982, **23**, 2679.

27 P. M. Henry, *Adv. Organomet. Chem.*, 1975, **13**, 363.

28 J. K. Stille and R. Divakaruni, *J. Am. Chem. Soc.*, 1978, **100**, 1303.

29 J. P. Collman and L. S. Hegedus, *Principles and Applications of Organotransition Metal Chemistry*, University Science Books, 1980, Ch. 12.

30 B. Åkermark, B. C. Söderberg, and S. S. Hall, *Organometallics*, 1987, **6**, 2608; B. C. Söderberg, B. Åkermark, Y.-H. Chen, and S. S. Hall, *J. Org. Chem.*, 1990, **55**, 1344.

31 J. Tsuji, I. Shimizu, H. Suzuki, and Y. Naito, *J. Am. Chem. Soc.*, 1979, **101**, 5070.

32 I. Shimizu, Y. Naito, and J. Tsuji, *Tetrahedron Lett.*, 1980, **21**, 487.

33 N. Cohen, *Acc. Chem. Res.*, 1976, **9**, 412.

34 J. Tsuji, Y. Kobayashi, and T. Takahashi, *Tetrahedron Lett.*, 1980, **21**, 483.

35 T. Takahashi, H. Ueno, M. Miyazawa, and J. Tsuji, *Tetrahedron Lett.*, 1985, **26**, 4463.

36 T. Takahashi, K. Kasuga, M. Takahashi, and J. Tsuji, *J. Am. Chem. Soc.*, 1979, **101**, 5072.

37 W. E. Walker, R. M. Manyik, K. E. Atkins, and M. L. Farmer, *Tetrahedron Lett.*, 1970, 3817; S. Takahashi, T. Shibano, and N. Hagihara, *Tetrahedron Lett.*, 1967, 2451.

38 J. Tsuji, I. Shimizu, and K. Yamamoto, *Tetrahedron Lett.*, 1976, 2975.

39 J. Tsuji, T. Yamada, and I. Shimizu, *J. Org. Chem.*, 1980, **45**, 5209.

40 E. Wenkert, B. L. Buckwalter, A. A. Craveireo, E. L. Sanchez, and S. S. Sathe, *J. Am. Chem. Soc.*, 1978, **100**, 1267.

41 K. Iseki, M. Yamazaki, M. Shibasaki, and S. Ikegami, *Tetrahedron Lett.*, 1981, **37**, 4411.

42 J. Tsuji, I. Shimizu, and Y. Kobayashi, *Israel J. Chem.*, 1984, **24**, 153.

43 J. Tsuji, Y. Kobayashi, and I. Shimizu, *Tetrahedron Lett.*, 1979, 39.

44 F. Sato, H. Watanabe, Y. Tanaka, T. Yamaji, and M. Sato, *Tetrahedron Lett.*, 1983, **24**, 1041.

45 G. T. Rodeheaver and D. F. Hunt, *J. Chem. Soc., Chem. Commun.*, 1971, 818.

46 T. Yanami, M. Miyashita, and A. Yoshikoshi, *J. Chem. Soc., Chem. Commun.*, 1979, 525.

47 F. J. McQuillin and D. G. Parker, *J. Chem. Soc., Perkin Trans. 1*, 1975, 2092.

48 B. R. James and G. L. Rempel, *Canad. J. Chem.*, 1968, **46**, 571; H. Mimoun, M. M. P. Machirant, and I. S. de Roch, *J. Am. Chem. Soc.*, 1978, **100**, 5437.

49 W. Kitching, Z. Rappoport, S. Winstein, and W. G. Young, *J. Am. Chem. Soc.*, 1966, **88**, 2054.

50 P. M. Henry, *J. Org. Chem.*, 1967, **32**, 2575; P. M. Henry, *J. Org. Chem.*, 1973, **38**, 1681.

51 M. Green, R. N. Haszeldine, and J. Lindley, *J. Organomet. Chem.*, 1966, **6**, 107.

52 S. Wolfe and P. G. C. Campbell, *J. Am. Chem. Soc.*, 1971, **93**, 1497.

53 A. Heumann and B. Åkermark, *Angew. Chem. Int. Ed. Engl.*, 1984, **23**, 453.

54 H. O. House, L. J. Czuba, M. Gall, and H. D. Olmstead, *J. Org. Chem.*, 1969, **34**, 2324.

55 Y. Ito, T. Hirao, and T. Saegusa, *J. Org. Chem.*, 1978, **43**, 1011.

56 B. M. Trost, Y. Nishimura, K. Yamamoto, and S. S. McElvain, *J. Am. Chem. Soc.*, 1979, **101**, 1328.

57 R. E. Ireland, J. D. Godfrey, and S. Thaisrivongs, *J. Am. Chem. Soc.*, 1981, **103**, 2446.

58 P. A. Wender and J. C. Lechleiter, *J. Am. Chem. Soc.*, 1980, **102**, 6340.

59 J. M. Reuter, A. Sinha, and R. G. Salomon, *J. Org. Chem.*, 1978, **43**, 2438.

60 Y. Ito, H. Aoyama, T. Hirao, A. Mochizuki, and T. Saegusa, *J. Am. Chem. Soc.*, 1979, **101**, 494.

61 Y. Ito, H. Aoyama, and T. Saegusa, *J. Am. Chem. Soc.*, 1980, **102**, 4519.

62 J. Tsuji, I. Minami, I. Shimizu, and H. Kataoka, *Chem. Lett.*, 1984, 1133.

63 J. Tsuji, I. Minami, and I. Shimizu, *Tetrahedron Lett.*, 1983, **24**, 5635.

64 E. Mincione, G. Ortaggi, and A. Sirna, *Synthesis*, 1977, 773.

65 S.-I. Murahashi, T. Tsumiyana, and Y. Mitsue, *Chem. Lett.*, 1984, 1419.

Chapter 5

1 B. Åkermark, J.-E. Bäckvall, K. Siirala-Hansén, K. Sjöberg, and K. Zetterberg, *Tetrahedron Lett.*, 1974, 1363; see also *J. Organomet. Chem.*, 1974, **72**, 127.

2 G. Palaro, A. de Renzi, and R. Palumbo, *J. Chem. Soc., Chem. Commun.*, 1967, 1150.

3 J.-E. Bäckvall, E. E. Bjorkman, S. E. Bystrom, and A Solladie-Cavallo, *Tetrahedron Lett.*, 1982, 943.

4 L. S. Hegedus, B. Åkermark, K. Zetterberg, and L. F. Olsson, *J. Am. Chem. Soc.*, 1984, **106**, 7122.

5 R. A. Holton and R. A. Kjonaas, *J. Am. Chem. Soc.*, 1977, **99**, 4177.

6 R. A. Holton and R. A. Kjonaas, *J. Organomet. Chem.*, 1977, **142**, C15.

7 R. A. Holton, *J. Am. Chem. Soc.*, 1977, **99**, 8083.

8 J. K. Stille and R. A. Morgan, *J. Am. Chem. Soc.*, 1966, **88**, 5135.

9 R. C. Larock, K. Takagi, S. S. Hershberger, and M. A. Mitchell, *Tetrahedron Lett.*, 1981, 5231.

10 R. C. Larock and S. Babu, *Tetrahedron Lett.*, 1985, **26**, 2763.

11 L. S. Hegedus, R. E. Williams, M. A. McGuire, and T. Hayashi, *J. Am. Chem. Soc.*, 1980, **102**, 4973.

12 M. A. Andrews, T. C.-T. Chang, C.-W. Cheng, L. V. Kapustay, K. P. Kelly, and M. J. Zweifel, *Organometallics*, 1984, **3**, 1479.

13 H. Takahashi and J. Tsuji, *J. Am. Chem. Soc.*, 1968, **90**, 2387.

14 P. M. Henry, M. Davies, G. Ferguson, S. Phillips, and R. Restivo, *J. Chem. Soc., Chem. Commun.*, 1974, 112.

15 A. Heumann, M. Reglier, and B. Waegell, *Tetrahedron Lett.*, 1983, **24**, 1971.

16 L. S. Hegedus, G. F. Allen, and E. L. Waterman, *J. Am. Chem. Soc.*, 1976, **98**, 2674.

17 A. Kasahara, T. Izumi, and M. Ooshima, *Bull. Chem. Soc. Jpn.*, 1974, **47**, 2526.

18 C. Lambert, K. Utimoto, and H. Nozaki, *Tetrahedron Lett.*, 1984, **25**, 5323.

19 O. L. Chapman, M. R. Engel, J. P. Springer, and J. C. Clardy, *J. Am. Chem. Soc.*, 1971, **93**, 6696.

20 B. F. Hallam and P. L. Pauson, *J. Chem. Soc.*, 1956, 3030.

21 E. O. Fischer and K. Fichtel, *Chem. Ber.*, 1961, **94**, 1200.

22 M. L. H. Green and P. L. I. Nagy, *J. Chem. Soc.*, 1963, 189.

23 W. P. Giering and M. Rosenblum, *J. Chem. Soc., Chem. Commun.*, 1971, 441; A. Cutler, D. Ehntholt, W. P. Giering, P. Lennon, S. Raghu, A. Rosan, M. Rosenblum, J. Tancrede, and D. Wells, *J. Am. Chem. Soc.*, 1976, **98**, 3495; S. Samuels, S. R. Berryhill, and M. Rosenblum, *J. Organomet. Chem.*, 1979, **166**, C9.

24 A. Rosan, M. Rosenblum, and J. Tancrede, *J. Am. Chem. Soc.*, 1973, **95**, 3062; M. Rosenblum, *Acc. Chem. Res.*, 1974, **7**, 122.

25 M. Rosenblum, T. C. T. Chang, B. M. Foxman, S. B. Samuels, and C. Stockman, *Org. Syn. Today, Tomorrow, Proc. 3rd IUPAC Sym. Org. Synth.*, 1980, 47.

26 M. Rosenblum, A. Bucheister, T. C. T. Chang, M. Cohen, M. Marsi, S. B. Samuels, D. Scheck, N. Sofen, and J. C. Watkins, *Pure & Appl. Chem.*, 1984, **56**, 129.

27 P. Lennon, A. M. Rosan, and M. Rosenblum, *J. Am. Chem. Soc.*, 1977, **99**, 8426.

28 A. Rosan and M. Rosenblum, *J. Org. Chem.*, 1975, **40**, 3621.

29 L. Busetto, A. Palazzi, R. Ros, and U. Belluco, *J. Organomet. Chem.*, 1970, **25**, 207; P. Lennon, M. Madhavarao, A. Rosan, and M. Rosenblum, *J. Organomet. Chem.*, 1976, **108**, 93.

30 D. E. Laycock, J. Hartgerink, and M. C. Baird, *J. Org. Chem.*, 1980, **45**, 291.

31 W. P. Giering, M. Rosenblum, and J. Tancrede, *J. Am. Chem. Soc.*, 1972, **94**, 7170.

32 D. F. Marten and M. N. Akbari, *J. Organomet. Chem.*, 1987, **322**, 99.

33 D. L. Reger, C. J. Coleman, and P. J. McElligott, *J. Organomet. Chem.*, 1979, **171**, 73.

34 T. C. T. Chang, M. Rosenblum, and S. B. Samuels, *J. Am. Chem. Soc.*, 1980, **102**, 5930.

35 T. C. T. Chang and M. Rosenblum, *J. Org. Chem.*, 1981, **46**, 4103.

36 T. C. T. Chang, T. S. Coolbaugh, B. M. Foxman, M. Rosenblum, N. Simms, and C. Stockman, *Organometallics*, 1987, **6**, 2394.

37 T. C. T. Chang, B. M. Foxman, M. Rosenblum, and C. Stockman, *J. Am. Chem. Soc.*, 1981, **103**, 7361.

38 T. C. T. Chang and M. Rosenblum, *Tetrahedron Lett.*, 1983, **24**, 695.

39 T. C. T. Chang and M. Rosenblum, *Israel J. of Chem.*, 1984, **24**, 99.

40 M. Marsi and M. Rosenblum, *J. Am. Chem. Soc.*, 1984, **106**, 7264.

41 M. Rosenblum, M. M. Turnbull, and B. M. Foxman, *Organometallics*, 1986, **5**, 1062.

42 S. N. Anderson, C. W. Fong, and M. D. Johnson, *J. Chem. Soc., Chem. Commun.*, 1973, 163; K. M. Nicholas and M. Rosenblum, *J. Am. Chem. Soc.*, 1973, **95**, 4449.

43 A. J. Birch and I. D. Jenkins, in *Transition Metal Organometallics in Organic Synthesis*, ed. H. Alper, Academic Press, London, 1976, p. 8.

44 P. L. Bock, D. J. Boschetto, J. R. Rasmussen, J. P. Demers, and G. M. Whitesides, *J. Am. Chem. Soc.*, 1974, **96**, 2814; P. L. Bock and G. M. Whitesides, *J. Am. Chem. Soc.*, 1974, **96**, 2826.

45 T. C. Flood and D. L. Miles, *J. Organomet. Chem.*, 1977, **127**, 33.

46 B. W. Roberts and J. Wong, *J. Chem. Soc., Chem. Commun.*, 1977, 20.

47 D. J. Bates, M. Rosenblum, and S. B. Samuels, *J. Organomet. Chem.*, 1981, **209**, C55.

48 J. Benaim and A. L'Honore, *J. Organomet. Chem.*, 1981, **202**, C53.

49 D. M. T. Chan, T. B. Marder, D. Milstein, and N. J. Taylor, *J. Am. Chem. Soc.*, 1987, **109**, 6385.

50 D. L. Reger, K. A. Belmore, E. Mintz, and P. J. McElligott, *Organometallics*, 1984, **3**, 134.

51 M. Rosenblum and D. Scheck, *Organometallics*, 1982, **1**, 397; D. L. Reger, P. J. McElligott, N. G. Charles, E. A. H. Griffith, and E. L. Amma, *Organometallics*, 1982, **1**, 443.

52　A. Rosan, M. Rosenblum, and J. Tancrede, *J. Am. Chem. Soc.*, 1973, **95**, 3060; D. F. Marten, *J. Chem. Soc., Chem. Commun.*, 1980, 341.

53　M. H. Chisholm and H. C. Clark, *Inorganic Chem.*, 1971, **10**, 2557; H. C. Clark and H. Kurosawa, *Inorganic Chem.*, 1972, **11**, 1275; M. H. Chisholm and H. C. Clark, *J. Am. Chem. Soc.*, 1972, **94**, 1532.

54　S. R. Allen, P. K. Baker, S. G. Barnes, M. Bottrill, M. Green, A. G. Orpen, I. D. Williams, and A. J. Welch, *J. Chem. Soc., Dalton Trans.*, 1983, 927; M. Bottrill and M. Green, *J. Am. Chem. Soc.*, 1977, **99**, 5795.

55　R. F. Lockwood and K. M. Nicholas, *Tetrahedron Lett.*, 1977, 4163.

56　S. Padmanabhan and K. M. Nicholas, *Tetrahedron Lett.*, 1983, **24**, 2239.

57　H. D. Hodes and K. M. Nicholas, *Tetrahedron Lett.*, 1970, 4349.

58　J. E. O'Boyl and K. M. Nicholas, *Tetrahedron Lett.*, 1980, **21**, 1595.

59　K. M. Nicholas, M. Mulvaney, and M. Bayer, *J. Am. Chem. Soc.*, 1980, **102**, 2508.

60　M. Saha and K. M. Nicholas, *Israel J. Chem.*, 1984, **24**, 105.

61　S. Padmanabhan and K. M. Nicholas, *Tetrahedron Lett.*, 1982, **23**, 2555.

62　G. S. Mikaelian, A. S. Gybin, W. A. Smit, and R. Caple, *Tetrahedron Lett.*, 1985, **26**, 1269.

63　J. A. Marshall and W. Y. Gung, *Tetrahedron Lett.*, 1989, **30**, 309,

64　R. Tester, V. Vargese, A. M. Montana, M. Khan, and K. M. Nicholas, *J. Org. Chem.*, 1990, **55**, 186.

Chapter 6

1　B. M. Trost, *Acc. Chem. Res.*, 1980, **13**, 385.

2　J. Tsuji, *Pure Appl. Chem.*, 1982, **54**, 197.

3　B. M. Trost and P. E. Strege, *Tetrahedron Lett.*, 1974, 2603.

4　A. Kasahara, K. Tanaka, and K. Asamiya, *Bull. Chem. Soc. Jpn.*, 1967, **40**, 351.

5　B. M. Trost and P. E. Strege, *J. Am. Chem. Soc.*, 1975, **97**, 2534.

6　K. Dunne and F. J. McQuillin, *J. Chem. Soc. (C)*, 1970, 2200.

7　B. M. Trost, P. E. Strege, L. Weber, T. J. Fullerton, and T. J. Dietsche, *J. Am. Chem. Soc.*, 1978, **100**, 3407.

8　B. M. Trost, T. J. Dietsche, and T. J. Fullerton, *J. Org. Chem.*, 1974, **39**, 737.

9　M. Takahashi, H. Suzuki, Y. Moro-Oka, and T. Ikawa, *Chem. Lett.*, 1979, 53.

10　T. A. Albright, P. R. Clemens, R. P. Hughes, D. E. Hunton, and L. D. Margerum, *J. Am. Chem. Soc.*, 1982, **104**, 5369.

11　W. A. Donaldson, *Organometallics*, 1986, **5**, 223.

12　W. A. Donaldson, *J. Organomet. Chem.*, 1984, **269**, C25.

13　W. A. Donaldson, *Tetrahedron Lett.*, 1987, **43**, 2901.

14　W. T. Dent, R. Long, and A. J. Wilkinson, *J. Chem. Soc.*, 1964, 1585; M. Sakakibara, Y. Takahashi, S. Sakai, and Y. Ishi, *J. Chem Soc., Chem. Commun.*, 1969, 396.

15　K. Henderson, Ph.D. Thesis, University of Newcastle upon Tyne, 1977 (unpublished).

16　J. K. Nicholson, J. Powell, and B. L. Shaw, *J. Chem. Soc., Chem. Commun.*, 1966, 174.

17　Y. Inoue, J. Yamashita, and H. Hashimoto, *Synthesis*, 1984, 244.

18　B. M. Trost and T. J. Fullerton, *J. Am. Chem. Soc.*, 1973, **95**, 292

19　B. M. Trost, *Tetrahedron*, 1977, **33**, 2615

20　B. Åkermark, B. Krakenberger, S. Hansson, and A. Vitagliano, *Organometallics*, 1987, **6**, 620.

21　W. R. Jackson and J. U. G. Strauss, *Tetrahedron Lett.*, 1975, 2591.

22 B. M. Trost and L. Weber, *J. Org. Chem.*, 1975, **40**, 3617.

23 D. J. Colins, W. R. Jackson, and R. N. Timms, *Aust. J. Chem.*, 1977, **30**, 2167.

24 Y. Tamaru, M. Kagotani, R. Suzuki, and Z.-I. Yoshida, *J. Org. Chem.*, 1981, **46**, 3374.

25 B. M. Trost, L. Weber, P. Strege, T. J. Fullerton, and T. J. Dietsche, *J. Am. Chem. Soc.*, 1978, **100**, 3426.

26 B. M. Trost and S. J. Brickner, *J. Am. Chem. Soc.*, 1983, **105**, 568.

27 P. S. Manchand, H. S. Wong, and J. F. Blount, *J. Org. Chem.*, 1978, **43**, 4769.

28 E. Keinan and Z. Roth, *J. Org. Chem.*, 1983, **48**, 1769.

29 B. M. Trost and M.-H. Hung, *J. Am. Chem. Soc.*, 1984, **106**, 6837.

30 R. J. P. Corriu, N. Escudié, and C. Guerin, *J. Organomet. Chem.*, 1984, **271**, C7.

31 S. A. Godleski and E. B. Villhauer, *J. Org. Chem.*, 1984, **49**, 2246.

32 J. Tsuji, M. Yuhara, M. Minato, H. Yamada, F. Sato, and Y. Kobayashi, *Tetrahedron Lett.*, 1988, **29**, 343.

33 B. M. Trost and F. W. Gowland, *J. Org. Chem.*, 1979, **44**, 3448.

34 J. Zhu and X. Lu, *Tetrahedron Lett.*, 1987, **28**, 1897.

35 Y. Kitagawa, A. Itoh, S. Hashimoto, H. Yamamoto, and H. Nozaki, *J. Am. Chem. Soc.*, 1977, **99**, 3864.

36 B. M. Trost and T. R. Verhoeven, *J. Am. Chem. Soc.*, 1977, **99**, 3867.

37 B. M. Trost and T. R. Verhoeven, *Tetrahedron Lett.*, 1978, 2275.

38 B. M. Trost, in *Second SCI–RSC Medicinal Chem. Sym.*, Special Publication 50, ed. J. C. Emmett, The Royal Society of Chemistry, 1983, 162.

39 J. Tsuji, Y. Kobayashi, H. Kataoka, and T. Takahashi, *Tetrahedron Lett.*, 1980, 1475.

40 S. A. Godleski and R. S. Valpey, *J. Org. Chem.*, 1982, **47**, 381.

41 B. M. Trost and A. Tenaglia, *Tetrahedron Lett.*, 1988, **29**, 2927.

42 A. K. Kende, B. Roth, and P. J. Sanfilippo, *J. Am. Chem. Soc.*, 1982, **104**, 1784.

43 A. S. Kende, R. A. Battista, and S. B. Sandoval, *Tetrahedron Lett.*, 1984, **25**, 1341.

44 B. M. Trost and L. Weber, *J. Am. Chem. Soc.*, 1975, **97**, 1611.

45 D. J. Collins, W. R. Jackson, and R. N. Timms, *Tetrahedron Lett.*, 1976, 495.

46 B. M. Trost and T. R. Verhoeven, *J. Org. Chem.*, 1976, **41**, 3215.

47 B. M. Trost, T. R. Verhoeven, and J. M. Fortunak, *Tetrahedron Lett.*, 1979, 2301.

48 B. M. Trost and E. Keinan, *J. Am. Chem. Soc.*, 1978, **100**, 7779.

49 T. Hayashi, T. Hagihara, M. Konishi, and M. Kumada, *J. Am. Chem. Soc.*, 1983, **105**, 7767.

50 J.-C. Fiaud and J.-Y. Legros, *J. Org. Chem.*, 1987, **52**, 1907.

51 J.-E. Bäckvall, R. E. Nordberg, and J. Vägberg, *Tetrahedron Lett.*, 1983, **24**, 411.

52 R. E. Nordberg and J.-E. Bäckvall, *J. Organomet. Chem.*, 1985, **285**, C24.

53 S. A. Godleski, K. B. Gundlach, H. Y. Ho, E. Keinan, and F. Frolow, *Organometallics*, 1984, **3**, 21.

54 E. Keinan and N. Greenspoon, *Tetrahedron Lett.*, 1982, 241.

55 J.-E. Bäckvall, R. E. Nordberg, E. E. Björkman, and C. Moberg, *J. Chem. Soc., Chem. Commun.*, 1980, 943.

56 H. Matsushita and E.-I. Negishi, *J. Chem. Soc., Chem. Commun.*, 1982, 160.

57 J. S. Temple, M. Riediker, and J. Schwartz, *J. Am. Chem. Soc.*, 1982, **104**, 1310.

58 B. M. Trost and T. R. Verhoeven, *J. Am. Chem. Soc.*, 1976, **98**, 630; B. M. Trost and T. R. Verhoeven, *J. Am. Chem. Soc.*, 1978, **100**, 3435.

59 M. Riediker and J. Schwartz, *Tetrahedron Lett.*, 1981, 4655.

60 Y. Hayasi, M. Riediker, J. S. Temple, and J. Schwartz, *Tetrahedron Lett.*, 1981, **22**, 2629.

61 K. Yamamoto, R. Deguchi, Y. Ogimura, and J. Tsuji, *Chem. Lett.*, 1984, 1657.

62 T. Takahashi, Y. Jinbo, K. Kitamura, and J. Tsuji, *Tetrahedron Lett.*, 1984, **25**, 5921.

63 B. M. Trost and T. P. Klun, *J. Am. Chem. Soc.*, 1981, **103**, 1864.

64 B. M. Trost and T. P. Klun, *J. Org. Chem.*, 1980, **45**, 4256.

65 B. M. Trost and T. J. Dietsche, *J. Am. Chem. Soc.*, 1973, **95**, 8200.

66 J. C. Fiaud and J. V. Legros, *Tetrahedron Lett.*, 1988, **29**, 2959.

67 B. M. Trost and P. E. Strege, *J. Am. Chem. Soc.*, 1977, **99**, 1649.

68 B. Bosnich and P. B. Mackenzie, *Pure Appl. Chem.*, 1982, **54**, 189.

69 P. B. Mackenzie, J. Whelan, and B. Bosnich, *J. Am. Chem. Soc.*, 1985, **107**, 2046;
    P. A. Auburn, P. B. Mackenzie, and B. Bosnich, *J. Am. Chem. Soc.*, 1985, **107**, 2033.

70 B. M. Trost and D. J. Murphy, *Organometallics*, 1985, **4**, 1143.

71 J. C. Fiaud and J. L. Malleron, *Tetrahedron Lett.*, 1981, **23**, 1399.

72 T. Hosokawa, S. Miyagi, S.-I. Murahashi, and A. Sonoda, *J. Chem. Soc., Chem. Commun.*, 1978, 687.

73 T. Hayashi, A. Yamamoto, and Y. Ito, *Tetrahedron Lett.*, 1988, **29**, 99, 669

74 J. P. Genêt and F. Colobert, *Tetrahedron Lett.*, 1985, **26**, 2779.

75 T Hayashi, A Yamamoto, and Y. Ito, *Chem. Lett.*, 1987, 177.

76 J.-C. Fiaud and J.-L. Malleron, *J. Chem. Soc., Chem. Commun.*, 1981, 1159.

77 B. M. Trost and E. Keinan, *Tetrahedron Lett.*, 1980, 2591.

78 H. H. Baer and Z. S. Hanna, *Carbohydrate Res.*, 1980, **78**, C11.

79 B. M. Trost, G.-H. Kuo, and T. Benneche, *J. Am. Chem. Soc.*, 1988, **110**, 621.

80 M. Brakta, F. Le Borgne, and D. Sinou, *J. Carbohydr. Chem.*, 1987, **6** (2), 307.

81 J. P. Genêt, M. Balabane, J.-E. Bäckvall, and J.-E. Nyström, *Tetrahedron Lett.*, 1983, **24**, 2745; J.-E. Bäckvall, R. E. Nordberg, and J.-E. Nyström, *Tetrahedron Lett.*, 1982, 1617.

82 S. E. Byström, R. Aslanian, and J.-E. Bäckvall, *Tetrahedron Lett.*, 1985, **26**, 1749.

83 J.-E. Bäckvall, S. E. Byström, and R. E. Nordberg, *J. Org. Chem.*, 1984, **49**, 4619.

84 J.-E. Bäckvall and J. O. Vågberg, *J. Org. Chem.*, 1988, **53**, 5695; J.-E. Bäckvall and A. Gogoll, *Tetrahedron Lett.*, 1988, **29**, 2243.

85 J.-E. Bäckvall, Z. D. Renko, and S. E. Byström, *Tetrahedron Lett.*, 1987, **28**, 4199.

86 M. Souchet, M. Baillargé, and F. Le Goffic, *Tetrahedron Lett.*, 1988, **29**, 191.

87 J.-E. Bäckvall and A. Gogoll, *J. Chem. Soc., Chem. Commun.*, 1987, 1237.

88 D. Ferroud, J. P. Genêt, and R. Kiolle (in part), *Tetrahedron Lett.*, 1986, **27**, 23.

89 J. P. Genêt, J. Uziel, and S. Juge, *Tetrahedron Lett.*, 1988, **29**, 4559.

90 M. Nikaido, R. Aslanian, F. Scavo, P. Helquist, B. Åkermark, and J.-E. Bäckvall, *J. Org. Chem.*, 1984, **49**, 4738.

91 Y. Tamaru, Y. Yamada, M. Kagotani, H. Ochiai, E. Nakajo, R. Suzuki, and Z.-I. Yoshida, *J. Org. Chem.*, 1983, **48**, 4669.

92 T. Hayashi, A. Yamamoto, T. Iwata, and Y. Ito, *J. Chem. Soc., Chem. Commun.*, 1987, 398.

93 T. Tabuchi, J. Inanaga, and M. Yamaguchi, *Tetrahedron Lett.*, 1987, **28**, 215.

94 Y. Tsukahara, H. Kinoshita, K. Inomata, and H. Kotake, *Bull. Chem. Soc. Jpn.*, 1984, **57**, 3013.

95 Y. Inoue, M. Taguchi, M. Toyofuku, and H. Hashimoto, *Bull. Chem. Soc. Jpn.*, 1984, **57**, 3021.

96 V. I. Ognyanov and M. Hesse, *Synthesis*, 1985, 645.

97 J.-P. Haudegond, Y. Chauvin, and D. Commereuc, *J. Org. Chem.*, 1979, **44**, 3063.

98 J. P. Genêt and S. Grisoni, *Tetrahedron Lett.*, 1986, **27**, 4165.

99 J. P. Genêt and S. Grisoni, *Tetrahedron Lett.*, 1988, **29**, 4543.

100 J. Godschafx and J. K. Stille, *Tetrahedron Lett.*, 1980, **21**, 2599; A. Goliaszewski and J. Schwartz, *Organometallics*, 1985, **4**, 417; B. M. Trost and E. Keinan, *Tetrahedron Lett.*, 1980, **21**, 2595.

101 M. Moreno-Mañas, M. Prat, J. Ribas, and A. Virgili, *Tetrahedron Lett.*, 1988, **29**, 581.

102 E.-I. Negishi, S. Chatterjee, and H. Matsushita, *Tetrahedron Lett.*, 1981, **22**, 3737.

103 T. Hayashi, M. Konishi, K.-I. Yokota, and M. Kumada, *J. Organomet. Chem.*, 1985, **285**, 359.

104 J. Tsuji, *Tetrahedron Lett.*, 1986, **42**, 4361.

105 J. Tsuji, I. Shimizu, I. Minami, Y. Ohashi, T. Sugiura, and K. Takahashi, *J. Org. Chem.*, 1985, **50**, 1523.

106 X. Lu and Z. Ni, *Synthesis*, 1987, 66.

107 N. Ono, I. Hamamoto, and A. Kaji, *J. Chem. Soc., Chem. Commun.*, 1982, 821.

108 R. Tamura, M. Kato, K. Saegusa, M. Kakihana, and D. Oda, *J. Org. Chem.*, 1987, **52**, 4121.

109 K. Hiroi, R. Kitayama, and S. Sato, *J. Chem. Soc., Chem. Commun.*, 1984, 303.

110 B. M. Trost and T. A. Runge, *J. Am. Chem. Soc.*, 1981, **103**, 7550.

111 B. M. Trost and T. A. Runge, *J. Am. Chem. Soc.*, 1981, **103**, 7559.

112 B. M. Trost, T. A. Runge, and L. N. Jungheim, *J. Am. Chem. Soc.*, 1980, **102**, 2840.

113 B. M. Trost and T. A. Runge, *J. Am. Chem. Soc.*, 1981, **103**, 2485.

114 T. Tsuda, Y. Chujo, S.-I. Nishi, K. Tawara, and T. Saegusa, *J. Am. Chem. Soc.*, 1980, **102**, 6381.

115 I. Shimizu, T. Yamada, and J. Tsuji, *Tetrahedron Lett.*, 1980, **21**, 3199.

116 I. Shimizu, I. Minami, and J. Tsuji, *Tetrahedron Lett.*, 1983, **24**, 1797; J. Tsuji, I. Minami, and I. Shimizu, *Tetrahedron Lett.*, 1983, **24**, 5639.

117 I. Minami, M. Nisar, M. Yuhara, I. Shimizu, and J. Tsuji, *Synthesis*, 1987, **11**, 992.

118 J. Tsuji, T. Yamada, I. Minami, M. Yuhara M. Nisar,, and I. Shimizu, *J. Org. Chem.*, 1987, **52**, 2988.

119 J. Tsuji, and I. Minami, *Acc. Chem. Res.*, 1987, **20**, 140.

120 I. Minami and J. Tsuji, *Tetrahedron Lett.*, 1987, **43**, 3903.

121 B. M. Trost and G. A. Molander, *J. Am. Chem. Soc.*, 1981, **103**, 5969.

122 B. M. Trost and A. Tenaglia, *Tetrahedron Lett.*, 1988, **29**, 2931.

123 R. C. Larock and S. K. Stolz-Dunn, *Tetrahedron Lett.*, 1988, **29**, 5069.

124 T. Takahashi, H. Kataoka, and J. Tsuji, *J. Am. Chem. Soc.*, 1983, **105**, 147.

125 T. Takahashi, A. Ootake, and J. Tsuji, *Tetrahedron Lett.*, 1984, **25**, 1921.

126 B. M. Trost and T. N. Nanninga, *J. Am. Chem. Soc.*, 1985, **107**, 1293.

127 B. M. Trost, J. T. Hane, and P. Metz, *Tetrahedron Lett.*, 1986, **27**, 5695.

128 I. Shimizu, M. Oshima, M. Nisar, and J. Tsuji, *Chem. Lett.*, 1986, 1775.

129 T. Fujinami, T. Suzuki, M. Kamiya, S.-I. Fukuzawa, and S. Sakai, *Chem. Lett.*, 1985, 199; B. M. Trost, J. K. Lynch, and S. R. Angle, *Tetrahedron Lett.*, 1987, **28**, 375.

130 I. Minami, M. Yuhara, and J. Tsuji, *Tetrahedron Lett.*, 1987, **28**, 629.

131 Y. Morizawa, K. Oshima, and H. Nozaki, *Israel J. Chem.*, 1984, **24**, 149.

132 K. Burgess, *Tetrahedron Lett.*, 1985, **26**, 3049.

133 K. Burgess, *J. Org. Chem.*, 1987, **52**, 2046.

134 K. Yamamoto, T. Ishida, and J. Tsuji, *Chem. Lett.*, 1987, 1157.

135 J. Muzart, P. Pale, and J.-P. Pete, *J. Chem. Soc., Chem. Commun.*, 1981, 668.

136 J. Y. Satoh and C. A. Horiuchi, *Bull. Chem. Soc. Jpn.*, 1981, **54**, 625.

137 V. Spyridon, M. Paraskewas, and A. Danopoulos, *Chemiker-Zeitung*, 1980, **104**, 238.

138 J. Tsuji, I. Minami, and I. Shimizu, *Tetrahedron Lett.*, 1984, **25**, 2791.

139 M. Donati and F. Conti, *Tetrahedron Lett.*, 1966, 4953.

140 R. K. Haynes, W. R. Jackson, and A. Stragalinou, *Aust. J. Chem.*, 1980, **33**, 1537.

141 J. Tsuji, T. Yamakawa, M. Kaito, and T. Mandai, *Tetrahedron Lett.*, 1978, 2075.

142 R. O. Hutchins, K. Learn, and R. P. Fulton, *Tetrahedron Lett.*, 1980, **21**, 27.

143 S. Suzuki, Y. Fujita, and T. Nishida, *Tetrahedron Lett.*, 1983, **24**, 5737.

144 B. M. Trost and J. M. Fortunak, *J. Am. Chem. Soc.*, 1980, **102**, 2841.

145 J. Tsuji, M. Nisar, and I. Minami, *Chem. Lett.*, 1987, 23.

146 A. Carpita, F. Bonaccorsi, and R. Rossi, *Tetrahedron Lett.*, 1984, **25**, 5193.

Chapter 7

1   T. H. Whitesides, R. W. Arhart, and R. W. Slaven, *J. Am. Chem. Soc.*, 1973, **95**, 5792.

2   A. J. Birch and A. J. Pearson, *J. Chem. Soc., Perkin Trans. 1*, 1976, 954.

3   G. S. Silverman, S. Strickland, and K. M. Nicholas, *Organometallics*, 1986, **5**, 2117.

4   J. W. Dieter, Z. Li, and K. M. Nicholas, *Tetrahedron Lett.*, 1987, **28**, 5415.

5   D. H. Gibson and R. L. Vonnahme, *J. Organomet. Chem.*, 1974, **70**, C33.

6   A. J. Pearson, *Aust. J. Chem.*, 1976, **29**, 1841.

7   D. N. Cox and R. Roulet, *Organometallics*, 1986, **5**, 1886.

8   R. L. Roustan, J. Y. Mérour, and F. Houlihan, *Tetrahedron Lett.*, 1979, 3721; B. Zhou and Y. Xu, *J. Org. Chem.*, 1988, **53**, 4419.

9   S. J. Ladoulis and K. M. Nicholas, *J. Organomet. Chem.*, 1985, **285**, C13.

10 J. Tsuji, I. Minami, and I. Shimizu, *Tetrahedron Lett.*, 1984, **25**, 5157; J.-C. Fiaud and L. Arib-Zouioueche, *J. Chem. Soc., Chem. Commun.*, 1986, 390; D. J. Krysan and P. B. Mackenzie, *J. Am. Chem. Soc.*, 1988, **110**, 6273.

11 B. M. Trost and M.-H. Hung, *J. Am. Chem. Soc.*, 1983, **105**, 7757; B. M. Trost and M. Lautens, *J. Am. Chem. Soc.*, 1983, **105**, 3343.

12 B. M. Trost and M. Lautens, *J. Am. Chem. Soc.*, 1987, **109**, 1469.

13 B. M. Trost and M. Lautens, *Tetrahedron*, 1987, **43**, 4817.

14 Y. Masuyama, K. Otake, and Y. Kurusu, *Bull. Chem. Soc. Jpn.*, 1987, **60**, 1527.

15 B. M. Trost, G. B. Tometzki, and M.-H. Hung, *J. Am. Chem. Soc.*, 1987, **109**, 2176.

16 B. M. Trost, M. Lautens, and B. Peterson, *Tetrahedron Lett.*, 1983, **24**, 4525.

17 R. J. Blade and J. E. Robinson, *Tetrahedron Lett.*, 1986, **27**, 3209.

18 M. F. Semmelhack and J. W. Herndon, *Organometallics*, 1983, **2**, 363.

19 M. F. Semmelhack, J. W. Herndon, and J. P. Springer, *J. Am. Chem. Soc.*, 1983, **105**, 2497.

20 D. J. Darensbourg and M. Y. Darensbourg, *Inorg. Chem.*, 1970, **9**, 1691.

21 S. E. Thomas, *J. Chem. Soc., Chem. Commun.*, 1987, 226; H. Kitahara, Y. Tozawa, S. Fujita, A. Tajiri, N. Morita, and T. Asao, *Bull. Chem. Soc. Jpn.*, 1988, **61**, 3362.

22 D. Rakshit and S. E. Thomas, *J. Organomet. Chem.*, 1987, **333**, C3.

23 N. W. Alcock, T. N. Danks, C. J. Richards, and S. E. Thomas, *J. Chem. Soc., Chem. Commun.*, 1989, 21; L. Hill, C. J. Richards, and S. E. Thomas, *J. Chem. Soc., Chem. Commun.*, 1990, 1085.

24 J. W. Faller, H. H. Murray, D. L. White, and K. H. Chao, *Organometallics*, 1983, **2**, 400.

25 A. J. Pearson and M. N. I. Khan, *J. Am. Chem. Soc.*, 1984, **106**, 1872.

26 A. J. Pearson and M. N. I. Khan, *Tetrahedron Lett.*, 1985, **26**, 1407; A. J. Pearson, M. N. I. Khan, J. C. Clardy, and H. Cun-heng, *J. Am. Chem. Soc.*, 1985, **107**, 2748.

27 M. Green, S. Greenfield, and M. Kersting, *J. Chem. Soc., Chem. Commun.*, 1985, 18.

28 M. J. Hynes, M. F. T. Mahon, and P. McArdle, *J. Organomet. Chem.*, 1987, **320**, C44.

29 M. Crocker, M. Green, C. E. Morton, K. R. Nagle, and A. G. Orpen, *J. Chem. Soc., Dalton Trans.*, 1985, 2145.

30 L. S. Barinelli and K. M. Nicholas, *J. Org. Chem.*, 1988, **53**, 2114.

31 A. J. Birch, P. E. Cross, J. Lewis, and D. A. White, *Chem. & Ind. (London)*, 1964, 838.

32 A. J. Birch, P. E. Cross, J. Lewis, D. A. White, and S. B. Wild, *J. Chem. Soc. (A)*, 1968, 332.

33 E. O. Fischer and R. D. Fischer, *Angew. Chem.*, 1960, **72**, 919.

34 A. J. Birch, B. M. R. Bandara, K. Chamberlain, B. Chauncy, P. Dahler, A. I. Day, I. D. Jenkins, L. F. Kelly, T.-C. Khor, G. Kretschmer, A. J. Liepa, A. S. Narula, W. D. Raverty, E. Rizzardo, C. Sell, G. R. Stephenson, D. J. Thompson, and D. H. Williamson, *Tetrahedron Supplement 9*, 1981, **37**, 289.

35 A. J. Birch and L. F. Kelly, *J. Organomet. Chem.*, 1985, **285**, 267.

36 A. J. Birch and K. B. Chamberlain, *Org. Synth.*, 1977, **57**, 107.

37 A. J. Birch and I. D. Jenkins, *Tetrahedron Lett.*, 1975, 119.

38 I. M. Palotai, W. J. Ross, G. R. Stephenson, and D. E. Tupper, *J. Organomet. Chem.*, 1989, **364**, C11.

39 D. A. Owen and G. R. Stephenson, unpublished results; D. A. Owen, G. R. Stephenson, H. Finch, and S. Swanson, *Tetrahedron Lett.*, 1990, **31**, 3401.

40 A. J. Birch, K. B. Chamberlain, M. A. Haas, and D. J. Thompson, *J. Chem. Soc., Perkin Trans. 1*, 1973, 1882.

41 A. J. Birch and A. J. Pearson, *J. Chem. Soc., Perkin Trans. 1*, 1978, 638.

42 F. Effenberger and M. Keil, *Chem. Ber.*, 1982, **115**, 1113.

43 A. J. Birch and M. A. Haas, *J. Chem. Soc. (C)*, 1971, 2465.

44 A. J. Birch, L. F. Kelly, and D. J. Thompson, *J. Chem. Soc., Perkin Trans. 1*, 1981, 1006.

45 A. J. Birch, B. Chauncy, L. F. Kelly, and D. J. Thompson, *J. Organomet. Chem.*, 1985, **286**, 37.

46 H. Curtis, B. F. G. Johnson, and G. R. Stephenson, *J. Chem. Soc., Dalton Trans.*, 1985, 1723.

47 I. M. Palotai, M.Sc. Thesis, University of East Anglia, 1985 (unpublished).

48 D. A. Owen, G. R. Stephenson, H. Finch, and S. Swanson, *J. Organomet. Chem.*, 1990, **395**, C5.

49 A. J. Birch and B. M. R. Bandara, *Tetrahedron Lett.*, 1980, 3499.

50 R. P. Alexander and G. R. Stephenson, *J. Chem. Soc., Dalton Trans.*, 1987, 885.

51 C. R. Jablonski and T. S. Sorensen, *Can. J. Chem.*, 1974, **52**, 2085.

52 B. F. G. Johnson, J. Lewis, D. G. Parker, and S. R. Postle, *J. Chem. Soc., Dalton Trans.*, 1977, 794; B. F. G. Johnson, J. Lewis, and G. R. Stephenson, *Tetrahedron Lett.*, 1980, **21**, 1995.

53 A. J. Birch, D. Bogsanyi, and L. F. Kelly, *J. Organomet. Chem.*, 1981, **214**, C39.

54 A. J. Birch and A. J. Pearson, *Tetrahedron Lett.*, 1975, 2379.

55 A. J. Pearson, *Aust. J. Chem.*, 1976, **29**, 1101; A. J. Pearson, *Aust. J. Chem.*, 1977, **30**, 345.

56 A. Pelter, K. J. Gould, and L. A. P. Kane-Maguire, *J. Chem. Soc., Chem. Commun.*, 1974, 1029.

57 L. F. Kelly, A. S. Narula, and A. J. Birch, *Tetrahedron Lett.*, 1979, 4107.

58 A. J. Birch, L. F. Kelly, and A. S. Narula, *Tetrahedron*, 1982, **38**, 1813.

59 A. J. Birch, A. S. Narula, P. Dahler, G. R. Stephenson, and L. F. Kelly, *Tetrahedron Lett.*, 1980, **21**, 979.

60 L. F. Kelly, A. S. Narula, and A. J. Birch, *Tetrahedron Lett.*, 1980, **21**, 871; L. F. Kelly, P. Dahler, A. S. Narula, and A. J. Birch, *Tetrahedron Lett.*, 1981, 1433.

61 B. M. R. Bandara, A. J. Birch, and T.-C. Khor, *Tetrahedron Lett.*, 1980, 3625.

62 P. A. Grieco and S. D. Larsen, *J. Org. Chem.*, 1986, **51**, 3553.

63 A. J. Pearson and J. Yoon, *Tetrahedron Lett.*, 1985, **26**, 2399.

64 A. J. Pearson, S. L. Kole, and B. Chen, *J. Am. Chem. Soc.*, 1983, **105**, 4483; A. J. Pearson, S. L. Kole, and T. Ray, *J. Am. Chem. Soc.*, 1984, **106**, 6060.

65 A. J. Pearson and I. C. Richards, *Tetrahedron Lett.*, 1983, **24**, 2465.

66 A. J. Birch and K. B. Chamberlain, *Org. Synth.*, 1977, **57**, 16.

67 B. F. G. Johnson, J. Lewis, and D. G. Parker, *J. Organomet. Chem.*, 1977, **141**, 139.

68 R. P. Alexander and G. R. Stephenson, *J. Organomet. Chem.*, 1986, **299**, C1; R. P. Alexander, T. D. James, and G. R. Stephenson, *J. Chem. Soc., Dalton Trans.*, 1987, 2013.

69 F. Franke and I. D. Jenkins, *Aust. J. Chem.*, 1978, **31**, 595.

70 D. J. Thompson, *J. Organomet. Chem.*, 1976, **108**, 381.

71 Y. Shvo and E. Hazum, *J. Chem. Soc., Chem. Commun.*, 1974, 336.

72 A. J. Pearson, *J. Chem. Soc., Perkin Trans. 1*, 1977, 2069.

73 A. J. Birch, W. D. Raverty, and G. R. Stephenson, *J. Org. Chem.*, 1981, **46**, 5166.

74 G. R. Stephenson, *J. Chem. Soc., Perkin Trans. 1*, 1982, 2449.

75 J. M. Kern, D. Martina, and M. P. Heitz, *Tetrahedron Lett.*, 1985, **26**, 737.

76 M. Franck-Neumann, M. P. Heitz, and D. Martina, *Tetrahedron Lett.*, 1983, 1615.

77 M. Franck-Neumann, D. Martina, and F. Brion, *Angew. Chem. Int. Ed. Engl.*, 1978, **17**, 690.

78 A. J. Birch, A. J. Liepa, and G. R. Stephenson, *Tetrahedron Lett.*, 1979, **20**, 3565.

79 A. J. Birch, P. Dahler, A. S. Narula, and G. R. Stephenson, *Tetrahedron Lett.*, 1980, 3817.

80 A. J. Pearson, I. C. Richards, and D. V. Gardner, *J. Chem. Soc., Chem. Commun.*, 1982, 807.

81 F. Hossner, M.Sc. Thesis, University of East Anglia, 1987 (unpublished); F. Hossner and M. Voyle, *J. Organomet. Chem.*, 1988, **347**, 365.

82 A. J. Pearson, *J. Chem. Soc., Perkin Trans. 1*, 1978, 495.

83 A. J. Pearson, E. Mincione, M. Chandler, and P. R. Raithby, *J. Chem. Soc., Perkin Trans. 1*, 1980, 2774; E. Mincione, A. J. Pearson, P. Bovicelli, M. Chandler, and G. C. Heywood, *Tetrahedron Lett.*, 1981, 2929; A. J. Pearson and T. Ray, *Tetrahedron Lett.*, 1985, **26**, 2981.

84 E. Mincione, P. Bovicelli, S. Cerrini, and D. Lamba, *Heterocycles*, 1985, **23**, 1607.

85 A. J. Pearson and M. K. O'Brien, *Tetrahedron Lett.*, 1988, **29**, 869.

86 M. Chandler, E. Mincione, and P. J. Parsons, *J. Chem. Soc., Chem. Commun.*, 1985, 1233.

87 A. J. Pearson, *J. Chem. Soc., Perkin Trans. 1*, 1980, 400.

88 A. J. Pearson, P. Ham, and D. C. Rees, *J. Chem. Soc., Perkin Trans. 1*, 1982, 489; A. J. Pearson and P. Ham, *J. Chem. Soc., Perkin Trans. 1*, 1983, 1421.

89 A. J. Pearson and T. R. Perrior, *J. Organomet. Chem.*, 1985, **285**, 253; H.-J. Knölker, R. Boese, and K. Hartmann, *Angew. Chem. Int. Ed. Engl.*, 1989, **28**, 1678.

90 A. Nakamura and M. Tsutsui, *J. Med. Chem.*, 1963, **6**, 796.

91 G. Evans, B. F. G. Johnson, and J. Lewis, *J. Organomet. Chem.*, 1975, **102**, 507.

92 H. Alper and C.-C. Huang, *J. Organomet. Chem.*, 1973, **50**, 213.

93 A. J. Birch, L. F. Kelly, and A. J. Liepa, *Tetrahedron Lett.*, 1985, **26**, 501.

94 J. E. Mahler and R. Pettit, *J. Am. Chem. Soc.*, 1963, **85**, 3955.

95 R. Grée, M. Laabassi, P. Mosset, and R. Carrie, *Tetrahedron Lett.*, 1985, **26**, 2317.

96 J. Morey, D. Grée, P. Mosset, L. Toupet, and R. Grée, *Tetrahedron Lett.*, 1987, **28**, 2959.

97 J. P. Lellouche, P. Breton, J. P. Beaucourt, L. Troupet, and R. Grée, *Tetrahedron Lett.*, 1988, **29**, 2449.

98 K. Nunn, P. Mosset, R. Grée, and R. W. Saalfrank, *Angew. Chem. Int. Ed. Engl.*, 1988, **27**, 1188.

99 W. A. Donaldson and M. Ramaswamy, *Tetrahedron Lett.*, 1988, **29**, 1343; W. A. Donaldson and M. Ramaswamy, *Tetrahedron Lett.*, 1989, **30**, 1343.

100 A. J. Pearson and T. Ray, *Tetrahedron*, 1985, 5765.

101 M. F. Semmelhack and J. Park, *J. Am. Chem. Soc.*, 1987, **109**, 935.

102 T. G. Bonner, K. A. Holder, and P. Powell, *J. Organomet. Chem.*, 1974, **77**, C37.

103 R. S. Bayoud, E. R. Biehl, and P. C. Reeves, *J. Organomet. Chem.*, 1978, **150**, 75; B. F. G. Johnson, J. Lewis, D. G. Parker, and S. R. Postle, *J. Chem. Soc., Dalton Trans.*, 1977, 794.

104 R. S. Bayoud, E. R. Biehl, and P. C. Reeves, *J. Organomet. Chem.*, 1979, **174**, 297.

105 G. Maglio and R. Palumbo, *J. Organomet. Chem.*, 1974, **76**, 367.

106 T. H. Whitesides and J. P. Neilan, *J. Am. Chem. Soc.*, 1976, **98**, 63.

107 J. R. Bleeke and M. K. Hays, *Organometallics*, 1987, **6**, 1367.

108 T. S. Sorensen and C. R. Jablonski, *J. Organomet. Chem.*, 1970, **25**, C62.

109 C. P. Lillya and R. A. Sahatjian, *J. Organomet. Chem.*, 1970, **25**, C67.

110 N. A. Clinton and C. P. Lillya, *J. Am. Chem. Soc.*, 1970, **92**, 3065; T. G. Bonner, K. A. Holder, P. Powell, and E. Styles, *J. Organomet. Chem.*, 1977, **131**, 105.

111 J. E. Mahler, D. H. Gibson, and R. Pettit, *J. Am. Chem. Soc.*, 1963, **85**, 3959.

112 M. Uemura, T. Minami, Y. Yamashita, K. Hiyoshi, and Y. Hayashi, *Tetrahedron Lett.*, 1987, **28**, 641.

113 B. R. Bonazza, C. P. Lillya, and G. Scholes, *Organometallics*, 1982, **1**, 137.

114 B. F. G. Johnson, J. Lewis, D. G. Parker, P. R. Raithby, and G. M. Sheldrick, *J. Organomet. Chem.*, 1978, **150**, 115.

115 B. M. R. Bandara, W. D. Raverty, and A. J. Birch, *J. Chem. Soc., Perkin Trans. 1*, 1982, 1745.

116 F. Birencwaig, H. Shamai, and Y. Shvo, *Tetrahedron Lett.*, 1979, 2947.

117 J. A. S. Howell, B. F. G. Johnson, P. L. Josty, and J. Lewis, *J. Organomet. Chem.*, 1972, **39**, 329.

118 D. Fărcaşiu and G. Marino, *J. Organomet. Chem.*, 1983, **253**, 243.

119 M. Slupczynski, I. Wolszczak, and P. Kosztolowicz, *Inorg. Chim. Acta.*, 1979, **33**, L97.

120 W. R. Roth and M. D. Meier, *Tetrahedron Lett.*, 1967, 2053; B. F. G. Johnson, J. Lewis, and D. J. Thompson, *Tetrahedron Lett.*, 1974, 3789; P. Eilbracht and R. Jelitte, *Chem. Ber.*, 1985, **118**, 1983.

121 A. G. M. Barrett, D. H. R. Barton, and G. Johnson, *J. Chem. Soc., Perkin Trans. 1*, 1978, 1014.

122 Y. Shvo and E. Hazum, *J. Chem. Soc., Chem. Commun.*, 1975, 829.

123 J. Rodrigues, P. Brun, and B. Waegell, *Tetrahedron Lett.*, 1986, **27**, 835.

124 S. Sarel, R. Ben-Shoshan, and B. Kirson, *Israel J. Chem.*, 1972, **10**, 787; T. H.
     Whitesides and R. W. Slaven, *J. Organomet. Chem.*, 1974, **67**, 99.

125 K. S. Suslick, P. F. Schubert, and J. W. Goodale, *J. Am. Chem. Soc.*, 1981, **103**, 7342.

126 S. V. Ley, C. M. R. Low, and A. D. White, *J. Organomet. Chem.*, 1986, **302**, C13.

127 A. J. Birch and L. F. Kelly, *J. Organomet. Chem.*, 1985, **286**, C5; B. F. G. Johnson, J.
     Lewis, G. R. Stephenson, and E. J. S. Vichi, *J. Chem. Soc., Dalton Trans.*, 1978, 369.

128 E. D. Sternberg and K. P. C. Vollhardt, *J. Am. Chem. Soc.*, 1980, **102**, 4839; E. D.
     Sternberg and K. P. C. Vollhardt, *J. Org. Chem.*, 1984, **49**, 1564.

129 D. Astruc, P. Michaud, A. M. Madonik, J. Y. Saillard, and R. Hoffmann, *Nouveau J.
     Chim.*, 1985, **9**, 41.

130 B. F. G. Johnson, J. Lewis, T. W. Matheson, I. E. Ryder, and M. V. Twigg, *J. Chem. Soc.,
     Chem. Commun.*, 1974, 269; J. Ashley-Smith, D. V. Hope, B. F. G. Johnson, J. Lewis, and
     I. E. Ryder, *J. Organomet. Chem.*, 1974, **82**, 257.

131 P. L. Pauson and K. H. Todd, *J. Chem. Soc. C*, 1970, 2638.

132 M. F. Semmelhack, *Ann. N.Y. Acad. Sci.*, 1977, **295**, 36; G. Jaouen, *Ann. N.Y. Acad.
     Sci.*, 1977, **295**, 59.

133 C. A. L. Mahaffy and P. L. Pauson, *Inorg. Synth.*, 1979, **19**, 154.

134 M. F. Semmelhack, H. T. Hall, M. Yoshifuji, and G. Clark, *J. Am. Chem. Soc.*, 1975, **97**,
     1247.

135 M. F. Semmelhack and H. T. Hall, *J. Am. Chem. Soc.*, 1974, **96**, 7091; 7092.

136 M. F. Semmelhack, G. R. Clark, J. L. Garcia, J. J. Harrison, Y. Thebtaranonth, W. Wulff,
     and A. Yamashita, *Tetrahedron*, 1981, **37**, 3957.

137 F. Rose-Munch, E. Rose, and A. Semra, *J. Chem. Soc., Chem. Commun.*, 1986, 1551;
     R. Khourzom, F. Rose-Munch, and E. Rose, *Tetrahedron Lett.*, 1990, **31**, 2011.

138 C. Baldoli, P. Del Buttero, E. Licandro, and S. Maiorana, *Synthesis*, 1988, 344.

139 R. G. Sutherland, A. S. Abd-El-Aziz, A. Piórko, and C. C. Lee, *Synth. Commun.*, 1987, **17**,
     393.

140 A. S. Abd-El-Aziz, C. C. Lee, A. Piórko, and R. G. Sutherland, *Synth. Commun.*, 1988, **18**,
     291.

141 U. S. Gill, *Inorg. Chim. Acta*, 1986, **114**, L25.

142 C. Baldoli, P. Del Buttero, S. Maiorana, and A. Papagni, *J. Chem. Soc., Chem. Commun.*,
     1985, 1181.

143 R. G. Sutherland, R. L. Chowdhury, A. Piórko, and C. C. Lee, *J. Org. Chem.*, 1987, **52**,
     4618.

144 P. L. Pauson and J. A. Segal, *J. Chem. Soc., Dalton Trans.*, 1975, 1677.

145 A. J. Pearson, P. R. Bruhn, and S.-Y. Hsu, *J. Org. Chem.*, 1986, **51**, 2137; A. J. Pearson,
     S.-H. Lee, and F. Gouzoules, *J. Chem. Soc., Perkin Trans. 1*, 1990, 2251.

146 M. F. Semmelhack, J. L. Garcia, D. Cortes, R. Farina, R. Hong, and B. K. Carpenter,
     Organometallics, 1983, **2**, 467.

147 M. F. Semmelhack, G. R. Clark, R. Farina, and M. Saeman, *J. Am. Chem. Soc.*, 1979,
     101, 217.

148 E. P. Kündig, *Pure Appl. Chem.*, 1985, **57**, 1855.

149 V. Desobry and E. P. Kündig, *Helv. Chim. Acta.*, 1981, **64**, 1288.

150 V. Desobry, Ph.D. Thesis No. 2042, University of Geneva, 1982 (unpublished).

151 E. P. Kündig, V. Desobry, and D. P. Simmons, *J. Am. Chem. Soc.*, 1983, **105**, 6962.

152 Y. K. Chung, H. S. Choi, D. A. Sweigert, and N. G. Connelly, *J. Am. Chem. Soc.*, 1982, **104**, 4245.

153 Y. K. Chung, P. G. Willard, and D. A. Sweigert, *Organometallics*, 1982, **1**, 1053.

154 R. P. Alexander and G. R. Stephenson, *J. Organomet. Chem.*, 1986, **314**, C73.

155 A. J. Pearson and I. C. Richards, *J. Organomet. Chem.*, 1983, **258**, C41.

156 H. Kunzer and M. Thiel, *Tetrahedron Lett.*, 1988, **29**, 1135.

157 E. P. Kündig and D. P. Simmons, *J. Chem. Soc., Chem. Commun.*, 1983, 1320.

158 M. F. Semmelhack and A. Yamashita, *J. Am. Chem. Soc.*, 1980, **102**, 5924.

159 G. Jaouen, B. Caro, and J.-Y. Le Bihan, *C. R. Acad. Sc. Paris*, 1972, **274C**, 902; A. Meyer and G. Jaouen, *J. Chem. Soc., Chem. Commun.*, 1974, 787; B. Caro and G. Jaouen, *J. Chem. Soc., Chem. Commun.*, 1976, 655; B. Caro, E. Gentric, D. Grandjean, and G. Jaouen, *Tetrahedron Lett.*, 1978, 3009.

160 W. R. Jackson and T. R. B. Mitchell, *J. Chem. Soc., B*, 1969, 1228.

161 G. Jaouen, in *Transition Metal Organometallics in Organic Synthesis* Vol. 2, ed. H. Alper, Academic Press, 1978, pp. 88–93; B. Caro and G. Jaouen, *J. Organomet. Chem.*, 1982, **228**, 87.

162 M. Uemura, T. Minami, K. Isobe, T. Kobayashi, and Y. Hayashi, *Tetrahedron Lett.*, 1986, **27**, 967; M. Uemura, T. Minami, and Y. Hayashi, *Tetrahedron Lett.*, 1988, **29**, 6271.

163 J. Brocard, L. Pelinski, and J. Lebibi, *J. Organomet. Chem.*, 1987, **337**, C47.

164 M. A. Boudeville and H. des Abbayes, *Tetrahedron Lett.*, 1975, 2727.

165 M. Uemura, K. Isobe, and Y. Hayashi, *Chem. Lett.*, 1985, 91.

166 M. F. Semmelhack, *J. Organomet. Chem. Library*, 1976, **1**, 361.

167 R. J. Card and W. S. Trahanovsky, *J. Org. Chem.*, 1980, **45**, 2555.

168 M. D. Rausch, G. A. Moser, and W. A. Lee, *Synth. React. Inorg. Met.-Org. Chem.*, 1979, **9**, 357.

169 R. J. Card and W. S. Trahanovsky, *J. Org. Chem.*, 1980, **45**, 2560.

170 J. P. Gilday and D. A. Widdowson, *J. Chem. Soc., Chem. Commun.*, 1986, 1235.

171 J. P. Gilday and D. A. Widdowson, *Tetrahedron Lett.*, 1986, **27**, 5525.

172 N. F. Masters and D. A. Widdowson, *J. Chem. Soc., Chem. Commun.*, 1983, 955.

173 M. Fukui, T. Ikeda, and T. Oishi, *Tetrahedron Lett.*, 1982, **23**, 1605.

174 P. M. Treichel and R. U. Kirss, *Organometallics*, 1987, **6**, 249.

175 M. Uemura, N. Nishikawa, and Y. Hayashi, *Tetrahedron Lett.*, 1980, **21**, 2069; M. Uemura, S. Tokuyama, and T. Sakan, *Chem. Lett.*, 1975, 1195.

176 M. Uemura, N. Nishikawa, K. Take, M. Ohnishi, K. Hirotsu, T. Higuchi, and Y. Hayashi, *J. Org. Chem.*, 1983, **48**, 2349.

177 J. C. Gill, B. A. Marples, and J. R. Traynor, *Tetrahedron Lett.*, 1987, **28**, 2643.

178 R. M. Moriarty, S. G. Engerer, O. Prakash, I. Prakash, U. S. Gill, and W. A. Freeman, *J. Chem. Soc., Chem. Commun.*, 1985, 1715.

179 P. J. Beswick, S. J. Leach, N. F. Masters, and D. A. Widdowson, *J. Chem. Soc., Chem. Commun.*, 1984, 46.

180 G. Jaouen, A. Meyer, and G. Simonneaux, *J. Chem. Soc., Chem. Commun.*, 1975, 813.

181 G. Simonneaux and G. Jaouen, *Tetrahedron*, 1979, **35**, 2249.

182 V. N. Kalinin, N. I. Udalov, and A. V. Vsatov, *Bull. Acad. Sci. USSR*, 1987, **36**, 1550.

183 J. Brocard, J. Lebibi, and D. Couturier, *J. Chem. Soc., Chem. Commun.*, 1981, 1264.

184 G. Jaouen, S. Top, A. Laconi, D. Couturier, and J. Brocard, *J. Am. Chem. Soc.*, 1984, **106**, 2207.

185 J. Blagg, S. G. Davies, N. J. Holman, C. A. Laughton, and B. E. Mobbs, *J. Chem. Soc., Perkin Trans. 1*, 1986, 1581.

186 J. Blagg and S. G. Davies, *J. Chem. Soc., Chem. Commun.*, 1985, 653.

187 J. Blagg and S. G. Davies, *Tetrahedron*, 1988, **44**, 4463.

188 J. Blagg, S. G. Davies, and B. E. Mobbs, *J. Chem. Soc., Chem. Commun.*, 1985, 620.

189 J. Blagg, S. J. Coote, S. G. Davies, and B. E. Mobbs, *J. Chem. Soc., Perkin Trans. 1*, 1986, 2257.

190 J. Blagg, S. J. Coote, S. G. Davies, D. Middlemass, and A. Naylor, *J. Chem. Soc., Perkin Trans. 1*, 1987, 689.

191 P. D. Baird, J. Blagg, S. G. Davies, and K. H. Sutton, *Tetrahedron*, 1988, **44**, 171.

192 M. F. Semmelhack, W. Seafort, and L. Keller, *J. Am. Chem. Soc.*, 1980, **102**, 6584.

193 Y. L. Chiu, A. E. G. Sant'ana, and J. H. P. Utley, *Tetrahedron Lett.*, 1987, **28**, 1349.

194 J. D. Holmes, D. A. K. Jones, and R. Petit, *J. Organomet. Chem.*, 1965, **4**, 324; R. S. Bly, R. A. Mateer, K.-K. Tse, and R. L. Veazey, *J. Org. Chem.*, 1973, **38**, 1518; D. Seyferth and C. S. Eschbach, *J. Organomet. Chem.*, 1975, **94**, C5.

195 S. Top and G. Jaouen, *J. Organomet. Chem.*, 1980, **197**, 199.

196 M. T. Reetz and M. Sauerwald, *Tetrahedron Lett.*, 1983, **24**, 2837.

197 G. Jaouen and A. Meyer, *J. Am. Chem. Soc.*, 1975, **97**, 4667.

198 M. Uemura, T. Kobayashi, T. Minami, and Y. Hayashi, *Tetrahedron Lett.*, 1986, **27**, 2479; M. Uemura, T. Kobayashi, K. Isobe, T. Minami, and Y. Hayashi, *J. Org. Chem.*, 1986, **51**, 2859.

199 J. Brocard, J. Lebibi, L. Pelinski, and M. Mahmoudi, *Tetrahedron Lett.*, 1986, **27**, 6325.

200 M. Uemura, T. Minami, and Y. Hayashi, *J. Am. Chem. Soc.*, 1987, **109**, 5277.

201 M. Uemura, K. Isobe, K. Take, and Y. Hayashi, *J. Org. Chem.*, 1983, **48**, 3855.

202 M. Uemura, T. Minami, and Y. Hayashi, *J. Chem. Soc., Chem. Commun.*, 1984, 1193.

203 M. Ghavshou and D. A. Widdowson, *J. Chem. Soc., Perkin Trans. 1*, 1983, 3065.

204 P. J. Dickens, A. M. Z. Slawin, D. A. Widdowson, and D. J. Williams, *Tetrahedron Lett.*, 1988, **29**, 103.

205 A. J. Birch, K. B. Chamberlain, and D. J. Thompson, *J. Chem. Soc., Perkin Trans. 1*, 1973, 1900.

206 C. W. Ong and A. J. Pearson, *Tetrahedron Lett.*, 1980, **21**, 2349.

207 A. J. Pearson, *J. Chem. Soc., Chem. Commun.*, 1980, 488.

208 A. J. Pearson and M. Chandler, *Tetrahedron Lett.*, 1980, **21**, 3933.

209 A. J. Pearson, S. L. Kole, and J. Yoon, *Organometallics*, 1986, **5**, 2075.

210 P. W. Howard, G. R. Stephenson, and S. C. Taylor, *J. Organomet. Chem.*, 1988, **339**, C5; R. W. Ashworth and G. A. Berchtold, *J. Am. Chem. Soc.*, 1977, **99**, 5200.

211 L. A. Paquette, R. G. Daniels, and R. Gleiter, *Organometallics*, 1984, **3**, 560.

212 W. Fink, *Helv. Chim. Acta*, 1976, **59**, 276.

213 D. Mandon, L. Toupet, and D. Astruc, *J. Am. Chem. Soc.*, 1986, **108**, 1320.

214 A. J. Pearson and H. S. Bansal, *Tetrahedron Lett.*, 1986, **27**, 283.

215 D. Astruc, *Acc. Chem. Res.*, 1986, **19**, 377; D. Astruc, *Tetrahedron*, 1983, **39**, 4027.

216 Y.-H. Lai, W. Tam, and K. P. C. Vollhardt, *J. Organomet. Chem.*, 1981, **216**, 97.

217 D. Jones, L. Pratt, and G. Wilkinson, *J. Chem. Soc.*, 1962, 4458; S. L. Grundy and P. M. Maitlis, *J. Chem. Soc., Chem. Commun.*, 1982, 379.

218 M. Brookhart and A. Lukacs, *J. Am. Chem. Soc.*, 1984, **106**, 4161.

219 Y. K. Chung, E. D. Honig, W. T. Robinson, D. A. Sweigart, N. G. Connelly, and S. D. Ittel, *Organometallics*, 1983, **2**, 1479.

220 Y. K. Chung, D. A. Swiegart, N. G. Connelly, and J. B. Sheridan, *J. Am. Chem. Soc.*, 1985, **107**, 2388; W. D. Meng and G. R. Stephenson, unpublished results.

221 A. Efraty, D. Liebman, J. Sikora, and D. Z. Denney, *Inorg. Chem.*, 1976, **15**, 886.

222 J. T. Bamberg and R. G. Bergman, *J. Am. Chem. Soc.*, 1977, **99**, 3173.

223 N. G. Connelly and R. L. Kelly, *J. Chem. Soc., Dalton Trans.*, 1974, 2334.

224 A. J. Pearson, P. R. Bruhn, and I. C. Richards, *Israel J. Chem.*, 1984, **24**, 93.

225 E. D. Honig and D. A. Sweigart, *J. Chem. Soc., Chem. Commun.*, 1986, 691.

226 S. G. Davies, M. L. H. Green, and D. M. P. Mingos, *Tetrahedron*, 1978, **34**, 3047.

227 G. R. John, L. A. P. Kane-Maguire, and R. Kanitz, *J. Organomet. Chem.*, 1986, **312**, C21.

228 M. A. Hashmi, J. D. Munro, P. L. Pauson, and J. M. Williamson, *J. Chem. Soc. (A)*, 1967, 240.

229 P. Sautet, O. Eisenstein, and K. M. Nicholas, *Organometallics*, 1987, **6**, 1845.

230 A. J. Birch and G. R. Stephenson, *J. Organomet. Chem.*, 1981, **218**, 91; A. J. Deeming, S. S. Ullah, A. J. Domingos, B. F. G. Johnson, and J. Lewis, *J. Chem. Soc., Dalton Trans.*, 1974, 2093; R. Edwards, J. A. S. Howell, B. F. G. Johnson, and J. Lewis, *J. Chem. Soc., Dalton Trans.*, 1974, 2105; A. L. Burrows, B. F. G. Johnson, J. Lewis, and D. G. Parker, *J. Organomet. Chem.*, 1980, **194**, C11; R. J. H. Cowles, B. F. G. Johnson, P. L. Josty, and J. Lewis, *J. Chem. Soc., Chem. Commun.*, 1969, 392; E. G. Bryan, A. L. Burrows, B. F. G. Johnson, J. Lewis, and G. M. Schiavon, *J. Organomet. Chem.*, 1977, **129**, C19.

231 A. Eisenstadt, *J. Organomet. Chem.*, 1973, **60**, 335; A. Eisenstadt, *J. Organomet. Chem.*, 1976, **113**, 147; A. D. Charles, P. Divers, B. F. G. Johnson, K. D. Karlin, J. Lewis, A. V. Rivera, and G. M. Sheldrick, *J. Organomet. Chem.*, 1977, **128**, C31.

232 A. Rosan, M. Rosenblum, and J. Tancrede. *J. Am. Chem. Soc.*, 1973, **95**, 3062; A. Rosan and M. Rosenblum, *J. Organomet. Chem.*, 1974, **80**, 103.

233 D. W. Clack, M. Monshi, and L. A. P. Kane-Maguire, *J. Organomet. Chem.*, 1976, **107**, C40.

234 D. W. Clack, M. Monshi, and L. A. P. Kane-Maguire, *J. Organomet. Chem.*, 1976, **120**, C25; R. Hoffmann and P. Hofmann, *J. Am. Chem. Soc.*, 1976, **98**, 598; D. W. Clack and K. D. Warren, *J. Organomet. Chem.*, 1978, **162**, 83; D. W. Clack and L. A. P. Kane-Maguire, *J. Organomet. Chem.*, 1979, **174**, 199.

235 P. A. Dobosh, D. G. Gresham, C. P. Lillya, and E. S. Magyar, *Inorg. Chem.*, 1976, **15**, 2311.

236 A. J. Pearson, *Metallo-organic Chemistry*, Wiley Interscience, 1985, p. 291.

237 A. J. Pearson and P. R. Raithby, *J. Chem. Soc., Dalton Trans.*, 1981, 884.

238 O. Eisenstein, W. M. Butler, and A. J. Pearson, *Organometallics*, 1984, **3**, 1150.

239 M. F. Semmelhack, G. R. Clark, R. Farina, and M. Saeman, *J. Am. Chem. Soc.*, 1979, **101**, 217; R. A. Kok and M. B. Hall, *J. Am. Chem. Soc.*, 1985, **107**, 2599.

240 T. A. Albright and B. K. Carpenter, *Inorg. Chem.*, 1980, **19**, 3092; A. Solladie-Cavallo and G. Wipff, *Tetrahedron Lett.*, 1980, **21**, 3047; J. C. Boutonnet, L. Mordenti, E. Rose, O. le Martret, and G. Precigoux, *J. Organomet. Chem.*, 1981, **221**, 147.

241 E. Elian and R. Hoffmann, *Inorg. Chem.*, 1975, **14**, 1058.

242 J. Brocard and J. Lebibi, *J. Organomet. Chem.*, 1987, **320**, 295.

243 O. Eisenstein and R. Hoffmann, *J. Am. Chem. Soc.*, 1980, **102**, 6148.

244 G. P. Randall, G. R. Stephenson, and E. J. T. Chrystal, *J. Organomet. Chem.*, 1988, **353**, C47.

245 L. A. P. Kane-Maguire, E. D. Honig, and D. A. Sweigart, *Chem. Rev.*, 1984, **84**, 525.

246 H. S. Choi and D. A. Sweigart, *Organometallics*, 1982, **1**, 60.

247 C. D. Johnson and B. Stratton, *J. Chem. Soc., Perkin Trans. 2*, 1988, 1903.

248 L. A. P. Kane-Maguire, E. D. Honig, and D. A. Sweigart, *J. Chem. Soc., Chem. Commun.*, 1984, 345.

249 H. Brunner, *Adv. Organomet. Chem.*, 1980, **18**, 151; H. Brunner, *Acc. Chem. Res.*, 1979, 250.

250 J. A. S. Howell and M. J. Thomas, *J. Chem. Soc., Dalton Trans.*, 1983, 1401.

251 K. Schlögl, *Topics in Stereochemistry*, Vol. 1, ed. N. L. Allinger and E. L. Eliel, Wiley Interscience, 1967, p. 68.

252 G. Paiaro, R. Palumbo, A. Musco, and A. Panunzi, *Tetrahedron Lett.*, 1965, 1067; A. Mandelbaum, Z. Neuwirth, and M. Cais, *Inorg. Chem.*, 1963, **2**, 902; R. Riemschneider and W. Herrmann, *Annalen*, 1961, **648**, 68; L. Westman and K. L. Rinehart, Jr., *Acta Chim. Scand.*, 1962, **16**, 1199.

253 V. Schurig, *Tetrahedron Lett.*, 1984, **25**, 2739.

254 A. J. Birch and B. M. R. Bandara, *Tetrahedron Lett.*, 1980, **21**, 2981.

255 B. M. R. Bandara, A. J. Birch, and L. F. Kelly, *J. Org. Chem.*, 1984, **49**, 2496.

256 A. J. Birch, W. D. Raverty, and G. R. Stephenson, *J. Chem. Soc., Chem. Commun.*, 1980, 857.

257 A. J. Birch, L. F. Kelly, and D. V. Weerasuria, *J. Org. Chem.*, 1988, **53**, 278.

258 B. M. R. Bandara, A. J. Birch, L. F. Kelly, and T. C. Khor, *Tetrahedron Lett.*, 1983, **24**, 2491.

259 J. G. Atton, L. A. P. Kane-Maguire, P. A. Williams, and G. R. Stephenson, *J. Organomet. Chem.*, 1982, **232**, C5; J. G. Atton, D. J. Evans, L. A. P. Kane-Maguire, and G. R. Stephenson, *J. Chem. Soc., Chem. Commun.*, 1984, 1246.

260 D. J. Evans, L. A. P. Kane-Maguire, and S. B. Wild, *J. Organomet. Chem.*, 1982, **232**, C9.

261 A. J. Birch and L. F. Kelly, *J. Org. Chem.*, 1985, **50**, 712.

262 A. J. Birch and G. R. Stephenson, *Tetrahedron Lett.*, 1981, **22**, 779.

263 A. J. Birch, W. D. Raverty, and G. R. Stephenson, *Organometallics*, 1984, **3**, 1075.

264 G. R. Stephenson, *Aust. J. Chem.*, 1981, **34**, 2339.

265 A. J. Birch, *Ann. N.Y. Acad. Sci.*, 1980, **333**, 110.

266 P. W. Howard, G. R. Stephenson, and S. C. Taylor, *J. Chem. Soc., Chem. Commun.*, 1988, 1603.

267 G. R. Stephenson, *Aust. J. Chem.*, 1982, **35**, 1939.

268 S. Litman, A. Gedanken, Z. Goldschmidt, and Y. Bakal, *J. Chem. Soc., Chem. Commun.*, 1978, 983.

269 A. Tajiri, N. Morita, T. Asao, and M. Hatano, *Angew. Chem. Int. Ed. Engl.*, 1985, **24**, 329.

270 N. W. Alcock, D. H. G. Crout, C. M. Henderson, and S. E. Thomas, *J. Chem. Soc., Chem. Commun.*, 1988, 746.

271 J. A. S. Howell and M. J. Thomas, *J. Organomet. Chem.*, 1983, **247**, C21; J. A. S. Howell and M. J. Thomas, *Organometallics*, 1985, **4**, 1054.

272 A. Meyer and G. Jaouen, *J. Chem. Soc., Chem. Commun.*, 1974, 787.

273 S. G. Davies and C. L. Goodfellow, *J. Chem. Soc., Perkin Commun.*, 1989, 192.

274 S. Top, G. Jaouen, J. Gillois, C. Baldoli, and S. Maiorana, *J. Chem. Soc., Chem. Commun.*, 1988, 1284.

275 J. Gillois, D. Buisson, R. Azerad, and G. Jaouen, *J. Chem. Soc., Chem. Commun.*, 1988, 1224.

276 G. Jaouen and R. Dabard, *Tetrahedron Lett.*, 1971, 1015.

277 G. Jaouen, in *Transition Metal Organometallics in Organic Synthesis*, Vol. 2, ed. H. Alper, Academic Press, 1978, p. 90.

278 G. Jaouen and A. Meyer, *Tetrahedron Lett.*, 1976, 3547; A. Meyer and O. Hofer, *J. Am. Chem. Soc.*, 1980, **102**, 4410.

279 S. Top, G. Jaouen, and M. J. McGlinchey, *J. Chem. Soc., Chem. Commun.*, 1980, 1110.

280 R. M. Moriarty, S. C. Engerer, O. Prakash, I. Prakash, U. S. Gill, and W. A. Freeman, *J. Org. Chem.*, 1987, **52**, 153.

281 H. Falk, K. Schlogl, and W. Steyrer, *Monatsch. Chem.*, 1966, **97**, 1029.

282 K. R. Stewart, S. G. Levine, and J. Bordner, *J. Org. Chem.*, 1984, **49**, 4082.

283 J. Blagg, S. G. Davies, C. L. Goodfellow, and K. H. Sutton, *J. Chem. Soc., Perkin Trans. 1*, 1987, 1805.

284 A. Solladie-Cavallo and J. Suffert, *Tetrahedron Lett.*, 1985, **26**, 429.

285 A. Solladie-Cavallo and E. Tsamo, *J. Organomet. Chem.*, 1979, **172**, 165.

286 M. Franck-Neumann, *Pure Appl. Chem.*, 1983, **55**, 1715; A. Monpert, J. Martelli, R. Grée, and R. Carrié, *Nouveau J. Chim.*, 1983, **7**, 345; M. Franck-Neumann, D. Martina, and M. P. Heitz, *Tetrahedron Lett.*, 1982, **23**, 3493.

287 A. Monpert, J. Martelli, R. Grée, and R. Carrié, *Tetrahedron Lett.*, 1981, **22**, 1961.

288 M. Laabassi and R. Grée, *Tetrahedron Lett.*, 1988, **29**, 611.

289 S. Boulaajaj, T. Le Gall, M. Vaultier, R. Grée, L. Toupet, and R. Carrié, *Tetrahedron Lett.*, 1987, **28**, 1761.

290 O. Jaenicke, R. C. Kerber, P. Kirsch, E. A. Koerner von Gustorf, and R. Rumin, *J. Organomet. Chem.*, 1980, **187**, 361.

291 H. W. Whitlock, Jr. and Y. N. Chuah, *J. Am. Chem. Soc.*, 1965, **87**, 3605; H. W. Whitlock, Jr. and R. L. Markezich, *J. Am. Chem. Soc.*, 1971, **93**, 2590; H. W. Whitlock, Jr., C. Reich, and W. P. Woessner, *J. Am. Chem. Soc.*, 1971, **93**, 2483; R. H. Grubbs and R. A. Grey, *J. Chem. Soc., Chem. Commun.*, 1973, 76.

292 J. W. Faller, M. E. Thomsen, and M. J. Mattina, *J. Am. Chem. Soc.*, 1971, **93**, 2642; H. Felkin, M. Joly-Goudket, and S. G. Davies, *Tetrahedron Lett.*, 1981, 1157; B. Bosnich and P. B. Mackenzie, *Pure. Appl. Chem.*, 1982, **54**, 189.

293 T. H. Whitesides and J. P. Neilan, *J. Am. Chem. Soc.*, 1976, **98**, 63.

294 I. M. Palotai, G. R. Stephenson, and L. A. P. Kane-Maguire, *J. Organomet. Chem.*, 1987, **319**, C5.

295 A. J. Pearson, M. Zettler, and A. A. Pinkerton, *J. Chem. Soc., Chem. Commun.*, 1987, 264.

296 K. E. Hine, B. F. G. Johnson, and J. Lewis, *J. Chem. Soc., Dalton Trans.*, 1976, 1702.

297 A. J. Pearson and M. W. Zettler, *J. Chem. Soc., Chem. Commun.*, 1987, 1243.

298 J. W. Faller and K.-H. Chao, *J. Am. Chem. Soc.*, 1983, **105**, 3893.

299 A. J. Pearson, S. L. Blystone, and B. A. Roden, *Tetrahedron Lett.*, 1987, **28**, 2459; A. J. Pearson and J. Yoon, *J. Chem. Soc., Chem. Commun.*, 1986, 1467; A. J. Pearson, S. L. Blystone, H. Nar, A. A. Pinkerton, B. A. Roden, and J. Yoon, *J. Am. Chem. Soc.*, 1989, **111**, 134.

300 G. R. Stephenson, R. P. Alexander, C. Morley, and P. W. Howard, *Phil. Trans. R. Soc. Lond. A*, 1988, **326**, 545.

301 D. Gentric, J.-Y. le Bihan, M.-C. Senechal-Tocquer, D. Senechal, and B. Caro, *Tetrahedron Lett.*, 1986, **27**, 3849.

302 A. Ceccon, A. Gambaro, and A. Venzo, *J. Chem. Soc., Chem. Commun.*, 1985, 540.

303 A. N. Nesmeyanov, N. A. Ustynyuk, L. N. Novikova, T. N. Rybina, Y. A. Ustynyuk, Y. F. Oprunenko, and O. I. Trifonova, *J. Organomet. Chem.*, 1980, **184**, 63.

304 M. Oda, N. Morita, and T. Asao, *Chem. Lett.*, 1981, 397.

305 H. Maltz and B. A. Kell, *J. Chem. Soc., Chem. Commun.*, 1971, 1390; M. Moll, P. Würstl, H. Behrens, and P. Merbach, *Z. Naturforsch.*, 1978, **33b**, 1304.

306 M. Airoldi, G. Barbera, G. Deganello, and G. Gennaro, *Organometallics*, 1987, **6**, 398.

307 M. F. Semmelhack and E. J. Fewkes, *Tetrahedron Lett.*, 1987, **28**, 1497.

308 A. J. Pearson, M. S. Holden, and R. D. Simpson, *Tetrahedron Lett.*, 1986, **27**, 4121.

309 M. Franck-Neumann, D. Martina, and M.-P. Heitz, *J. Organomet. Chem.*, 1986, **315**, 59.

310 J. E. Ellis and E. A. Flom, *J. Organomet. Chem.*, 1975, **99**, 263.

311 J. A. Gladysz, G. M. Williams, W. Tam, D. L. Johnson, D. W. Parker, and J. C. Selover, *Inorg. Chem.*, 1979, **18**, 553.

312 T. S. Piper and G. Wilkinson, *J. Inorg. Nucl. Chem.*, 1956, **3**, 104.

313 J. E. Ellis, R. A. Faltynek, and S. G. Hentges, *J. Organomet. Chem.*, 1976, **120**, 389; R. J. Kinney, W. D. Jones, and R. G. Bergman, *J. Am. Chem. Soc.*, 1978, **100**, 7902.

314 B. A. Kelsey and J. E. Ellis, *J. Chem. Soc., Chem. Commun.*, 1986, 331; B. A. Kelsey and J. E. Ellis, *J. Am. Chem. Soc.*, 1986, **108**, 1344.

315 A. Davison and D. L. Reger, *J. Organomet. Chem.*, 1970, **23**, 491; R. J. Bernhardt, M. A. Wilmoth, J. J. Weers, D. M. LaBrush, D. P. Eyman, and J. C. Huffman, *Organometallics*, 1986, **5**, 883; P. K. Rush, S. K. Noh, and M. Brookhart, *Organometallics*, 1986, **5**, 1745.

316 W. P. Henry and R. D. Rieke, *J. Am. Chem. Soc.*, 1983, **105**, 6314; R. D. Rieke, W. P. Henry, and J. S. Arney, *Inorg. Chem.*, 1987, **26**, 420.

317 V. S. Leong and N. J. Cooper, *Organometallics*, 1987, **6**, 2000.

318 N. E. Murr and J. D. Payne, *J. Chem. Soc., Chem. Commun.*, 1985, 162.

319 A. N. Nesmeyanov and I. I. Kritskaya, *J. Organomet. Chem.*, 1968, **14**, 387.

320 R. D. Rieke, K. P. Daruwala, and M. W. Forkner, *J. Org. Chem.*, 1989, **54**, 24.

321 L. D. Schulte and R. D. Rieke, *Tetrahedron Lett.*, 1988, **29**, 5483.

322 D. R. Falkowski, D. F. Hunt, C. P. Lillya, and M. D. Rausch, *J. Am. Chem. Soc.*, 1967, **89**, 6387; D. A. T. Young, J. R. Holmes, and H. D. Kaesz, *J. Am. Chem. Soc.*, 1969, **91**, 6968; T. H. Whitesides and R. W. Arhart, *J. Am. Chem. Soc.*, 1971, **93**, 5296; M. Brookhart, E. R. Davis, and D. L. Harris, *J. Am. Chem. Soc.*, 1972, **94**, 7853; T. H. Whitesides, R. W. Arhart, and R. W. Slaven, *J. Am. Chem. Soc.*, 1973, **95**, 5792; M. Brookhart and D. L. Harris, *Inorg. Chem.*, 1974, **13**, 1540; T. H. Whitesides and R. W. Arhart, *Inorg. Chem.*, 1975, **14**, 209; M. Brookhart, T. H. Whitesides, and J. M. Crockett, *Inorg. Chem.*, 1976, **15**, 1550.

323 B. F. G. Johnson, J. Lewis, A. W. Parkins, and G. L. P. Randall, *J. Chem. Soc., Chem. Commun.*, 1969, 595; E. O. Greaves, G. R. Knox, P. L. Pauson, and S. Toma, *J. Chem. Soc., Chem. Commun.*, 1974, 257; A. D. U. Hardy and G. A. Sim, *J. Chem. Soc., Dalton Trans.*, 1972, 2305; A. N. Nesmeyanov, L. V. Rybin, N. T. Gubenko, M. I. Rybinskaya, and P. V. Petrovskii, *J. Organomet. Chem.*, 1974, **71**, 271; R. E. Graf and C. P. Lillya, *J. Organomet. Chem.*, 1976, **122**, 377; B. F. G. Johnson, J. Lewis, and D. G. Parker, *J. Organomet. Chem.*, 1977, **141**, 319.

324 M. Franck-Neumann, M. Sedrati, and M. Mokhi, *Tetrahedron Lett.*, 1986, **27**, 3861; M. Franck-Neumann, M. Sedrati, and M. Mokhi, *Angew. Chem. Int. Ed. Engl.*, 1986, **25**, 1131.

325 M. Franck-Neumann, F. Brion, and D. Martina, *Tetrahedron Lett.*, 1978, 5033; A. J. Pearson, *Transition Met. Chem.*, 1981, **6**, 67.

326 G. M. Williams and D. E. Rudisill, *Tetrahedron Lett.*, 1986, **27**, 3465.

Chapter 8

1 J. F. Normant, *Synthesis*, 1972, 63; A. E. Jukes, *Adv. Organomet. Chem.*, 1978, **12**, 215; H. O. House, *Acc. Chem. Res.*, 1976, **9**, 59; J. F. Normant, *Pure Appl. Chem.*, 1978, **50**, 709.

2 R. J. K. Taylor, *Synthesis*, 1985, 364.

3 E. J. Corey and D. J. Beames, *J. Am. Chem. Soc.*, 1972, **94**, 7210.

4 G. H. Posner, C. E. Whitten, and J. J. Sterling, *J. Am. Chem. Soc.*, 1973, **95**, 7788; G. H. Posner, D. J. Brunelle, and L. Sinoway, *Synthesis*, 1974, 662.

5 R. Adams, W. Reifschneider, and A. Ferretti, *Org. Synth.*, 1962, **42**, 22.

6 H. O. House, C. Y. Chu, J. M. Wilkins, and M. J. Umen, *J. Org. Chem.*, 1975, **40**, 1460.

7 B. H. Lipshutz, R. S. Wilhelm, and D. M. Floyd, *J. Am. Chem. Soc.*, 1981, **103**, 7672.

8 C. R. Johnson and D. S. Dhanoa, *J. Org. Chem.*, 1987, **52**, 1885.

9 S. H. Bertz, C. P. Gibson, and G. Dabbagh, *Tetrahedron Lett.*, 1987, **28**, 4251.

10 F. DiNinno, E. V. Linek, and B. G. Christensen, *J. Am. Chem. Soc.*, 1979, **101**, 2210.

11 E. J. Corey, J. R. Cashman, T. M. Eckrich, and D. R. Corey, *J. Am. Chem. Soc.*, 1985, **107**, 713.

12 H. Gilman and J. M. Straley, *Rec. Trav. Chim. Pays-Bas*, 1936, **55**, 821.

13 H. Gilman, R. G. Jones, and L. A. Woods, *J. Org. Chem.*, 1952, **17**, 1630; H. O. House, W. L. Respess, and G. M. Whitesides, *J. Org. Chem.*, 1966, **31**, 3128.

14 S. R. Krauss and S. G. Smith, *J. Am. Chem. Soc.*, 1981, **103**, 141; B. H. Lipshutz, J. A. Kozlowski, and C. M. Breneman, *J. Am. Chem. Soc.*, 1985, **107**, 3197.

15 B. H. Lipshutz, R. S. Wilhelm, and J. A. Kozlowski, *Tetrahedron*, 1984, **40**, 5005; B. H. Lipshutz, D. A. Parker, S. L. Nguyen, K. E. McCarthy, J. C. Barton, S. E. Whitney, and H. Kotsuki, *Tetrahedron*, 1986, **42**, 2873.

16 B. H. Lipshutz, R. S. Wilhelm, and J. A. Kozlowski, *J. Org. Chem.*, 1984, **49**, 3938.

17 B. H. Lipshutz, M. Koerner, and D. A. Parker, *Tetrahedron Lett.*, 1987, **28**, 945.

18 B. H. Lipshutz, R. S. Wilhelm, J. A. Koslowski, and D. A. Parker, *J. Org. Chem.*, 1984, **49**, 3928.

19 B. H. Lipshutz, E. L. Ellsworth, J. R. Behling, and A. L. Campbell, *Tetrahedron Lett.*, 1988, **29**, 893; T. N. Majid and P. Knochel, *Tetrahedron Lett.*, 1990, **31**, 4413.

20 P. G. Edwards, R. W. Gellert, M. W. Marks, and R. Bau, *J. Am. Chem. Soc.*, 1982, **104**, 2072; C. Eaborn, P. B. Hitchcock, J. D. Smith, and A. C. Sullivan, *J. Organomet. Chem.*, 1984, **263**, C23.

21 H. Hope, D. Oram, and P. P. Power, *J. Am. Chem. Soc.*, 1984, **106**, 1149.

22 R. G. Carlson and E. G. Zey, *J. Org. Chem.*, 1972, **37**, 2468.

23 G. Stork and T. L. Macdonald, *J. Am. Chem. Soc.*, 1975, **97**, 1264.

24 J. H. Babler and T. R. Mortell, *Tetrahedron Lett.*, 1972, 669.

25 P. A. Wender and A. W. White, *J. Am. Chem. Soc.*, 1988, **110**, 2218.

26 G. Bozzato, J.-P. Bachmann, and M. Pesaro, *J. Chem. Soc., Chem. Commun.*, 1974, 1005.

27 M. Pesaro, G. Bozzato, and P. Schudel, *J. Chem. Soc., Chem. Commun.*, 1968, 1152.

28 J. A. Marshall and S. F. Brady, *Tetrahedron Lett.*, 1969, 1387.

29 R. Lohmar and W. Steglich, *Angew. Chem. Int. Ed. Engl.*, 1978, **17**, 450.

30 Y. Yamamoto, *Angew. Chem. Int. Ed. Engl.*, 1986, **25**, 947.

31 W. Oppolzer, P. Dudfield, T. Stevenson, and T. Godel, *Helv. Chim. Acta*, 1985, **68**, 212.

32 R. K. Dieter and M. Tokles, *J. Am. Chem. Soc.*, 1987, **109**, 2040.

33 T. Takahashi, H. Yamada, and J. Tsuji, *J. Am. Chem. Soc.*, 1981, **103**, 5259.

34 L. A. Pacquette and Y. K. Han, *J. Am. Chem. Soc.*, 1981, **103**, 1835.

35 S. Danishefsky, K. Vaughan, R. Gadwood, and K. Tsuzuki, *J. Am. Chem. Soc.*, 1981, **103**, 4136.

36 C. M. Lentz and G. H. Posner, *Tetrahedron Lett.*, 1978, 3769; G. H. Posner, M. J. Chapdelaine, and C. M. Lentz, *J. Org. Chem.*, 1979, **44**, 3661.

37 G. H. Posner, C. E. Whitten, J. J. Sterling, and D. J. Brunelle, *Tetrahedron Lett.*, 1974, 2591; G. H. Posner, J. J. Sterling, C. E. Whitten, C. M. Lentz, and D. J. Brunelle, *J. Am. Chem. Soc.*, 1975, **97**, 107.

38 R. K. Boeckman, Jr., *J. Org. Chem.*, 1973, **38**, 4450.

39 T. Tanaka, S. Kurozumi, T. Toru, M. Kobayashi, S. Miura, and S. Ishimoto, *Tetrahedron Lett.*, 1975, 1535; K. K. Heng and R. A. J. Smith, *Tetrahedron*, 1979, **35**, 425.

40 R. G. Salomon and M. F. Salomon, *J. Org. Chem.*, 1975, **40**, 1488; T. Toru, S. Kurozumi, T. Tanaka, S. Miura, M. Kobayashi, and S. Ishimoto, *Tetrahedron Lett.*, 1976, 4087.

41 F. Näf, R. Decorzant, and W. Thommen, *Helv. Chim. Acta*, 1975, **58**, 1808.

42 J. E. McMurry and S. J. Isser, *J. Am. Chem. Soc.*, 1972, **94**, 7132.

43 C.-T. Hsu, N.-Y. Wang, L. H. Latimer, and C. J. Sih, *J. Am. Chem. Soc.*, 1983, **105**, 593.

44 W. K. Bornack, S. S. Bhagwat, J. Ponton, and P. Helquist, *J. Am. Chem. Soc.*, 1981, **103**, 4647.

45 S. Danishefsky, M. Kahn, and M. Silvestri, *Tetrahedron Lett.*, 1982, **23**, 1419.

46 E. Nakamura, S. Matsuzawa, Y. Horiguchi, and I. Kuwajima, *Tetrahedron Lett.*, 1986, **27**, 4029.

47 Y. Horiguchi, S. Matsuzawa, E. Nakamura, and I. Kuwajima, *Tetrahedron Lett.*, 1986, **27**, 4025.

48 C. R. Johnson and T. J. Marren, *Tetrahedron Lett.*, 1987, **28**, 27.

49 T. Shono, Y. Matsumara, and S. Kashimura, *J. Org. Chem.*, 1981, **46**, 3719; S. Bernasconi, P. Gariboldi, G. Jommi, S. Montanari, and M. Sisti, *J. Chem. Soc., Perkin Trans. 1*, 1981, 2394.

50 R. Davis and K. G. Untch, *J. Org. Chem.*, 1979, **44**, 3755.

51 T. Tanaka, N. Okamura, K. Bannai, A. Hazato, S. Sugiura, K. Manabe, and S. Kurozumi, *Tetrahedron Lett.*, 1985, **26**, 5575.

52 M. Shibasaki, K. Iseki, and S. Ikegami, *Chem. Lett.*, 1979, 1299.

53 S. B. Bowlus and J. A. Katzenellenbogen, *Tetrahedron Lett.*, 1973, 1277; S. B. Bowlus and J. A. Katzenellenbogen, *J. Org. Chem.*, 1973, **38**, 2733.

54 E. J. Corey and J. A. Katzenellenbogen, *J. Am. Chem. Soc.*, 1969, **91**, 1851.

55 A. Alexakis, G. Cahiez, and J. F. Normant, *J. Organomet. Chem.*, 1979, **177**, 293.

56 J. F. Normant and A. Alexakis, *Synthesis*, 1981, 841.

57 H. Westmijze, H. Kleijn, and P. Vermeer, *Tetrahedron Lett.*, 1977, 2023.

58 H. Westmijze, J. Meijer, H. J. T. Bos, and P. Vermeer, *Rec. Trav. Chim. Pays-Bas*, 1976, **95**, 304.

59 A. Marfat, P. R. McGuirk, and P. Helquist, *Tetrahedron Lett.*, 1978, 1363.

60 A. Alexakis, J. Normant, and J. Villieras, *J. Organomet. Chem.*, 1975, **96**, 471.

61 A. Alexakis, C. Chuit, M. Commerçon-Bourgain, J. P. Foulon, N. Jabri, P. Mangeney, and J. F. Normant, *Pure Appl. Chem.*, 1984, **56**, 91.

62 A. Alexakis, A. Commerçon, J. Villieras, and J. F. Normant, *Tetrahedron Lett.*, 1976, 2313.

63 P. Rona and P. Crabbe, *J. Am. Chem. Soc.*, 1968, **90**, 4733; J. L. Luche, E. Barreiro, J. M. Dollat, and P. Crabbe, *Tetrahedron Lett.*, 1975, 4615; W. H. Pirkle and C. W. Boeder, *J. Org. Chem.*, 1978, **43**, 1950.

64 I. Marek, P. Mangeney, A. Alexakis, and J. F. Normant, *Tetrahedron Lett.*, 1986, **27**, 5499.

65 R. Baudouy and J. Gore, *J. Chem. Res. (S)*, 1981, 278.

66 A. Alexakis, G. Cahiez, and J. F. Normant, *Tetrahedron*, 1969, **36**, 1961.

67 A. Alexakis, G. Cahiez, and J. F. Normant, *Tetrahedron Lett.*, 1978, 2027.

68 M. Furber, R. J. K. Taylor, and S. C. Burford, *Tetrahedron Lett.*, 1985, **26**, 3285.

69 N. Jabri, A. Alexakis, and J. F. Normant, *Bull. Soc. Chim. Fr.*, 1983, II-321; N. Jabri, A. Alexakis, and J. F. Normant, *Bull. Soc. Chim. Fr.*, 1983, II-332.

70 G. M. Whitesides and C. P. Casey, *J. Am. Chem. Soc.*, 1966, **88**, 4541.

71 G. Büchi and J. A. Carlson, *J. Am. Chem. Soc.*, 1968, **90**, 5336.

72 G. M. Whitesides, J. SanFilippo, Jr., C. P. Casey, and E. J. Panek, *J. Am. Chem. Soc.*, 1967, **89**, 5302; G. Linstrumelle, J. K. Krieger, and G. M. Whitesides, *Org. Synth.*, 1976, **55**, 103.

73 E. J. Corey, K. Achiwa, and J. A. Katzenellenbogen, *J. Am. Chem. Soc.*, 1969, **91**, 4318.

74 G. H. Posner, G. L. Loomis, and H. S. Sawaya, *Tetrahedron Lett.*, 1975, 1373.

75 J. A. Marshall and J. A. Ruth, *J. Org. Chem.*, 1974, **39**, 1971.

76 K. Kitatani, T. Hiyama, and H. Nozaki, *J. Am. Chem. Soc.*, 1976, **98**, 2362.

77 J. Klein and R. Levene, *J. Am. Chem. Soc.*, 1972, **94**, 2520; J. P. Marino and D. M. Floyd, *J. Am. Chem. Soc.*, 1974, **96**, 7138.

78 E. J. Corey and I. Kuwajima, *Tetrahedron Lett.*, 1972, 487.

79 R. E. Atkinson, R. F. Curtis, and J. A. Taylor, *J. Chem. Soc. (C)*, 1967, 578; R. E. Atkinson, R. F. Curtis, D. M. Jones, and J. A. Taylor, *J. Chem. Soc. (C)*, 1969, 2173.

80 D. W. Knight and G. Pattenden, *J. Chem. Soc., Chem. Commun.*, 1974, 188.

81 K. C. Nicolaou and S. E. Webber, *J. Am. Chem. Soc.*, 1984, **106**, 5734.

82 E. J. Corey and N. W. Boaz, *Tetrahedron Lett.*, 1984, **25**, 3059; E. J. Corey and N. W. Boaz, *Tetrahedron Lett.*, 1984, **25**, 3063.

83 R. J. Cave, C. C. Howard, G. Klinkert, R. F. Newton, D. P. Reynolds, A. H. Wadsworth, and S. M. Roberts, *J. Chem. Soc., Perkin Trans. 1*, 1979, 2954.

84 S. Masamune, C. U. Kim, K. E. Wilson, G. O. Spessard, P. E. Georghiou, and G. S. Bates, *J. Am. Chem. Soc.*, 1975, **97**, 3512.

85 R. J. Anderson, *J. Am. Chem. Soc.*, 1970, **92**, 4978; R. W. Herr and C. R. Johnson, *J. Am. Chem. Soc.*, 1970, **92**, 4979.

86 J. M. Marino, R. Fernandez de la Pradilla, and E. Laborde, *J. Org. Chem.*, 1987, **52**, 4898.

87 J. A. Marshall and J. D. Trometer, *Tetrahedron Lett.*, 1987, **28**, 4985.

88 A. Alexakis, D. Jachiet, and J. F. Normant, *Tetrahedron Lett.*, 1986, **27**, 5607.

89 A. Alexakis and D. Jachiet, *Tetrahedron Lett.*, 1988, **29**, 217.

90 M. J. Chong, D. R. Cyr, and E. K. Mar, *Tetrahedron Lett.*, 1987, **28**, 5009.

91 M. J. Kurth and M. A. Abreo, *Tetrahedron Lett.*, 1987, **28**, 5631.

92 M. J. Eis and B. Ganem, *Tetrahedron Lett.*, 1985, **26**, 1153.

93 T. Sato, T. Kawara, M. Kawashima, and T. Fujisawa, *Chem. Lett.*, 1980, 571; T. Sato, M. Kawashima, and T. Fujisawa, *Tetrahedron Lett.*, 1981, **22**, 2375.

94 A. Ghribi, A. Alexakis, and J. F. Normant, *Tetrahedron Lett.*, 1984, **25**, 3075; A. Ghribi,

A. Alexakis, and J. F. Normant, *Tetrahedron Lett.*, 1984, **25**, 3079; A. Ghribi, A. Alexakis, and J. F. Normant, *Tetrahedron Lett.*, 1984, **25**, 3083; Y. Yamamoto, *Angew. Chem. Int. Ed. Engl.*, 1986, **25**, 947; A. Alexakis, P. Mangeney, A. Ghribi, D. Jachiet, and J. F. Normant, *Phil. Trans. R. Soc. Lond. A*, 1988, **326**, 557.

95 R. M. Magid, *Tetrahedron*, 1980, **36**, 1901.

96 P. Rona, L. Tokes, J. Tremble, and P. Crabbe, *J. Chem. Soc., Chem. Commun.*, 1969, 43.

97 R. J. Anderson, C. A. Henrick, and J. B. Siddall, *J. Am. Chem. Soc.*, 1970, **92**, 735; E. E. van Tamelen and J. P. McCormick, *J. Am. Chem. Soc.*, 1970, **92**, 737; R. J. Anderson, C. A. Henrick, J. B. Siddall, and R. Zurfluh, *J. Am. Chem. Soc.*, 1972, **94**, 5379.

98 P. Crabbe, J.-M. Dollat, J. Gallina, J.-L. Luche, E. Verlarde, M. L. Maddox, and L. Tokes, *J. Chem. Soc., Perkin Trans. 1*, 1978, 730.

99 J. Levisalles, M. Rudler-Chauvin, and H. Rudler, *J. Organomet. Chem.*, 1977, **136**, 103.

100 H. L. Goering and V. D. Singleton, Jr., *J. Am. Chem. Soc.*, 1976, **98**, 7854.

101 C. R. Johnson and G. A. Dutra, *J. Am. Chem. Soc.*, 1973, **95**, 7777; C. R. Johnson and G. A. Dutra, *J. Am. Chem. Soc.*, 1973, **95**, 7783.

102 E. J. Corey and J. Mann, *J. Am. Chem. Soc.*, 1973, **95**, 6832.

103 C. Descoins, C. A. Henrick, and J. B. Siddall, *Tetrahedron Lett.*, 1972, 3777.

104 Y. Yamamoto and K. Maruyama, *J. Organomet. Chem.*, 1978, **156**, C9.

105 Y. Tanigawa, H. Kanamaru, A. Sonoda, and S. I. Murahashi, *J. Am. Chem. Soc.*, 1977, **99**, 2361; Y. Yamamoto, H. Yatagai, K. Maruyama, A. Sonoda, and S. I. Murahashi, *J. Am. Chem. Soc.*, 1977, **99**, 5652; Y. Tanigawa, H. Ohta, A. Sonoda, and S. I. Murahashi, *J. Am. Chem. Soc.*, 1978, **100**, 4610.

106 R. E. Ireland, C. A. Lipinski, C. J. Kowalski, J. W. Tilley, and D. M. Walba, *J. Am. Chem. Soc.*, 1974, **96**, 3333.

107 C. Chuit, H. Flekin, C. Frajerman, G. Roussi, and G. Swierczewski, *J. Organomet. Chem.*, 1977, **127**, 371; H. Felkin, E. Jampel-Costa, and G. Swierczewski, *J. Organomet. Chem.*, 1977, **134**, 265.

108 B. L. Buckwalter, I. R. Burfitt, H. Felkin, M. Joly-Goudket, K. Naemura, M. F. Salomon, E. Wenkert, and P. M. Wovkulich, *J. Am. Chem. Soc.*, 1978, **100**, 6445.

109 G. Consiglio, F. Morandini, and O. Piccolo, *J. Am. Chem. Soc.*, 1981, **103**, 1846.

110 P. C. Wailes and H. Weigold, *J. Organomet. Chem.*, 1970, **24**, 405; P. C. Wailes, H. Weigold, and A. P. Bell, *J. Organomet. Chem.*, 1971, **27**, 373; P. C. Wailes, H. Weigold, and A. P. Bell, *J. Organomet. Chem.*, 1972, **43**, C32.

111 D. W. Hart and J. Schwartz, *J. Am. Chem. Soc.*, 1974, **96**, 8115.

112 T. Yoshida and E. Negishi, *J. Am. Chem. Soc.*, 1981, **103**, 4985; J. A. Miller and E. Negishi, *Tetrahedron Lett.*, 1984, **25**, 5863.

113 E. Negishi and T. Takahashi, *Aldrichimica Acta*, 1985, **18**, 31.

114 J. Schwartz, M. J. Loots, and H. Kosugi, *J. Am. Chem. Soc.*, 1980, **102**, 1333.

115 M. J. Loots and J. Schwartz, *Tetrahedron Lett.*, 1978, 4381; J. Schwartz and Y. Hayashi, *Tetrahedron Lett.*, 1980, **21**, 1497.

116 J. Schwartz, D. B. Carr, R. T. Hansen, and F. M. Dayrit, *J. Org. Chem.*, 1980, **45**, 3053.

117 A. E. Greene, J.-P. Lansard, J.-L. Luche, and C. Petrier, *J. Org. Chem.*, 1984, **49**, 931.

118 N. Genco, D. Marten, S. Raghu, and M. Rosenblum, *J. Am. Chem. Soc.*, 1976, **98**, 848.

119 Y. Yamamoto and K. Maruyama, *Tetrahedron Lett.*, 1981, **22**, 2895; K. Mashima, H. Yasuda, K. Asami, and A. Nakamura, *Chem. Lett.*, 1983, 219.

120 H. Yasuda, Y. Kajihara, K. Mashima, K. Nagasuna, and A. Nakamura, *Chem. Lett.*, 1981, 671.

121 B. Weidmann, C. D. Maycock, and D. Seebach, *Helv. Chim. Acta*, 1981, **64**, 1552; M. T. Reetz, R. Steinbach, J. Westermann, R. Urz, B. Wenderoth, and R. Peter, *Angew. Chem., Int. Ed. Engl.*, 1982, **21**, 135.

122 B. Weidmann and D. Seebach, *Angew. Chem., Int. Ed. Engl.*, 1983, **22**, 31.

123 C. H. Heathcock, J. J. Doney, and R. G. Bergman, *Pure Appl. Chem.*, 1985, **57**, 1789; G. Cahiez, A. Masuda, D. Bernard, and J. F. Normant, *Tetrahedron Lett.*, 1976, 3155.

124 G. Cahiez, D. Bernard, and J. F. Normant, *Synthesis*, 1977, 130.

125 G. Cahiez, *Tetrahedron Lett.*, 1981, **22**, 1239.

Chapter 9

1 C. Narayana and M. Periasamy, *Synthesis*, 1985, 253.

2 E. Negishi and T. Takahashi, *Aldrichimica Acta*, 1985, **18**, 31.

3 T. H. Coffield, J. Kozikowski, and R. D. Closson, *J. Org. Chem.*, 1957, **22**, 598.

4 R. J. Mawby, F. Basolo, and R. G. Pearson, *J. Am. Chem. Soc.*, 1964, **86**, 3994; W. D. Bannister, M. Green, and R. N. Haszeldine, *J. Chem. Soc., Chem. Commun.*, 1965, 54.

5 R. W. Johnson and R. G. Pearson, *J. Chem. Soc., Chem. Commun.*, 1970, 986.

6 H. Brunner, B. Hammer, I. Bernal, and M. Draux, *Organometallics*, 1983, **2**, 1595.

7 T. C. Flood, K. D. Campbell, H. H. Downs, and S. Nakanishi, *Organometallics*, 1983, **2**, 1590.

8 D. T. Thompson, in *Catalysis and Chemical Processes*, ed. R. Pearce and W. R. Patterson, Leonord Hill, 1981, Ch. 8, p. 164; B. Cornils, in *New Syntheses with Carbon Monoxide*, ed. J. Falbe, Springer-Verlag, 1980, p. 1–225.

9 O. Roelen, US Patent, 1947, 2 327 066.

10 R. F. Heck and D. S. Breslow, *J. Am. Chem. Soc.*, 1961, **83**, 4023.

11 P. Pino, S. Pucci, and F. Paicenti, *Chem. Ind. (London)*, 1963, 294.

12 A. G. Inventa, NL Patent, 1964, 298/834; Fr Patent, 1964, 1 371 085; Jefferson Chem. Co. Inc. (R. M. Gipson), US Patent, 1976, 3 954 877; A Spencer, *J. Organomet. Chem.*, 1977, **124**, 85; L. I. Zakharkin, V. V. Guseva, V. V. Kaverin, and V. P. Yurlev, *J. Org. Chem. USSR*, 1987, **23**, 1201.

13 P. F. Beal, M. A. Rebenstorf, and J. E. Pike, *J. Am. Chem. Soc.*, 1959, **81**, 1231; A. L. Nussbaum, T. L. Popper, E. P. Oliveto, S. Friedman, and I. Wender, *J. Am. Chem. Soc.*, 1959, **81**, 1228.

14 A. Rosenthal and D. Abson, *Can. J. Chem.*, 1964, **42**, 1811; A. Rosenthal and H. J. Koch, *Can. J. Chem.*, 1965, **43**, 1375; H. Paulsen and K. Todt, *Adv. Carbohydr. Chem.*, 1968, **23**, 115.

15 F. Piacenti, M. Bianchi, E. Benedetti, and P. Frediani, *J. Organomet. Chem.*, 1970, **23**, 257; E. R. Tucci, *I & EC Product Res. & Dev.*, 1968, **7**, 32.

16 L. H. Slaugh and R. D. Mullineaux, US Patent, 1966, 3 239 569; C. R. Greene and R. E. Meeker, US Patent, 1966, 3 274 263.

17 D. Evans, J. A. Osborn, and G. Wilkinson, *J. Chem. Soc. (A)*, 1968, 3133; D. Evans, J. A. Osborn, and G. Wilkinson, *Inorg. Synth.*, 1966, **11**, 99.

18 R. Fowler, H. Connor, and R. A. Baehe, *Hydrocarbon Processing*, 1976, 247; E. A. V. Brewester, *Chem. Engineering*, 1976, 90.

19 A. Stefani, G. Consiglio, C. Botteghi, and P. Pino, *J. Am. Chem. Soc.*, 1973, **95**, 6504.

20 F. J. McQuillin and D. G. Parker, *J. Chem. Soc., Perkin Trans. 1*, 1975, 2092.

21 C. Botteghi, G. Consiglio, G. Ceccarelli, and A. Stefani, *J. Org. Chem.*, 1972, **37**, 1835.

22 W. Hoffmann and H. Siegel, *Tetrahedron Lett.*, 1975, 533.

23 J. Schwartz and J. B. Cannon, *J. Am. Chem. Soc.*, 1974, **96**, 4721.

24 P. G. M. Wuts, M. L. Obrzut, and P. A. Thompson, *Tetrahedron Lett.*, 1984, **25**, 4051.

25 B. Cornil in *New Syntheses with Carbon Monoxide*, ed. J. Falbe, Springer-Verlag, 1980, Ch. 1.

26 C. W. Bird, *Transition Metal Intermediates in Organic Synthesis*, Academic Press, 1967, Ch. 6.

27 P. Pino, G. Consiglio, C. Botteghi, and C. Salomon, *Advan. Chem. Ser.*, 1974, **132**, 295; M. Tanaka, Y. Ikeda, and I. Ogata, *Chem. Lett.*, 1975, 1115.

28 C. Botteghi, M. Branca, G. Micera, F. Piacenti, and G. Menchi, *Chim. Ind. (Milan)*, 1978, **60**, 16.

29 C. Botteghi, M. Branca, and A. Saba, *J. Organomet. Chem.*, 1980, **184**, C17.

30 G. Consiglio and P. Pino, *Top. Curr. Chem.*, 1982, **105**, 77.

31 J. K. Stille and D. E. James, *J. Organomet. Chem.*, 1976, **108**, 401; S. Brewis and P. R. Hughes, *J. Chem. Soc., Chem. Commun.*, 1966, 6.

32 J. K. Stille, D. E. James, and L. F. Hines, *J. Am. Chem. Soc.*, 1973, **95**, 5062.

33 J. K. Stille and R. Divakaruni, *J. Org. Chem.*, 1979, **44**, 3474.

34 R. F. Heck, *J. Am. Chem. Soc.*, 1972, **94**, 2712.

35 J. G. Buchanan, A. R. Edgar, M. J. Power, and C. T. Shanks, *J. Chem. Soc., Perkin Trans. 1*, 1979, 225.

36 R. C. Larock, *J. Org. Chem.*, 1975, **40**, 3237.

37 M. Mori, K. Chiba, M. Okita, and Y. Ban, *J. Chem. Soc., Chem. Commun.*, 1979, 698.

38 N. D. Trieu, C. J. Elsevier, and K. Vrieze, *J. Organomet. Chem.*, 1987, **325**, C23.

39 T. F. Murray, V. Varma, and J. R. Norton, *J. Chem. Soc., Chem. Commun.*, 1976, 907; J. R. Norton, K. E. Shenton, and J. Schwartz, *Tetrahedron Lett.*, 1975, 51; T. F. Murray, V. Varma, and J. R. Norton, *J. Org. Chem.*, 1978, **43**, 353.

40 T. F. Murray, E. G. Samsel, V. Varma, and J. R. Norton, *J. Am. Chem. Soc.*, 1981, **103**, 7520.

41 A. Cowell and J. K. Stille, *J. Am. Chem. Soc.*, 1980, **102**, 4193.

42 T. Takahashi, H. Ikeda, and J. Tsuji, *Tetrahedron Lett.*, 1981, **22**, 1363.

43 E. Negishi and J. M. Tour, *Tetrahedron Lett.*, 1986, **27**, 4869; E. Negishi, G. Wu, and J. M. Tour, *Tetrahedron Lett.*, 1988, **29**, 6745.

44 R. Jaouhari, P. H. Dixneuf, and S. Lecolier, *Tetrahedron Lett.*, 1986, **27**, 6315; T. Jintoku, H. Taniguchi, and Y. Fujiwara, *Chem. Lett.*, 1987, 1159.

45 D. Forster, *Adv. Organomet. Chem.*, 1979, **17**, 255; F. E. Paulik and J. F. Roth, *J. Chem. Soc., Chem. Commun.*, 1968, 1578.

46 J. F. Roth, J. H. Craddock, A. Hershman, and F. E. Paulik, *Chem. Tech.*, 1971, 600.

47 M. F. Semmelhack and C. Bodurow, *J. Am. Chem. Soc.*, 1984, **106**, 1496.

48 M. F. Semmelhack, C. Bodurow, and M. Baum, *Tetrahedron Lett.*, 1984, **25**, 3171.

49 Y. Tamaru, T. Kobayashi, S. Kawamura, H. Ochiai, M. Hojo, and Z. Yoshida, *Tetrahedron Lett.*, 1985, **26**, 3207.

50 N. Meyer and D. Seebach, *Angew. Chem. Int. Ed. Engl.*, 1978, **17**, 521.

51 D. Lathbury, P. Vernon, and T. Gallagher, *Tetrahedron Lett.*, 1986, **27**, 6009.

52 Y. Tamaru, M. Hojo, and Z. Yoshida, *Tetrahedron Lett.*, 1987, **28**, 325; Y. Tamura, M. Hojo, H. Higashimura, and Z. Yoshida, *J. Am. Chem. Soc.*, 1988, **110**, 3994.

53 H. Alper and D. Leonard, *J. Chem. Soc., Chem. Commun.*, 1985, 511.

54 E. G. Samsel and J. R. Norton, *J. Am. Chem. Soc.*, 1984, **106**, 5505.

55 S. I. Murahashi, Y. Imada, and K. Nishimura, *J. Chem. Soc., Chem. Commun.*, 1988, 1578.

56 I. Rhee, M. Ryang, and S. Tsutsumi, *J. Organomet. Chem.*, 1967, **9**, 361.

57 R. Aumann and J. Knecht, *Chem. Ber.*, 1976, **109**, 174.

58 V. Heil, B. F. G. Johnson, J. Lewis, and D. J. Thompson, *J. Chem. Soc., Chem. Commun.*, 1974, 270.

59 B. F. G. Johnson, J. Lewis, and D. J. Thompson, *Tetrahedron Lett.*, 1974, 3789.

60 B. F. G. Johnson, K. D. Karlin, and J. Lewis, *J. Organomet. Chem.*, 1978, **145**, C23.

61 P. Eilbracht, R. Jelitte, and L. Walz, *Chem. Ber.*, 1984, **117**, 3473; P. Eilbracht, R. Jelitte, and P. Trabold, *Chem. Ber.*, 1986, **119**, 169.

62 R. Victor, R. Ben-Shoshan, and S. Sarel, *Tetrahedron Lett.*, 1970, 4253; R. Aumann, *J. Am. Chem. Soc.*, 1974, **96**, 2631.

63 G. D. Annis, E. M. Hebblethwaite, and S. V. Ley, *J. Chem. Soc., Chem. Commun.*, 1980, 297; R. Aumann and H. Averbeck, *J. Organomet. Chem.*, 1975, **85**, C4; G. D. Annis and S. V. Ley, *J. Chem. Soc., Chem. Commun.*, 1977, 581.

64 R. Aumann, H. Ring, C. Krüger, and R. Goddard, *Chem. Ber.*, 1979, **112**, 3644.

65 G. D. Annis, S. V. Ley, C. R. Self, and R. Sivaramakrishnan, *J. Chem. Soc., Perkin Trans. 1*, 1981, 270.

66 G. D. Annis, E. M. Hebblethwaite, S. T. Hodgson, A. M. Horton, D. M. Hollinshead, S. V. Ley, C. R. Self, and R. Sivaramakrishnan, *Proceedings of the Second SCI-RSC Medicinal Chemistry Symposium*, Special Publication No. 50, ed. J. C. Emmett, 1983, p. 148.

67 S. V. Ley, *Phil. Trans. R. Soc. Lond. A*, 1988, **326**, 633.

68 R. G. Finke and T. N. Sorrell, *Org. Synth.*, 1979, **59**, 102.

69 J. P. Collman, *Acc. Chem. Res.*, 1975, **8**, 342.

70 J. Y. Merour, J. L. Roustan, C. Charrier, J. Collin, and J. Benaim, *J. Organomet. Chem.*, 1973, **51**, C24.

71 J. E. McMurry, A. Andrus, G. M. Ksander, J. H. Musser, and M. A. Johnson, *J. Am. Chem. Soc.*, 1979, **101**, 1330.

72 J. P. Collman, R. G. Finke, J. N. Cawse, and J. I. Brauman, *J. Am. Chem. Soc.*, 1978, **100**, 4766; C. M. Lukehart, *Adv. Organomet. Chem.*, 1986, **25**, 45; C. M. Lukehart, J. B. Meyers, Jr., and B. J. Sweetman, *J. Organomet. Chem.*, 1986, **316**, 319.

73 Y. Kimura, Y. Tomita, S. Nakanishi, and Y. Otsuji, *Chem. Lett.*, 1979, 321.

74 H. Alper and H. des Abbayes, *J. Organomet. Chem.*, 1977, **134**, C11.

75 L. Cassar and M. Foa, *J. Organomet. Chem.*, 1977, **134**, C15; H. Alper, *Adv. Organomet. Chem.*, 1981, **19**, 183.

76 P. L. Pauson, *Tetrahedron*, 1985, **41**, 5855.

77 L. Daalman, R. F. Newton, P. L. Pauson, R. G. Taylor, and A. Wadsworth, *J. Chem. Res. (S)*, 1984, 344.

78 P. Magnus and D. P. Becker, *J. Am. Chem. Soc.*, 1987, **109**, 7495.

79 P. Magnus, C. Exon, and P. Albaugh-Robertson, *Tetrahedron*, 1985, **41**, 5861.

80 E. Negishi, S. J. Holmes, J. M. Tour, and J. A. Miller, *J. Am. Chem. Soc.*, 1985, **107**, 2568.

81 E. Negishi, D. R. Swanson, F. E. Cederbaum, and T. Takahashi, *Tetrahedron Lett.*, 1987, **28**, 917.

82 R. Aumann and H.-J. Weidenhaupt, *Chem. Ber.*, 1987, **120**, 23.

83 I. R. Butler, W. R. Cullen, W. E. Lindsell, P. N. Preston, and S. J. Rettig, *J. Chem. Soc., Chem. Commun.*, 1987, 439.

84 R. Aumann, H. D. Melchers, and H. J. Weidenhaupt, *Chem. Ber.*, 1987, **120**, 17.

85 E. Carceller, V. Centellas, A. Moyano, M. A. Pericas, and F. Serratosa, *Tetrahedron Lett.*, 1985, **26**, 2475.

86 P. Magnus and L. M. Principe, *Tetrahedron Lett.*, 1985, **26**, 4851.

87 F. Camps, J. Coll, J. M. Moreto, and J. Torras, *Tetrahedron Lett.*, 1987, **28**, 4745.

88 N. E. Schore and M. J. Knudsen, *J. Org. Chem.*, 1987, **52**, 569.

89 M. E. Krafft, *J. Am. Chem. Soc.*, 1988, **110**, 968.

90 M. E. Krafft, *Tetrahedron Lett.*, 1988, **29**, 999.

91 H. J. Jaffer and P. L. Pauson, *J. Chem. Res. (S)*, 1983, 244.

92 S. L. Schreiber, T. Sammakia, and W. E. Crowe, *J. Am. Chem. Soc.*, 1986, **108**, 3128.

93 N. E. Schore and S. D. Najdi, *J. Org. Chem.*, 1987, **52**, 5296.

94 P. W. Jolly and G. Wilke, *The Organic Chemistry of Nickel*, Vol. II, Academic Press, 1975, Chapter VI; A. Mullen in *New Syntheses with Carbon Monoxide*, ed. J. Falbe, Springer-Verlag, 1980, Chapter 3.

95 W. Reppe, *Justus Liebigs Ann. Chem*, 1953, **582**, 1; E. R. H. Jones, T. Y. Shen, and M. C. Whiting, *J. Chem. Soc.*, 1950, 230.

96 E. J. Corey and K. Achiwa, *Tetrahedron Lett.*, 1970, 2245.

97 R. F. Heck, *J. Am. Chem. Soc.*, 1963, **85**, 2013; J. Tsuji, J. Kiji, S. Imamura, and M. Morikawa, *J. Am. Chem. Soc.*, 1964, **86**, 4350; S. Brewis and P. R. Hughes, *J. Chem. Soc., Chem. Commun.*, 1965, 157.

98 M. F. Semmelhack and S. J. Brickner, *J. Am. Chem. Soc.*, 1981, **103**, 3945.

99 N. T. Byrom, R. Grigg, and B. Kongkathip, *J. Chem. Soc., Chem. Commun.*, 1976, 216; J. Tsuji, Y. Mori, and M. Hara, *Tetrahedron*, 1972, **28**, 3721.

100 S. Hosaka and J. Tsuji, *Tetrahedron*, 1971, **27**, 3821; C. Bordenca and W. E. Marsico, *Tetrahedron Lett.*, 1967, 1541.

101 R. Aumann and H. Ring, *Angew. Chem. Int. Ed. Engl.*, 1977, **16**, 50; Y. Seki, A. Hidaka, S. Murai, and N. Sonoda, *Angew. Chem. Int. Ed. Engl.*, 1977, **16**, 174.

102 Y. Seki, S. Murai, I. Yamamoto, and N. Sonoda, *Angew. Chem. Int. Ed. Engl.*, 1977, **16**, 789; S. Murai and N. Sonoda, *Angew. Chem. Int. Ed. Engl.*, 1979, **18**, 837.

103 J. L. Graff and M. G. Romanelli, *J. Chem. Soc., Chem. Commun.*, 1987, 337.

104 P. Eilbracht, E. Balss, and M. Acker, *Tetrahedron Lett.*, 1984, **25**, 1131.

105 R. Baker, R. C. Cookson, and J. R. Vinson, *J. Chem. Soc., Chem. Commun.*, 1974, 515.

106 K. Tamao, K. Kobayashi, and Y. Ito, *J. Am. Chem. Soc.*, 1988, **110**, 1286.

107 J. W. Suggs, *J. Am. Chem. Soc.*, 1979, **101**, 489.

108 K. H. Theopold, P. N. Becker, and R. G. Bergman, *J. Am. Chem. Soc.*, 1982, **104**, 5250.

109 L. S. Liebeskind, M. E. Welker, and V. Goedken, *J. Am. Chem. Soc.*, 1984, **106**, 441.

110 N. Aktogu, H. Felkin, and S. G. Davies, *J. Chem. Soc., Chem. Commun.*, 1982, 1303.

111 S. G. Davies, Royal Society Discussion Meeting, 24/25 Feb. 1988; S. G. Davies, G. Bashiardes, R. P. Beckett, S. J. Coote, I. M. Dordor-Hedgecock, C. L. Goodfellow, G. L. Gravatt, J. P. McNally, and M. Whittaker, *Phil. Trans. R. Soc. Lond. A*, 1988, **326**, 619.

112 S. G. Davies, I. M. Dordor-Hedgecock, K. H. Sutton, J. C. Walker, C. Bourne, R. H. Jones, and K. Prout, *J. Chem. Soc., Chem. Commun.*, 1986, 607.

113 G. Bashiardes and S. G. Davies, *Tetrahedron Lett.*, 1987, **28**, 5563.

114 G. Bashiardes, S. P. Collingwood, S. G. Davies, and S. C. Preston, *J. Chem. Soc., Perkin Trans. 1*, 1989, 1162; S. P. Collingwood, S. G. Davies, and S. C. Preston, *Tetrahedron Lett.*, 1990, **31**, 4067.

115 G. J. Baird, J. A. Bandy, S. G. Davies, and K. Prout, *J. Chem. Soc., Chem. Commun.*, 1983, 1202; S. G. Davies and J. C. Walker, *J. Chem. Soc., Chem. Commun.*, 1986, 495.

116 L. S. Liebeskind, R. W. Fengl, and M. E. Welker, *Tetrahedron Lett.*, 1985, **26**, 3075.

117 S. G. Davies, R. J. C. Easton, K. H. Sutton, J. C. Walker, and R. H. Jones, *J. Chem. Soc., Perkin Trans. 1*, 1987, 489.

118 S. G. Davies, I. M. Dordor, and P. Warner, *J. Chem. Soc., Chem. Commun.*, 1984, 956; S. G. Davies, I. M. Dordor-Hedgecock, P. Warner, R. H. Jones, and K. Prout, *J. Organomet. Chem.*, 1985, **285**, 213.

119 S. L. Brown, S. G. Davies, P. Warner, R. H. Jones, and K. Prout, *J. Chem. Soc., Chem. Commun.*, 1985, 1446; S. G. Davies and P. Warner, *Tetrahedron Lett.*, 1985, **26**, 4815.

120 L. S. Liebeskind and M. E. Welker, *Tetrahedron Lett.*, 1984, **25**, 4341.

121 S. G. Davies, I. M. Dordor-Hedgecock, and P. Warner, *Tetrahedron Lett.*, 1985, **26**, 2125.

122 J. I. Seeman and S. G. Davies, *J. Chem. Soc., Chem. Commun.*, 1984, 1019; S. G. Davies, and J. I. Seeman, *Tetrahedron Lett.*, 1984, **25**, 1845; S. G. Davies, J. I. Seeman, and I. H. Williams, *Tetrahedron Lett.*, 1986, **27**, 619; S. L. Brown, S. G. Davies, D. F. Foster, J. I. Seeman, and P. Warner, *Tetrahedron Lett.*, 1986, **27**, 623; B. K. Blackburn, S. G. Davies, and M. Whittaker, *J. Chem. Soc., Chem. Commun.*, 1987, 1344; S. G. Davies, I. M. Dordor-Hedgecock, K. H. Sutton, and M. Whittaker, *J. Am. Chem. Soc.*, 1987, **109**, 5711.

123 S. G. Davies and J. C. Walker, *J. Chem. Soc., Chem. Commun.*, 1985, 209; S. G. Davies, I. M. Dordor-Hedgecock, K. H. Sutton, and J. C. Walker, *Tetrahedron*, 1986, **42**, 5123.

124 P. W. Amber and S. G. Davies, *Tetrahedron Lett.*, 1988, **29**, 6979 and 6983.

125 K. Broadley and S. G. Davies, *Tetrahedron Lett.*, 1984, **25**, 1743.

126 S. G. Davies, I. M. Dorder-Hedgecock, K. H. Sutton, and J. C. Walker, *Tetrahedron Lett.*, 1986, **27**, 3787.

127 S. G. Davies, I. M. Dorder-Hedgecock, J. C. Walker, and P. Warner, *Tetrahedron Lett.*, 1984, **25**, 2709.

128 S. G. Davies and M. Wills, *J. Chem. Soc., Chem. Commun.*, 1987, 1647.

129 E. R. Burkhardt, J. J. Doney, R. G. Bergman, and C. H. Heathcock, *J. Am. Chem. Soc.*, 1987, **109**, 2022.

130 F. Francalanci and M. Foa, *J. Electroanalytical Chem.*, 1982, **232**, 59.

131 M. Tanaka, T. Kobayashi, and T. Sakakura, *J. Chem. Soc., Chem. Commun.*, 1985, 837.

132 F. Ozawa, N. Kawasaki, T. Yamamoto, and A. Yamamoto, *Chem. Lett.*, 1985, 567.

133 F. Ozawa, H. Soyama, H. Yanagihara, I. Aoyama, H. Takino, K. Izawa, T. Yamamoto, and A. Yamamoto, *J. Am. Chem. Soc.*, 1985, **107**, 3235.

134 J.-T. Chen and A. Sen, *J. Am. Chem. Soc.*, 1984, **106**, 1506.

135 F. Ozawa, N. Kawasaki, H. Okamoto, T. Yamamoto, and A. Yamamoto, *Organometallics*, 1987, **6**, 1640.

136 H. Alper, H. Arzoumanian, J.-F. Petrignani, and M. Saldana-Maldonado, *J. Chem. Soc., Chem. Commun.*, 1985, 340.

137 S.-I. Murahashi, Y. Mitsue, and K. Ike, *J. Chem. Soc., Chem. Commun.*, 1987, 125.

138 P. DeShong, D. R. Sidler, and G. A. Slough, *Tetrahedron Lett.*, 1987, **28**, 2233.

139 P. DeShong, G. A. Slough, and A. L. Rheingold, *Tetrahedron Lett.*, 1987, **28**, 2229.

140 J. L. Vidal and W. E. Walker, *Inorg. Chem.*, 1980, **19**, 896; R. L. Pruett, *Ann. N.Y. Acad. Sci.*, 1977, **295**, 239.

141 D. G. Parker, R. Pearce, and D. W. Prest, *J. Chem. Soc., Chem. Commun.*, 1982, 1193.

142 J. A. Osborn, F. H. Jardine, J. F. Young, and G. Wilkinson, *J. Chem. Soc. A*, 1966, 1711.

143 J. Tsuji and K. Ohno, *Tetrahedron Lett.*, 1965, 3969.

144 H. M. Walborsky and L. E. Allen, *Tetrahedron Lett.*, 1970, 823.

145 K. Ohno and J. Tsuji, *J. Am. Chem. Soc.*, 1968, **90**, 99.

146 R. J. Anderson, R. P. Hanzlik, K. B. Sharpless, E. E. van Tamelen, and R. B. Clayton, *J. Chem. Soc., Chem. Commun.*, 1969, 53.

147 Y. Shimizu, H. Mitsuhashi, and E. Caspi, *Tetrahedron Lett.*, 1966, 4113; M. Sergent, M. Mongrain, and P. Deslongchamps, *Can. J. Chem.*, 1972, **50**, 336.

148 K. Heusler, *Helv. Chim. Acta*, 1972, **55**, 388.

149 D. E. Iley and B. Fraser-Reid, *J. Am. Chem. Soc.*, 1975, **97**, 2563; R. W. Fries and J. K. Stille, *Synth. React. Inorg. Met.-Org. Chem.*, 1971, **1**, 295; D. H. Doughty and L. H. Pignolet, *J. Am. Chem. Soc.*, 1978, **100**, 7083.

150 P. D. Hobbs and P. D. Magnus, *J. Chem. Soc., Chem. Commun.*, 1974, 856.

151 D. J. Dawson and R. E. Ireland, *Tetrahedron Lett.*, 1968, 1899.

152 B. M. Trost and M. Preckel, *J. Am. Chem. Soc.*, 1973, **95**, 7862.

153 K. Sakai and O. Oda, *Tetrahedron Lett.*, 1972, 4375.

154 J. Blum, *Tetrahedron Lett.*, 1966, 1605; J. Blum, H. Rosenman, and E. D. Bergmann, *J. Org. Chem.*, 1968, **33**, 1928.

155 S. Friedman, S. R. Harris, and I. Wender, *Ind. Eng. Chem. Prod. Res. Dev.*, 1970, **9**, 347.

156 B. M. Trost and F. Chen, *Tetrahedron Lett.*, 1971, 2603.

157 K. Hori, M. Ando, N. Takaishi, and Y. Inamoto, *Tetrahedron Lett.*, 1986, **27**, 4615.

Chapter 10

1 A. D. English and T. Herskovitz, *J. Am. Chem. Soc.*, 1976, **99**, 1645.

2 S. D. Ittel, C. A. Tolman, A. D. English, and J. P. Jesson, *J. Am. Chem. Soc.*, 1978, **100**, 7577.

3 T. Tsuda, Y. Chujo, and T. Saegusa, *J. Chem. Soc., Chem. Commun.*, 1976, 415.

4 T. Tsuda, Y. Chujo, and T. Saegusa, *J. Am. Chem. Soc.*, 1978, **100**, 630.

5 D. J. Darensbourg and G. Grötsch, *J. Am. Chem. Soc.*, 1985, **107**, 7473.

6 A. Musco, *J. Chem. Soc., Perkin Trans. 1*, 1980, 693.

7 A. Behr and G. von Ilsemann, *J. Organomet. Chem.*, 1984, **276**, C77.

8 M. Santi and M. Marchi, *J. Organomet. Chem.*, 1979, **182**, 117.

9 T. Fujinami, T. Suzuki, M. Kamiya, S. Fukuzawa, and S. Sakai, *Chem. Lett.*, 1985, 199.

10 T. Tsuda, R. Sumiya, and T. Saegusa, *Synth. Commun.*, 1987, **17**, 147.

11 T. Tsuda, S. Morikawa, R. Sumiya, and T. Saegusa, *J. Org. Chem.*, 1988, **53**, 3140; T. Tsuda, S. Morikawa, and T. Saegusa, *J. Chem. Soc., Chem. Commun.*, 1989, 9.

12 J. F. Young, J. A. Osbourne, F. H. Jardine, and G. Wilkinson, *J. Chem. Soc., Chem. Commun.*, 1965, 131; M. A. Bennett and P. A. Longstaff, *Chem. & Ind. (London)*, 1965, 846; R. S. Coffey and J. B. Smith, UK Patent, 1965, 1 121 642.

13 J. P. Candlin and A. R. Oldham, *Discuss. Faraday Soc.*, 1968, **46**, 60.

14 A. J. Birch and K. A. M. Walker, *Aust. J. Chem.*, 1971, **24**, 513.

15 J. F. Biellmann and H. Liesenfelt, *Bull. Soc. Chim. Fr.*, 1966, 4029.

16 B. Zeeh, G. Jones, and C. Djerassi, *Chem. Ber.*, 1967, **100**, 3204.

17 A. R. Pinder, *Tetrahedron Lett.*, 1970, 413; H. C. Odom and A. R. Pinder, *J. Chem. Soc., Perkin Trans. 1*, 1972, 2193.

18 M. Brown and L. W. Piszkiewicz, *J. Org. Chem.*, 1967, **32**, 2013.

19 A. J. Birch and K. A. M. Walker, *J. Chem. Soc. C*, 1966, 1894.

20 F. H. Jardine and G. Wilkinson, *J. Chem. Soc. C*, 1967, 270; J. R. Shapley, R. R. Schrock, and J. A. Osborn, *J. Am. Chem. Soc.*, 1969, **91**, 2816; W. S. Knowles, M. J. Sabacky, and B. D. Vineyard, *Ann. N.Y. Acad. Sci.*, 1970, **172**, 232.

21 J. F. Biellman and M. J. Jung, *J. Am. Chem. Soc.*, 1968, **90**, 1673.

22 C. H. Heathcock and S. R. Poulter, *Tetrahedron Lett.*, 1969, 2755.

23 A. Tanaka, R. Tanaka, H. Uda, and A. Yoshikoshi, *J. Chem. Soc., Perkin Trans. 1*, 1972, 1721.

24 A. J. Birch and K. A. M. Walker, *Tetrahedron Lett.*, 1967, 3457.

25 M. Pardhasaradhi and G. S. Sidhu, *Tetrahedron Lett.*, 1972, 4201.

26 C. Bianchini, A. Meli, M. Peruzzini, F. Vizza, Y. Fujiwara, T. Jintoko, and H. Taniguchi, *J. Chem. Soc., Chem. Commun.*, 1988, 299.

27 R. H. Crabtree and D. G. Hamilton, *Adv. Organomet. Chem.*, 1988, **28**, 299.

28 R. H. Crabtree and D. G. Hamilton, *J. Am. Chem. Soc.*, 1986, **108**, 3124.

29 R. H. Crabtree, M. Lavin, and L. Bonneviot, *J. Am. Chem. Soc.*, 1986, **108**, 4032.

30 R. Stern and L. Sajus, *Tetrahedron Lett.*, 1968, 6313.

31 J. M. Brown, A. E. Derome, and S. A. Hall, *Tetrahedron*, 1985, **41**, 4647.

32 P. S. Hallman, D. Evans, J. A. Osborn, and G. Wilkinson, *J. Chem. Soc., Chem. Commun.*, 1967, 305.

33 P. S. Hallman, B. R. McGarvey, and G. Wilkinson, *J. Chem. Soc. A*, 1968, 3143.

34 I. Jardine and F. J. McQuillin, *Tetrahedron Lett.*, 1968, 5189.

35 E. F. Litvin, A. K. Freidlin, and K. K. Karimov, *Izv. Akad. Nauk. SSSR, Ser. Khim*, 1972, 1853.

36 A. Micsono and I. Ogata, *Bull. Chem. Soc. Jpn.*, 1967, **40**, 2718.

37 K. M. Ho, M.-C. Chan, and T. Y. Luh, *Tetrahedron Lett.*, 1986, **27**, 5383.

38 A. S. Hussey and Y. Takeuchi, *J. Org. Chem.*, 1970, **35**, 643.

39 D. F. Fahey, *J. Org. Chem.*, 1973, **38**, 3343.

40 W. Strohmeier and W. Rehder-Stirnweiss, *Z. Naturforschg.*, 1969, **24b**, 1219.

41 G. Dolcetti, *Inorg. Nucl. Chem. Lett.*, 1973, **9**, 705.

42 I. Jardine and F. J. McQuillin, *Tetrahedron Lett.*, 1968, 4871.

43 W. M. Moreau and K. Weiss, *Nature*, 1965, **208**, 1203.

44 F. H. Jardine, J. A. Osborn, G. Wilkinson, and J. F. Young, *Chem. Ind. (London)*, 1965, 560.

45 H. Greenfield, J. H. Wotiz, and I. Wender, *J. Org. Chem.*, 157, **22**, 542.

46 B. de Vries, *Koninkl. Ned. Akad. Wetenschap., Proc. Ser. B*, 1960, **63**, 443; A. F. Mabrouk, H. J. Dutton, and J. C. Cowan, *J. Am. Chem. Soc.*, 1964, **41**, 53; A. F. Mabrouk, E. Selke, W. K. Rohwedde, and H. J. Dutton, *J. Am. Oil Chem. Soc.*, 1965, **42**, 432; T. Takagi, *Nippon Kagaku Zasshi*, 1966, **87**, 600.

47 M. Murakami, K. Suzuki, and J.-W. Kang, *Nippon Kagaku Zasshi*, 1962, **83**, 1226.

48 J. Kwiatek, I. L. Mador, and J. K. Seyler, *Adv. Chem. Ser.*, 1963, **37**, 201.

49 J. Kwiatek, I. L. Mador, and J. K. Seyler, *J. Am. Chem. Soc.*, 1962, **84**, 304.

50 D. L. Reger, M. M. Habib, and D. J. Fauth, *J. Org. Chem.*, 1980, **45**, 3860; D. L. Reger, M. M. Habib, and D. J. Fauth, *Tetrahedron Lett.*, 1979, 115.

51 A. Fischli and P. M. Müller, *Helv. Chim. Acta*, 1980, **63**, 529, and 1619.

52 M. N. Ricroch and A. Gaudemer, *J. Organomet. Chem.*, 1974, **67**, 119.

53 R. Peyer and C. Weymuth, unpublished results cited in *Modern Synthetic Methods 1983*, ed. R. Scheffolf, Verlag Sauerländer, 1983.

54 I. Amer, T. Bravdo, J. Blum, and K. P. C. Vollhardt, *Tetrahedron Lett.*, 1987, **28**, 1321.

55 C. Larpent, R. Dabard, and H. Patin, *Tetrahedron Lett.*, 1987, **28**, 2507.

56 C. Masters, *Homogeneous Transition Metal Catalysis*, Chapman and Hall, 1981, pp. 24–8; B. R. James in *Comprehensive Organometallic Chemistry*, Vol. 8, ed. G. Wilkinson, Pergamon, 1982, pp. 290–9.

57 B. R. James, in *Comprehensive Organometallic Chemistry*, Vol. 8, ed. G. Wilkinson, Pergamon, 1982.

58 A. J. Birch and D. H. Williamson, *Organic Reactions*, 1976, **24**, 1.

59 B. R. James, *Homogeneous Hydrogenation*, Wiley, 1973; M. Nogradi, *Stereoselective Synthesis*, VCH, 1986, Ch. 2.

60 T. Yoshida, T. Okano, and S. Otsuka, *J. Chem. Soc., Chem. Commun.*, 1979, 870.

61 J. M. Brown, *Angew. Chem. Int. Ed. Engl.*, 1987, **26**, 190.

62 A. Nakamura and S. Otsuka, *Tetrahedron Lett.*, 1973, 4529.

63 E. N. Frankel, E. Selke, and C. A. Glass, *J. Am. Chem. Soc.*, 1968, **90**, 2446.

64 M. Cais, E. N. Frankel, and A. Rejoan, *Tetrahedron Lett.*, 1968, 1919.

65 T. Funabiki, M. Mohri, and K. Tarama, *J. Chem. Soc., Dalton Trans.*, 1973, 1813.

66 J. R. Tucker and D. P. Riley, *J. Organomet. Chem.*, 1985, **279**, 49.

67 M. Shibasaki and M. Sodeoka, *Tetrahedron Lett.*, 1985, **26**, 3491.

68 E. L. Muetterties and F. J. Hirsekorn, *J. Am. Chem. Soc.*, 1974, **96**, 4063; J. W. Johnson and E. L. Muetterties, *J. Am. Chem. Soc.*, 1977, **99**, 7395; L. S. Stuhl, M. Rakowski DuBois, F. J. Hirsekorn, J. R. Bleeke, A. E. Stevens, and E. L. Muetterties, *J. Am. Chem. Soc.*, 1978, **100**, 2405; M. A. Bennett, T.-N. Huang, and T. W. Turney, *J. Chem. Soc., Chem. Commun.*, 1979, 312; M. J. Russel, C. White, and P. M. Maitlis, *J. Chem. Soc., Chem. Commun.*, 1977, 427; K. J. Klabunde, B. B. Anderson, M. Bader, and L. J. Radonovich, *J. Am. Chem. Soc.*, 1978, **100**, 1313; M. S. Cohen, J. G. Noltes, and G. van Koten, U.S. Patent, 1980, 4 207 213.

69 C. J. Love and F. J. McQuillin, *J. Chem. Soc., Perkin Trans. 1*, 1973, 2509.

70 I. Wender, H. Greenfield, and M. Orchin, *J. Am. Chem. Soc.*, 1951, **73**, 2656.

71 R. Ercoli and R. E. Torregrosa, *Chim. Ind.* (Milan), 1958, **40**, 552.

72 L. K. Freidlin, V. Z. Sharf, V. N. Krutii, and S. I. Shcherbakova, *Zh. Org. Khim.*, 1972, **8**, 979.

73 Y. Ohgo, S. Takeuchi, and J. Yoshimura, *Bull. Chem. Soc. Jpn.*, 1971, **44**, 283.

74 Y. M. Y. Haddad, H. B. Henbest, J. Husbands, and T. R. B. Mitchell, *Proc. Chem. Soc.*, 1964, 361.

75 E. L. Eliel, T. W. Doyle, R. O. Hutchins, and E. C. Gilbert, *Org. Synth.*, 1970, **50**, 13.

76 H. B. Henbest and T. R. B. Mitchell, *J. Chem. Soc. C*, 1970, 785.

77 M. Hanaoka, N. Ogawa, and Y. Arata, *Tetrahedron Lett.*, 1973, 2355.

78 P. A. Browne and D. N. Kirk, *J. Chem. Soc., C*, 1969, 1653.

79 M. Murakami and J.-W. Kang, *Bull. Chem. Soc. Jpn.*, 1963, **36**, 763; M. Murakami and J.-W. Kang, *Bull. Chem. Soc. Jpn.*, 1962, **35**, 1243.

80 M. Murakami, K. Suzuki, M. Fujishige, and J.-W. Kang, *J. Chem. Soc. Jpn.*, 1964, **85**, 235.

81 L. Markó and J. Bakos, *J. Organomet. Chem.*, 1974, **81**, 411.

82 J. Halpern, in *Asymmetric Syntheses*, Vol. 5, ed. J. D. Morrison, Academic Press, 1985, Ch. 2, p. 41.

83 B. Bosnich, (ed.), *Asymmetric Catalysis*, Martin Nijhoff, 1986.

84 R. Noyori, T. Ohkuma, and M. Kitamura, *J. Am. Chem. Soc.*, 1987, **109**, 5856.

85 M. Kitamura, T. Ohkuma, S. Inoue, N. Sayo, H. Kumobayashi, S. Akutagawa, T. Ohta, H. Takaya, and R. Noyori, *J. Am. Chem. Soc.*, 1988, **110**, 629.

86 T. Hayashi, N. Kawamura, and Y. Ito, *J. Am. Chem. Soc.*, 1987, **109**, 7876.

87 B. D. Zwick, A. M. Arif, A. T. Patton, and J. A. Gladysz, *Angew. Chem. Int. Ed. Engl.*, 1987, **26**, 910.

88 W. S. Knowles, M. J. Sabacky, and B. D. Vineyard, *J. Chem. Soc., Chem. Commun.*, 1972, 10; W. S. Knowles, M. J. Sabacky, and B. D. Vineyard, *Chem. Technol.*, 1972, 590; W. S. Knowles, M. J. Sabacky, and B. D. Vineyard, *Ann. N.Y. Acad. Sci.*, 1973, **214**, 119.

89 H. Brunner, J. Wachter, J. Schmidbauer, G. M. Sheldrick, and P. G. Jones, *Angew. Chem. Int. Ed. Engl.*, 1986, **25**, 371.

90 P. Kvintovics, B. R. James, and B. Heil, *J. Chem. Soc., Chem. Commun.*, 1986, 1810.

91 T. Yamagishi, S. Ikeda, M. Yatagai, M. Yamaguchi, and M. Hida, *J. Chem. Soc., Perkin Trans. 1*, 1988, 1787.

92 Y. Ohgo, S. Takeuchi, and J. Yoshimura, *Bull. Chem. Soc. Jpn.*, 1971, **44**, 583.

93 V. Massonneau, P. Le Maux, and G. Simonneaux, *J. Organomet. Chem.*, 1987, **327**, 269; T. Ohta, H. Takaya, M. Kitamura, K. Nagai, and R. Noyori, *J. Org. Chem.*, 1987, **52**, 3174.

94 H. Kawano, Y. Ishii, T. Ikariya, M. Saburi, S. Yoshikawa, Y. Uchida, and H. Kumobayashi, *Tetrahedron Lett.*, 1987, **28**, 1905.

95 J. M. Brown and I. Cutting, *J. Chem. Soc., Chem. Commun.*, 1985, 578.

96 D. A. Evans, M. M. Morrissey, and R. L. Dow, *Tetrahedron Lett.*, 1985, **26**, 6005.

97 B. Bosnich, (ed.), *Asymmetric Catalysis*, Martinus Nijhoff, 1986, pp. 24–8.

98 R. H. Crabtree, J. M. Mihelcic, and J. M. Quirk, *J. Am. Chem. Soc.*, 1979, **101**, 7738; R. H. Crabtree, M. F. Mellea, J. M. Mihelcic, and J. M. Quirk, *J. Am. Chem. Soc.*, 1982, **104**, 107; J. W. Faller and H. Felkin, *Organometallics*, 1985, **4**, 1488.

99 D. Baudry, M. Ephritikhine, H. Felkin, and J. Zakrzewski, *Tetrahedron Lett.*, 1984, **25**, 1283.

100 R. H. Crabtree, E. M. Holt, M. Lavin, and S. M. Morehouse, *Inorg. Chem.*, 1985, **24**, 1986.

101 J.-Y. Saillard and R. Hoffmann, *J. Am. Chem. Soc.*, 1984, **106**, 2006.

102 R. B. Cracknell, A. G. Orpen, and J. L. Spencer, *J. Chem. Soc., Chem. Commun.*, 1984, 326; G. F. Schmidt and M. Brookhart, *J. Am. Chem. Soc.*, 1985, **107**, 1443; R. B. Cracknell, A. G. Orpen, and J. L. Spencer, *J. Chem. Soc., Chem. Commun.*, 1986, 1005.

103 J. J. La Placa and J. A. Ibers, *Inorg. Chem.*, 1965, **4**, 778; D. M. Roe, P. M. Bailey, K. Moseley, and P. M. Maitlis, *J. Chem. Soc., Chem. Commun.*, 1972, 1273.

104 J. J. Turner, J. K. Burdett, R. N. Perutz, and M. Poliakoff, *Pure Appl. Chem.*, 1977, **49**, 271.

105 J. M. Buchanan, J. M. Stryler, and R. G. Bergman, *J. Am. Chem. Soc.*, 1986, **108**, 1537.

106 D. E. Webster, *Adv. Organomet. Chem.*, 1977, **15**, 147.

107 P. L. Watson, *J. Am. Chem. Soc.*, 1983, **105**, 6491; P. L. Watson, *J. Chem. Soc., Chem. Commun.*, 1983, 276.

108 P. Svoboda, M. Čapka, J. Hetflejš, and V. Chvalovský, *Collct. Czech. Chem. Commun.*, 1972, **37**, 1585.

109 K. Takeshita, Y. Seki, K. Kawamoto, S. Murai, and N. Sonoda, *J. Chem. Soc., Chem. Commun.*, 1983, 1195.

110 A. J. Chalk, *J. Organomet. Chem.*, 1970, **21**, 207.

111 Y. Hori, T.-A. Mitsudo, and Y. Watanabe, *Bull. Chem. Soc. Jpn.*, 1988, **61**, 3011.

112 R. N. Perutz, personal communication, S. B. Duckett and R. N. Perutz, poster at the MICRA 88 symposium in Sheffield (1988).

113 J.-E. Bäckvall and O. S. Andell, *J. Chem. Soc., Chem. Commun.*, 1984, 260.

114 G. D. Fallon, N. J. Fitzmaurice, W. R. Jackson, and P. Perlmutter, *J. Chem. Soc., Chem. Commun.*, 1985, 4.

115 K. Narasaka, T. Yamada, and H. Minamikawa, *Chem. Lett.*, 1987, 2073.

116 H. X. Zhang, F. Guibé, and G. Balavoine, *Tetrahedron Lett.*, 1988, **29**, 619.

117 T. Hayashi and K. Kabeta, *Tetrahedron Lett.*, 1985, **26**, 3023.

118 I. Ojima, K. Yamamoto, and M. Kumada, *Aspects of Homogeneous Catalysis*, D. Reidel, Dordrecht, 1977, Ch. 3, p. 186; T. Hayashi, K. Yamamoto, and M. Kumada, *Tetrahedron Lett.*, 1974, 4405.

119 I. Ojima, T. Kogure, and M. Kumagai, *J. Org. Chem.*, 1977, **42**, 1671; I. Ojima, T. Kogure, and Y. Nagai, *Tetrahedron Lett.*, 1974, 1889.

120 I. Ojima, T. Kogure, and Y. Nagai, *Chem. Lett.*, 1973, 541; I. Ojima and Y. Nagai, *Chem. Lett.*, 1974, 223.

121 W. Dumont, J.-C. Poulin, T.-P. Dang, and H. B. Kagan, *J. Am. Chem. Soc.*, 1973, **95**, 8295.

122 N. Langlois, T.-P. Dang, and H. B. Kagan, *Tetrahedron Lett.*, 1973, 4865.

123 E. G. Samsel and J. R. Norton, *J. Am. Chem. Soc.*, 1984, **106**, 5505.

124 D. L. Thorn and R. Hoffmann, *J. Am. Chem. Soc.*, 1978, **100**, 2079.

125 B. M. Trost and R. Braslau, *Tetrahedron Lett.*, 1988, **29**, 1231.

126 B. M. Trost and S.-F. Chen, *J. Am. Chem. Soc.*, 1986, **108**, 6053.

127 B. M. Trost and J. M. Tour, *J. Am. Chem. Soc.*, 1987, **109**, 5268.

128 R. Van Helden and G. Verberg, *Rec. Trav. Chim. Pays-Bas*, 1965, **84**, 1263.

129 F. R. S. Clark, R. O. C. Norman, C. B. Thomas, and J. S. Willson, *J. Chem. Soc., Perkin Trans. 1*, 1974, 1289.

130 C. H. Bushweller, *Tetrahedron Lett.*, 1968, 6123.

131 D. R. Bryant, J. E. McKeon, and B. C. Ream, *Tetrahedron Lett.*, 1968, 3371.

132 R. A. Kretchmer and R. Glowinski, *J. Org. Chem.*, 1976, **41**, 2661.

133 G. Van Koten, J. T. B. H. Jastrzebski, and J. G. Noltes, *J. Chem. Soc., Chem. Commun.*, 1977, 203.

134 F. E. Ziegler, I. Chliwner, K. W. Fowler, S. J. Kanfer, S. J. Kuo, and N. D. Sinha, *J. Am. Chem. Soc.*, 1980, **102**, 790.

135 J. Yamashita, Y. Inoue, T. Kondo, and H. Hashimoto, *Chem. Lett.*, 1986, 407.

136 A. C. Cope and E. C. Friedrich, *J. Am. Chem. Soc.*, 1968, **90**, 909.

137 R. A. Holton, *Tetrahedron Lett.*, 1977, 355.

138 A. D. Ryabov, *Synthesis*, 1985, 233.

139 T. Itahara, *J. Org. Chem.*, 1985, **50**, 5272.

140 B. M. Trost, S. A. Godleski, and J. P. Genêt, *J. Am. Chem. Soc.*, 1978, **100**, 3930.

141 B. M. Trost, S. A. Godleski, and J. L. Belletire, *J. Org. Chem.*, 1979, **44**, 2052.

142 Y. Murakami, Y. Yogoyama, and T. Aoki, *Heterocycles*, 1984, **22**, 1493.

143 B. M. Trost and J. P. Genêt, *J. Am. Chem. Soc.*, 1976, **98**, 8516.

144 R. Grigg, T. R. B. Mitchell, and A. Ramasubbu, *J. Chem. Soc., Chem. Commun.*, 1980, 27.

145 R. Grigg, T. R. B. Mitchell, and A. Ramasubbu, *J. Chem. Soc., Chem. Commun.*, 1979, 669.

146 B. M. Trost and J. Y. L. Chung, *J. Am. Chem. Soc.*, 1985, **107**, 4586.

147 B. M. Trost and M. Lautens, *J. Am. Chem. Soc.*, 1985, **107**, 1781.

148 B. M. Trost and F. Rise, *J. Am. Chem. Soc.*, 1987, **109**, 3161.

149 R. Grigg, P. Stevenson, and T. Worakun, *Tetrahedron*, 1988, **44**, 4967.

150 W. A. Nugent and J. C. Calabrese, *J. Am. Chem. Soc.*, 1984, **106**, 6422.

151 S. L. Buchwald, A. Sayers, B. T. Watson, and J. C. Dewan, *Tetrahedron Lett.*, 1987, **28**, 3245.

152 R. A. Raphael, *Acetylenic Compounds in Organic Synthesis*, Butterworths, London, 1955; F. Straus, *Annalen*, 1905, **342**, 190.

153 L. Carlton and G. Read, *J. Chem. Soc., Perkin Trans. 1*, 1978, 1631.

154 C. W. Bird, *Transition Metal Intermediates in Organic Synthesis*, Academic Press, New York, 1967.

155 P. M. Maitlis, *The Organic Chemistry of Palladium*, Vol. 1, Academic Press, New York and London, 1971.

156 T. Hosokawa and I. Moritani, *Tetrahedron Lett.*, 1969, 3021.

157 T. Hosokawa, I. Moritani, and S. Nishioka, *Tetrahedron Lett.*, 1969, 3833.

158 L. Malatesta, G. Santarella, L. Vallarino, and F. Zingales, *Angew. Chem.*, 1960, **72**, 34.

159 H. Sakurai, K. Hirama, Y. Nakadaira, and C. Kabuto, *Chem. Lett.*, 1988, 485.

160 H. Sakurai, K. Hirama, Y. Nakadaira, and C. Kabuto, *J. Am. Chem. Soc.*, 1987, **109**, 6880.

161 H. Reinheimer, J. Moffat, and P. M. Maitlis, *J. Am. Chem. Soc.*, 1970, **92**, 2285.

162 K. P. C. Vollhardt and R. G. Bergman, *J. Am. Chem. Soc.*, 1974, **96**, 4996; W. G. L. Aalbersberg, A. J. Barkovich, R. L. Funk, R. L. Hillard III, and K. P. C. Vollhardt, *J. Am. Chem. Soc.*, 1975, **97**, 5600; R. L. Funk and K. P. C. Vollhardt, *J. Chem. Soc., Chem. Commun.*, 1976, 833.

163 R. L. Funk and K. P. C. Vollhardt, *J. Am. Chem. Soc.*, 1977, **99**, 5483; R. L. Funk and K. P. C. Vollhardt, *J. Am. Chem. Soc.*, 1979, **101**, 215.

164 R. L. Funk and K. P. C. Vollhardt, *J. Am. Chem. Soc.*, 1980, **102**, 5253.

165 E. D. Sternberg and K. P. C. Vollhardt, *J. Org. Chem.*, 1984, **49**, 1574.

166 K. P. C. Vollhardt, *Pure Appl. Chem.*, 1985, **57**, 1819.

167 K. P. C. Vollhardt, *Angew. Chem. Int. Ed. Engl.*, 1984, **23**, 539.

168 R. L. Hillard III, C. A. Parnell, and K. P. C. Vollhardt, *Tetrahedron*, 1983, **39**, 905.

169 C. J. Saward and K. P. C. Vollhardt, *Tetrahedron Lett.*, 1975, 4539.

170 B. C. Berris, Y.-H. Lai, and K. P. C. Vollhardt, *J. Chem. Soc., Chem. Commun.*, 1982, 953.

171 R. Diercks and K. P. C. Vollhardt, *J. Am. Chem. Soc.*, 1986, **108**, 3150.

172 R. Grigg, R. Scott, and P. Stevenson, *Tetrahedron Lett.*, 1982, **23**, 2691.

173 H. A. Dieck and R. F. Heck, *J. Am. Chem. Soc.*, 1974, **96**, 1133.

174 M. Mori, Y. Kubo, and Y. Ban, *Tetrahedron Lett.*, 1985, **26**, 1519.

175 S. J. Tremont and H. U. Rahman, *J. Am. Chem. Soc.*, 1984, **106**, 5759.

176 C. Sahlberg and L. Gawell, poster at the fourth IUPAC symposium on organometallic chemistry directed towards organic synthesis, Vancouver, 1987.

177 R. Grigg, P. Stevenson, and T. Worakun, *J. Chem. Soc., Chem. Commun.*, 1984, 1073.

178 F. E. Ziegler, U. R. Chakraborty, and R. B. Weisenfeld, *Tetrahedron*, 1981, **37**, 4035.

179 W. J. Scott, M. R. Pena, K. Swärd, S. J. Stoessel, and J. K. Stille, *J. Org. Chem.*, 1985, **50**, 2302.

180 J. R. Luly and H. Rapoport, *J. Org. Chem.*, 1984, **49**, 1671.

181 B. A. Patel, L.-C. Kao, N. A. Cortese, J. V. Minkiewicz, and R. F. Heck, *J. Org. Chem.*, 1979, **44**, 918.

182 L. Shi, C. K. Narula, K. T. Mak, L. Kao, Y. Xu, and R. F. Heck, *J. Org. Chem.*, 1983, **48**, 3894.

183 R. F. Heck, *Org. Reactions*, 1982, **27**, 345.

184 R. F. Heck and J. P. Nolley, Jr., *J. Org. Chem.*, 1972, **37**, 2320; B. A. Patel, J. F. Dickerson, and R. F. Heck, *J. Org. Chem.*, 1978, **43**, 5018.

185 N. Miyaura, I. Ishiyama, H. Sasaki, M. Ishikawa, M. Satoh, and A. Suzuki, *J. Am. Chem. Soc.*, 1989, **111**, 314.

186 R. C. Larock and B. E. Baker, *Tetrahedron Lett.*, 1988, **29**, 905.

187 P. C. Amos and D. A. Whiting, *J. Chem. Soc., Chem. Commun.*, 1987, 510.

188 K. Karabelas and A. Hallberg, *Tetrahedron Lett.*, 1985, **26**, 3131; Y. Hatanaka and T. Hiyama, *J. Org. Chem.*, 1988, **53**, 918.

189 N. A. Bumagin, I. G. Bumagina, and I. P. Beletskaya, *Dokl. Chem. (Engl. Trans.)*, 1984, **274**, 39.

190 K. Kikukawa, K. Ikenaga, F. Wada, and T. Matsuda, *Tetrahedron Lett.*, 1984, **25**, 5789.

191 J. B. Melpolder and R. F. Heck, *J. Org. Chem.*, 1976, **41**, 265; A. J. Chalk and S. A. Magennis, *J. Org. Chem.*, 1976, **41**, 273.

192 Y. Tamaru, Y. Yamada, and Z.-I. Yoshida, *Tetrahedron*, 1979, **35**, 329.

193 W. C. Frank, Y. C. Kim, and R. F. Heck, *J. Org. Chem.*, 1978, **43**, 2947.

194 H. Horino and N. Inoue, *J. Chem. Soc., Chem. Commun.*, 1976, 500.

195 S. Cacchi, P. G. Ciattini, E. Morera, and G. Ortar, *Tetrahedron Lett.*, 1987, **28**, 3039.

196 M. M. Abelman, T. Oh, and L. E. Overman, *J. Org. Chem.*, 1987, **52**, 4130.

197 G. E. Stokker, *Tetrahedron Lett.*, 1987, **28**, 3179.

198 S. G. Davies, B. E. Mobbs, and C. J. Goodwin, *J. Chem. Soc., Perkin Trans. 1*, 1987, 2597.

199 M. M. Abelman and L. E. Overman, *J. Am. Chem. Soc.*, 1988, **110**, 2328.

200 T. R. Bailey, *Tetrahedron Lett.*, 1986, **27**, 4407.

201 J. Stavenuiter, M. Hamzink, R. van der Hulst, G. Zommer, G. Westra, and E. Kriek, *Heterocycles*, 1987, **26**, 2711.

202 W. J. Scott, *J. Chem. Soc., Chem. Commun.*, 1987, 1755.

203 J. M. Clough, I. S. Mann, and D. A. Widdowson, *Tetrahedron Lett.*, 1987, **28**, 2645.

204 A. Dondoni, M. Fogagnolo, G. Fantin, A. Medici, and P. Pedrini, *Tetrahedron Lett.*, 1986, **27**, 5269.

205 M. Satoh, N. Miyaura, and A. Suzuki, *Chem. Lett.*, 1986, 1329.

206 S. Kanemoto, S. Matsubara, K. Oshima, K. Utimoto, and H. Nozaki, *Chem. Lett.*, 1987, 5.

207 N. Miyaura, K. Yamada, and A. Suzuki, *Tetrahedron Lett.*, 1979, 3437.

208 N. Miyaura, K. Yamada, H. Suginome, and A. Suzuki, *J. Am. Chem. Soc.*, 1985, **107**, 972.

209 R. Rossi, A. Carpita, and M. G. Quirici, *Tetrahedron*, 1981, **37**, 2617.

210 W. J. Scott and J. K. Stille, *J. Am. Chem. Soc.*, 1986, **108**, 3033.

211 J. K. Stille and M. Tanaka, *J. Am. Chem. Soc.*, 1987, **109**, 3785.

212 J. K. Stille and B. L. Groh, *J. Am. Chem. Soc.*, 1987, **109**, 813.

213 J. K. Stille and J. H. Simpson, *J. Am. Chem. Soc.*, 1987, **109**, 2138.

214 N. Miyaura, M. Satoh, and A. Suzuki, *Tetrahedron Lett.*, 1986, **27**, 3745.

215 Y. Kobayashi, T. Shimazaki, and F. Sato, *Tetrahedron Lett.*, 1987, **28**, 5849.

216 V. Nair, D. W. Powell, and S. C. Suri, *Synth. Commun.*, 1987, **17**, 1897.

217 E. J. Corey and M. F. Semmelhack, *J. Am. Chem. Soc.*, 1967, **89**, 2755; L. S. Hegedus and R. K. Stiverson, *J. Am. Chem. Soc.*, 1974, **96**, 3250; P. W. Jolly and G. Wilke, *The Organic Chemistry of Nickel*, Vol. II, Academic Press, 1975.

218 D. C. Billington, *Chem. Soc. Rev.*, 1985, **14**, 93.

219 J. A. Marshall and P. G. M. Wuts, *J. Am. Chem. Soc.*, 1978, **100**, 1627.

220 K. Sato, S. Inoue, and K. Saito, *J. Chem. Soc., Perkin Trans. 1*, 1973, 2289.

221 K. Sato, S. Inoue, and R. Yamaguchi, *J. Org. Chem.*, 1972, **37**, 1889; S. Inoue, R. Yamaguchi, K. Saito, and K. Sato, *Bull. Chem. Soc. Jpn.*, 1974, **47**, 3098.

222 S. Inoue, K. Saito, K. Kato, S. Nozaki, and K. Sato, *J. Chem. Soc., Perkin Trans. 1*, 1974, 2097.

223 E. J. Corey and E. Hamanaka, *J. Am. Chem. Soc.*, 1964, **86**, 1641; E. J. Corey and E. K. W. Wat, *J. Am. Chem. Soc.*, 1967, **89**, 2757; E. J. Corey and E. Hamanaka, *J. Am. Chem. Soc.*, 1967, **89**, 2758.

224 M. F. Semmelhack, *Org. React.*, 1972, **19**, 115.

225 L. Crombie, G. Kneen, G. Pattenden, and D. Whybrow, *J. Chem. Soc., Perkin Trans. 1*, 1980, 1711.

226 M. F. Semmelhack and S. J. Brickner, *J. Am. Chem. Soc.*, 1981, **103**, 3945.

227 H. P. Dang and G. Linstrumelle, *Tetrahedron Lett.*, 1978, 191.

228 V. Fiandanese, G. Marchese, F. Naso, and L. Ronzini, *Synthesis*, 1987, 1034.

229 K. Fugami, K. Oshima, and K. Utimoto, *Chem. Lett.*, 1987, 2203.

230 P. Kocieński, N. J. Dixon, and S. Wadman, *Tetrahedron Lett.*, 1988, **29**, 2353.

231 P. Kocieński, S. Wadman, and K. Cooper, *Tetrahedron Lett.*, 1988, **29**, 2357.

232 L. Crombie, A. J. W. Hobbs, M. A. Horsham, and R. J. Blade, *Tetrahedron Lett.*, 1987, **28**, 4875.

233 T. L. Gilchrist and R. J. Summersell, *Tetrahedron Lett.*, 1987, **28**, 1469.

234 R. McCague, *Tetrahedron Lett.*, 1987, **28**, 701; M. I. Al-Hassan, *J. Coll. Sci. King Saud Univ.*, 1987, **18**, 103.

235 A. Minato, K. Suzuki, and K. Tamao, *J. Am. Chem. Soc.*, 1987, **109**, 1257.

236 B. P. Andreini, M. Benetti, A. Carpita, and R. Rossi, *Tetrahedron*, 1987, **43**, 4591.

237 K. Sonogashira, Y. Tohda, and N. Hagihara, *Tetrahedron Lett.*, 1975, 4467; R. Rossi, A. Carpita, M. G. Quirici, and M. L. Gaudenzi, *Tetrahedron*, 1982, **38**, 631.

238 A. Carpita and R. Rossi, *Tetrahedron Lett.*, 1986, **27**, 4351.

239 K. C. Nicolaou and S. E. Webber, *J. Am. Chem. Soc.*, 1984, **106**, 5734.

240 L. Castedo, A. Mouriño, and L. A. Sarandeses, *Tetrahedron Lett.*, 1986, **27**, 1523.

241 E.-I. Negishi, T. Takahashi, S. Baba, D. E. Van Horn, and N. Okukado, *J. Am. Chem. Soc.*, 1987, **109**, 2393.

242 M. Uno, K. Seto, W. Ueda, M. Masuda, and S. Takahashi, *Synthesis*, 1985, 506; M. Uno, T. Takahashi, and S. Takahashi, *J. Chem. Soc., Chem. Commun.*, 1987, 785.

243 T. Migita, T. Shimizu, Y. Asami, J. Shiobara, Y. Kato, and M. Kosugi, *Bull. Chem. Soc. Jpn.*, 1980, **53**, 1385.

244 D. L. Boger and J. S. Panek, *Tetrahedron Lett.*, 1984, **25**, 3175.

245 A. S. Bell, D. A. Roberts, and K. S. Ruddock, *Tetrahedron Lett.*, 1988, **29**, 5013.

246 J. W. Tilley and S. Zawoiski, *J. Org. Chem.*, 1988, **53**, 386.

247 K. S. Petrakis and T. L. Nagabhushan, *J. Am. Chem. Soc.*, 1987, **109**, 2833.

248 M. A. Ciufolini and M. E. Browne, *Tetrahedron Lett.*, 1987, **28**, 171.

249 L. Crombie, M. A. Horsham, and R. J. Blade, *Tetrahedron Lett.*, 1987, **28**, 4879.

250 T. Sakamoto, H. Nagata, Y. Kondo, M. Shiraiwa, and H. Yamanaka, *Chem. Pharm. Bull.*, 1987, **35**, 823.

251 G. Just and R. Singh, *Tetrahedron Lett.*, 1987, **28**, 5981.

252 M. I. Al-Hassan, *Synthesis*, 1987, 816.

253 Y. Hatanaka and T. Hiyama, *Tetrahedron Lett.*, 1988, **29**, 97.

254 G. Dumartin, M. Pereyre, and J. P. Quintard, *Tetrahedron Lett.*, 1987, **28**, 3935.

255 B. M. Trost and R. Walchli, *J. Am. Chem. Soc.*, 1987, **109**, 3487.

256 J. K. Stille, *Angew. Chem. Int. Ed. Engl.*, 1986, **25**, 508; T. N. Mitchell, *J. Organomet. Chem.*, 1986, **304**, 1.

257 P. L. Castle and D. A. Widdowson, *Tetrahedron Lett.*, 1986, **27**, 6013.

258 A. Pelter and M. Rowlands, *Tetrahedron Lett.*, 1987, **28**, 1203; A. Pelter, M. Rowlands, and G. Clements, *Synthesis*, 1987, 49.

259 D. Djahanbini, B. Cazes, and J. Gore, *Tetrahedron*, 1987, **43**, 3441.

260 M. Kosugi, K. Ohashi, K. Akuzawa, T. Kawazoe, H. Sano, and T. Migita, *Chem. Lett.*, 1987, 1237.

261 K. Isomura, N. Okada, M. Saruwatari, H. Yamasaki, and H. Taniguchi, *Chem. Lett.*, 1985, 385.

262 A. S. Kende, B. Roth, P. J. Sanfilippo, and T. J. Blacklock, *J. Am. Chem. Soc.*, 1982, **104**, 5808.

263 S. Danishefsky, K. Vaughan, R. C. Gadwood, and K. Tsuzuki, *J. Am. Chem. Soc.*, 1980, **102**, 4262; S. Danishefsky, K. Vaughan, R. Gadwood, and K. Tsuzuki, *J. Am. Chem. Soc.*, 1981, **103**, 4136.

264 D. Milstein and J. K. Stille, *J. Am. Chem. Soc.*, 1978, **100**, 3636.

265 K. Hiroto, Y. Isobe, Y. Kitade, and Y. Maki, *Synthesis*, 1987, 495.

266 J. L. van der Baan and F. Bickelhaupt, *Tetrahedron Lett.*, 1986, **27**, 6267.

267 L. E. Overman and E. J. Jacobsen, *J. Am. Chem. Soc.*, 1982, **104**, 7225.

268 L. E. Overman and F. M. Knoll, *Tetrahedron Lett.*, 1979, 321: A. C. Oehlschlager, P. Mishra, and S. Dhami, *Can. J. Chem.*, 1984, **62**, 791.

269 L. E. Overman, *Acc. Chem. Res.*, 1980, **13**, 218; T. G. Schenck and B. Bosnich, *J. Am. Chem. Soc.*, 1985, **107**, 2058.

270 Y. Tamaru, M. Kagotani, and Z. Yoshida, *Tetrahedron Lett.*, 1981, **22**, 4245.

271 J. Yamada and Y. Yamamoto, *J. Chem. Soc., Chem. Commun.*, 1987, 1302.

272 A. F. Renaldo, *Synth. Commun.*, 1987, **17**, 1823.

273 R. A. Holton and K. J. Natalie, *Tetrahedron Lett.*, 1981, **22**, 267.

274 P. W. Clark, H. J. Dyke, S. F. Dyke, and G. Perry, *J. Organomet. Chem.*, 1983, **253**, 399.

275 P. Heimbach, P. W. Jolly, and G. Wilke, *Adv. Organomet. Chem.*, 1970, **8**, 29.

276 J. Tsuji, *Acc. Chem. Res.*, 1973, **6**, 8.

277 A. Tenaglia, P. Brun, and B. Waegell, *Tetrahedron Lett.*, 1985, **26**, 3571.

278 T. Mandai, H. Yasuda, M. Kaito, J. Tsuji, R. Yamaoka, and H. Fukami, *Tetrahedron*, 1979, **35**, 309.

279 J. Tsuji, T. Yamakawa, and T. Mandai, *Tetrahedron Lett.*, 1979, 3741.

280 L. Crombie, S. H. Harper, and F. C. Newman, *J. Chem. Soc.*, 1956, 3963.

281 R. N. Fakhretdinov, R. N. Khusnitdinov, I. B. Abdranmanov, and U. M. Dzhemislev, *J. Org. Chem. USSR*, 1984, **20**, 1002.

282 G. Hata, K. Takahashi, and A. Miyake, *J. Org. Chem.*, 1971, **36**, 2116.

283 J. Tsuji, H. Nagashima, T. Takahashi, and K. Masaoka, *Tetrahedron Lett.*, 1977, 1917.

284 T. Mitsuyasa and J. Tsuji, *Tetrahedron*, 1974, **30**, 831.

285 J. Tsuji and T. Mandai, *Tetrahedron Lett.*, 1977, 3285.

286 T. Takahashi, S. Hashiguchi, K. Kasuga, and J. Tsuji, *J. Am. Chem. Soc.*, 1978, **100**, 7424.

287 J. Tsuji, T. Yamakawa, and T. Mandai, *Tetrahedron Lett.*, 1978, 565.

288 S. Takahashi, T. Shibano, and N. Hagihara, *Tetrahedron Lett.*, 1967, 2451.

289 T. Takahashi, K. Kasuga, M. Takahashi, and J. Tsuji, *J. Am. Chem. Soc.*, 1979, **101**, 5072.

290 J. Tsuji, I. Shimizu, H. Suzuki, and Y. Naito, *J. Am. Chem. Soc.*, 1979, **101**, 5070.

291 J. Tsuji, K. Masaoka, and T. Takahashi, *Tetrahedron Lett.*, 1977, 2267.

292 M. Hidai, H. Ishiwatari, H. Yagi, E. Tanaka, K. Onozawa, and Y. Uchida, *J. Chem. Soc., Chem. Commun.*, 1975, 170.

293 D. Neibecker, M. Touma, and I. Tkatchenko, *Synthesis*, 1984, 1023.

294 J.-P. Haudegond, Y. Chauvin, and D. Commereuc, *J. Org. Chem.*, 1979, **44**, 3063.

295 J. Tsuji, Y. Mori, and M. Hara, *Tetrahedron*, 1972, **28**, 3721.

296 J. E. Mahler, D. H. Gibson, and R. Pettit, *J. Am. Chem. Soc.*, 1963, **85**, 3959.

297 A. J. Birch, I. D. Jenkins, and A. J. Liepa, *Tetrahedron Lett.*, 1975, 1723.

298 M. A. Hashmi, J. D. Munro, P. L. Pauson, and J. M. Williamson, *J. Chem. Soc. A*, 1967, 240; S.-I. Sasaoka, T. Yamamoto, H. Kinoshita, K. Inomata, and H. Kotake, *Chem. Lett.*, 1985, 315.

299 A. J. Pearson, Y.-S. Chen, M. L. Daroux, A. A. Tanaka, and M. Zettler, *J. Chem. Soc., Chem. Commun.*, 1987, 155.

300 J. Muzart and J.-P. Pete, *J. Chem. Soc., Chem. Commun.*, 1980, 257.

301 S. Torii, H. Tanaka, and K. Morisaki, *Tetrahedron Lett.*, 1985, **26**, 1655.

302 R. D. Ernst, H. Ma, G. Sergeson, T. Zahn, and M. L. Ziegler, *Organometallics*, 1987, **6**, 848; N. A. Vol'kenau, P. V. Petrovskii, L. S. Shilovtseva, and D. N. Kravstov, *J. Organomet. Chem.*, 1986, **303**, 121.

303 B. F. G. Johnson, J. Lewis, D. G. Parker, and S. R. Postle, *J. Chem. Soc., Dalton Trans.*, 1977, 794.

304 A. J. Birch and G. R. Stephenson, *J. Organomet. Chem.*, 1981, **218**, 91.

305 R. S. Sapienza, P. E. Riley, R. E. Davis, and R. Pettit, *J. Organomet. Chem.*, 1976, **121**, C35.

Chapter 11

1 F. D. Mango, *Coord. Chem. Rev.*, 1975, **15**, 109; R. G. Pearson, *Chem. Br.*, 1976, **12**, 160.

2 Z. Goldschmidt and H. E. Gottlieb, *J. Organomet. Chem.*, 1987, **329**, 391.

3 Z. Goldschmidt, H. E. Gottlieb, and D. Cohen, *J. Organomet. Chem.*, 1985, **294**, 219; Z. Goldschmidt, H. E. Gottlieb, E. Genizi, D. Cohen, and I. Goldberg, *J. Organomet. Chem.*, 1986, **301**, 337.

4 P. L. Pruitt, E. R. Biehl, and P. C. Reeves, *J. Organomet. Chem.*, 1977, **134**, 37.

5 M. Bottrill, R. Goddard, M. Green, R. P. Hughes, M. K. Lloyd, B. Lewis, and P. Woodward, *J. Chem. Soc., Chem. Commun.,* 1975, 253; R. C. Kerber in *The Organic Chemistry of Iron,* Vol. 2, ed. E. A. Koerner von Gustorf, F.-W. Grevels, and I. Fischer, Academic Press, 1981, p. 28.

6 G. N. Schrauzer and P. Glockner, *Chem. Ber.,* 1964, **97**, 2451; W. Brenner, P. Heimbach, H. Hey, E. W. Müller, and G. Wilke, *Justus Liebigs Ann. Chem.,* 1969, **727**, 161; P. Heimbach and G. Wilke, *Justus Liebigs Ann. Chem.,* 1969, **727**, 183; W. Brenner, P. Heimbach, and G. Wilke, *Justus Liebigs Ann. Chem.,* 1969, **727**, 194; P. J. Garratt and M. Wyatt, *J. Chem. Soc., Chem. Commun.,* 1974, 251.

7 P. Vioget, P. Vogel, and R. Roulet, *Angew. Chem. Int. Ed. Engl.,* 1982, **21**, 430.

8 F.-W. Grevels and K. Schneider, *Angew. Chem. Int. Ed. Engl.,* 1981, **20**, 410.

9 J. P. Genêt and J. Ficini, *Tetrahedron Lett.,* 1979, 1499; P. Heimbach, *Angew. Chem. Int. Ed. Engl.,* 1973, **121**, 975.

10 R. Davis, M. Green, and R. P. Hughes, *J. Chem. Soc., Chem. Commun.,* 1975, 405.

11 C. G. Kreiter, E. Michels, and H. Kurz, *J. Organomet. Chem.,* 1982, **232**, 249.

12 S. Ozkar, *Doga, Seri A.,* 1980, **4**, 31; J. H. Rigby and H. S. Ateeq, *J. Am. Chem. Soc.,* 1990, **112**, 6442.

13 C. G. Kreiter and H. Kurz, *Chem. Ber.,* 1983, **116**, 1494.

14 E. Michels and C. G. Kreiter, *J. Organomet. Chem.,* 1983, **252**, C1.

15 E. Michels, W. S. Sheldrick, and C. G. Kreiter, *Chem. Ber.,* 1985, **118**, 964.

16 C. G. Kreiter and E. Michels, *J. Organomet. Chem.,* 1986, **312**, 59.

17 P. A. Wender and N. C. Ihle, *Tetrahedron Lett.,* 1987, **28**, 2451.

18 H. Nagashima, H. Matsuda, and K. Itoh, *J. Organomet. Chem.,* 1983, **258**, C15.

19 K. Itoh, K. Mukai, H. Nagashima, and H. Nishiyama, *Chem. Lett.,* 1983, 499.

20 D. G. Bourner, L. Brammer, M. Green, G. Moran, A. G. Orpen, C. Reeve, and C. J. Schaverien, *J. Chem. Soc., Chem. Commun.,* 1985, 1409.

21 I. Matsuda, M. Shibata, S. Sato, and Y. Izumi, *Tetrahedron Lett.,* 1987, **28**, 3361.

22 M. Green, S. Heathcock, and D. C. Wood, *J. Chem. Soc., Dalton Trans.,* 1973, 1564.

23 M. Green, S. M. Heathcock, T. W. Turney, and D. M. P. Mingos, *J. Chem. Soc., Dalton Trans.,* 1977, 204.

24 S. K. Chopra, G. Moran, and P. McArdle, *J. Organomet. Chem.,* 1981, **214**, C36.

25 Z. Goldschmidt, H. E. Gottlieb, and A. Almadhoun, *J. Organomet. Chem.,* 1987, **326**, 405.

26 D. M. P. Mingos, *J. Chem. Soc., Dalton Trans.,* 1977, 20, 26 and 31.

27 D. Cunningham, P. McArdle, H. Sherlock, B. F. G. Johnson, and J. Lewis, *J. Chem. Soc., Dalton Trans.,* 1977, 2340.

28 R. E. Davis, T. A. Dodds, T.-H. Hseu, J. C. Wagnon, T. Devon, J. Tancrede, J. S. McKennis, and R. Pettit, *J. Am. Chem. Soc.,* 1974, **96**, 7562; R. Gandolfi and L. Toma, *Tetrahedron,* 1980, **36**, 935.

29 S. K. Chopra, M. J. Hynes, and P. McArdle, *J. Chem. Soc., Dalton Trans.,* 1981, 586.

30 Z. Goldschmidt and S. Antebi, *J. Organomet. Chem.,* 1983, **259**, 119.

31 Z. Goldschmidt, S. Antebi, D. Cohen, and I. Goldberg, *J. Organomet. Chem.,* 1984, **273**, 347.

32 L. A. Paquette, S. V. Ley, M. J. Broadhurst, J. M. Truesdell, J. Fayos, and J. Clardy, *Tetrahedron Lett.,* 1973, 2943; L. A. Paquette, S. V. Ley, S. Maiorana, D. F. Schneider, M. J. Broadhurst, and R. A. Boggs, *J. Am. Chem. Soc.,* 1975, **97**, 4658.

33 A. P. Humphries and S. A. R. Knox, *J. Chem. Soc., Dalton Trans.*, 1978, 1514; R. Goddard and P. Woodward, *J. Chem. Soc., Dalton Trans.*, 1979, 661.

34 M. E. Wright, *Organometallics*, 1983, **2**, 558.

35 R. S. Glass and W. W. McConnell, *Organometallics*, 1984, **3**, 1630.

36 S. R. Su and A. Wojcicki, *Inorg. Chim. Acta.*, 1974, **8**, 55.

37 W. P. Giering and M. Rosenblum, *J. Am. Chem. Soc.*, 1971, **93**, 5299.

38 W. Kitching and C. W. Fong, *Organometal. Chem. Rev. A*, 1970, **5**, 281.

39 W. P. Giering, S. Raghu, M. Rosenblum, A. Cutler, D. Ehntholt, and R. W. Fish, *J. Am. Chem. Soc.*, 1972, **94**, 8251.

40 T. W. Leung, G. G. Christoph, J. Gallucci, and A. Wojcicki, *Organometallics*, 1986, **5**, 846.

41 R. Baker, C. M. Exon, V. B. Rao, and R. W. Turner, *J. Chem. Soc., Perkin Trans. 1*, 1982, 295.

42 T. S. Abram, R. Baker, C. M. Exon, V. B. Rao, and R. W. Turner, *J. Chem. Soc., Perkin Trans. 1*, 1982, 301.

43 A. Cutler, D. Ehntholt, W. P. Giering, P. Lennon, S. Raghu, A. Rosan, M. Rosenblum, J. Tancrede, and D. Wells, *J. Am. Chem. Soc.*, 1976, **98**, 3495.

44 A. Bucheister, P. Klemarczyk, and M. Rosenblum, *Organometallics*, 1982, **1**, 1679.

45 R. Baker, V. B. Rao, and E. Erdik, *J. Organomet. Chem.*, 1983, **243**, 451.

46 T. S. Abram, R. Baker, C. M. Exon, and V. B. Rao, *J. Chem. Soc., Perkin Trans. 1*, 1982, 285.

47 R. Baker, R. B. Keen, M. D. Morris, and R. W. Turner, *J. Chem. Soc., Chem. Commun.*, 1984, 987.

48 N. Genco, D. Marten, S. Raghu, and M. Rosenblum, *J. Am. Chem. Soc.*, 1976, **98**, 848.

49 J. C. Watkins and M. Rosenblum, *Tetrahedron Lett.*, 1984, **25**, 2097; J. C. Watkins and M. Rosenblum, *Tetrahedron Lett.*, 1985, **26**, 3531.

50 B. M. Trost, *Angew. Chem. Int. Ed. Engl.*, 1986, **25**, 1.

51 B. M. Trost and D. M. T. Chan, *J. Am. Chem. Soc.*, 1979, **101**, 6329 and 6432.

52 I. Shimizu, Y. Ohashi, and J. Tsuji, *Tetrahedron Lett.*, 1984, **25**, 5183.

53 B. M. Trost and D. M. T. Chan, *J. Am. Chem. Soc.*, 1981, **103**, 5972.

54 B. M. Trost, T. N. Nanninga, and T. Satoh, *J. Am. Chem. Soc.*, 1985, **107**, 721.

55 D. G. Cleary and L. A. Paquette, *Synth. Commun.*, 1987, **17**, 497.

56 B. M. Trost and P. R. Seoane, *J. Am. Chem. Soc.*, 1987, **109**, 615.

57 R. Noyori, T. Odagi, and H. Takaya, *J. Am. Chem. Soc.*, 1970, **92**, 5780; P. Binger and P. Bentz, *J. Organomet. Chem.*, 1981, **221**, C33; P. Binger, A. Brinkmann, and P. Wedemann, *Chem. Ber.*, 1983, **116**, 2920; P. Binger and P. Wedemann, *Tetrahedron Lett.*, 1983, **24**, 5847; P. Binger, Q.-H. Lü, and P. Wedemann, *Angew. Chem. Int. Ed. Engl.*, 1985, **24**, 316.

58 B. M. Trost and M. L. Miller, *J. Am. Chem. Soc.*, 1988, **110**, 3687.

59 B. M. Trost and D. M. T. Chan, *J. Am. Chem. Soc.*, 1980, **102**, 6359; B. M. Trost and D. M. T. Chan, *J. Am. Chem. Soc.*, 1983, **105**, 2326; D. J. Gordon, R. F. Fenske, T. N. Nanninga, and B. M. Trost, *J. Am. Chem. Soc.*, 1981, **103**, 5974.

60 B. M. Trost and P. Renaut, *J. Am. Chem. Soc.*, 1982, **104**, 6668; B. M. Trost and T. N. Nanninga, *J. Am. Chem. Soc.*, 1985, **107**, 1075.

61 B. M. Trost and T. N. Nanninga, *J. Am. Chem. Soc.*, 1985, **107**, 1293.

62 B. M. Trost, S. A. King, and T. N. Nanninga, *Chem. Lett.*, 1987, 15.

63 B. M. Trost, J. Lynch, P. Renunt, and D. H. Steinman, *J. Am. Chem. Soc.*, 1986, **108**, 284.

64 R. Baker and R. B. Keen, *J. Organomet. Chem.*, 1985, **285**, 419.

65 B. M. Trost and D. M. T. Chan, *J. Am. Chem. Soc.*, 1983, **105**, 2315.

66 B. M. Trost and P. J. Bonk, *J. Am. Chem. Soc.*, 1985, **107**, 1778.

67 J. van der Louw, J. L. van der Baan, H. Stichter, G. J. J. Out, F. Bickelhaupt, and G. W. Klumpp, *Tetrahedron Lett.*, 1988, **29**, 3579.

68 B. M. Trost and D. M. T. Chan, *J. Am. Chem. Soc.*, 1982, **104**, 3733.

69 B. M. Trost, T. N. Nanninga, and D. M. T. Chan, *Organometallics*, 1982, **1**, 1543; B. M. Trost and D. T. MacPherson, *J. Am. Chem. Soc.*, 1987, **109**, 3483.

70 B. M. Trost and S. Schneider, *Angew. Chem. Int. Ed. Engl.*, 1989, **28**, 213.

71 B. M. Trost and P. R. Seoane, *J. Am. Chem. Soc.*, 1987, **109**, 615.

72 Y. Huang and X. Lu, *Tetrahedron Lett.*, 1987, **28**, 6219; Y. Huang and X. Lu, *Tetrahedron Lett.*, 1988, **29**, 5663.

73 G. A. Molander and D. C. Shubert, *Tetrahedron Lett.*, 1986, **27**, 787.

74 P. Breuilles and D. Uguen, *Tetrahedron Lett.*, 1988, **29**, 201.

75 A. Yamamoto, Y. Ito, and T. Hayashi, *Tetrahedron Lett.*, 1989, **30**, 375.

76 I. Shimizu, Y. Ohashi, and J. Tsuji, *Tetrahedron Lett.*, 1985, **26**, 3825.

77 R. Chidgey and H. M. R. Hoffmann, *Tetrahedron Lett.*, 1977, 2633; R. Noyori, *Acc. Chem. Res.*, 1979, **12**, 61; G. Fierz, R. Chidgey, and H. M. R. Hoffmann, *Angew. Chem. Int. Ed. Engl.*, 1974, **13**, 410; J. Mann, *Tetrahedron*, 1986, **42**, 4611.

78 H. M. R. Hoffmann, *Angew. Chem. Int. Ed. Engl.*, 1972, **11**, 324.

79 T. Sato, R. Ito, Y. Hayakawa, and R. Noyori, *Tetrahedron Lett.*, 1978, 1829; R. Noyori, T. Sato, and Y. Hayakawa, *J. Am. Chem. Soc.*, 1978, **100**, 2561.

80 H. Takaya, Y. Hayakawa, S. Makino, and R. Noyori, *J. Am. Chem. Soc.*, 1978, **100**, 1765, 1778.

81 Y. Hayakawa, M. Sakai, and R. Noyori, *Chem. Lett.*, 1974, 509; R. Noyori, S. Makino, T. Okita, and Y. Hayakawa, *J. Org. Chem.*, 1975, **40**, 806.

82 Y. Hayakawa, F. Shimizu, and R. Noyori, *Tetrahedron Lett.*, 1978, 993.

83 R. Noyori, M. Nishizawa, F. Shimizu, Y. Hayakawa, K. Maruoka, S. Hashimoto, H. Yamamoto, and H. Nozaki, *J. Am. Chem. Soc.*, 1979, **101**, 220.

84 B. E. La Belle, M. J. Knudsen, M. M. Olmstead, H. Hope, M. D. Yanuck, and N. E. Schore, *J. Org. Chem.*, 1985, **50**, 5215.

85 R. Noyori, F. Shimizu, K. Fukuta, H. Takaya, and Y. Hayakawa, *J. Am. Chem. Soc.*, 1977, **99**, 5196.

86 A. S. Narula, *Tetrahedron Lett.*, 1979, 1921.

87 D. Savoia, E. Tagliavini, C. Trombini, and A. Umani-Ronchi, *J. Org. Chem.*, 1982, **47**, 876.

88 T. Ishizu, K. Harano, M. Yasuda, and K. Kanematsu, *J. Org. Chem.*, 1981, **46**, 3630; T. Ishizu, K. Harano, M. Yasuda, and K. Kanematsu, *Tetrahedron Lett.*, 1981, **22**, 1601.

89 T. Ishizu, M. Mori, and K. Kanematsu, *J. Org. Chem.*, 1981, **46**, 526.

Chapter 12

1 S. D. Burke and P. A. Grieco, *Org. React.*, 1979, **26**, 361.

2 A. Demonceau, A. F. Noels, A. J. Hubert, and P. Teyssié, *J. Chem. Soc., Chem. Commun.*, 1981, 688.

3 H. Fischer, J. Schmid, and R. Märkl, *J. Chem. Soc., Chem. Commun.*, 1985, 572.

4 A. J. Anciaux, A. Demonceau, A. F. Noels, A. J. Hubert, R. Warin, and P. Teyssié, *J. Org. Chem.*, 1981, **46**, 873; D. F. Taber and E. H. Petty, *J. Org. Chem.*, 1982, **47**, 4808.

5 D. F. Taber, E. H. Petty, and K. Raman, *J. Am. Chem. Soc.*, 1985, **107**, 196.

6 H. Ledon, G. Linstrumelle, and S. Julia, *Tetrahedron Lett.*, 1973, 25.

7 D. F. Taber, K. Raman, and M. D. Gaul, *J. Org. Chem.*, 1987, **52**, 28.

8 P. Brown and R. Southgate, *Tetrahedron Lett.*, 1986, **27**, 247.

9 D. Häbich and W. Hartwig, *Tetrahedron Lett.*, 1987, **28**, 781.

10 M. P. Moyer, P. L. Feldman, and H. Rapoport, *J. Org. Chem.*, 1985, **50**, 5223.

11 J. C. Heslin, C. J. Moody, A. M. Z. Slawin, and D. J. Williams, *Tetrahedron Lett.*, 1986, **27**, 1403.

12 C. J. Moody and R. J. Taylor, *Tetrahedron Lett.*, 1987, **28**, 5351.

13 H. J. Monteiro, *Tetrahedron Lett.*, 1987, **28**, 3459.

14 B. Corbel, D. Hernot, J.-P. Haelters, and G. Sturtz, *Tetrahedron Lett.*, 1987, **28**, 6605.

15 G. Stork and K. Nakatani, *Tetrahedron Lett.*, 1988, **29**, 2283.

16 M. Hrytsak and T. Durst, *J. Chem. Soc., Chem. Commun.*, 1987, 1150.

17 M. C. Pirung and J. A. Werner, *J. Am. Chem. Soc.*, 1986, **108**, 6060.

18 H. M. L. Davies and L. van T. Crisco, *Tetrahedron Lett.*, 1987, **28**, 371; A. Padwa and P. D. Stull, *Tetrahedron Lett.*, 1987, **28**, 5407; F. Kido, S. C. Sinha, T. Abiko and A. Yoshikoshi, *Tetrahedron Lett.*, 1989, **30**, 1575.

19 M. E. Maier and K. Evertz, *Tetrahedron Lett.*, 1988, **29**, 1677.

20 K. A. M. Kremer, P. Helquist, and R. C. Kerber, *J. Am. Chem. Soc.*, 1981, **103**, 1862; K. A. M. Kremer and P. Helquist, *J. Organomet. Chem.*, 1985, **285**, 231; R. S. Iyer, G.-H. Kuo, and P. Helquist, *J. Org. Chem.*, 1985, **50**, 5898; C. Knors, G.-H. Kuo, J. W. Lauher, C. Eigenbrot, and P. Helquist, *Organometallics*, 1987, **6**, 988.

21 A. Wienand and H. U. Reissig, *Tetrahedron Lett.*, 1988, **29**, 2315.

22 C. P. Casey, N. W. Vollendorf, and K. J. Haller, *J. Am. Chem. Soc.*, 1984, **106**, 3754; A. Parlier and H. Rudler, *J. Chem. Soc., Chem. Commun.*, 1986, 514.

23 M. Suda, *Synthesis*, 1981, 714; D. F. Taber, J. C. Amedio, Jr., and R. G. Sherrill, *J. Org. Chem.*, 1986, **51**, 3382.

24 R. G. Salomon and J. K. Kochi, *J. Am. Chem. Soc.*, 1973, **95**, 3300.

25 R. G. Salomon, M. F. Salomon, and T. R. Heyne, *J. Org. Chem.*, 1975, **40**, 756.

26 A. J. Hubert, A. F. Noels, A. J. Anciaux, and P. Teyssié, *Synthesis*, 1976, 600.

27 A. J. Anciaux, A. J. Hubert, A. F. Noels, N. Petiniot, and P. Teyssié, *J. Org. Chem.*, 1980, **45**, 695.

28 N. Petiniot, A. J. Anciaux, A. F. Noels, A. J. Hubert, and P. Teyssié, *Tetrahedron Lett.*, 1978, 1239.

29 J. Adams and M. Belley, *Tetrahedron Lett.*, 1986, **27**, 2075.

30 A. Padwa, T. J. Wisnieff, and E. J. Walsh, *J. Org. Chem.*, 1986, **51**, 5036.

31 H. M. L. Davies, H. D. Smith, and O. Korkor, *Tetrahedron Lett.*, 1987, **28**, 1853.

32 H. M. L. Davies, D. M. Clark, D. B. Alligood, and G. R. Eiband, *Tetrahedron*, 1987, **43**, 4265.

33 M. Elliot, A. W. Farnham, N. F. James, P. H. Needham, A. Pulman, and J. H. Stevenson, *Nature*, 1973, **246**, 169.

34 A. J. Anciaux, A. Demonceau, A. F. Noels, R. Warin, A. J. Hubert, and P. Teyssié, *Tetrahedron*, 1983, **39**, 2169.

35 D. Holland, D. A. Laidler, and D. J. Milner, *J. Mol. Catal.*, 1981, **11**, 119.

36 M. F. Dull and P. G. Abend, *J. Am. Chem. Soc.*, 1959, **81**, 2588.

37 P. Dowd, C. Kaufman, and Y. H. Paik, *Tetrahedron Lett.*, 1985, **26**, 2283.

38 T. Aratani, Y. Yoneyoshi, and T. Nagase, *Tetrahedron Lett.*, 1982, **23**, 685.

39 H. Fritschi, Y. Leutenegger, and A. Pfaltz, *Angew. Chem. Int. Ed. Engl.*, 1986, **25**, 1005.

40 A. Nakamura, *Pure Appl. Chem.*, 1978, **50**, 37.

41 M. Brookhart, D. Timmers, J. R. Tucker, and G. D. Williams, *J. Am. Chem. Soc.*, 1983, **105**, 6721.

42 M. Brookhart, Y. Liu, and R. C. Buck, *J. Am. Chem. Soc.*, 1988, **110**, 2337.

43 R. R. Schrock, *J. Am. Chem. Soc.*, 1976, **98**, 5399.

44 F. N. Tebbe, G. W. Parshall, and G. S. Reddy, *J. Am. Chem. Soc.*, 1978, **100**, 3611.

45 S. H. Pine, R. J. Pettit, G. D. Geib, S. G. Cruz, C. H. Gallego, T. Tijerina, and R. D. Pine, *J. Org. Chem.*, 1985, **50**, 1212.

46 J. B. Lee, K. C. Ott, and R. H. Grubbs, *J. Am. Chem. Soc.*, 1982, **104**, 7491.

47 D. A. Straus and R. H. Grubbs, *Organometallics*, 1982, **1**, 1658.

48 K. H. Dötz, *Angew. Chem. Int. Ed. Engl.*, 1984, **23**, 587.

49 K. A. Brown-Wensley, S. L. Buchwald, L. Cannizzo, L. Clawson, S. Ho, D. Meinhardt, J. R. Stille, D. Straus, and R. H. Grubbs, *Pure Appl. Chem.*, 1983, **55**, 1733.

50 R. E. Ireland, private communication cited in ref. 49.

51 J. R. Stille and R. H. Grubbs, *J. Am. Chem. Soc.*, 1983, **105**, 1664; T.-S. Chou and B.-S. Huang, *Tetrahedron Lett.*, 1983, **24**, 2169.

52 J. M. Hawkins and R. H. Grubbs, *J. Am. Chem. Soc.*, 1988, **110**, 2821.

53 S. L. Buchwald and R. H. Grubbs, *J. Am. Chem. Soc.*, 1983, **105**, 5490.

54 F. W. Hartner, Jr., J. Schwartz, and S. M. Clift, *J. Am. Chem. Soc.*, 1983, **105**, 640.

55 S. M. Clift and J. Schwartz, *J. Am. Chem. Soc.*, 1984, **106**, 8300.

56 T. Okazoe, K. Takai, K. Oshima, and K. Utimoto, *J. Org. Chem.*, 1987, **52**, 4410.

57 W. A. Kinney, M. J. Coghlan, and L. A. Paquette, *J. Am. Chem. Soc.*, 1984, **106**, 6868.

58 R. L. Banks and G. C. Bailey, *Ind. Eng. Chem.*, *Prod. Res. Dev.*, 1964, **3**, 170.

59 N. Calderon, H. Y. Chen, and K. W. Scott, *Tetrahedron Lett.*, 1967, 3327.

60 N. Calderon, E. A. Ofstead, and W. A. Judy, *Angew. Chem. Int. Ed. Engl.*, 1976, **15**, 401; T. J. Katz, *Adv. Organomet. Chem.*, 1977, **16**, 283; R. H. Grubbs, *Prog. Inorg. Chem.*, 1978, **24**, 1; N. Calderon, J. P. Lawrence, and E. A. Ofstead, *Adv. Organomet. Chem.*, 1979, **17**, 449.

61 R. R. Schrock, *J. Am. Chem. Soc.*, 1974, **96**, 6796; H. C. Foley, L. M. Strubinger, T. S. Targos, and G. L. Geoffroy, *J. Am. Chem. Soc.*, 1983, **105**, 3064; K. Angermund, F.-W. Grevels, C. Krüger, and V. Skibbe, *Angew. Chem. Int. Ed. Engl.*, 1984, **23**, 904.

62 J.-L. Herisson and Y. Chauvin, *Makromol. Chem.*, 1970, **141**, 161; G. Wilke, *Pure Appl. Chem.*, 1978, **50**, 677; J. Kress and J. A. Osborn, *J. Am. Chem. Soc.*, 1983, **105**, 6346.

63 O. Eisenstein, R. Hoffmann, and A. R. Rossi, *J. Am. Chem. Soc.*, 1981, **103**, 5582.

64 R. J. Haines and G. J. Leigh, *Chem. Soc. Rev.*, 1975, **4**, 155.

65 R. L. Banks, *Chemtech*, 1979, 494; R. L. Banks, *Proceedings of the Award Symposium on Olefin Polymerisation and Disproportionation (Metathesis)* in the joint meeting of the American Chemical Society and Chemical Society of Japan, Honolulu, 1979, p. 399.

66 R. S. Bauer, P. W. Glockner, W. Kelm, and R. F. Mason, US Patent, 1972, 3 647 915; A. J. Berger, US Patent, 1973, 3 726 938; P. A. Verbrugge and G. J. Helszwolf, US Patent, 1973, 3 776 975; UK Patent, 1972, 1 416 317.

67 P. D. Montgomery, R. N. Moore, and W. R. Knox, US Patent, 1976, 3 965 206.

68 P. Chevalier, D. Sinou, G. Descotes, R. Mutin, and J. Basset, *J. Organomet. Chem.*, 1976, **113**, 1.

69 L. G. Wideman, *J. Org. Chem.*, 1968, **33**, 4541.

70 J.-P. Laval, A. Lattes, R. Mutin, and J. M. Basset, *J. Chem. Soc., Chem. Commun.*, 1977, 502.

71 W. B. Hughes, *Ann. N.Y. Acad. Sci.*, 1977, **295**, 271.

72 J. Tsuji and S. Hashiguchi, *J. Organomet. Chem.*, 1981, **218**, 69.

73 S. R. Wilson and D. E. Schalk, *J. Org. Chem.*, 1976, **41**, 3928.

74 P. B. van Dam, M. C. Mittelmeijer, and C. Boelhouwer, *J. Am. Oil Chem. Soc.*, 1974, **51**, 389.

75 P. B. van Dam, M. C. Mittelmeijer, and C. Boelhouwer, *J. Chem. Soc., Chem. Commun.*, 1972, 1221.

76 F.-W. Küpper and R. Streck, *Z. Naturforsch. Teil B*, 1976, **316**, 1256.

77 E. Dalcanale, F. Casagrande, T. Martinengo, and F. Montanari, *J. Chem. Res (S)*, 1985, 294.

78 J. R. Stille and R. H. Grubbs, *J. Am. Chem. Soc.*, 1986, **108**, 855.

79 T. R. Howard, J. B. Lee, and R. H. Grubbs, *J. Am. Chem. Soc.*, 1980, **102**, 6876; K. C. Ott and R. H. Grubbs, *J. Am. Chem. Soc.*, 1981, **103**, 5922; D. A. Straus and R. H. Grubbs, *J. Mol. Catal.*, 1985, **28**, 9.

80 C. P. Casey, N. L. Hornung, and W. P. Kosar, *J. Am. Chem. Soc.*, 1987, **109**, 4908.

81 W. D. Wulff, P.-C. Tang, K.-S. Chan, J. S. McCallum, D. C. Yang, and S. R. Gibertson, *Tetrahedron*, 1985, **41**, 5813.

82 E. O. Fischer and A. Maasböl, *Angew. Chem. Int. Ed. Engl.*, 1964, **3**, 580.

83 K. H. Dötz, *Pure Appl. Chem.*, 1983, **55**, 1689.

84 K. H. Dötz, *Angew. Chem. Int. Ed. Engl.*, 1975, **14**, 644.

85 W. D. Wulff, P.-C. Tang, and J. S. McCallum, *J. Am. Chem. Soc.*, 1981, **103**, 7677.

86 W. D. Wulff, K.-S. Chan, and P.-C. Tang, *J. Org. Chem.*, 1984, **49**, 2293.

87 K. H. Dötz and R. Dietz, *Chem. Ber.*, 1978, **111**, 2517.

88 W. D. Wulff, J. S. McCallum, and F.-A. Kunng, *J. Am. Chem. Soc.*, 1988, **110**, 7419.

89 A. Yamashita and A. Toy, *Tetrahedron Lett.*, 1986, **27**, 3471.

90 K. H. Dötz, R. Dietz, D. Kappenstein, D. Neugebauer, and U. Schubert, *Chem. Ber.*, 1979, **112**, 3682.

91 W. D. Wulff, S. R. Gilbertson, and J. P. Springer, *J. Am. Chem. Soc.*, 1986, **108**, 520.

92 K. H. Dötz and R. Dietz, *J. Organomet. Chem.*, 1978, **157**, C55; A. Yamashita, T. A. Scahill, and A. Toy, *Tetrahedron Lett.*, 1985, **26**, 2969.

93 A. Yamashita, *Tetrahedron Lett.*, 1986, **27**, 5915; W. Flitsch, J. Lauterwein, and W. Micke, *Tetrahedron Lett.*, 1989, **30**, 1633.

94 K. H. Dötz and M. Popall, *Chem. Ber.*, 1988, **121**, 665; K. H. Dötz and M. Popall, *Angew. Chem. Int. Ed. Engl.*, 1987, **26**, 1158; W. D. Wulff and Y.-C. Xu, *J. Am. Chem. Soc.*, 1988, **110**, 2312.

95 A. Yamashita, *J. Am. Chem. Soc.*, 1985, **107**, 5823.

96 K. H. Dötz, W. Kuhn, and K. Ackerman, *Z. Naturforsch. Teil B*, 1983, **38**, 1351.

97 M. F. Semmelhack and J. J. Bozell, *Tetrahedron Lett.*, 1982, **23**, 2931.

98 K. H. Dötz, W. Sturm, M. Popall, and J. Riede, *J. Organomet. Chem.*, 1984, **277**, 267; K. H. Dötz and M. Popall, *Tetrahedron*, 1985, **41**, 5797.

99 D. L. Boger and I. C. Jacobson, *J. Org. Chem.*, 1990, **55**, 1919.

100 W. D. Wulff and P.-C. Tang, *J. Am. Chem. Soc.*, 1984, **106**, 434.

101 J. King, P. Quayle, and J. F. Malone, *Tetrahedron Lett.*, 1990, **31**, 5221.

102 W. D. Wulff and Y.-C. Xu, *Tetrahedron Lett.*, 1988, **29**, 415.

103 G. A. Peterson, F.-A. Kunng, J. S. McCallum, and W. D. Wulff, *Tetrahedron Lett.*, 1987, **28**, 1381.

104 W. D. Wulff, R. W. Kaesler, G. A. Peterson, and P.-C. Tang, *J. Am. Chem. Soc.*, 1985, **107**, 1060.

105 T. M. Sivavec and T. J. Katz, *Tetrahedron Lett.*, 1985, **26**, 2159.

106 W. D. Wulff and R. W. Kaesler, *Organometallics*, 1985, **4**, 1461.

107 P.-C. Tang and W. D. Wulff, *J. Am. Chem. Soc.*, 1984, **106**, 1132.

108 M. F. Semmelhack, R. Tamura, W. Schnatter, and J. Springer, *J. Am. Chem. Soc.*, 1984, **106**, 5363.

109 C. P. Casey, R. A. Boggs, and R. L. Anderson, *J. Am. Chem. Soc.*, 1972, **94**, 8947; C. P. Casey and W. R. Brunsvold, *J. Organomet. Chem.*, 1975, **102**, 175.

110 M. Rudler-Chauvin and H. Rudler, *J. Organomet, Chem.*, 1981, **212**, 203.

111 W. D. Wulff and S. R. Gilbertson, *J. Am. Chem. Soc.*, 1985, **107**, 503.

Chapter 13

1 K. M. Nicholas, *J. Am. Chem. Soc.*, 1975, **97**, 3254.

2 M. L. H. Green and G. Wilkinson, *J. Chem. Soc.*, 1958, 4314.

3 Y. Ishii, T. Kagayama, A. Inada, and M. Ogawa, *Bull. Chem. Soc. Jpn.*, 1983, **56**, 2861.

4 A. Sanders and W. P. Giering, *J. Am. Chem. Soc.*, 1975, **97**, 919.

5 D. Seyferth and A. T. Wehman, *J. Am. Chem. Soc.*, 1970, **92**, 5520.

6 H. Kappeler and J. Wild, *Ger. Offen.*, 1969, 1 801 661 (C.A. 1969, **71**, 61600).

7 D. V. Banthorpe, H. Fitton, and J. Lewis, *J. Chem. Soc., Perkin Trans. 1*, 1973, 2051.

8 G. A. Taylor, *J. Chem. Soc., Perkin Trans. 1*, 1979, 1716.

9 A. J. Birch and A. J. Pearson, *J. Chem. Soc., Chem. Commun.*, 1976, 601.

10 A. A. L. Gunatilaka and A. F. Mateos, *J. Chem. Soc., Perkin Trans. 1*, 1979, 935; A. Nakamara and M. Tsutsur, *J. Med. Chem.*, 1963, **6**, 796.

11 G. Evans, B. F. G. Johnson, and J. Lewis, *J. Organomet. Chem.*, 1975, **102**, 507.

12 P. E. Cross, Ph.D. Thesis, University of Manchester, 1966 (unpublished).

13 D. H. R. Barton, A. A. L. Gunatilaka, T. Nakanishi, H. Patin, D. V. Widdowson, and B. R. Worth, *J. Chem. Soc., Perkin Trans. 1*, 1976, 821.

14 A. J. Birch, B. M. R. Bandara, K. Chamberlain, B. Chauncy, P. Dahler, A. I. Day, I. D. Jenkins, L. F. Kelly, T.-C. Khor, G. Kretschmer, A. J. Liepa, A. S. Narula, W. D. Raverty, E. Rizzardo, C. Sell, G. R. Stephenson, D. J. Thompson, and D. H. Williamson, *Tetrahedron*, 1981, **37**, *Suppl. 1*, 289.

15 K. M. Nicholas and R. Pettit, *Tetrahedron Lett.*, 1971, 3475; C. H. Mauldin, E. R. Biehl, and P. C. Reeves, *Tetrahedron Lett.*, 1972, 2955.

16 A. J. Birch, L. F. Kelly, and A. J. Liepa, *Tetrahedron Lett.*, 1985, **26**, 501; A. J. Birch and H. Fitton, *Aust. J. Chem.*, 1969, **22**, 971.

17 D. L. Reger and A. Gabrielli, *J. Am. Chem. Soc.*, 1975, **97**, 4421.

18 T. Ishizu, K. Harano, N. Hori, M. Yasuda, and K. Kanematsu, *Tetrahedron*, 1983, **39**, 1281.

19 H. Alper and S. Amaratunga, *Tetrahedron Lett.*, 1980, **21**, 1589.

20 B. F. G. Johnson, J. Lewis, P. McArdle, and G. L. P. Randall, *J. Chem. Soc., Dalton Trans.*, 1972, 456; B. F. G. Johnson, J. Lewis, and G. L. P. Randall, *J. Chem. Soc., Chem.*

Commun., 1969, 1273; B. F. G. Johnson, J. Lewis, and G. L. P. Randall, *J. Chem. Soc. (A)*, 1971, 422.

21 D. Lloyd, *Carbocyclic Non-benzenoid Aromatic Compounds*, Elsevier, Amsterdam, 1966, pp. 135 and 144.

22 M. Franck-Neumann, F. Brion, and D. Martina, *Tetrahedron Lett.*, 1978, 5033.

23 G. R. Knox and I. G. Thom, *J. Chem. Soc., Chem. Commun.*, 1981, 373.

24 A. J. Pearson and P. R. Raithby, *J. Chem. Soc., Perkin Trans. 1*, 1980, 395.

25 A. J. Pearson and C. W. Ong, *J. Am. Chem. Soc.*, 1981, **103**, 6686; A. J. Pearson and C. W. Ong, *Tetrahedron Lett.*, 1980, **21**, 4641.

26 A. J. Pearson and Y.-S. Chen, *J. Org. Chem.*, 1986, **51**, 1939; A. J. Pearson and M. K. O'Brien, *J. Org. Chem.*, 1989, **54**, 4663.

27 D. H. R. Barton and H. Patin, *J. Chem. Soc., Perkin Trans. 1*, 1976, 829.

28 D. Martina and F. Brion, *Tetrahedron Lett.*, 1982, **23**, 865.

29 A. Hafner, W. von Philipsborn, and A. Salzer, *Angew. Chem. Int. Ed. Engl.*, 1985, **24**, 126.

30 A. J. Pearson, *J. Chem. Soc., Perkin Trans. 1*, 1979, 1255; S. A. Gardner and M. D. Rausch, *J. Organomet. Chem.*, 1973, **56**, 365.

31 A. J. Pearson and M. Chandler, *J. Organomet. Chem.*, 1980, **202**, 175.

32 A. J. Birch and D. H. Williamson, *J. Chem. Soc., Perkin Trans. 1*, 1973, 1892; B. M. R. Bandara, A. J. Birch, B. Chauncy, and L. F. Kelly, *J. Organomet. Chem.*, 1981, **208**, C31.

33 A. J. Birch and G. R. Stephenson, *J. Organomet. Chem.*, 1981, **218**, 91.

34 B. F. G. Johnson, J. Lewis, D. G. Parker, and G. R. Stephenson, *J. Organomet. Chem.*, 1981, **204**, 221.

35 A. G. M. Barrett, D. H. R. Barton, and G. Johnson, *J. Chem. Soc., Perkin Trans. 1*, 1978, 1014.

36 A. Salzer and W. von Philipsborn, *J. Organomet. Chem.*, 1979, **170**, 63.

37 A. J. Pearson, *Acc. Chem. Res.*, 1980, **13**, 463.

38 A. J. Birch, K. B. Chamberlain, and D. J. Thompson, *J. Chem. Soc., Perkin Trans. 1*, 1973, 1900; H.-J. Knölker, M. Bauermeister, D. Bläser, R. Boese, and J. B. Pannek, *Angew. Chem. Int. Ed. Engl.*, 1989, **28**, 223; H. J. Knölker and M. Bauermeister, *J. Chem. Soc., Chem. Commun.*, 1989, 1468.

39 E. J. Corey and G. Moinet, *J. Am. Chem. Soc.*, 1973, **95**, 7185.

40 G. P. Randall and G. R. Stephenson, unpublished results.

41 E. Mincione, P. Bovicelli, M. Chandler, and A. R. D. Jacono, *Heterocycles*, 1985, **23**, 75; B. F. G. Johnson, J. Lewis, D. G. Parker, and G. R. Stephenson, *J. Organomet. Chem.*, 1980, **194**, C14.

42 B. F. G. Johnson, J. Lewis, D. G. Parker, and G. R. Stephenson, *J. Organomet. Chem.*, 1980, **197**, 67.

43 B. F. G. Johnson, J. Lewis, D. G. Parker, and G. R. Stephenson, *J. Organomet. Chem.*, 1980, **194**, C14.

44 B. F. G. Johnson, J. Lewis, D. G. Parker, P. R. Raithby, and G. M. Sheldrick, *J. Organomet. Chem.*, 1978, **150**, 115.

45 R. P. Alexander, C. Morley, and G. R. Stephenson, *J. Chem. Soc., Perkin Trans. 1*, 1988, 2069.

46 A. J. Pearson, P. Ham, and D. C. Rees, *Tetrahedron Lett.*, 1980, **21**, 4637.

47 A. J. Birch, P. Dahler, A. S. Narula, and G. R. Stephenson, *Tetrahedron Lett.*, 1980, **21**, 3817.

48 M. Chandler and A. J. Pearson, *Tetrahedron Lett.*, 1983, **24**, 5781; A. J. Pearson and I. C. Richards, *Tetrahedron Lett.*, 1983, 2465.

49 M. E. Wright, *Organometallics*, 1983, **2**, 558; M. E. Wright, J. F. Hoover, G. O. Nelson, C. P. Scott, and R. S. Glass, *J. Org. Chem.*, 1984, **49**, 3059.

50 V. A. Mironov, E. V. Sobolev, and A. N. Elizarova, *Tetrahedron*, 1963, **19**, 1939; S. McLean and P. Haynes, *Tetrahedron*, 1965, **21**, 2329.

51 J. Tsuji, Y. Kobayashi, and I. Shimizu, *Tetrahedron Lett.*, 1979, 39.

52 Y. Wakatsuki, S. Nozakura, and S. Murahashi, *Bull. Chem. Soc. Jpn.*, 1969, **42**, 273.

53 A. J. Birch, P. E. Cross, J. Lewis, D. A. White, and S. B. Wild, *J. Chem. Soc. A*, 1968, 332.

54 G. Jaouen, B. F. G. Johnson, and J. Lewis, *J. Organomet. Chem.*, 1982, **231**, C21.

55 A. Sanders and W. P. Giering, *J. Am. Chem. Soc.*, 1975, **97**, 919; S. A. Gardner and M. D. Rausch, *J. Organomet. Chem.*, 1973, **56**, 365; M. Rosenblum, B. North, D. Wells, and W. P. Giering, *J. Am. Chem. Soc.*, 1972, **94**, 1239.

56 L. Watts, J. D. Fitzpatrick, and R. Pettit, *J. Am. Chem. Soc.*, 1965, **87**, 3253; R. Pettit, *J. Organomet. Chem.*, 1975, **100**, 205; L. Watts, J. D. Fitzpatrick, and R. Pettit, *J. Am. Chem. Soc.*, 1966, **88**, 623.

57 M. Görlitz and H. Günther, *Tetrahedron*, 1969, **25**, 4467; E. Ciganek, *J. Am. Chem. Soc.*, 1971, **93**, 2207.

58 W. von E. Doering and D. W. Wiley, *Tetrahedron*, 1960, **11**, 183.

59 M. A. Ogliaruso, M. G. Romanelli, and E. I. Becker, *Chem. Rev.*, 1965, **65**, 261.

60 W. R. Roth and J. D. Meier, *Tetrahedron Lett.*, 1967, 2053; R. Victor and R. Ben-Shoshan, *J. Organomet. Chem.*, 1974, **80**, C1.

61 W. Grimme and H. G. Köser, *J. Am. Chem. Soc.*, 1981, **103**, 5919.

62 B. F. G. Johnson, J. Lewis, P. McArdle, and G. L. P. Randall, *J. Chem. Soc., Dalton Trans.*, 1972, 2076.

63 E. Weiss, W. Hübel, and R. Merényi, *Chem. Ber.*, 1962, **95**, 1155; E. Weiss, R. Merényi, and W. Hübel, *Chem. Ber.*, 1962, **95**, 1170; E. R. F. Gesing, J. P. Tane, and K. P. C. Vollhardt, *Angew. Chem. Int. Ed. Engl.*, 1980, **19**, 1023.

64 N. Morita, M. Oda, and T. Asao, *Tetrahedron Lett.*, 1984, **25**, 5419.

65 H. C. Longuet-Higgins and L. E. Orgel, *J. Chem. Soc.*, 1956, 1969.

66 G. Amiet, K. Nicholas, and R. Pettit, *J. Chem. Soc., Chem. Commun.*, 1970, 161.

67 G. D. Burt and R. Pettit, *J. Chem. Soc., Chem. Commun.*, 1965, 517; L. Watts, J. D. Fitzpatrick, and R. Pettit, *J. Am. Chem. Soc.*, 1965, **87**, 3253.

68 J. C. Barborak, L. Watts, and R. Pettit, *J. Am. Chem. Soc.*, 1966, **88**, 1328.

69 J. Rebek, Jr., and F. Gaviña, *J. Am. Chem. Soc.*, 1975, **97**, 3453.

70 J. D. Fitzpatrick, L. Watts, G. F. Emerson, and R. Pettit, *J. Am. Chem. Soc.*, 1965, **87**, 3254.

71 R. G. Amiet and R. Pettit, *J. Am. Chem. Soc.*, 1968, **90**, 1059.

72 J. Agar, F. Kaplan, and B. W. Roberts, *J. Org. Chem.*, 1974, **39**, 3451.

73 M. D. Rausch and A. V. Grossi, *J. Chem. Soc., Chem. Commun.*, 1978, 401; M. Rosenblum and C. Gatsonis, *J. Am. Chem. Soc.*, 1967, **89**, 5074; M. Rosenblum and B. North, *J. Am. Chem. Soc.*, 1968, **90**, 1060.

74 W. Pritschins and W. Grimme, *Tetrahedron Lett.*, 1982, **23**, 1151.

75 K. K. Joshi, *J. Chem. Soc. (A)*, 1966, 594.

76 M. Franck-Neumann, D. Martina, and F. Brion, *Angew. Chem. Int. Ed. Engl.*, 1981, 864.

77 L. A. Paquette, J. M. Photis, and G. D. Ewing, *J. Am. Chem. Soc.*, 1975, **97**, 3538; B. F. G. Johnson, J. Lewis, and D. Wege, *J. Chem. Soc., Dalton Trans.*, 1976, 1874; M.

Brookhardt, G. W. Koszalka, G. O. Nelson, G. Scholes, and R. A. Watson, *J. Am. Chem. Soc.*, 1976, 8155; C. R. Graham, G. Scholes, and M. Brookhardt, *J. Am. Chem. Soc.*, 1977, **99**, 1180.

78 E. J. Corey and J. W. Suggs, *J. Org. Chem.*, 1973, **38**, 3224.

79 E. J. Corey and J. W. Suggs, *Tetrahedron Lett.*, 1975, 3775.

80 P. A. Gent and R. Gigg, *J. Chem. Soc., Chem. Commun.*, 1974, 277; P. A. Gent and R. Gigg, *J. Chem. Soc., Perkin Trans. 1*, 1974, 1835.

81 B. Moreau, S. Lavielle, and A. Marquet, *Tetrahedron Lett.*, 1977, 2591.

82 B. C. Laguzza and B. Ganem, *Tetrahedron Lett.*, 1981, **22**, 1483.

83 J. Tsuji, *Pure Appl. Chem.*, 1986, **58**, 869; I. Minami, Y. Ohashi, I. Shimizu, and J. Tsuji, *Tetrahedron Lett.*, 1985, **26**, 2449; H. Kinoshita, K. Inomata, T. Kameda, and H. Kotake, *Chem. Lett.*, 1985, 515.

84 I. Minami, M. Yuhara, and J. Tsuji, *Tetrahedron Lett.*, 1987, **28**, 2737.

85 T. Yamada, K. Goto, Y. Mitsuda, and J. Tsuji, *Tetrahedron Lett.*, 1987, **28**, 4557.

86 J. Tsuji and T. Yamakawa, *Tetrahedron Lett.*, 1979, 613.

87 F. DiNinno, D. A. Muthard, R. W. Ratcliffe, and B. G. Christensen, *Tetrahedron Lett.*, 1982, **23**, 3535.

88 H. Mastalerz, *J. Org. Chem.*, 1984, **49**, 4092.

89 O. Dangles, F. Guibé, G. Balavoine, S. Lavielle, and A. Marquet, *J. Org. Chem.*, 1987, **52**, 4984.

90 H. X. Zhang, F. Guibé, and G. Balavoine, *Tetrahedron Lett.*, 1988, **29**, 623.

91 H. X. Zhang, F. Guibé, and G. Balavoine, *Tetrahedron Lett.*, 1988, **29**, 619.

92 Y. Hayakawa, S. Wakabayashi, T. Nobori, and R. Noyori, *Tetrahedron Lett.*, 1987, **28**, 2259.

Chapter 14

1 N. T. Byrom, R. Grigg, and B. Kongkathip, *J. Chem. Soc., Chem. Commun.*, 1976, 216.

2 J. Tsuji, Y. Mori, and M. Hara, *Tetrahedron*, 1972, **28**, 3721.

3 A. C. Oehlschlager and B. D. Johnston, *J. Org. Chem.*, 1987, **52**, 940.

4 K. Mori and Y.-B. Seu, *Tetrahedron*, 1985, **41**, 3429.

5 S. Hatakeyama, K. Sakurai, and S. Takano, *J. Chem. Soc., Chem. Commun.*, 1985, 1759.

6 P. Mangeney, A. Alexakis, and J. Normant, *Tetrahedron Lett.*, 1987, **28**, 2363.

7 J. Meinwald, A. M. Chalmers, T. E. Pliske, and T. Eisner, *J. Chem. Soc., Chem. Commun.*, 1969, 86.

8 J. Tsuji, H. Kataoka, and Y. Kobayashi, *Tetrahedron Lett.*, 1981, **22**, 2575.

9 B. M. Trost, M. J. Bogdanowicz, W. J. Frazee, and T. N. Salzmann, *J. Am. Chem. Soc.*, 1978, **100**, 5512.

10 B. Bosnich and P. B. Mackenzie, *Pure Appl. Chem.*, 1982, **54**, 189.

11 K. Jankowski, *Experimentia*, 1973, **29**, 1334.

12 P. Mangeney, N. Langlois, C. Leroy, C. Riche, and Y. Langlois, *J. Org. Chem.*, 1982, **47**, 4261; P. Mangeney, R. Z. Andriamialisoa, N. Langlois, Y. Langlois, and P. Potier, *J. Am. Chem. Soc.*, 1979, **101**, 2243; J. P. Kutney, T. Honda, P. M. Kazmaier, N. J. Lewis, and B. R. Worth, *Helv. Chim. Acta*, 1980, **63**, 366.

13 A. I. Scott, F. Gueritte, and S.-L. Lee, *J. Am. Chem. Soc.*, 1978, **100**, 6253.

14 B. M. Trost and A. G. Romero, *J. Org. Chem.*, 1986, **51**, 2332.

15 R. A. Holton, *J. Am. Chem. Soc.*, 1977, **99**, 8083.

16 R. A. Holton, *Tetrahedron Lett.*, 1977, 355; R. A. Holton and R. A. Kjonaas, *J.*

Organomet. Chem., 1977, **133**, C5; R. A. Holton and R. A. Kjonaas, *J. Organomet. Chem.*, 1977, **133**, C5; R. A. Holton and R. A. Kjonaas, *J. Am. Chem. Soc.*, 1977, **99**, 4177.

17 R. B. Moffett, *Org. Synth.*, Collected Vol IV, Wiley, New York, 1963, p. 238.

18 E. J. Corey, S. M. Albonico, U. Koelliker, T. K. Schaaf, and R. K. Varma, *J. Am. Chem. Soc.*, 1971, **93**, 1491.

19 R. Noyori, I. Tomino, M. Yamada, and M. Nishizawa, *J. Am. Chem. Soc.*, 1984, **106**, 6717.

20 G. A. Cordell, *The Alkaloids*, Vol. 17, Ed. R. H. F. Manske and R. G. A. Rodrigo, Academic Press, New York, 1979, p. 199.

21 M. Pinar, W. von Philipsborn, W. Vetter, and H. Schmid, *Helv. Chim. Acta*, 1962, **45**, 2260.

22 A. J. Pearson, D. C. Rees, and C. W. Thornber, *J. Chem. Soc., Perkin Trans. 1*, 1983, 619.

23 G. Stork and J. E. Dolfini, *J. Am. Chem. Soc.*, 1963, **85**, 2872.

24 A. J. Pearson, P. Ham, C. W. Ong, T. R. Perrior, and D. C. Rees, *J. Chem. Soc., Perkin Trans. 1*, 1982, 1527.

25 A. J. Pearson, T. R. Perrior, and D. C. Rees, *J. Organomet. Chem.*, 1982, **226**, C39.

26 H.-J. Liu and H. K. Lai, *Can. J. Chem.*, 1979, **57**, 2522.

27 A. J. Pearson and D. C. Rees, *Tetrahedron Lett.*, 1980, **21**, 3037.

28 A. J. Pearson and D. C. Rees, *J. Am. Chem. Soc.*, 1982, **104**, 1118.

29 J. H. Cardllina II, R. E. Moore, E. V. Arnold, and J. Clardy, *J. Org. Chem.*, 1979, **44**, 4039.

30 A. M. Horton and S. V. Ley, *J. Organomet. Chem.*, 1985, **285**, C17.

31 K. Matsuo, T. Kinuta, and K. Tanaka, *Chem. Pharm. Bull.*, 1981, **29**, 3047; K. Matsuo and K. Tanaka, *Chem. Pharm. Bull.*, 1981, **29**, 3070.

32 J. C. van Meter, M. Dann, and N. Bohonos, *Antimicrobial Agents Annual, 1960*, Plenum Press, New York, 1961, p. 77.

33 G. A. Ellestad, M. P. Kunstmann, H. A. Whaley, and E. L. Patterson, *J. Am. Chem. Soc.*, 1968, **90**, 1325; Y. Iwai, A. Kora, Y. Takahashi, T. Hayashi, J. Awaya, R. Masuma, R. Oiwa, and S. Omura, *J. Antibiot.*, 1978, 31, 959.

34 A. Ichihara, M. Ubukata, H. Oikawa, K. Murakami, and S. Sakamura, *Tetrahedron Lett.*, 1980, **21**, 4469.

35 M. F. Semmelhack, J. J. Bozell, T. Sato, W. Wulff, E. Spiess, and A. Zask, *J. Am. Chem. Soc.*, 1982, **104**, 5850.

36 E. O. Fischer and A. Maasböl, *Chem. Ber.*, 1967, **100**, 2445.

37 M. F. Semmelhack and A. Zask, *J. Am. Chem. Soc.*, 1983, **105**, 2034.

38 H. Gilman and J. F. Nobis, *J. Am. Chem. Soc.*, 1950, **72**, 2629.

39 S. J. Selikson and D. S. Watt, *J. Org. Chem.*, 1975, **40**, 267.

40 M. C. Croudace and N. E. Schore, *J. Org. Chem.*, 1981, **46**, 5357; N. E. Schore and M. C. Croudace, *J. Org. Chem.*, 1981, **46**, 5436.

41 S. Takahashi, H. Naganawa, H. Iinuma, T. Takita, K. Maeda, and H. Umezawa, *Tetrahedron Lett.*, 1971, 1955.

42 C. Exon and P. Magnus, *J. Am. Chem. Soc.*, 1983, **105**, 2477.

43 B. M. Trost and D. P. Curran, *J. Am. Chem. Soc.*, 1981, **103**, 7380.

44 G. S. Mikaelian, A. S. Gybin, W. A. Smit, and R. Caple, *Tetrahedron Lett.*, 1985, **26**, 1269; W. A. Smit, A. A. Schegolev, A. S. Gybin, G. S. Mikaelian, and R. Caple, *Synthesis*, 1984, 887.

45 W. A. Smit, A. S. Gybin, A. S. Shashkov, Y. T. Strychov, L. G. Kyz'mina, G. S. Mikaelian, R. Caple, and E. D. Swanson, *Tetrahedron Lett.*, 1986, **27**, 1241.

46 S. O. Simonian, W. A. Smit, A. S. Gybin, A. S. Shashkov, G. S. Mikaelian, V. A. Tarasov, I. I. Ibragimov, R. Caple, and D. E. Froen, *Tetrahedron Lett.*, 1986, **27**, 1245.

47 M. Saha, B. Bagby, and K. M. Nicholas, *Tetrahedron Lett.*, 1986, **27**, 915.

48 M. Franck-Neumann, D. Martina, and F. Brion, *Angew. Chem. Int. Ed. Engl.*, 1978, **17**, 690.

49 J. Golik, G. Dubay, G. Groenewold, H. Kawaguchi, M. Konishi, B. Krishnan, H. Ohkuma, K. Saitoh, and T. W. Doyle, *J. Am. Chem. Soc.*, 1987, **109**, 3462.

50 M. D. Lee, T. S. Dunne, C. C. Chang, G. A. Ellestad, M. M. Siegel, G. O. Morton, W. J. McGahren, and D. B. Borders, *J. Am. Chem. Soc.*, 1987, **109**, 3466.

51 P. Magnus, R. T. Lewis, and J. C. Huffman, *J. Am. Chem. Soc.*, 1988, **110**, 6921; P. Magnus, H. Annoura, and J. Harling, *J. Org. Chem.*, 1990, **55**, 1709.

52 S. L. Schreiber and L. L. Kiessling, *J. Am. Chem. Soc.*, 1988, **110**, 631.

53 R. P. Beckett and S. G. Davies, *J. Chem. Soc., Chem. Commun.*, 1988, 160.

54 G. J. Hanson, J. S. Baran, and T. Lindberg, *Tetrahedron Lett.*, 1986, **27**, 3577.

Chapter 15

1 J. L. Davidson and P. N. Preston, *Adv. Heterocyclic Chem.*, 1982, **30**, 319.

2 A. G. M. Barrett and M. A. Sturgess, *Tetrahedron*, 1988, **44**, 5615.

3 H. Alper, F. Urso, and D. J. H. Smith, *J. Am. Chem. Soc.*, 1983, **105**, 6737.

4 S. T. Hodgson, D. M. Hollinshead, and S. V. Ley, *Tetrahedron*, 1985, **41**, 5871.

5 L. S. Hegedus, L. M. Schultze, J. Toro, and C. Yijun, *Tetrahedron*, 1985, **41**, 5833.

6 K. Broadley and S. G. Davies, *Tetrahedron Lett.*, 1984, **25**, 1743.

7 L. S. Liebeskind and M. E. Welker, *Tetrahedron Lett.*, 1985, **26**, 3079.

8 P. K. Wong, M. Madhavarao, D. F. Marten, and M. Rosenblum, *J. Am. Chem. Soc.*, 1977, **99**, 2823.

9 S. R. Berryhill and M. Rosenblum, *J. Org. Chem.*, 1980, **45**, 1984.

10 S. R. Berryhill, T. Price, and M. Rosenblum, *J. Org. Chem.*, 1983, **48**, 158.

11 M. Mori, K. Chiba, M. Okita, and Y. Ban., *J. Chem. Soc., Chem. Commun.*, 1979, 698.

12 S. J. Brickner, J. I. Gaikema, J. T. Torrado, L. J. Greenfield, and D. A. Ulanowicz, *Tetrahedron Lett.*, 1988, **29**, 5601.

13 W. Chamchaang and A. R. Pinhas, *J. Chem. Soc., Chem. Commun.*, 1988, 710.

14 H. Alper and C. P. Mahatantila, *Organometallics*, 1982, **1**, 70.

15 Y. Becker, A. Eisenstadt, and Y. Shvo, *J. Organomet. Chem.*, 1978, **155**, 63.

16 F. DiNinno, E. V. Linek, and B. G. Christensen, *J. Am. Chem. Soc.*, 1979, **101**, 2210; Y. Becker, A. Eisenstadt, and Y. Shvo, *Tetrahedron*, 1978, **34**, 799.

17 G. D. Annis, E. M. Hebblethwaite, S. T. Hodgson, D. M. Hollinshead, and S. V. Ley, *J. Chem. Soc., Perkin Trans. 1*, 1983, 2851.

18 T. Kametani, T. Nagahara, and M. Ihara, *J. Chem. Soc., Perkin Trans. 1*, 1981, 3048.

19 M. A. McGuire and L. S. Hegedus, *J. Am. Chem. Soc.*, 1982, **104**, 5538; L. S. Hegedus, M. A. McGuire, L. M. Schultze, C. Yijun, and O. P. Anderson, *J. Am. Chem. Soc.*, 1984, **106**, 2680.

20 C. Borel, L. S. Hegedus, J. Krebs, and Y. Satoh, *J. Am. Chem. Soc.*, 1987, **109**, 1101.

21 A. G. M. Barrett, C. P. Brock, and M. A. Sturgess, *Organometallics*, 1985, **4**, 1903.

22 M. I. Bruce and A. G. Swincer, *Adv. Organomet. Chem.*, 1983, **22**, 59.

23 B. E. Boland-Lussier, M. R. Churchill, R. P. Hughes, and A. L. Rheingold, *Organometallics*, 1982, **1**, 628.

24 A. G. M. Barrett and M. A. Sturgess, *Tetrahedron Lett.*, 1986, **27**, 3811.

25 A. G. M. Barrett and M. A. Sturgess, *J. Org. Chem.*, 1987, **52**, 3940.

26 P. Brown and R. Southgate, *Tetrahedron Lett.*, 1986, **27**, 247.

27 R. J. Ponsford and R. Southgate, *J. Chem. Soc., Chem. Commun.*, 1979, 846.

28 L. S. Liebeskind, M. E. Welker, and R. W. Fengl, *J. Am. Chem. Soc.*, 1986, **108**, 6328.

29 I. Ojima and H. B. Kwon, *Chem. Lett.*, 1985, 1327.

30 M. Mori, N. Kanda, and Y. Ban, *J. Chem. Soc., Chem. Commun.*, 1986, 1375.

31 G. Georg and T. Durst, *J. Org. Chem.*, 1983, **48**, 2092.

32 L. S. Hegedus, G. F. Allen, and E. L. Waterman, *J. Am. Chem. Soc.*, 1976, **98**, 2674; L. S. Hegedus, G. F. Allen, J. J. Bozell, and E. L. Waterman, *J. Am. Chem. Soc.*, 1978, **100**, 5800.

33 L. S. Hegedus, G. F. Allen, and D. J. Olsen, *J. Am. Chem. Soc.*, 1980, **102**, 3583.

34 A. Kasahara and T. Saito, *Chem. & Ind.*, 1975, 745.

35 L. S. Hegedus, T. A. Mulhern, and H. Asada, *J. Am. Chem. Soc.*, 1986, **108**, 6224.

36 H. Sheng, S. Lin, and Y. Huang, *Tetrahedron Lett.*, 1986, **27**, 4893; H. Sheng, S. Lin, and Y. Huang, *Synthesis*, 1987, 1022.

37 B. M. Trost and J. P. Genêt, *J. Am. Chem. Soc.*, 1976, **98**, 8516.

38 B. M. Trost, S. A. Godleski, and J. L. Belletire, *J. Org. Chem.*, 1979, **44**, 2052.

39 B. M. Trost, S. A. Godleski, and J. P. Genêt, *J. Am. Chem. Soc.*, 1978, **100**, 3930.

40 R. C. Larock, L. W. Harrison, and M. H. Hsu, *J. Org. Chem.*, 1984, **49**, 3662.

41 T. Hayashi, A. Yamamoto, and Y. Ito, *Tetrahedron Lett.*, 1987, **28**, 4837.

42 J.-E. Bäckvall, P. G. Andersson, and J. O. Vägberg, *Tetrahedron Lett.*, 1989, **30**, 137.

43 R. P. Houghton, M. Voyle, and R. Price, *J. Chem. Soc., Perkin Trans 1*, 1984, 925.

44 M. F. Semmelhack, W. Wulff, and J. L. Garcia, *J. Organomet. Chem.*, 1982, **240**, C5.

45 A. P. Kozikowski and K. Isobe, *J. Chem. Soc., Chem. Commun.*, 1978, 1076.

46 R. M. Moriarty, Y.-Y. Ku, and L. Guo, *J. Chem. Soc., Chem. Commun.*, 1988, 1621.

47 J. P. Kutney, M. Noda, and B. R. Worth, *Heterocycles*, 1979, **12**, 1269.

48 J. P. Kutney, T. C. W. Mak, D. Mostowicz, J. Trotter, and B. R. Worth, *Heterocycles*, 1979, **12**, 1517.

49 G. Nechvatal, D. A. Widdowson, and D. J. Williams, *J. Chem. Soc., Chem. Commun.*, 1981, 1260.

50 P. J. Beswick, C. S. Greenwood, T. J. Mowlem, G. Nechvatal, and D. A. Widdowson, *Tetrahedron*, 1988, **44**, 7325.

51 M. L. Bruce, *Angew. Chem. Int. Ed. Engl.*, 1977, **16**, 73.

52 J. M. Thompson and R. F. Heck, *J. Org. Chem.*, 1975, **40**, 2667; H. Takahashi and J. Tsuji, *J. Organomet. Chem.*, 1967, **10**, 511.

53 R. C. Davis, T. J. Grinter, D. Leaver, and R. M. O'Neil, *Tetrahedron Lett.*, 1979, 3339.

54 F. Maassarani, M. Pfeffer, and G. Le Borgne, *Organometallics*, 1987, **6**, 2029 and 2043.

55 S. E. Diamond, A. Szalkiewicz, and F. Mares, *J. Am. Chem. Soc.*, 1979, **101**, 490.

56 S. Murahashi, *J. Am. Chem. Soc.*, 1955, **77**, 6403.

57 R. D. O'Sullivan and A. W. Parkins, *J. Chem. Soc., Chem. Commun.*, 1984, 1165.

58 W. D. Jones and W. P. Kosar, *J. Am. Chem. Soc.*, 1986, **108**, 5640.

59 A. Kasahara, T. Izumi, S. Murakami, H. Yanai, and M. Takatori, *Bull. Chem. Soc. Jpn.*, 1986, **59**, 927.

60 P. Martin, *Helv. Chim. Acta*, 1984, **67**, 1647.

61 M. Mori, S. Kudo, and Y. Ban, *J. Chem. Soc., Perkin Trans. 1*, 1979, 771.

62 M. F. Semmelhack, R. D. Stauffer, and T. D. Rogerson, *Tetrahedron Lett.*, 1973, 4519.

63 W. C. Frank, Y. C. Kim, and R. F. Heck. *J. Org. Chem.*, 1978, **43**, 2947.

64 A. J. Birch, A. J. Liepa, and G. R. Stephenson, *Tetrahedron Lett.*, 1979, 3565; A. J. Birch, A. J. Liepa, and G. R. Stephenson, *J. Chem. Soc., Perkin Trans. 1*, 1982, 713; G. R. Stephenson, *J. Organomet. Chem.*, 1985, **286**, C41.

65 H.-J. Knölker and M. Bauermister, *J. Chem. Soc., Chem. Commun.*, 1990, 664.

66 T. N. Danks and S. E. Thomas, *Tetrahedron Lett.*, 1988, **29**, 1425; *J. Chem. Soc., Perkin Trans* 1990, 761.

67 J. P. Kutney, M. Noda, N. G. Lewis, B. Monteiro, D. Mostowicz, and B. R. Worth, *Heterocycles*, 1981, **16**, 1469.

68 W. P. Henry and R. P. Hughes, *J. Am. Chem. Soc.*, 1986, **108**, 7862.

69 H. Alper, J. E. Prickett, and S. Wollowitz, *J. Am. Chem. Soc.*, 1977, **99**, 4330.

70 K. Isomura, K. Uto, and H. Taniguchi, *J. Chem. Soc., Chem. Commun.*, 1977, 664.

71 F. Bellamy, *J. Chem. Soc., Chem. Commun.*, 1978, 998.

72 H. Alper, J. E. Prickett, and S. Wollowitz, *J. Am. Chem. Soc.*, 1977, **99**, 4330; A. Inada, H. Heimgartner, and H. Schmid, *Tetrahedron Lett.*, 1979, 2983.

73 H. Alper and J. E. Prickett, *J. Chem. Soc., Chem. Commun.*, 1976, 191.

74 D. Roberto and H. Alper, *J. Chem. Soc., Chem. Commun.*, 1987, 212.

75 T. Kobayashi and M. Nitta, *Chem. Lett.*, 1983, 1233.

76 M. Nitta and T. Kobayashi, *Heterocycles*, 1985, **23**, 339.

77 S. Nakanishi, Y. Shirai, K. Takahashi, and Y. Otsuji, *Chem. Lett.*, 1981, 869; S. Nakanishi, J. Nantaku, and Y. Otsuji, *Chem. Lett.*, 1983, 341.

78 T. L. Gilchrist, E. E. Nunn, and C. W. Rees, *J. Chem. Soc., Perkin Trans. 1*, 1974, 1262.

79 T. Kobayashi and M. Nitta, *Bull. Chem. Soc. Jpn.*, 1985, **58**, 152.

80 B. Potthoff and E. Breitmaier, *Synthesis*, 1986, 584.

81 H. Bönnemann and R. Brinkmann, *Synthesis*, 1975, 600.

82 Y. Wakatsuki and H. Yamazaki, *Synthesis*, 1976, 26.

83 H. Bönnemann, *Angew. Chem. Int. Ed. Engl.*, 1985, **24**, 248; H. Bönnemann and W. Brijoux, *Bull. Soc. Chim. Belg.*, 1985, **94**, 635.

84 H. Bönnemann, W. Brijoux, R. Brinkmann, and W. Meurers, *Helv. Chim. Acta*, 1984, **67**, 1616.

85 G. E. Herberich and G. Greiss, *Chem. Ber.*, 1972, **105**, 3413.

86 A. Naiman and K. P. C. Vollhardt, *Angew. Chem. Int. Ed. Engl.*, 1977, **16**, 708.

87 D. J. Brien, A. Naiman, and K. P. C. Vollhardt, *J. Chem. Soc., Chem. Commun.*, 1982, 133.

88 R. E. Geiger, M. Lalonde, H. Stoller, and K. Schleich, *Helv. Chim. Acta*, 1984, **67**, 1274.

89 P. Hong and H. Yamazaki, *Tetrahedron Lett.*, 1977, 1333; P. Hong and H. Yamazaki, *Synthesis*, 1977, 50.

90 E. R. F. Gesing, U. Groth, and K. P. C. Vollhardt, *Synthesis*, 1984, 351.

91 D. B. Grotjahn and K. P. C. Vollhardt, *J. Am. Chem. Soc.*, 1986, **108**, 2091.

92 K. Jonas, E. Deffense, and D. Habermann, *Angew. Chem. Int. Ed. Engl.*, 1983, **22**, 716; E. H. Bray, C. Hoogzand, W. Hübel, U. Krüerke, R. Merenyl, and E. Weiss, *Proc. 6th Int. Conf. Coord. Chem.*, 1961, 190.

93 M. Jautelat and K. Ley, *Synthesis*, 1970, 593.

94 Y. Ohshiro, K. Kinugasa, T. Minami, and T. Agawa, *J. Org. Chem.*, 1970, **35**, 2136.

95   K. Kinugasa and T. Agawa, *J. Organomet. Chem.*, 1973, **51**, 329.

96   H. Hoberg and B. W. Oster, *J. Organomet. Chem.*, 1983, **252**, 359.

97   T. Ban, K. Nagai, Y. Miyamoto, K. Harano, M. Yasuda, and K. Kanematsu, *J. Org. Chem.*, 1982, **47**, 110.

98   Z. Goldschmidt, S. Antebi, and I. Goldberg, *J. Organomet. Chem.*, 1984, **260**, 105.

99   C. W. Jefford and W. Johncock, *Helv. Chim. Acta*, 1983, **66**, 2666.

100   C. W. Jefford, T. Kubota, and A. Zaslona, *Helv. Chim. Acta*, 1986, **69**, 2048.

101   R. Connell, F. Scavo, P. Helquist, and B. Åkermark, *Tetrahedron Lett.*, 1986, **27**, 5559.

102   R. Aumann, H. Heinen, C. Krüger, and Y.-H. Tsay, *Chem. Ber.*, 1986, **199**, 3141.

103   R. Aumann and H. Heinen, *Chem. Ber.*, 1986, **119**, 3801.

104   K. S. Chan and W. D. Wulff, *J. Am. Chem. Soc.*, 1986, **108**, 5229.

105   W. D. Wulff and D. C. Yang, *J. Am. Chem. Soc.*, 1983, **105**, 6726; W. D. Wulff and D. C. Yang, *J. Am. Chem. Soc.*, 1984, **106**, 7565; K. H. Dötz and W. Kuhn, *J. Organomet. Chem.*, 1985, **286**, C23.

106   F. R. Kreissl, E. O. Fischer, and C. G. Kreiter, *J. Organomet. Chem.*, 1973, **57**, C9.

107   K. S. Chan and W. D. Wulff, *J. Am. Chem. Soc.*, 1986, **108**, 5229.

108   W. D. Wulff, S. R. Gibertson, and J. P. Springer, *J. Am. Chem. Soc.*, 1986, **108**, 520.

109   S. Murahashi and S. Horiie, *J. Am. Chem. Soc.*, 1956, **78**, 4816.

110   A. Rosenthal, R. F. Astbury, and A. Hubscher, *J. Org. Chem.*, 1958, **23**, 1037.

111   Y. Ohshiro and T. Hirao, *Heterocycles*, 1984, **22**, 859.

112   R. Millini, H. Kisch, and C. Kruger, *Z. Naturforsch. Teil B.*, 1985, **40**, 187.

113   C. B. Argo and J. T. Sharp, *J. Chem. Soc., Perkin Trans. 1*, 1984, 1581.

114   T. Tsuda, R. Sumiya, and T. Saegusa, *Synth. Commun.*, 1987, **17**, 147.

115   M. Mori, K. Chiba, and Y. Ban, *Tetrahedron Lett.*, 1977, 1037.

116   R. C. Larock and S. Babu, *Tetrahedron Lett.*, 1987, **28**, 5291.

117   M. Mori and Y. Ban, *Tetrahedron Lett.*, 1976, 1803.

118   J. R. Norton, K. E. Shenton, and J. Schwartz, *Tetrahedron Lett.*, 1975, 51; T. F. Murray, E. G. Samsel, V. Varma, and J. R. Norton, *J. Am. Chem. Soc.*, 1981, **103**, 7520.

119   M. F. Semmelhack and S. J. Brickner, *J. Am. Chem. Soc.*, 1981, **103**, 3945.

120   M. F. Semmelhack and S. J. Brickner, *J. Org. Chem.*, 1981, **46**, 1723.

Index

584   *Index*